Engineering Differential Equations

Bill Goodwine

Engineering Differential Equations

Theory and Applications

Springer

Bill Goodwine
Department of Aerospace and
    Mechanical Engineering
University of Notre Dame
Notre Dame, Indiana 46556
USA
jgoodwin@nd.edu

MATLAB® is a registered trademark of The MathWorks, Inc.

For MATLAB product information, please contact:
The MathWorks Inc.
3 Apple Hill Drive
Natick, MA, 01760-2098 USA
Tel: 508-647-7000
Fax: 508-647-7001
E-mail: info@mathworks.com
Web: www.mathworks.com

ISBN 978-1-4899-8167-7   ISBN 978-1-4419-7919-3 (eBook)
DOI 10.1007/978-1-4419-7919-3
Springer New York Dordrecht Heidelberg London

Mathematics Subject Classification Codes (2010): 34-01, 34H05, 35B05, 34B24, 93-01, 65-01,
                                                 70J10, 70J25, 80-01.

Printed on acid-free paper

Springer is part of Springer Science+Business Media (www.springer.com)

To Amy, Bridget, Carolyn and Katie.

Preface

This book is intended for engineering undergraduate students, particularly aerospace and mechanical engineers and students in other disciplines concerned with system modeling, analysis, and control. It is intended to be a relatively comprehensive treatment of engineering undergraduate differential equations as well as two primary applications thereof: linear vibrations and classical feedback control. This material is traditionally separated into different courses in undergraduate engineering curricula, however, consistent with the theme of this book, the current trend to optimize and streamline curricula results in many programs combining courses where there is a common underlying theoretical basis. Specifically this book was developed from the materials presented in a two-course, required, junior-level sequence of courses that I developed and have taught in the Department of Aerospace and Mechanical Engineering at the University of Notre Dame over the past several years. The rationale behind the selection, arrangement, and relationship of the content of book has four primary facets.

The first facet relates to the role of mathematical analysis in modern engineering. The modern reality, especially in industry but also in academic settings, is that sophisticated software packages enabled by fast computing are starting to play a dominant role in engineering analysis. Hence, to some extent, the skill of being able to solve problems "by hand" is being displaced by computer simulation. This is not an argument for not covering what has been traditionally been the subject of engineering analysis courses, however, it can be taken as a justification for a slightly altered focus toward fundamental understanding versus problem solving by hand. Because the algorithmic aspects of many problem-solving methods are increasingly "hidden" in software, a fundamental understanding of the relationship between the attributes of a differential equation and the nature of its solution is critical; for few things are as dangerous as an engineer who places complete faith in the output of a computer.

This point may be best represented in the language of educational objectives. Perhaps the most common is due to Bloom [7], which categorizes cognitive processes in a hierarchical manner. From the lowest to highest process there are: knowledge, comprehension, application, analysis, synthesis, and evaluation. Despite the label of "engineering analysis" many homework problems given to students fall within the application process; that is, they are asked to apply a particular solution or analysis method to a given problem. Then, through repeated exposure to a subject on the application level, it is hoped that most students develop the ability (inductively through experience) to become competent at the higher levels, at which point higher-level synthesis or evaluation problems may be addressed.

With the application level being more and more automated, an increased focus in courses on the higher cognitive levels is necessary for the students to remain competent. I cannot argue that being able to "do" problems is not a necessary skill, and it is one that certainly has not been removed from this book. However, focusing on higher-level cognitive processes is what is going to serve students best. At a minimum, it will allow them to be distinguishable from a computer, which is also able to solve differential equations [27, 55]. By combining the mathematics and the application in the same course, the full range of the theoretical mathematics can be exercised through the engineering applications for which the students will ultimately be accountable.

The second facet is pedagogical. There are several nonstandard features of the content and presentation in this book that should be highlighted. First, there is an abundance of detailed examples. These are present, not to serve as a template from which students can copy the procedure to solve a problem, but rather a recognition of the fact that, although traditionally mathematics and related application fields are taught in a deductive manner, *inductive learning* actually "promotes deeper learning and longer retention of information" [15, 16, 33]. Thus, one way to consider this abundance of examples is that they replace, to some extent, the more direct application-oriented homework problems. Second, material is sometimes covered or named in a nonstandard manner purely to promote a deeper understanding of the ultimate result, but which otherwise does not directly help one "use" the result or is otherwise nonstandard. Examples of this are found throughout the text. A superficial example would be naming the procedure normally referred to as "integrating factors" for first-order equations "variation of parameters" because it shares a common derivation with that method as typically applied to higher-order equations. A deeper example would be in the study of frequency response methods for feedback control, where emphasis is placed on the fact that what is plotted in Bode plots relates to the harmonically forced steady-state solution of the system, when the usual use of the plot for stability under unity feedback is essentially unrelated to that. Finally, an example of including a whole nonstandard section is related to Taylor series methods for numerical methods. It is included purely as a setup for the Runge–Kutta method to facilitate developing a deeper understanding by the students. This is particularly important in such a case. When the end result is just a formula to be used, without the proper development, many students would be inclined perhaps simply to be content to "use" the formula, in which case the more modern approach would be simply to bring the `ode45()` function in MATLAB® to the attention of the students.

The third facet is that this text provides a means for a streamlined and efficient treatment of material normally covered in three courses. In the author's program it resulted in combining three courses into two courses with little lost in terms of content coverage. With the ever-broadening scope of engineering programs to include more, for example, biological sciences and design content, this would provide a means to allocate credit hours to such content without overly substantially cutting into the traditional engineering science material.

The fourth facet relates to the motivation engineering students have for studying mathematics. Ultimately engineering students study mathematics in order to be able to solve problems that are of importance to them. Although it is certainly legitimate for engineering students to have courses solely focused on the mathematics followed by the applications, it has been my experience that engineering students approach the mathematical subjects with much greater interest and enthusiasm when they have an application immediately at hand. Assessment from the sequence of courses I teach verifies this conclusion.

Content

This book covers what is normally covered in undergraduate engineering differential equations, vibrations, and controls courses. Less emphasis is placed on "recipes" or enumerated "procedures" to solve problems than is usual, although such content is not completely missing. There are plenty of problems that ask the student to simply solve some differential equations, however, quite a few of them are, for the reasons outlined above, deeper.

In addition to the combination of subjects unified by the content of the book, there are a couple of additional unique features related to content. The final chapter on nonlinear systems is perhaps longer than what is typically covered in engineering courses. Also, in the appendix there are many computer programs that were used to solve the example problems. These are presented in both the C programming language as well as in FORTRAN. The former is included because it is still widely used. The latter is, perhaps somewhat uniquely, still used in the aerospace industry but, more important, is fairly transparent in syntax, so that a student who is not proficient in programming can still easily determine what the program is doing.

It is important to note that the chapters are not of approximately uniform length. This is particularly important for an instructor making a course syllabus from the table of contents to note.

A Web page has been created for this book that contains:

- Some media content such as movies illustrating problem solutions that are amenable to such a presentation
- Source code for computer programs
- Additional exercises
- Errata

The URL is: http://controls.ame.nd.edu/engdiffeq/

Prerequisites

The student is assumed to have a good background in calculus (at least through multivariable calculus) and linear algebra. A dynamics course would be useful, but the basic mechanics from the typical undergraduate engineering physics sequence seems to suffice. A basic exposure to circuit analysis along the lines of the content typical in introductory physics courses would also be helpful. Finally, a good introduction to computer programming would be very useful, but not necessarily required.

Chapter Dependencies

The book is organized in what I consider the most logical order for a fundamental treatment of the subject matter. Even if a chapter does not explicitly depend on a previous one, the general progression of understanding and sophistication that would be developed when the chapters are covered in order was carefully considered.

However, curricular realities may prevent covering the chapters in order. Although it is not ideal, it would be possible to treat some of the material out of order. Specifically, the order of the following chapters may be altered without an extreme disruption in the logical flow of the material.

- Chapter 5 considers variable-coefficient ordinary differential equations. The method used, assuming a power series, is sufficiently different from the methods that precede the chapter that it could really be considered at any point.
- Chapters 8 through 10 cover Laplace transforms and control applications. It would be possible to treat these chapters as an independent unit.
- Chapter 11 considers the simplest linear partial differential equations using the separation of variables method. Hence, as long as Chapter 3 or the equivalent has been covered, it should be possible to cover this material.
- Chapter 12 considers numerical methods, and hence really only requires an understanding of Taylor series.

In the text, these chapters do occasionally refer back to earlier chapters, however, the dependence is typically one of pedagogy, rather than of theoretical necessity. These references could typically be treated in a lecture with a relatively quick aside. In my own case, for example, because of the structure of our curriculum, I cover partial differential equations and numerical methods in the first semester of the two-semester sequence.

Acknowledgments

I would like to acknowledge the valuable help I have received from several of my colleagues. Michael M. Stanišić provided many insightful comments and corrections to several chapters of a draft of this book. I would also like to thank Mary

Frandsen for reviewing and suggesting improvements to the section on music in Chapter 11 and Robert A. Howland for reviewing and suggesting improvements to some of the mechanics material in Chapter 1. I would also like to thank the approximately 300 undergraduate students who used draft versions of this text in the courses I teach and pointed out many typographical errors, substantive errors, confusingly written parts, and so on.

Finally, I would like to thank my wife, Amy, for her patience and support throughout the process of writing this book. Especially during the very long time I thought this project was "95% finished," she was unwaveringly supportive.

And now on to business.

Notre Dame, Indiana   *Bill Goodwine*
June 2010

Contents

Chapter 1
Introduction and Preliminaries

This chapter presents an assortment of material, covered in varying degrees of detail, which is needed as background material for the rest of this book. Students with a strong background in mathematics can probably skim some of the sections in this chapter. Section 1.1 discusses why differential equations are important to study in engineering. Section 1.2 discusses functions in mathematics, which is important not only because functions are solutions to differential equations but also because it helps define the types of variables that appear in differential equations. Implicit functions are also reviewed. The most important sections in this chapter are Sections 1.3 and 1.4 because they present new material that is used throughout the entire book related to the types of differential equations and their solutions. Section 1.5 reviews concepts from mechanical and electronic systems, including the properties of the common elements that make up simple electromechanical engineering systems. Finally, Section 1.6 presents the most basic method to use a computer to determine an approximate solution to an initial value problem for a given ordinary differential equation. More advanced numerical methods are considered in Chapter 12, but this introductory material is presented so that, even in the earliest chapters, students can check their work using a computer-solving method.

1.1 The Engineering Utility of Differential Equations

Nearly all the fundamental principles that govern physical processes of engineering interest are described by differential equations. Hence, it is fair to say that the ability to analyze, solve, and understand differential equations is fundamentally important for engineers. This is particularly important in *design*, because in design an engineer is usually tasked with choosing the parameters of a system. Those parameters, then, are parameters in a differential equation describing the system, and hence insight into the nature of the dependence of the solutions to that differential equation on the parameters is critical to making design decisions.

B. Goodwine, *Engineering Differential Equations: Theory and Applications*,
DOI 10.1007/978-1-4419-7919-3_1, © Springer Science+Business Media, LLC 2011

This book is intended to make differential equations more accessible to engineering students by presenting and developing some application areas in parallel with the presentation of the mathematics. This is done sometimes by way of simply using the application as a motivational problem, and other times by fully developing the application material. The main two application areas in this book are mechanical vibrations and classical feedback control theory.

Additionally, there is an emphasis on analyzing the solutions to each problem, for example, instead of the "answer" being simply a mathematical expression, an analysis of that answer is often required. For example, the "answer" to the differential equation that is a simple model of an automotive suspension is

$$x_p(t) = h \sqrt{\frac{1 + \left(2\zeta\frac{\omega}{\omega_n}\right)^2}{\left(1 - \frac{\omega^2}{\omega_n^2}\right)^2 + \left(2\zeta\frac{\omega}{\omega_n}\right)^2}} \cos\left(\omega t + \hat{\phi} + \psi\right).$$

Such an equation is only really useful in the context of designing the suspension and, for example, a more useful question may include determining speed of the vehicle, which is related to the frequency, ω, at which $x_p(t)$ (how much the passengers in the automobile are being shaken) has the greatest magnitude.

1.2 Sets, Relations, and Functions

Most engineering students have a pretty decent grip on the idea of a function, and functions are important in this book because the solution to a differential equation is a function. This section first deals briefly with *sets* inasmuch as functions are relationships between sets. Then functions and implicit functions are defined as well. Implicit functions arise in this book because they are the natural representation of a solution to certain differential equations, particularly those considered in Section 2.5.1, dealing with so-called separable first-order differential equations, which is followed by the definition of multivariable functions and a review of their calculus.

1.2.1 Sets

Without getting bogged down in the nuances of basic set theory, we consider a *set* to be a collection of *elements*.[1] We assume that there is a way for us to determine

---

[1] More precisely, a collection of elements is a *class* and a set is a certain kind of class. A reader interested in the distinction is referred to [24].

whether an element is in a set[2] and whether two elements are the same. Many sets have common names. The two sets we are most concerned with are the set of real numbers, denoted by $\mathbb{R}$, and the set of complex numbers, denoted by $\mathbb{C}$. We often deal with particular subsets, the most common of which are *intervals of* $\mathbb{R}$, such as the *closed interval*

$$[a,b] = \{x \in \mathbb{R} \, | \, a \leq x \leq b\},$$

that is, real numbers that are either a, b, or between[3] a and b, or

$$(a,b] = \{x \in \mathbb{R} \, | \, a < x \leq b\},$$

that is, real numbers that are between a and b or are b. An *open interval* is an interval of the form

$$(a,b) = \{x \in \mathbb{R} \, | \, a < x < b\},$$

where the term "open" connotes the fact that the interval does not include its boundary or endpoints. As is the usual convention, a parenthesis indicates that the boundary point is not included in the set, and a square brace indicates that the boundary point is included.

Sometimes we put more than one set together to make a new set. A common way in which this is done is called the *Cartesian product*.

Definition 1.1. Let $\mathcal{D}_1, \mathcal{D}_2, \ldots, \mathcal{D}_n$ be sets. The *Cartesian product* of $\mathcal{D}_1$, $\mathcal{D}_2$, ..., $\mathcal{D}_n$, is the set

$$\mathcal{D}_1 \times \mathcal{D}_2 \times \cdots \times \mathcal{D}_n = \{(x_1, x_2, \ldots, x_n) \, | \, x_1 \in \mathcal{D}_1, x_2 \in \mathcal{D}_2, \cdots, x_n \in \mathcal{D}_n\}.$$

If the sets $\mathcal{D}_1, \mathcal{D}_2, \ldots, \mathcal{D}_n$ are the same, then we use the notation

$$\mathcal{D}_1 \times \mathcal{D}_2 \times \cdots \times \mathcal{D}_n = \mathcal{D}^n,$$

and elements of $\mathcal{D}^n$ are called called *n-tuples*. Elements of $\mathcal{D}_1 \times \mathcal{D}_2 \times \cdots \times \mathcal{D}_n$ are *ordered* which means that

$$(x_1, x_2, \ldots, x_n) = (y_1, y_2, \ldots, y_n)$$

if and only if $x_1 = y_1$, $x_2 = y_2$, $\cdots$, and $x_n = y_n$.

An example of the way the Cartesian product is used is when vectors in Euclidean space are used to represent something.

---

[2] Whether an element is a member of a set is not even necessarily an either–or proposition. *Fuzzy logic* is the branch of logic and mathematics where set theory is generalized to include the notion of partial set membership. In this book an element is either in a set or it is not in the set. In contrast, in fuzzy logic an element may be partially in a set. A classic example of a fuzzy set is the set of "warm days." It is natural to think of some days as "kind of" warm, which is represented in fuzzy logic by being partially in the set of warm days and partially not in it. There is a vast literature on fuzzy logic and the interested reader is referred to the original paper [56].

[3] For an element to be between two others, we must be able to *order* the set and also to be able to determine the order.

Example 1.1. An example of a Cartesian product is the set of vectors in three-dimensional Euclidean space. To specify a point in space, a set of three basis vectors is needed, and the point is then represented by its component along each of these three basis vectors. In this book we write

$$\xi = \begin{bmatrix} x_1 \\ x_2 \\ x_3 \end{bmatrix}$$

to represent the point. The set to which this point belongs is $\mathbb{R} \times \mathbb{R} \times \mathbb{R} = \mathbb{R}^3$.

1.2.2 Relations and Functions

In this book, a *relation* between elements of sets may be defined by an equation or a set of equations. Elements of the sets satisfy the relation if they satisfy the equation. A special kind of relation is a function.

Definition 1.2. Given two sets, $\mathcal{D}$ and $\mathcal{R}$, if, for each element of $x \in \mathcal{D}$ there is a rule that assigns one and only one element of $y \in \mathcal{R}$ then we say that y is a *function* of x. The set $\mathcal{D}$ is called the *domain* and the set $\mathcal{R}$ the *range*.

The variable x denoting an element of the domain is called the *independent variable* and the variable y denoting the elements of the range is called the *dependent variable*. It is common to write $y = f(x)$ to indicate that y is a function of x, where f is a name for the function. Two functions, f and g are equal if they have the same domain and range and $f(x) = g(x)$ for every element x of the domain.

Note that it is necessary to specify which set is the domain and which is the range. Of course, we do not usually bother to do that and it is normally clear from the context which set is the domain and which is the range. We often indirectly specify the domain and range by saying that a function is *from* the domain *to* the range.

Example 1.2. If $s = \mu + i\omega$, the equation

$$r = \|s\| = \sqrt{\mu^2 + \omega^2}$$

defines a function from the complex numbers to the real numbers (the complex numbers are the domain and the real numbers are the range) because there is one and only one real number for each complex number that satisfies the equation. The equation does not define a function from the real numbers to the complex numbers because for most real numbers r, there are many complex numbers with $\|s\| = r$.

So far we have been considering functions between two sets. Of course, functions may exist between multiple sets, which is manifested in the case where the dependent variable depends upon more than one independent variable. In such a case, the dependent variable is a function of the independent variables if, for each possible combination of the independent variables, there corresponds only one value of the dependent variable. Solutions to partial differential equations are multivariable functions.

Definition 1.3. If, given $m+1$ sets, $\mathcal{D}_1, \mathcal{D}_2, \ldots, \mathcal{D}_m$, and $\mathcal{R}$, and elements $x_1 \in \mathcal{D}_1, x_2 \in \mathcal{D}_2, \ldots, x_m \in \mathcal{D}_m$, there corresponds one and only one element of $y \in \mathcal{R}$, then we say that y is a *function* (or *multivariable function* of $x_1, x_2, \ldots, x_m$). The variables $x_1, x_2, \ldots, x_m$ are called the *independent variables* and the variable y is called the *dependent variable*. Using the Cartesian product, the domain is given by $\mathcal{D} = \mathcal{D}_1 \times \mathcal{D}_2 \times \cdots \times \mathcal{D}_m$ and the function is a function from $\mathcal{D}$ to $\mathcal{R}$. It is common to write $y = f(x_1, x_2, \ldots, x_m)$ to indicate that y is a function of $x_1, x_2, \ldots, x_m$.

Example 1.3. For $r \in \mathbb{R}$ and $(x, y) \in \mathbb{R} \times \mathbb{R}$, the equation

$$r = \sqrt{x^2 + y^2}$$

defines a function inasmuch as there is only one r for any specified values for x and y.

1.2.3 The Derivative

The derivative is given by the usual limit definition.

Definition 1.4. Let $x(t)$ be a function with the single independent variable t. The *derivative of x with respect to t* is defined by

$$\frac{dx}{dt}(t) = \lim_{\Delta t \to 0} \frac{x(t + \Delta t) - x(t)}{\Delta t}.$$

The usual interpretation of the derivative is that it is the rate of change of the function with respect to the independent variable. If graphed, it is the slope of the curve of $x(t)$. If the function depends on more than one independent variable, then we must consider the partial derivative.

Definition 1.5. Let $x(t_1, \ldots, t_n)$ be a function with independent variables $t_1, \ldots, t_n$. The *partial derivative of x with respect to t_m* is defined by

$$\frac{\partial x}{\partial t_m}(t_1, \ldots, t_n) = \lim_{\Delta t \to 0} \frac{x(t_1, \ldots, t_m + \Delta t, \ldots, t_n) - x(t_1, \ldots, t_m, \ldots, t_n)}{\Delta t}.$$

This book uses practically all the usual notational means to represent derivatives. Which one is used typically depends on the conventional notation used by various application areas. In particular, because it can be difficult to interpret an equation with many parentheses, we often use a "subscript" notation to indicate the values at which a derivative function is evaluated instead of following the function name by parentheses, that is,

$$\frac{df}{dx}\bigg|_{x=x_0} = \frac{df}{dx}(x_0).$$

In cases where it is obvious what the independent variable is, it may be omitted, for example,

$$m\ddot{x} + b\dot{x} + kx = 0$$

may be written instead of

$$m\ddot{x}(t) + b\dot{x}(t) + kx(t) = 0.$$

1.2.4 Implicit Functions

So far things are simple: given an element of the domain, if we have a way to determine one and only one element of the range, then we have a function. In some cases, however, it naturally arises that for a function of more than one variable, we are interested not so much in what element of the range corresponds to elements of the domain, but rather in the relationship among the elements of the domain that correspond to one particular element in the range. A circle is a typical example.

Example 1.4. Returning to Example 1.3, consider the set of points that satisfy

$$x^2 + y^2 = 1. \tag{1.1}$$

A plot of all points that satisfy this equation is illustrated in Figure 1.1.

In Example 1.3 we had a function of two variables, and in Example 1.4 we studied the set of points that satisfy $x^2 + y^2 = 1$. This second example defines a *relation*, which is more general than a function. Two points $x \in \mathbb{R}$ and $y \in \mathbb{R}$ satisfy the relation if they satisfy Equation (1.1). Mathematically a relation is defined to be a subset of the domain. For purposes of this book we consider them to be the subset of the domain that satisfies some equation, such as $f(x,y) = 1$, or $f(x,y) \geq 2$.

It is logical in the second example to study the relationship between x and y beyond simply asking whether they satisfy the relation. By referring to Figure 1.1, it is clear that x and y are not related by a function because for any $x \in (-1, 1)$ there are two values for y that satisfy the relation (and vice–versa). In the next example, we show that it is possible to make the relationship between x and y that satisfies $f(x,y) = 1$ into a function, at least for a limited domain or range.

Example 1.5. Consider the set of points that satisfy

$$x^2 + y^2 = 1. \tag{1.2}$$

One way to make this relation into a function from one of the independent variables to the other independent variable is to appropriately restrict the domain and range.

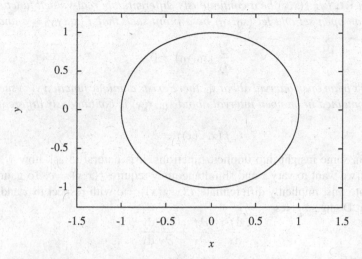

Fig. 1.1 A plot of the subset of points in $\mathbb{R}^2$ that satisfy $x^2 + y^2 = 1$.

It is clear from Figure 1.1 that, at most, the domain must be limited at least to the interval $\mathcal{D} = [-1, 1]$. With respect to the range, it must also be restricted so that only the top or bottom half of the circle is included in the range.

So, in this example, Equation (1.2) defines a function $y = f(x)$ if we restrict the domain to be

$$\mathcal{D} = \{x \in \mathbb{R} \mid -1 \leq x \leq 1\}$$

and specify either $y = \sqrt{1-x^2}$ or $y = -\sqrt{1-x^2}$, which corresponds to either the top or bottom half of the circle, respectively.

Do not infer from Example 1.5 that it will always be the case that an equation that defines an implicit function can be "solved" for one of the variables. In fact doing so will typically be difficult.

In Example 1.5 we were able to solve for y in terms of x for some region of the domain. Motivated by this we define an *implicit function* as follows.

Definition 1.6. The equation $f(x, y) = c$ where c is a constant, defines an implicit function if and only if there exists a function $g(x)$ such that

$$f(x, g(x)) = c.$$

A more natural way to write this is to write $g(x) = y(x)$; that is, the y variable is actually a function of x.

The following theorem gives a sufficient condition for the existence of an implicit function.

Theorem 1.1. *Let $f(x,y)$ be a continuously differentiable real-valued function defined on an open set and let (x_0,y_0) be a point such that $f(x_0,y_0) = c$ and such that*

$$\frac{\partial f}{\partial y}(x_0,y_0) \neq 0.$$

Then, for x in an open interval about x_0, there exists a unique function $y(x)$ such that $y(x)$ is contained in an open interval about y_0, $y(x)$ is continuously differentiable, and

$$f(x,y(x)) = c.$$

To gain some insight into implicit functions, it is natural to ask how y should change if we want to vary x and simultaneously require $f(x,y) = c$. To gain some insight into this, implicitly differentiate $f(x,y(x)) = c$ with respect to x and solve for dy/dx. Doing so gives

$$\frac{df}{dx} = \frac{\partial f}{\partial x} + \frac{\partial f}{\partial y}\frac{dy}{dx}.$$

Because $f(x,y(x)) = c$, then $df/dx = 0$, and hence

$$\frac{dy}{dx} = -\frac{\frac{\partial f}{\partial x}}{\frac{\partial f}{\partial y}}. \tag{1.3}$$

Intuitively, in order to determine how y should change as a function of x, we need the denominator on the right-hand side of Equation (1.3) to be nonzero and points where the denominator is zero to correspond exactly to points where the theorem does not guarantee the existence of $y(x)$.[4]

1.3 Types of Differential Equations

This section provides the basic definitions necessary to categorize a differential equation so that the appropriate solution method can be identified. The solution methods developed subsequently are only applicable to certain types of differential equations; hence, it is critical from the beginning to be able to properly categorize them. Before that, however, we must first consider exactly what a "differential equation" is.

Definition 1.7. Let $x(t_1,t_2,\ldots,t_m)$ be a function of the m independent variables $t_1,t_2,\ldots,t_m$. A *differential equation* is an equation that contains at least one derivative (of any order) of $x(t_1,\ldots,t_m)$.

Example 1.6. The equation

$$\frac{1}{t}\ddot{x}(t) = 3$$

---

[4] Note that this does not prove the theorem; on the contrary, we assumed that y was a function of x when we implicitly differentiated the equation.

is a differential equation with dependent variable x and independent variable t.

Sometimes we have to consider a set of differential equations, which is called a *system of differential equations*.

Definition 1.8. Let each function in the set of functions $\{x_1, x_2, \ldots, x_n\}$ be a function of the m independent variables $t_1, t_2, \ldots, t_m$. A *system of differential equations* is a set of equations where each equation contains at least one derivative of each of the functions in the set $\{x_1, x_2, \ldots, x_n\}$.

Example 1.7. The set of equations describing the projectile motion of a particle with mass m is

$$m\ddot{x} = 0$$
$$m\ddot{y} = -mg,$$

where g is the acceleration due to gravity. This is a system of two differential equations with dependent variables x and y and one independent variable t.

In general, because they can be determined from fundamental scientific principles, the differential equation governing a system is known, but the solution is unknown. For example, in Example 1.7, the differential equations follow from basic principles from mechanics. "Solving" a differential equation amounts to determining the function (dependent variable) of the independent variable that satisfies the differential equation.

Several chapters in this book deal with solution methods for differential equations, and which solution method works depends, not surprisingly, upon the characteristics of the differential equation. The characteristics that are used in this book to determine the correct solution method are:

- Whether the differential equation is an ordinary or partial differential equation
- What the order of the differential equation is
- Whether the differential equation is linear or nonlinear
- If the equation is linear, whether the equation is homogeneous
- If the equation is linear, whether the equation has constant or variable coefficients

The next few subsections present the definitions of these characteristics.

1.3.1 Ordinary Versus Partial Differential Equations

In a differential equation, if the dependent variable is a function of only one independent variable, then the differential equation is an *ordinary differential equation*. If the dependent variable depends on more than one independent variable and the equation contains derivatives with respect to more than one of the independent variables, then the differential equation is a *partial differential equation*. Generally it is trivial to distinguish between ordinary and partial differential equations because the derivatives are notationally different.

Example 1.8. The equation describing a mass–spring–damper system, studied in complete detail in Chapter 4, under the influence of a harmonic forcing function given by

$$\ddot{x}(t) + 3\dot{x}(t) + 5x(t) = \sin(t)$$

is an ordinary differential equation with independent variable t and dependent variable x.

Example 1.9. The equation that described the motion of a vibrating string

$$\frac{\partial^2 u}{\partial t^2}(x,t) = \frac{\partial^2 u}{\partial x^2}(x,t),$$

where $u(x,t)$ gives the displacement of the string at position x and time t, is a partial differential equation.

A system of differential equations is ordinary if each of the dependent variables is a function of one and the same independent variable. Generally speaking if there are partial derivative signs in the equation it is a partial differential equation and if there are only ordinary derivative operators (ds) or "dots" ($\dot{x}$) or "primes," (y') then the equation is ordinary.

1.3.2 The Order of a Differential Equation

The *order* of a differential equation is the order of the highest derivative in the equation.

Example 1.10. The equation

$$\sin(t) + x(t) + \ddot{x}(t) = 35\dot{x}(t)\cos(t)$$

is second-order.

Example 1.11. The wave equation,

$$\frac{\partial^2 u}{\partial x^2} = \frac{\partial^2 u}{\partial t^2}$$

is second-order.

Remark 1.1. This text only considers nth-order ordinary differential equation that may be written in the form

$$\frac{d^n x}{dt^n} = f\left(\frac{d^{n-1}x}{dt^{n-1}}, \frac{d^{n-2}x}{dt^{n-2}}, \ldots, x, t\right); \qquad (1.4)$$

that is, the equation can be solved for the highest derivative of the dependent variable.

So, this text considers differential equations of the form

$$m\ddot{x}(t) + b\dot{x}(t) + kx(t) = \cos(t),$$

but not of the form

$$m\left(\ddot{x}(t) + \dot{x}^2(t)\right) + b\dot{x}(t) + kx(t) = \cos(t),$$

because the latter cannot be written in the form of Equation (1.4).

1.3.3 Linear Versus Nonlinear Differential Equations

This is perhaps the most important distinction of all. With the exception of some first-order equations and other very specific examples, nonlinear differential equations do not have any known solution techniques; in contrast, linear differential equations have some very nice properties and are easily solved. A differential equation is *linear* if all the terms in the equation are linear in the dependent variable and its derivatives; otherwise, it is *nonlinear*.

Considering first an nth-order ordinary differential equation with independent variable t and dependent variable x, if the equation can be put in the form

$$\boxed{f_n(t)\frac{\mathrm{d}^n x}{\mathrm{d}t^n}(t) + f_{n-1}(t)\frac{\mathrm{d}^{n-1}x}{\mathrm{d}t^{n-1}}(t) + \cdots + f_1(t)\frac{\mathrm{d}x}{\mathrm{d}t}(t) + f_0(t)x(t) = g(t)} \qquad (1.5)$$

it is linear.

Remark 1.2. The functions $f_i(t)$ and $g(t)$ do not have to be linear functions of t in order for the equation to be linear. Only linearity in the dependent variable matters.

Extending this to the partial differential equation case is straightforward. The equation is linear if all the terms containing the dependent variable or any of its derivatives appear linearly in the equation; otherwise, it is nonlinear.

Considering an nth-order partial differential equation with independent variables x and t and dependent variable u, if the equation can be put in the form

$$\sum_{i,j,i+j\leq n} f_{i,j}(x,t)\frac{\partial^{i+j}u}{\partial x^i \partial t^j}(x,t) = g(x,t) \qquad (1.6)$$

it is linear.

Example 1.12. The differential equations listed in Table 1.1 are linear or nonlinear as indicated.

Differential Equation	Linear or Nonlinear
$\ddot{x}(t) + t^2\sin(t)x(t) = 5t$	linear
$\ddot{x}(t) + t^2\sin(t)x^2(t) = 5t$	nonlinear
$\ddot{x}(t) + t^2\dot{x}(t)x(t) = 5t$	nonlinear
$\ddot{x}(t) + t^2\sin(t)\sin(x(t)) = 5t$	nonlinear
$\ddot{x}(t) + t^2\sin(t)x(t) = 5tx(t)$	linear
$\ddot{x}(t) + 2t\dot{x}(t) = 5x(t)$	linear
$\ddot{x}(t) + 2\dot{x}(t) = 5x(t)$	linear
$\ddot{x}(t) + 2x(t)t = 5x(t)$	linear
$\ddot{x}(t) + 2x(t)\dot{x}(t) = 5x(t)$	nonlinear
$\ddot{x}(t) + 2x(t) = 5\sin(t)$	linear
$\ddot{x}(t) + 2x(t) = 5\sin(t)\dot{x}(t)$	linear
$\ddot{x}(t) + 2x(t) = 5\sin(t)\sin(\dot{x}(t))$	nonlinear
$\dfrac{\partial^2 u}{\partial x^2}(x,t) = \dfrac{\partial u}{\partial t}(x,t)$	linear
$\dfrac{\partial^2 u}{\partial x^2}(x,t) = u(x,t)\dfrac{\partial u}{\partial t}(x,t)$	nonlinear
$\dfrac{\partial^2 u}{\partial x^2}(x,t) = \dfrac{\partial u}{\partial t}(x,t)\dfrac{\partial u}{\partial t}(x,t)$	nonlinear
$\dfrac{\partial^2 u}{\partial x^2}(x,t) = \dfrac{\partial u}{\partial t}(x,t) + x$	linear
$\dfrac{\partial^2 u}{\partial x^2}(x,t) = x\dfrac{\partial u}{\partial t}(x,t) + u(x,t)$	linear

Table 1.1 Linear and nonlinear differential equations

1.3.4 Homogeneous Versus Inhomogeneous Linear Differential Equations

If any of the nonzero terms of a linear differential equation are not a function of the dependent variable, then the equation is inhomogeneous; otherwise, it is homogeneous. The "terms" of a differential equation are the elements of the equation that are on either side of the equality and that are combined by addition or subtraction. In the form of Equation (1.5), a linear ordinary differential equation is homogeneous if $g(t) = 0$; otherwise, it is inhomogeneous.

Example 1.13. The linear differential equations listed in Table 1.2 with dependent variable x and independent variable t are homogeneous or inhomogeneous as indicated.

Differential Equation	Homogeneous or inhomogeneous
$\ddot{x}(t) + t^2 \sin(t)x(t) = 5t$	inhomogeneous
$\ddot{x}(t) + t^2 \sin(t)x(t) = 5tx(t)$	homogeneous
$\dfrac{\partial^2 u}{\partial x^2}(x,t) = \dfrac{\partial u}{\partial t}(x,t)$	homogeneous
$\dfrac{\partial^2 u}{\partial x^2}(x,t) = \dfrac{\partial u}{\partial t}(x,t) + \sin(x)$	inhomogeneous

Table 1.2 Homogeneous and inhomogeneous linear differential equations

1.3.5 Constant-Coefficient Versus Variable-Coefficient Linear Differential Equations

If all the functions $f_i(t)$, $i \in \{1,\ldots,n\}$ in Equation (1.5) are constants, then the linear ordinary differential equation is *constant-coefficient;* otherwise, it is *variable-coefficient.* Note that if the equation is inhomogeneous, then there may be terms that are functions of the independent variable, but if they are not coefficients of the dependent variable it will still be a constant-coefficient differential equation. Especially in control theory and in dynamical systems, constant-coefficient equations are often referred to as *time invariant.*

Example 1.14. The linear differential equations listed in Table 1.3 with dependent variable x and independent variable t are either constant- or variable-coefficient as indicated.

Differential Equation	Constant- or Variable-Coefficient
$\ddot{x}(t) + t^2 \sin(t)x(t) = 5t$	variable-coefficient
$\ddot{x}(t) + t^2 \sin(t)x(t) = 5tx(t)$	variable-coefficient
$\ddot{x}(t) + 2\dot{x}(t) = 5x(t)$	constant-coefficient
$\ddot{x}(t) + 2\dot{x}(t)t = 5x(t)$	variable-coefficient
$\ddot{x}(t) + 2x(t) = 5\sin(t)$	constant-coefficient
$\dfrac{\partial^2 u}{\partial x^2}(x,t) = \dfrac{\partial u}{\partial t}(x,t)$	constant-coefficient
$\dfrac{\partial^2 u}{\partial x^2}(x,t) = \dfrac{\partial u}{\partial t}(x,t) + x$	constant-coefficient
$\dfrac{\partial^2 u}{\partial x^2}(x,t) = x\dfrac{\partial u}{\partial t}(x,t) + u(x,t)$	variable-coefficient

Table 1.3 Constant- and variable-coefficient linear differential equations

1.3.6 Types of Linear Second-Order Partial Differential Equations

By referring to Equation (1.6), a second-order linear partial differential equation may be written as

$$f_{2,0}(x,t)\frac{\partial^2 u}{\partial x^2}(x,t) + f_{1,1}(x,t)\frac{\partial^2 u}{\partial x \partial t}(x,t) + f_{0,2}(x,t)\frac{\partial^2 u}{\partial t^2}(x,t)$$
$$+ f_{1,0}(x,t)\frac{\partial u}{\partial x}(x,t) + f_{0,1}(x,t)\frac{\partial u}{\partial t}(x,t) + f_{0,0}(x,t)u(x,t) = g(x,t). \tag{1.7}$$

If

1. $(f_{1,1}(x,t))^2 - 4f_{2,0}(x,t)f_{0,2}(x,t) = 0$, then the equation is called *parabolic*,
2. $(f_{1,1}(x,t))^2 - 4f_{2,0}(x,t)f_{0,2}(x,t) > 0$, then the equation is called *hyperbolic*,
3. $(f_{1,1}(x,t))^2 - 4f_{2,0}(x,t)f_{0,2}(x,t) < 0$, then the equation is called *elliptic*.

The functions $f_{i,j}(x,t)$ depend on the values of the independent variables, therefore whether the equation is parabolic, hyperbolic, or elliptic may change depending on the values of the independent variables. In the case of constant coefficients, the equation will be of the same type throughout the domain.

Example 1.15. The one-dimensional heat conduction equation

$$\frac{\partial^2 u}{\partial x^2}(x,t) = \frac{\partial u}{\partial t}(x,t)$$

is parabolic. In general, parabolic equations describe diffusionlike processes.

Example 1.16. The wave equation

$$\frac{\partial^2 u}{\partial x^2}(x,t) = \frac{\partial^2 u}{\partial t^2}(x,t),$$

which describes a vibrating string, is hyperbolic. In general, hyperbolic equations describe vibrating and wavelike motions.

Example 1.17. Laplace's equation

$$\frac{\partial^2 u}{\partial x^2}(x,y) + \frac{\partial^2 u}{\partial y^2}(x,y) = 0,$$

which describes the steady-state temperature distribution in a plate is elliptic. In general, elliptic equations describe steady-state phenomena.

1.4 Solutions of Differential Equations

This section deals with solutions of differential equations. There are, in fact, several different types of solutions and distinguishing among them is important, not only for fundamentally understanding the subject, but also for avoiding frustration subsequently when "solving" problems so that the right type of solution is actually obtained. Another issue that arises in the study of differential equations relates to whether a solution even exists to a given differential equation, and if one does exist, whether it is unique.

1.4.1 Types of Solutions to Differential Equations

The projectile motion problem, introduced previously in Example 1.7 is used here to motivate the need to consider different "types" of solutions to a differential equation. The differential equations were given as

$$m\ddot{x} = 0$$
$$m\ddot{y} = -mg. \tag{1.8}$$

Thinking of this problem as a cannon shooting a cannonball, these differential equations are a consequence of the laws of mechanics, that is, the sum of the forces on the ball is proportional to the acceleration.

It can be easily verified by substituting them into the differential equations that the two functions

$$x(t) = x_0 + \dot{x}_0 t$$
$$y(t) = y_0 + \dot{y}_0 t - \frac{1}{2}gt^2 \tag{1.9}$$

satisfy the differential equations, for any constants x_0, y_0, $\dot{x}_0$, and $\dot{y}_0$. These functions describe the trajectory of the ball. Clearly, the path of the ball depends not only on the mechanics governing projectile motion, but also on the location of the cannon, the angle from the ground at which it is pointed θ, and the velocity of the ball when it leaves the cannon. These factors appear in the solution in the form of the four constants, where x_0 and y_0 represent the position of the cannon and $\dot{x}_0$ and $\dot{y}_0$ may be computed from the angle and initial velocity of the cannonball. Figure 1.2 illustrates three solutions starting from the same position, $(x_0, y_0) = (0, 0)$ and the same initial velocity, but with different angles.

The first type of solution that we consider is a function that, when substituted for the dependent variable, satisfies the differential equation.

Definition 1.9. An *explicit solution* (usually just called "a *solution*") of a differential equation is a function that satisfies the differential equation.

The set of functions in Equation (1.9) is the solution to the set of equations in Equation (1.8). There are constants in the solution to the projectile motion equations that could take on any constant values. We are also interested in the case where there are no arbitrary constants.

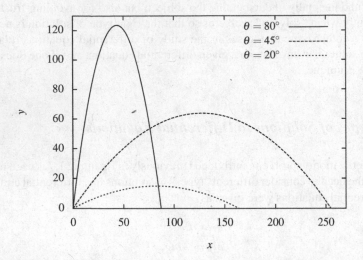

Fig. 1.2 Three paths corresponding to the solution to Equation (1.8) with different cannon angles.

Definition 1.10. A *particular solution* of a differential equation is a function that satisfies the differential equation, but contains no arbitrary constants.

Each of the three curves illustrated in Figure 1.2 represents a different particular solution. In the case where the solution contains arbitrary constants, then it is natural to ask when a solution has enough arbitrary constants to represent every possible solution to a differential equation.

Definition 1.11. The *general solution* of a differential equation is a solution from which every particular solution may be obtained by an appropriate choice of values for arbitrary constants.

It should be apparent that it will typically be very difficult to know whether a given solution is a general solution, even if it contains many arbitrary constants. The way to think of the general solution to the cannonball problem would be to ask if there is an expression for the solution that describes any possible motion of the cannonball. At this point we do not have the tools to prove it, but it is the case that the functions in Equation (1.9) represent every possible solution to the differential equations in Equation (1.8). This is clearly very valuable because there is no way the ball can move that is not represented by the general solution. The data used to determine the arbitrary constants in a general solution to determine a specific particular solution are called either the *initial conditions* or *boundary conditions*.

The distinction between initial conditions and boundary conditions is that initial conditions all are specified at the same point in time (more precisely, for the same value of the independent variable). Boundary conditions may be specified either at different times, or more commonly in the case where an independent variable represents a spatial dimension, specified at different points in space. Generally a problem comprised of a differential equation and initial conditions is called an *initial value problem* and a problem comprised of a differential equation and boundary values is called a *boundary value problem.*

For the cannonball example, an initial value problem would correspond to specifying the location and angle of the cannon and the initial velocity of the ball, and the solution would be the trajectory that the ball follows. We would expect a solution to such an initial value problem to be unique, meaning that once the initial conditions are established, there is one and only one trajectory that the cannonball follows. In the case where a function satisfies both the differential equation and the initial conditions, the function is said to be the solution of the initial value problem.

In contrast, an example of a solution to the boundary value problem would correspond to finding a solution that connects the initial location of the cannon and a location of a target. In contrast to the initial value problem, this problem may have many solutions. Figure 1.3 illustrates multiple solutions that start at the same location and pass through the same target location. A solution that satisfies the differential equation and the boundary conditions is said to be a solution of the boundary value problem. It is not necessarily always the case that there are many solutions to a boundary value problem and one unique solution to an initial value problem, but for most of the types of equations considered in this book, that is indeed the case.

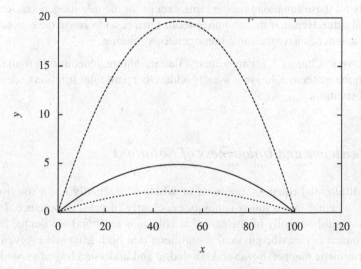

Fig. 1.3 Three projectile paths, each of which satisfies the same boundary values.

Finally, there is a definition that arises mainly because it is part of a procedure to solve linear ordinary differential equations.

Definition 1.12. A *homogeneous solution* is the general solution to an ordinary, linear, homogeneous differential equation. If the equation is inhomogeneous, the homogeneous solution is the general solution obtained by setting the inhomogeneous term to zero.

If the equation is inhomogeneous, then a subscript h is used to designate the fact that the solution is the homogeneous solution as opposed to the particular or general solution. For the types of differential equations considered in this book where we deal with homogeneous solutions, an nth-order differential equation has n different[5] homogeneous solutions. An alternative common name for homogeneous solutions is *complementary functions*.

Example 1.18. The function $x_h(t) = c_1 \sin t + c_2 \cos t$ is the homogeneous solution to all of the following differential equations,

$$\ddot{x} + x = \sin t$$
$$\ddot{x} + x = \cos t$$
$$\ddot{x} + x = t$$
$$\ddot{x} + x = e^t$$
$$\ddot{x} + x = \frac{\sin t + 35 \cos t}{e^t + 6}$$
$$2\ddot{x} + 2x = 5.$$

Obviously all these equations are the same except for the inhomogeneous term (the right-hand side). Hence, if the inhomogeneous term is set to zeros, the equations are the same and hence have the same homogeneous solution.

As shown in Chapter 3, for an ordinary, linear, inhomogeneous differential equation, the homogeneous solution is usually added to a particular solution to determine a general solution.

1.4.2 Existence and Uniqueness of Solutions

Given a differential equation, the issue of whether it actually has a solution, and if it does, whether that solution is unique, is clearly of great importance. In engineering it would normally be the case that solutions exist and are unique because they are meant to describe physical phenomena that both exist and evolve in time in a deterministic manner; however, knowledge and understanding of existence and uniqueness theorems are still important for engineers because they provide insight regarding the accuracy of the model of the physical phenomenon that the differential equation is supposed to represent.

Although this topic is obviously an important one, it is not directly addressed in this book. One reason was mentioned above: as engineers we deal with physical phenomena that will be modeled by equations that we would expect to have

---

[5] What exactly constitutes "different" solutions is a bit subtle. In fact, we require *linearly independent* solutions, which are considered in detail in Section 3.2.2.

unique solutions. Furthermore, because much of the focus in this book is on solution methods, such methods obviously will only work for equations that have solutions. Hence, at the beginning of each chapter, the type of differential equation that is considered is stated with precision and, in most cases, a theorem relating to the existence and uniqueness of solutions to the equation is presented in Appendix D.

1.5 A Few Fundamental Principles from Science

Differential equations arise in engineering because the fundamental laws governing many physical processes are known relationships between various quantities and their derivatives. Hence, the fundamental law is known, and is often quite simple such as Newton's second law, $F = ma$; however, the ultimate consequences of this law may be quite complicated. This section reviews a few fundamental laws of science, some of which are the foundation that gives rise to differential equations that have engineering importance.

1.5.1 Units

In order for numeric descriptions of quantities to be meaningful, a system of units must be employed. As is conventional this book uses the following as the base units for the seven base quantities, and all other units are derived from these. These base units are

1. The *meter*, m, which is a unit for the base quantity of *length*
2. The *second*, s, which is a unit for the base quantity of *time*
3. The *kilogram*, kg, which is a unit for the base quantity of *mass*
4. The *ampere*, A, which is a unit for the base quantity of *electric current*
5. The *kelvin*, K, which is a unit for the base quantity of *thermodynamic temperature*
6. The *mole*, mol, which is a unit for the base quantity of the *amount of substance*
7. The *candela*, cd, which is a unit for the base quantity of the *luminous intensity* of light

See [37] for further information. Units derived from these base units that are used in this book are presented in Table 1.4. For completeness, the usual prefixes used for different orders of magnitude are presented in Table 1.5.

Remark 1.3. Most of the examples in this book do not include units in order to emphasize the mathematics. However, engineering requires units, and hence some of the exercises require that all the terms be labeled with the appropriate units.

The calculus operations of differentiation and integration change the units of a function in an intuitive manner. By the definition of the derivative,

Derived Quantity	Name	Symbol	Base Units	Other Units
Area	square meter		m^2	
Volume	cubic meter		m^3	
Plane angle	radian	rad	m/m	
Speed, velocity	meters per second		m/s	
Angular velocity	radians per second		1/s	rad/s
Acceleration	meters per second squared		m/s^2	
Mass density	kilograms per cubic meter		kg/m^3	
Frequency	hertz	Hz	1/s	
Force	newton	N	$(kg \cdot m)/s^2$	
Moment	newton meter		$(kg \cdot m^2)/s^2$	$N \cdot m$
Energy, work	joule	J	$(kg \cdot m^2)/s^2$	$N \cdot m$
Power	watt	W	$(kg \cdot m^2)/s^3$	J/s
Electric charge	coulomb	C	$A \cdot s$	
Electric potential	volt	V	$(kg \cdot m^2)/(s^3 \cdot A)$	W/A
Electric capacitance	farad	F	$(A^2 \cdot s^4)/(kg \cdot m^2)$	C/V
Electric resistance	ohm	Ω	$(kg \cdot m^2)/(s^3 \cdot A^2)$	V/A
Electric inductance	henry	H	$(kg \cdot m^2)/(s^2 \cdot A^2)$	
Heat capacity	joules per kelvin		$(kg \cdot m^2)/(s^2 \cdot K)$	J/K
Thermal conductivity	watt per meter kelvin		$(kg \cdot m)/(s^3 \cdot K)$	$W/(m \cdot K)$

Table 1.4 Some derived units based upon the seven base units in the SI system (adapted from [37])

Magnitude	Name	Symbol	Magnitude	Name	Symbol
10^{24}	yotta	Y	10^{-1}	deci	d
10^{21}	zetta	Z	10^{-2}	centi	c
10^{18}	exa	E	10^{-3}	milli	m
10^{15}	peta	P	10^{-6}	micro	μ
10^{12}	tera	T	10^{-9}	nano	n
10^{9}	giga	G	10^{-12}	pico	p
10^{6}	mega	M	10^{-15}	femto	f
10^{3}	kilo	k	10^{-18}	atto	a
10^{2}	hecto	h	10^{-21}	zepto	z
10^{1}	deka	da	10^{-24}	yocto	y

Table 1.5 Standard prefixes corresponding to different orders of magnitude [37]

$$\frac{df}{dt}(t) = \lim_{\Delta t \to 0} \frac{f(t + \Delta t) - f(t)}{\Delta t}$$

the units of a derivative of a function are the units of that function divided by the units of the independent variable with respect to which it is being differentiated.

Example 1.19. If $x(t)$ has units of meters, then $\dot{x}(t)$ will have units of meters divided by seconds.

Example 1.20. If $u(x,t)$ has units of kelvin, then $\partial^2 u/\partial x^2 (x,t)$ will have units of kelvin divided by meters squared.

Conversely, the units of the integral of a function are the units of that function multiplied by the units of the independent variable with respect to which it is being integrated.

Example 1.21. If $f(t)$ has units of meters and t has units of seconds, then $\int f(t)dt$ has units of meters times seconds.

1.5.2 Mechanical Systems

In this section we consider some basic ways to determine the equations of motion for mechanical systems. This text is not intended to be a mechanics book; however, it is important to consider the manner in which differential equations arise. Keep in mind that the point of this section is only the means to determine the right equations. The rest of the book is about how to solve them.

This section is intended as a summary of basic results from Newtonian dynamics and is far from complete. An interested reader is referred to, for example, [34] for a comprehensive introductory treatment, [23] for an intermediate treatment, or to [11, 19, 43, 51] for a more advanced treatment, and [1, 4] for very advanced treatments.

In *The Principia*, [39, 40], Isaac Newton states the following three laws of motion.

Law 1.1. *Every body preserves in its state of rest, or of uniform motion in a right line, unless it is compelled to change that state by forces impressed thereon.*[6]

The modern expression of this law is *conservation of momentum*.

Law 1.2. *The alteration of motion is ever proportional to the motive forces impressed; and is made in the direction of the right line in which that force is impressed.*[7]

This gives rise to the familiar "force equals mass times acceleration" rule.

Law 1.3. *To every action there is always opposed an equal reaction: or the mutual actions of two bodies upon each other are always equal, and directed to contrary parts.*[8]

In other words, forces occur in equal and opposite pairs. If you push on a body, the force you exert is exactly the same as the force that the body exerts on you. This law plays a critical role in the development of rigid body mechanics.

---

[6] As originally published, it states *"Lex I: Corpus omne perseverare in statu suo quiescendi vel movendi uniformiter in directum, nisi quatenus a viribus impressis cogitur statum illum mutare."*

[7] *"Lex II: Mutationem motus proportionalem esse vi motrici impressae, et fieri secundum lineam rectam qua vis illa imprimitur."*

[8] *"Lex III: Actioni contrariam semper et qualem esse reactionem: sive corporum duorum actiones in se mutuo semper esse quales et in partes contrarias dirigi."*

1.5.2.1 Application of Newton's Laws to the Motion of Particles

Newton's first law speaks of a "body." We need to make a distinction between particles and rigid bodies. A *particle* is an object that generally has a finite mass, but has no appreciable physical extent compared to its range of motion and negligible rotational motion. In such a case it is valid to assume that the mass is concentrated at a point. A *rigid body* is a collection of particles where the distance between any two particles remains fixed. Unless otherwise indicated, all vectors describing physical systems are with respect to an *inertial coordinate system,* which is a coordinate system that is not rotating and has an origin that is not accelerating.[9]

In this section, bold variables represent vector quantities, as is the convention in mechanics. We consider a collections of N particles, each with a mass m_i, position vector with respect to the origin of some inertial coordinate system $\mathbf{x}_i$, and velocity $\dot{\mathbf{x}}_i = \mathbf{v}_i$.

Definition 1.13. The *linear momentum* $\mathbf{p}_i$ of a particle of mass m_i with velocity $\mathbf{v}_i$ measured relative to an inertial coordinate system is given by

$$\mathbf{p}_i(t) = m_i \mathbf{v}_i(t). \tag{1.10}$$

Newton's second law states that if $\mathbf{F}_i$ represents the vector sum of the forces acting on the ith particle; then

$$\frac{\mathrm{d}\,(m_i \mathbf{v}_i)}{\mathrm{d}t}(t) = \mathbf{F}_i(t), \tag{1.11}$$

and in the case where m_i is constant,

$$\boxed{m_i \frac{\mathrm{d}^2 \mathbf{x}_i}{\mathrm{d}t^2}(t) = \mathbf{F}_i(t).} \tag{1.12}$$

Because it turns out to be very useful in the case of rotational motion, we now take the cross-product of a vector from some fixed point with each side of Equations (1.10)–(1.12).

Definition 1.14. The *angular momentum* about the origin $\mathbf{h}_i(t)$ of a particle of mass m_i with velocity $\mathbf{v}_i(t)$ about the origin is given by

$$\mathbf{h}_i(t) = \mathbf{x}_i(t) \times \mathbf{p}_i(t) = m_i\,(\mathbf{x}_i(t) \times \mathbf{v}_i(t)), \tag{1.13}$$

where $\times$ is the usual cross-product in $\mathbb{R}^3$.

---

[9] Exactly how to determine whether a coordinate system is inertial is not an easy thing. Generally, however, on the earth if it is not accelerating with respect to the surface of the earth, then it is approximately inertial unless the acceleration is extremely small or the extent of motion is large. Sometimes an inertial frame is defined to be one in which Newton's laws hold; however, this is not of much use if our purpose is to apply Newton's laws! Appealing to Einstein's general theory of relativity gives a complete answer, but is beyond the scope of this text.

For the rest of this development, we assume that the mass of each particle is constant and that, unless otherwise indicated, the angular momentum is computed about the origin. Computing the derivative of angular momentum about the origin with respect to time gives

$$\frac{d\mathbf{h}_i}{dt}(t) = \frac{d}{dt}\left(\mathbf{x}_i(t) \times \mathbf{p}_i(t)\right) = \left(\frac{d\mathbf{x}_i}{dt}(t) \times \mathbf{p}_i(t)\right) + \left(\mathbf{x}_i(t) \times \frac{d\mathbf{p}_i}{dt}(t)\right)$$

$$= \left(\mathbf{v}_i(t) \times m_i\mathbf{v}_i(t)\right) + \left(\mathbf{x}_i(t) \times \mathbf{F}_i(t)\right)$$

$$= \mathbf{x}_i(t) \times \mathbf{F}_i(t). \tag{1.14}$$

Putting together Equations (1.13) and (1.14):

$$\boxed{\mathbf{h}_i(t) = m_i\left(\mathbf{x}_i(t) \times \mathbf{v}_i(t)\right) \quad \text{and} \quad \frac{d\mathbf{h}_i}{dt}(t) = \mathbf{x}_i(t) \times \mathbf{F}_i(t).} \tag{1.15}$$

Equation (1.15) is the usual, "the rate of change of angular momentum of a particle about the origin is equal to the sum of the moments about the origin."

In order to extend Newton's laws to rigid bodies, we must consider the application of them to a collection of particles. The first thing to consider is the *center of mass*, defined by

$$\mathbf{x}_{\text{com}} = \frac{\sum_i m_i\mathbf{x}_i}{\sum_i m_i} = \frac{\sum_i m_i\mathbf{x}_i}{m},$$

where $\sum_i$ is the sum over all the particles in the system; that is, $\sum_i = \sum_{i=1}^N$ and $m = \sum_i m_i$ is the total mass of the system.

Summing each side of Equation (1.11) over all the particles and using the definition of the center of mass we have

$$\sum_i \frac{d(m_i\mathbf{v}_i)}{dt}(t) = \sum_i \mathbf{F}_i(t) \quad \Longrightarrow \quad m\frac{d^2\mathbf{x}_{\text{com}}}{dt^2}(t) = \sum_i \mathbf{F}_i(t). \tag{1.16}$$

This equation expresses Newton's law for a system of particles in terms of the acceleration of the center of mass and the applied forces. If all we are concerned about is the center of mass, this would be useful; however, it does not necessarily provide any information about any individual particle in the system.

Summing Equations (1.13) and (1.15) over all the particles gives

$$\mathbf{h}(t) = \sum_i \left[m_i\left(\mathbf{x}_i(t) \times \mathbf{v}_i(t)\right)\right] \quad \text{and} \quad \frac{d\mathbf{h}}{dt}(t) = \sum_i \left(\mathbf{x}_i(t) \times \mathbf{F}_i(t)\right). \tag{1.17}$$

This expression gives the rate of change of the angular momentum for the collection of particles that only depends on the applied forces and their location. As with Equation (1.16), it does not necessarily provide any information about any individual particle, but does provide information about the whole system.

Observe that Equations (1.12), (1.15)–(1.17) always hold, and are nothing more than expressions of Newton's second law. Mechanics would be simple if there were not too many particles and the forces that each particle was subjected to were known. Unfortunately, for engineering systems complications arise because

- Most systems are rigid bodies, which are a collection of particles which are exerting forces on each other.
- Most engineering systems involve constrained motion, and the forces that enforce the constraints may be difficult to determine.

Rigid bodies are considered next, followed by considering certain special types of constrained motions.

1.5.2.2 Application of Newton's Laws to a Rigid Body

A *rigid body* is a system of particles where the particles are constrained by internal forces to remain a fixed distance from each other. We need to do two things to extend the considerations of the previous section to rigid bodies. First, there are internal constraint forces that keep the body rigid. Second, because most rigid bodies are considered a continuum of material, the summations of the previous section have to be replaced by integrals.

Let $\mathbf{F}_{ij}$ denote the force to which particle j is subjected by particle i, which will be aligned with the vector connecting the two particles, $\mathbf{x}_i - \mathbf{x}_j$. Newton's third law states that $\mathbf{F}_{ji} = \mathbf{F}_{ij}$. The internal forces all cancel, therefore the equations are the same as before:

$$m\frac{d^2\mathbf{x}_{com}}{dt^2}(t) = \sum_i \mathbf{F}_i(t) \ \text{ and } \ \frac{d\mathbf{h}}{dt}(t) = \sum_i \left(\mathbf{x}_i(t) \times \mathbf{F}_i(t)\right), \quad (1.18)$$

where it is emphasized that $\mathbf{F}_i(t)$ are the external forces only.

The fact that the distance between points in a rigid body remains fixed allows a useful reformulation of the angular momentum equation. Let $\mathbf{r}_i$ denote the vector from the center of mass to particle i, so that $\mathbf{x}_i = \mathbf{x}_{com} + \mathbf{r}_i$ and $\mathbf{v}_i = \mathbf{v}_{com} + \dot{\mathbf{r}}_i$. Note that because the $\mathbf{r}_i$ are measured with respect to the center of mass, they are not measured with respect to a inertial coordinate system and we cannot directly express Newton's law using them. However, we can use them within an expression of Newton's law that is with respect to an inertial coordinate system.

Using curly braces to denote the extent of each summation, substituting this into the definition of angular momentum for a system of particles gives

$$\mathbf{h} = \sum_i \{m_i [\mathbf{x}_i \times \mathbf{v}_i]\} = \sum_i \{m_i [(\mathbf{x}_{\text{com}} + \mathbf{r}_i) \times (\mathbf{v}_{\text{com}} + \dot{\mathbf{r}}_i)]\}$$

$$= \sum_i \{m_i [(\mathbf{x}_{\text{com}} \times \mathbf{v}_{\text{com}}) + (\mathbf{x}_{\text{com}} \times \dot{\mathbf{r}}_i) + (\mathbf{r}_i \times \mathbf{v}_{\text{com}}) + (\mathbf{r}_i \times \dot{\mathbf{r}}_i)]\}$$

$$= (m(\mathbf{x}_{\text{com}} \times \mathbf{v}_{\text{com}})) + \left(\mathbf{x}_{\text{com}} \times \sum_i \{m_i \dot{\mathbf{r}}_i\}\right) + \left(\sum_i \{m_i \mathbf{r}_i\} \times \mathbf{v}_{\text{com}}\right)$$

$$+ \left(\sum_i \{m_i \mathbf{r}_i \times \dot{\mathbf{r}}_i\}\right).$$

In as much as $\sum_i m_i \mathbf{r}_i = 0$ (Exercise 1.15), the middle two terms are zero, and hence

$$\mathbf{h} = m(\mathbf{x}_{\text{com}} \times \mathbf{v}_{\text{com}}) + \sum_i (m_i \mathbf{r}_i \times \dot{\mathbf{r}}_i). \tag{1.19}$$

The interpretation of Equation (1.19) is that the angular momentum can be decomposed into two parts, one that is the related to the total mass and center of mass position and velocity, and one that is related to the motion of the body about its center of mass.

The last thing to incorporate is the fact that rigid bodies are often modeled as continua, so the sums must be replaced by integrals. For the center of mass, the computation is straightforward,

$$\mathbf{x}_{\text{com}} = \frac{\int_B dm}{\int_B \mathbf{x} dm} = \frac{1}{m} \int_B \mathbf{x} dm,$$

where the subscript B on the integral denotes that the integral is over the volume of the body.

For rigid body rotational motion, the sums in Equation (1.19) must be integrated through the whole body, which requires a fairly general formulation. In this book, we only consider rigid body rotations constrained in a plane. Consider the planar rigid body illustrated in Figure 1.4. A coordinate frame B is affixed to the body, and the angle that the x_b-axis of the body frame makes with respect to the inertial x-axis is θ.

Define the vector ω to be out of the plane with magnitude $\dot{\theta}$. It is left as an exercise (Exercise 1.14) to show that the magnitude of $\mathbf{r}_i \times \dot{\mathbf{r}}_i$ is given by

$$\|\mathbf{r}_i \times \dot{\mathbf{r}}_i\| = \|\omega\| \|\mathbf{r}_i\|^2 \tag{1.20}$$

and the direction is out of the plane. Replacing the sum in Equation (1.19) with an integral and assuming the mass per unit area of the body is given by ρ gives

$$\mathbf{h} = m(\mathbf{x}_{\text{com}} \times \mathbf{v}_{\text{com}}) + \dot{\theta} \int_B \rho \|\mathbf{r}\|^2 dA.$$

The quantity

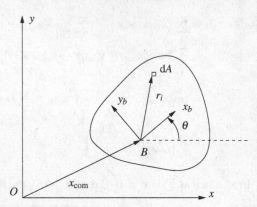

Fig. 1.4 Planar rigid body.

$$I_{\text{com}} = \int_{\mathcal{B}} \rho \, \|\mathbf{r}\|^2 \, dA,$$

is called the *mass moment of inertia* of the body about its center of mass, which finally gives

$$\mathbf{h} = m \left(\mathbf{x}_{\text{com}} \times \mathbf{v}_{\text{com}}\right) + I_{\text{com}} \dot{\theta}. \tag{1.21}$$

The time rate of change of **h** is still given by Equation (1.18), but it may be the case that it is convenient to express the location of the force in terms of the center of mass, which gives

$$\frac{d\mathbf{h}}{dt} = \mathbf{x}_{\text{com}} \times \sum_i \mathbf{F}_i + \sum_i \left(\mathbf{r}_i \times \mathbf{F}_i\right), \tag{1.22}$$

where the sum over i is over the points of application of the forces $\mathbf{F}_i$.

1.5.2.3 Examples of the Application of Newton's Laws

Most of the use of Newton's laws in this book are concerned with the special case of *rectilinear motion*, which is motion along a straight line. This case is nice because the equations of motion reduce to a scalar differential equation and the application of Newton's law is simply to write $F = ma$ in the relevant direction. The student is cautioned to be cognizant of how restrictive this case actually is and to exercise care in applying Newton's law in the appropriate form, in particular Equation (1.11), when the motion is not necessarily rectilinear.

Example 1.22. Consider a particle of mass m constrained to move along the x-axis and subjected to an applied force $\mathbf{F}(t)$ as is illustrated in Figure 1.5. The force $\mathbf{F}(t)$ may have both a magnitude and orientation that changes with time. Assume that the constraint is frictionless.

Let

$$\mathbf{x}(t) = \begin{bmatrix} x(t) \\ y(t) \end{bmatrix} \quad \text{and} \quad \mathbf{F}(t) = \begin{bmatrix} F_x(t) \\ F_y(t) \end{bmatrix}$$

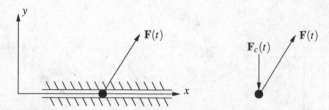

Fig. 1.5 System for Example 1.22.

denote the position of the particle and the two components of the applied force, respectively. A *free body diagram*[10] of the particle is illustrated on the right of Figure 1.5. There are two forces acting on the particle: the applied force $\mathbf{F}(t)$ and some unknown constraint force $\mathbf{F}_c(t)$. The constraint is frictionless, thus $\mathbf{F}_c(t)$ must be purely in the y-direction with no component in the x-direction, so we may write

$$\mathbf{F}_c(t) = \begin{bmatrix} 0 \\ -F_c(t) \end{bmatrix}.$$

For this particle, Equation (1.12) is of the form

$$m\frac{d^2\mathbf{x}}{dt^2}(t) = \mathbf{F}(t) + \mathbf{F}_c(t), \implies m\frac{d^2}{dt^2}\begin{bmatrix} x(t) \\ y(t) \end{bmatrix} = \begin{bmatrix} F_x(t) \\ F_y(t) \end{bmatrix} + \begin{bmatrix} 0 \\ -F_c(t) \end{bmatrix} = \begin{bmatrix} F_x(t) \\ F_y(t) - F_c(t) \end{bmatrix}$$

which is equivalent to the two scalar equations

$$m\ddot{x}(t) = F_x(t) \tag{1.23}$$
$$m\ddot{y}(t) = F_y(t) - F_c(t). \tag{1.24}$$

Because the motion is constrained to be only in the x-direction, $\ddot{y} = 0$ and Equation (1.24) reduces to $F_y(t) - F_c(t) = 0$.

Observe that if the point of interest in the problem were only the motion in the x-direction, we could have easily determined Equation (1.23) by only considering the forces in the x-direction. In such a case there is no need to even determine Equation (1.24). This nice form of the equations occurred not only because the direction of motion and constraint force were orthogonal, but because they were in constant directions, which is a consequence of the motion being rectilinear.

The following example illustrates that things become much more complicated when the motion is not rectilinear.

Example 1.23. Consider a particle constrained to move along a curve described by $y = f(x)$ as illustrated in Figure 1.6 subjected to a known external force $\mathbf{F}(t)$. This may be thought of as a bead moving along a curved frictionless wire.

---

[10] A free body diagram is a representation of the particle isolated from the environment wherein all the forces acting on the body are illustrated.

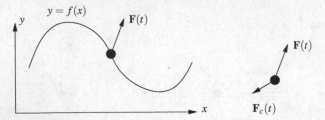

Fig. 1.6 System for Example 1.23.

The particle must obey Newton's second law; hence,

$$m\frac{d^2\mathbf{x}}{dt^2}(t) = \mathbf{F}(t) + \mathbf{F}_c(t), \tag{1.25}$$

where $\mathbf{F}_c(t)$ is the constraint force between the bead and wire. Because the constraint is frictionless, the constraint force must always be perpendicular to the wire.

This seems like two equations for the two components of $\mathbf{x}(t)$. However, those two components are not actually the unknowns. Because the particle must stay on the wire, if we know $x(t)$, then we know $y(t)$ because $y = f(x)$. So, only one of the two components is really not known. The other unknown is the magnitude of $\mathbf{F}_c(t)$. Inasmuch as $\mathbf{F}_c(t)$ must be orthogonal to the wire at the location of the bead, we know its direction at any location $\mathbf{x}(t)$, but not its magnitude. The slope of $y = f(x)$ is given by the derivative at x; then the vector

$$\mathbf{t}(x) = \begin{bmatrix} 1 \\ \frac{df}{dx}\big|_x \end{bmatrix}$$

is in the direction tangent to the curve. Computing a normal vector, $n(x)$ such that $n(x) \cdot t(x) = 0$, we have that

$$\mathbf{n}(x) = \begin{bmatrix} -\frac{df}{dx}\big|_x \\ 1 \end{bmatrix}$$

is normal to the curve at the point x. Hence the constraint force is in the direction of this normal vector, but not necessarily of the same magnitude. It is a vector that has the same direction as the normal, but not the same magnitude; that is,

$$\mathbf{F}_c(t) = F_c \begin{bmatrix} -\frac{df}{dx}\big|_{x(t)} \\ 1 \end{bmatrix}, \tag{1.26}$$

where the scalar F_c scales the constraint force vector to give it the appropriate magnitude. Note that F_c itself is not the magnitude because the vector part in Equation (1.26) does not have unit magnitude, in general.

If we write

$$\mathbf{x}(t) = \begin{bmatrix} x(t) \\ y(t) \end{bmatrix}$$

because $y(t) = f(x(t))$, then the chain rule for differentiation gives

$$\dot{y}(t) = \frac{\mathrm{d}f}{\mathrm{d}x}\bigg|_{x(t)} \dot{x}(t)$$

and then the chain rule and product rule for differentiation give

$$\ddot{y}(t) = \frac{\mathrm{d}^2 f}{\mathrm{d}x^2}\bigg|_{x(t)} \dot{x}^2(t) + \frac{\mathrm{d}f}{\mathrm{d}x}\bigg|_{x(t)} \ddot{x}(t). \tag{1.27}$$

Now we can substitute Equation (1.27) into the second component of Equation (1.25) and substitute Equation (1.26) into Equation (1.25), solve one of them for $F_c(t)$, and substitute into the other to result in a single differential equation in dependent variable x and independent variable t. Substituting Equation (1.26) into Equation (1.25) and writing it in components gives

$$m\ddot{x}(t) = F_x(t) - F_c(t) \frac{\mathrm{d}f}{\mathrm{d}x}\bigg|_{x(t)} \tag{1.28}$$

$$m\ddot{y}(t) = F_y(t) + F_c(t).$$

Hence

$$F_c(t) = m\ddot{y}(t) - F_y(t)$$

and substituting from Equation (1.27) gives

$$F_c(t) = m\left(\frac{\mathrm{d}^2 f}{\mathrm{d}x^2}\bigg|_{x(t)} \dot{x}^2(t) + \frac{\mathrm{d}f}{\mathrm{d}x}\bigg|_{x(t)} \ddot{x}(t) \right) - F_y(t).$$

Finally, substituting for $F_c(t)$ in Equation (1.28) gives

$$m\ddot{x}(t) = F_x(t) - \left[m\left(\frac{\mathrm{d}^2 f}{\mathrm{d}x^2}\bigg|_{x(t)} \dot{x}^2(t) + \frac{\mathrm{d}f}{\mathrm{d}x}\bigg|_{x(t)} \ddot{x}(t) \right) - F_y(t) \right] \left[\frac{\mathrm{d}f}{\mathrm{d}x}\bigg|_{x(t)} \right], \tag{1.29}$$

which is an ordinary second-order nonlinear differential equation with dependent variable x and independent variable t.

The point of Example 1.23 was to demonstrate that the case of nonrectilinear motion is rather involved. There are other ways to approach the problem, which in some cases may be more efficient, but there is basically no way to just be able to write the equations as one equation that is of the form $m\ddot{x} = F$, where F is simply the applied forces projected onto the direction of motion. This is because, in general, the motion of the particle, the applied force, and the constraint force have components in both coordinate directions. Fortunately, for most problems in this book rectilinear motion is what is considered and hence, it is usually relatively easy to write $F = ma$ in the correct direction.

The following example shows the application of Newton's laws to the motion of a particle that is constrained to move in a circle.

Example 1.24. Consider a particle with mass m constrained to move along a frictionless circular hoop with radius r under the influence of gravity, as illustrated in Figure 1.7. Determine a differential equation that describes the motion of the particle using θ as the dependent variable and t as the independent variable.

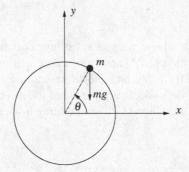

Fig. 1.7 Hoop for Example 1.24.

The particle is constrained to move along the hoop therefore the magnitude of the velocity is

$$\|\mathbf{v}(t)\| = r\left|\dot{\theta}(t)\right|.$$

The angular momentum about the center of the hoop is

$$\mathbf{h}(t) = m\left(\mathbf{r}(t) \times \mathbf{v}(t)\right).$$

We could determine the components of $\mathbf{v}$ and $\mathbf{r}$ as a function of θ. However, it is easier to observe that because of the geometry of the hoop, $\mathbf{v}$ will always be orthogonal to $\mathbf{r}$, and $\mathbf{h}$ about the center of the hoop and will always be orthogonal to the plane of the hoop. Hence, we can let h be the scalar that represents the magnitude of the angular momentum of the particle along the axis through the center of the hoop and orthogonal to and out of the plane of the hoop and write

$$h(t) = mr^2(t)\dot{\theta}(t),$$

and because r is constant,

$$\frac{dh}{dt}(t) = mr^2\ddot{\theta}(t).$$

To determine the motion of the particle, we use Equation (1.15), so we need to compute $\mathbf{r} \times \mathbf{F}$. This cross-product is also along the axis orthogonal to the plane of the hoop, so we may write

$$\|\mathbf{r}(t) \times \mathbf{F}(t)\| = -rmg\cos\theta(t).$$

By Equation (1.15)

$$mr^2\ddot{\theta}(t) = -rmg\cos\theta(t)$$

or

$$\ddot{\theta}(t) = -\frac{g}{r}\cos\theta(t), \tag{1.30}$$

which is a second-order, nonlinear, ordinary differential equation. Equation (1.30) is Equation (1.15) expressed in terms of the variable θ. Because the particle's motion is constrained to move in a circle, determining the equation of motion for the system was more efficiently expressed in a formulation based on angular momentum.

The final example is a planar rigid body that requires the use of the rigid body formulations of Newton's laws.

Example 1.25. This example considers a simple model of a hovercraft. Consider the body of the hovercraft to have uniformly distributed mass and be rectangular, with a length of 2 m and width of 3 m and total mass of 12 kg, as illustrated in Figure 1.8. Assume that the body is subjected to a force $\mathbf{F}$ with magnitude F that is applied a distance $1/2$ m from the back of the hovercraft at the centerline with an angle ψ with respect to the centerline. The purpose of the problem is to determine the equations of motion for the hovercraft.

Because it is common to do so, the dependence of the dependent variables on the independent variable t, are dropped. It is important to be cognizant of which variables depend on time and which are constant.

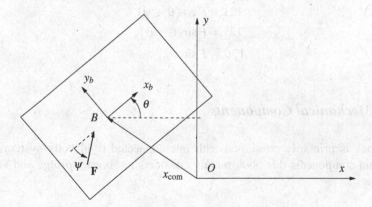

Fig. 1.8 Hovercraft model for Example 1.25.

This is a planar rigid body, so the differential equations describing its motion are given by Equations (1.18), (1.21), and (1.22). For the derivative of the angular velocity we could use either the right equation in Equations (1.18) or Equation (1.22). The quantities in each of these equations are as follows:

$$\mathbf{x}_{com} = \begin{bmatrix} x \\ y \end{bmatrix}, \quad \mathbf{F} = F\begin{bmatrix} \cos(\theta+\psi) \\ \sin(\theta+\psi) \end{bmatrix}, \quad \mathbf{v}_{com} = \begin{bmatrix} \dot{x} \\ \dot{y} \end{bmatrix}, \quad \mathbf{r} = -\frac{1}{2}\begin{bmatrix} \cos\theta \\ \sin\theta \end{bmatrix}.$$

Substituting the relevant terms into Equation (1.21) and computing the cross-products gives $\|\mathbf{h}\|$,

$$\|\mathbf{h}\| = 12\,(x\dot{y} - y\dot{x}) + \frac{13}{2}\dot{\theta},$$

and hence

$$\|\dot{\mathbf{h}}\| = 12\,(x\ddot{y} - y\ddot{x}) + \frac{13}{2}\ddot{\theta}.$$

Computing the right-hand side of Equation (1.22) gives

$$\|\dot{\mathbf{h}}\| = F\,(x\sin(\theta + \psi) - y\cos(\theta + \psi)) - \frac{F}{2}\sin\psi.$$

Equation (1.18) gives

$$12\ddot{x} = F\cos(\theta + \psi)$$
$$12\ddot{y} = F\sin(\theta + \psi)$$

and equating the two $\|\dot{\mathbf{h}}\|$ equations gives

$$12\,(x\ddot{y} - y\ddot{x}) + 13\ddot{\theta} = F\,(x\sin(\theta + \psi) - y\cos(\theta + \psi)) + \frac{F}{2}\sin\psi,$$

and finally substituting for $\ddot{x}$ and $\ddot{y}$ from the first two equations into the third gives what is finally a pretty simple answer:

$$12\ddot{x} = F\cos(\theta + \psi)$$
$$12\ddot{y} = F\sin(\theta + \psi) \tag{1.31}$$
$$13\ddot{\theta} = F\sin\psi.$$

1.5.3 Mechanical Components

This book is primarily concerned with interconnected rigid body systems. The two main components this book mainly considers are linear *springs* and *viscous dampers*.

1.5.3.1 Springs

An ideal linear spring is a mechanical device that requires a force to extend it which is proportional to the amount of extension. Mathematically,

$$f_s(t) = kx(t),$$

where f_s is the force required to extend the length of the spring, x is the amount by which the length of the spring has been extended, and k is the *spring constant*, which is a characteristic of the spring. The force f_s and extension x must be defined in a manner so that they have the same sign when a positive force and positive extension are in the same direction. Negative extension is compression, and the equation still holds. The relationship is illustrated in Figure 1.9 where the unstretched length of the spring is l.

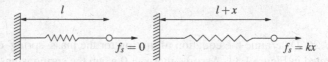

Fig. 1.9 The relationship between force and extension of an ideal spring.

Throughout this book we make an important assumption regarding the reference point from which spring displacements are measured.

Assumption 1.1. *Unless stated otherwise, any variable that represents the extension or compression of a spring is assumed to have a value of zero when the spring is at an equilibrium. If there is no gravity, then the variable is zero when the spring is unstretched. If there is gravity acting on a mass that is supported by the spring, then the variable is zero when the spring is stretched by an amount that results in a force equal to the weight of the mass.*

Another important assumption is that, unless otherwise specified, the mass of the spring itself may be neglected.

1.5.3.2 Viscous Dampers

A viscous damper[11] is a mechanical device that requires a force to extend it which is proportional to the rate at which it is being extended. A common example of such a device is an automobile shock absorber. Mathematically,

$$f_d(t) = b\dot{x}(t),$$

where f_d is the force required to extend the damper, $\dot{x}$ is the rate at which the damper is being extended, and b is the *damper constant*, which is a characteristic of the damper. The force f_d and the rate of extension $\dot{x}$ must be defined in a manner so that they have the same sign when a positive force and positive rate of extension are in the same direction. Negative extension is compression, and the equation still holds. The common schematic representation of a viscous damper as well as the relationship between force and rate of displacement is illustrated in Figure 1.10. Note that for an ideal damper, the force is independent of the length of the damper.

---

[11] Another common term used to refer to these devices is *viscous dashpot*.

Unless otherwise specified, throughout this book we assume that the mass of the damper itself is negligible, and hence may be omitted from any model.

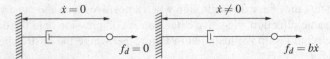

Fig. 1.10 The relationship between force and the rate of extension of an ideal viscous damper.

Example 1.26. Determine the equation of motion for the mass–spring–damper system illustrated in Figure 1.11. Assume that $x = 0$ when the spring is unstretched.

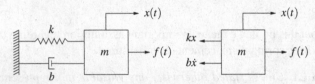

Fig. 1.11 Mechanical system with a mass, spring, and damper and its free body diagram for Example 1.26.

A free body diagram of the mass is illustrated on the right in Figure 1.11. So, because the acceleration of the mass is equal to $\ddot{x}$, we have

$$m\ddot{x}(t) = f(t) - kx(t) - b\dot{x}(t),$$

which is usually expressed in the form

$$m\ddot{x}(t) + b\dot{x}(t) + kx(t) = f(t).$$

1.5.4 Kirchhoff's Laws

Kirchhoff's laws provide the scientific principles that are used to derive the equations describing electrical circuits.

Law 1.4. Kirchhoff's voltage law (KVL) *states that the sum of the voltage drops around any closed loop in a circuit is zero.*

Law 1.5. Kirchhoff's current law (KCL) *states that the sum of the currents into any point in a circuit is zero.*

Law 1.4 is basically conservation of energy and Law 1.5 is basically conservation of charge. These laws apply to circuits containing electrical components, which are defined next.

1.5.5 Electronic Components

There are many types of electronic components, and properly modeling some of them is necessary in this book. In particular, we consider resistors, capacitors, inductors, voltage sources, current sources, direct current motors ("dc motors"), and operational amplifiers ("op-amps").

1.5.5.1 Resistors

In an ideal *resistor*, the voltage drop across the resistor is proportional to the current passing through the resistor. The constant of proportionality is called the *resistance*, is represented by the symbol R and has units of ohms. The equation describing this property is

$$v_R(t) = i(t)R, \tag{1.32}$$

where v_R is the voltage across the resistor, i is the current passing through it, and R is the resistance of the resistor. The typical schematic representation of a resistor is illustrated on the left in Figure 1.12.

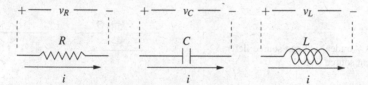

Fig. 1.12 Representation of an ideal resistor, capacitor, and inductor.

1.5.5.2 Capacitors

In an ideal *capacitor*, the time rate of change of the voltage across the capacitor is proportional to the current through it. The constant of proportionality is called the capacitance, is represented by the symbol C, and has units of farads. The equation describing it is given by

$$i(t) = C\frac{dv_C}{dt}(t).$$

This should make sense inasmuch as charge will not flow through the capacitor; the effect of current flow will be the accumulation of charge on the plates of the capacitor, which results in a change in voltage across the plates. The schematic representation of a capacitor is illustrated in the center in Figure 1.12.

1.5.5.3 Inductors

In an ideal *inductor*, the voltage drop across the inductor is proportional to the time rate of change of the current through it. The constant of proportionality is called the inductance, is represented by the symbol L and has units of henrys. The equation governing it is

$$v_L(t) = L\frac{di}{dt}(t).$$

The schematic representation of an inductor is illustrated on the right in Figure 1.12.

1.5.5.4 Voltage Source

An ideal *voltage source* supplies a specified voltage that is independent of the current that the circuit draws. A schematic illustration of an ideal voltage source is illustrated on the left in Figure 1.13. The voltage $v(t)$ is specified, whereas the current through the voltage source i is determined by the circuit to which it is attached. Of course, a real voltage source cannot maintain a specified voltage if it would require a very high current, for example, in a short circuit.

Fig. 1.13 An ideal voltage source (left) and current source (right).

1.5.5.5 Current Source

An ideal *current source* supplies a specified current that is independent of the terminal voltage across the source. Its schematic representation is illustrated on the right in Figure 1.13.

1.5.5.6 Direct Current Motors

The schematic representation for a direct current motor ("dc motor") is illustrated in Figure 1.14. The two idealized properties of a dc motor we need in this book relate the output torque of the motor to the current flowing through it and the voltage drop across the motor to the angular velocity of the shaft of the motor. Mathematically,

$$\tau(t) = k_\tau i(t)$$
$$v_m(t) = k_e \dot{\theta}(t),$$

where i is the current through the motor, τ is the torque produced by the motor, v_m is the voltage drop across the motor, $\dot{\theta}$ is the angular velocity of the shaft of the motor, and k_τ and k_e are the *torque* and *back emf* proportionality constants of the motor.

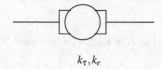

Fig. 1.14 Schematic representation of a direct current motor.

k_τ, k_e

1.5.5.7 Operational Amplifier

An operational amplifier ("op-amp") scales an input voltage difference by an amount called the *gain*.[12] The mathematical description is

$$v_{out}(t) = kv_{in}(t)$$

where v_{in} is the potential difference across the two input pins and k is the *open loop gain*. An ideal op-amp has infinite input impedance, which means that no current flows across the input pins. A schematic representation of an op-amp is illustrated in Figure 1.15.

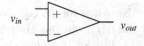

Fig. 1.15 Operational amplifier schematic.

v_{in} v_{out}

The application of these laws is illustrated with a simple example.

Example 1.27. Consider the circuit illustrated in Figure 1.16. Let v_R, v_L, v_C, and v_m denote the voltage across the resistor, inductor, capacitor, and motor, respectively, and let i_1, i_2, and i_3 be the current through the resistor, motor, and inductor, respectively.

The differential equations describing the circuit are determined by using Kirchhoff's laws and the properties of the components in the circuit. Kirchhoff's voltage law around the left loop in the circuit[13] gives

$$v_{in}(t) = v_R(t) + v_L(t) + v_C(t), \tag{1.33}$$

and around the right loop gives

---

[12] Sometimes the gain is specifically called the *open-loop gain* to distinguish it from the closed-loop gain. The closed-loop gain of an op-amp is discussed in Chapter 9.

[13] The signs are not indicated in the figure. A reader that is unsure should make a note of them.

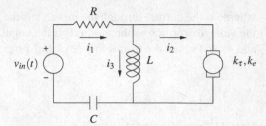

Fig. 1.16 Circuit for Example 1.27.

$$v_L(t) = v_m(t). \tag{1.34}$$

Kirchhoff's current law at the top center node gives

$$i_1(t) = i_2(t) + i_3(t). \tag{1.35}$$

The definitions of the components give

$$v_R(t) = i_1(t)R \tag{1.36}$$

$$v_L(t) = L\frac{di_3}{dt}(t) \tag{1.37}$$

$$C\frac{dv_C}{dt}(t) = i_1(t) \tag{1.38}$$

$$v_m(t) = k_e\frac{d\theta}{dt}(t). \tag{1.39}$$

Note that for the equation for the capacitor, the current through the capacitor must be $i_1(t)$, which is given by the current law applied to the bottom center node.

Finally, if the mass moment of inertia of the shaft of the motor is J, then

$$J\frac{d^2\theta}{dt^2}(t) = k_\tau i_2(t). \tag{1.40}$$

Equations (1.33) through (1.40) are a set of algebraic and differential equations describing the circuit and motor. It makes physical sense that if $v_{in}(t)$ and initial conditions were specified, then unique solutions would exist for all the other variables. Hence, one check on whether these equations completely represent the system is to compare the number of variables to the number of equations. There are nine $(v_{in}, v_R, v_C, v_L, v_m, i_1, i_2, i_3, \theta)$ variables that depend on time, and there are eight equations. Hence, if one is specified, then there should be a solution for all of the rest.[14]

---

[14] Note that in order for consistency between the number of variables and equations to be meaningful, the equations and variables must be independent. For example, v_R could be replaced with $v_R = v_1 + v_2$ everywhere and it would seem there are two variables instead of one. Similarly, a duplicate equation could be added by multiplying one of the equations by a constant. Normally, for simple systems such features can be determined by inspection. In cases where the system is too complicated to determine this by inspection, either linear algebraic approaches such as reducing

The way to combine them to simplify things and reduce the number of variables is not always straightforward because they are not all algebraic equations, so this subject is deferred until Chapter 8. Leaving this as a long list of equations may not be the most satisfying, but is where things stand at this point.

1.6 Introduction to Numerical Methods

Because it is a sad, but true, fact that most differential equations cannot be solved using methods in this book (and any other book, for that matter) methods that use computers to determine approximate solutions are extremely important. This section considers *Euler's method* for solving initial value problems for ordinary differential equations, which is the most basic, and perhaps most common, method to use a computer to determine an approximate solution to a differential equation. Chapter 12 considers more advanced topics on numerical methods including more sophisticated methods for initial value problems for ordinary and partial differential equations.

As should be clear subsequently there are two major shortcomings to resorting to numerical techniques. First, only explicit solutions may be obtained; that is, general solutions that can be used for any initial conditions generally cannot be determined using numerical methods. Therefore, if the initial conditions to a problem change, the entire method must be used again. It is not simply a matter of computing different coefficients within a solution. Secondly, the "answer" is only an approximate answer and is in the form of tabulated data. If a more accurate solution is required, then more computer resources must be allocated to the problem and if an expression of the solution in terms of elementary functions such as sine, cosine, the exponential, and so on, is required, the method is not appropriate. Even with these two caveats, however, numerical methods are extremely useful and commonplace in engineering.

1.6.1 Euler's Method

Consider an ordinary, first-order differential equation of the form

$$\dot{x} = f(x(t),t), \tag{1.41}$$

and assume that either we do not know how to solve it, or we are too lazy to solve it by hand. In order to derive an algorithm to determine an approximate solution, recall the definition of the derivative from calculus

---

the system to echelon form can be applied or problems will arise when trying to solve the system indicating that there is a problem with the set of equations describing the system.

$$\dot{x}(t) = \frac{dx}{dt}(t) = \lim_{\Delta t \to 0}\left(\frac{x(t+\Delta t)-x(t)}{\Delta t}\right). \tag{1.42}$$

Another way to interpret this equation is that, if the limit exists and Δt is small, then

$$\dot{x}(t) \approx \frac{x(t+\Delta t)-x(t)}{\Delta t}.$$

Keep in mind that the typical scenario is that the differential equation is known; that is, $f(x,t)$ in Equation (1.41) is known and the solution, $x(t)$ is unknown. This is in contrast to the usual use of Equation (1.42) in calculus where $x(t)$ is known and the derivative is unknown.

Now, assume that an initial condition is known as well, so that

$$x(t_0) = x_0 \tag{1.43}$$

has been specified. So, what is known is $f(x,t)$ in Equation (1.41), the initial condition in Equation (1.43), and also the definition of the derivative in Equation (1.42). Now, at $t = t_0$, the approximate derivative is given by

$$\dot{x}(t_0) \approx \frac{x(t_0+\Delta t)-x(t_0)}{\Delta t}.$$

For a specified Δt, everything in the preceding equation is known except $x(t_0+\Delta t)$, so it can be solved for $x(t_0+\Delta t)$ as

$$x(t_0+\Delta t) \approx \dot{x}(t_0)\Delta t + x(t_0)$$

or, from Equation (1.41),

$$x(t_0+\Delta t) \approx f(x(t_0),t_0)\Delta t + x(t_0). \tag{1.44}$$

In words, if $x(t_0)$ is known and the differential equation, $\dot{x} = f(x,t)$, is known, then an approximation for $x(t+\Delta t)$ is given by Equation (1.44). Also, given normal convergence properties, it will be the case that as Δt gets smaller, the approximation will be more accurate. The final piece of the puzzle is to note that once $x(t+\Delta t)$ is computed, $x(t+2\Delta t)$ can be computed from Equation (1.44) by substituting the value for $x(t+\Delta t)$ for $x(t_0)$ and $t_0+\Delta t$ for t_0 in the right-hand side of Equation (1.44), that is,

$$x(t+2\Delta t) \approx f(x(t_0+\Delta t),t_0+\Delta t)\Delta t + x(t_0+\Delta t),$$

and by recursion, then

$$\boxed{x(t+n\Delta t) \approx f(x(t_0+(n-1)\Delta t),t_0+(n-1)\Delta t)\Delta t + x(t_0+(n-1)\Delta t).}$$
$$\tag{1.45}$$

Equation (1.45) is "the answer." The algorithm to implement it for a given Δt is called Euler's method, and is as follows.

1. Let $x(t_0) = x_0$ and let $n = 0$.
2. Let $n = n + 1$.
3. Let $x(t + n\Delta t) = f\big(x(t_0 + (n-1)\Delta t), t_0 + (n-1)\Delta t\big)\Delta t + x\big(t_0 + (n-1)\Delta t\big)$.
4. If $n\Delta t$ is less than the time to which the approximate solution is needed, return to step 2.

Example 1.28. Determine an approximate numerical solution to

$$\dot{x} = \sin 2t, \tag{1.46}$$

where $x(0) = 3$. It is useful to compare the approximate solution with an exact solution. The exact solution, which can be verified by differentiating it and substituting into Equation (1.46), is

$$x(t) = \frac{7}{2} - \frac{1}{2}\cos 2t.$$

The method to determine this exact solution is covered subsequently in Section 2.5.1.

In this example, $t_0 = 0$, $x_0 = 3$, and $f(x, t) = \sin 2t$. Picking $\Delta t = 0.5$ (a discussion on how to choose Δt appears subsequently), the first 20 steps of the algorithm are presented in Table 1.6. The last column is the exact solution, which is included for comparison. A plot of the approximate numerical solution and the exact solution are illustrated in Figure 1.17.

t	n	$x(t)$	$f(x,t)$	$x(t+\Delta t)$	$\frac{7}{2} - \frac{1}{2}\cos 2(t+\Delta t)$
0.000000	0	3.000000	0.000000	3.000000	3.229849
0.500000	1	3.000000	0.841471	3.420735	3.708073
1.000000	2	3.420735	0.909297	3.875384	3.994996
1.500000	3	3.875384	0.141120	3.945944	3.826822
2.000000	4	3.945944	-0.756802	3.567543	3.358169
2.500000	5	3.567543	-0.958924	3.088081	3.019915
3.000000	6	3.088081	-0.279415	2.948373	3.123049
3.500000	7	2.948373	0.656987	3.276866	3.572750
4.000000	8	3.276866	0.989358	3.771545	3.955565
4.500000	9	3.771545	0.412118	3.977605	3.919536
5.000000	10	3.977605	-0.544021	3.705594	3.497787
5.500000	11	3.705594	-0.999990	3.205599	3.078073
6.000000	12	3.205599	-0.536573	2.937312	3.046277
6.500000	13	2.937312	0.420167	3.147396	3.431631
7.000000	14	3.147396	0.990607	3.642699	3.879844
7.500000	15	3.642699	0.650288	3.967844	3.978830
8.000000	16	3.967844	-0.287903	3.823892	3.637582
8.500000	17	3.823892	-0.961397	3.343193	3.169842
9.000000	18	3.343193	-0.750987	2.967700	3.005648
9.500000	19	2.967700	0.149877	3.042638	3.295959

Table 1.6 Tabulated data for Example 1.28

Note that the numerical solution is approximate in two ways. First, in between the times $n\Delta t$, the solution can only be interpolated. In Figure 1.17 the interpolation is linear, as is the default in many graphics packages. However, keep in mind that

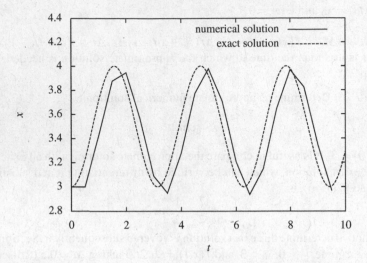

Fig. 1.17 Approximate and exact solutions for Example 1.28 with $\Delta t = 0.5$.

the only data about the solution we really have is at a finite list of points in time. Second, even for the exact times $n\Delta t$, the solution still does not exactly match the exact solution. This is due to the fact that each computation for $x(t_0 + n\Delta t)$ is only an approximation. Thus, the only point where we know the answer is correct is point $t = t_0$.

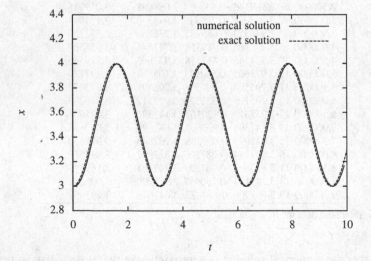

Fig. 1.18 Approximate and exact solutions for Example 1.28 with $\Delta t = 0.1$.

Decreasing the time step to $\Delta t = 0.1$ gives the result illustrated in Figure 1.18. Note that decreasing the step size by a factor of 5 greatly improves the accuracy of the approximate solution. A code listing using the C programming language is included in Appendix E.1.0.1. A code listing using FORTRAN is included in Appendix E.2.0.16.

1.6.2 Determining an Appropriate Step Size

A more detailed and theoretically rigorous analysis of the types of errors introduced by numerical methods is considered in Chapter 12. At this point a heuristic approach is used to ensure accuracy of the approximate solution, which is simply to continue to reduce the step size by a certain factor (say by a factor of two, or perhaps even by an order of magnitude) until the answer seems to have converged to a fixed solution. This is best illustrated by means of an example.

Example 1.29. Find an approximate solution to

$$\dot{x} = 75x(1-x) \tag{1.47}$$

$$x(-1) = \frac{1}{1+e^{75}} \tag{1.48}$$

using Euler's method on the time interval $-1 \leq t \leq 1$. The solution to this problem is simply implementing Euler's method using

$$t_0 = -1$$

$$x_0 = \frac{1}{1+e^{75}}$$

$$f(x,t) = 75x(1-x).$$

Figure 1.19 illustrates the solution for a variety of values for Δt. Note that Δt must be quite small before the solution converges. A code listing using C is included in Appendix E.1.0.2. A code listing using FORTRAN is included in Appendix E.2.0.17.

Euler's method may be represented by the following algorithm.

Algorithm 1.1 (Euler's Method).

1. Let $x(t_0) = x_0$ and $n = 0$.
2. Let $n = n + 1$.
3. Let $x(t + n\Delta t) = f(x(t_0 + (n-1)\Delta t), t_0 + (n-1)\Delta t)\Delta t + x(t_0 + (n-1)\Delta t)$.
4. If $n\Delta t$ is less than the time to which the approximate solution is needed, return to step 2.
5. If only one approximate solution has been obtained, then reduce Δt and return to step 1.

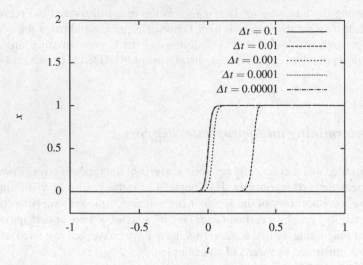

Fig. 1.19 Approximate solutions for Example 1.29 using various Δt values.

6. *Compare the solution obtained to the previous one and if the solutions differ significantly, then reduce Δt and return to step 1.*

1.6.3 Numerical Methods for Higher-Order Differential Equations

The development so far is limited to ordinary, first-order differential equations. This section extends the approach to higher-order ordinary differential equations by using a straightforward reformulation of the problem to convert it into a system of first-order equations, which is illustrated by means of an example.

Example 1.30. Find an approximate numerical solution to

$$\ddot{x} + \sin(t)\dot{x} + \cos(t)x = e^{-5t} \tag{1.49}$$

where $x(0) = 2$ and $\dot{x}(0) = 5$.

The main idea is the following. Consider the change of variables

$$x_1(t) = x(t)$$
$$x_2(t) = \dot{x}(t).$$

Then the following equations are equivalent

$$\ddot{x} + \sin(t)\dot{x} + \cos(t)x = e^{-5t} \quad \Longleftrightarrow \quad \begin{bmatrix} \dot{x}_1 \\ \dot{x}_2 \end{bmatrix} = \begin{bmatrix} x_2 \\ e^{-5t} - \sin(t)x_2 - \cos(t)x_1 \end{bmatrix}.$$
$$\tag{1.50}$$

This is because the second line of the right-hand equation is determined by solving Equation (1.49) for $\ddot{x}$ and recognizing that $\ddot{x} = \dot{x}_2$ because $x_2 = \dot{x}$. The initial value problems are equivalent as well if $x_1(0) = 2$ and $x_2(0) = 5$.

Observe that in general terms, the right-hand formulation of Equation (1.50) is of the form

$$\dot{x}_1 = f_1(x_1, x_2, t)$$
$$\dot{x}_2 = f_2(x_1, x_2, t).$$

Hence, for this case Euler's method, expressed in Equation (1.45), has the simple reformulation of

$$x_1(t + n\Delta t) \approx f_1\big(x_1(t + (n-1)\Delta t), x_2(t + (n-1)\Delta t), t + (n-1)\Delta t\big)\Delta t$$
$$+ x_1(t + (n-1)\Delta t)$$
$$x_2(t + n\Delta t) \approx f_2\big(x_1(t + (n-1)\Delta t), x_2(t + (n-1)\Delta t), t + (n-1)\Delta t\big)\Delta t$$
$$+ x_2(t + (n-1)\Delta t),$$

or using the particular equations of this example

$$x_1(t + x\Delta t) \approx x_2\big(t + (n-1)\Delta t\big)\Delta t + x_1\big(t + (n-1)\Delta t\big)$$
$$x_2(t + x\Delta t) \approx \Big[e^{-5(t+(n-1)\Delta t)} - \sin(t + (n-1)\Delta t)x_2(t + (n-1)\Delta t)$$
$$- \cos(t + (n-1)\Delta t)x_1(t + (n-1)\Delta t)\Big]\Delta t$$
$$+ x_2\big(t + (n-1)\Delta t\big).$$

Inasmuch as this is notationally a bit cumbersome, it may be easier to refer to the example code in the appendix. A code listing using the C programming language is included in Appendix E.1.0.3. A code listing using FORTRAN is included in Appendix E.2.0.18. A plot of the solution for $\Delta t = 0.02$ and $\Delta t = 0.01$ is illustrated in Figure 1.20.

1.6.4 Numerical Packages

The ode series[15] of functions in MATLAB provide the basic functionality for solving initial value problems for ordinary differential equations. Perhaps the most common of these is ode45(), the usage of which is outlined here. This function used the fourth-order Runge–Kutta method, the details of which are included in Chapter 12.

---

[15] The functions include, ode43(), ode23(), ode113(), ode15s(), ode23s(), ode23t(), ode23tb(), ode15i() which provide functionality using a variety of solution methods applicable to a variety of differential equations.

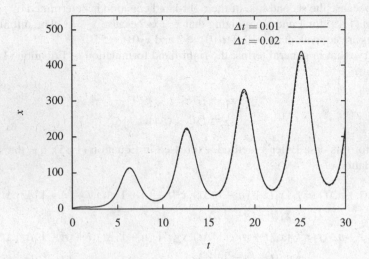

Fig. 1.20 Solution for Example 1.30.

The reader is cautioned that although such computer software packages are very common and used widely, they are essentially "black boxes" and a user must exercise caution when relying upon the output. It is, frankly, rather easy to find differential equations where these functions fail to produce an accurate approximation.

The basic usage of ode45() is

```
>> [T,Y] = ODE45(ODEFUN,TSPAN,Y0,OPTIONS)
```

where

T

is the time vector,

Y

is the solution vector (or matrix),

ODEFUN

is a function that provides the derivative information (the right-hand side of the equation),

Y0

is the initial condition, and

OPTIONS

is a list of optional parameters sent to the solver. The following example illustrates its basic use.

Example 1.31. Use MATLAB to determine an approximate numerical solution to the set of equations from Example 1.30.

The file "secondorder.m" contains the following.

```
function xdot = secondorder(t,x)
  xdot = zeros(2,1);
  xdot(1) = x(2);
  xdot(2) = exp(-5.0*t) - sin(t)*x(2) - cos(t)*x(1);
```

and in the command window

```
>> [t,y] = ode45(@secondorder,[0 30],[2 5]);
>> plot(t,y(:,1));
>> xlabel('t');
>> ylabel('x(t)');
```

The output of these commands would be very similar to the solution plotted in Figure 1.20.

Octave[16] is free software that has many features similar to MATLAB. The main function for computing approximate solutions for ordinary differential equations in Octave is lsode(). Similar to MATLAB, a user must exercise caution to not necessarily assume the answer provided is a good approximation.

The basic usage is

```
y = lsode("f",x0,t)
```

where

```
f
```

is a function (which must be defined in a file with the same name),

```
x0
```

is the initial condition and

```
t
```

is a time vector. The following example illustrates its use.

Example 1.32. Use Octave to determine a solution to the set of equations from example 1.30.

The file "secondorder.m" contains the following.

```
function xdot = secondorder(x,t)
  xdot = zeros(2,1);
  xdot(1) = x(2);
  xdot(2) = exp(-5.0*t) - sin(t)*x(2) - cos(t)*x(1);
endfunction
```

---

[16] For more information visit http://www.octave.org.

Within the Octave command line interface, the steps

```
octave:1> t = linspace(0,30,10000);
octave:2> y = lsode("secondorder",[2;5],t);
octave:3> plot(t,y(:,1),';;')
octave:4> xlabel('t')
octave:5> ylabel('x(t)')
```

produce a solution vector called y. The output of these commands would be very similar to the solution plotted in Figure 1.20.

1.7 Exercises

1.1. Let $\mathcal{A} = (-10, 10)$ and $\mathcal{B} = \mathbb{R}$. For $x \in \mathcal{A}$ and $y \in \mathcal{B}$ which of the following equations define a function from $\mathcal{A}$ to $\mathcal{B}$?

1. $y = 2x$.
2. $y = 1/x$.
3. $y = \sin x$.
4. $x = \sin y$.
5. $y = \sin^{-1} x$.
6. $y = 0$.
7. $y = \begin{cases} 0, & x > 0 \\ -1, & x < 0. \\ -24, & x = 0 \end{cases}$
8. $y = (x - 2)/(x - 6)$.

For each of the equations that you determined were not functions, is it possible to alter the domain or range to make them functions?

1.2. Let $\mathcal{C}$ be the set composed of some vehicles in a parking lot and let $\mathcal{M} = \mathbb{R}$. If you had a means to determine the mass of each automobile, would that define a function from $\mathcal{C}$ to $\mathcal{M}$? Would it necessarily define a function from $\mathcal{M}$ to $\mathcal{C}$? Explain your answer.

1.3. Let $\mathcal{D} = \mathcal{R} = \mathbb{R}$. For $x, y \in \mathbb{R}$, does $y = 1/x$ define a function? Does it define a function from $\mathcal{R}$ to $\mathcal{D}$?

1.4. Use the limit definition of the derivative from Definition 1.4 to compute the derivative with respect to t of the following functions.

1. $f(t) = t^2$.
2. $f(t) = \sin t$. *Hint:* You may use the fact that $\lim_{t \to 0} \sin(t)/t = 1$.

1.5. The *ideal gas law* relates the pressure, volume, temperature, and amount of a gas by the well-known equation

$$pV = nRT.$$

Apply the implicit function theorem to this equation to determine the domain of values of V so that for a fixed temperature T there exists a function which gives the pressure for any value of the volume in that domain. Does the result from the implicit function theorem make physical sense?

1.6. For each of the following equations, use implicit differentiation to determine dx/dt and the intervals of values for t so that the equation $f(x,t) = 0$ defines x as an implicit function of t.

1. $t^2 + 5t + 4x^2 + 3x + 21 = 0$.
2. $\sin(2t - 4x) = 12x$.
3. $\sqrt{6x - 3t} = t + 4x$.
4. $e^{2t+3x} = 5$.
5. $\left(x^2 + t^2\right)^2 - x^2 + t^2 = 0$.

On the (t,x)-plane, plot the values of x and t that satisfy the equation. You may use a computer program for this. Then sketch $x(t)$ versus t by hand.

1.7. Theorem 1.1 says that $y(x)$ is a unique function when the conditions of the theorem are satisfied. However, for the circle example, $x^2 + y^2 = c$, for any value of $x < \sqrt{c}$ there are *two* values of y on the circle, given in Example 1.5, which would seem to indicate that function is not unique. Explain how this is consistent with the theorem.

1.8. Classify each of the following differential equations according to whether it is

- Ordinary or partial
- Linear or nonlinear

If it is linear, indicate whether it is

- Constant- or variable-coefficient
- homogeneous or inhomogeneous.

Also determine

- Its order
- The dependent and independent variables

1. $5\ddot{x} + 6\dot{x} + \sin(t)x = \cos(t^2), x(0) = 1, \dot{x}(0) = \pi$.
2. $5\ddot{x} + 6\dot{x} + \sin(t)x = \cos(t), x(0) = 1, \dot{x}(0) = \pi$.
3. $5\ddot{x} + 6\dot{x} + \sin(t)x = \cos(x), x(0) = 1, \dot{x}(0) = \pi$.
4. $\cos(t)\dot{x} + e^t x = x^2, x(2) = e$.
5. $\cos(x)\dot{x} + e^t x = x^2, x(2) = e$.
6. $\cos(t)\dot{x} + e^t x = x, x(7) = e$.
7. $\cos(t)\cos(\dot{x}) + e^t x = x, x(7) = e$.
8. $\cos(t)\dot{x} + e^t x = 2, x(-6) = e$.
9. $\dot{x} + e^\pi x = 2, x(0) = 1$.

10. $\dot{x} + e^t x = 2, x(0) = 1.$

11. $\dot{x} + e^x x = 2, x(0) = 1.$

12. $2\ddot{x} + 19\dot{x} + 24x = 0, x(0) = 1, \dot{x}(0) = 0.$

13. $2\frac{\partial^2 \zeta}{\partial \gamma^2} + 19\frac{\partial \zeta}{\partial \alpha} + 24\zeta = \gamma^2 + \alpha^2.$

14. $6\ddot{x} + 23\dot{x} + t^3 x^2 = 0, x(0) = 1, \dot{x}(0) = 0.$

15. $6\ddot{x} + 23\dot{x} + x^3 = \sin(t^2), x(0) = 1, \dot{x}(0) = 0.$

16. $2\frac{d^2 \xi}{d\eta^2} + 19\frac{d\xi}{d\eta} + 25\xi = 0, \xi(0) = 1, \frac{d\xi}{d\eta}(0) = 0.$

17. $\pi\ddot{x} + e\dot{x} + x = \sin(t), x(0) = 1, \dot{x}(0) = 0.$

18. $2\frac{d^2 \zeta}{d\gamma^2} + 19\frac{d\zeta}{d\gamma} + \gamma 24\zeta = 0, \zeta(0) = 1, \frac{d\zeta}{d\gamma}(0) = 0.$

19. $2\frac{d^2 \zeta}{d\gamma^2} + 19\frac{d\zeta}{d\gamma} + \gamma 24\zeta = \sin(\gamma), \zeta(0) = 1, \frac{d\zeta}{d\gamma}(0) = 0.$

20. $2\frac{d^2 \zeta}{d\gamma^2} + 19\frac{d\zeta}{d\gamma} + \gamma 24\zeta = \sin(\zeta), \zeta(0) = 1, \frac{d\zeta}{d\gamma}(0) = 0.$

21. $2\frac{d^2 \zeta}{d\gamma^2} + 19\frac{d\zeta}{d\gamma} + 24\zeta = \gamma, \zeta(0) = 1, \dot{\zeta}(0) = 0.$

1.9. Is the partial differential equation

$$\frac{\partial u}{\partial t}(x,t) = \alpha \frac{\partial^2 u}{\partial x^2}(x,t) - \beta u(x,t),$$

where α and β are real constants, elliptic, parabolic or hyperbolic?

1.10. Is the *telegraph equation*

$$\frac{\partial^2 u}{\partial t^2}(x,t) = \alpha \frac{\partial^2 u}{\partial x^2}(x,t) - \beta \frac{\partial u}{\partial t}(x,t) - \gamma u(x,t),$$

where α, β, and γ are real positive constants, elliptic, parabolic or hyperbolic?

When modeling a telegraph, $u(x,t)$ represents the voltage at time t at location x in a wire. What are the units for α, β, and γ?

1.11. Is the *interior Dirichlet problem* for a circle

$$\frac{\partial^2 u}{\partial r^2}(r,\theta) + \frac{1}{r}\frac{\partial u}{\partial r}(r,\theta) + \frac{1}{r^2}\frac{\partial^2 u}{\partial \theta^2}(r,\theta) = 0$$

elliptic, parabolic, or hyperbolic?

1.12. Determine whether each of the given functions, $x(t)$ is a solution to the associated differential equation.

1. $x(t) = \cos(t), \ddot{x} + x = 0.$
2. $x(t) = \sin(t), \ddot{x} + x = 0.$
3. $x(t) = 3\cos(t) + 2\sin(t), \ddot{x} + x = 0.$
4. $x(t) = \cos 2t, \ddot{x} + x = 0.$
5. $x(t) = c_1 t + c_2/t, t^2\ddot{x} + t\dot{x} - x = 0, t \neq 0.$
6. $x(t) = 3t + 2/t, t^2\ddot{x} + t\dot{x} - x = 0, t \neq 0.$

7. $x(t) = t + 1/t + t/2 \ln t, \; t^2\ddot{x} + t\dot{x} - x = t, \; t \neq 0.$
8. $x(t) = \cos 2t, \; x^2\ddot{x} + \dot{x}/x = -4\cos^3(2t) - 2\tan 2t.$
9. $x(t) = 3\cos 2t, \; x^2\ddot{x} + \dot{x}/x = -4\cos^3(2t) - 2\tan 2t.$
10. $x(t) = t + 1/t + 5t/2 \ln t, \; t^2\ddot{x} + t\dot{x} - x = t, \; t \neq 0.$

1.13. Consider three particles that are free to move in the plane. Let the positions of the particles be

$$\mathbf{x}_1 = \begin{bmatrix} 0 \\ 2 \end{bmatrix}, \quad \mathbf{x}_2 = \begin{bmatrix} 2 \\ 3 \end{bmatrix}, \quad \mathbf{x}_3 = \begin{bmatrix} -3 \\ 1 \end{bmatrix},$$

and the velocities be

$$\mathbf{v}_1 = \begin{bmatrix} 1 \\ 1 \end{bmatrix}, \quad \mathbf{v}_2 = \begin{bmatrix} -1 \\ -1 \end{bmatrix}, \quad \mathbf{v}_3 = \begin{bmatrix} 2 \\ 0 \end{bmatrix}.$$

1. Compute the linear momentum of the system.
2. Compute the angular momentum of the system about the origin.
3. Compute the time rate of change of the linear momentum of the system if a force

$$\mathbf{F}_1 = \begin{bmatrix} 3 \\ -2 \end{bmatrix}$$

   is applied to the first particle.
4. What is the time rate of change of the linear momentum if the force is applied to the second particle? Does changing the particle to which the force is applied matter for $d\mathbf{p}/dt$?
5. Compute the time rate of change of the angular momentum about the origin of the system if $\mathbf{F}_1$ is applied to the first particle.
6. What is the time rate of change of the angular momentum about the origin if $\mathbf{F}_1$ is applied to the second particle? Does changing the particle to which the force is applied matter for $d\mathbf{p}/dt$? Explain why this is the same or different from the case for linear momentum when the particle to which the force was applied was changed.
7. Consider a second force

$$\mathbf{F}_2 = \begin{bmatrix} -3 \\ 2 \end{bmatrix}.$$

Let $\mathbf{F}_1$ be applied to the first particle and $\mathbf{F}_2$ be applied to the second particle. What is the time rate of change of the linear momentum of the system? What is the time rate of change of the angular momentum of the system? What can you conclude from the answers?

1.14. Derive Equation (1.20).

1.15. Use the definition of the center of mass of a system of particles to show that $\sum_i m_i \mathbf{r}_i = 0$. Explain why this makes sense.

1.16. Write a computer program that solves for the motion of the particle in Example 1.24 with $m = 1$ and $r = 2$, $\theta(0) = 1$ and $\dot{\theta}(0) = 0$.

1. Plot θ versus t.
2. By considering the total energy for the particle, determine what $\dot{\theta}(0)$ should be so that the particle goes clockwise from $\theta(0) = 1$ and stops at the top. Verify your prediction by changing your initial conditions in the computer program and plotting θ versus t.
3. Is it possible to determine an initial velocity starting from $\theta(0) = 1$ such that the particle moves around the hoop more than one time and then comes to rest at the top? If so, verify this with the computer program.

1.17. Referring to Example 1.23, determine the differential equation with dependent variable x and independent variable t describing the motion of a particle with a mass of 3 kg, constrained to move along the curve $y = \sin(x)$ subjected to an applied force

$$\mathbf{F} = \begin{bmatrix} \cos(t) \\ 3 \end{bmatrix}.$$

Write a computer program to solve this equation starting at $t = 0$ if $x(0) = 0$ and $\dot{x}(0) = 5$. Plot the solution in a manner that clearly illustrates the nature of the solution. Label all the axes and include appropriate units.

1.18. Referring to Example 1.23, determine the differential equation with dependent variable x and independent variable t describing the motion of a particle of mass 3, constrained to move along the curve $y = \cos(x) - x$ where the mass is subjected to gravity in the negative y-direction. Write a computer program to solve this differential equation starting at $t = 0$ with $x(0) = 0$ and $\dot{x}(0) = 1$.

If this is a roller coaster, at what time and position, if any, does the roller coaster leave the track? *Hint:* Keep track of the constraint force and if it ever has a y-component that is negative.

1.19. Consider a body of mass m sliding along an incline plane with angle α under the influence of gravity as is illustrated on the left in Figure 1.21. Assume the contact between the mass and plane is frictionless and determine a differential equation with dependent variable x and independent variable t.

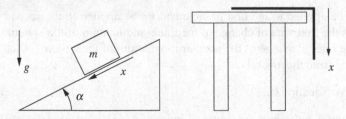

Fig. 1.21 System for Exercises 1.19 (left) 1.20 (right).

1.20. Consider a uniform flexible rope of length l and total mass ρl, as is illustrated on the right in Figure 1.21. Assume that the length of rope hanging off the end of the table is x and that the coefficient of friction (both dynamic and static) between the rope and table is μ. Determine a differential equation with dependent variable x and independent variable t that describes the motion of the rope. What are the units for each term in the equation?

1.21. Consider a mass suspended from a string under the influence of gravity, as illustrated on the left in Figure 1.22. An identical string hangs from the bottom. If a force is applied at the bottom of the bottom string which starts with $F = 0$ and is slowly increased, which string will break first, the top or bottom string? If the force is very rapidly increased, which one will break first? Explain your answers. This problem was adapted from [20]. If you have trouble with the problem, you may want to do Exercise 1.28 first.

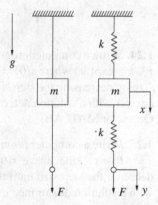

Fig. 1.22 System for Exercise 1.21

1.22. For each of the circuits illustrated in Figure 1.23, determine a set of equations that, if v_{in} were specified, would be sufficient to uniquely determine v_{out}.

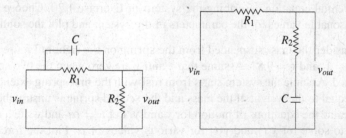

Fig. 1.23 Circuits for Exercise 1.22.

1.23. This problem investigates series and parallel elements in mechanical and electrical systems. Match the following mathematical relationship with the corresponding system in Figure 1.24. Justify your answer by deriving the relationship for each figure based on first principles, such as Newton's laws, Kirchhoff's laws, or the mathematical representation of the properties of the components of the system:

$$f = (k_1 + k_2)x \qquad\qquad f = \frac{1}{1/k_1 + 1/k_2}x$$

$$f = (b_1 + b_2)\dot{x} \qquad\qquad f = \frac{1}{1/b_1 + 1/b_2}\dot{x}$$

$$v = i(R_1 + R_2) \qquad\qquad v = i\frac{1}{1/R_1 + 1/R_2}$$

$$\frac{dv}{dt} = i\frac{1}{C_1 + C_2} \qquad\qquad \frac{dv}{dt} = i\left(\frac{1}{C_1} + \frac{1}{C_2}\right)$$

$$v = \frac{di}{dt}(L_1 + L_2) \qquad\qquad v = \frac{di}{dt}\frac{1}{1/L_1 + 1/L_2}.$$

1.24. Write a computer program to determine an approximate numerical solution to $\dot{x} + x = \exp(3t)$ where $x(0) = 1$ using Euler's method. Determine an appropriate step size by decreasing the step size until the solution seems to converge. Compare your answer with a solution determined using a numerical computation package such as Octave or MATLAB.

1.25. Write a computer program to determine an approximate numerical solution to $\dot{x} = (t^2 - x^2)\sin x$ where $x(0) = -1$ using Euler's method. Be sure to continue to decrease the step size until the solution seems to converge. Compare your answer with a solution determined using MATLAB or Octave.

1.26. Write a computer program to determine an approximate numerical solution to $\ddot{x} + t\dot{x} + 2x = 0$, where $x(0) = 3$ and $\dot{x}(0) = -2$ using Euler's method. Be sure to continue to decrease the step size until the solution seems to converge. Compare your answer with a solution determined using MATLAB or Octave.

1.27. Write a computer program to determine an approximate numerical solution to the differential equation describing the system in Exercise 1.20. Choose your one set of reasonable values for the parameters in the system and plot the solution.

1.28. Consider the mass suspended from the spring on the right in Figure 1.22. Let $m = 1$, $k = 1$, and $g = 9.81$. Assume that x and y are zero when the two springs are unstretched. Assume the system starts from rest with the first spring extended by an amount equal to the weight of the mass and the second spring is unstretched.

Determine the equations of motion for x and y. Let $F = \alpha t$ and write a computer program to solve for $x(t)$ and $y(t)$ for various values of α. Plot the force in each spring versus time for different values of α. Use your results to provide an answer to Exercise 1.21. *Hint:* The force in the top spring is kx and the force in the bottom spring is $k(y - x)$.

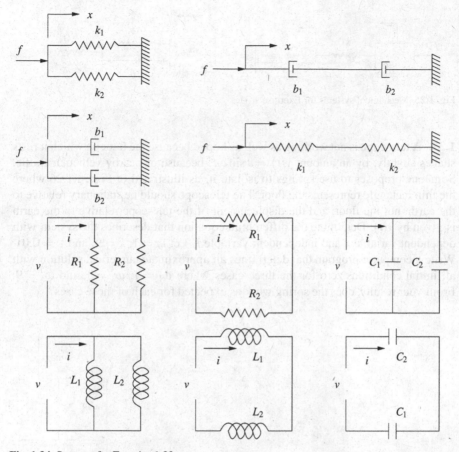

Fig. 1.24 Systems for Exercise 1.23.

1.29. The *exhaust velocity* is the velocity, relative to the engine, at which air and spent fuel leave a jet or rocket. Explain why a rocket can attain a velocity greater than the exhaust velocity whereas a jet can not.

1.30. Use the ode45() function in MATLAB to determine an approximate numerical solution to the initial value problem in Example 1.29. If it does not give an answer that appears to be close to that illustrated in Figure 1.19, investigate the options for ode45() or other MATLAB functions to obtain a good answer.

1.31. Determine the equations of motion for the systems in Figure 1.25. Write a computer program that determines an approximate numerical solution for each where $k_1 = 1$, $k_2 = 2$, $b = 0.25$, $m = 1$, $f = \sin(t)$, $x(0) = 0$, $\dot{x}(0) = 0$, and $y(0) = 0$. Plot x versus t and y versus t starting from $t = 0$ for a sufficient interval of time to illustrate the nature of the solution.

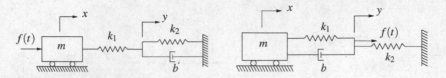

Fig. 1.25 Mechanical systems for Exercise 1.31.

1.32. A telescope is not working as well as hoped because the floor on which it rests shakes slightly, by an amount $y(t) = \varepsilon \sin \omega t$, because of nearby vehicular traffic. Someone proposes to use springs to isolate it, as illustrated in Figure 1.26 where the thin rectangle represents the floor. The telescope should be stationary relative to the earth, not the floor, and the displacement of the telescope relative to the earth is given by $x(t)$. Determine the differential equation that describes the system with dependent variable x and independent variable t. Let $m = 1$, $k = 9$, and $\varepsilon = 0.01$. Write a computer program that determines an approximate numerical solution with all initial conditions zero for the three cases where $\omega_1 = 1$, $\omega_2 = 3$, and $\omega_3 = 9$. From your results, does the spring work as expected for each of those cases?

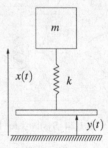

Fig. 1.26 System for Exercise 1.32.

Chapter 2
First-Order Ordinary Differential Equations

First-order ordinary differential equations have some rather special properties, which result for the most part because they can only contain a limited number of terms. In fact, all linear first-order ordinary differential equations can easily be solved. This is in contrast to higher-order ordinary differential equations that become much more difficult to solve when, for example, they contain variable coefficients. Furthermore, some methods exist to solve a pretty large class of nonlinear first-order ordinary differential equations, which is generally not the case for higher-order nonlinear equations.

This chapter considers methods to solve first-order ordinary differential equations of the form

$$\frac{\mathrm{d}x}{\mathrm{d}t}(t) = f(x,t). \tag{2.1}$$

If the first-order ordinary differential equation that must be solved is not of the form of Equation (2.1) it must be transformed into that form, and in the process care must be taken that solutions are neither gained nor lost. Theorem D.1 gives conditions for existence and uniqueness of solutions to problems involving equations in the form of Equation (2.1). For the purposes of this chapter, note that as long as $f(x,t)$ is infinitely differentiable in both x and t, then the theorem is satisfied. All the equations considered in this chapter have the properties necessary for existence and uniqueness of solutions.

Even if a reader is familiar with ordinary first-order differential equations, this chapter should not be skipped. Detailed coverage of the method of undetermined coefficients and solutions to constant-coefficient linear equations of any order is in this chapter and serves as a basis for material in subsequent chapters.

2.1 Motivational Examples

The first example of a first-order differential equation comes from heat transfer.

B. Goodwine, *Engineering Differential Equations: Theory and Applications*,
DOI 10.1007/978-1-4419-7919-3_2, © Springer Science+Business Media, LLC 2011

Example 2.1. Consider the problem of determining the temperature of an object placed in an oven (or conversely, a refrigerator). If the inside of the oven is at temperature T_a, and is constant, and the initial temperature of the body is $T(0)$, we want to determine $T(t)$.

Although a complete exposition of heat transfer requires an entire course, a couple of relevant concepts can be introduced here. First, temperature can be considered as a measure of the amount of thermal energy that a body contains. Second, heat transfer, then, is a measure of how much energy is transferred between systems in a given amount of time. Let q denote the rate of heat transfer. The units for q are energy per unit time J/s or watts W.

Considering an energy balance on the body, we have that the rate of change of the internal energy of the body must be equal to the rate of energy transfer into (or out of) the body from the surrounding air. A basic result from heat transfer is that the heat transfer from a surrounding fluid to a body is given by

$$q(t) = hA\left(T_a - T(t)\right), \tag{2.2}$$

where A is the surface area of the body and h is the *convection heat transfer coefficient* which will have units of $W \cdot K/m^2$. Equation (2.2) should make perfect sense. The rate at which energy is transferred from the body to the fluid, or vice versa is proportional to the difference in their temperatures and the amount of area over which it may occur.

Inasmuch as temperature is a measure of the amount of thermal energy contained in the body, the rate of change of temperature should be proportional to the rate at which energy is transferred into the body. This is true, and in particular,

$$q(t) = \rho V c_p \frac{dT}{dt}(t), \tag{2.3}$$

where ρ is the density of the body, V is the volume, and c_p is the *specific heat* of the material, which has units of $J \cdot K/kg$.

Conservation of energy requires that the rate of heat transfer into the body must equal the rate of change of its internal energy, thus Equations (2.2) and (2.3) must be equal, so we have

$$hA\left(T_a - T(t)\right) = \rho V c_p \dot{T}(t).$$

If we let $\theta(t) = T(t) - T_a$, then

$$-hA\theta(t) = \rho V c \dot{\theta}(t)$$

or

$$\dot{\theta}(t) + \frac{hA}{\rho V c}\theta(t) = 0. \tag{2.4}$$

Usually, this equation is written in the form

$$\dot{\theta}(t) + \frac{1}{RC}\theta(t) = 0, \tag{2.5}$$

where R is the resistance to convective heat transfer and C is called the lumped thermal capacitance.[1] Equation (2.5) is a linear, first-order, ordinary, constant-coefficient, homogeneous differential equation.

The next section outlines how to solve various forms of first-order equations. As it turns out, there are multiple ways to solve Equation (2.5), and in particular, the two different methods from Section 2.3 may be used to solve this problem.

The next examples come from the field of bioengineering, but first we need to consider some basic reaction rate concepts. The *Michaelis–Menton equation* describes many physiological processes: among other things, biological process catalyzed by enzymes and protein facilitated diffusion of substances into or out of cells. The form of the equation is

$$v_o = \frac{v_{\max}[s]}{k_m + [s]} \tag{2.6}$$

where v_o is the reaction rate or uptake rate, $[s]$ is the concentration of some substrate, and $v_{\max}$ and k_m are constants that depend upon the particular process under consideration. A plot of v_o versus $[s]$ for various values of $v_{\max}$ and k_m is illustrated in Figures 2.1 and 2.2.

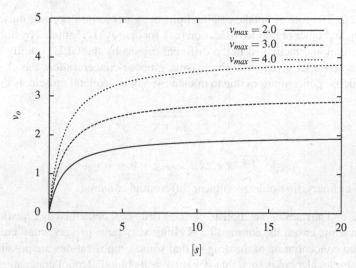

Fig. 2.1 Reaction rate for various $v_{\max}$ and $k_m = 1.0$.

---

[1] A careful reader, or one with a background in heat transfer, will recognize that fact that when we use $T(t)$ to represent the temperature of the body, it is implicitly assuming that the temperature distribution in the body is uniform. This is intuitively appropriate in some cases, and is rigorously justified by considering the *Boit number*, which is defined as the dimensionless quantity $Bi = hk/L$, where k is the *thermal conductivity* of the body and L is a characteristic length of the body. When $Bi \ll 1$, then the approach taken in this example problem, which is called the *lumped capacitance method*, is a justified approximation. See [25] for a complete exposition.

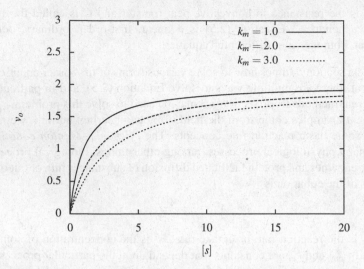

Fig. 2.2 Reaction rate for various k_m and $v_{max} = 2.0$.

Example 2.2. The rate of uptake of blood plasma glucose into skeletal muscle, the brain, liver, and other organs for oxidation (use for energy) is regulated by hormones such as insulin and facilitated in the different organs by the GLUT family of proteins. Thus if we let g represent the plasma glucose concentration, the change in plasma glucose concentrations due to uptake by, say in, skeletal muscle, is given by

$$\dot{g} = -\frac{v_{max}g}{k_m + g}$$

or

$$k_m\dot{g} + g\dot{g} + v_{max}g = 0. \tag{2.7}$$

This is an ordinary, first-order, nonlinear differential equation.

Example 2.3. The rates of metabolism of many drugs are described by Equation (2.6) as well. In some cases, the constant k_m is either very large or very small compared to the blood concentration of the drug so that some simplifications are possible.

For example, alcohol is such that if x represents blood alcohol concentrations, x is always much larger than k_m. In this case the denominator of Equation (2.6) can be approximated by $k_m + x \approx x$, and then the equation describing the blood alcohol concentration as a function of time is

$$\dot{x} = -v_{max}. \tag{2.8}$$

This is an ordinary, first-order, constant-coefficient, inhomogeneous, linear differential equation.

Example 2.4. For other drugs, cocaine is an illicit example but there are many pharmaceutical examples; metabolism is such that the constant k_m is very large compared to the drug concentration levels. In that case, the denominator of Equation (2.6) can be approximated simply by k_m ($k_m + x \approx k_m$) and the blood drug concentration as a function of time is given by

$$k_m \dot{x} = -v_{\max} x \tag{2.9}$$

which is an ordinary, first-order, constant-coefficient, homogeneous, linear differential equation.

2.2 Homogeneous Constant-Coefficient Linear First-Order Ordinary Differential Equations

Because it is the case that the coefficients of the dependent variable terms in engineering differential equations are often parameters that describe the physical properties of a system, and it is also often the case that such parameters are constant (mass, thermal capacitance, etc.), it is thus often the case that differential equations in engineering have constant coefficients. This section presents a method to solve ordinary, first-order, constant-coefficient, linear differential equations.

The following fact regarding ordinary, constant-coefficient, linear, homogeneous differential equations of any order and worthy of repeated emphasis.

If you remember anything from differential equations, remember the following: ordinary, linear, constant-coefficient, homogeneous differential equations of any order have exponential solutions. To emphasize the fact, let us make it a theorem.

Theorem 2.1. *Ordinary, linear, constant-coefficient, homogeneous differential equations with dependent variable x and independent variable t have solutions of the form $x = ce^{\lambda t}$ where c and λ are constants.*

Proof. Consider an nth-order, ordinary, linear, constant-coefficient, homogeneous differential equation of the form

$$\alpha_n \frac{d^n x}{dt^n} + \alpha_{n-1} \frac{d^{n-1} x}{dt^{n-1}} + \cdots + \alpha_1 \frac{dx}{dt} + \alpha_0 x = 0. \tag{2.10}$$

To verify the form of the solution, substitute $x = ce^{\lambda t}$ into Equation (2.10):

$$\alpha_n \lambda^n ce^{\lambda t} + \alpha_{n-1} \lambda^{n-1} ce^{\lambda t} + \cdots + \alpha_1 \lambda ce^{\lambda t} + \alpha_0 e^{\lambda t} = 0.$$

Note that $c = 0$ results in $x(t) = 0$, which is a solution to Equation (2.10), but only satisfies the initial value problem where $x(0) = 0$. For the case where $c \neq 0$, because $ce^{\lambda t}$ is never zero, it is legitimate to divide each side of the equation by it which gives

$$\alpha_n \lambda^n + \alpha_{n-1} \lambda^{n-1} + \cdots + \alpha_1 \lambda + \alpha_0 = 0 \tag{2.11}$$

which is an nth-order polynomial in λ. Because, by the fundamental theorem of algebra, Equation (2.11) has n solutions, there may be, in fact, up to n different solutions of the form $x = e^{\lambda t}$. □

Remark 2.1. The fact bears repeating: ordinary, linear, constant-coefficient, homogeneous differential equations of any order have exponential solutions.

Armed with this knowledge, we now consider solutions to ordinary, first-order, constant-coefficient, linear differential equations. This provides the general solution to Equation (2.9) (because it is already homogeneous) as well as the homogeneous solution to Equations (2.5) and (2.8). It does nothing for us for Equation (2.7) because it is not linear.

Any ordinary, first-order, homogeneous, linear, constant-coefficient differential equation can be written as

$$\dot{x} + \alpha x = 0.$$

Note that there is no restriction that α may not be zero. Assuming a solution of the form

$$x(t) = ce^{\lambda t}$$

and substituting gives

$$\lambda = -\alpha$$

or

$$x(t) = ce^{-\alpha t}. \tag{2.12}$$

This is the general solution because, by Theorem D.1, the solution is unique for a given initial condition, and the constant c in Equation (2.12) may be used to satisfy any initial condition.

Example 2.5. Returning to Example 2.4, we have a general solution of the form

$$x(t) = ce^{-v_{max}t/k_m}. \tag{2.13}$$

To determine c, we would have to know the initial blood concentration of the drug. Assuming $x(0) = x_0$, substituting $t = 0$ into Equation (2.13) gives $c = x_0$, so the solution to

$$k_m\dot{x} = -v_{max}x$$

with $x(0) = x_0$, is

$$x(t) = x_0 e^{-v_{max}t/k_m}.$$

Remark 2.2. It is worth memorizing that the solution to

$$\dot{x} + \alpha x = 0,$$

where $x(0) = x_0$ is

$$\boxed{x(t) = x_0 e^{-\alpha t}.} \tag{2.14}$$

2.3 Inhomogeneous Constant-Coefficient Linear First-Order Ordinary Differential Equations

Now we consider the same case as in the previous section but where the equation is inhomogeneous. Two solution methods are presented. The first is easier, but only works when the inhomogeneous term is in a particular class of functions, and the second is computationally a bit harder, but will always work. Both approaches require that a homogeneous solution be known, so the first order of business is to determine the homogeneous solution in the form of Equation (2.12) (not in the form of Equation (2.14)) as outlined in the previous section.

2.3.1 Undetermined Coefficients

The idea behind undetermined coefficients is relatively simple, as is illustrated by the following example. The approach has two components. First homogeneous and particular solutions are determined separately and then combined for the solution (this is mathematically justified after the example). Second, a specific form of the particular solution is assumed, which is then substituted into the differential equation which gives rise to equations for some undetermined coefficients in the particular solution.

Example 2.6. Solve

$$\dot{x} + 3x = \sin 2t \qquad (2.15)$$

where $x(0) = 1$. This is an ordinary, first-order, linear, constant-coefficient, inhomogeneous differential equation. From Equation (2.12), the homogeneous solution (the solution to $\dot{x} + 3x = 0$) is

$$x_h(t) = ce^{-3t},$$

where c is an arbitrary real number.

To determine the particular solution, consider the following logic. We seek a function $x(t)$ such that if we take its derivative and add it to three times itself we obtain the function $\sin 2t$. A moment's reflection results in the conclusion that the only sorts of functions that can be combined with their derivative to obtain a sine function are sines and cosines that are a function of the same argument. So, it is logical to assume that the particular solution is of the form

$$x_p(t) = c_1 \cos 2t + c_2 \sin 2t,$$

where c_1 and c_2 are coefficients that are yet to be determined, that is, the undetermined coefficients. The manner in which to compute the undetermined coefficients should be obvious: substitute x_p into the differential equation to see if equations for c_1 and c_2 can be derived. So, because $\dot{x}_p(t) = -2c_1 \sin 2t + 2c_2 \cos 2t$, and substituting gives

$$\dot{x} + 3x = (-2c_1 \sin 2t + 2c_2 \cos 2t) + 3(c_1 \cos 2t + c_2 \sin 2t)$$
$$= (2c_2 + 3c_1) \cos 2t + (-2c_1 + 3c_2) \sin 2t$$
$$= \sin 2t,$$

where the last $\sin 2t$ term is the inhomogeneous term from Equation (2.15). The second and third lines of the above equation must be true for all time, therefore

$$3c_1 + 2c_2 = 0$$
$$-2c_1 + 3c_2 = 1,$$

which gives $c_1 = -2/13$ and $c_2 = 3/13$, so the particular solution is

$$x_p(t) = -\frac{2}{13} \cos 2t + \frac{3}{13} \sin 2t.$$

The final task is to ensure that the initial condition is satisfied; that is, $x(0) = 1$. Note the following facts.

1. The particular solution satisfies Equation (2.15) but does not satisfy the initial condition.
2. The homogeneous solution does not satisfy the differential equation in Equation (2.15), but does have a coefficient that has not yet been specified which perhaps may be used in some way to satisfy the initial condition.

Now observe that inasmuch as x_h is a homogeneous solution, by definition when it is substituted into Equation (2.15) the result will be zero. So, *because the equation is linear*, it may be added to the particular solution and the sum will still satisfy the differential equation. In particular, using $x = x_h + x_p$ and substituting gives

$$\dot{x} + 3x = (\dot{x}_h + \dot{x}_p) + 3(x_h + x_p)$$
$$= (\dot{x}_h + 3x_h) + (\dot{x}_p + 3x_p)$$
$$= 0 + (\dot{x}_p + 3x_p)$$
$$= \sin 2t.$$

Because $x = x_h + x_p$ satisfies Equation (2.15) and also contains a coefficient that has not yet been specified (the c in x_h), evaluating $x(0)$ and setting it equal to the initial condition gives an equation for c. So,

$$x(0) = x_h(0) + x_p(0)$$
$$= c - \frac{2}{13}.$$

The initial condition was $x(0) = 1$, thus $c = 15/13$ and the solution to the differential equation is

$$x(t) = \frac{15}{13} e^{-3t} - \frac{2}{13} \cos 2t + \frac{3}{13} \sin 2t.$$

At first glance, the main idea behind the undetermined coefficients approach may seem to be simply educated guesswork. However, the method is actually guaranteed to work if the right conditions are met. Insight into the method is obtained by observing that certain functions have only a finite number of linearly independent derivatives.[2]

Example 2.7. Returning to Example 2.6, we computed that if

$$x(t) = c_1 \cos 2t + c_2 \sin 2t,$$

then

$$\dot{x}(t) + 3x(t) = (3c_1 + 2c_2) \cos 2t + (-2c_1 + 3c_2) \sin 2t.$$

The critical observation is that we started with a function of the form

$$x(t) = c_1 \cos 2t + c_2 \sin 2t,$$

and after substituting it into the differential equation obtained a function of the form

$$x(t) = k_1 \cos 2t + k_2 \sin 2t.$$

Specifically, a linear combination of the function $x(t)$ and its derivative, which results when it is substituted into the differential equation, is exactly the same form as the original function albeit with different coefficients.

As the following theorem shows that if the inhomogeneous term $g(t)$ is such that it only has a finite number of linearly independent derivatives, then, assuming a solution that is a linear combination of $g(t)$ and its derivatives will always lead to a set of equations that will give a solution for the undetermined coefficients. First we need to define what it means for functions to be linearly independent.

Definition 2.1. A set of functions, $\{f_1(t), \ldots, f_n(t)\}$ is *linearly dependent* on an interval $\mathcal{I} = (t_0, t_1)$ if there exists a set of constants, $c_1, \ldots, c_n$ that are not all zero such that

$$c_1 f_1(t) + c_2 f_2(t) + \cdots + c_n f_n(t) = 0, \quad t \in \mathcal{I}. \tag{2.16}$$

If the functions are not linearly dependent, then they are *linearly independent*.

A necessary condition for linear dependence is easy to construct. Differentiating Equation (2.16) $n - 1$ times gives the system of algebraic equations

---

[2] By specifying "only a finite number" of linearly independent derivatives, we mean that the largest set of derivatives of the function that is linearly independent does not contain an infinite number of elements.

$$c_1 f_1(t) + c_2 f_2(t) + \cdots + c_n f_n(t) = 0$$

$$c_1 \frac{df_1}{dt}(t) + c_2 \frac{df_2}{dt}(t) + \cdots + c_n \frac{df_n}{dt}(t) = 0$$

$$c_1 \frac{d^2 f_1}{dt^2}(t) + c_2 \frac{d^2 f_2}{dt^2}(t) + \cdots + c_n \frac{d^2 f_n}{dt^2}(t) = 0$$

$$\vdots = \vdots$$

$$c_1 \frac{d^{n-1} f_1}{dt^{n-1}}(t) + c_2 \frac{d^{n-1} f_2}{dt^{n-1}}(t) + \cdots + c_n \frac{d^{n-1} f_n}{dt^{n-1}}(t) = 0$$

which may be written as

$$\begin{bmatrix} f_1(t) & f_2(t) & \cdots & f_n(t) \\ \frac{df_1}{dt}(t) & \frac{df_2}{dt}(t) & \cdots & \frac{df_n}{dt}(t) \\ \frac{d^2 f_1}{dt^2}(t) & \frac{d^2 f_2}{dt^2}(t) & \cdots & \frac{d^2 f_n}{dt^2}(t) \\ \vdots & \vdots & \ddots & \vdots \\ \frac{d^{n-1} f_1}{dt^{n-1}}(t) & \frac{d^{n-1} f_2}{dt^{n-1}}(t) & \cdots & \frac{d^{n-1} f_n}{dt^{n-1}}(t) \end{bmatrix} \begin{bmatrix} c_1 \\ c_2 \\ c_3 \\ \vdots \\ c_n \end{bmatrix} = \begin{bmatrix} 0 \\ 0 \\ 0 \\ \vdots \\ 0 \end{bmatrix}.$$

From basic linear algebra, in order for this to have a nonzero solution, the determinant of the matrix must be zero. Hence, if the determinant is nonzero, then the set of functions is not linearly dependent; that is, the set of functions is linearly independent. This determinant is called the *Wronskian*.

Definition 2.2. Given n functions, $f_1(t), f_2(t), \ldots, f_n(t)$ define the *Wronskian, W* as the following determinant

$$W(f_1, f_2, \ldots, f_n)(t) = \begin{vmatrix} f_1(t) & f_2(t) & \cdots & f_n(t) \\ \frac{df_1}{dt}(t) & \frac{df_2}{dt}(t) & \cdots & \frac{df_n}{dt}(t) \\ \vdots & \vdots & \ddots & \vdots \\ \frac{d^{n-1} f_1}{dt^{n-1}}(t) & \frac{d^{n-1} f_2}{dt^{n-1}}(t) & \cdots & \frac{d^{n-1} f_n}{dt^{n-1}}(t) \end{vmatrix}.$$

Be careful about the logic. If the Wronskian is nonzero, then the set of functions is linearly independent. If the Wronskian is zero, it does not necessarily mean that the set of functions is linearly dependent. As is emphasized in the first few examples, the main utility of knowing that a set of functions is linearly independent is that it justifies equating coefficients of the functions on either side of an equality.

Theorem 2.2. *An nth-order, linear, ordinary, constant-coefficient, inhomogeneous differential equation of the form*

$$\alpha_n \frac{d^n x}{dt^n}(t) + \alpha_{n-1} \frac{d^{n-1} x}{dt^{n-1}}(t) + \cdots + \alpha_1 \frac{dx}{dt}(t) + \alpha_0 x(t) = g(t), \qquad (2.17)$$

where $g(t)$ has only a finite number of linearly independent derivatives has a particular solution of the form

$$x_p(t) = c_0 g(t) + c_1 \frac{\mathrm{d}g}{\mathrm{d}t}(t) + c_2 \frac{\mathrm{d}^2 g}{\mathrm{d}t^2}(t) + \cdots + c_m \frac{\mathrm{d}^m g}{\mathrm{d}t^m}(t)$$

where m is the maximum number of linearly independent derivatives and there does not exist a combination of coefficients, c_i, where not all the c_i are zero such that $x_p(t)$ is a homogeneous solution to Equation (2.17).

Proof. Consider the vector space

$$V = \left\{ c_0 g(t) + c_1 \frac{\mathrm{d}g}{\mathrm{d}t}(t) + \cdots + c_m \frac{\mathrm{d}^m g}{\mathrm{d}t^m}(t) \,\middle|\, c_i \in \mathbb{R}, i \in \{1, \ldots, m\} \right\}.$$

The functions $g, \mathrm{d}g/\mathrm{d}t, \ldots, \mathrm{d}^m g/\mathrm{d}t^m$ are the basis elements for V and the operator $\mathrm{d}/\mathrm{d}t$ is a linear operator on V. Consequently

$$\mathrm{D} = \alpha_0 + \alpha_1 \frac{\mathrm{d}}{\mathrm{d}t} + \cdots + \alpha_m \frac{\mathrm{d}^m}{\mathrm{d}t^m}$$

is also a linear operator on V. The null space of D contains only the zero function because by assumption no element of V is a homogeneous solution to Equation (2.17). This implies that the set of functions $\mathrm{D}g, \mathrm{D}(\mathrm{d}g/\mathrm{d}t), \ldots, \mathrm{D}(\mathrm{d}^m g/\mathrm{d}t^m)$ also is a basis for V. Hence,

$$\mathrm{D}x_p(t) = c_0 \mathrm{D}g(t) + \cdots + c_m \mathrm{D}\frac{\mathrm{d}^m g}{\mathrm{d}t^m}(t) = g(t)$$

is satisfied by a unique set of coefficients.   □

In general, computing the derivatives of $g(t)$ and at each time checking whether the set is linearly independent is the general approach. However, the number of functions that are commonly encountered in engineering that have this property is somewhat limited and the functions along with what to assume for a particular solution are listed in Table 2.1.

If the inhomogeneous term, $g(t)$ is	Then assume for $x_p(t)$
$\hat{c}\cos \omega t$	$c_1 \cos \omega t + c_2 \sin \omega t$
$\hat{c}\sin \omega t$	$c_1 \cos \omega t + c_2 \sin \omega t$
$\hat{c}e^{\lambda t}$	$ce^{\lambda t}$
$\hat{c}_n t^n + \cdots + \hat{c}_1 t + \hat{c}_0$	$c_n t^n + \cdots + c_1 t + c_0$
sum of above terms	sum of corresponding terms
product of above terms	product of corresponding terms

Table 2.1 Forms to assume for x_p depending on the inhomogeneous term $g(t)$

Example 2.8. Determine the particular solution to the ordinary, first-order, linear, constant-coefficient, inhomogeneous differential equation

$$3\dot{x} + 6x = 9e^t.$$

Because $g(t) = e^t$ does not have any linearly independent derivatives, assume $x_p(t) = c_0 e^t$. Then $\dot{x}_p(t) = c_1 e^t$ and substituting gives

$$3c_1 e^t + 6c_1 e^t = 9e^t \implies c_1 = 1.$$

Hence

$$x_p(t) = e^t.$$

Example 2.9. Determine the particular solution to

$$\dot{x} - x = \cos t + e^{-t}. \tag{2.18}$$

Computing the first few derivatives of $g(t)$ gives

$$g(t) = \cos t + e^{-t}$$
$$\frac{dg}{dt}(t) = -\sin t - e^{-t}$$
$$\frac{d^2 g}{dt^2}(t) = -\cos t + e^{-t}$$
$$\frac{d^3 g}{dt^3}(t) = \sin t - e^{-t}.$$

Either by computing the Wronskian or by simply observing that the third derivative is equal to -1 times the sum of $g(t)$ and its first two derivatives, shows that the particular solution is of the form

$$x_p(t) = c_0 \left(\cos t + e^{-t}\right) + c_1 \left(-\sin t - e^{-t}\right) + c_2 \left(-\cos t + e^{-t}\right)$$

as long as there is no set of constants which make it a homogeneous solution. Because the homogeneous solution is $x_h(t) = e^t$, no combination of coefficients makes $x_p(t)$ a homogeneous solution.

Substituting x_p into Equation (2.18) gives

$$\left[c_0 \left(-\sin t - e^{-t}\right) + c_1 \left(-\cos t + e^{-t}\right) + c_2 \left(\sin t - e^{-t}\right)\right]$$
$$- \left[c_0 \left(\cos t + e^{-t}\right) + c_1 \left(-\sin t - e^{-t}\right) + c_2 \left(-\cos t + e^{-t}\right)\right] = \cos t + e^{-t},$$

where the first term in square braces is $\dot{x}_p$ and the second term in square braces is x_p. As pointed out, the third term in the $\dot{x}_p$ term can be expressed as a sum of the other terms, so

$$\cos t + e^{-t} = \left[c_0 \left(-\sin t - e^{-t}\right) + c_1 \left(-\cos t + e^{-t}\right)\right.$$
$$\left. + c_2 \left(-\cos t - e^{-t} + \sin t + e^{-t} + \cos t - e^{-t}\right)\right]$$
$$- \left[c_0 \left(\cos t + e^{-t}\right) + c_1 \left(-\sin t - e^{-t}\right) + c_2 \left(-\cos t + e^{-t}\right)\right].$$

Equating the coefficients of $g(t)$, $\dot{g}(t)$, and $\ddot{g}(t)$ gives the system of equations

$$-c_2 - c_0 = 1$$
$$c_0 - c_2 - c_1 = 0$$
$$c_1 - c_2 - c_2 = 0,$$

which gives

$$c_0 = -\frac{3}{4}, \quad c_1 = -\frac{1}{2}, \quad c_2 = -\frac{1}{4}$$

or

$$x_p(t) = -\frac{3}{4}\left(\cos t + e^{-t}\right) - \frac{1}{2}\left(-\sin t - e^{-t}\right) - \frac{1}{4}\left(-\cos t + e^{-t}\right).$$

2.3.2 Complication: When the Assumed Solution Contains a Homogeneous Solution

By the proof of Theorem 2.2, the method is not guaranteed to work if the function $g(t)$ is not in the range of the operator D. This can happen if D is not full rank, which will be the case if there exists a linear combination of $g(t)$ and its derivatives that is a homogeneous solution of Equation (2.17). If there is a homogeneous solution of this form, then the question is whether $g(t)$ is in the range of D, and if it is not, what can be done about it? First, consider an example.

Example 2.10. Use the method of undetermined coefficients to determine the general solution to

$$\dot{x} + 3x = e^{-3t} + \sin 2t.$$

Referring to Table 2.1, it is logical to assume

$$x_p(t) = c_1 e^{-3t} + c_2 \sin 2t + c_3 \cos 2t.$$

Differentiating and substituting gives

$$\left(-3c_1 e^{-3t} + 2c_2 \cos 2t - 2c_3 \sin 2t\right) + 3\left(c_1 e^{-3t} + c_2 \sin 2t + c_3 \cos 2t\right)$$
$$= e^{-3t} + \sin 2t.$$

Equating coefficients of e^{-3t}, $\sin 2t$ and $\cos 2t$, respectively, gives the following set of equations

$$-3c_1 + 3c_1 = 1$$
$$-2c_3 + 3c_2 = 1$$
$$2c_2 + 3c_3 = 0.$$

Note that the first equation is $0 = 1$; that is, there does not exist any c_1 that will satisfy the equations, and hence the assumed form for the particular solution is incorrect.

This problem is due to the fact that e^{-3t} is, in addition to being a component of the inhomogeneous term, a homogeneous solution to the differential equation. When it is substituted into the differential equation it must evaluate to zero, by definition.

To determine a method to deal with this case, first consider

$$\dot{x} + \alpha x = e^{-\alpha t}. \tag{2.19}$$

Table 2.1 would indicate to choose $x_p(t) = ce^{-\alpha t}$; however, this is also the homogeneous solution. Using a technique that actually foreshadows the method of variation of parameters presented subsequently, assume a particular solution of the form

$$x_p(t) = \mu(t)e^{-\alpha t},$$

substitute into Equation (2.19) and use the result to (one hopes) determine $\mu(t)$.[3] Differentiating $x_p(t)$ and substituting gives

$$\left(\dot{\mu}(t)e^{-\alpha t} - \alpha\mu(t)e^{-\alpha t}\right) + \alpha\mu(t)e^{-\alpha t} = e^{-\alpha t},$$

which simplifies to

$$\dot{\mu}(t) = 1$$

or

$$\mu(t) = t + c.$$

Hence,

$$x_p(t) = (t + c)e^{-\alpha t}.$$

Note that inasmuch as the term $ce^{-\alpha t}$ is actually a homogeneous solution, it is not necessary to add it to the particular solution at this stage in the process of determining the solution as it will be added to it subsequently anyway. So the simplest form for the particular solution is

$$x_p(t) = te^{-\alpha t}.$$

Hence, when the assumed form of the particular solution is also the homogeneous solution to the differential equation, the approach is to multiply the assumed form by the independent variable.

Example 2.11. Continuing from Example 2.10, instead of assuming

$$x_p(t) = c_1 e^{-3t} + c_2 \sin 2t + c_3 \cos 2t$$

---

[3] Assuming a solution of the form of an unknown function times a homogeneous solution is somewhat common and should make some sense that it perhaps might work. Given the relationship between a homogeneous solution and the differential equation, it is definitely plausible that it could be used in combination with other functions to generate a different type of solution.

assume

$$x_p(t) = c_1 t e^{-3t} + c_2 \sin 2t + c_3 \cos 2t.$$

Differentiating and substituting gives

$$\left(c_1 e^{-3t} - 3c_1 t e^{-3t} + 2c_2 \cos 2t - 2c_3 \sin 2t\right)$$
$$+ 3\left(c_1 t e^{-3t} + c_2 \sin 2t + c_3 \cos 2t\right) = e^{-3t} + \sin 2t.$$

Collecting terms now gives

$$c_1 = 1$$
$$-2c_3 + 3c_2 = 1$$
$$2c_2 + 3c_3 = 0$$

which has the solution

$$c_1 = 1, \quad c_2 = \frac{3}{13}, \quad c_3 = -\frac{2}{13},$$

and hence

$$x_p(t) = t e^{-3t} + \frac{3}{13} \sin 2t - \frac{2}{13} \cos 2t,$$

and the general solution is

$$x(t) = x_h(t) + x_p(t) = c e^{-3t} + t e^{-3t} + \frac{3}{13} \sin 2t - \frac{2}{13} \cos 2t.$$

The next example illustrates that it may be the case that the homogeneous solution is "hidden" in some linear combination of $g(t)$ and some of its derivatives.

Example 2.12. Consider

$$\dot{x} + x = t e^{-t}. \tag{2.20}$$

Computing the first two derivatives of $g(t)$ gives

$$\frac{dg}{dt} = e^{-t} - t e^{-t}, \quad \frac{d^2 g}{dt^2} = -2e^{-t} + t e^{-t}.$$

By inspection, we may think that the second derivative is possibly a linear combination of the first two. In fact it is because

$$-1g(t) - 2\frac{dg}{dt}(t) = -t e^{-t} - 2e^{-t} + 2t e^{-t} = -2e^{-t} + t e^{-t}.$$

The fact that the set $\{g, dg/dt\}$ is linearly independent is verified by the Wronskian

$$\begin{vmatrix} t e^{-t} & e^{-t} - t e^{-t} \\ e^{-t} - t e^{-t} & -2e^{-t} + t e^{-t} \end{vmatrix} = -e^{-2t} \neq 0.$$

Assuming a solution of the form

$$x_p(t) = c_0 t e^{-t} + c_1 \left(e^{-t} - t e^{-t} \right)$$

gives

$$\dot{x}_p(t) = c_0 \left(e^{-t} - t e^{-t} \right) + c_1 \left(-2 e^{-t} + t e^{-t} \right).$$

Substituting into Equation (2.20) gives

$$\left[c_0 \left(e^{-t} - t e^{-t} \right) + c_1 \left(-2 e^{-t} + t e^{-t} \right) \right] + \left[c_0 t e^{-t} + c_1 \left(e^{-t} - t e^{-t} \right) \right] = (c_0 - c_1) e^{-t},$$

and hence there is no combination of c_0 and c_1 that will allow x_p to satisfy the differential equation because we want this to equal $g(t) = t e^{-t}$. If, instead, the assumed form of the particular solution is multiplied by t,

$$x_p(t) = t \left(c_0 t e^{-t} + c_1 \left(e^{-t} - t e^{-t} \right) \right),$$

then

$$\dot{x}_p(t) = \left(c_0 t e^{-t} + c_1 \left(e^{-t} - t e^{-t} \right) \right) + t \left(c_0 \left(e^{-t} - t e^{-t} \right) + c_1 \left(-2 e^{-t} + t e^{-t} \right) \right).$$

Substituting into the differential equation and simplifying gives

$$\begin{aligned}
\dot{x}_p + x_p = & \left(c_0 t e^{-t} + c_1 \left(e^{-t} - t e^{-t} \right) \right) \\
& + t \left(c_0 \left(e^{-t} - t e^{-t} \right) + c_1 \left(-2 e^{-t} + t e^{-t} \right) \right) \\
& + t \left(c_0 t e^{-t} + c_1 \left(e^{-t} - t e^{-t} \right) \right),
\end{aligned}$$

which, after a tedious bit of work, simplifies to

$$\dot{x}_p + x_p = 2 (c_0 - c_1) t e^{-t} + c_1 e^{-t}.$$

Equating coefficients with

$$g(t) = t e^{-t}$$

gives

$$c_0 = \frac{1}{2}, \quad c_1 = 0.$$

Hence

$$x_p(t) = \frac{1}{2} t^2 e^{-t}.$$

One way to think of multiplying by t if the inhomogeneous term is a homogeneous solution is that because of the way the product rule for differentiation works, it is "plugging the hole" in the assumed form of the particular solution caused by the homogeneous solution being part of it. It may arise that the inhomogeneous term contains some terms that combine to be a homogeneous solution and some terms that do not. In such a case it would be incorrect to multiply the terms that are not part of the homogeneous solution by t. In such a case there are two approaches.

- It will always work to assume

$$x_p(t) = \left(c_1 g(t) + c_2 \frac{dg}{dt}(t) + \cdots c_m \frac{d^m g}{dt}(t) \right)$$
$$+ t \left(d_1 g(t) + d_2 \frac{dg}{dt}(t) + \cdots d_m \frac{d^m g}{dt}(t) \right),$$

even if there is not a homogeneous solution in the first term. In such a case, all the d_i coefficients will be zero. If there are some terms that combine to be homogeneous solutions and some that do not, then some of the c_i and some of the d_i coefficients will be not zero.

- Although the above approach is nice in that it will always work, it is more work to compute all the coefficients. A smarter approach is to try to identify which terms are combining to make a homogeneous solution, and multiply only those by t.

In general, other than being more work, it is not wrong to assume more terms in the particular solution. It will just work out that the coefficients must be zero. If, after substituting an assumed form for the particular solution into the differential equation, it is not possible to determine one or more of the coefficients, it is not possible to solve for them, it generally is due to the fact that they are combining as a homogeneous solution. Clearly, it is advisable to always compute the homogeneous solution first, so that if the homogeneous solution appears explicitly in the assumed form for the particular solution they can be multiplied by the independent variable right away.

The following example illustrates both approaches.

Example 2.13. Determine the general solution to

$$\dot{x} + 3x = e^{-3t} + \sin 2t.$$

According to Table 2.1, we should assume

$$x_p(t) = c_1 e^{-3t} + c_2 \sin 2t + c_3 \cos 2t.$$

It should be apparent that this will not work because the exponential is also a homogeneous solution. So, one approach would be to assume

$$x_p(t) = \left(c_1 e^{-3t} + c_2 \sin 2t + c_3 \cos 2t \right) + t \left(d_1 e^{-3t} + d_2 \sin 2t + d_3 \cos 2t \right).$$

In this case, it will work out that $d_2 = d_3 = 0$ and c_1 will be arbitrary. Because it is a lot of work to deal with six equations and six coefficients, a more insightful assumption for the particular solution would be

$$x_p(t) = c_1 t e^{-3t} + c_2 \sin 2t + c_3 \cos 2t,$$

where only the problematic term is multiplied by t.

2.3.3 Variation of Parameters

This method will always work for linear first-order ordinary differential equations. As long as one is willing to evaluate the integrals required, it will yield the solution.

The idea behind the variation of parameters method is that if a homogeneous solution for a differential equation is known, denoted by x_h, then assume a solution of the form $x(t) = \mu(t)x_h(t)$. Substituting the assumed form of the solution into the differential equation will yield an equation for μ that, if it can be solved, will give the solution. Unlike the method for undetermined coefficients, this method will work for a variable-coefficient equation as well, but this section limits the coverage to the constant-coefficient case. Also unlike the case for undetermined coefficients, no special form of the inhomogeneous term is necessary.

Consider the ordinary, first-order, linear, constant-coefficient, inhomogeneous differential equation

$$\dot{x} + \alpha x = g(t),$$

where $x(t_0) = x_0$. From before, $x_h(t) = ce^{-\alpha t}$. Assume $x(t) = c\mu(t)e^{-\alpha t}$. Substituting into the differential equation gives

$$c\dot{\mu}(t)e^{-\alpha t} - c\mu(t)\alpha e^{-\alpha t} + \alpha c\mu(t)e^{-\alpha t} = c\dot{\mu}(t)e^{-\alpha t} = g(t).$$

Hence

$$\dot{\mu}(t) = \frac{1}{c}e^{\alpha t}g(t)$$

which can be directly integrated. So

$$\mu(t) - \mu(t_0) = \int_{t_0}^{t} \frac{1}{c}e^{\alpha s}g(s)ds$$

or

$$\mu(t) = \int_{t_0}^{t} \frac{1}{c}e^{\alpha s}g(s)ds + \mu(t_0),$$

where $\mu(t_0)$ is arbitrary. So

$$x(t) = \mu(t)ce^{-\alpha t}$$
$$= ce^{-\alpha t}\int_{t_0}^{t} \frac{1}{c}e^{\alpha s}g(s)ds + \mu(t_0)ce^{-\alpha t}$$
$$= e^{-\alpha t}\int_{t_0}^{t} e^{\alpha s}g(s)ds + c_1 e^{-\alpha t},$$

where $c_1 = \mu(t_0)c$. Evaluating $x(t_0)$ gives

$$x(t_0) = e^{-\alpha t}\int_{t_0}^{t_0} e^{\alpha s}g(s)ds + c_1 e^{-\alpha t_0} = c_1 e^{-\alpha t_0} = x_0.$$

Thus $c_1 = x_0 e^{\alpha t_0}$ and

$$x(t) = e^{-\alpha t} \int_{t_0}^{t} e^{\alpha s} g(s) ds + x_0 e^{\alpha t_0} e^{-\alpha t}.$$ (2.21)

Remark 2.3. If the initial condition were not specified and a general solution were desired, the integral in the above method would become an indefinite integral and a constant of integration would be necessary. It is left as an exercise to prove that the general solution to the ordinary, first-order, linear, constant-coefficient, inhomogeneous differential equation

$$\dot{x} + \alpha x = g(t)$$ (2.22)

is

$$x(t) = e^{-\alpha t} \int e^{\alpha t} g(t) dt + c e^{-\alpha t}.$$ (2.23)

2.4 Variable-Coefficient Linear First-Order Ordinary Differential Equations: Variation of Parameters

The same procedure as above may be used in the case of ordinary, first-order, linear, variable-coefficient, differential equations (regardless of whether it is homogeneous or inhomogeneous). Consider the initial value problem

$$\dot{x} + h(t)x = g(t)$$ (2.24)
$$x(t_0) = x_0.$$ (2.25)

The procedure is the same as before: find a homogeneous solution, $x_h(t)$, assume the solution of the form $x(t) = \mu(t)x_h(t)$, substitute to determine an equation for $\mu(t)$, and if possible, solve for $\mu(t)$. The first task is to determine the homogeneous solution, which is not simply $x_h(t) = ce^{\lambda t}$ in the case of a variable-coefficient equation.

First consider the corresponding homogeneous equation

$$\frac{dx_h}{dt}(t) + h(t)x_h(t) = 0.$$

Rearranging gives

$$\frac{1}{x_h(t)} \frac{dx_h}{dt}(t) = -h(t).$$

Integrating each side with respect to t gives

$$\int \frac{1}{x_h(t)} \frac{dx_h}{dt}(t) dt = \int \frac{d}{dt} \left(\ln(x_h(t)) \right) dt = \ln(x_h(t)) + c = -\int h(t) dt.$$

Hence

$$x_h(t) = k e^{-\int h(t) dt},$$ (2.26)

where $k = -e^{-c}$.

Remark 2.4. This procedure to find the homogeneous solution is a special case of the method for separable equations considered subsequently in Section 2.5.1.

Now armed with the homogeneous solution, assume a solution of the form

$$x(t) = \mu(t)x_h(t) = \mu(t)ke^{-\int h(t)dt}.$$

Substituting gives

$$
\begin{aligned}
(\dot{\mu}(t)x_h + \mu(t)\dot{x}_h) + h(t)\left(\mu(t)x_h\right) &= \left(\dot{\mu}(t)ke^{-\int h(t)dt} - \mu(t)h(t)ke^{-\int h(t)dt}\right) \\
&\quad + h(t)\left(\mu(t)ke^{-\int h(t)dt}\right) \\
&= \dot{\mu}(t)ke^{-\int h(t)dt} \\
&= g(t).
\end{aligned}
$$

Hence

$$\dot{\mu}(t) = \frac{1}{k}g(t)e^{\int h(t)dt} \implies \mu(t) = \int\left(\frac{1}{k}g(t)e^{\int h(t)dt}\right)dt + c.$$

and

$$x(t) = \left(\int\left(\frac{1}{k}g(t)e^{\int h(t)dt}\right)dt + c\right)\left(ke^{-\int h(t)dt}\right)$$

or

$$\boxed{x(t) = \left(\int\left(g(t)e^{\int h(t)dt}\right)dt + c\right)\left(e^{-\int h(t)dt}\right).} \tag{2.27}$$

Remark 2.5. Even though the integrals in Equation (2.27) are indefinite, when you are evaluating them you should not include the constants of integration because they were included in the derivation of the solution.

Remark 2.6. Occasionally it is convenient to combine arbitrary constants but not change the name of the variable, as was done in Equation (2.27). The constant k was distributed across both terms in the left side of the equation, so the constant term c is now actually ck; however, because both c and k are arbitrary, it is most convenient just to keep the variable name as c.

At this point it is worth observing that Equation (2.27) is the solution to Equation (2.24). The only possible complication is that sometimes the integrals may not have a closed-form solution, or may simply be difficult to evaluate.

Example 2.14. Determine the general solution to

$$\dot{x} + \frac{3}{t}x = \sin t.$$

This equation is of the form of Equation (2.24), so the general solution is given by Equation (2.27) where $h(t) = 3/t$ and $g(t) = \sin t$. Substituting into the solution gives

$$x(t) = \left(\int \left(g(t) e^{\int h(t) dt} \right) dt + c \right) \left(e^{-\int h(t) dt} \right)$$

$$= \left(\int \left((\sin t) e^{\int \frac{3}{t} dt} \right) dt + c \right) \left(e^{-\int \frac{3}{t} dt} \right)$$

$$= \left(\int \left((\sin t) e^{3 \ln t} \right) dt + c \right) \left(e^{-3 \ln t} \right)$$

$$= \left(\int t^3 \sin t \, dt + c \right) \frac{1}{t^3}$$

$$= \frac{1}{t^3} \left[3 \left(t^2 - 2 \right) \sin t - t \left(t^2 - 6 \right) \cos t + c \right].$$

For most people, completing the last step without an integral table or computer program would probably be quite difficult.

2.5 Ordinary First-Order Nonlinear Differential Equations

Unfortunately, it is generally the case that nonlinear differential equations are difficult to solve and often do not even have solutions that can be expressed in terms of elementary functions. In the case of first-order equations, however, there is one case in which a solution may be obtained, and that case is the so-called exact equation. Before presenting the theory and method of exact equations, the next section presents a simplified special case of exact equations, namely, separable equations.

2.5.1 Separable Equations

A notationally simplistic, yet nonetheless useful, description of the idea behind separable equations is that if it is possible to put all the terms that are a function of the dependent variable on one side of the equation and all the terms that are a function of the independent variable on the other side of the equation the equation is separable. In such a case, both sides may be directly integrated.

Example 2.15. Find the general solution to

$$(x+1)\left(t^2 + 5t + 3\right) = x\dot{x}.$$

This may be rearranged as

$$t^2 + 5t + 3 = \frac{x}{x+1} \frac{dx}{dt}$$

and each side may be integrated with respect to t

$$\int t^2 + 5t + 3dt = \int \frac{x(t)}{x(t)+1} \frac{dx}{dt}(t)dt.$$

Recall from calculus the substitution rule for integration, namely,

$$\int_{t_0}^{t} f(x(s)) \frac{dx}{ds}(s)ds = \int_{x(t_0)}^{x(t)} f(x)dx.$$

Using this fact,

$$\int t^2 + 5t + 3dt = \int \frac{x(t)}{x(t)+1} \frac{dx(t)}{dt} dt = \int \frac{x}{x+1} dx,$$

so

$$\frac{t^3}{3} + \frac{5t^2}{2} + 3t = x(t) - \ln(x(t)+1) + c.$$

Note that one problem is that the solution $x(t)$ may be, as is the case in this example, only determined as an implicit function of the dependent variable.

The preceding example was rather precise and in practice the approach is a bit more informal. In words, the simplest way to approach the problem is to notationally treat $\dot{x}$ as dx/dt and try to manipulate the equation so that all the x terms are on one side of the equation along with the dx term and all the t terms are on the other side with the dt term. This casual use of notational convenience works correctly in this case, however, it is important to recognize that what is actually going on is an integration by substitution on the x side of the equation. Another example illustrates this point and completes the treatment of separable equations. It also illustrates the slight variation in the approach when the problem is an initial value problem rather than finding a general solution, the only difference being that data are now available to make the integrals definite integrals.

Example 2.16. Determine the solution to

$$\dot{x} + \sin(t)x = 0,$$

where $x(1) = 2$. Note this can perhaps be more easily solved by directly using Equation (2.27) with $g(t) = 0$; however, just for the fun of it, this example solves it by recognizing it is separable.

A bit of manipulation gives

$$\frac{dx}{dt} + \sin(t)x = 0 \quad \Longleftrightarrow \quad \frac{dx}{x} = -\sin(t)dt,$$

so

$$\int_{x(t_0)}^{x(t)} \frac{1}{x} dx = -\int_{t_0}^{t} \sin(s)ds$$

or

$$\int_{2}^{x} (t) \frac{1}{x} dx = \int_{1}^{t} \sin(t)dt \quad \Longleftrightarrow \quad \ln x - \ln 2 = \cos t - \cos 1,$$

which gives, upon taking the exponential of each side

$$x(t) = 2e^{\cos t - \cos 1}.$$

2.5.2 Exact Equations

Although actually using it is another matter, the idea behind exact equations is actually quite simple. Consider a function $\psi(x(t), t)$ (as usual, t is the independent variable and x is the dependent variable) and consider the level sets of ψ; namely, $\psi(x(t), t) = c$. Differentiating ψ constrained to the level set with respect to t gives

$$\frac{d\psi}{dt} = \frac{\partial \psi}{\partial x}\frac{dx}{dt} + \frac{\partial \psi}{\partial t} = 0.$$

Note that this is of the form

$$f(x,t)\dot{x} + g(x,t) = 0, \tag{2.28}$$

where f and g are functions of both the independent variable t and the dependent variable x. If a differential equation just so happens to be of the form of Equation (2.28) such that there exists a $\psi(x(t), t)$ such that $\partial \psi/\partial x = f(x,t)$ and $\partial \psi/\partial t = g(x,t)$, then solving Equation (2.28) is simply a matter of determining ψ and setting $\psi(x,t) = c$ for the general solution. The correct value of c is determined from the initial condition in the case of the initial value problem.

Because the order of differentiating the partial derivatives does not matter, that is,

$$\frac{\partial^2 \psi}{\partial x \partial t} = \frac{\partial^2 \psi}{\partial t \partial x}$$

and because

$$\frac{\partial \psi}{\partial x} = f(x,t), \quad \frac{\partial \psi}{\partial t} = g(x,t)$$

the following are equivalent

$$\frac{\partial^2 \psi}{\partial x \partial t} = \frac{\partial^2 \psi}{\partial t \partial x} \quad \Longleftrightarrow \quad \frac{\partial f}{\partial t} = \frac{\partial g}{\partial x}.$$

In other words, this proves the following theorem.

Theorem 2.3. *For the ordinary, first-order differential equation*

$$f(x,t)\dot{x} + g(x,t) = 0, \tag{2.29}$$

if

$$\frac{\partial f}{\partial t} = \frac{\partial g}{\partial x}$$

then there exists a function $\psi(x(t),t)$ such that

$$\frac{\partial \psi}{\partial x} = f(x,t) \quad and \quad \frac{\partial \psi}{\partial t} = g(x,t).$$

The general solution to Equation (2.29) is given implicitly by

$$\psi(x(t),t) = c.$$

So far, so good, but although the theory is nice and tidy, there are still two practical problems. First, the solution is only given implicitly by ψ. Second, we still need to determine a way to find ψ. The first problem is inherent in the method and is unavoidable. The second problem is addressed subsequently. First, we have an example.

Example 2.17. Consider

$$2x\dot{x} = -2t - 1.$$

In this case $f(x,t) = 2x$ and $g(x,t) = 2t + 1$.

$$\frac{\partial f}{\partial t} = 0, \quad \frac{\partial g}{\partial x} = 0,$$

the equation is exact.[4] Note that

$$\psi(x,t) = x^2 + t^2 + t$$

is such that

$$\dot{\psi} = 0 \quad \Longleftrightarrow \quad 2x\dot{x} + 2t + 1 = 0,$$

so

$$x^2 + t^2 + t = c$$

gives the solution $x(t)$ implicitly.

Determining $\psi(x,t)$ is actually rather straightforward. Inasmuch as

$$\frac{\partial \psi}{\partial x} = f(x,t)$$

then

$$\psi(x,t) = \int f(x,t)\mathrm{d}x + h(t),$$

and

$$g(x,t) = \frac{\partial \psi}{\partial t} = \frac{\partial}{\partial t}\left(\int f(x,t)\mathrm{d}x + h(t) \right) = \frac{\partial}{\partial t}\left(\int f(x,t)\mathrm{d}x \right) + \dot{h}(t).$$

Thus,

---

[4] Observe that it is also separable, but that is not the fact we exploit to solve it.

$$h(t) = \int \left(g(x,t) - \frac{\partial}{\partial t} \left(\int f(x,t)\mathrm{d}x \right) \right) \mathrm{d}t$$

and the general solution is given by

$$\psi(x,t) = \int f(x,t)\mathrm{d}x + \int \left(g(x,t) - \frac{\partial}{\partial t} \left(\int f(x,t)\mathrm{d}x \right) \right) \mathrm{d}t = c.$$

2.5.3 Integrating Factors

From the preceding section it may seem that whether a first-order differential equation is exact is simply a matter of luck. In fact, it is possible to convert any first-order equation of the form

$$f(x,t)\dot{x} + g(x,t) = 0 \qquad (2.30)$$

that has a solution of the form

$$\psi(x,t) = c$$

into one that is exact. Unfortunately, doing this generally involves solving a partial differential equation. To see this, if there exists a solution to Equation (2.30) of the form $\psi(x,t) = c$, then, as before

$$\frac{\mathrm{d}\psi}{\mathrm{d}t} = 0 \quad \Longrightarrow \quad \frac{\partial \psi}{\partial x}\dot{x} + \frac{\partial \psi}{\partial t} = 0. \qquad (2.31)$$

In Section 2.5.2 we simply equated $\partial \psi/\partial x$ with f and $\partial \psi/\partial t$ with g, and used the fact that mixed partials were equal as the basis for Theorem 2.3. But this is too restrictive because we could multiply Equation (2.30) by some function $\mu(x,t)$, and as long as it is not zero and is defined, then the solution $x(t)$ is the same. To determine such a function, $\mu(x,t)$, solve both Equation (2.30) and the equation on the right in Equation (2.31) for $\dot{x}$, because x is what ultimately interests us. Doing so and equating them gives

$$\frac{g}{f} = \frac{\frac{\partial \psi}{\partial t}}{\frac{\partial \psi}{\partial x}}$$

or

$$\frac{1}{g(x,t)} \frac{\partial \psi}{\partial t}(x,t) = \frac{1}{f(x,t)} \frac{\partial \psi}{\partial x}(x,t).$$

If we set

$$\mu(x,t) = \frac{1}{g(x,t)} \frac{\partial \psi}{\partial t}(x,t) = \frac{1}{f(x,t)} \frac{\partial \psi}{\partial x}(x,t),$$

then

$$\frac{\partial \psi}{\partial x}(x,t) = \mu(x,t)f(x,t), \qquad \frac{\partial \psi}{\partial t}(x,t) = \mu(x,t)g(x,t),$$

which is exactly what is needed for

$$\mu(x,t)f(x,t)\dot{x}(t) + \mu(x,t)g(x,t) = 0 \qquad (2.32)$$

to be exact. So, even if Equation (2.30) is not exact, Equation (2.32) will be as long as we can find $\mu(x,t)$.

These equations were based on computations using the solution $\psi(x,t)$, which is not known. To find an equation for $\mu(x,t)$, expand the equation it must satisfy from Theorem 2.3, which is

$$\frac{\partial(\mu f)}{\partial t} - \frac{\partial(\mu g)}{\partial x} = 0,$$

which, expanding using the product rule gives the partial differential equation for μ

$$f(x,t)\frac{\partial \mu}{\partial t}(x,t) + \mu(x,t)\frac{\partial f}{\partial t}(x,t) - g(x,t)\frac{\partial \mu}{\partial x}(x,t) - \mu(x,t)\frac{\partial g}{\partial x}(x,t) = 0. \quad (2.33)$$

This, unfortunately, is not easy to solve, except in certain special cases.

The following example illustrates the fact that an integrating factor works to make an equation that is not exact into one that is exact. It does not show, however, how to find the integrating factor.

Example 2.18. Consider

$$\left(\frac{1+\sin t}{x+1}\right)\frac{dx}{dt} + \frac{x\cos t}{x+1} = 0. \qquad (2.34)$$

This is not exact because

$$\frac{\partial f}{\partial t} = \frac{\cos t}{x+1} \neq \frac{\partial g}{\partial x} = \frac{\cos t}{(x+1)^2}.$$

However, if we multiply Equation (2.34) by $\mu = x+1$, then

$$(1+\sin t)\frac{dx}{dt} + x\cos t = 0$$

is exact because

$$\frac{\partial}{\partial t}(1+\sin t) = \cos t$$

and

$$\frac{\partial}{\partial x}(x\cos t) = \cos t.$$

Doing the necessary computations to find the solution gives

$$x\sin t + x = c.$$

Because finding the integrating factor involves solving a partial differential equation, most approaches depend on special cases or iterative guesswork. A good review of the special cases, such as when the integrating factor only depends on x or t but not both, and how to exploit them are given in [44].

2.6 Summary

Ordinary first-order differential equations are solved using the following methods.

- If the equation is linear, constant-coefficient, and homogeneous, then assuming a solution of the form $x(t) = ce^{\lambda t}$ is probably the easiest method.
- If the equation is linear, variable-coefficient, and homogeneous, then using Equation (2.26) is probably the easiest method.
- If the equation is linear, constant-coefficient, and inhomogeneous with an inhomogeneous term of the form given in Table 2.1, then the method of undetermined coefficients outlined in Section 2.3.1 is probably the easiest.
- If the equation is linear, constant-coefficient, and inhomogeneous the method of variation of parameters with a solution given by Equation (2.23) will work. If the inhomogeneous term is not given in Table 2.1 then this is probably the easiest method.
- If the equation is linear, variable-coefficient, and inhomogeneous the method of variation of parameters with a solution given by Equation (2.27) will work.
- If the equation is nonlinear, first check if it is separable; if it is not, then check if it is exact. If it is not exact, attempt to determine an integrating factor.

2.7 Exercises

2.1. Based on Theorem D.1, which of the following differential equations are guaranteed to have solutions that exist and are unique?

1. $\dot{x} = x$ where $x(0) = 0$.
2. $\dot{x}^2 = x$ where $x(0) = 0$.
3. $\dot{x} = \begin{cases} -1, & x \geq 0, \\ 1, & x < 0, \end{cases}$
   where $x(0) = 0$.
4. $\dot{x} = \begin{cases} -1, & t \geq 5, \\ 1, & t < 5, \end{cases}$
   where $x(0) = 0$.
5. $\dot{x} = x^2$ where $x(0) = 0$.
6. $\dot{x} = x^{1/2}$ where $x(0) = 0$.
7. $\dot{x} = -|x|$ where $x(0) = 0$.
8. $\dot{x} = \sqrt{x^2 + 9}$ where $x(0) = 0$.

2.2. Determine the solution to $\dot{x} = \alpha x$ where $x(0) = 1$. On the same graph, sketch the solution for $\alpha = -1$, $\alpha = 0$, and $\alpha = 1$.

2.3. In dead organic matter, the C^{14} isotope decays at a rate proportional to the amount of it that is present. Furthermore, it takes approximately 5600 years for half of the original amount present to decay.

1. If $x(0)$ denotes the amount present when the organism is alive, determine a differential equation that describes the amount of the C^{14} isotope present if $x(t)$ represents the amount present after time t elapses after the organism dies.
2. In contrast to C^{14}, the C^{12} isotope does not decay and the ratio of C^{12} to C^{14} is constant while an organism is alive. Hence, one should be able to compare the ratio of the two isotopes in a dead specimen to that of a live specimen. Determine how many years have elapsed if the ratio of the amount of C^{14} to C^{12} is 30% of the original value.

Do not look up the formula for half-life and exponential decay problems. The point is to derive the equation in order to relate it to the problem, and then to solve it.

2.4. Consider the first-order, linear, variable-coefficient, homogeneous ordinary differential equation

$$\dot{x} + tx = 0.$$

Does assuming a solution of the form $x(t) = e^{\lambda t}$ where λ is a constant work? Why or why not?

2.5. Consider the first-order, nonlinear, ordinary differential equation $\dot{x} + x^2 = 0$. Does assuming a solution of the form $x(t) = e^{\lambda t}$ work? Why or why not?

2.6. As part of a fabrication process, you encounter the following scenario. A vat contains 100 liters of water. In error someone pours 100 grams of a chemical into the vat instead of the correct amount, which is 50 grams. To correct this condition, a stopper is removed from the bottom of the vat allowing 1 liter of the mixture to flow out each minute. At the same time, 1 liter of fresh water per minute is pumped into the vat and the mixture is kept uniform by constant stirring.

1. Show that if $x(t)$ represents the number of grams of chemical in the solution at time t, the equation governing x is

$$\frac{dx}{dt} = -\frac{x}{100},$$

   where $x(0) = 100$. How long will it take for the mixture to contain the desired amount of chemical?
2. Determine the equation governing $x(t)$ if the amount of water in the vat is W liters, the rate at which the mixture flows out is F liters/minute (and the same amount of fresh water is added), and the amount of the chemical initially added is C grams.

2.7. Determine the general solution to Equation (2.5).

1. Determine the temperature of a body for which $R = 1$ and $C = 10$ if it is initially at $100°$ and is plunged into a medium held at a constant temperature of $T_a = 20°$.
2. If two hot objects with equal masses are dropped into the ocean, which will cool faster, the object that has the shape of a sphere or the object that has the shape of a cube? Justify your answer by referring to Equation (2.4).

2.8. The rate by which people are infected by the zombie plague is proportional to the number of people already infected. Let x denote the number of people infected.[5]

1. What is the differential equation describing the number of people infected? Denote the proportionality constant by k. What are the units for k in the differential equation? What is the general solution to this equation? What are the units for k in the solution?
2. If at time $t = 0$ there are 100,000 people infected and at time $t = 1$ (the next day) there are 150,000 people infected, what is the numerical value of k?
3. For the value of k determined in the previous part, if at time $t = 0$, one person is infected, how long will it take for the zombie plague to infect every person on earth?

2.9. Assume that the rate of loss of a volume of a substance, such as dry ice or a moth ball, due to evaporation is proportional to its surface area.

1. If the substance is in the shape of a sphere, determine the differential equation describing the radius of the ball and solve it to find the radius as a function of time.
2. If the substance is in the shape of a cube, determine the differential equation describing the length of an edge of the cube and solve it to find the length of the edge as a function of time.
3. Use the answers from the previous two parts to determine which shape would be better for a given quantity of material if it is desired for it to take as long as possible to evaporate.

2.10. Use undetermined coefficients to determine the general solution to the following first-order ordinary differential equations.

1. $\dot{x} + x = \cos t$.
2. $\dot{x} + x = \cos 2t$.
3. $\dot{x} + x = \cos t + 2\sin t$.
4. $\dot{x} + 5x = \cos t + 2\sin t$.
5. $5\dot{x} + x = \cos t + 2\sin t$.
6. $\dot{x} + 3x = t^2 + 2t + 1$.
7. $\dot{x} + 3x = t^2 + 2t$.
8. $\dot{x} + 3x = 3t^2$.
9. $\dot{x} + 3x = 3t^2 + \cos 2t$.
10. $\dot{x} + 2x = e^{-3t}$.
11. $\dot{x} + 2x = e^{-2t}$.
12. $\dot{x} + 2x = 2e^{-2t}$.
13. $\dot{x} + 2x = 2e^{-2t} + \cos t$.
14. $\dot{x} + 2x = 2e^{-2t} + \cos t + t^3$.

2.11. Show that the set of functions $\{t^0, t^1, t^2, t^3, t^4, \ldots, t^n\}$ is linearly independent.

---

[5] In this problem, let x be a real number and do not restrict it to be an integer.

2.12. From Example 2.13, substitute all three particular solutions into the differential equation to verify the conclusions from that example.

2.13. Determine the general solution to

$$\dot{x} + x/t = \cos 5t.$$

2.14. Use two different methods to determine the general solution to $\dot{x} + x = \sin 5t$. Also, find the solution if $x(0) = 0$.

2.15. Use two different methods to determine the general solution to

$$\dot{x} + 5x = e^{-5t}.$$

Also, find the solution if $x(0) = 1$ and plot the solution versus time for a length of time that is appropriate to demonstrate the qualitative nature of the solution.

2.16. Determine the solution to

$$t\dot{x} + 2x = t^2 - t + 1$$

where $x(1) = 1/2$ and $t > 0$.

2.17. Prove that Equation (2.23) is the solution to Equation (2.22).

2.18. Determine the general solution to

$$\dot{x} + t^2 x = 0$$

using two different methods.

2.19. You are in desperate need to determine (as in make up), by hand, 100 different exact first-order differential equations in less than one hour. What would be a good way to do that? Determine 10 different exact first-order ordinary differential equations using your method.

2.20. Determine the general solution to

$$(2x + 1)\dot{x} = 3t^2.$$

If necessary, you may express the solution as an implicit function.

2.21. Use two different methods to determine the general solution to

$$3t^2 \dot{x} + 6tx + 5 = 0.$$

2.22. Prove that all separable first-order ordinary differential equations are exact. In other words, show that separable first-order differential equations are a special case of exact first-order ordinary differential equations.

2.23. A special type of nonlinear first-order ordinary differential equation that can be converted into one that is separable is called a *homogeneous equation*.[6] A function $f(x,y)$ is *homogeneous of order n* if it can be written $f(x,y) = x^n g(u)$ where $u = y/x$. A first-order ordinary differential equation is homogeneous if it can be written as

$$P(x,y) + Q(x,y)\frac{dx}{dy} = 0,$$

where $P(x,y)$ and $Q(x,y)$ are homogeneous of the same order and hence may be written $P(x,y) = x^n p(u)$ and $Q(x,y) = x^n q(u)$. Because $y = ux$,

$$\frac{dy}{dx} = x\frac{du}{dx} + u$$

then the differential equation is of the form

$$x^n p(u) + x^n q(u)\left(x\frac{du}{dx} + u\right) = 0$$

which is separable, can be solved for u as an implicit function of x, and then $u = y/x$ may be substituted to obtain the answer. Alternatively, the substitution $u = x/y$ may be used and the variable x eliminated. Which is better for any particular problem requires some experience or trial and error.

Verify that each of the following equations is homogeneous, and use this fact to solve them. Some of the integrals may be tricky, so resorting to a table or symbolic mathematics computer package may be necessary.

1. Show that $y^3 + 2x^2 y = c$ implicitly defines the general solution $y(x)$ to $2xy + (x^2 + y^2)\,dy/dx = 0$ by using the substitution $u = x/y$ and eliminating x.
2. Show that $\tan^{-1}(y/x) - 1/2\ln(x^2 + y^2) = c$ is the general solution to $(x+y) + (y-x)\,dy/dx = 0$ by using the substitution $u = y/x$.

2.24. An interesting class of nonlinear first-order ordinary differential equations arises in so-called *trajectory problems*. Given a family of curves, the problem is to find an orthogonal family of curves that are orthogonal at every point to the family of curves. If $y_f(x)$ is the given family of curves, then the orthogonal family will have a slope of $-1/y'_f(x)$ at any point $(x, y_f(x))$. If we denote the orthogonal family by $y_o(x)$, then the orthogonality condition requires

$$\frac{dy_f}{dx}(x)\frac{dy_o}{dx}(x) = -1.$$

For example, consider the family of curves given by $y_f(x) = cx^5$, which is illustrated in Figure 2.3 for various values of c. The slope is given by $dy_f/dx(x) = 5cx^4$. Hence, for a given point (x,y), the value of c is $c = y/x^5$ so the slope at a given point (x,y) is given by $dy_f/dx = 5y/x$. So, finally, a curve orthogonal to $y_f(x)$ at the point (x,y) satisfies

---

[6] This is a homogeneous equation, not to be confused with a homogeneous solution.

$$\frac{dy_o}{dx}(x) = -\frac{1}{5}\frac{x}{y}.$$

Determine the general solution to this equation. Plot it for various values of the parameter that appears in the general solution and also plot $y_f(x)$ for various values of c. Are the curves orthogonal at every point? *Hint:* The orthogonal curves should be ellipses.

1. Repeat this problem for $y_f(x) = cx^4$.
2. Repeat this problem for $y_f(x) = cx$.

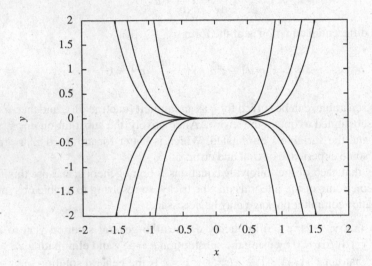

Fig. 2.3 Family of curves for Exercise 2.24.

2.25. For each of the following first-order, ordinary differential equations, indicate

- If the equation is linear
- If the equation is not linear but it is separable
- If the equation is neither linear nor separable but it is exact
- If the equation is neither linear, separable nor exact.

1. $\dot{x} + 5tx = \cosh t$.
2. $\dot{x}\sqrt{x^3}\cos t + 35t^2 = t$.
3. $\dot{x}(2t^2x + t) + 2tx^2 = -x$.
4. $\dot{x} + x = \cosh t$.
5. $\dot{x}(1 - x^2) = t$.
6. $(2tx^2 + 2x) = -\dot{x}(2t^2x + 2t)$.
7. $(2tx^2 + 2x) = \dot{x}(2t^2x + 2t)$.
8. $t^2x^3 + t(1 + x^2)\dot{x} = 0$.

9. $\dot{x}(\cos t + t) - x\sin t + x = 0$.
10. $\dot{x} + (\cos t)x = 0$.

2.26. Determine the general solution to each of the differential equations in Exercise 2.25 using the first method from the list that is applicable. If it is not linear and not exact, then determine an approximate numerical solution. You may choose your own initial conditions in such a case. Be careful not to pick an initial condition that is a singularity, for example, using $x = 0$ or $t = 0$ if that causes a term in the equation to be undefined.

2.27. Show that

$$\left(2t^2 + 3x\sin^2 t\right)\frac{dx}{dt} + 2x\left(t + x\sin t \cos t\right) = 0$$

is not exact, but when multiplied by $\mu(x,t) = x$, it is exact. Find the solution. Leaving the solution in implicit form is fine.

2.28. Show that

$$\left(2x^2t^2 + 3x^3\sin^2 t\right)\frac{dx}{dt} + \left(2x^3t + 2x^4\sin t \cos t\right) = 0$$

is not exact, but when multiplied by $\mu(x,t) = 1/x$, it is exact. Find the solution. Leaving the solution in implicit form is fine.

2.29. One special case where it is possible to determine an integrating factor is when it only depends on t. In such a case, Equation (2.33) reduces to

$$f(x,t)\frac{d\mu}{dt}(t) + \mu(t)\frac{\partial f}{\partial t}(x,t) - \mu(t)\frac{\partial g}{\partial x}(x,t) = 0$$

which gives

$$\frac{1}{\mu(t)}\frac{d\mu}{dt}(t) = \frac{\frac{\partial g}{\partial x}(x,t) - \frac{\partial f}{\partial t}(x,t)}{f(x,t)}. \tag{2.35}$$

1. Show that if we additionally have that the right-hand side of Equation (2.35) is only a function of t, then $\mu(t)$ is given by

$$\mu(t) = \exp\left(\int \frac{\frac{\partial g}{\partial x}(x) - \frac{\partial f}{\partial t}(x)}{f(x)}dx\right).$$

2. Show that $(e^t - \sin x) + \cos x(dx/dt) = 0$ is not exact, but that the above method to determine an integrating factor applies. Use that to find the solution.

2.30. For each of the first-order differential equations listed in Problem 1.8, determine which, if any, of the following solution methods apply based upon what has been covered in this book so far.

1. Assuming exponential solutions

2. Undetermined coefficients
3. Variation of parameters
4. Using the fact that the equation is separable
5. Using the fact that the equation is exact
6. Determining an approximate numerical solution

It may be the case that no method, one method, or more than one method may apply.

2.31. For relatively high velocities[7] the drag due to the motion of the body through air is proportional to the square of the velocity of the body. Hence, if the direction of positive velocity is down, Newton's law on the body can be represented as

$$m\frac{dv}{dt}(t) = mg - kv^2(t). \tag{2.36}$$

1. If a body falls from a sufficiently high altitude, it will reach its *terminal velocity* which is the velocity at which it will stop accelerating. Determine an expression for the terminal velocity, v_{term} from Equation (2.36).
2. Determine the general solution to Equation (2.36).
3. For re-entry of a 5000 kg space vehicle into the atmosphere, it is determined experimentally that the terminal velocity at low altitude is 300 kilometers per hour. If the velocity at time $t = 0$ is 600 kilometers per hour, determine $v(t)$ of the vehicle and plot it versus time. Plot the deceleration (g-force) experienced by the payload versus time.[8]

2.32. You work for the ACME parachute company. A person and parachute weigh 192 lb. Assume that a safe landing velocity is 16 ft/sec and that air resistance is proportional to the square of the velocity, equaling 1/2 lb for each square foot of cross-sectional area of the parachute when it is moving at 20 ft/sec. What must the cross-sectional area of the parachute be in order for the paratrooper to land safely?[9]

---

[7] More precisely, for certain geometries with Reynolds' numbers between approximately 10^3 and 10^5 [52].

[8] In this problem you are using a constant value for k. For a real vehicle re-entering the atmosphere, due to variation in the density of the atmosphere, k varies significantly.

[9] This problem is adapted from [50].

Chapter 3
Second-Order Linear Constant-Coefficient Ordinary Differential Equations

This chapter considers ordinary, linear, second-order differential equations and the main focus is on constant-coefficient equations of the form

$$\frac{d^2x}{dt^2}(t) + \alpha_1 \frac{dx}{dt}(t) + \alpha_0 x(t) = g(t). \tag{3.1}$$

A general solution to Equation (3.1) is constructed by a linear combination of two linearly independent homogeneous solutions and one particular solution. In order to know that the constructed solution is in fact the general solution, we need a theorem concerning the existence and uniqueness of solutions for these equations.

Theorem 3.1. *If the function $g(t)$ is continuous on an open interval $t \in (t_1, t_2)$, then there exists one and only one function $x(t)$ satisfying Equation (3.1) and the given initial conditions*

$$x(t_0) = x_0, \quad \frac{dx}{dt}(t_0) = \dot{x}_0.$$

This theorem[1] is important because it allows us to know when we have computed a general solution, which is useful because if we can do that, we know we have every possible solution to the differential equation. If we conclude something based on an analysis of a solution to an equation, we want to be guaranteed that it represents all the possible behaviors of the system.

3.1 Introduction

Second-order equations arise quite frequently in engineering. In mechanical and aerospace engineering, in particular, they arise often in the context of the study of vibrations. First let us consider a few prototypical example problems to help motivate the importance of second-order ordinary differential equations as well as to illustrate their apparent importance.

---

[1] For a proof, see [50], Lesson 65.

B. Goodwine, *Engineering Differential Equations: Theory and Applications*,
DOI 10.1007/978-1-4419-7919-3_3, © Springer Science+Business Media, LLC 2011

Example 3.1. Consider the very simple mechanical system illustrated on the left in Figure 3.1. The scenario modeled by the problem is that a mass is attached to a moving base by a spring (on the left) and viscous damper (on the right). The base is moving with a specified motion $z(t)$. The question is what is the resulting motion of the mass given a specified motion of the base?

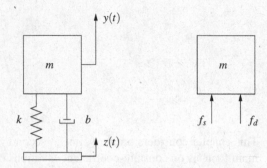

Fig. 3.1 Mechanical system for Example 3.1.

A free body diagram of the mass is illustrated on the right in Figure 3.1, where f_s and f_d are the forces that the spring and damper exert on the mass, respectively. Assume that y and z are measured from a configuration where the spring is unstretched; that is, at $y = 0$ and $z = 0$ the spring is unstretched. In that case, $f_s = k(z-y)$ and $f_d = b(\dot{z} - \dot{y})$.

Newton's law gives

$$m\ddot{y} = f_s + f_d = k(z-y) + b(\dot{z} - \dot{y})$$

or rearranging

$$m\ddot{y} + b\dot{y} + ky = kz + b\dot{z}.$$

This is an ordinary, second-order, linear, constant-coefficient, inhomogeneous differential equation. Remember that $z(t)$ is assumed to be known, so, for example, if $z(t) = Z\sin \omega t$, then

$$m\ddot{y} + b\dot{y} + ky = kZ\sin \omega t + bZ\omega \cos \omega t.$$

For such a system, important and interesting questions may be the following.

1. Given $y(0)$ and $\dot{y}(0)$ what is the resulting motion of the mass $y(t)$?
2. What is the magnitude of the resulting motion of the mass as a function of the magnitude of the base motion Z?
3. How does the magnitude of the motion of the mass change as Z or ω are changed?
4. Given either or both Z and ω, what are good choices for k and b so that the magnitude of either or both the motion or acceleration of the mass is minimized? This is basically designing a suspension or vibration absorber.

3.2 Theory of Linear Homogeneous Equations

In the study of mechanical vibrations, the starting point is usually the case of free, undamped vibrations. This case is illustrated by the following example and is the starting point for the study of homogeneous equations.

Example 3.2. Consider the mass–spring system illustrated in Figure 3.2. Assume for present purposes that there is no gravitational force acting on the spring and that $x = 0$ when the spring is unstretched. The only force on the mass is due to the spring and the equation of motion is

$$m\ddot{x} + kx = 0. \tag{3.2}$$

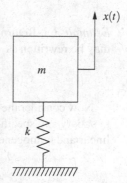

Fig. 3.2 Mass spring system for Example 3.2.

From Theorem 2.1, this ordinary, second-order, homogeneous, linear, constant-coefficient differential equation must have solutions of the form

$$x(t) = ce^{\lambda t}.$$

Substituting this into the differential equation gives

$$m\lambda^2 ce^{\lambda t} + kce^{\lambda t} = 0.$$

Inasmuch as $e^{\lambda t}$ is never zero and assuming that $c \neq 0$, we have the *characteristic equation*

$$m\lambda^2 + k = 0$$

or

$$\lambda = \pm\sqrt{-\frac{k}{m}}.$$

Real spring constants and masses have only positive values, therefore

$$\lambda = \pm i\sqrt{\frac{k}{m}}.$$

Let $\omega_n = \sqrt{k/m}$ denote the *natural frequency* of the system. Using this notation, there are two possible solutions

$$x_1(t) = e^{i\omega_n t}, \quad x_2(t) = e^{-i\omega_n t}.$$

The fact that these two functions are indeed solutions to the differential equation may be verified by direct substitution.

At this point it would behoove the reader to read Appendix A which gives a several-page review of complex variable theory. The one fact that is necessary and repeated here is the definition of *Euler's formula*, which relates the exponential of a complex number to trigonometric functions. Namely,

$$e^{(\mu + i\omega)t} = e^{\mu t}\left(\cos \omega t + i \sin \omega t\right).$$

Example 3.3. Returning to Example 3.2 and using Euler's formula, the two solutions may be rewritten as

$$x_1(t) = \cos \omega_n t + i \sin \omega_n t, \quad x_2(t) = \cos \omega_n t - i \sin \omega_n t.$$

Now consider the question: when will it be possible to combine the two solutions to satisfy any specified initial conditions? First, note that because Equation (3.2) is linear and homogeneous, the solution

$$x(t) = c_1 x_1(t) + c_2 x_2(t)$$

also satisfies the differential equation. This may be verified by direct substitution using either form of the solutions. Using the sine and cosine form, then

$$\dot{x}_1(t) = -\omega_n \sin \omega_n t + i\omega_n \cos \omega_n t$$
$$\dot{x}_2(t) = -\omega_n \sin \omega_n t - i\omega_n \cos \omega_n t$$
$$\ddot{x}_1(t) = -\omega_n^2 \cos \omega_n t - i\omega_n^2 \sin \omega_n t$$
$$\ddot{x}_2(t) = -\omega_n^2 \cos \omega_n t + i\omega_n^2 \sin \omega_n t$$

and substituting into Equation (3.2) and using the fact that $\omega_n^2 = k/m$ gives

$$
\begin{aligned}
m\ddot{x} + kx &= m\left(c_1 \ddot{x}_1 + c_2 \ddot{x}_2\right) + k\left(c_1 x_1 + c_2 x_2\right) \\
&= mc_1 \left(-\omega_n^2 \cos \omega_n t - i\omega_n^2 \sin \omega_n t\right) + mc_2 \left(-\omega_n^2 \cos \omega_n t + i\omega_n^2 \sin \omega_n t\right) \\
&\quad + kc_1 \left(\cos \omega_n t + i \sin \omega_n t\right) + kc_2 \left(\cos \omega_n t - i \sin \omega_n t\right) \\
&= -c_1 k \cos \omega_n t - ic_1 k \sin \omega_n t - c_2 k \cos \omega_n t + ic_2 k \sin \omega_n t \\
&\quad + kc_1 \cos \omega_n t + ikc_1 \sin \omega_n t + kc_2 \cos \omega_n t - ikc_2 \sin \omega_n t \\
&= 0.
\end{aligned}
$$

The fact that a linear combination of the two solutions of Equation (3.2) also satisfies the equation is a particular example of the *principle of superposition*.

3.2.1 The Principle of Superposition

The principle of superposition states that any linear combination of solutions to an ordinary, linear, homogeneous differential equation is also a solution. An example is the final computation in Example 3.3. The following theorem proves the principle for the second-order case. Proving it for nth-order equations is left as an exercise.

Theorem 3.2. *Let the functions $x_1(t)$ and $x_2(t)$ each satisfy the ordinary, second-order, linear, homogeneous differential equation*

$$f_2(t)\ddot{x}(t) + f_1(t)\dot{x}(t) + f_0(t)x = 0. \tag{3.3}$$

Then any linear combination of $x_1(t)$ and $x_2(t)$, that is,

$$x(t) = c_1 x_1(t) + c_2 x_2(t),$$

also satisfies Equation (3.3).

Proof. The proof is simply by direct substitution.

$$
\begin{aligned}
f_2(t)\ddot{x}(t) + f_1(t)\dot{x}(t) + f_0(t)x(t) &= f_2(t)\left(c_1\ddot{x}_1(t) + c_2\ddot{x}_2(t)\right) \\
&\quad + f_1(t)\left(c_1\dot{x}_1(t) + c_2\dot{x}_2(t)\right) \\
&\quad + f_0(t)\left(c_1 x_1(t) + c_2 x_2(t)\right) \\
&= c_1\left(f_2(t)\ddot{x}_1(t) + f_1(t)\dot{x}_1(t) + f_0(t)x_1(t)\right) \\
&\quad + c_2\left(f_2(t)\ddot{x}_2(t) + f_1(t)\dot{x}_2(t) + f_0(t)x_2(t)\right) \\
&= 0 + 0.
\end{aligned}
$$

$\square$

Note that the principle of superposition does not require that the equation have constant coefficients.

Example 3.4. Returning to Example 3.3, at this point it has been shown that the two functions

$$x_1(t) = \cos\omega_n t + \mathrm{i}\sin\omega_n t, \quad x_2(t) = \cos\omega_n t - \mathrm{i}\sin\omega_n t$$

are solutions to Equation (3.2) and that any linear combination

$$x(t) = c_1 x_1(t) + c_2 x_2(t)$$

is also a solution.

Because this text is also concerned with the solutions and analysis of vibration problems, an alternative simpler form of the combination of the two solutions would be nice. To achieve this, it is possible to simply rearrange the linear combination as

$$x(t) = c_1 x_1(t) + c_2 x_2(t)$$
$$= c_1 \left(\cos \omega_n t + i \sin \omega_n t \right) + c_2 \left(\cos \omega_n t - i \sin \omega_n t \right)$$
$$= (c_1 + c_2) \cos \omega_n t + i (c_1 - c_2) \sin \omega_n t$$
$$= \hat{c}_1 \cos \omega_n t + \hat{c}_2 \sin \omega_n t,$$

where

$$\hat{c}_1 = c_1 + c_2, \quad \hat{c}_2 = i(c_1 - c_2).$$

Note also, by direct substitution, it may be verified that the two functions

$$\hat{x}_1(t) = \sin \omega_n t, \quad \hat{x}_2(t) = \cos \omega_n t$$

are also solutions to Equation (3.2). This is an otherwise unremarkable fact, but they are generally a more convenient representation of two homogeneous solutions than the forms containing imaginary terms because most engineering problems are only concerned with real solutions.

3.2.2 Linear Independence

Now consider the question of determining when it will be the case that any initial conditions can be satisfied by appropriately determining the two unspecified coefficients in the various forms of the solutions above.

Example 3.5. Adding initial conditions to the problem statement corresponding to the system illustrated in Figure 3.2 gives

$$m\ddot{x} + kx = 0,$$

where $x(0) = x_0$, $\dot{x}(0) = \dot{x}_0$. The examples above showed that

$$x(t) = c_1 \hat{x}_1(t) + c_2 \hat{x}_2(t) = c_1 \cos \omega_n t + c_2 \sin \omega_n t$$

is a solution to the differential equation. Now to determine the values for c_1 and c_2 that satisfy the initial conditions, simply evaluate $x(0)$ and $\dot{x}(0)$ and set them equal to x_0 and $\dot{x}_0$, respectively. Because $x(0) = c_1$ and $\dot{x}(0) = c_2 \omega_n$, then

$$c_1 = x_0, \quad c_2 = \frac{\dot{x}_0}{\omega_n}.$$

The main point of this example is the fact that regardless of the values given for x_0 and $\dot{x}_0$, there are values for c_1 and c_2 that satisfy the differential equation as well as the initial conditions and the solution is

$$x(t) = x_0 \cos \omega_n t + \frac{\dot{x}_0}{\omega_n} \sin \omega_n t.$$

Now consider the more general question: given two solutions, $x_1(t)$ and $x_2(t)$ of an ordinary, second-order, linear, homogeneous differential equation, when will it be the case that the two coefficients in the general solution

$$x(t) = c_1 x_1(t) + c_2 x_2(t)$$

may be used to satisfy any given initial conditions?

Stated a bit more mathematically, consider the initial value problem

$$\ddot{x} + p(t)\dot{x} + q(t)x = 0,$$

$x(t_0) = x_0$, $\dot{x}(t_0) = \dot{x}_0$, and assume that $x_1(t)$ and $x_2(t)$ are solutions. From the principle of superposition in Theorem 3.2, because the equation is ordinary, linear, and homogeneous, then

$$x(t) = c_1 x_1(t) + c_2 x_2(t)$$

also satisfies the differential equation.

Now solving $x(t_0) = x_0$ and $\dot{x}(t_0) = \dot{x}_0$ for c_1 and c_2 gives

$$x(t_0) = c_1 x_1(t_0) + c_2 x_2(t_0) = x_0$$
$$\dot{x}(t_0) = c_1 \dot{x}_1(t_0) + c_2 \dot{x}_2(t_0) = \dot{x}_0$$

which yields

$$c_1 = \frac{\dot{x}_0 x_2(t_0) - x_0 \dot{x}_2(t_0)}{x_1(t_0)\dot{x}_2(t_0) - x_2(t_0)\dot{x}_1(t_0)}, \quad c_2 = \frac{\dot{x}_0 x_1(t_0) - x_0 \dot{x}_1(t_0)}{x_1(t_0)\dot{x}_2(t_0) - x_2(t_0)\dot{x}_1(t_0)}, \quad (3.4)$$

so the only time there will be a problem with solving for the coefficients is when the denominator is equal to zero (note that both denominators are equal). Observe that the denominator is the Wronskian for the two functions, $x_1(t)$ and $x_2(t)$,

$$W(x_1, x_2)(t) = \begin{vmatrix} x_1(t) & x_2(t) \\ \dot{x}_1(t) & \dot{x}_2(t) \end{vmatrix} = x_1(t)\dot{x}_2(t) - x_2(t)\dot{x}_1(t).$$

Thus, if the Wronskian is nonzero at the time where the initial conditions are specified, we may use the two solutions to satisfy any initial conditions. If the Wronskian is nonzero over an interval, then the two functions form a *fundamental set of solutions* on that interval. Due to the uniqueness of solutions for the types of equations we are considering, we have the following theorem.

Theorem 3.3. *If $x_1(t)$ and $x_2(t)$ satisfy*

$$\ddot{x} + \alpha_1 \dot{x} + \alpha_0 x = 0,$$

and if

$$W(x_1), x_2)(t) = x_1(t)\dot{x}_2(t) - x_2(t)\dot{x}_1(t) \neq 0,$$

then

$$x(t) = c_1 x_1(t) + c_2 x_2(t)$$

is the general solution to the initial value problem

$$\ddot{x} + \alpha_1 \dot{x} + \alpha_0 x = 0,$$

$x(t_0) = x_0$, $\dot{x}(t_0) = \dot{x}_0$, *where c_1 and c_2 are given by Equation (3.4).*[2]

So, if the goal is to solve an ordinary, linear, homogeneous, second-order initial value problem, then it will not suffice to find any two homogeneous solutions to combine, but two linearly independent solutions. Fortunately, be illustrated subsequently, the methods developed in the next few sections all generate linearly independent solutions, so obsessive attention to this detail is not always necessary.

3.3 Constant-Coefficient Homogeneous Equations

From Theorem 2.1 it is clear that ordinary, linear, constant-coefficient, homogeneous, second-order differential equations have solutions of the form $x(t) = e^{\lambda t}$. However, in contrast to the case of first-order equations of this type, there will generally be two solutions for λ which complicates matters somewhat, as illustrated by the following example.

Example 3.6. Determine the general solution to

$$\ddot{x} + 5\dot{x} + 6x = 0. \tag{3.5}$$

Assuming

$$x(t) = e^{\lambda t}$$

and substituting gives

$$\ddot{x} + 5\dot{x} + 6x = \lambda^2 e^{\lambda t} + 5\lambda e^{\lambda t} + 6e^{\lambda t} = 0$$

and inasmuch as $e^{\lambda t}$ is never equal to zero,

$$\lambda^2 + 5\lambda + 6 = 0.$$

Using the quadratic formula (or simply factoring, as is possible in this case) gives $\lambda = -2$ or $\lambda = -3$, so

$$x_1(t) = e^{-2t}, \quad x_2(t) = e^{-3t}$$

both satisfy Equation (3.5) and

$$x(t) = c_1 e^{-2t} + c_2 e^{-3t}$$

is a general solution as long as the Wronskian, $W(x_1, x_2)(t)$ is nonzero. Checking the Wronskian gives

---

[2] In fact, a similar theorem holds for the variable-coefficient case.

$$W(x_1, x_2)(t) = \begin{vmatrix} e^{-2t} & e^{-3t} \\ -2e^{-2t} & -3e^{-3t} \end{vmatrix} = -3e^{-5t} + 2e^{-5t} = -e^{-5t} \neq 0,$$

so the initial conditions may be specified at any time t.

Now, it should be clear that assuming exponential solutions for second-order equations of this type will result in a quadratic characteristic equation which, in general, will have two roots. Because a quadratic equation may have distinct roots, a complex conjugate pair of roots, or a repeated root, and each case results in a solution of a different type, each case must be considered separately.

As a recurring example throughout the investigation of the three possible cases, consider the system mass–spring–damper illustrated in Figure 3.3 which has the equation of motion

$$m\ddot{x} + b\dot{x} + kx = 0. \tag{3.6}$$

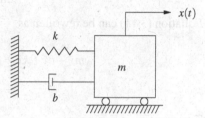

Fig. 3.3 Mechanical system described by Equation (3.6).

Assuming $x(t) = e^{\lambda t}$ results in the characteristic equation

$$m\lambda^2 + b\lambda + k = 0$$

which has roots

$$\lambda_1 = \frac{-b + \sqrt{b^2 - 4mk}}{2m}, \qquad \lambda_2 = \frac{-b - \sqrt{b^2 - 4mk}}{2m}.$$

The roots, λ_1 and λ_2 are either

- Real and distinct (when $b^2 - 4mk > 0$)
- A complex conjugate pair (when $b^2 - 4mk < 0$)
- Repeated (when $b^2 - 4mk = 0$)

As is already clear, the solutions to Equation (3.6) involve the parameters m, b, and k. However, there exists a standard canonical form for such equations. This form is valuable to know both because it is a standard formulation for second-order problems and also because it simplifies notation for the problem.

3.3.0.1 Canonical Form for Second-Order Systems

Consider

$$m\ddot{x} + b\dot{x} + kx = 0$$

and the following definitions

$$\zeta = \frac{b}{2\sqrt{mk}}$$

$$\omega_d = \omega_n\sqrt{1-\zeta^2} = \sqrt{\frac{k}{m}}\sqrt{1 - \frac{b^2}{4mk}} = \sqrt{\frac{k}{m}\frac{4mk-b^2}{4mk}} = \frac{\sqrt{4mk-b^2}}{2m}.$$

These definitions hold for any values of m, b, and k, but are generally only used when the three parameters have positive values.

The first term ζ is called *the damping ratio* and the second term ω_d is called *the damped natural frequency*. Observing that

$$\frac{b}{2m} = \zeta\omega_n,$$

Equation (3.11) can be rewritten as

$$m\ddot{x} + b\dot{x} + kx = m\left(\ddot{x} + \frac{b}{m}\dot{x} + \frac{k}{m}x\right)$$
$$= m\left(\ddot{x} + 2\zeta\omega_n\dot{x} + \omega_n^2 x\right).$$

The equation is homogeneous and $m \neq 0$, thus the two differential equations

$$m\ddot{x} + b\dot{x} + kx = 0 \quad \Longleftrightarrow \quad \ddot{x} + 2\zeta\omega_n\dot{x} + \omega_n^2 x = 0,$$

are equivalent, meaning that they have the same solutions. In this case, the characteristic equation is

$$\lambda^2 + 2\zeta\omega_n\lambda + \omega_n^2 = 0$$

which gives

$$\lambda = \frac{-2\zeta\omega_n \pm \sqrt{4\zeta^2\omega_n^2 - 4\omega_n^2}}{2}$$
$$= -\zeta\omega_n \pm \omega_n\sqrt{\zeta^2 - 1}.$$

The three cases corresponding to distinct, real roots, complex conjugate roots, and repeated roots correspond to the cases where $\zeta > 1$, $\zeta < 1$, and $\zeta = 1$ respectively. The "simplification" is in that the roots only contain two parameters, ω_n and ζ instead of the three parameters, m, b, and k.

3.3.1 Distinct Real Roots

In the case that the quadratic equation has distinct real roots, as was illustrated in Example 3.6, the two solutions

$$x_1(t) = e^{\lambda_1 t}, \quad x_2(t) = e^{\lambda_2 t},$$

where λ_1 and λ_2 are the roots of the characteristic equation, will both satisfy the differential equation and by the principle of superposition the linear combination

$$x(t) = c_1 x_1(t) + c_2 x_2(t) \tag{3.7}$$

will also satisfy it. Furthermore, in this case where $\lambda_1 \neq \lambda_2$ the Wronskian is always nonzero. This fact is illustrated by the direct computation,

$$W(x_1, x_2)(t) = \begin{vmatrix} e^{\lambda_1 t} & e^{\lambda_2 t} \\ \lambda_1 e^{\lambda_1 t} & \lambda_2 e^{\lambda_2 t} \end{vmatrix} = \lambda_2 e^{(\lambda_1 + \lambda_2)t} - \lambda_1 e^{(\lambda_1 + \lambda_2)t} = (\lambda_2 - \lambda_1) e^{(\lambda_1 + \lambda_2)t} \neq 0.$$
$$\tag{3.8}$$

Therefore any initial conditions specified at any t may be satisfied by a choice of c_1 and c_2, and this fact, in combination with Theorem 3.1, implies Equation (3.7) is the general solution.

To emphasize its importance, the above results are restated in the form of a theorem.

Theorem 3.4. *For an ordinary, second-order, linear, constant-coefficient, homogeneous differential equation, if the roots of the corresponding characteristic equation are real and distinct, denoted by λ_1 and λ_2, then the functions*

$$x_1(t) = e^{\lambda_1 t}, \quad x_2(t) = e^{\lambda_2 t}$$

both are solutions to the differential equation. Furthermore $x_1(t)$ and $x_2(t)$ are linearly independent and therefore

$$x(t) = c_1 e^{\lambda_1 t} + c_2 e^{\lambda_2 t}$$

is a general solution of the differential equation.

In the case of the mass–spring–damper system illustrated in Figure 3.3 and described by Equation (3.6) the general solution will be

$$x(t) = c_1 e^{\left(\left(-b + \sqrt{b^2 - 4mk}\right)/(2m)\right)t} + c_2 e^{\left(\left(-b - \sqrt{b^2 - 4mk}\right)/(2m)\right)t}$$
$$= c_1 e^{\left(-\zeta \omega_n + \omega_n \sqrt{\zeta^2 - 1}\right)t} + c_2 e^{\left(-\zeta \omega_n - \omega_n \sqrt{\zeta^2 - 1}\right)t}.$$

Note that this solution corresponds to a linear combination of two decaying exponentials.

3.3.2 *Complex Roots*

In the case of the mass–spring–damper problem where $b^2 - 4mk < 0$ or $\zeta < 1$, the sign of the term inside the square root will be negative and hence the two roots will be a complex conjugate pair given by

$$\lambda_1 = \frac{-b + i\sqrt{4mk - b^2}}{2m} = -\zeta\omega_n + i\omega_n\sqrt{1 - \zeta^2}$$

$$\lambda_2 = \frac{-b - i\sqrt{4mk - b^2}}{2m} = -\zeta\omega_n - i\omega_n\sqrt{1 - \zeta^2}.$$

Using Euler's formula, the two solutions

$$\hat{x}_1(t) = \hat{c}_1 e^{\lambda_1 t}, \quad \hat{x}_2(t) = \hat{c}_2 e^{\lambda_2 t},$$

where, using the canonical formulation

$$\hat{x}_1(t) = \hat{c}_1 e^{-\zeta\omega_n t} \left(\cos\left(\omega_n\sqrt{1 - \zeta^2}t\right) + i\sin\left(\omega_n\sqrt{1 - \zeta^2}t\right) \right)$$

$$= \hat{c}_1 e^{-\zeta\omega_n t} \left(\cos\omega_d t + i\sin\omega_d t \right)$$

$$\hat{x}_2(t) = \hat{c}_2 e^{-\zeta\omega_n t} \left(\cos\left(\omega_n\sqrt{1 - \zeta^2}t\right) - i\sin\left(\omega_n\sqrt{1 - \zeta^2}t\right) \right)$$

$$= \hat{c}_2 e^{-\zeta\omega_n t} \left(\cos\omega_d t - i\sin\omega_d t \right),$$

and, following the procedure outlined in Example 3.4 and defining

$$c_1 = \hat{c}_1 + \hat{c}_2, \quad c_2 = i(\hat{c}_1 - \hat{c}_2), \tag{3.9}$$

then the two solutions

$$x_1(t) = e^{-\zeta\omega_n t} \cos\omega_d t, \quad x_2(t) = e^{-\zeta\omega_n t} \sin\omega_d t \tag{3.10}$$

may be added in a linear combination

$$x(t) = c_1 e^{-\zeta\omega_n t} \cos\omega_d t + c_2 e^{-\zeta\omega_n t} \sin\omega_d t \tag{3.11}$$

to form a solution.

There is nothing wrong with repeating the Wronskian computation for this case; however, it is worth noting that the computation in Equation (3.8) is valid for the case where the λs are complex as well. Also, because the combination of solutions expressed by the constants in Equation (3.9) is full rank, the sine and cosine combination of the solutions will be linearly independent as well. However, just to complete the picture, the detailed Wronskian computation is as follows.

$$W(x_1,x_2)(t) = \begin{vmatrix} e^{-\zeta\omega_n t}\cos\omega_d t & e^{-\zeta\omega_n t}\sin\omega_d t \\ -e^{-\zeta\omega_n t}(\omega_d\sin\omega_d t + \zeta\omega_n\cos\omega_d t) & e^{-\zeta\omega_n t}(\omega_d\cos\omega_d t - \zeta\omega_n\sin\omega_d t) \end{vmatrix}$$

$$= \omega_d e^{-2\zeta\omega_n t} \neq 0.$$

So, the above proves the following theorem.

Theorem 3.5. *For an ordinary, second-order, linear, constant-coefficient, homogeneous differential equation, if the roots of the corresponding characteristic equation are a complex conjugate pair, denoted by λ_1 and λ_2, then the functions*

$$x_1(t) = e^{\lambda_1 t}, \quad x_2(t) = e^{\lambda_2 t}$$

both are solutions to the differential equation. Furthermore $x_1(t)$ and $x_2(t)$ are linearly independent and therefore

$$x(t) = c_1 e^{\lambda_1 t} + c_2 e^{\lambda_2 t}$$

is a general solution of the differential equation.

Using the sine and cosine formulation gives the following corollary to Theorem 3.5.

Corollary 3.1. *Equivalently, if the two roots to the characteristic polynomial are denoted by*

$$\lambda_1 = -\zeta\omega_n + i\omega_n\sqrt{1-\zeta^2}, \quad \lambda_2 = -\zeta\omega_n - i\omega_n\sqrt{1-\zeta^2}$$

then the functions

$$x_1(t) = e^{-\zeta\omega_n t}\cos\omega_d t, \quad x_2(t) = e^{-\zeta\omega_n t}\sin\omega_d t$$

both are solutions to the differential equation. Furthermore $x_1(t)$ and $x_2(t)$ are linearly independent and therefore

$$x(t) = c_1 e^{-\zeta\omega_n t}\cos\omega_d t + c_2 e^{-\zeta\omega_n t}\sin\omega_d t$$

is a general solution of the differential equation.

An example may be helpful at this point.

Example 3.7. Determine a general solution to

$$\ddot{x} + 2\dot{x} + 5x = 0.$$

Just for fun, let us solve this two ways.

1. Assuming a solution of the form $x(t) = e^{\lambda t}$ gives the characteristic equation

$$\lambda^2 + 2\lambda + 5 = 0 \quad \Longrightarrow \quad \lambda = -1 \pm 2i.$$

Immediately we can write either

$$x(t) = c_1 e^{(-1-2i)t} + c_2 e^{(-1+2i)t}$$

or

$$x(t) = c_1 e^{-t} \cos 2t + c_2 e^{-t} \sin 2t.$$

2. Alternatively, using the definition of ω_n, ζ and ω_d,

$$\omega_n = \sqrt{\frac{k}{m}} = \sqrt{5}$$

$$\zeta = \frac{b}{2\sqrt{mk}} = \frac{1}{\sqrt{5}} = \frac{\sqrt{5}}{5}$$

$$\omega_d = \omega_n \sqrt{1 - \zeta^2} = \sqrt{5}\sqrt{1 - \frac{1}{5}} = 2.$$

and substituting into Equation (3.11) gives

$$x(t) = c_1 e^{-t} \cos 2t + c_2 e^{-t} \sin 2t.$$

3.3.3 Repeated Roots

Now consider the case when $b^2 = 4mk$ or, equivalently, $\zeta = 1$. In this case, $\lambda = -\zeta \omega_n = -\omega_n$, and at this point there is only one solution

$$x_1(t) = e^{-\omega_n t}. \tag{3.12}$$

To find another solution, assume a solution of the form

$$x_2(t) = \mu(t) x_1(t),$$

and substitute to see if it determines $\mu(t)$. Computing

$$\dot{x}_2 = \mu \dot{x}_1 + \dot{\mu} x_1, \quad \ddot{x}_2 = 2\dot{\mu}\dot{x}_1 + \mu\ddot{x}_1 + \ddot{\mu}x_1$$

and substituting into

$$\ddot{x} + 2\zeta \omega_n \dot{x} + \omega_n^2 x = 0$$

gives

$$(2\dot{\mu}\dot{x}_1 + \mu\ddot{x}_1 + \ddot{\mu}x_1) + 2\zeta\omega_n(\mu\dot{x}_1 + \dot{\mu}x_1) + \omega_n^2\mu x_1 =$$
$$\mu\left(\ddot{x}_1 + 2\zeta\omega_n\dot{x}_1 + \omega_n^2 x_1\right) + (2\dot{\mu}\dot{x}_1 + \ddot{\mu}x_1 + 2\zeta\omega_n\dot{\mu}x_1) =$$
$$\ddot{\mu}x_1 + 2\dot{\mu}\left(\dot{x}_1 + \omega_n x_1\right) =$$
$$\ddot{\mu}x_1 =$$
$$\ddot{\mu} = 0 \quad \Longrightarrow \quad \mu(t) = t + c.$$

Note that in the second line the term in the left pair of parentheses is zero because x_1 is a solution to the homogeneous equation. In the third line the term in parentheses is zero due to the form of $x_1(t)$ from Equation (3.12). Finally, because c is arbitrary, it may be zero and hence, finally,

$$x_2(t) = tx_1(t) = te^{\lambda t} = te^{-\omega_n t}.$$

So, the two solutions

$$x_1(t) = e^{\lambda t} = e^{\omega_n t}, \quad x_2(t) = te^{\lambda t} = te^{\omega_n t}$$

both satisfy the differential equation.

A direct computation with the Wronskian shows they are linearly independent, which is left as an exercise.

So, the above proves the following theorem.

Theorem 3.6. *For an ordinary, second-order, linear, constant-coefficient, homogeneous differential equation, if the roots of the corresponding characteristic equation, are equal (i.e., the roots are repeated, and are denoted by λ), then the following two functions*

$$x_1(t) = e^{\lambda t}, \quad x_2(t) = te^{\lambda t}$$

both are solutions to the differential equation. Furthermore $x_1(t)$ and $x_2(t)$ are linearly independent and therefore

$$x(t) = c_1 e^{\lambda_1 t} + c_2 te^{\lambda_2 t}$$

is a general solution of the differential equation.

An example follows.

Example 3.8. Find a general solution to

$$\ddot{x} + 4\dot{x} + 4x = 0.$$

The corresponding characteristic equation is

$$\lambda^2 + 4\lambda + 4 = 0.$$

Hence, $\lambda = -2$ is the repeated solution. Therefore

$$x(t) = c_1 e^{-2t} + c_2 te^{-2t}$$

is the general solution.

3.4 Inhomogeneous Equations

The two methods for solving inhomogeneous, second-order, ordinary, linear differential equations go by the same name and are essentially equivalent in approach to the methods outlined in Chapter 2 for inhomogeneous first-order equations, namely, the method of undetermined coefficients and the method of variation of parameters.

3.4.1 The Method of Undetermined Coefficients Constant-Coefficient Differential Equations

The method of undetermined coefficients is essentially the same as was presented for first-order equations in Section 2.3.1. Thus this section limits the presentation to a few examples.

Example 3.9. Find the general solution to

$$m\ddot{x} + kx = F\cos\omega t.$$

From Examples 3.2 through 3.4, the homogeneous solution is

$$x_h(t) = c_1\cos\omega_n t + c_2\sin\omega_n t,$$

where, as usual, $\omega_n = \sqrt{k/m}$. Although not necessary to simply find the solution, this example will work with the normal form

$$\ddot{x} + \omega_n^2 x = \frac{F}{m}\cos\omega t.$$

Referring to Table 2.1, as long as $\omega \neq \omega_n$, then a correct assumption for the form of the particular solution is

$$x_p(t) = A\cos\omega t + B\sin\omega t.$$

Skipping the gory details, differentiating $x_p(t)$, and substituting into the differential equation gives

$$A = \frac{F}{m(\omega_n^2 - \omega^2)}, \quad B = 0,$$

so the entire solution is

$$x(t) = c_1 \cos \omega_n t + c_2 \sin \omega_n t + \frac{F}{m(\omega_n^2 - \omega^2)} \cos \omega t$$

$$= c_1 \cos \omega_n t + c_2 \sin \omega_n t + \frac{F}{k\left(1 - \left(\frac{\omega}{\omega_n}\right)^2\right)} \cos \omega t.$$

The case where $\omega = \omega_n$ is referred to as *resonance*, and is further explored in the next chapter. At this point it suffices to note that it corresponds to the case where the initially assumed form of the particular solution is the same as a homogeneous solution; hence, the assumed form of $x_p(t)$ must be multiplied by t, which results in a solution with a magnitude that increases linearly with time, as is illustrated by the following example.

Example 3.10. Solve

$$\ddot{x} + x = \cos t,$$

where $x(0) = 0$ and $\dot{x}(0) = 0$, which corresponds, for example, to $m = 1$, $k = 1$ and hence, $\omega_n = 1$. Assuming homogeneous solutions of the form $x_h(t) = e^{\lambda t}$ gives $\lambda = \pm i$, so

$$x_h(t) = c_1 \cos t + c_2 \sin t$$

is a homogeneous solution. Note that due to the inhomogeneous term $\cos t$, one may be inclined to assume $x_p(t) = A \cos t + B \sin t$; however, because $\sin t$ and $\cos t$ are homogeneous solutions, then the appropriate particular solution is

$$x_p(t) = t(A \cos t + B \sin t).$$

Differentiating twice, substituting, equating coefficients of $\sin t$ and $\cos t$ and solving for A and B gives $A = 0$ and $B = 1/2$; hence

$$x(t) = c_1 \cos t + c_2 \sin t + \frac{t}{2} \sin t.$$

Evaluating the initial conditions gives $c_1 = c_2 = 0$. Hence

$$x(t) = \frac{t}{2} \sin t$$

is the solution of the initial value problem. A plot of this solution is illustrated in Figure 3.4 and is an illustration of the phenomenon of *resonance*. Note that the solution grows unbounded.

Example 3.11. Find the general solution to

$$\ddot{x} + \dot{x} + 4x = t \sin 2t.$$

Assuming

$$x_h(t) = e^{\lambda t}$$

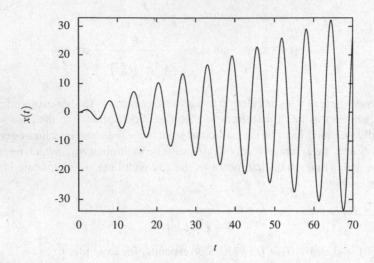

Fig. 3.4 Resonance response of solution to Example 3.10.

gives the characteristic equation

$$\lambda^2 + \lambda + 4 = 0$$

so

$$\lambda = \frac{-1 \pm \sqrt{1 - 16}}{2} = -\frac{1}{2} \pm \frac{\sqrt{15}}{2}i.$$

Hence,

$$x_h(t) = e^{-t/2}\left(c_1 \sin\frac{\sqrt{15}}{2}t + c_2\cos\frac{\sqrt{15}}{2}t\right).$$

The inhomogeneous term is of the product of a polynomial in t and $\sin 2t$, therefore we must assume a solution that contains the product of all the corresponding linearly independent derivatives. Hence, assume

$$x_p(t) = At\sin 2t + Bt\cos 2t + C\sin 2t + D\cos 2t.$$

Differentiating gives

$$\dot{x}_p(t) = A\sin 2t + 2At\cos 2t + B\cos 2t - 2Bt\sin 2t + 2C\cos 2t - 2D\sin 2t$$
$$= -2Bt\sin 2t + 2At\cos 2t + (A - 2D)\sin 2t + (B + 2C)\cos 2t$$

and differentiating again gives

$$\ddot{x}_p(t) = -2B\sin 2t - 4Bt\cos 2t + 2A\cos 2t - 4At\sin 2t$$
$$+ 2(A - 2D)\cos 2t - 2(B + 2C)\sin 2t$$
$$= -4At\sin 2t - 4Bt\cos 2t - 4(B + C)\sin 2t + 4(A - D)\cos 2t.$$

Substituting into the differential equation gives

$$[-4At\sin 2t - 4Bt\cos 2t - 4(B + C)\sin 2t + 4(A - D)\cos 2t]$$
$$+ [-2Bt\sin 2t + 2At\cos 2t + (A - 2D)\sin 2t + (B + 2C)\cos 2t]$$
$$+ 4[At\sin 2t + Bt\cos 2t + C\sin 2t + D\cos 2t] = t\sin 2t$$

and equating the coefficients of $t\sin 2t$, $t\cos 2t$, $\sin 2t$, and $\cos 2t$, respectively, gives
the following set of equations

$$-4A - 2B + 4A = 1$$
$$-4B + 2A + 4B = 0$$
$$-4(B + C) + (A - 2D) + 4C = 0$$
$$4(A - D) + (B + 2C) + 4D = 0.$$

From the first two equations, $A = 0$ and $B = -1/2$. Substituting this into the third
equation gives

$$2 - 2D = 0$$

so $D = 1$. From the last equation, $C = 1/4$. Hence

$$x_p(t) = -\frac{1}{2}t\cos 2t + +\frac{1}{4}\sin 2t + \cos 2t$$

and the general solution is

$$x(t) = x_h(t) + x_p(t)$$
$$= e^{-\frac{1}{2}t}\left(c_1\sin\frac{\sqrt{15}}{2}t + c_2\cos\frac{\sqrt{15}}{2}t\right) - \frac{1}{2}t\cos 2t + \frac{1}{4}\sin 2t + \cos 2t.$$

3.4.2 Method of Variation of Parameters for Constant or Variable-Coefficient Equations

Recall in Section 2.3.3 the method of variation of parameters was used to find solutions to ordinary, first-order, linear, inhomogeneous differential equations (either
constant or variable-coefficient). The same approach may be used in the case of
second-order equations; however, due to the second-order nature of the problem,
the computations involved become a bit more algebraically complex. Nevertheless,
proceed, as before, and consider the ordinary, second-order, linear, inhomogeneous

differential equation

$$\ddot{x}(t) + p(t)\dot{x}(t) + q(t)x(t) = f(t) \tag{3.13}$$

and assume a particular solution of the form

$$x_p(t) = \mu_1(t)x_1(t) + \mu_2(t)x_2(t), \tag{3.14}$$

where $x_1(t)$ and $x_2(t)$ are homogeneous solutions to Equation (3.13). The approach is (one hopes) obvious: substitute $x_p(t)$ into Equation (3.13) to see if equations for $\mu_1(t)$ and $\mu_2(t)$ may be obtained. So, proceeding thusly, and dropping the explicit dependence on t

$$\dot{x}_p = \dot{\mu}_1 x_1 + \mu_1 \dot{x}_1 + \dot{\mu}_2 x_2 + \mu_2 \dot{x}_2$$
$$\ddot{x}_p = \ddot{\mu}_1 x_1 + 2\dot{\mu}_1 \dot{x}_1 + \mu_1 \ddot{x}_1 + \ddot{\mu}_2 x_2 + 2\dot{\mu}_2 \dot{x}_2 + \mu_2 \ddot{x}_2$$

and substituting into Equation (3.13) gives

$$\ddot{x} + p\dot{x} + qx = (\ddot{\mu}_1 x_1 + 2\dot{\mu}_1 \dot{x}_1 + \mu_1 \ddot{x}_1 + \ddot{\mu}_2 x_2 + 2\dot{\mu}_2 \dot{x}_2 + \mu_2 \ddot{x}_2)$$
$$+ p(\dot{\mu}_1 x_1 + \mu_1 \dot{x}_1 + \dot{\mu}_2 x_2 + \mu_2 \dot{x}_2)$$
$$+ q(\mu_1 x_1 + \mu_2 x_2)$$
$$= f.$$

Rearranging a bit gives

$$\ddot{x} + p\dot{x} + qx = \mu_1 (\ddot{x}_1 + p\dot{x}_1 + qx_1) + \mu_2 (\ddot{x}_2 + p\dot{x}_2 + qx_2)$$
$$+ (\ddot{\mu}_1 x_1 + 2\dot{\mu}_1 \dot{x}_1 + \ddot{\mu}_2 x_2 + 2\dot{\mu}_2 \dot{x}_2) + p(\dot{\mu}_1 x_1 + \dot{\mu}_2 x_2)$$
$$= f,$$

and noting that because x_1 and x_2 are homogeneous solutions, the terms in the parentheses multiplying μ_1 and μ_2 in the first line are zero, the equation reduces to

$$(\ddot{\mu}_1 x_1 + 2\dot{\mu}_1 \dot{x}_1 + \ddot{\mu}_2 x_2 + 2\dot{\mu}_2 \dot{x}_2) + p(\dot{\mu}_1 x_1 + \dot{\mu}_2 x_2) = f. \tag{3.15}$$

At this point, there is one equation for two unknown functions, $\mu_1(t)$ and $\mu_2(t)$; furthermore, it is second-order, so at first glance it may seem not much progress has been made inasmuch as one second-order equation (Equation (3.13)) has been replaced with another one (Equation (3.15)). However, because it is one equation with two unknowns, the system is underdetermined, and we have the freedom to choose another independent equation. So, let us try to make the term in the left set of parentheses zero. Note that if we choose (with much foresight)

$$\dot{\mu}_1 x_1 + \dot{\mu}_2 x_2 = 0 \tag{3.16}$$

then its derivative must also be zero, so

$$\ddot{\mu}_1 x_1 + \dot{\mu}_1 \dot{x}_1 + \ddot{\mu}_2 x_2 + \dot{\mu}_2 \dot{x}_2 = 0.$$

In light of this, Equation (3.15) reduces to

$$\dot{\mu}_1 \dot{x}_1 + \dot{\mu}_2 \dot{x}_2 = f. \tag{3.17}$$

Solving Equations (3.16) and (3.17) for $\dot{\mu}_1$ and $\dot{\mu}_2$ gives

$$\dot{\mu}_1(t) = -\frac{x_2(t)f(t)}{x_1(t)\dot{x}_2(t) - \dot{x}_1(t)x_2(t)} \tag{3.18}$$

$$\dot{\mu}_2(t) = \frac{x_1(t)f(t)}{x_1(t)\dot{x}_2(t) - \dot{x}_1(t)x_2(t)}. \tag{3.19}$$

Thus, if $x_1(t)$ and $x_2(t)$ are known, everything on the right-hand sides of the above equations is known and $\mu_1(t)$ and $\mu_2(t)$ may be determined by integration. Hence

$$\mu_1(t) = -\int \frac{x_2(t)f(t)}{x_1(t)\dot{x}_2(t) - \dot{x}_1(t)x_2(t)} dt + c_1$$

$$\mu_2(t) = \int \frac{x_1(t)f(t)}{x_1(t)\dot{x}_2(t) - \dot{x}_1(t)x_2(t)} dt + c_2,$$

where c_1 and c_2 are the integration constants and are arbitrary. Substituting this into the original assumed form of the solution, Equation (3.14) gives

$$x_p(t) = -x_1(t) \left(\int \frac{x_2(t)f(t)}{x_1(t)\dot{x}_2(t) - \dot{x}_1(t)x_2(t)} dt + c_1 \right)$$

$$+ x_2(t) \left(\int \frac{x_1(t)f(t)}{x_1(t)\dot{x}_2(t) - \dot{x}_1(t)x_2(t)} dt + c_2 \right).$$

Note the following.

1. Because the denominator in each integrand must be nonzero, $x_1(t)$ and $x_2(t)$ must be linearly independent.
2. Because $x_p(t)$ has a linear combination of the two homogeneous solutions contained in it, it is actually the complete solution.

Hence the final answer is

$$\boxed{\begin{aligned} x(t) = c_1 x_1(t) + c_2 x_2(t) - x_1(t) \int \frac{x_2(t)f(t)}{x_1(t)\dot{x}_2(t) - \dot{x}_1(t)x_2(t)} dt \\ + x_2(t) \int \frac{x_1(t)f(t)}{x_1(t)\dot{x}_2(t) - \dot{x}_1(t)x_2(t)} dt. \end{aligned}} \tag{3.20}$$

To illustrate the use of the method, consider some examples.

Example 3.12. Find the general solution to

$$\ddot{x} + x = \frac{1}{\cos t},$$

where $x(0) = 0$ and $\dot{x}(0) = 0$. Note that the method of undetermined coefficients cannot be used for this problem because the inhomogeneous term is not of the appropriate form. For the homogeneous solution, there are complex roots, $\lambda = \pm i$; hence,

$$x_1(t) = \cos t, \quad x_2(t) = \sin t..$$

Note that $W(x_1, x_2)(t) = 1$. Substituting into Equation (3.20) gives

$$x(t) = c_1 \cos t + c_2 \sin t - \cos t \int \frac{\sin t}{\cos t} \, dt + \sin t \int \frac{\cos t}{\cos t} \, dt$$
$$= c_1 \cos t + c_2 \sin t + \cos t \ln(\cos t) + t \sin t.$$

Evaluating the initial conditions gives

$$x(0) = 0 \quad \Longrightarrow \quad c_1 = 0$$

and

$$\dot{x}(0) = 0 \quad \Longrightarrow \quad c_2 = 0.$$

The following example repeats Example 3.10 using variation of parameters, illustrating the fact that variation of parameters may be used for all the types of second-order equations that may be solved using undetermined coefficients. Using undetermined coefficients is probably usually easier, however, inasmuch as it requires solving some algebraic equations rather than evaluating some integrals.

Example 3.13. Solve

$$\ddot{x} + x = \cos t,$$

where $x(0) = 0$ and $\dot{x}(0) = 0$. The homogeneous solutions are

$$x_1(t) = \cos t, \quad x_2(t) = \sin t.$$

A quick computation shows that $W(x_1, x_2) = 1$. Hence

$$x(t) = c_1 \cos t + c_2 \sin t - \cos t \int \sin t \cos t \, dt + \sin t \int \cos^2 t \, dt.$$

Aside 3.1. *As a quick reminder of what you should already know from calculus, these integrals are worked out in detail. For the first one, note that if $u = \sin t$ then $du/dt = \cos t$; hence, by substitution*

$$\int \sin t \cos t \, dt = \int u \frac{du}{dt} \, dt = \int \frac{d}{dt}\left(\frac{u^2}{2}\right) dt = \frac{u^2}{2} + c = \frac{1}{2}\sin^2 t + c.$$

Of course, mentally most people just "cancel" the dt terms on the right hand side of the first line and skip right to $\int u \, du$, which is the familiar substitution rule in the above process.

For the second integral, integrating by parts[3] gives

---

[3] To remember integration by parts, simply integrate the product rule, that is,

$$\int \cos^2 t dt = \cos t \sin t + c + \int \sin^2 t dt = \cos t \sin t + c \int \left(1 - \cos^2 t\right) dt$$

$$= \cos t \sin t + c + t - \int \cos^2 t dt,$$

hence

$$\int \cos^2 t dt = \frac{t}{2} + \frac{\cos t \sin t}{2} + c.$$

So, returning to the example, the general solution is

$$x(t) = c_1 \cos t + c_2 \sin t - \frac{1}{2} \cos t \sin^2 t + \frac{1}{2} t \sin t + \frac{1}{2} \sin^2 t \cos t$$

$$= c_1 \cos t + c_2 \sin t + \frac{1}{2} t \sin t.$$

Applying the initial conditions gives $c_1 = 0$ and $c_2 = 0$, so

$$x(t) = \frac{1}{2} t \sin t,$$

which is the same answer as before.

3.5 Summary

1. For ordinary, second-order, linear, constant-coefficient, homogeneous differential equations, solutions are of the form $e^{\lambda t}$. Substituting this into the differential equation gives the characteristic equation, which will have either distinct and real roots, a pair of complex conjugate roots, or repeated roots. When the equation is in canonical form

$$\ddot{x} + 2\zeta \omega_n \dot{x} + \omega_n^2 x = 0$$

the roots are

$$\lambda_1 = -\zeta \omega_n + \omega_n \sqrt{\zeta^2 - 1}, \quad \lambda_2 = -\zeta \omega_n - \omega_n \sqrt{\zeta^2 - 1}.$$

a. If the roots are real and distinct, then $\zeta > 1$ and the general solution is

$$x(t) = c_1 e^{\left(-\zeta \omega_n + \omega_n \sqrt{\zeta^2 - 1}\right)t} + c_2 e^{\left(-\zeta \omega_n - \omega_n \sqrt{\zeta^2 - 1}\right)t}.$$

$$\int \frac{d}{dt} (uv) dt = \int \left(u \frac{dv}{dt} + v \frac{du}{dt} \right) dt$$

which, following the substitution rules gives the usual formula

$$\int u dv = uv - \int v du.$$

b. If the roots are a complex conjugate pair, then $0 < \zeta < 1$ and the general solution is

$$x_1(t) = c_1 e^{-\zeta \omega_n t} \cos \omega_d t + c_2 e^{-\zeta \omega_n t} \sin \omega_d t,$$

where $\omega_d = \omega_n \sqrt{1 - \zeta^2}$.

c. If the roots are repeated, then $\zeta = 1$ and the general solution is

$$x(t) = c_1 e^{-\omega_n t} + c_2 t e^{-\omega_n t}.$$

2. For ordinary, second-order, linear, constant-coefficient, inhomogeneous differential equations use

   a. Undetermined coefficients if the inhomogeneous term is sums or products of polynomials, sines, cosines or exponentials
   b. Variation of parameters if the inhomogeneous term is not of that form

   Variation of parameters works for any form of inhomogeneous term, but is generally more difficult than undetermined coefficients. For both methods, two homogeneous solutions are also needed.

3. For ordinary, second-order, linear, variable-coefficient, inhomogeneous differential equations, the method of variation of parameters works. However, two linearly independent homogeneous solutions are required for the method, and at least at this point, you do not have any method to find them!

3.6 Exercises

3.1. Assume that $x_1(t)$ and $x_2(t)$ are (individually) solutions to the following ordinary, second-order differential equations. For which of the following is the linear combination

$$x(t) = c_1 x_1(t) + c_2 x_2(t)$$

also a solution?

1. $\ddot{x} + 5\dot{x} + 4x = 0$.
2. $\ddot{x} + \sin t \dot{x} + 4x = 0$.
3. $\ddot{x} + 4\dot{x}x = 0$.
4. $\ddot{x} + 5\dot{x} + 4x = t$.

What are the differences between the equations for which $x(t)$ is a solution and $x(t)$ is not a solution?

3.2. Prove the following theorem regarding the principle of superposition for ordinary, linear, nth-order, homogeneous differential equations.

Theorem 3.7. *Let the functions $x_1(t), \ldots, x_n(t)$ each satisfy the ordinary, nth-order, linear, homogeneous differential equation*

$$f_n(t)\frac{d^n x}{dt^n} + f_{n-1}(t)\frac{d^{n-1}x}{dt^{n-1}} + \cdots + f_1(t)\frac{dx}{dt} + f_0(t)x = 0. \qquad (3.21)$$

Then any linear combination of $x_1(t), \ldots, x_n(t)$, that is,

$$x(t) = c_1 x_1(t) + \cdots + c_n x_n(t),$$

also satisfies Equation (3.21).

3.3. Show that[4] $x_1(t) = \sqrt{t}$ and $x_2(t) = 1/t$ are solutions to

$$2t^2\ddot{x} + 3t\dot{x} - x = 0 \qquad (3.22)$$

on the interval $0 < t < \infty$.

1. What happens to the Wronskian, $W(x_1, x_2)(t)$ as $t \to 0$?
2. Show that $x(t) = c_1 x_1(t) + c_2 x_2(t)$ is the general solution to Equation (3.22) for $0 < t < \infty$.
3. Determine the solution to Equation (3.22) if $x(1) = 2$ and $\dot{x}(1) = 1$.

3.4. In the case of repeated roots of the characteristic equation

$$\lambda^2 + 2\zeta\omega_n\lambda + \omega_n^2 = 0,$$

prove the following two facts.

1. If the root is repeated, the value of the root is $\lambda = -\omega_n$.
2. If the root is repeated, then the two solutions

$$x_1(t) = e^{-\omega t}, \quad x_2(t) = te^{-\omega t}$$

are linearly independent.

3.5. Determine the solution to $6\ddot{x} - 5\dot{x} + x = 0$, where $x(0) = 4$ and $\dot{x}(0) = 0$.

3.6. Determine the solution to $\ddot{x} + 4\dot{x} + 5x = 0$, where $x(0) = 1$ and $\dot{x}(0) = 0$.

3.7. Determine the general solution to $25\ddot{x} - 20\dot{x} + 4x = 0$.

3.8. This chapter mainly deals with constant-coefficient second-order ordinary differential equations. However, there is one class of variable-coefficient equations that is easy to solve. The equation

$$t^2\ddot{x} + \alpha t\dot{x} + \beta x = 0 \qquad (3.23)$$

is called *Euler's equation*. Show that $x(t) = t^\lambda$ is a solution.

1. Are there usually two solutions to Euler's equation? If so, are they linearly independent? If they are not linearly independent everywhere, on what intervals do they form a fundamental set of solutions?

---

[4] This is from [9].

2. Determine the general solution to

$$t^2\ddot{x} + 4t\dot{x} + 2x = 0.$$

3.9. Show that the set of functions $\{e^{\omega_n t}, te^{\omega_n t}\}$ is linearly independent.

3.10. In the case of distinct real roots where the solution is a linear combination of two exponentials, that is,

$$x(t) = c_1 e^{\lambda_1 t} + c_2 e^{\lambda_2 t},$$

it may seem initially that it is only possible for the solutions to decay or grow. However, due to the linear combination it is possible for the slope of the solution to change once. Let $\lambda_1 = -1$ and $\lambda_2 = -2$ and plot $x(t)$ for the following combinations of c_1 and c_2.

1. $c_1 = 1$ and $c_2 = 1$.
2. $c_1 = -1$ and $c_2 = 1$.
3. $c_1 = 1$ and $c_2 = -1$.
4. $c_1 = -1$ and $c_2 = -1$.

Explain in words why the characteristics of each solution make sense.

3.11. The solution to $2\ddot{x} + \dot{x} + 10x = 0$ with $x(0) = 1$ and $\dot{x}(0) = 0$ is illustrated in Figure 3.5. By referring to one of the forms of the solution given in Section 3.3, without solving the equation sketch what the solution will look like if

1. The coefficient of $\ddot{x}$ is increased
2. The coefficient of $\ddot{x}$ is decreased
3. The coefficient of $\dot{x}$ is increased a little
4. The coefficient of $\dot{x}$ is increased a lot
5. The coefficient of $\dot{x}$ is decreased
6. The coefficient of x is increased
7. The coefficient of x is decreased

Verify your predictions by solving the equation and plotting the solution. Using a computer package is acceptable. Insight from problems of this type is very useful in designing feedback controllers in Chapter 10.

3.12. For

$$a\ddot{x} + b\dot{x} + cx = 0$$

prove that if $b/a < 0$ or $c/a < 0$ then one or both of the homogeneous solutions is unstable; that is, as t gets large, the magnitude of the solution gets large. If one of the two homogeneous solutions blows up, is it mathematically possible for a linear combination to remain bounded? Is it practically possible if the equation represents a real system for the solution to remain bounded?

3.13. Determine the solution to $\ddot{x} + 4x = t^2 + 3e^t$, where $x(0) = 0$ and $\dot{x}(0) = 0$.

3.14. Determine the general solution to $\ddot{x} + 4\dot{x} + 8x = \sin 2t$.

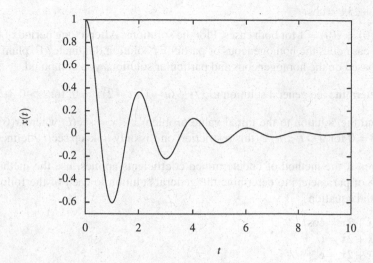

Fig. 3.5 Response of system for Exercise 3.11.

3.15. Consider $\ddot{x} + 2\dot{x} + 2x = e^{-t}\sin t + 1$.

1. Determine the general solution.
2. Sketch the solution if $x(0) = \dot{x}(0) = 0$.

3.16. Find the general solution of $\ddot{x} + 4\dot{x} + 4x = 4t^2 + 6e^t$.

3.17. Use the method of undetermined coefficients to find the general solution to

$$\ddot{x} + 4\dot{x} + 4x = e^{-2t}.$$

3.18. Use the method of undetermined coefficients to determine the general solution to the following equations.

1. $\ddot{x} + 3\dot{x} + 2x = \cos 2t$.
2. $\ddot{x} + 3\dot{x} + 2x = t^3$.
3. $\ddot{x} + 3\dot{x} + 2x = e^{-2t}$.
4. $\ddot{x} + 2x = \cos 2t$.
5. $\ddot{x} + 4x = \cos 2t$.
6. $\ddot{x} + 4x = \cos 2t + e^{-2t}$.
7. $\ddot{x} + 4x = \cos 2t + e^{-2t} + t^2$.
8. $\ddot{x} + 3\dot{x} + 2x = e^{-t}$.
9. $\ddot{x} + 3\dot{x} + 2x = e^{-t} + 1$.
10. $\ddot{x} + 3\dot{x} + 2x = t\cos 2t$.
11. $\ddot{x} + 3\dot{x} + 2x = t^2\cos 2t$.

3.19. Use the method of undetermined coefficients to solve

1. $\ddot{x} + \dot{x} + 25x = \cos t$

2. $\ddot{x} - \dot{x} + 25x = \cos t$

where $x(0) = \dot{x}(0) = 1$ for both cases. Plot the solutions. After a long period of time, for each case does the homogeneous or particular solution dominate? Explain your answer based on the homogeneous and particular solutions you computed.

3.20. Determine the general solution to $x/t + 6t + (\ln t - 2)\dot{x} = 0$, for $t > 0$.

3.21. Find the solution to the initial value problem $\ddot{x} + x = \sec(t)$ where $x(0) = 1$ and $\dot{x}(0) = 0$ for $0 \le t < \pi/2$ (this restriction on t is only to keep $\sec(t)$ defined).

3.22. Even if the method of undetermined coefficients applies, use the method of variation of parameters to determine the general solution to each of the following differential equations.

1. $\ddot{x} + 3\dot{x} + 2x = \cos t$.
2. $\ddot{x} + 3\dot{x} + 2x = t^3$.
3. $\ddot{x} + 3\dot{x} + 2x = e^{-2t}$.
4. $\ddot{x} + 2x = \cos 2t$.
5. $\ddot{x} + 4x = \cos 2t$.

3.23. Determine the general solution to $\ddot{x} - 2t\dot{x} + x = \sec t$.

3.24. Determine the solution to $\dot{x}/x^2 = 1 - 2t$, where $x(0) = -1/6$.

3.25. One nice thing about the method of variation of parameters is that it works for variable-coefficient problems also. For each of the following differential equations[5]

- Verify that $x_1(t)$ and $x_2(t)$ are homogeneous solutions.
- Determine the interval(s) of t for which they are linearly independent and hence form a fundamental set of solutions
- Determine the general solution

1. $x_1(t) = 1 + t, x_2(t) = e^t$, for $t\ddot{x} - (1+t)\dot{x} + x = t^2 e^{2t}$.
2. $x_1(t) = t^{-1/2}\sin t, x_2(t) = t^{-1/2}\cos t$, for $t^2\ddot{x} + t\dot{x} + (t^2 - 1/4)x = 3t^{3/2}\sin t$.
3. $x_1(t) = e^t, x_2(t) = t$, for $(1-t)\ddot{x} + t\dot{x} - x = 2(t-1)^2 e^{-t}$.

3.26. For each of the second-order differential equations listed in Problem 1.8, determine which, if any, of the following solution methods apply based upon what has been covered in this book so far.

1. Assuming exponential solutions
2. Undetermined coefficients
3. Variation of parameters
4. Using the fact that the equation is separable
5. Using the fact that the equation is exact
6. Determining an approximate numerical solution

---

[5] Adapted from [8].

It may be the case that no method, one method, or more than one method may apply.

3.27. Plot the solution to $\ddot{x} + \dot{x} + x = \cos 2.8t + \cos 3.0t$, where $x(0) = 2$ and $\dot{x}(0) = 5$. You may use any method you want, including writing a computer program to determine an approximate numerical solution, but plot the whole solution, not just the steady-state solution. Explain the various features of the problem, namely

1. What is happening between 0 and 10 seconds
2. What is happening between 10 and 60 seconds

3.28. For each of the following differential equations, which of the following methods may be used to find the solution?

1. Undetermined coefficients
2. Variation of parameters
3. A numerical approach (Euler's method)
4. None of the above

Which would be the best method to use and why?

1. $\ddot{x} + e^1 \dot{x} + \sin(5)x = (\sin t)/t$.
2. $t\ddot{x} + te^\pi \dot{x} + t\sin(5)x = \sin t$.
3. $t^2\ddot{x} + t\dot{x} - x = t\sin t$, for $t > 0$,
   where you know that $x_1(t) = t$ and $x_2(t) = 1/t$ are solutions to $t^2\ddot{x} + t\dot{x} - x = 0$.
4. $\ddot{x} + x\dot{x} + 5x = 0$.
5. $\frac{\partial^2 \phi}{\partial x^2} = \frac{\partial^2 \phi}{\partial t^2}$.
6. $\ddot{x} - 3\dot{x} + x = 2e^{2t}\sin t$.
7. $\ddot{x} + e^t \dot{x} + \sin(5)x = 0$.
8. $t\ddot{x} + te^t \dot{x} + t\sin(5)x = \sin t$.
9. $\ddot{x} + 3\dot{x} + \pi x = e^{23t}t^4 + t^2\sin(t)$.
10. $\ddot{x} + 3\dot{x} + \pi x = (e^{23t})/t^4 + t^2\sin(t)$.

3.29. Table 3.1 contains 27 differential equations and Figure 3.6 contains plots of 27 different solutions. Each plot has three solutions, except the ones in the last row which have two and one solution, respectively. The plots are the solution $x(t)$ versus t. Match each equation with the corresponding plot. It is possible to do this by solving only six equations! In order to clearly communicate your answer, sketch each of the plots by hand and indicate which equation goes with which solution. On your sketch, indicate the feature(s) of the solution that were the basis for your conclusion.

3.30. Determine the solution to $t + x\dot{x}e^t = 0$, where $x(0) = 1$.

	A	B	C
1	$\ddot{x} + 8\dot{x} + 4x = \sin t$ $x(0) = 1, \quad \dot{x}(0) = 0$	$\dot{x} = -5x$ $x(0) = 1$	$\ddot{x} + \dot{x} + 4x = 0$ $x(0) = 1, \quad \dot{x}(0) = 0$
2	$\dot{x} + 3x = 1$ $x(0) = 1$	$\ddot{x} + x = 0$ $x(0) = 1, \quad \dot{x}(0) = 0$	$\ddot{x} + \frac{1}{2}\dot{x} + 4x = 0$ $x(0) = 1, \quad \dot{x}(0) = 0$
3	$x\dot{x}e^{2t} - t = 0$ $x(0) = 1$	$\ddot{x} + 8\dot{x} + 4x = \sin 2t$ $x(0) = 1, \quad \dot{x}(0) = 0$	$\dot{x} + x = 1$ $x(0) = 1$
4	$\dot{x} - 0.1x = 0$ $x(0) = 1$	$\dot{x} = -5x + 1$ $x(0) = 1$	$\dot{x} + (t - 0.1)x = 0$ $x(0) = 1$
5	$\ddot{x} + 4x = \sin t$ $x(0) = 1, \quad \dot{x}(0) = 0$	$\dot{x} + (t - 1)x = 0$ $x(0) = 1$	$x\dot{x}e^{3t} - t = 0$ $x(0) = 1$
6	$\dot{x} = 0.5x$ $x(0) = 1$	$\ddot{x} + 4x = \sin 2t$ $x(0) = 1, \quad \dot{x}(0) = 0$	$\ddot{x} + 3x = 0$ $x(0) = 1, \quad \dot{x}(0) = 0$
7	$\dot{x} - x = 0$ $x(0) = 1$	$\dot{x} + x = 0$ $x(0) = 1$	$\ddot{x} + 2\dot{x} + 4x = 0$ $x(0) = 1, \quad \dot{x}(0) = 0$
8	$x\dot{x}e^{t} - t = 0$ $x(0) = 1$	$\ddot{x} + 4x = \sin 1.9t$ $x(0) = 1, \quad \dot{x}(0) = 0$	$\ddot{x} + 2x = 0$ $x(0) = 1, \quad \dot{x}(0) = 0$
9	$\dot{x} + (t - 0.5)x = 0$ $x(0) = 1$	$\dot{x} + 3x = 0$ $x(0) = 1$	$\ddot{x} + 0.2\dot{x} - x + x^3 = 0.3\sin t$ $x(0) = 1, \quad \dot{x}(0) = 0$

Table 3.1 Differential equations for Exercise 3.29

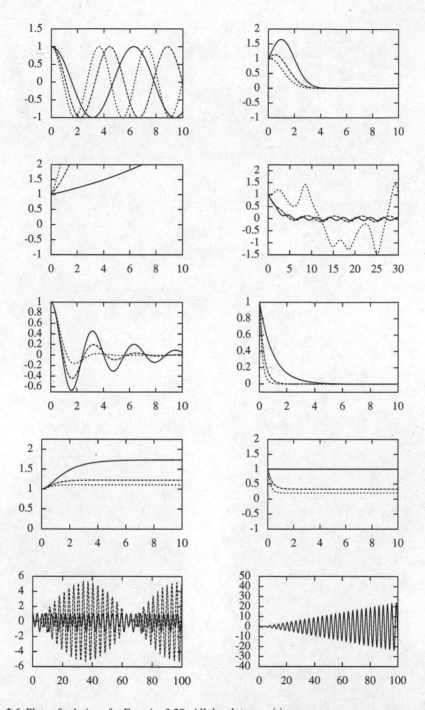

Fig. 3.6 Plots of solutions for Exercise 3.29. All the plots are $x(t)$ versus t.

Chapter 4
Single Degree of Freedom Vibrations

This chapter presents applications of second-order, ordinary, constant-coefficient differential equations. The primary applications in mechanical engineering and related fields is that of vibrations analysis. Additionally, because a second-order system is a canonical system for the design of some feedback controllers, this material serves as important background for the design of feedback controllers, which are considered in Chapter 10.

The study of single degree of freedom vibrations considers the analysis of problems of the type illustrated in Figure 4.1 and described by

$$m\ddot{x} + b\dot{x} + kx = f(t), \tag{4.1}$$

where $f(t)$ is an applied force. The term "single" refers to the fact that the system has only one degree of freedom. This is in contrast with a multiple degree of freedom system, an example of which is illustrated in Figure 6.1. This type of problem is generally categorized according to whether it is

- Free or forced
- Damped or undamped

Sections 4.1 through 4.4 consider each of the four possible permutations of these cases.

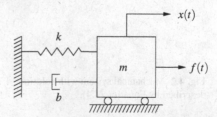

Fig. 4.1 Mechanical system described by Equation (4.1).

This chapter is a complete study and analysis of the solutions to

B. Goodwine, *Engineering Differential Equations: Theory and Applications*, DOI 10.1007/978-1-4419-7919-3_4, © Springer Science+Business Media, LLC 2011

$$m\ddot{x} + b\dot{x} + kx = f(t) \quad \Longleftrightarrow \quad \ddot{x} + 2\zeta\omega_n\dot{x} + \omega_n^2 x = \frac{f(t)}{m}. \tag{4.2}$$

The system is *free* if it is unforced; that is, $f(t) = 0$; otherwise it is *forced*. The system is *undamped* if $b = 0$ (equivalently $\zeta = 0$); otherwise, it is damped. Although somewhat scattered throughout the example problems in Chapter 3, the quantities of major importance in this chapter that have already been introduced include the following.

1. The *natural frequency*: $\omega_n = \sqrt{k/m} > 0$
2. The *damping ratio*: $\zeta = b/\left(2\sqrt{km}\right) > 0$
3. The *damped natural frequency*: $\omega_d = \omega_n\sqrt{1 - \zeta^2}$ (only relevant for $0 < \zeta < 1$)

4.1 Free Undamped Oscillations

This problem has been completely solved in Section 3.4.1 and particularly in Example 3.9. Free and undamped implies that in Figure 4.1 $b = 0$ and $f(t) = 0$, or equivalently, that the system is as illustrated in Figure 4.2, so the equation of motion reduces to

$$m\ddot{x} + kx = 0 \quad \Longleftrightarrow \quad \ddot{x} + \omega_n^2 x = 0,$$

which, as presented previously, has a general solution

$$x(t) = c_1\cos\omega_n t + c_2\sin\omega_n t.$$

If the initial conditions are specified as

$$x(0) = x_0, \quad \dot{x}(0) = \dot{x}_0,$$

then the solution is

$$x(t) = x_0\cos\omega_n t + \frac{\dot{x}_0}{\omega_n}\sin\omega_n t. \tag{4.3}$$

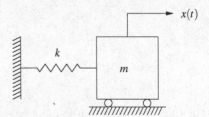

Fig. 4.2 Mechanical system with solution described by Equation (4.4).

This equation is relatively simple to interpret and plot, however, it can be made even simpler to analyze if the sine and cosine terms are combined. In particular, equate the solution in Equation (4.3) with a phase-shifted cosine function

$$x_0 \cos \omega_n t + \frac{\dot{x}_0}{\omega_n} \sin \omega_n t = c \cos(\omega_n t + \phi)$$
$$= c(\cos \phi \cos \omega_n t - \sin \phi \sin \omega_n t).$$

Inasmuch as the set of functions $\{\sin \omega_n t, \cos \omega_n t\}$ is linearly independent, the coefficients of those functions must be equal. Hence,

$$c \cos \phi = x_0, \quad c \sin \phi = \frac{\dot{x}_0}{\omega_n}.$$

Normalizing the right hand side of each to have an absolute value less than or equal to one and to provide a common value for the constant c gives

$$c \cos \phi = \sqrt{x_0^2 + \frac{\dot{x}_0^2}{\omega_n^2}} \, \frac{x_0}{\sqrt{x_0^2 + \frac{\dot{x}_0^2}{\omega_n^2}}}, \quad c \sin \phi = \sqrt{x_0^2 + \frac{\dot{x}_0^2}{\omega_n^2}} \, \frac{\dot{x}_0}{\omega_n \sqrt{x_0^2 + \frac{\dot{x}_0^2}{\omega_n^2}}}.$$

Hence, solving for c and ϕ gives

$$c = \sqrt{x_0^2 + \left(\frac{\dot{x}_0}{\omega_n}\right)^2}, \quad \phi = \tan^{-1}\left(-\frac{\dot{x}_0}{\omega_n x_0}\right),$$

so an equivalent representation of the solution is

$$\boxed{x(t) = \sqrt{x_0^2 + \left(\frac{\dot{x}_0}{\omega_n}\right)^2} \cos(\omega_n t + \phi).} \tag{4.4}$$

Remark 4.1. The function commonly denoted by $\tan^{-1}$ normally can not distinguish between quadrants in the plane. Throughout this text the arc tangent function is always considered to be the one that is able to distinguish the quadrants.[1] In particular, $\tan^{-1}$ is used to denote the angle ϕ that satisfies

$$\sin \phi = \frac{-\dot{x}_0/\omega_n}{\sqrt{x_0^2 + \left(\frac{\dot{x}_0}{\omega_n}\right)^2}}, \quad \cos \phi = \frac{x_0}{\sqrt{x_0^2 + \left(\frac{\dot{x}_0}{\omega_n}\right)^2}}.$$

Example 4.1. Figure 4.3 is a plot of the solution to

$$\ddot{x} + \omega_n x = 0,$$

where $x(0) = 1$ and $\dot{x}(0) = 1$ for $\omega_n = 1, 2$, and 3.

As is obvious from the form of the solution in Example 4.1, the solution is a harmonic with a constant amplitude. As the natural frequency increases, the fre-

---

[1] Numerical computational packages and programming languages often use the function `atan2()` to denote this function.

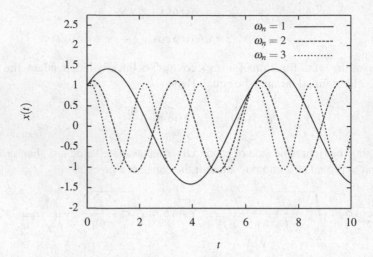

Fig. 4.3 Solutions to system in Example 4.1.

quency of the response increases. Also due to the $\dot{x}_0/\omega_n$ term in the amplitude of the response, as ω_n increases, the amplitude of the response decreases.

From the example and an analysis of the form of the solution in Equation (4.4), one may conclude the following regarding the response of an undamped, free, single degree of freedom system.

1. If ω_n increases, the frequency of the response will increase.
2. If k increases, the frequency of the response will increase.
3. If m increases, the frequency of the response will decrease.
4. If $|x_0|$ increases, the magnitude of the response will increase.
5. If $|\dot{x}_0|$ increases, the magnitude of the response will increase.
6. If $\dot{x}_0 \neq 0$ and ω_n increases, the magnitude of the response will decrease.

4.2 Harmonically Forced Undamped Vibrations

Now the problem considered in the previous section is modified to add a forcing function acting on the mass as illustrated in Figure 4.4. The most common scenario is the case when the forcing function, $f(t)$ is a harmonic function (i.e., sines, cosines, or combinations thereof).

Consider the case when $f(t) = F \cos \omega t$, that is, a harmonic function of magnitude F and frequency ω. Note that there are now two frequencies appearing in the problem; namely, the natural frequency $\omega_n = \sqrt{k/m}$ and the frequency of the forcing function ω. In general, they are not the same and care must be taken to observe the subscript or absence thereof. The equation of motion for this system is

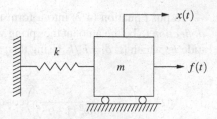

Fig. 4.4 Single degree of freedom, undamped, forced oscillator.

$$m\ddot{x} + kx = F\cos\omega t \quad \Longleftrightarrow \quad \ddot{x} + \omega_n^2 x = \frac{F}{m}\cos\omega t. \tag{4.5}$$

Clearly, this is an ordinary, second-order, constant-coefficient, linear, inhomogeneous differential equation; furthermore, due to the form of the inhomogeneous term, the method of undetermined coefficients is probably the most expedient solution method. From Section 4.1, the homogeneous solution is

$$x_h(t) = c_1\cos\omega_n t + c_2\sin\omega_n t.$$

Hence, for undetermined coefficients, assume

$$x_p(t) = A\cos\omega t + B\sin\omega t,$$

as long as $\omega \neq \omega_n$! The special case where $\omega = \omega_n$, which requires

$$x_p(t) = t\left(A\cos\omega t + B\sin\omega t\right),$$

is considered subsequently.

Differentiating x_p and substituting gives

$$A = \frac{F}{m(\omega_n^2 - \omega^2)}, \quad B = 0,$$

so a general solution to Equation (4.5) is

$$x(t) = c_1\cos\omega_n t + c_2\sin\omega_n t + \frac{F}{m(\omega_n^2 - \omega^2)}\cos\omega t. \tag{4.6}$$

If the initial conditions are specified as $x(0) = x_0$ and $\dot{x}(0) = \dot{x}_0$, then a quick calculation gives

$$c_1 = x_0 - \frac{F}{m(\omega_n^2 - \omega^2)}, \quad c_2 = \frac{\dot{x}_0}{\omega_n},$$

and hence the solution to the initial value problem is

$$x(t) = \left(x_0 - \frac{F}{m(\omega_n^2 - \omega^2)}\right)\cos\omega_n t + \frac{\dot{x}_0}{\omega_n}\sin\omega_n t + \frac{F}{m(\omega_n^2 - \omega^2)}\cos\omega t. \tag{4.7}$$

To put Equation (4.7) into a form more amenable to analysis, define the *static deflection* to be the amount the spring would displace due to a static force of magnitude F, which is, $\delta = F/k$. Using this, and defining the frequency ratio as $r = \omega/\omega_n$ gives

$$x(t) = \left(x_0 - \frac{F}{m\omega_n^2\left(1 - r^2\right)} \right) \cos\omega_n t + \frac{\dot{x}_0}{\omega_n}\sin\omega_n t + \frac{F}{m\omega_n^2\left(1 - r^2\right)}\cos\omega t,$$

or

$$\boxed{x(t) = \left(x_0 - \frac{\delta}{1 - r^2} \right)\cos\omega_n t + \frac{\dot{x}}{\omega_n}\sin\omega_n t + \frac{\delta}{1 - r^2}\cos\omega t.} \qquad (4.8)$$

This solution can be considered in two parts. The first two terms depend on the natural frequency and are from the homogeneous solution. The third term depends on the forcing frequency and is from the particular solution. The effect of the magnitude and frequency of the forcing also appears in the coefficient $\delta/\left(1 - r^2\right)$. If the magnitude of the force is increased, δ increases proportionally and the two corresponding coefficients increase in magnitude.

If the frequency of the forcing is changed, the frequency of the $\cos\omega t$ terms correspondingly changes which changes the frequency of that component of the solution. However, the magnitude of two of the coefficients also changes. The amount by which the frequency affects these coefficients is defined by the *magnification factor*

$$M = \frac{1}{1 - \left(\frac{\omega}{\omega_n}\right)^2} = \frac{1}{1 - r^2}.$$

The magnification factor is the amount by which the static deflection is either amplified or attenuated in the solution and is a function of the ratio between the forcing frequency of the system and the natural frequency. A plot of the magnification factor versus frequency ratio is illustrated in Figure 4.5. Note that the case where $r = 1$ is seemingly problematic; however, recall that is the case where $\omega = \omega_n$, which has a different solution. Also observe that for frequency ratios greater than one, the magnification ratio is negative, which represents the fact that the particular solution is out of phase with the forcing function.

Note that the solution, in Equation (4.8) depends upon

1. The natural frequency ω_n,
2. The forcing frequency ω,
3. The static deflection δ, and,
4. The initial conditions x_0 and $\dot{x}_0$.

Note, however, that the initial conditions as well as the static deflection simply scale individual terms of the solution. Therefore, the most interesting feature of the solution is its dependence on the forcing and natural frequencies, which are explored in the following example.

Example 4.2. Plot the solution for

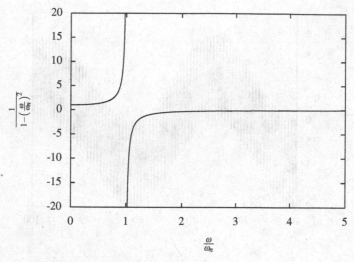

Fig. 4.5 Magnification factor versus frequency ratio.

$$\ddot{x} + \omega_n^2 x = \frac{F}{m} \cos \omega t$$

where $x(0) = 0$, $\dot{x}(0) = 0$ and $\omega \ll \omega_n$; that is, the forcing frequency is much smaller than the natural frequency. With zero initial conditions, the solution is

$$x(t) = \frac{\delta}{1 - r^2} \left(\cos \omega t - \cos \omega_n t \right)$$

and if $\omega \ll \omega_n$, $r \approx 0$; hence,

$$x(t) \approx \delta \left(\cos \omega t - \cos \omega_n t \right).$$

Thus, the solution will vary in magnitude between 0 and 2δ depending upon whether ω and ω_n are in phase or out of phase.

A plot of the solution where $\delta = 1$, $\omega = 0.1$, and $\omega_n = 5$ is illustrated in Figure 4.6. Note that because the two frequencies are well separated, the solution is clearly the superposition of two cosine functions, one relatively fast and the other relatively slow.

Example 4.3. Now consider the other extreme where $\omega \gg \omega_n$; that is, the system is forced at a frequency that is much greater than the natural frequency. In this case, the frequency ratio will become very large and the coefficient of the solution

$$\frac{\delta}{1 - r^2} \approx -\frac{\delta}{r^2}$$

will be very small.

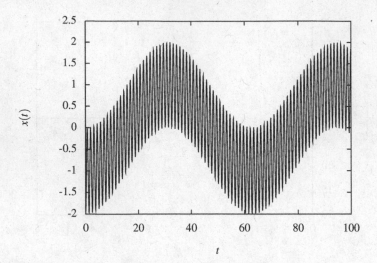

Fig. 4.6 Harmonically forced, undamped solution where $\omega \ll \omega_n$.

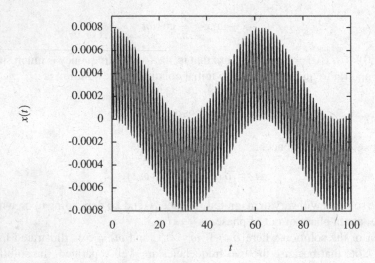

Fig. 4.7 Harmonically forced, undamped solution where $\omega_n \ll \omega$.

A plot of the solution where $\delta = 1$, $\omega = 5$, and $\omega_n = 0.1$ is illustrated in Figure 4.7. At first glance this appears similar to the response when $\omega \ll \omega_n$; however, note the scale on the graph. The response is still the sum of two cosine functions but the magnitude of the response is much smaller than in the case where $\omega \ll \omega_n$

Example 4.4. Yet another interesting feature of this solution is apparent when one considers the relative phase between the forcing function $F \cos \omega t$ and the response of the system. Just for the fun of it, let us assume that

$$x(0) = -\frac{\delta}{1 - r^2}, \quad \dot{x}(0) = 0.$$

The initial conditions were picked so that the terms in the solution due to the homogeneous solutions are zero and the complete solution is the same as the particular solution; namely,

$$x(t) = \frac{\delta}{1 - r^2} \cos \omega t.$$

Recall that the forcing function is

$$f(t) = F \cos \omega t.$$

The response $x(t)$ and forcing function $f(t)$ are plotted together for the two cases where $\omega < \omega_n$ ($r = 0.5$) and $\omega > \omega_n$ ($r = 1.5$) in Figures 4.8 and 4.9, respectively. In both figures, $\delta = 1$.

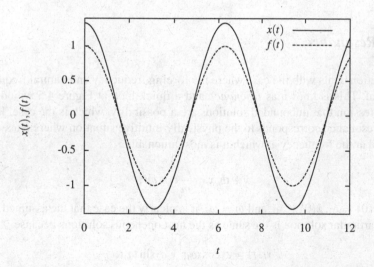

Fig. 4.8 Forcing function and particular solution in phase ($\omega < \omega_n$).

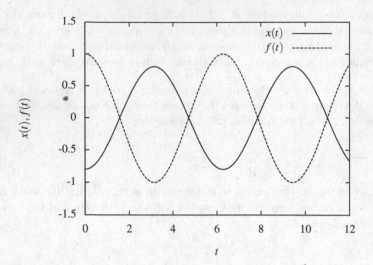

Fig. 4.9 Forcing function and particular solution out of phase ($\omega > \omega_n$).

The interesting feature of these solutions is that the response of the system is in phase with the forcing function when $\omega < \omega_n$ and out of phase with the forcing function when $\omega > \omega_n$. The latter is the somewhat counterintuitive case when the force is always directed in the opposite direction of the velocity of the mass.

4.2.1 Resonance

This section deals with the case where the forcing frequency and natural frequency are equal. This is known as *resonance* and a quick look at Figure 4.5 would give the impression that unbounded solutions are a possibility, which is the case. In addition, resonance corresponds to the physically intuitive situation wherein a system is forced at the frequency at which it is most amenable.

For

$$\ddot{x} + \omega_n^2 x = \frac{F}{m} \cos \omega t, \tag{4.9}$$

where $x(0) = x_0$, $\dot{x}(0) = \dot{x}_0$, and $\omega = \omega_n$, it is clearly the case that the assumed form of the particular solution is the same as the homogeneous solutions because

$$x_h(t) = c_1 \cos \omega_n t + c_2 \sin \omega_n t.$$

So, the correct assumption is

$$x_p(t) = t \left(A \cos \omega_n t + B \sin \omega_n t \right).$$

Skipping the mundane details of substituting and equating coefficients, the solution is

$$x(t) = x_0 \cos \omega_n t + \frac{\dot{x}_0}{\omega_n} \sin \omega_n t + \frac{\delta \omega_n t}{2} \sin \omega_n t. \qquad (4.10)$$

The part of the solution that is the particular solution (the second $\sin \omega_n t$ term) is multiplied by t, thus it grows linearly in time. A specific example follows, but the general point that the solution grows with time is the fundamentally important point regarding resonance.

Example 4.5. Solve

$$\ddot{x} + x = \cos t,$$

where $x(0) = 0$, $\dot{x}(0) = 0$ and plot the solution versus time.

This equation is exactly of the form of Equation (4.9) with $\omega_n = F = m = 1$, therefore simply substituting those values into Equation (4.10) gives the solution

$$x(t) = \frac{t}{2} \sin t,$$

which is plotted in Figure 4.10.

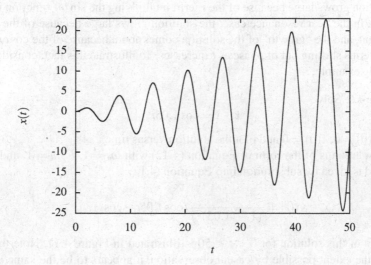

Fig. 4.10 Solution for Example 4.5.

4.2.2 Near Resonance

Obviously in physical situations, it is impossible to have exactly $\omega = \omega_n$, so the question regarding the nature of the solution when $\omega \approx \omega_n$ and its relationship to the resonance solution naturally arises.

Consider

$$\ddot{x} + \omega_n^2 x = \frac{F}{m}\cos\omega t, \qquad (4.11)$$

where $x(0) = x_0$, $\dot{x}(0) = \dot{x}_0$, and $\omega \approx \omega_n$. Because $\omega \neq \omega_n$, the solution is not from Equation (4.10) but rather is from Equation (4.7),

$$x(t) = \left(x_0 - \frac{F}{m(\omega_n^2 - \omega^2)}\right)\cos\omega_n t + \frac{\dot{x}_0}{\omega_n}\sin\omega_n t + \frac{F}{m(\omega_n^2 - \omega^2)}\cos\omega t.$$

If $\omega \approx \omega_n$, then the two coefficients with $(\omega_n - \omega)$ in the denominator will be very large. Rewriting the solution by grouping those two terms gives

$$x(t) = x_0\cos\omega_n t + \frac{\dot{x}_0}{\omega_n}\sin\omega_n t + \frac{F}{m(\omega_n^2 - \omega^2)}\left(\cos\omega t - \cos\omega_n t\right). \qquad (4.12)$$

Note that the terms in this solution and in the resonance solution in Equation (4.10) that depend on the initial conditions are identical. In the resonance case, the solution grows large because of the t term multiplying the $\sin\omega t$ function in the solution. In the near resonance case, the solution grows large because of the large coefficient, and the "growth" of the solution comes about because of the $\cos\omega t$ and $\cos\omega_n t$ terms shifting out of phase as t increases. To illustrate this fact, consider the following example.

Example 4.6. Solve

$$\ddot{x} + x = \cos 1.05t,$$

where $x(0) = 0$, $\dot{x}(0) = 0$ and plot the solution versus time.

The solution is of the form of Equation (4.12) with $\omega_n = 1$, $F/m = 1$, and $\omega = 1.05$, and is given by substitution into Equation (4.12)

$$x(t) = \frac{1}{(1 - (1.05)^2)}\left(\cos 1.05t - \cos t\right).$$

A plot of this solution for $0 < t < 50$ is illustrated in Figure 4.11. Note that, at least to the extent possible by casual observation, it appears to be the same as the solution illustrated for resonance in Figure 4.10.

Plotting the solution for a longer period of time, $0 < t < 500$, as illustrated in Figure 4.12, highlights the main difference. The solution grows because the cosine terms slowly go out of phase as time increases, therefore they eventually must go back in phase, resulting in a decrease in magnitude of the solution, which is in contrast to the resonance solution which always grows with time. This phenomenon is called *beating*.

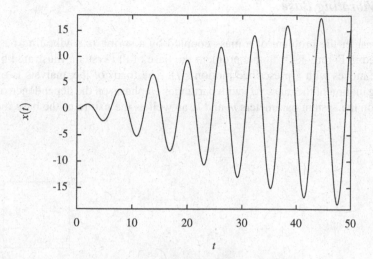

Fig. 4.11 Solution for Example 4.6 near resonance for $0 < t < 50$.

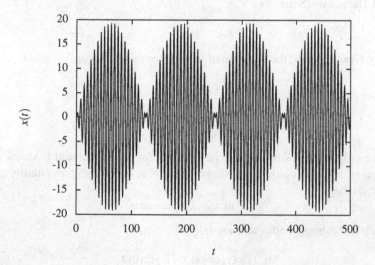

Fig. 4.12 Solution for Example 4.6 for $0 < t < 500$.

4.2.3 Vibrating Base

Now consider the problem of a mass coupled by a spring to a vibrating base, as illustrated in Figure 4.13. In this problem the base of the system, illustrated by the thin bar, moves with a prescribed motion $z(t)$. The focus of the analysis is on the resulting motion of the mass $x(t)$ with particular emphasis on the dependence of this motion on the system parameters m and k, as well as the nature of the base motion $z(t)$.

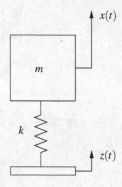

Fig. 4.13 Undamped vibrating base system.

Using Newton's law, the equation of motion for this system is

$$m\ddot{x} + kx = kz(t).$$

or

$$\ddot{x} + \omega_n^2 x = \omega_n^2 z(t).$$

Thus, the only variables of concern in the problem are the natural frequency and the nature of $z(t)$. For simplicity, assume that $z(t)$ is harmonic, particularly, $z(t) = Z\cos \omega t$, so that

$$\ddot{x} + \omega_n^2 x = Z\omega_n^2 \cos \omega t.$$

Clearly, the homogeneous solution is

$$x_h(t) = c_1\cos \omega_n t + c_2\sin \omega_n t.$$

Assuming that $\omega \neq \omega_n$ and

$$x_p(t) = A\cos \omega t + B\sin \omega t,$$

substituting and equating coefficients gives

$$x_p(t) = \frac{Z\omega_n^2}{\omega_n^2 - \omega^2}\cos \omega t$$

so that

$$x(t) = c_1 \cos \omega_n t + c_2 \sin \omega_n t + \frac{Z}{1 - r^2} \cos \omega t,$$

where, as before, $r = \omega/\omega_n$.

The coefficient of the part of the solution which is the particular solution is exactly of the form of Equation (4.8), therefore the analysis is exactly the same as the undamped, forced oscillation case, but where the static deflection is replaced by the magnitude of the base motion. In this case, the magnification factor,

$$M = \frac{1}{1 - r^2}$$

has an even more direct interpretation in that it is the magnification of the base motion in the response of the mass motion. Referring back to Figure 4.5, M has a value of 1 at $r = 0$, increases to an unbounded value at $r = 1$, decreases to $M = 1$ at $r = \sqrt{2}$ and asymptotically approaches zero as r gets large. Hence, the resonance analysis is similar to that of the simple forced case. The smallest magnification occurs at very high frequencies and for frequency ratios greater than one, the motion of the base and mass are out of phase.

4.3 Free Damped Vibrations

This section considers the case of damped oscillations with no forcing function, that is, the solution to

$$\ddot{x} + 2\zeta \omega_n \dot{x} + \omega_n^2 x = 0$$

where $x(0) = x_0$, $\dot{x}(0) = \dot{x}_0$ and $\zeta \neq 0$.

Because this is a constant-coefficient, linear, homogeneous, second-order ordinary differential equation, it has exponential solutions. The resulting characteristic equation is

$$\lambda^2 + 2\zeta \omega_n \lambda + \omega_n^2 = 0$$

with roots

$$\lambda = -\zeta \omega_n \pm \omega_n \sqrt{\zeta^2 - 1}.$$

The nature of the solution clearly depends upon whether ζ is less than one, equal to one, or greater than one.

4.3.1 Damping Ratio Greater than One

In this case, the solution is

$$x(t) = c_1 e^{\left(-\zeta \omega_n + \omega_n \sqrt{\zeta^2 - 1}\right)t} + c_2 e^{\left(-\zeta \omega_n - \omega_n \sqrt{\zeta^2 - 1}\right)t}$$

and evaluating the initial conditions gives

$$x(0) = c_1 + c_2 = x_0$$

and

$$\dot{x}(0) = \left(-\zeta\omega_n + \omega_n\sqrt{\zeta^2 - 1} \right) c_1 + \left(-\zeta\omega_n - \omega_n\sqrt{\zeta^2 - 1} \right) c_2 = \dot{x}_0$$

which gives

$$x(t) = \frac{\dot{x}_0 + x_0\omega_n\left(\zeta + \sqrt{\zeta^2 - 1} \right)}{2\omega_n\sqrt{\zeta^2 - 1}} e^{\left(-\zeta\omega_n + \omega_n\sqrt{\zeta^2 - 1} \right)t}$$
$$- \frac{\dot{x}_0 - x_0\omega_n\left(\zeta + \sqrt{\zeta^2 - 1} \right)}{2\omega_n\sqrt{\zeta^2 - 1}} e^{\left(-\zeta\omega_n - \omega_n\sqrt{\zeta^2 - 1} \right)t}$$

Figure 4.14 illustrates the response for $\omega_n = 1$, $x(0) = 1$, and $\dot{x}(0) = 0$ for various values of ζ.

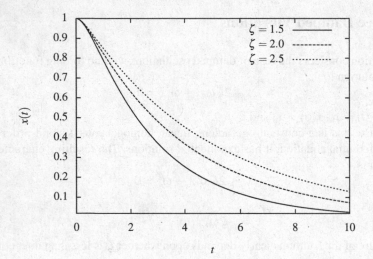

Fig. 4.14 Solution of second-order system for various values of ζ.

4.3.2 Damping Ratio Equal to One

When the damping ratio is equal to one there are repeated roots of the characteristic equation

$$\lambda = -\omega_n$$

so the general solution to the homogeneous equation is

$$x(t) = c_1 e^{-\omega_n t} + c_2 t e^{-\omega_n t}.$$

Inasmuch as the exponential decays faster than t grows, this solution approaches zero as t gets large.

4.3.3 Damping Ratio Less than One

When the damping ratio is less than one, the characteristic equation has complex roots

$$\lambda = -\zeta \omega_n \pm \omega_n \sqrt{\zeta^2 - 1} = -\zeta \omega_n \pm i\omega_n \sqrt{1 - \zeta^2}$$

so the general solution to the differential equation is

$$x(t) = e^{-\zeta \omega_n t} \left(c_1 \cos \omega_n \sqrt{1 - \zeta^2} t + c_2 \sin \omega_n \sqrt{1 - \zeta^2} t \right). \qquad (4.13)$$

Figure 4.15 illustrates the response for $\omega_n = 1$, $x(0) = 1$, and $\dot{x}(0) = 0$ for various values of ζ.

4.4 Harmonically Forced Damped Vibrations

In this section we consider the system illustrated in Figure 4.1 with the equation of motion given by Equation (4.2) where the applied force $f(t)$ is assumed to be harmonic. For the rest of this section we assume the forcing function is of the form

$$f(t) = F \cos \omega t.$$

Confirming the details that the solution for a sine function of a combination of sines and cosines is left as an exercise.

For the system

$$\ddot{x} + 2\zeta \omega_n x + \omega_n^2 x = \frac{F}{m} \cos \omega t \qquad (4.14)$$

Section 4.3 provides all the possible cases for the homogeneous solution.

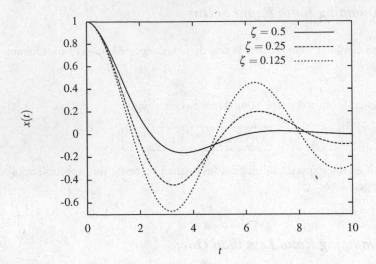

Fig. 4.15 Solution of second-order system for various values of ζ.

Using the method of undetermined coefficients, and consequently assuming a particular solution of the form

$$x_p(t) = A\cos\omega t + B\sin\omega t$$

gives

$$\dot{x}_p = -A\omega\sin\omega t + B\omega\cos\omega t, \quad \ddot{x}_p = -A\omega^2\cos\omega t - B\omega^2\sin\omega t$$

and substituting into Equation (4.14) gives

$$\left(-A\omega^2\cos\omega t - B\omega^2\sin\omega t\right) + 2\zeta\omega_n\left(-A\omega\sin\omega t + B\omega\cos\omega t\right)$$
$$+ \omega_n^2\left(A\cos\omega t + B\sin\omega t\right) = \frac{F}{m}\cos\omega t.$$

A bit of algebra gives A and B so that

$$x_p(t) = \frac{F}{m}\left(\frac{\omega_n^2 - \omega^2}{\left(\omega_n^2 - \omega^2\right)^2 + \left(2\zeta\omega\omega_n\right)^2}\cos\omega t + \frac{2\zeta\omega\omega_n}{\left(\omega_n^2 - \omega^2\right)^2 + \left(2\zeta\omega\omega_n\right)^2}\sin\omega t\right).$$
$$(4.15)$$

Observe that as long as $\zeta \neq 0$ the solution given by Equation (4.15) is correct, even in the case of resonance when $\omega = \omega_n$. Furthermore, when $\zeta = 0$ this reduces to the undamped forced solution as long as $\omega \neq \omega_n$.

As before, we convert the solution to the form of a single trigonometric function with a phase shift; for example,

$$x_p(t) = c\cos(\omega t + \phi),$$

where the magnitude of the response c and the phase shift ϕ must be determined. Note that, reverting back to the coefficients A and B

$$x_p(t) = A\cos\omega t + B\sin\omega t$$
$$= \sqrt{A^2 + B^2}\left(\frac{A}{\sqrt{A^2 + B^2}}\cos\omega t + \frac{B}{\sqrt{A^2 + B^2}}\sin\omega t\right).$$

The coefficients of the sine and cosine terms must have values in the interval $[-1, 1]$, thus we can write this as

$$x_p(t) = c\left(\cos\phi\cos\omega t - \sin\phi\sin\omega t\right) = c\cos(\omega t + \phi),$$

where

$$c = \sqrt{A^2 + B^2}, \quad \phi = \tan^{-1}\left(-\frac{B}{A}\right),$$

or, using the actual expressions for A and B

$$c = \frac{F}{m}\sqrt{\frac{1}{(\omega_n^2 - \omega^2)^2 + (2\zeta\omega_n\omega)^2}}, \quad \phi = \tan^{-1}\left(-\frac{2\zeta\omega\omega_n}{\omega_n^2 - \omega^2}\right).$$

So, that bit of work resulted in

$$x_p(t) = \frac{F}{m}\sqrt{\frac{1}{(\omega_n^2 - \omega^2)^2 + (2\zeta\omega_n\omega)^2}}\cos(\omega t + \phi),$$

where ϕ is given as above.

A final step, that may not be obvious a priori is to factor an ω_n^2 out of the denominator of c, which gives

$$x_p(t) = \frac{F}{\omega_n^2 m}\sqrt{\frac{1}{\left(1 - \frac{\omega^2}{\omega_n^2}\right)^2 + \left(2\zeta\frac{\omega}{\omega_n}\right)^2}}\cos(\omega t + \phi).$$

Note that

$$\frac{F}{\omega_n^2 m} = \frac{F}{k} = \delta,$$

which gives

$$\boxed{x_p(t) = \delta\sqrt{\frac{1}{\left(1 - \frac{\omega^2}{\omega_n^2}\right)^2 + \left(2\zeta\frac{\omega}{\omega_n}\right)^2}}\cos(\omega t + \phi)}$$

which is a complete description of the particular solution in terms of the static deflection, the ratio of the forcing frequency to the natural frequency and the damping ratio. The quantity

$$M = \sqrt{\frac{1}{\left(1 - \frac{\omega^2}{\omega_n^2}\right)^2 + \left(2\zeta\frac{\omega}{\omega_n}\right)^2}}$$

can be interpreted to represent the amount that the static deflection is amplified or attenuated in the response of the system due to the frequency of the forcing function. The phase shift can similarly be expressed in terms of the frequency ratio by dividing the numerator and denominator by ω_n^2 giving

$$\phi = \tan^{-1}\left(-\frac{2\zeta\frac{\omega}{\omega_n}}{1 - \frac{\omega^2}{\omega_n^2}}\right).$$

We may gain insight into the nature of the response by considering the nature of the dependence of M and ϕ on the frequency ratio ω/ω_n and the damping ratio, ζ. A plot of M as a function of the frequency ratio for different damping ratios is illustrated in Figure 4.16 and a plot of the phase shift ϕ as a function of the frequency ratio for different damping ratios is illustrated in Figure 4.17.

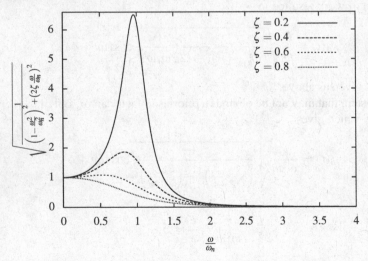

Fig. 4.16 Magnification of static deflection for various damping ratios versus frequency ratio.

Up to now this section has considered only the particular solution to Equation (4.14). However, as long as the damping ratio is not zero, the homogeneous solution will decay and the above analysis of M and ϕ are appropriate to consider

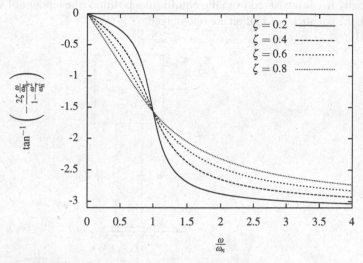

Fig. 4.17 Phase shift between response and forcing function for various damping ratios versus frequency ratio.

for the steady-state solution, that is, after the transient response represented by the homogeneous solution has become negligible.

4.4.1 Resonance

Resonance when the damping ratio is greater than zero does not require a different solution method. If damping is light, however, the magnitude of the response may be large; however, unlike the undamped case, it does not grow unbounded.

4.4.2 Vibrating Base

In this section we consider an example that is illustrative of the operation of a suspension system.

Example 4.7. Consider the system illustrated in Figure 4.18. Assume that a vehicle is driving over a road with constant velocity v and that the surface of the road is such that the center of the wheel follows a sinusoidal path with wavelength λ and height h. Determine the magnitude of the steady-state motion of the car body (the mass) as well as the force transmitted to the mass as a function of the velocity of the vehicle. For this problem we assume there is no gravity and that x is measured from the unstretched position of the spring. It is left as an exercise to show that if there is gravity if x is measured from the equilibrium position (the amount of static deflection), the equations of motion are unchanged.

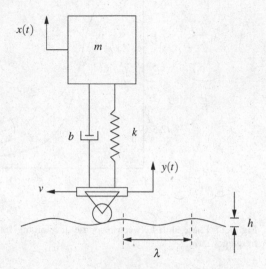

Fig. 4.18 Model suspension system.

The first task is to determine the vertical motion of the wheel. Because the velocity of the wheel is v, the horizontal position of the wheel at time t is given by vt. Because the wavelength of the oscillations of the road surface is λ, that means that the argument to the sine function will need to go from zero to 2π in the amount of time it takes the vehicle to travel the distance λ. Hence, the time to travel λ is $T = \lambda/v$ is the period and the vertical motion of the wheel is given by

$$y(t) = h\sin\left(\frac{2\pi}{T}t + \hat{\phi}\right) = h\sin\left(\frac{2\pi vt}{\lambda} + \hat{\phi}\right) \qquad (4.16)$$

where $\hat{\phi}$ is some unknown phase angle (we do not know where on the road the car was at $t = 0$, and we show subsequently that if all we care about is the steady-state behavior, then it does not matter).

To use Newton's law to derive the equations of motion of the system, we must draw a free body diagram of the mass, as illustrated in Figure 4.19. The only forces acting on the mass are from the damper and spring in the suspension. The force of the spring is proportional to the amount it is compressed, which in this case is $y(t) - x(t)$. Similarly, the force from the damper is proportional to the rate at which it is being compressed, which is $\dot{y}(t) - \dot{x}(t)$.

Hence, using Newton's law, we have

$$b(\dot{y} - \dot{x}) + k(y - x) = m\ddot{x},$$

or, substituting for $y(t)$ from Equation (4.16) and rearranging gives

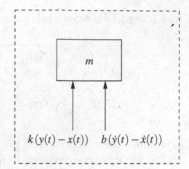

Fig. 4.19 Free body diagram for mass in suspension problem.

$$m\ddot{x} + b\dot{x} + kx = \frac{2\pi vhb}{\lambda}\cos\left(\frac{2\pi vt}{\lambda} + \hat{\phi}\right) + kh\sin\left(\frac{2\pi vt}{\lambda} + \hat{\phi}\right).$$

Following the procedure used several times previously, we can rewrite the right-hand side to transform the equation into

$$m\ddot{x} + b\dot{x} + kx = \sqrt{\left(\frac{2\pi vhb}{\lambda}\right)^2 + (kh)^2}\cos\left(\frac{2\pi vt}{\lambda} + \hat{\phi} + \overline{\phi}\right),$$

where

$$\overline{\phi} = \tan^{-1}\left(-\frac{2\pi vb}{\lambda k}\right).$$

Dividing both sides by m and letting $\phi = \hat{\phi} + \overline{\phi}$ gives

$$\ddot{x} + 2\zeta\omega_n\dot{x} + \omega_n^2 x = \sqrt{\left(\frac{2\pi vhb}{\lambda m}\right)^2 + \left(\frac{kh}{m}\right)^2}\cos\left(\frac{2\pi vt}{\lambda} + \phi\right).$$

Finally, to simplify writing it, let

$$\omega = \frac{2\pi v}{\lambda}$$

(so ω is just proportional to v) which gives

$$\ddot{x} + 2\zeta\omega_n\dot{x} + \omega_n^2 x = \frac{h}{m}\sqrt{(\omega b)^2 + k^2}\cos\left(\omega t + \phi\right).$$

It may be tempting to think that we have already solved this problem inasmuch as this looks a lot like Equation (4.14); however, there is one critical distinction. In Equation (4.14) the coefficient of the forcing term was constant. In this problem ω appears in the coefficient of the forcing term. Let us see what effect, if any, this has.

Assuming a particular solution of the form

$$x_p(t) = A\cos(\omega t + \phi) + B\sin(\omega t + \phi)$$

and doing all the usual work gives

$$A = \frac{h}{m}\sqrt{(\omega b)^2 + k^2}\left(\frac{\omega_n^2 - \omega^2}{(\omega_n^2 - \omega^2)^2 + (2\zeta\omega_n\omega)^2}\right)$$

$$B = \frac{h}{m}\sqrt{(\omega b)^2 + k^2}\left(\frac{2\zeta\omega_n\omega}{(\omega_n^2 - \omega^2)^2 + (2\zeta\omega_n\omega)^2}\right).$$

Examining the coefficients of the fractions and noting the k and m, one might be inclined to try to convert those to ω_n and get the b term expressed somehow as ζ. In fact,

$$\frac{h}{m}\sqrt{(\omega b)^2 + k^2} = h\sqrt{\left(\frac{\omega b}{m}\right)^2 + \left(\frac{k}{m}\right)^2} = h\sqrt{(2\zeta\omega_n\omega)^2 + (\omega_n^2)^2}$$

$$= h\omega_n^2\sqrt{\left(2\zeta\frac{\omega}{\omega_n}\right)^2 + 1}.$$

Dividing the numerator of both A and B by ω_n^2 and the denominator by ω_n^4, and while we are at it, computing $\sqrt{A^2 + B^2}$ gives the final answer

$$x_p(t) = h\sqrt{\frac{1 + \left(2\zeta\frac{\omega}{\omega_n}\right)^2}{\left(1 - \frac{\omega^2}{\omega_n^2}\right)^2 + \left(2\zeta\frac{\omega}{\omega_n}\right)^2}}\cos(\omega t + \phi + \psi),$$

where

$$\psi = \tan^{-1}\left(-\frac{2\zeta\frac{\omega}{\omega_n}}{1 - \left(\frac{\omega^2}{\omega_n^2}\right)}\right).$$

Note that the magnitude of the variation in the road height h is scaled by the term in the square root. In other words, the magnitude of the oscillation of the mass is the magnitude of the oscillation of the road times a factor that we call the *displacement transmissibility*. The displacement transmissibility tells how much the oscillation of the road is transmitted to result in an oscillation of the mass. Plotting the displacement transmissibility as a function of the frequency ratio for various damping ratios is probably a good idea, so it appears in Figure 4.20.

Note that Figures 4.16 and 4.20 are not identical. In particular, in the latter case all the curves have the value of one at a frequency ratio of $\sqrt{2}$. Also, for high frequency ratios, corresponding to high velocities, a low damping ratio is preferable. This is in contrast to the magnitude factor for an applied force where it is always the case that a larger damping ratio produces a smaller magnitude response, as should be apparent from Figure 4.16.

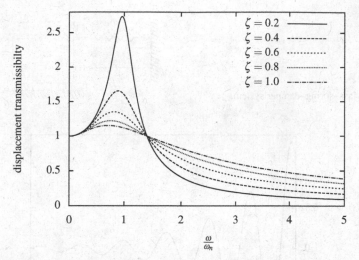

Fig. 4.20 Displacement transmissibility as a function of frequency ratio and damping ratio.

4.5 System Identification

This section considers the problem of *system identification* which is the problem of determining the differential equation(s) governing a system based upon experimental data rather than first principles. In principle, it should always be possible to use first principles to determine the governing equations for a given system; however, in practice this is not always the case. First, many engineering systems may simply be too complicated to reduce to a collection of interconnected systems that can be individually modeled. Second, even if the components may be individually modeled, the interaction among them may not be. Finally, even if both of the above are possible, the approximations involved in modeling each individual component may combine in a manner that make the overall model a poor representation of the actual system. Hence, if it is the case that some data are available regarding how the system behaves, it makes sense to use these data to either validate the given model or as a basis for modeling the system.

Consider the problem of modeling the system illustrated in Figure 4.21 where it is the case that the parameters for the model, m, k and b are not known, but what is known is that the system responds in a particular manner as illustrated in Figure 4.22.

The system is governed by the differential equation

$$m\ddot{x} + b\dot{x} + kx = 0, \tag{4.17}$$

therefore at first it may seem like a simple matter to find m, b, and k to give a response that looks like what is in the figure. In fact, attempting to do so by trial and

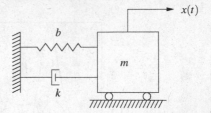

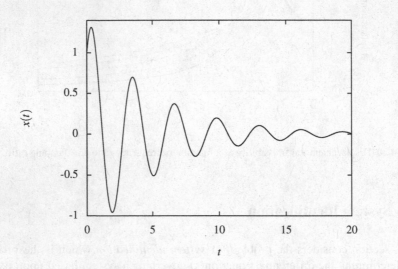

Fig. 4.21 Mass–spring–damper system.

Fig. 4.22 Response of a second-order system.

error is not too difficult. However, because the system is simple enough, we may as well make the effort to at least be a bit more sophisticated about it in order to save the time involved in a trial-and-error method and to gain some insight into the problem at hand.

First, note that there will actually be an infinite number of sets of values for m, b, and k that give the same response. This is because of the fact that if $x(t)$ satisfied Equation (4.17) it will also satisfy a scaled version of the equation such as

$$\alpha m\ddot{x} + \alpha b\dot{x} + \alpha k x = 0,$$

or, in particular, it will also satisfy

$$\ddot{x} + 2\zeta\omega_n\dot{x} + \omega_n^2 x = 0.$$

Because we may arbitrarily scale the equation without changing the solution, it seems reasonable to conclude that we may only find at most two of the three parameters. In fact this is the case, as outlined subsequently, and hence it makes sense to attempt to find the natural frequency ω_n and the damping ratio ζ which are the parameters in the canonical form of the second-order linear oscillation equation. Of course, once these two parameters are determined, it will be possible to use their definitions to find all the possible combinations of m, b, and k that are equivalent. Furthermore, if one of the parameters can be determined using an independent method, then the unique set of three parameters may be determined.

The system is characterized by decaying oscillations, thus we know that $0 < \zeta < 1$. Hence, the form of the solution to

$$\ddot{x} + 2\zeta \omega_n \dot{x} + \omega_n^2 x = 0,$$

where $x(0) = x_0$ and $\dot{x}(0) = \dot{x}_0$ that we need is given by Equation (4.13) which is

$$x(t) = e^{-\zeta \omega_n t} \left(c_1 \cos \omega_n \sqrt{1 - \zeta^2} t + c_2 \sin \omega_n \sqrt{1 - \zeta^2} t \right)$$
$$= e^{-\zeta \omega_n t} \left(c_1 \cos \omega_d t + c_2 \sin \omega_d t \right). \qquad (4.18)$$

Inspecting Equation (4.18) indicates that it should be straightforward to determine ω_d from simply inspecting the period of oscillation in the figure, as is illustrated in Figure 4.23. If the period is given by T, then the relationship between the frequency and period is simply $\omega_d T = 2\pi$, which gives

$$\omega_d = \frac{2\pi}{T}. \qquad (4.19)$$

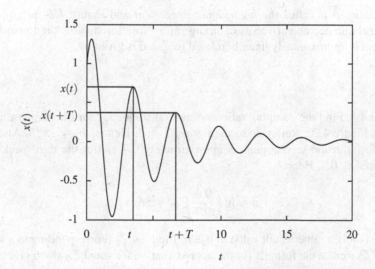

Fig. 4.23 Response of a second-order system.

Another quantity that is easy to determine from the response of the system is the ratio of the magnitudes of two successive peaks in the response. Using Equation (4.18) we have

$$\frac{x(t+T)}{x(t)} = \frac{e^{-\zeta \omega_n(t+T)}\left(c_1 \cos \omega_d (t+T) + c_2 \sin \omega_d (t+T)\right)}{e^{-\zeta \omega_n t}\left(c_1 \cos \omega_d t + c_2 \sin \omega_d t\right)}.$$

However, since the period of oscillation is T, then the sine and cosine terms in the ratio are the same, so

$$\frac{x(t+T)}{x(t)} = \frac{e^{-\zeta \omega_n(t+T)}}{e^{-\zeta \omega_n t}} = e^{-\zeta \omega_n T} = e^{(-\zeta \omega_d T)/\sqrt{1-\zeta^2}} = e^{(-2\pi\zeta)/\sqrt{1-\zeta^2}}. \quad (4.20)$$

Hence, the ratio of the magnitude of two successive peaks is a function of the damping ratio only. Simply reading the values of two successive peaks, computing their ratio, and then solving Equation (4.20) for the damping ratio is all that is necessary. Observe that in the previous computations t was not specified; hence, it does not matter which peaks are used as long as they are successive peaks.

Because the study of linear oscillations is a classical subject, we take it one step further to make the presentation consistent with the usual treatment. Taking the natural logarithm of both sides of Equation (4.20) gives

$$\hat{\delta} = \ln\left(\frac{x(t+T)}{x(t)}\right) = \ln x(t+T) - \ln x(t) = \frac{-2\pi\zeta}{\sqrt{1-\zeta^2}}.$$

This quantity $\hat{\delta}$ is called the *logarithmic decrement* and Figure 4.24 is a plot of the logarithmic decrement versus damping ratio. Note for small ζ the logarithmic decrement is approximately linearly related to ζ and is given by

$$\hat{\delta} \approx -2\pi\zeta.$$

Example 4.8. Find the damping ratio and natural frequency for the response illustrated in Figure 4.22. Referring to the figure, $T \approx 3$. Hence, $\omega_d \approx 2\pi/3$. Also the value of $x(t)$ at the second peak is approximately 0.7 and at the third peak it is approximately 0.4. Hence

$$\hat{\delta} \approx \ln\left(\frac{0.4}{0.7}\right) = -.56.$$

Because that is a rather small value to use in Figure 4.24 (corresponding to a small value of ζ) we use the formula for the approximation for small ζ, which gives

$$\zeta \approx -\frac{\delta}{2\pi} = -\frac{-.56}{2\pi} = 0.089.$$

Using this value gives

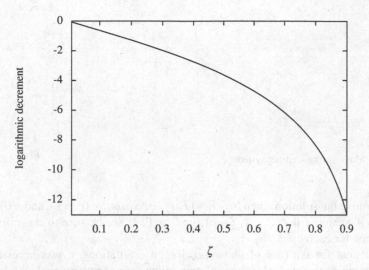

Fig. 4.24 Plot of the logarithmic decrement versus damping ratio.

$$\omega_n = \frac{\omega_d}{\sqrt{1-\zeta^2}} = 2.1.$$

In fact, the plot was generated using $\zeta = 0.1$ and $\omega_n = 2$, so the approximations involved in reading the values from the graphs and for the linear approximation for the relationship between the damping ratio and logarithmic decrement were really quite good.

4.6 Exercises

Several of the following exercises refer to the mass–spring–damper system illustrated in Figure 4.25. Unless otherwise indicated, assume that there is no gravity and that $x = 0$ at the unstretched position of the spring.

4.1. Consider the system illustrated in Figure 4.25. Assume that $b = 0$ and use either undetermined coefficients or variation of parameters to determine the solution to $\ddot{x} + \omega_n^2 x = F/m \sin \omega t$ where $x(0) = x_0$ and $\dot{x}(0) = \dot{x}_0$. Does it matter whether $\omega = \omega_n$? If so, be sure to consider both cases.

4.2. Write a computer program to determine an approximate numerical solution to the system in Problem 4.1 for the case where $m = 1$, $k = 4$, $F = 1$, and $\omega = 1.99$ or $\omega = 2.0$. Plot the solution for each case on the same graph and explain any significant phenomena that you observe.

4.3. Consider the system illustrated in Figure 4.25.

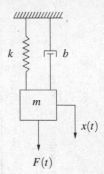

Fig. 4.25 Mass–spring–damper system.

1. Determine the solution when $b \neq 0$ and $F(t) = F \sin \omega t$, $x(0) = x_0$, and $\dot{x}(0) = \dot{x}_0$. Does it matter if $0 < \zeta < 1$, $\zeta = 1$, or $\zeta > 1$? If so, be sure to determine the solution for each case.
2. Recall that for the case of undamped forced oscillations, it was necessary to determine a separate form of the solution in the case of resonance. Is the form of the solution determined in Part 1 the same if $\omega = \omega_n$ or $\omega = \omega_d$? If so, be sure to determine those solutions as well.
3. Determine the magnification factor for the steady-state solution in Part 1 and plot it for $\zeta = 0.2, 0.4, 0.6$, and 0.8 versus ω/ω_n.
4. Determine the phase shift between the forcing function and the steady-state response in Part 1 and plot it for $\zeta = 0.2, 0.4, 0.6$, and 0.8. Be sure to indicate what form you assumed for the particular solution, for example, $\cos(\omega t + \phi)$ or $\sin(\omega t - \phi)$, and so on, because the phase may be different depending on the form of the solution you used.

4.4. Use the figures you plotted for Problem 4.3 to determine good approximations to the steady-state solutions for the following equations.

1. $\ddot{x} + 2\dot{x} + 25x = 3 \sin 2t$.
2. $\ddot{x} + 2\dot{x} + 25x = 3 \sin 5t$.
3. $\ddot{x} + 2\dot{x} + 25x = 3 \sin 10t$.
4. $\ddot{x} + 4\dot{x} + 25x = 3 \sin 10t$.
5. $\ddot{x} + 4\dot{x} + 49x = 3 \sin 10t$.
6. $\ddot{x} + 2\dot{x} + 36x = 6 \sin 20t$.

4.5. Consider the system illustrated in Figure 4.25. If $0 < \zeta < 1$ and

$$F(t) = F_c \cos \omega_c t + F_s \sin \omega_s t$$

is it possible to combine your answer from Exercise 4.3 and the solution in Equation (4.15) to obtain the steady-state solution, or is it necessary to work out the whole thing again? In either case, provide the answer and justify it.

4.6. Consider

$$\ddot{x} + 4\dot{x} + 16x = \cos 4t + \cos 4.2t.$$

If we are only interested in the steady-state response, is it valid to write

$$x_{ss} = \delta_1 M_1 \cos(\omega t + \phi_1) + \delta_2 M_2 \cos(\omega t + \phi_2),$$

where M_1, M_2, ϕ_1, and ϕ_2 are determined from the appropriate graphs? How would you determine δ_1 and δ_2? Demonstrate whether it works by picking some initial conditions and writing a computer program to determine an approximate numerical solution and comparing it to the combination of the approximate steady-state solutions determined from the graphs.

4.7. Consider the system illustrated in Figure 4.25 and let $m = 1$, $b = 1$, $k = 1$, and $F(t) = 3\cos 2t$. Use Figures 4.16 and 4.17 to determine a good approximation for the steady-state response of the system. Will the magnitude of the steady state response increase or decrease if the forcing frequency $\omega = 2$ is increased?

4.8. Write a computer program to determine an approximate numerical solution to the system in Part 1 of Problem 4.4 with $x(0) = 0$ and $\dot{x}(0) = 0$. Plot the approximate solution as well as the solution determined in Problem 4.4 and compare the results. Explain any significant differences.

4.9. Consider the system illustrated in Figure 4.25 and assume that there is gravity.

1. Determine the equation of motion for the system when $x = 0$ at the unstretched position of the spring.
2. Determine the equation of motion for the system when $x = 0$ at the equilibrium position. In other words, $x = 0$ at the position when the spring is stretched by an amount due to the weight of the mass.

4.10. Consider the system illustrated in Figure 4.25 with $F(t) = 0$ (damped, unforced). Let $\omega_n = 1$, $x(0) = 1$, $\dot{x}(0) = 0$, and plot the solution for $\zeta = 0.0, 0.2, 0.4$, 0.6, 0.8, and 1.0 for $t = 0$ to $t = 10$. Plot all the solutions on the same plot. Explain the effect of increasing the damping ratio.

4.11. Write a computer program to determine an approximate numerical solution for

$$m\ddot{x} + b\dot{x} + kx = F\sin\omega t,$$

when $\omega_n = 2$, $\zeta = 0.3$, $m = 1$, $\omega = 1.5$, $F = 5$, $x(0) = 1$, $\dot{x}(0) = 1$.

1. On the same plot, plot the numerical solution and the solution for these values substituted into the closed-form solution from Exercise 4.3.
2. Vary the step size for the numerical solution to determine the largest step size that gives a reasonable approximation to the exact solution.
3. Use the figures from Exercise 4.3 to determine a good approximation for the steady-state solution and plot the solution on the same graph. At approximately what time does the transient solution decay sufficiently so that the steady-state solution is approximately equal to the exact solution?

4.12. Figure 4.20 plots the magnitude of the steady-state oscillation of a mass subjected to a vibrating base. For some applications, such as an automotive suspension, the magnitude of the response is not the critical factor, but rather the net force to which the mass is subjected.

1. Determine an expression for the force to which the mass illustrated in Figure 4.18 is subjected.
2. Manipulate the expression for the force so that it is in the form of

$$f = -khM_f\cos(\omega t + \phi).$$

Explain the interpretation of the term M_f, which is called the *force transmissibility*. Plot M_f as a function of ω/ω_n for various damping ratios to make a plot similar to Figure 4.20.

4.13. Use Figure 4.20 to determine the magnitude of the motion of the mass in Figure 4.18 if $k = 2$, $m = 2$, $b = 1$, $h = 0.25$. Plot the magnitude of the motion versus ω.

4.14. The system illustrated in Figure 4.26 is comprised of a board of length l with uniform mass per unit length of ρ. It sits on top of two counter-rotating cylinders, which are rotating with an angular velocity of ω in the directions indicated and are separated by a distance $l/2$. You may assume that ω is large enough that the points of contact between the board and cylinders are always slipping and that the coefficient of dynamic friction between the board and cylinders is μ.

If the board is not centered on the two cylinders, then the normal force will be greater on the side to which it is displaced because more of the weight of the board will be supported on that side. For example, in the figure, the normal force between the board and cylinder will be greater at the left cylinder, and this will cause a net force to the right.

Let x denote the distance that the center of the board is displaced from the center of the two cylinders, as is illustrated in the figure, and determine the differential equation for this system. Solve it if $x(0) = l/8$ and $\dot{x}(0) = 0$.

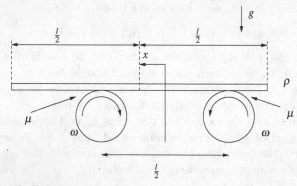

Fig. 4.26 System for Exercise 4.14.

4.15. Determine and plot the total energy versus time for an undamped mass–spring system with $m = 1$, $k = 9$ subjected to a harmonic forcing function, $f(t) = \cos 3t$. Determine and plot the total energy versus time for an undamped mass–spring system with $m = 1$, $k = 8$ subjected to a harmonic forcing function, $f(t) = \cos 3t$. If the energy is different, explain why the same forcing function acting on the same mass can do a different amount of work in each case.

4.16. Find the motion of the mass illustrated in Figure 4.25 if $m = 1$, $b = 2$, $k = 1$, $x(0) = 0$, $\dot{x}(0)m = 0$, and $F(t) = \cos t + 4t$.

4.17. The free response of a second-order system is illustrated in Figure 4.27. Determine the natural frequency and the damping ratio.

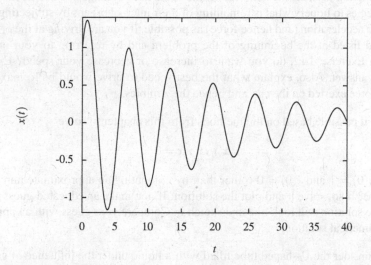

Fig. 4.27 System response for Exercise 4.17.

4.18. Add gravity to the suspension problem from Section 4.4.2 and determine the velocity v, if any, at which the wheel would leave the road.

4.19. You are driving down a sinusoidally bumpy road in a vehicle that can be modeled like the system illustrated in Figure 4.18.

- The distance between the peaks of the bumps is $\lambda = 10$ m.
- The mass of the vehicle and everything in/on it is $m = 1000$ kg.
- The spring constant in the suspension is $k = 9000$ N/m.
- The constant for the shock absorbers is $b = 2400$ N s/m.
- The speed you are driving is $60/(2\pi)$ m/s.

There are zombies all over the outside of the car that you need to shake off the car by making the magnitude of the shaking as great as possible.

1. By referring to Figure 4.20, do you want to speed up, slow down, or maintain your speed? Explain your answer, perhaps by reproducing the figure and annotating it with data from this problem.
2. By what factor are you able to maximize the magnitude of the shaking by changing your velocity, for example, two times greater, 10 times greater, no times greater, and so on?
3. As is well known, the zombies are tenacious and after driving down the road for a while the shock absorbers are now trashed so b is decreasing. If you are driving at the original speed, will the decreasing b increase or decrease the magnitude of the shaking? If you are driving at the speed that always maximizes the magnitude of the shaking, will the decreasing b increase or decrease the magnitude of the shaking? Explain your answer.
4. Attempting to shake the zombies off the car is not working. You decide to switch strategies to liquefy what is remaining of their internal organs by subjecting to as much acceleration (and hence force) as possible. If you are driving at the original speed listed at the beginning of the problem and by referring to your answer from Exercise 4.12, do you want to increase or decrease your speed? Explain your answer. Also, explain what the best speed to drive would be to maximize the force exerted on the car, and hence the zombies.

4.20. Is it possible based on the methods from this chapter to solve

$$\ddot{x} + t\dot{x} + x = 0$$

where $x(0) = 1$ and $\dot{x}(0) = 0$ (other than by computing an approximate numerical solution)? If so, solve it and plot the solution. If not, make an educated guess about what the solution will look like and sketch it. Compare your guess with an approximate numerical solution.

4.21. Consider the U-shaped tube filled with a liquid under the influence of gravity illustrated in Figure 4.28. Assume the fluid is inviscid[2] and has a density ρ and that the tube has a cross-sectional area of A. If the fluid is displaced down by an amount x on one side and up by an amount x on the other side, what is the total force on the body of fluid? Determine and solve the equation of motion.

4.22. Determine the equations of motion and determine the homogeneous solution for the systems illustrated in Figure 4.29.

4.23. Consider the system illustrated in Figure 4.30. Assume that $\theta \ll 1$ so that the amount that the spring is extended is equal to $l_2\theta$. Determine the equation of motion for this system.

---

[2] An inviscid fluid has no viscosity. For the purposes of this problem, that means we may ignore any force resisting the flow of the fluid in the tube.

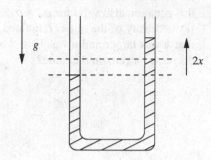

Fig. 4.28 System for Exercise 4.21.

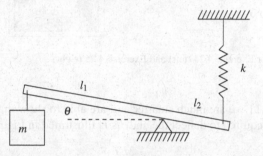

Fig. 4.29 Mechanical systems for Exercise 4.22.

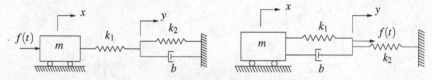

Fig. 4.30 System for Exercise 4.23.

4.24. Consider the system illustrated on the left in Figure 4.31. Assume the pulleys are light with negligible inertia. Determine and solve the equation of motion.

4.25. Consider a solid block floating in a vat of liquid as is illustrated on the right in Figure 4.31. Assume the height of the block is h and the widths are a (so if you look from the top it is a square. Recall from elementary fluid mechanics, the buoyant force exerted on a submerged object is equal to the amount of fluid it displaces. Assume the density of the liquid is ρ_f, the density of the object is ρ_o, and that the object does not rotate.[3]

1. If $\rho_o < \rho_f$, determine the equilibrium value for x.

---

[3] Extra credit: would the object, in fact, stay oriented in the manner illustrated?

2. If the object were displaced downward from its equilibrium position slightly and let go, it would bob in the liquid. Determine the equation of motion for this. Is this equation also valid for $\rho_o \geq \rho_f$?

3. If the density of the object remained the same, but the geometry were altered so that h was larger and a was less, would the frequency of the bobbing increase, decrease, or remain the same?

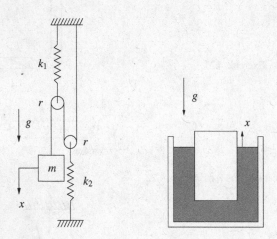

Fig. 4.31 System for Exercise 4.24 (left) and Exercise 4.25 (right).

4.26. In Chapter 11 we are much smarter and are able to show that for a cantilever beam, the force required to deflect the end, as is illustrated in Figure 4.32 is

$$F = \frac{3EI}{L^3}x,$$

where E is the modulus of elasticity, which is a property of the material used, I is the area moment of inertia of the cross-section of the beam, and L is the length.

If a mass m, is attached to the end of the beam, what is the equation of motion for the vertical motion of the beam? If the length is doubled, what is the change in frequency of the free vibration of the beam?

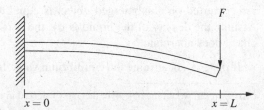

Fig. 4.32 Cantilever beam subjected to a force at the end.

4.27. For most of this chapter, the forcing function was of the form $f(t) = F \cos \omega t$, where F and ω were considered independently. There is an important class of problems in which they are related, however, which is when the force is caused by some imbalance in rotating machinery.

Consider the mass constrained to rotate about the origin at a fixed distance r at a fixed angular velocity ω. Assume this models an eccentricity of a motor, which is attached to a mass–spring–damper system illustrated in Figure 4.33.

1. Show that the vertical force exerted on the mass by the motor is given by $f(t) = m_e r \omega^2 \cos \omega t$.
2. Compute the steady-state solution for this system.
3. Plot the ratio of the magnitude of the steady-state response to the amount of eccentricity versus frequency ratio; that is, plot $|x_p/r|$ versus ω/ω_n. Explain why the low frequency, resonant frequency, and high frequency parts of the graph make sense. Compare and contrast your plot with Figure 4.5.

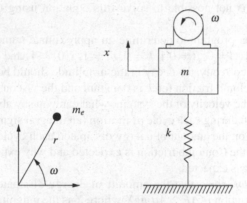

Fig. 4.33 Rotating system with eccentricity for Exercise 4.27.

4.28. Referring to Exercises 4.26 and 4.27, assume an electric motor is mounted at the end of a cantilever beam as illustrated in Figure 4.34. Assume the modulus of elasticity of the beam is $E = 200$ GPa, the area moment of inertia is $I = 1/4$ m^4, the length is $L = 1$ m, and the imbalance in the motor is $rm = 1$ kg $\cdot$ cm.

1. Determine the resonant frequency of the system.
2. Determine the magnitude of the motion of the beam if the speed of the motor is $\omega = 1000$ rpm (note the units).

4.29. Determine the solution to $\ddot{x} + 1/5\dot{x} + x = 5 \cos 4t$, $x(0) = 1$, and $\dot{x}(0) = -1$.

1. On the same plot, plot the particular solution and the whole solution versus t.
2. Identify the part of the solution you would identify as the *transient response* and the part you would identify as the *steady-state response*.

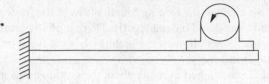

Fig. 4.34 Motor mounted on a cantilever beam for Exercise 4.28.

3. Explain why the solution is not the same as the steady-state response during the transient response.

4.30. Damping is particularly difficult to properly model and viscous damping is often assumed and this problem compares viscous damping to Coulomb friction. Consider the system illustrated in Figure 4.35 where the coefficients of static and dynamic friction between the block and surface are equal and are given by μ.

1. Draw a free body diagram for the mass and determine the equation of motion. Explain why it is not possible to solve this equation using the methods from Chapter 3.
2. Write a computer program to determine an approximate numerical solution for the case where $f(t) = 0$, $\mu = 0.1$, $k = 9$, $m = 1$, $x(0) = 1$, and $\dot{x}(0) = -1$. If you do this problem correctly, the decay in the amplitude should be linear.
3. Because the Coulomb friction force is constant and the viscous damping force is proportional to the velocity of the system, what can you say about the amount of energy dissipated during each cycle of motion (an integral sign should be in your answer)? Based on the amount of energy dissipated each cycle, explain why the linear decay for the Coulomb friction is expected and why exponential decay for viscous damping is expected.
4. If $f(t) = F\cos\omega t$, show that the amount of energy dissipated over one cycle with Coulomb friction is $\Delta E = 4\mu mgX$ where X is the magnitude of the motion, and that the amount of energy dissipated over one cycle with viscous damping is $\Delta E = \pi b_{eq}\omega X^2$. Equating these two and solving for b_{eq} should give a system with viscous damping that approximates the one with Coulomb friction. Compute b_{eq} for the system you solved in Part 2, and plot the solution to the system with the equivalent viscous damping on the same plot as the system with Coulomb damping. How good is the approximation?

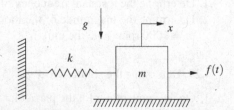

Fig. 4.35 System subjected to Coulomb friction for Exercise 4.30.

Chapter 5
Variable-Coefficient Linear Ordinary Differential Equations

This chapter presents the the use of power series to determine solutions to differential equations. Although the method applies to a variable-coefficient linear equation of any order, the primary use of such solutions is for second-order variable-coefficient linear ordinary differential equations. The basic approach is to assume a solution of the form

$$x(t) = a_0 + a_1 (t - t_0) + a_2 (t - t_0)^2 + a_3 (t - t_0)^3 + \cdots$$
$$= \sum_{n=0}^{\infty} a_n (t - t_0)^n$$

and then substitute it into the differential equation to determine the coefficients, which is a relatively straightforward procedure. More complicated situations may require assuming a solution of the form

$$x(t) = (t - t_0)^m \left(a_0 + a_1 (t - t_0) + a_2 (t - t_0)^2 + a_3 (t - t_0)^3 + \cdots \right)$$
$$= \sum_{n=0}^{\infty} a_n (t - t_0)^{n+m}$$

and then substituting it into the differential equation to determine m as well as the coefficients a_i. Whether the series that is obtained by determining the coefficients converges to the actual solution is, unfortunately, a more difficult matter. Finally and interestingly, quite a few famous differential equations (famous enough to have names) are second-order, variable-coefficient equations. This chapter presents the equation and solutions for some of them, and contains mainly mathematical examples. The context in which most engineering uses for this subject arise appears later in Chapter 11.

B. Goodwine, *Engineering Differential Equations: Theory and Applications*,
DOI 10.1007/978-1-4419-7919-3_5, © Springer Science+Business Media, LLC 2011

5.1 Motivational Examples

This section presents two examples intended to illustrate two things. First, they illustrate the manner in which the coefficients in the power series solution are obtained. Second, they illustrate that whether and in what manner the series converges to the solution is not something that is obvious a priori.

Example 5.1. Determine the solution to

$$\ddot{x} + bt\dot{x} + x = 0$$

where $x(0) = 2$ and $\dot{x}(0) = 5$. Note that because this is a second-order variable-coefficient problem, none of the methods in the previous chapters can be used to determine a solution to it.

If we assume

$$x(t) = a_0 + a_1 t + a_2 t^2 + a_3 t^3 + \cdots = \sum_{n=0}^{\infty} a_n t^n$$

then differentiating the series termwise[1]

$$\dot{x}(t) = a_1 + 2a_2 t + 3a_3 t^2 + \cdots = \sum_{n=1}^{\infty} a_n n t^{n-1}$$

and

$$\ddot{x}(t) = 2a_2 + 6a_3 t + 12a_4 t^2 + \cdots = \sum_{n=2}^{\infty} a_n n(n-1) t^{n-2}.$$

Substituting into the differential equation gives

$$\left(\sum_{n=2}^{\infty} a_n n(n-1) t^{n-2} \right) + bt \left(\sum_{n=1}^{\infty} a_n n t^{n-1} \right) + \left(\sum_{n=0}^{\infty} a_n t^n \right) = 0. \qquad (5.1)$$

The procedure is to equate the coefficients of the different powers of t in the equation. Recall from Exercise 2.11 that the set of functions $\{1, t, t^2, t^3, \ldots, t^n\}$ is linearly independent. Hence, the only way for an equation of the form

$$\alpha_0 + \alpha_1 t + \alpha_2 t^2 + \alpha_3 t^3 + \cdots = 0$$

to hold for all t is for each coefficient α_n to be zero. In order to determine an expression for the coefficients of the different powers of t, it is desirable to shift the exponent in the first sum by 2, which, along with distributing the t in the second term gives

---

[1] This is valid only in the interior of domains where the series converges. Convergence is considered subsequently in this chapter.

$$\left(\sum_{n=0}^{\infty} a_{n+2}(n+2)(n+1)t^n\right) + b\left(\sum_{n=1}^{\infty} a_n n t^n\right) + \left(\sum_{n=0}^{\infty} a_n t^n\right) = 0.$$

To see that the first summation is identical to what it was in Equation (5.1), let $m = n - 2$ and substitute into the expression (alternatively, write out the first few terms). Because m is an index, it may be changed back to n.

Collecting the powers of t and observing that the middle sum starts at 1, gives

$$2a_2 + a_0 + \sum_{n=1}^{\infty} [(n+1)(n+1)a_{n+2} + b n a_n + a_n]t^n = 0,$$

where the first two terms are from writing the $n = 0$ term separately from the first and third series. From this we obtain

$$a_2 = -\frac{1}{2}a_0$$

and

$$a_{n+2} = -\frac{bn+1}{(n+2)(n+1)}a_n. \tag{5.2}$$

Equation (5.2) is called a *recurrence relation* and can be used to determine every term in the solution if a_0 and a_1 are known. The series solution partial sums[2] including different numbers of terms are illustrated along with an accurate solution (determined numerically) in Figure 5.1 where $b = 0.1$. The notation P_N in the figure indicates how many terms were included in the partial sum. Specifically, P_N means that the partial sum

$$P_N = \sum_{n=0}^{N} a_n t^n$$

was used.

In Example 5.1, by comparing the partial sums to the numerical solution, it appears that as more terms are included in the partial sum, the partial sum from the series becomes a better approximation to the real solution for longer periods of time. However, note that at this point it is not guaranteed that the series will converge to the solution. Intuitively it would seem correct to assume that more terms will guarantee a solution that converges for a larger interval of t, but as illustrated in the next example, this is not always the case.

Example 5.2. Determine the solution to

$$\ddot{x}(t) + \frac{1}{1+t^2}x(t) = 0, \tag{5.3}$$

where $x(0) = 1$ and $\dot{x}(0) = 5$ by assuming a power series solution about $t = 0$.

---

[2] The *partial sum to N* of a series is the sum of the first N terms of the series.

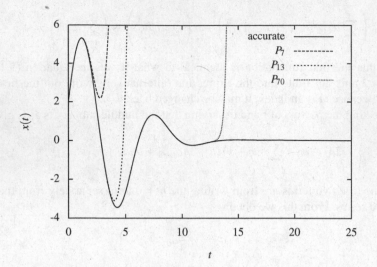

Fig. 5.1 Accurate and partial sum solutions for Example 5.1.

If we leave the coefficient function, $1/\left(1+t^2\right)$ as it is, it will prevent us from isolating the coefficients of different powers of t because of the denominator. One approach would be to expand $1/\left(1+t^2\right)$ in a Taylor series about $t=0$. Alternatively, and more simply, we can multiply the differential equation by $1+t^2$:

$$\left(1+t^2\right)\ddot{x}+x=0. \tag{5.4}$$

Assuming

$$x(t) = \sum_{n=0}^{\infty} a_n t^n$$

which, as before, gives

$$\ddot{x}(t) = \sum_{n=2}^{\infty} n\left(n-1\right)a_n t^{n-2} = \sum_{n=0}^{\infty} \left(n+1\right)\left(n+1\right)a_{n+2}t^n \tag{5.5}$$

and substituting, gives

$$\left(1+t^2\right)\sum_{n=2}^{\infty} n\left(n-1\right)a_n t^{n-2} + \sum_{n=0}^{\infty} a_n t^n = 0.$$

Distributing the $\left(1+t^2\right)$ term and using the second form of $\ddot{x}(t)$ in Equation (5.5) when multiplying by the 1 and the first form when multiplying by the t^2 gives

$$\sum_{n=2}^{\infty} n\left(n-1\right)a_n t^n + \sum_{n=0}^{\infty} \left(n+2\right)\left(n+1\right)a_{n+2}t^n + \sum_{n=0}^{\infty} a_n t^n = 0.$$

Collecting all the terms into one sum, which requires writing out the first two terms in the second and third series gives

$$2a_2 + 6a_3t + a_0 + a_1t + \sum_{n=2}^{\infty} \left[(n(n-1)+1)a_n + (n+2)(n+1)a_{n+2} \right] t^n =$$

$$(a_0 + 2a_2) + (a_1 + 6a_3)t^1 + \sum_{n=2}^{\infty} \left[(n(n-1)+1)a_n + (n+2)(n+1)a_{n+2} \right] t^n = 0.$$

The coefficient of each power of t must be zero, thus we have

$$a_2 = -\frac{1}{2}a_0$$

$$a_3 = -\frac{1}{6}a_1$$

$$a_{n+2} = -\frac{n(n+1)+1}{(n+2)(n+1)}a_n, \quad n \geq 2.$$

A plot illustrating an accurate numerical solution and several partial sums of the power series solution are illustrated in Figure 5.2. As is apparent from the figure, the partial sums provide a good approximate solution for $t < 1$ only. No matter how many terms are included in the partial sum, it diverges from the accurate solution once $t \approx 1$.

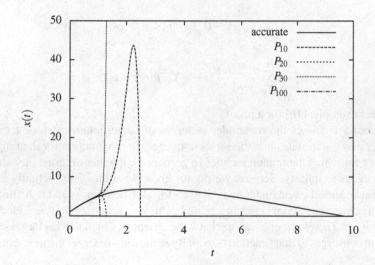

Fig. 5.2 Accurate and partial sum solutions for Example 5.2.

5.2 Convergence: Real Rational Functions

For a variable-coefficient linear differential equation of the form

$$\ddot{x} + f_1(t)\dot{x} + f_0(t)x = g(t), \tag{5.6}$$

as shown in Section 5.3, it turns out that whether an assumed power series solution to the differential equation converges to the actual solution depends on the extent to which the Taylor series expansions for the functions $f_1(t)$, $f_0(t)$, and $g(t)$ converge to those functions. The reason the series solutions for Example 5.2 did not converge to the real solution beyond a very limited range of t was precisely due to the fact that the Taylor series for $1/\left(1+t^2\right)$ only had a limited range of convergence. It makes some intuitive sense that attempting to formulate a series solution to a differential equation will only work for ranges of the independent variable for which terms in the differential equation have convergent Taylor series representations.

Hence, we need a means to determine when a Taylor series expansion for a function $f(t)$ actually converges to $f(t)$, and the most fundamental answer is provided by Taylor's theorem.

Theorem 5.1. *Define the nth Taylor remainder by*

$$R_n(t) = \frac{f^{(n+1)}(\hat{t})}{(n+1)!}(t-t_0)^{n+1}, \ \ where \ |\hat{t}-t_0| < |t-t_0|.$$

If

$$\lim_{n\to\infty} R_n(t) = 0 \ for \ |t_0 - t| < \rho$$

then

$$f(t) = \sum_{n=0}^{\infty} \frac{f^{(n)}(t_0)}{n!}(t-t_0)^n \ for \ |t-t_0| < \rho.$$

See, for example, [10] for a proof.

Theorem 5.1 defines the remainder in terms of some unknown value $\hat{t}$ that is between t and t_0 and states that the series converges to the function for the range of values of t for which the remainder goes to zero as the number of terms included in the series goes to infinity. Because we do not know what this value actually is, we need to make sure the remainder goes to zero for all $\hat{t}$ between t_0 and t. A function that has a convergent Taylor series in a neighborhood of points about t_0 is called *real analytic* at t_0.[3] Therefore, one way to check the extent to which the Taylor series for a function converges to that function is to analyze the remainder, as the next example illustrates.

Example 5.3. The Taylor series for the function

---

[3] The beautiful subject of analyticity is beyond the scope of this text. However, suffice it to say that a student is encouraged to take a course in complex analysis where this topic is considered in detail. Surprisingly, in the case of a complex-valued function of a complex variable, tests for analyticity are much simpler than for the case of a real-valued function.

$$f(t) = \cos(\omega t)$$

about any t_0 converges for all t. The remainder formula is given as above by

$$R_n(t) = \frac{\omega^{n+1} \hat{f}^{(n+1)}(\hat{t})}{(n+1)!}(t-t_0)^{n+1},$$

where $\hat{f}(t)$ is $\pm\cos\omega t$ or $\pm\sin\omega t$. Because $|\hat{f}(t)| \leq 1$, the remainder satisfies the bound

$$|R_n(t)| \leq \left|\frac{\omega^{n+1}}{(n+1)!}(t-t_0)^{n+1}\right|.$$

For any t, inasmuch as the factorial in the denominator grows faster than the powers of the fixed numbers ω and $(t-t_0)$,

$$\lim_{n\to\infty} R_n = 0.$$

Hence, we know that the Taylor series for $\cos\omega t$ about $t = 0$ converges, that is

$$\cos(\omega t) = \sum_{n=0}^{\infty} \frac{\omega^n(-1)^n}{(2n)!}t^{2n}.$$

The next example shows that the Taylor series for the function $f_0(t)$ in Example 5.2 does not converge for all t, and, in fact, converges for exactly the same range of t for which the solution seems to converge to the accurate solution to the differential equation as was illustrated in Figure 5.2. This example also illustrates the fact that determining the range of convergence based only on the remainder may be difficult because determining the form of the remainder term for any n may be difficult.

Example 5.4. Compute the Taylor series for

$$f(t) = \frac{1}{1+t^2}. \tag{5.7}$$

The first few derivatives are

$$\frac{df}{dt}(t) = \frac{2t}{(1+t^2)^2}$$

$$\frac{d^2f}{dt^2}(t) = \frac{8t^2}{(1+t^2)^3} - \frac{2}{(1+t^2)^2}$$

$$\frac{d^3f}{dt^3}(t) = \frac{48t^3}{(1+t^2)^4} + \frac{24t}{(1+t^2)^3}.$$

Hence, the Taylor series about $t = 0$ is given by

$$\frac{1}{1+t^2} = 1 - t^2 + t^4 - t^6 + t^8 + \cdots. \tag{5.8}$$

A plot of the function $f(t)$ and the partial sum including different numbers of terms in the Taylor series is illustrated in Figure 5.3. By examining the function in Equation (5.7), clearly

$$\lim_{t \to \pm\infty} |f(t)| = 0,$$

so in Figure 5.3 the actual graph of the function is the one with the outer tails approaching zero. As more and more terms are included in the Taylor series, the approximation becomes better, but only for values of $t \in (-1, 1)$.

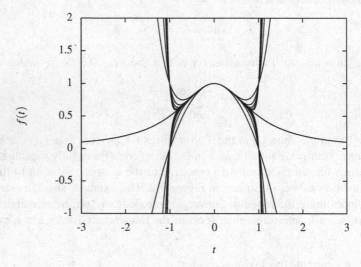

Fig. 5.3 The function $f(t) = 1/(1+t^2)$ and partial sums approximating it.

The main point from Example 5.4 is that even for functions that seem perfectly normal, a series approximation may not converge for it for all values of the independent variable, and attempts to increase the accuracy of the approximation by including more terms in the series will be futile outside the range of convergence.

If one is to apply a standard test for convergence of the Taylor series for the function in Equation (5.7) given by Equation (5.8), such as the ratio test[4] it would

---

[4] The ratio test is: for the series $\sum_{n=0}^{\infty} s_n$, if $s_n \neq 0$ for $n \geq 1$, suppose

$$\lim_{n \to \infty} \left| \frac{s_{n+1}}{s_n} \right| = r.$$

If $r < 1$, then the series converges absolutely. If $r > 1$, then the series diverges. If $r = 1$, no conclusion can be drawn from this test alone.

be clear that the series converges for $|t| < 1$ and one would be tempted to conclude that all we need to do is check whether the series converges for some interval of t. However, unfortunately, even that is not necessarily enough because it is not the case that if a Taylor series converges for some range of values of the independent variable that it actually converges to that function. It may, unfortunately, converge to something else.[5] So, we need to be able to say more than simply that the series converges. The most basic way to do that is to consider the remainder term given in Theorem 5.1. If the remainder goes to zero, not only does the series converge, it converges to the function $f(t)$.

Interestingly, a more general study of convergence properties of a series approximation beyond analyzing the remainder term depends on the properties of the function where the dependent variable may be a complex number, and is fundamentally related to the property of analyticity. A general study of analyticity is beyond the scope of this text and an interested reader is referred to [10] for a complete exposition.

One final set of results is useful subsequently and is presented without a proof. It is limited to *rational functions*, which are ratios of polynomials. This is particularly useful because many of the important examples of variable-coefficient ordinary differential equations in engineering have rational functions as the variable coefficients.

Theorem 5.2. *The Taylor series for a rational function*

$$f(t) = \frac{n(t)}{d(t)}$$

($n(t)$ and $d(t)$ are polynomials) about the point $t = t_0$ converges for all $|t - t_0| < \rho$ where ρ is the distance in the complex plane from t_0 to the nearest zero of the polynomial $d(t)$.

The proof follows from a few facts from complex variable theory.

1. Polynomial functions are analytic everywhere.
2. If the polynomials $n(t)$ and $d(t)$ have real coefficients, then they will be real-valued when t is real.
3. The quotient of two analytic functions is analytic except where the denominator is zero.
4. If a function is analytic everywhere inside a circle of radius ρ in the complex plane centered at t_0, then the Taylor series for the function about t_0 converges to the function for every point in that circle.

---

[5] A classic example is Taylor series for the function

$$f(t) = \begin{cases} \exp\left(-1/t^2\right), & t \neq 0, \\ 0, & t = 0, \end{cases}$$

about $t = 0$. The function is smooth, and the Taylor series converges. However, it converges to the function that is zero everywhere, which is not equal to the function f.

Referring to the function in Example 5.4, note that if t can be complex, then the denominator is zero for $t = \pm i$, and the distance in the complex plane from the origin, where $t = 0$, to $\pm i$ is one, which is exactly the distance from $t_0 = 0$ to $t = 1$, which is the point at which the Taylor series graphed in Figure 5.3 approximations diverge from the function. A list of common functions with the associated Taylor series expansions and radius of convergence are listed in Table 5.2.

Function	Convergence Interval
$e^{\mu t} = \displaystyle\sum_{n=0}^{\infty} \frac{(\mu t)^n}{n!}$	$t \in (-\infty, \infty)$
$\cos \omega t = \displaystyle\sum_{n=0}^{\infty} (-1)^n \frac{(\omega t)^{2n}}{(2n)!}$	$t \in (-\infty, \infty)$
$\sin \omega t = \displaystyle\sum_{n=0}^{\infty} (-1)^n \frac{(\omega t)^{2n+1}}{(2n+1)!}$	$t \in (-\infty, \infty)$
$\arcsin t = \displaystyle\sum_{n=0}^{\infty} \frac{(2n)!}{4^n (n!)^2 (2n+1)} t^{2n+1}$	$|t| < 1$
$\arctan t = \displaystyle\sum_{n=1}^{\infty} \frac{(-1)^n}{2n+1} t^{2n+1}$	$|t| < 1$
$\dfrac{1}{1-t} = \displaystyle\sum_{n=0}^{\infty} t^n$	$|t| < 1$
$\dfrac{t}{(1-t)^2} = \displaystyle\sum_{n=0}^{\infty} n t^n$	$|t| < 1$
$t e^t = \displaystyle\sum_{n=0}^{\infty} n \frac{t^n}{n!}$	$|t| < 1$
$\sqrt{1+t} = \displaystyle\sum_{n=0}^{\infty} \frac{(-1)^n (2n)!}{(1-2n)(n!)^2 4^n} t^n$	$|t| < 1$

Table 5.1 Series expansions for common functions (with $0! = 1$) and the domain of convergence

In summary, we have considered three ways to check the radius of convergence for the functions that are the variable coefficients in the differential equation.

1. Determining an expression or bound for the remainder and showing that it goes to zero as n becomes large.
2. In the specific case of the ratio of two polynomials, the radius of convergence of the Taylor series to the function will be the distance from t_0 to the nearest zero of the denominator in the *complex* plane.
3. Checking if the function is in Table 5.2 and having faith that it is correct.

5.3 Series Solutions About an Ordinary Point

In Section 1.3.3 a linear differential equation was defined by Equation (1.5) to be of the form

$$f_n(t)\frac{d^n x}{dt^n}(t) + f_{n-1}(t)\frac{d^{n-1} x}{dt^{n-1}}(t) + \cdots + f_1(t)\frac{dx}{dt}(t) + f_0(t)x(t) = g(t). \qquad (5.9)$$

Although the methods generalize in exactly the manner expected to higher order, we restrict our attention to second-order equations of the form

$$\frac{d^2 x}{dt^2}(t) + f_1(t)\frac{dx}{dt}(t) + f_0(t)x(t) = g(t), \qquad (5.10)$$

because a surprisingly large number of important differential equations in engineering are of this form.

Examining Equation (5.9), one may reasonably (and correctly) conclude that points where $f_n(t) = 0$ are problematic. This makes intuitive sense because the order of the equation changes at those points. Also, based on Theorem 5.2, these points also define the limit of the radius of convergence of the Taylor series for $f_1(t)$, $f_0(t)$, and $g(t)$ in the case where they are rational functions. This section considers solutions of Equation (5.10) for values of t where $f_1(t)$, $f_0(t)$, and $g(t)$ have properties that guarantee a unique solution exists that may be expressed as a convergent series for some range of the independent variable.

Definition 5.1. For the second-order, linear, ordinary differential equation

$$\frac{d^2 x}{dt^2}(t) + f_1(t)\frac{dx}{dt}(t) + f_0(t)x(t) = g(t),$$

a point $t = t_0$ where $g(t)$, $f_1(t)$, and $f_0(t)$ are analytic is called an *ordinary point*.

The main result of this section is the following theorem that says that about an ordinary point of a second-order linear ordinary differential equation, a unique solution exists and the Taylor series for that solution converges about the ordinary point.

Theorem 5.3. *If each function $g(t)$, $f_0(t)$, and $f_1(t)$ in Equation (5.10) is analytic at $t = t_0$, and the Taylor series for each function converges for $|t - t_0| < \rho$, then there is a unique solution $x(t)$ of Equation (5.10) that is also analytic at $t = t_0$ satisfying the initial conditions*

$$x(t_0) = a_0, \quad \dot{x}(t_0) = a_1.$$

In other words, the solution has a Taylor series expansion in powers of $(t - t_0)$. Furthermore, the Taylor series converges for $|t - t_0| < \rho$.

For a proof of the existence part, see [36].

Remark 5.1. Note that the solution is only guaranteed to converge in the region of convergence for the coefficient functions. So do not conclude that if this theorem applies, the series solution converges everywhere.

Example 5.5. Determine a solution to

$$\ddot{x} + t^2 x = 0,$$

where $x(0) = 1$ and $\dot{x}(0) = -1$. Before solving this equation, note that it can be interpreted as a mass–spring system where the spring constant is equal to t^2. Hence, expecting solutions that oscillate with a frequency that increases with time would be reasonable. Also, because the function $f_0(t) = t^2$ is analytic everywhere, Theorem 5.3 guarantees a series solution that converges for all t.

Assuming the usual

$$x(t) = \sum_{n=0}^{\infty} a_n t^n \quad \Longrightarrow \quad \ddot{x}(t) = \sum_{n=2}^{\infty} n(n-1) a_n t^{n-2}$$

and substituting gives

$$\left(\sum_{n=2}^{\infty} n(n-1) a_n t^{n-2} \right) + t^2 \left(\sum_{n=0}^{\infty} a_n t^n \right)$$

$$= \left(\sum_{n=2}^{\infty} n(n-1) a_n t^{n-2} \right) + \left(\sum_{n=0}^{\infty} a_n t^{n+2} \right) = 0.$$

Shifting the index on the first sum by four gives

$$\left(\sum_{n=-2}^{\infty} (n+4)(n+3) a_{n+4} t^{n+2} \right) + \left(\sum_{n=0}^{\infty} a_n t^{n+2} \right)$$

$$= 2a_2 + 6a_3 t + \sum_{n=0}^{\infty} \left((n+4)(n+3) a_{n+4} + a_n \right) t^{n+2} = 0.$$

From the initial conditions and the first two terms in the above equation, we have

$$a_0 = 1, \quad a_1 = -1, \quad a_2 = 0, \quad a_3 = 0,$$

and from solving the coefficient in the series for a_{n+4},

$$a_{n+4} = \frac{1}{(n+4)(n+3)} a_n \quad n = 0, 1, 2, \ldots$$

gives the recurrence relation for the rest of the coefficients. A plot comparing an accurate numerical solution with several partial sums is illustrated in Figure 5.4.

There is nothing special about computing the series about the point $t = 0$, unless we are particularly interested in the features of the solution there. For comparison,

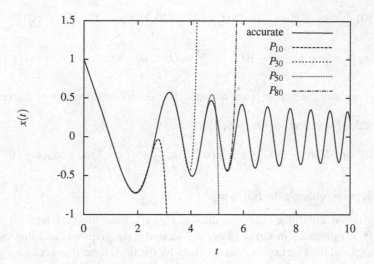

Fig. 5.4 Accurate solution and several partial sums for solution to Example 5.5.

the next example solves essentially the same problem as Example 5.2, but about $t = 10$ instead of $t = 0$.

Example 5.6. Determine the series solution to

$$\ddot{x} + \frac{5}{t^2 + 1} x = 0 \tag{5.11}$$

about $t = 10$.

Assuming a solution of the form

$$x(t) = \sum_{n=0}^{\infty} a_n (t - 10)^n, \tag{5.12}$$

observing that

$$t^2 + 1 = 101 + 20(t - 10) + (t - 10)^2,$$

and substituting gives

$$\left[101 + 20(t - 10) + (t - 10)^2 \right] \sum_{n=2}^{\infty} n(n - 1) a_n (t - 10)^{n-2} + 5 \sum_{n=0}^{\infty} a_n (t - 10)^n = 0.$$

Distributing the three terms before the second derivative term and shifting indices to make all the powers of $(t - 10)$ to be n gives

$$101 \sum_{n=0}^{\infty} (n+2)(n+1) a_{n+2} (t-10)^n + 20 \sum_{n=1}^{\infty} n(n+1) a_{n+1} (t-10)^n$$

$$+ \sum_{n=2}^{\infty} n(n-1) a_n (t-10)^n + 5 \sum_{n=0}^{\infty} a_n (t-10)^n = 0.$$

Explicitly writing the first terms of the sums that do not begin with $n = 2$ gives

$$(202a_2 + 5a_0) + (606a_3 + 40a_2 + 5a_1)(t-10)$$

$$+ \sum_{n=2}^{\infty} \left[101(n+2)(n+1) a_{n+2} + 20n(n+1) a_{n+1} + n(n-1) a_n + a_n \right] (t-10)^n = 0.$$

$$(5.13)$$

At this point, observe the following.

1. If the initial conditions are specified at $t = 10$, then the first term in Equation (5.13) gives a_2 in terms of a_0, the second term gives a_3, and the rest can be computed by a recursion relation given by the third term (the series).
2. Theorem 5.3 allows us to conclude that the series solution will converge in the range $t \in \left(10 - \sqrt{101}, 10 + \sqrt{101} \right)$.
3. It is not the case that $a_0 = x(0)$ and $a_1 = \dot{x}(0)$. If $t = 0$ is substituted into the assumed form of the solution (Equation (5.12)) every term in the series remains and must be evaluated.
4. If the initial conditions are given at a time other than $t = 10$, then we could include all the terms up to a point in the series and solve a system of algebraic equations to determine approximate values for the coefficients.

If the initial conditions are given at $t = 10$, such as $x(10) = 1$ and $\dot{x}(10) = -1$, then computing the coefficients is straightforward:

$$a_0 = 1, \quad a_1 = -1, \quad a_2 = -\frac{5}{202}, \quad a_3 = -\frac{1}{606}(5a_1 + 40a_2),$$

and the rest of the coefficients can be computed from

$$a_{n+2} = -\frac{n(n+1) a_{n+1} + n(n-1) a_n + a_n}{101(n+2)(n+1)}, \qquad (5.14)$$

for $n = 2, 3, 4, \ldots$ A graph of the solution determined numerically and several partial sums are illustrated in Figure 5.5.

5.4 Series Solutions About a Singular Point

Sometimes it is the case that a series solution about a point which is not an ordinary point is needed. Because singular points are less common when the independent variable is time and are more commonly associated with an independent *spatial*

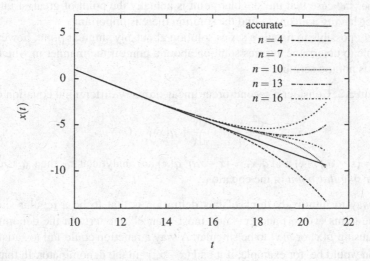

Fig. 5.5 Solution and several partial sums for Example 5.6.

variable, this section switches notation and has x represent the independent variable and y represent the dependent variable. A common scenario is when the second derivative term is multiplied by a function that can be zero at some points. An example would be

$$\left(1 - x^2\right) \frac{d^2 y}{dx^2}(x) + \sin(x) \frac{dy}{dx}(x) + y(x) = \cos(x),$$

where at $x = \pm 1$, $1 - x^2 = 0$. As a general matter, such points are clearly important because they correspond to situations in which the order of the equation changes. Also, if the equation is converted to the form of Equation (5.10),

$$\frac{d^2 y}{dx^2}(x) + \left(\frac{\sin(x)}{1 - x^2}\right) \frac{dy}{dx}(x) + \left(\frac{1}{1 - x^2}\right) y(x) = \frac{\cos(x)}{1 - x^2},$$

the points $x = \pm 1$ are the points at which the Taylor series for each of the functions

$$f_1(x) = \frac{\sin(x)}{1 - x^2}, \quad f_0(x) = \frac{1}{1 - x^2}, \quad g(x) = \frac{\cos(x)}{1 - x^2}$$

will not converge. Hence, we cannot make use of Theorem 5.3 to ensure we will find a unique solution at those points. Points where the coefficient of the highest derivative is equal to zero are called *singular points*. If it is not important to have a solution at the singular point, then using the method from the previous section and finding a series solution about a nearby ordinary point would suffice. However, it is sometimes the case that the singular point is actually the point of greatest interest, and hence having a means to find the solution there is important.

It is not possible to handle a series solution about any singular point, however, it is possible to determine a series solution about a point if the manner in which it is singular is not too serious.

Definition 5.2. Consider a second-order linear ordinary differential equation of the form

$$\frac{d^2y}{dx^2} + p(x)\frac{dy}{dx} + q(x)y = 0.$$

If $\hat{p}(x) = (x - x_0)\,p(x)$ and $\hat{q}(x) = (x - x_0)^2\,q(x)$ are analytic at x_0, then x_0 is called a *regular singular point* of the equation.

The way to intuitively think of this definition is that to be a regular singular point, the terms $(x - x_0)$ and $(x - x_0)^2$ must "cancel" the term in the denominator that is causing $p(x)$ or $q(x)$ to be singular. A way a function could fail to satisfy the definition would be, for example, if it had $(x - x_0)^5$ in the denominator. In that case the denominator would still go to zero even when multiplied by $(x - x_0)^2$ and hence not be analytic at $x = x_0$. Because $(x - x_0)^5$ goes to zero faster than $(x - x_0)^2$, a term in which it appears in the denominator will "blow up" faster, which is an indication of the singularity being, in some sense, worse.

If the usual approach outlined in the previous section is used to find a series solution, it will generally fail to work (see Exercise 5.16). Instead, the approach to find a series solution about a regular singular point is to assume a solution of the form

$$y(x) = (x - x_0)^m \sum_{n=0}^{\infty} a_n (x - x_0)^n = \sum_{n=0}^{\infty} a_n (x - x_0)^{n+m} \tag{5.15}$$

and, in addition to the coefficients a_n, determine m. Note that to fix values for m, we require that $a_0 \neq 0$.

Although the approach is simple ("find m and then all the coefficients a_i by substituting into the equation") the implementation is complicated by the fact that there are no less than five different possible outcomes. Before those can be outlined, we determine the formula used to determine m.

Assume that x_0 is a regular singular point of

$$\frac{d^2y}{dx^2}(x) + p(x)\frac{dy}{dx}(x) + q(x)y(x) = 0.$$

Multiplying this equation by $(x - x_0)^2$ and expanding $\hat{p}(x)$ and $\hat{q}(x)$ in a Taylor series, gives

$$(x-x_0)^2 \frac{d^2 y}{dx^2}(x) + (x-x_0)[(x-x_0)p(x)]\frac{dy}{dx}(x) + \left[(x-x_0)^2 q(x)\right] y(x)$$

$$= (x-x_0)^2 \frac{d^2 y}{dx^2}(x) + (x-x_0)[\hat{p}(x)]\frac{dy}{dx}(x) + [\hat{q}(x)]y(x)$$

$$= (x-x_0)^2 \frac{d^2 y}{dx^2}(x) + (x-x_0)\left[p_0 + p_1(x-x_0) + p_2(x-x_0)^2 + \cdots\right]\frac{dy}{dx}(x)$$

$$+ \left[q_0 + q_1(x-x_0) + q_2(x-x_0)^2 + \cdots\right]y(x)$$

$$= 0.$$

Assuming a solution of the form

$$y(x) = \sum_{n=0}^{\infty} a_n (x-x_0)^{n+m}$$

and substituting gives (Exercise 5.18) gives

$$[m(m-1) + p_0 m + q_0] a_0 (x-x_0)^m$$

$$+ [(m(m+1) + p_0(m+1) + q_0)a_1 + (p_1 m + q_1)a_0](x-x_0)^{m+1} \qquad (5.16)$$

$$+ \cdots = 0.$$

Remark 5.2. Three important conclusions are worth emphasizing.

1. Note that inasmuch as a_0 is assumed not to be zero, the term in the square brackets in the first line of Equation (5.16) must be zero, and hence determines the value(s) for m. It is quadratic, so there are two solutions and the values for m may be real and distinct, repeated real, or a complex conjugate pair. A complete investigation into the nature of series solutions about singular points requires considering all of those cases.

2. If the values for m are determined by the coefficient of $(x-x_0)^m$, then each term in the coefficient of $(x-x_0)^{m+1}$ is fixed except a_0 and a_1, so this gives a recursion relation. Unlike the examples considered previously for series solutions about ordinary points for second-order equations, this recursion relation is between coefficients with indices that differ only by one. The same is true for the coefficients for the higher powers in $(x-x_0)$. The consequence of this is that the value for a_0 determines the values for all the other a_i. In order to obtain a second linearly independent solution, the second value for m is used in the series. Hence, instead of alternating terms in the series forming the two solutions, the entire series is used twice, once for each value of m.

3. The different values for m are used to find the two solutions, therefore the case where m is repeated requires some work to find another solution. Less obvious is the case where there are distinct values for m, but they differ by an integer; this also requires additional work because when the ms differ by an integer, that effect is only to shift the powers of $(x-x_0)^{n+m}$ which may complicate matters significantly.

We do not delve into all the intricacies of this subject and work out every case. Good basic references are [8, 50], which provide all the answers, but not necessarily every derivation. A theorem presenting formulae for all the solutions appears in Appendix D.2. The remainder of this section presents a few examples

Euler's equation, which is of the form

$$x^2 \frac{d^2 y}{dx}(x) + x p_0 \frac{dy}{dx}(x) + q_0 y(x) = 0, \tag{5.17}$$

is an example of a variable-coefficient second-order ordinary differential equation where the series solution about a regular singular point simplifies dramatically, and is good to use to develop some intuition about the nature of the solutions to these types of problems.[6] Euler's equation is of the general form of a second order equation with a regular singular point, but where the functions $p(x)$ and $q(x)$ are constants.

Example 5.7. Determine a solution to

$$x^2 \frac{d^2 y}{dt^2} + x \frac{dy}{dt} - \frac{1}{4} y = 0 \tag{5.18}$$

valid in the neighborhood of $x = 0$. Note that the point $x_0 = 0$ is a regular singular point for this equation. Assuming

$$y(x) = x^m \sum_{n=0}^{\infty} a_n x^n = \sum_{n=0}^{\infty} a_n x^{n+m}$$

gives

$$\frac{dy}{dx}(x) = \sum_{n=0}^{\infty} (n+m) a_n x^{n+m-1}, \quad \frac{d^2 y}{dx^2}(x) = \sum_{n=0}^{\infty} (n+m)(n+m-1) a_n x^{n+m-2}.$$

Note that the sums start at $n = 0$, which is a consequence of the fact that m is not yet specified. Substituting gives

---

[6] A reader with a great memory will recall Euler's equation appeared earlier in this book. Exercise 3.8 presented the equation, but provided the form of the solution instead of deriving it, as we are doing here.

$$x^2 \sum_{n=0}^{\infty} (n+m)(n+m-1) a_n x^{n+m-2} + x \sum_{n=0}^{\infty} (n+m) a_n x^{n+m-1} - \frac{1}{4} \sum_{n=0}^{\infty} a_n x^{n+m} =$$

$$\sum_{n=0}^{\infty} (n+m)(n+m-1) a_n x^{n+m} + \sum_{n=0}^{\infty} (n+m) a_n x^{n+m} - \frac{1}{4} \sum_{n=0}^{\infty} a_n x^{n+m} =$$

$$\sum_{n=0}^{\infty} \left[(n+m)(n+m-1) + (n+m) - \frac{1}{4} \right] a_n x^{n+m} =$$

$$\sum_{n=0}^{\infty} \left[(n+m)^2 - \frac{1}{4} \right] a_n x^{n+m} = 0.$$

Observe that no recurrence relation results from this substitution. Hence, we can choose one value for n for which $a_n \neq 0$, but all the rest of the coefficients must be zero. For that value of n, the term in square brackets can be zero by an appropriate choice for m. Specifically, for whatever n is picked,

$$(n+m)^2 = \frac{1}{4} \implies n+m = \pm \frac{1}{2}.$$

There is a solution corresponding to the one term for each value of $n+m$, namely

$$y_1(x) = x^{1/2}, \quad y_2(x) = x^{-1/2}.$$

These do not come from two different terms in a recurrence relation, but rather from two different series collapsing into one term each. Computing the Wronskian gives

$$\begin{vmatrix} x^{1/2} & x^{-1/2} \\ \frac{1}{2} x^{-1/2} & -\frac{1}{2} x^{-3/2} \end{vmatrix} = -\frac{1}{x},$$

which shows that the two solutions are linearly independent for $-\infty < x < 0$ or $0 < x < \infty$, and hence

$$y(x) = c_1 \sqrt{x} + c_2 \frac{1}{\sqrt{x}} \tag{5.19}$$

is the general solution to Equation (5.18) for $x \neq 0$.

The complete solution to Euler's equation is given in Appendix D.2, Theorem D.2. It was the special structure of Euler's equation that led the power series solution to reduce to a combination of two powers in x. In the general case, the entire series is necessary, as is illustrated by the following example.

Remark 5.3. It is worth emphasizing that the solution to the differential equation in Example 5.7 is not defined at $x = 0$ if $c_2 \neq 0$. However, if the nature of the solution near $x = 0$ is important, then a series solution computed about a point other than zero cannot possibly capture this. This is because if the series is computed about an ordinary point near zero, it will be a polynomial with positive powers of x only and hence cannot possibly blow up at $x = 0$. The terms necessary for it to appropriately capture the nature of the singularity in the solution simply are not in the form of the series that was assumed for the solution. Exercise 5.21 compares two such solutions.

Example 5.8. Determine the general solution to

$$x^2 \frac{d^2 y}{dx^2} + x \left(x + \frac{1}{2} \right) \frac{dy}{dx} - \left(x^2 + \frac{1}{2} \right) y = 0$$

for values of x near zero.

The fact that $x = 0$ is a regular singular point is easily verified. Assuming

$$y(x) = \sum_{n=0}^{\infty} a_n x^{n+m}$$

and substituting gives

$$x^2 \sum_{n=0}^{\infty} (n+m)(n+m-1) a_n x^{n+m-2}$$

$$+ x \left(x + \frac{1}{2} \right) \sum_{n=0}^{\infty} (n+m) a_n x^{n+m-1} - \left(x^2 + \frac{1}{2} \right) \sum_{n=0}^{\infty} a_n x^{n+m} = 0.$$

Rearranging gives

$$\sum_{n=0}^{\infty} (n+m)(n+m-1) a_n x^{n+m} + \sum_{n=0}^{\infty} (n+m) a_n x^{n+m+1}$$

$$+ \frac{1}{2} \sum_{n=0}^{\infty} (n+m) a_n x^{n+m} - \sum_{n=0}^{\infty} a_n x^{n+m+2} - \frac{1}{2} \sum_{n=0}^{\infty} a_n x^{n+m} = 0.$$

Shifting the indices on the terms so that all the powers of x are $n+m$ gives

$$\sum_{n=0}^{\infty} (n+m)(n+m-1) a_n x^{n+m} + \sum_{n=1}^{\infty} (n+m-1) a_{n-1} x^{n+m}$$

$$+ \frac{1}{2} \sum_{n=0}^{\infty} (n+m) a_n x^{n+m} - \sum_{n=2}^{\infty} a_{n-2} x^{n+m} - \frac{1}{2} \sum_{n=0}^{\infty} a_n x^{n+m} = 0,$$

and writing out the $n = 0$ and $n = 1$ terms gives

$$m(m-1) a_0 x^m + (m+1) m a_1 x^{m+1} + m a_0 x^{m+1}$$

$$+ \frac{1}{2} m a_0 x^m + \frac{1}{2} (m+1) a_1 x^{m+1} - \frac{1}{2} a_0 x^m - \frac{1}{2} a_1 x^{m+1}$$

$$+ \sum_{n=2}^{\infty} \left\{ \left((n+m) \left(n+m-\frac{1}{2} \right) - \frac{1}{2} \right) a_n + (n+m-1) a_{n-1} - a_{n-2} \right\} x^{n+m} = 0.$$

Collecting terms in equal powers of x gives

$$\left(m(m-1)+\frac{1}{2}m-\frac{1}{2}\right)a_0 x^m + \left(ma_0 + \left[(m+1)\left(m+\frac{1}{2}\right)-\frac{1}{2}\right]a_1\right)x^{m+1}$$

$$+\sum_{n=2}^{\infty}\left\{\left[(n+m)\left(n+m-\frac{1}{2}\right)-\frac{1}{2}\right]a_n + (n+m-1)a_{n-1}-a_{n-2}\right\}x^{n+m}=0.$$

$$(5.20)$$

Setting the coefficient of $a_0 x^m$ in Equation (5.20) to zero gives

$$m(m-1)+\frac{1}{2}m-\frac{1}{2}=0$$

or

$$m^2 -\frac{1}{2}m-\frac{1}{2}=0, \implies m=1,-\frac{1}{2},$$

which are distinct and do not differ by an integer. We could have obtained these from Equation (5.16), but we needed the whole expansion anyway to construct the solution. Checking with Equation (5.16) verifies the two values for m.

Using $m=1$, the relationship between a_0 and a_1 given by the coefficient of x^{m+1} in Equation (5.20) is

$$a_1 = -\frac{2}{5}a_0,$$

and the recursion relation given by the series term in Equation (5.20), after some simplification, is

$$n\left(n+\frac{3}{2}\right)a_n = a_{n-2}-na_{n-1}.$$

So, for $n=2$,

$$7a_2 = a_1 - 2a_0 = \left(1+\frac{4}{5}\right)a_0, \implies a_2 = \frac{9}{35}a_0,$$

and so on for $n>2$. Hence, we have for the $m=1$ solution

$$y_1(x) = a_0 x\left(1-\frac{2}{5}x+\frac{9}{35}x^2+\cdots\right).$$

Note, the x outside the parentheses is the x^m term.

Following the same procedure, but for $m=-1/2$ gives

$$y_2(x) = \frac{a_0}{\sqrt{x}}\left(1-x+\frac{3}{2}x^2+\cdots\right).$$

The complete solution is a linear combination of y_1 and y_2, namely,

$$y(x) = c_1 x\left(1-\frac{2}{5}x+\frac{9}{35}x^2+\cdots\right) + \frac{c_2}{\sqrt{x}}\left(1-x+\frac{3}{2}x^2+\cdots\right).$$

5.5 A Collection of Famous Series Solutions

The following five examples are some famous differential equations and solutions. Many of these play a role in Sturm–Liouville boundary value problems considered in Chapter 11.

5.5.1 Airy Equation

One famous variable-coefficient differential equation is the *Airy equation*, which has an ordinary point at $x = 0$. This equation describes optical phenomena and is named after the astronomer George Airy.

The Airy equation is

$$\frac{d^2 y}{dx^2}(x) - xy(x) = 0.$$

Assume

$$y(x) = \sum_{n=0}^{\infty} a_n x^n.$$

As before

$$\frac{d^2 y}{dx^2}(x) = \sum_{n=2}^{\infty} n(n-1)a_n x^{n-1} = \sum_{n=0}^{\infty}(n+2)(n+1)a_{n+2} x^n$$

and substituting into the differential equation gives

$$\left(\sum_{n=0}^{\infty}(n+2)(n+1)a_{n+2} x^n \right) - x \left(\sum_{n=0}^{\infty} a_n x^n \right)$$

$$= \left(\sum_{n=0}^{\infty}(n+2)(n+1)a_{n+2} x^n \right) - \left(\sum_{n=0}^{\infty} a_n x^{n+1} \right) = 0.$$

In order to equate powers of x, shift the index of summation in the second sum by one; that is,

$$\sum_{n=0}^{\infty} a_n x^{n+1} = \sum_{n=1}^{\infty} a_{n-1} x^n,$$

so we have

$$\left(\sum_{n=0}^{\infty}(n+2)(n+1)a_{n+2} x^n \right) - \sum_{n=1}^{\infty} a_{n-1} x^n = 0.$$

The two series have the same powers in x, but start at different values of n. To handle this, simply write the first term of the first series by itself; that is,

$$(2)(1)a_2x^0 + \left(\sum_{n=1}^{\infty}(n+2)(n+1)a_{n+2}x^n\right) - \sum_{n=1}^{\infty}a_{n-1}x^n =$$

$$2a_2 + \sum_{n=1}^{\infty}[(n+2)(n+1)a_{n+2} - a_{n-1}]x^n = 0.$$

So, equating powers of x gives

$$a_2 = 0$$

and

$$a_{n+2} = \frac{1}{(n+2)(n+1)}a_{n-1}. \tag{5.21}$$

It is possible to show (see [8]) that the solution can be written as

$$y(x) = a_0\left(1 + \frac{1}{6}x^3 + \sum_{n=2}^{\infty}\frac{1}{(3n)(3n-1)(3n-3)(3n-4)\cdots(3)(2)}x^{3n}\right)$$

$$+ a_1\left(x + \frac{1}{12}x^4 + \sum_{n=2}^{\infty}\frac{1}{(3n+1)(3n)(3n-2)(3n-3)\cdots(4)(3)}x^{3n+1}\right).$$

Writing this solution as $y(x) = a_0y_1(x) + a_1y_2(x)$, Figures 5.6 and 5.7 illustrate $y_1(x)$ and $y_2(x)$ for partial sums including various numbers of terms. As one should expect by examining the differential equation, the solutions oscillate for $x < 0$ and grow exponentially for $x > 0$.[7]

5.5.2 Chebychev Equation

The differential equation

$$(1-x^2)\frac{d^2y}{dx^2} - x\frac{dy}{dx} + \lambda^2y = 0 \tag{5.22}$$

is called the *Chebychev equation*.[8] Dividing by $(1-x^2)$ to put Equation (5.22) into the form of Equation (5.10) makes it clear that $x_0 = 0$ is an ordinary point and a series solution about $x_0 = 0$ will converge to the solution to the differential equation for $x \in (-1, 1)$. Assuming a solution of the form

$$y(x) = \sum_{n=1}^{\infty}a_nx^n$$

---

[7] The reader is cautioned that the Airy function and Airy function of the second kind, as they are commonly defined and used in numerical computational packages, are not usually the same as these. They are, in fact, typically a linear combination of these two.

[8] Sometimes this name is spelled *Tchebycheff* or *Chebyshev*.

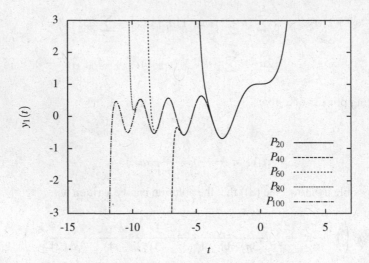

Fig. 5.6 Partial sum solutions for the Airy equation including various numbers of terms.

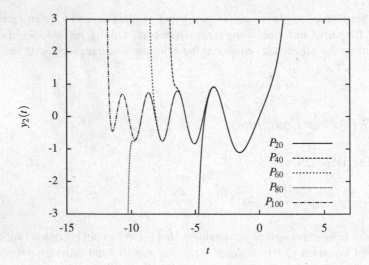

Fig. 5.7 Partial sum solutions for the Airy equation including various numbers of terms.

gives the recurrence relation

$$a_{n+1} = -\frac{(n-\lambda)(n+\lambda)}{(n+1)(n+2)}a_n.$$

As is common, then, $y(0) = a_0$ specifies all the values for a_n where n is even, and $dy/dx(0) = a_1$ specifies all the values for a_n where n is odd, and we can write the solution as

$$y(x) = a_0 S_e + a_1 S_o,$$

where S_e and S_o stand for the even and odd series with the even and odd powers of x, respectively.

Because of the $n - \lambda$ term in the numerator of the recurrence relation, when λ is a nonnegative integer, one of the two series S_e or S_o terminates at x^n; that is, it is a polynomial of finite degree that consequently globally converges. When properly normalized, these resulting polynomials are the *Chebychev polynomials*.

5.5.3 Hermite Equation

Another famous second-order variable-coefficient differential equation with an ordinary point at $x = 0$ is the *Hermite equation* given by

$$\frac{d^2y}{dx^2} - 2x\frac{dy}{dx} + \lambda y = 0. \tag{5.23}$$

Because $x = 0$ is an ordinary point, a series representation of the solution is given by

$$y(x) = \sum_{n=0}^{\infty} a_n x^n.$$

Upon substituting and equating coefficients of the powers of x, we obtain

$$a_2 = -\frac{\lambda}{2}a_0$$

and

$$a_{n+2} = \frac{2n - \lambda}{(n+2)(n+1)}a_n.$$

In the case where λ is a positive even integer, the series is finite and *Hermite polynomials* are the result. The details are left as an exercise (Exercise 5.9).

5.5.4 *Legendre Equation*

The Legendre equation is given by

$$\left(1-x^2\right)\frac{d^2y}{dx^2} - 2x\frac{dy}{dx} + \lambda\left(\lambda+1\right)y = 0. \tag{5.24}$$

As with the other equations, $x = 0$ is an ordinary point, and assuming a power series solution of the usual form results in a recurrence relation of the form

$$a_{n+1} = -\frac{(\lambda - n)(\lambda + n + 1)}{(n+1)(n+2)}a_n.$$

Similar to the case for Hermite's equation, if λ is a positive integer, then the series containing either the even or odd powers of x will terminate at the power of x^λ. When normalized in a particular way, the resulting polynomials are the *Legendre polynomials*.

5.5.5 *Bessel Equation*

The *Bessel equation* is

$$t^2\frac{d^2x}{dt^2} + t\frac{dx}{dt} + \left(t^2 - \lambda^2\right)x = 0. \tag{5.25}$$

The parameter λ may be a real or complex number. A couple of cases are considered in the exercises. In the special case where it is an integer, it is called the *order* of the equation. Because this is a second-order equation, two linearly independent solutions are necessary to determine a general solution. In the case when λ is an integer, the two solutions are given by

$$J_\lambda(t) = \sum_{n=1}^{\infty} \frac{(-1)^n}{2^{2n+\lambda}n!\,(n+\lambda)!}t^{2n+\lambda}$$

and

$$Y_\lambda(t) = \frac{J_r(t)\cos(\lambda\pi) - J_{-\lambda}(t)}{\sin(\lambda\pi)}.$$

The function $J_\lambda(t)$ is called the *Bessel function of the first kind* and the function $Y_\lambda(t)$ is called the *Bessel function of the second kind*. Figure 5.8 illustrates $J_\lambda(t)$ for various integer orders and Figure 5.9 illustrates $Y_\lambda(t)$ for various integer orders. In addition to being the linearly independent solutions to Equation (5.25), these two functions have some additional remarkable properties which are explored in Chapter 11.

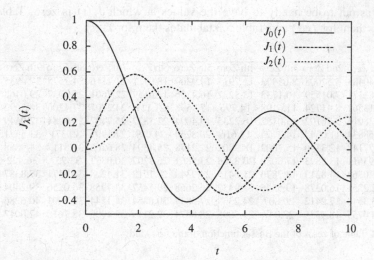

Fig. 5.8 Bessel functions of the first kind.

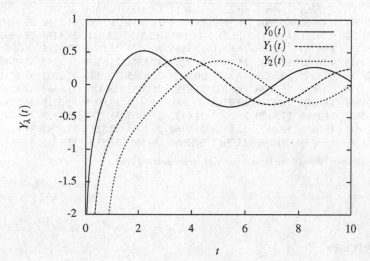

Fig. 5.9 Bessel functions of the second kind.

It turns out to be handy to have the values at which $J_\lambda(t)$ is zero. Table 5.2 tabulates them for $J_\lambda(t)$ and Table 5.3 tabulates them for $Y_\lambda(t)$.

Order	1st Zero	2nd Zero	3rd Zero	4th Zero	5th Zero	6th Zero	7th Zero	8th Zero	9th Zero	10th Zero
0	2.40483	5.52008	8.65373	11.7915	14.9309	18.0711	21.2116	24.3525	27.4935	30.6346
1	3.83171	7.01559	10.1735	13.3237	16.4706	19.6159	22.7601	25.9037	29.0468	32.1897
2	5.13562	8.41724	11.6198	14.796	17.9598	21.117	24.2701	27.4206	30.5692	33.7165
3	6.38016	9.76102	13.0152	16.2235	19.4094	22.5827	25.7482	28.9084	32.0649	35.2187
4	7.58834	11.0647	14.3725	17.616	20.8269	24.019	27.1991	30.371	33.5371	36.699
5	8.77148	12.3386	15.7002	18.9801	22.2178	25.4303	28.6266	31.8117	34.9888	38.1599
6	9.93611	13.5893	17.0038	20.3208	23.5861	26.8202	30.0337	33.233	36.422	39.6032
7	11.0864	14.8213	18.2876	21.6415	24.9349	28.1912	31.4228	34.6371	37.8387	41.0308
8	12.2251	16.0378	19.5545	22.9452	26.2668	29.5457	32.7958	36.0256	39.2404	42.4439
9	13.3543	17.2412	20.807	24.2339	27.5837	30.8854	34.1544	37.4001	40.6286	43.8438
10	14.4755	18.4335	22.047	25.5095	28.8874	32.2119	35.4999	38.7618	42.0042	45.2316 .

Table 5.2 Table of zeros of the Bessel function of the first kind

Order	1st Zero	2nd Zero	3rd Zero	4th Zero	5th Zero	6th Zero	7th Zero	8th Zero	9th Zero	10th Zero
0	0.893577	3.95768	7.08605	10.2223	13.3611	16.5009	19.6413	22.782	25.923	29.064
1	2.19714	5.42968	8.59601	11.7492	14.8974	18.0434	21.1881	24.3319	27.4753	30.6183
2	3.38424	6.79381	10.0235	13.21	16.379	19.539	22.694	25.8456	28.9951	32.143
3	4.52702	8.09755	11.3965	14.6231	17.8185	20.9973	24.1662	27.3288	30.487	33.642
4	5.64515	9.36162	12.7301	15.9996	19.2244	22.4248	25.6103	28.7859	31.9547	35.1185
5	6.74718	10.5972	14.0338	17.3471	20.6029	23.8265	27.0301	30.2203	33.4011	36.575
6	7.83774	11.811	15.3136	18.6707	21.9583	25.2062	28.429	31.6349	34.8286	38.0135
7	8.91961	13.0077	16.5739	19.9743	23.294	26.5668	29.8095	33.0318	36.2393	39.4358
8	9.99463	14.1904	17.8179	21.2609	24.6126	27.9105	31.1737	34.4129	37.6346	40.8434
9	11.0641	15.3613	19.0479	22.5328	25.9162	29.2394	32.5233	35.7797	39.0162	42.2376
10	12.1289	16.5223	20.266	23.7917	27.2066	30.555	33.8597	37.1336	40.3851	43.6195

Table 5.3 Table of zeros of the Bessel function of the second kind

5.6 Exercises

For any exercise that requires a series solution assume a series solution and substitute it into the differential equation to determine the recursion relation. Plot your solution for different numbers of terms in the partial sum. Your answer should include enough terms to satisfy you that the graph is an accurate representation of the solution.

5.1. Consider $\ddot{x} - 2t\dot{x} + 6x = 0$.

1. Is this one of the differential equations outlined in Section 5.5?
2. Is the point $x = 0$ an ordinary point or a singular point?

3. Assume a series solution and substitute to determine the coefficients.
4. If $x(0) = 1$, $\dot{x}(0) = 1$ and plot the solution for various numbers of terms. Compare your series solutions to a solution obtained numerically.
5. If $x(0) = 0$, $\dot{x}(0) = 1$ and plot the solution for various numbers of terms. Compare your series solutions to a solution obtained numerically.

5.2. Consider $\ddot{x} - 2t\dot{x} + 5x = 0$.

1. Is this one of the differential equations outlined in Section 5.5?
2. Is the point $x = 0$ an ordinary point or a singular point?
3. Assume a series solution and substitute to determine the coefficients.
4. If $x(0) = 1$, $\dot{x}(0) = 1$ and plot the solution for various numbers of terms. Compare your series solutions to a solution obtained numerically.
5. Did changing the 6 to a 5 have a large effect on the solution?

5.3. Consider $\ddot{x} + x = 0$.

1. Determine a power series solution about $t = 0$.
2. Determine the general solution using a method from Chapter 3.
3. Are the solutions the same?

5.4. Determine a partial sum of the series solution about $t = 0$ to

$$\ddot{x} - (t+1)\dot{x} + t^2 x = 0,$$

where $x(0) = 1$ and $\dot{x}(0) = 1$. Compare your partial sum approximations to a solution determined numerically.

5.5. Determine a partial sum of the series solution about $t = 0$ to

$$\ddot{x} + 3\left(t^2 + 1\right)\dot{x} + x = 0,$$

where $x(0) = 1$ and $\dot{x}(0) = -1$. Compare your partial sum approximations to a solution determined numerically.

5.6. Determine a series representation for two linearly independent solutions to Airy's equation

$$\frac{d^2 y}{dx^2} - 2xy = 0.$$

Plot them.

5.7. Determine a series solution to Airy's equation

$$\frac{d^2 y}{dx^2} + xy = 0,$$

where $y(0) = 1$ and $dy/dx(0) = -1$.

5.8. For the Chebychev equation:

1. Determine a series solution about $x = 0$ when $\lambda = 3$ and $y(0) = 1$ and $y'(0) = 0$.
2. Determine a series solution about $x = 0$ when $\lambda = 1/2$ and $y(0) = 0$ and $y'(0) = 1$.

For each case, plot your solution and compare it to an accurate approximate numerical solution.

5.9. For the Hermite equation:

1. Determine a series solution about $x = 0$ when $\lambda = 1$ and $y(0) = 1$ and $y'(0) = 0$.
2. Determine a series solution about $x = 0$ when $\lambda = 2$ and $y(0) = 0$ and $y'(0) = 1$.

For each case, plot your solution and compare it to an accurate approximate numerical solution.

5.10. For the Legendre equation:

1. Determine a series solution about $x = 0$ when $\lambda = 2$ and $y(0) = 1$ and $y'(0) = 0$.
2. Determine a series solution about $x = 0$ when $\lambda = 1/2$ and $y(0) = 0$ and $y'(0) = 1$.

For each case, plot your solution and compare it to an accurate approximate numerical solution.

5.11. Determine a partial sum of the series solution about $t = 0$ to

$$\ddot{x} + \dot{x} + (t + 2)x = 0,$$

where $x(0) = 1$ and $\dot{x}(0) = 1$. Compare your partial sum approximations to a solution determined numerically.

5.12. Determine a partial sum of the series solution about $t = 0$ to

$$\ddot{x} + \frac{t}{1 - t^2}\dot{x} - \frac{1}{1 - t^2}x = 0,$$

where $x(0) = 1$ and $\dot{x}(0) = 1$. Compare your partial sum approximations to a solution determined numerically. Based upon your analysis of the equation, what will be the range for t in which the series solution converges? Do your results represent this?

5.13. Consider

$$\ddot{x} + \frac{1}{t^2 + t2 + 2}\dot{x} + \frac{1}{t^2 + 4t + 8}x = 0.$$

1. For what interval of t would a series solution about $t = 0$ converge?
2. Determine the recursion relation for the series solution.
3. Compare the series solution for partial sums including different numbers of terms for the following two cases.

   a. $x(0) = 1$ and $\dot{x}(0) = -1$.
   b. $x(0) = 5$ and $\dot{x}(0) = 0$.

Do the initial conditions seem to affect the interval of convergence?

5.14. Determine a partial sum of the general series solution about $t = 0$ to

$$\ddot{x} + \sin(t)\dot{x} + 2e^t x = 0.$$

The answer should be in terms of a_0 and a_1 because it is the general solution. *Hint:* expand the sine and exponential functions in their Taylor series.

5.15. Determine the power series solution about $t = 1$ for

$$\ddot{x} + \frac{1}{t}\dot{x} - \frac{1}{t^2}x = 0,$$

where $x(0) = 1$ and $\dot{x}(0) = 1$.

5.16. Consider

$$x^2 \frac{d^2 y}{dt^2} + 2x^2 \frac{dy}{dt} + y = 0,$$

where $y(0) = 1$ and $dy/dx(0) = -1$. Assume a series solution of the form

$$y(x) = \sum_{n=0}^{\infty} a_n x^n.$$

Are you able to find coefficients that satisfy the differential equation as well as the initial conditions?

5.17. Determine a partial sum of the series solution about $t = 0$ to

$$\frac{d^3 x}{dt^3} + \frac{1}{t}\frac{dx}{dt} - \frac{1}{t^2}x = 0,$$

where $x(0) = 1$, $\dot{x}(0) = 0$, and $\ddot{x}(0) = 1$. Compare your partial sum approximations to a solution determined numerically. Based upon your analysis of the equation, what will be the range for t in which the series solution converges? Do your results represent this?

5.18. Verify Equation (5.16).

5.19. Find a series solution about $x = 0$ for the Bessel equation

$$x^2 \frac{d^2 y}{dx^2}(x) + x\frac{dy}{dx}(x) + \left(x^2 - \frac{1}{9} \right) y(x) = 0. \tag{5.26}$$

5.20. In the development of the method for series solutions about a regular singular point, we required that $a_0 \neq 0$. Explain why this was done.

5.21. Consider the solution to Euler's equation given in Example 5.7,

$$x^2 \frac{d^2 y}{dt^2} + x\frac{dy}{dt} - \frac{1}{4}y = 0$$

which was given by

$$y(x) = c_1\sqrt{x} + \frac{c_2}{\sqrt{x}}.$$

1. Let $c_1 = c_2 = 1$ and plot the solution from $0 < \varepsilon \ll 1$.
2. For these values of c_1 and c_2, compute $y(1)$ and $dy/dx(1)$. Show that $x = 1$ is an ordinary point for the equation. Compute the series solution about $x = 1$. Use a computer plotting package to plot a partial sum of the solution and compare the solution about $x = 1$ to $y(x) = c_1\sqrt{x} + c_2/\sqrt{x}$. How well does the series about $x = 1$ match the solution near $x = 0$?

5.22. For each of the following, compute a series solution about $x = 0$ and determine two linearly independent solutions.

1. $2x^2 \frac{d^2 y}{dx^2} + x\left(x + \frac{1}{2}\right)\frac{dy}{dx} + 2xy = 0$,
2. $2x^2 \frac{d^2 y}{dx^2} + 3x\frac{dy}{dx} + (3x - 1)y = 0$,
3. $2x\frac{d^2 y}{dx^2} + (x + 1)\frac{dy}{dx} + 3y = 0$.

5.23. Although this chapter focused on linear equations with variable coefficients, power series methods may also sometimes work for nonlinear equations. Assume a series solution about $t = 0$ and substitute it into $\dot{x} = t^2 + x^2$ to find the coefficients. Compare the series solution to an approximate numerical solution if $x(0) = 1$.

Chapter 6
Systems of First-Order Linear Constant-Coefficient Ordinary Differential Equations

This chapter considers systems of n first-order linear constant-coefficient ordinary differential equations, which arise from nth-order differential equations or from systems of coupled differential equations. As becomes readily apparent, the theoretical basis for solving such systems relies heavily upon matrix algebra theory. Appendix B contains a review of some of the more important concepts from linear algebra.

The types of equations considered in this chapter are systems of n first-order linear constant-coefficient ordinary differential equations of the form

$$\frac{d}{dt} \begin{bmatrix} \xi_1(t) \\ \xi_2(t) \\ \vdots \\ \xi_n(t) \end{bmatrix} = \begin{bmatrix} a_{11} & a_{12} & \cdots & a_{1n} \\ a_{21} & a_{22} & \cdots & a_{2n} \\ \vdots & \vdots & \ddots & \vdots \\ a_{n1} & a_{n2} & \cdots & a_{nn} \end{bmatrix} \begin{bmatrix} \xi_1(t) \\ \xi_2(t) \\ \vdots \\ \xi_n(t) \end{bmatrix} + \begin{bmatrix} g_1(t) \\ g_2(t) \\ \vdots \\ g_n(t) \end{bmatrix}$$

where all the elements of the matrix a_{ij} are constants. This is a system of n first-order linear constant-coefficient differential equations because each line may be put in the form of Equation (1.5) where the coefficients are constants. If all the functions $g_i(t)$ are zero, then the system is homogeneous.

6.1 Introduction

This section first gives an example which illustrates that equations of the type considered in this chapter are common, after which the foundation of the solution method for the homogeneous case is presented.

Example 6.1. Consider the mass–spring–damper system illustrated in Figure 6.1. As is the usual case, assume that x_1 and x_2 are the displacements of m_1 and m_2, respectively, measured in an inertial coordinate system where the values of x_1 and x_2 are zero when the springs are unstretched. Considering a free body diagram for

B. Goodwine, *Engineering Differential Equations: Theory and Applications*,
DOI 10.1007/978-1-4419-7919-3_6, © Springer Science+Business Media, LLC 2011

each mass illustrated in Figure 6.2 and applying Newton's law gives

$$m_1\ddot{x}_1 + (b_1 + b_2)\dot{x}_1 - b_2\dot{x}_2 + (k_1 + k_2)x_1 - k_2x_2 = 0$$
$$m_2\ddot{x}_2 - b_2\dot{x}_1 + b_2\dot{x}_2 - k_2x_1 + k_2x_2 = f(t).$$

(6.1)

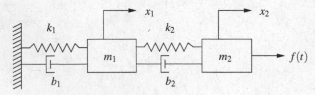

Fig. 6.1 Two degree of freedom mass–spring–damper system.

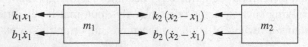

Fig. 6.2 Free body diagrams for masses in Figure 6.1.

These equations are coupled because x_1 appears in the x_2 equation and vice versa. A first inclination may be to try to solve one equation for one of either x_1 or x_2 and substitute into the other, but such an approach is problematic inasmuch as the equations involve the derivatives of the variables as well.

An insightful extrapolation of the method considered in Chapter 3 might lead one to attempt to solve the homogeneous problem first followed by some method for the particular solutions; indeed, this is fundamentally the approach we utilize. In fact, for the homogeneous case ($f(t) = 0$), that is,

$$m_1\ddot{x}_1 + (b_1 + b_2)\dot{x}_1 - b_2\dot{x}_2 + (k_1 + k_2)x_1 - k_2x_2 = 0$$
$$m_2\ddot{x}_2 - b_2\dot{x}_1 + b_2\dot{x}_2 - k_2x_1 + k_2x_2 = 0,$$

because the equations are linear, constant-coefficient, and homogeneous, a good guess may be to assume exponential solutions and substitute. In order to have a general approach that will work for all systems of linear, constant-coefficient, ordinary differential equations, we first convert the system into an equivalent system of first-order equations and the results are expressed in matrix form.

If we let

$$\xi_1 = x_1, \quad \xi_2 = \dot{x}_1, \quad \xi_3 = x_2, \quad \xi_4 = \dot{x}_2,$$

then

$$\frac{\mathrm{d}}{\mathrm{d}t} \begin{bmatrix} \xi_1 \\ \xi_2 \\ \xi_3 \\ \xi_4 \end{bmatrix} = \begin{bmatrix} \xi_2 \\ \frac{1}{m_1}\left(-b_1\xi_2 - k_1\xi_1 + k_2\xi_3 - k_2\xi_1 + b_2\xi_4 - b_2\xi_2\right) \\ \xi_3 \\ \frac{1}{m_2}\left(-k_2\xi_3 + k_2\xi_1 - b_2\xi_4 + b_4\xi_2\right) \end{bmatrix}.$$

Because this equation is linear in the ξ_is, it can be expressed as

$$\frac{\mathrm{d}}{\mathrm{d}t} \begin{bmatrix} \xi_1 \\ \xi_2 \\ \xi_3 \\ \xi_4 \end{bmatrix} = \begin{bmatrix} 0 & 1 & 0 & 0 \\ -\frac{k_1+k_2}{m_1} & -\frac{b_1+b_2}{m_1} & \frac{k_2}{m_1} & \frac{b_2}{m_1} \\ 0 & 0 & 0 & 1 \\ \frac{k_2}{m_2} & \frac{b_2}{m_2} & -\frac{k_2}{m_2} & -\frac{b_2}{m_2} \end{bmatrix} \begin{bmatrix} \xi_1 \\ \xi_2 \\ \xi_3 \\ \xi_4 \end{bmatrix}.$$

If we let

$$\xi = \begin{bmatrix} \xi_1 \\ \xi_2 \\ \xi_3 \\ \xi_4 \end{bmatrix}, \quad A = \begin{bmatrix} 0 & 1 & 0 & 0 \\ -\frac{k_1+k_2}{m_1} & -\frac{b_1+b_2}{m_1} & \frac{k_2}{m_1} & \frac{b_2}{m_1} \\ 0 & 0 & 0 & 1 \\ \frac{k_2}{m_2} & \frac{b_2}{m_2} & -\frac{k_2}{m_2} & -\frac{b_2}{m_2} \end{bmatrix}, \tag{6.2}$$

then this whole system can be expressed simply as

$$\dot{\xi} = A\xi. \tag{6.3}$$

Clearly, the nature of the solution to the vibrating masses depends on the properties of the matrix A because that is the only term in the equation where the parameters of the system are expressed. Exploiting the properties of A to solve this equation is our task at hand.

Now, the question arises as to the nature of the solution to Equation (6.3). Motivated by the results from Chapters 2 and 3, consider the possibility of a solution of the form

$$\xi(t) = \hat{\xi}e^{\lambda t},$$

where $\hat{\xi}$ is a constant vector. In full detail,

$$\xi(t) = \begin{bmatrix} \xi_1(t) \\ \xi_2(t) \\ \vdots \\ \xi_n(t) \end{bmatrix} = \begin{bmatrix} \hat{\xi}_1 \\ \hat{\xi}_2 \\ \vdots \\ \hat{\xi}_n \end{bmatrix} e^{\lambda t} = \begin{bmatrix} \hat{\xi}_1 e^{\lambda t} \\ \hat{\xi}_2 e^{\lambda t} \\ \vdots \\ \hat{\xi}_n e^{\lambda t} \end{bmatrix}.$$

Substituting this into Equation (6.3) gives

$$\lambda \begin{bmatrix} \hat{\xi}_1 e^{\lambda t} \\ \hat{\xi}_2 e^{\lambda t} \\ \vdots \\ \hat{\xi}_n e^{\lambda t} \end{bmatrix} = A \begin{bmatrix} \hat{\xi}_1 e^{\lambda t} \\ \hat{\xi}_2 e^{\lambda t} \\ \vdots \\ \hat{\xi}_n e^{\lambda t} \end{bmatrix}.$$

Inserting an identity matrix gives

$$\lambda \begin{bmatrix} 1 & 0 & \cdots & 0 \\ 0 & 1 & \cdots & 0 \\ \vdots & & \ddots & \vdots \\ 0 & 0 & \cdots & 1 \end{bmatrix} \begin{bmatrix} \hat{\xi}_1 e^{\lambda t} \\ \hat{\xi}_2 e^{\lambda t} \\ \vdots \\ \hat{\xi}_n e^{\lambda t} \end{bmatrix} = A \begin{bmatrix} \hat{\xi}_1 e^{\lambda t} \\ \hat{\xi}_2 e^{\lambda t} \\ \vdots \\ \hat{\xi}_n e^{\lambda t} \end{bmatrix}.$$

which can be rearranged to give

$$(A - \lambda I)\hat{\xi} = 0, \tag{6.4}$$

where the exponentials are canceled inasmuch as they are never zero. Recall from linear algebra that the values for λ that satisfy Equation (6.4) are the eigenvalues of the matrix A and the $\hat{\xi}$ that satisfy it are the corresponding eigenvectors of A. More importantly, what this shows is that solutions to Equation (6.3) are the product of the eigenvectors and exponentials of the eigenvalues of A.

Computing the eigenvalues and eigenvectors involves two steps. First, Equation (6.4) only has nonzero solutions for $\hat{\xi}$ if

$$\det(A - \lambda I) = 0.$$

If A is $n \times n$, this results in an nth-order polynomial for λ called the *characteristic polynomial*, and the roots of this polynomial are the eigenvalues. To determine the corresponding eigenvectors, substitute each of the eigenvalues into Equation (6.4) and solve for the eigenvector. Note that the eigenvalues and eigenvectors occur in pairs and that furthermore, eigenvectors are unique only up to multiplication by a scalar.

Because the system of equations is linear, if there are multiple pairs of eigenvalues and eigenvectors that satisfy Equation (6.4), a linear combination of the solutions is still a solution; that is,

$$\xi(t) = c_1 \hat{\xi}^1 e^{\lambda_1 t} + c_2 \hat{\xi}^2 e^{\lambda_2 t} + \cdots,$$

where each $\hat{\xi}^i$ corresponds to λ_i. If there are n components, then there are n initial conditions to be satisfied, so what we need to be able to satisfy any initial conditions are n linearly independent eigenvectors. Unfortunately, an $n \times n$ matrix does not always have n linearly independent eigenvectors, so, in such a case some more work is necessary, which, unsurprisingly, involves things such as multiplying solutions by t.

6.2 Converting to Systems of First-Order Differential Equations

Systems of first-order differential equations may arise naturally, but they are often the result of converting higher-order equations into that form. This has been addressed previously in Section 1.6.3 (and as subsequently in Section 12.5), so rather than present a formal procedure, an example should suffice.

Example 6.2. Consider the system of three ordinary differential equations

$$\frac{d^3x_1}{dt^3} + 4x_2 = x_3$$

$$\frac{dx_2}{dt} = x_1 \tag{6.5}$$

$$\frac{d^2x_3}{dt^2} + \frac{dx_3}{dt} = x_1 + x_2.$$

This is a system of linear, homogeneous, constant-coefficient, ordinary differential equations. Note that the highest-order derivative of x_1 is three, of x_2 is one, and of x_3 is two. If we let

$$\xi_1 = x_1, \quad \xi_2 = \frac{dx_1}{dt}, \quad \xi_3 = \frac{d^2x_1}{dt^2}, \quad \xi_4 = x_2, \quad \xi_5 = x_3, \quad \xi_6 = \frac{dx_3}{dt}, \tag{6.6}$$

then the system of equations in Equation (6.5) is equivalent to

$$\frac{d}{dt}\begin{bmatrix} \xi_1 \\ \xi_2 \\ \xi_3 \\ \xi_4 \\ \xi_5 \\ \xi_6 \end{bmatrix} = \begin{bmatrix} \xi_2 \\ \xi_3 \\ \xi_5 - 4\xi_4 \\ \xi_1 \\ \xi_6 \\ \xi_1 + \xi_4 - \xi_6 \end{bmatrix} = \begin{bmatrix} 0 & 1 & 0 & 0 & 0 & 0 \\ 0 & 0 & 1 & 0 & 0 & 0 \\ 0 & 0 & 0 & -4 & 1 & 0 \\ 1 & 0 & 0 & 0 & 0 & 0 \\ 0 & 0 & 0 & 0 & 0 & 1 \\ 1 & 0 & 0 & 1 & 0 & -1 \end{bmatrix}\begin{bmatrix} \xi_1 \\ \xi_2 \\ \xi_3 \\ \xi_4 \\ \xi_5 \\ \xi_6 \end{bmatrix}.$$

The $\dot{\xi}_1, \dot{\xi}_2, \dot{\xi}_4,$ and $\dot{\xi}_5$ components follow from the definitions in Equation (6.6) and the $\dot{\xi}_3, \dot{\xi}_4,$ and $\dot{\xi}_6$ components are determined by solving the original three differential equations in Equation (6.5) for d^3x_1/dt^3, dx_2/dt, and d^2x_3/dt^2, respectively.

6.3 Linearly Independent Full Set of Real Eigenvectors

This section deals with the case when $A \in \mathbb{R}^{n \times n}$ has a set of n linearly independent real eigenvectors. In this case, the general solution to $\dot{\xi} = A\xi$ is

$$\xi(t) = c_1\hat{\xi}^1 e^{\lambda_1 t} + c_2\hat{\xi}^2 e^{\lambda_2 t} + \cdots + c_n\hat{\xi}^n e^{\lambda_n t}. \tag{6.7}$$

This is the general solution because the n linearly independent eigenvectors allow any initial condition to be satisfied. This section deals with the case when the eigenvalues and eigenvectors are real and the following sections consider when they may be complex or when there may not be a full set of them.

The first issue, then, is to determine when an $n \times n$ matrix A will have n linearly independent real eigenvectors. The complete answer is determined from actually computing the eigenvectors; however, for any given problem, it would be nice to know early on if this is the type of problem we have. This book presents two results from linear algebra.

1. If the eigenvalues of a matrix are distinct, then the eigenvectors are linearly independent.
2. If the matrix is symmetric, the eigenvectors are linearly independent and also orthogonal, even if some of the eigenvalues are repeated.

These results only provide *sufficient* conditions for the existence of a set of n linearly independent eigenvectors. There may be cases where they are not satisfied, but there still is a full set of linearly independent eigenvectors.

6.3.1 Some Useful Results from Linear Algebra

This first theorem proves that eigenvectors corresponding to distinct eigenvalues are linearly independent.

Theorem 6.1. *Let $A \in \mathbb{R}^{n \times n}$. If A has n distinct, real eigenvalues, then it has a set of n linearly independent eigenvectors.*

Proof. Let $\lambda_1, \dots, \lambda_n$ denote the distinct eigenvalues of A; that is, $\lambda_i \neq \lambda_j$ if $i \neq j$ and let $\hat{\xi}^1, \dots, \hat{\xi}^n$ denote the corresponding eigenvectors. To show that the eigenvectors are linearly independent it suffices to show that

$$\alpha_1 \hat{\xi}^1 + \alpha_2 \hat{\xi}^2 + \cdots + \alpha_n \hat{\xi}^n = 0 \quad \Longleftrightarrow \quad \alpha_i = 0$$

for all i; that is, that is there is no linear combination of the eigenvectors that is zero.

First consider finding α_1 and α_2 such that

$$\alpha_1 \hat{\xi}^1 + \alpha_2 \hat{\xi}^2 = 0. \tag{6.8}$$

Multiply both sides of this equation by $(A - \lambda_2 I)$ (note the specific eigenvalue, λ_2):

$$\alpha_1 (A - \lambda_2 I) \hat{\xi}^1 + \alpha_2 (A - \lambda_2 I) \hat{\xi}^2 = 0$$

$$\alpha_1 \left(A \hat{\xi}^1 - \lambda_2 \hat{\xi}^1 \right) + 0 = 0$$

$$\alpha_1 \left(\lambda_1 \hat{\xi}^1 - \lambda_2 \hat{\xi}^1 \right) = 0$$

$$\alpha_1 (\lambda_1 - \lambda_2) \hat{\xi}^1 = 0.$$

Because $\lambda_1 \neq \lambda_2$ and $\hat{\xi}^1 \neq 0$, then $\alpha_1 = 0$. Hence by Equation (6.8), $\alpha_2 = 0$ and hence, by definition, the set $\left\{\hat{\xi}^1, \hat{\xi}^2\right\}$ is linearly independent.

Now proceed by induction and assume the set $\left\{\hat{\xi}^1, \hat{\xi}^2, \ldots, \hat{\xi}^i\right\}$ is linearly independent. Consider

$$\alpha_1 \hat{\xi}^1 + \alpha_2 \hat{\xi}^2 + \cdots + \alpha_i \hat{\xi}^i + \alpha_{i+1} \hat{\xi}^{i+1} = 0. \tag{6.9}$$

Multiplying both sides of the equation by $(A - \lambda_{i+1}I)$ gives

$$\alpha_1 (\lambda_1 - \lambda_{i+1}) \hat{\xi}^1 + \alpha_2 (\lambda_2 - \lambda_{i+1}) \hat{\xi}^2 + \cdots + \alpha_i (\lambda_i - \lambda_{i+1}) \hat{\xi}^i + 0 = 0.$$

The set $\left\{\hat{\xi}^1, \hat{\xi}^2, \ldots, \hat{\xi}^i\right\}$ is linearly independent, thus

$$\alpha_i = \alpha_2 = \cdots = \alpha_i = 0$$

and by Equation (6.9), $\alpha_{i+1} = 0$. Hence the set $\left\{\hat{\xi}^1, \hat{\xi}^2, \ldots, \hat{\xi}^i, \hat{\xi}^{i+1}\right\}$ is linearly independent and by induction, the set $\left\{\hat{\xi}^1, \hat{\xi}^2, \ldots, \hat{\xi}^n\right\}$ is linearly independent. $\quad\square$

The implication of this theorem for solving $\dot{\xi} = A\xi$ is that if all the eigenvalues of A are different, then simply substituting everything into Equation (6.7) gives the general solution.

Another useful theorem states that even if some of the eigenvalues are repeated, if the matrix is symmetric, then the eigenvectors are still linearly independent.

Theorem 6.2. *If $A \in \mathbb{R}^{n \times n}$ is symmetric (i.e., $A = A^T$), then following are true.*

1. *All the eigenvalues of A are real.*
2. *A has n linearly independent eigenvectors, regardless of the multiplicity of any eigenvalue.*
3. *Eigenvectors corresponding to different eigenvalues are orthogonal.*

The proof uses notation common in complex variable theory, which is

$$A^* = \overline{A}^T,$$

where A is transposed and if any of the elements are complex, then the complex conjugate is taken.[1]

Proof. 1. Assume $A = A^T$. Because

$$A\hat{\xi}^i = \lambda_i \hat{\xi}^i \quad \Longrightarrow \quad \left(\hat{\xi}^i\right)^* A\hat{\xi}^i = \lambda_i \left(\hat{\xi}^i\right)^* \hat{\xi}^i$$

the eigenvalue may be expressed as

---

[1] A matrix is called *Hermitian* if $A = A^*$.

$$\lambda_i = \frac{\left(\hat{\xi}^i\right)^* A \hat{\xi}^i}{\left(\hat{\xi}^i\right)^* \hat{\xi}^i}.$$

Note that the notation $\left(\hat{\xi}^i\right)^* \hat{\xi}^i$ is simply the dot product between the vector $\hat{\xi}^i$ and its complex conjugate. Then

$$\lambda_i^* = \left(\frac{\left(\hat{\xi}^i\right)^* A \hat{\xi}^i}{\left(\hat{\xi}^i\right)^* \hat{\xi}^i}\right)^* = \frac{\left(\left(\hat{\xi}^i\right)^* A \hat{\xi}^i\right)^*}{\left(\left(\hat{\xi}^i\right)^* \hat{\xi}^i\right)^*} = \frac{\left(\hat{\xi}^i\right)^* A^* \hat{\xi}^i}{\left(\hat{\xi}^i\right)^* \hat{\xi}^i} = \frac{\left(\hat{\xi}^i\right)^* A \hat{\xi}^i}{\left(\hat{\xi}^i\right)^* \hat{\xi}^i} = \lambda_i.$$

Because $\lambda_i = \lambda_i^*$, it must be real.

2. The proof of this part is beyond the scope of this book.
3. Let $\hat{\xi}^i$ be the eigenvector associated with eigenvalue λ_i and $\hat{\xi}^j$ be the eigenvector associated with eigenvalue λ_j. Because A is real and symmetric, it is Hermitian, and $A = A^*$, so

$$\left(A\hat{\xi}^i\right)^* \hat{\xi}^j = \left(\hat{\xi}^i\right)^* A^* \hat{\xi}^j = \left(\hat{\xi}^i\right)^* A \hat{\xi}^j = \lambda_j \left(\hat{\xi}^i\right)^* \hat{\xi}^j.$$

But we also have

$$\left(A\hat{\xi}^i\right)^* \hat{\xi}^j = \lambda_i^* \left(\hat{\xi}^i\right)^* \hat{\xi}^j.$$

Because these are equal, and $\lambda_i \neq \lambda_j$, then $\left(\hat{\xi}^i\right)^* \hat{\xi}^j = 0$; that is, they are orthogonal.

$\square$

This is a very useful theorem because if A is symmetric, then we are guaranteed to have n orthogonal eigenvectors, and orthogonal vectors are linearly independent. Nothing in the theorem required distinct eigenvalues, so this works even if some of the eigenvalues are repeated. Also, as we show subsequently, we often have to compute the inverse of some matrices, and when the eigenvectors are orthogonal this will save a lot of work. The next section illustrates the use of both of these theorems where they are applicable.

6.3.2 Solution Technique for $\dot{\xi} = A\xi$

The general solution to $\dot{\xi} = A\xi$ is a linear combination of n homogeneous solutions

$$\xi(t) = c_1 \hat{\xi}^1 e^{\lambda_1 t} + \cdots + c_n \hat{\xi}^n e^{\lambda_n t},$$

where the set of eigenvectors is linearly independent and the coefficients c_i may be used to satisfy specified initial conditions. The eigenvectors are linearly independent, thus any initial condition may be satisfied with the appropriate coefficients,

c_is. In particular, for a specified $\xi(0)$

$$\xi(0) = c_1 \hat{\xi}^1 + \cdots + c_n \hat{\xi}^n = \begin{bmatrix} \hat{\xi}^1 & \cdots & \hat{\xi}^n \end{bmatrix} \begin{bmatrix} c_1 \\ \vdots \\ c_n \end{bmatrix} \implies \begin{bmatrix} c_1 \\ \vdots \\ c_n \end{bmatrix} = \begin{bmatrix} \hat{\xi}^1 & \cdots & \hat{\xi}^n \end{bmatrix}^{-1} \xi(0),$$

although, as illustrated in the examples below, it is usually easiest just to solve for the coefficients using row reduction methods.

Example 6.3. Find the homogeneous solutions to

$$\dot{\xi} = A\xi, \tag{6.10}$$

where

$$A = \begin{bmatrix} 1 & 2 \\ 1 & 0 \end{bmatrix}, \quad \xi(0) = \begin{bmatrix} 1 \\ 0 \end{bmatrix}.$$

Note that A is not symmetric, so we cannot make use of Theorem 6.2. The computations for the eigenvalues of A are

$$\det(A - \lambda I) = \begin{vmatrix} 1 - \lambda & 2 \\ 1 & -\lambda \end{vmatrix} = -(1 - \lambda)\lambda - 2 = \lambda^2 - \lambda - 2 - 0,$$

so the eigenvalues are

$$\lambda_1 = 2, \quad \lambda_2 = -1.$$

These are real and distinct, so according to Theorem 6.1 the eigenvectors are guaranteed to be linearly independent.

Substituting each eigenvalue into $(A - \lambda I)\xi = 0$ gives

$$\begin{bmatrix} -1 & 2 \\ 1 & -2 \end{bmatrix} \begin{bmatrix} \hat{\xi}_1^1 \\ \hat{\xi}_2^1 \end{bmatrix} = \begin{bmatrix} 0 \\ 0 \end{bmatrix} \implies \hat{\xi}^1 = \begin{bmatrix} 2 \\ 1 \end{bmatrix}$$

$$\begin{bmatrix} 2 & 2 \\ 1 & 1 \end{bmatrix} \begin{bmatrix} \hat{\xi}_1^2 \\ \hat{\xi}_2^2 \end{bmatrix} = \begin{bmatrix} 0 \\ 0 \end{bmatrix} \implies \hat{\xi}^2 = \begin{bmatrix} 1 \\ -1 \end{bmatrix}.$$

Thus

$$\xi_1(t) = \begin{bmatrix} 2 \\ 1 \end{bmatrix} e^{2t}, \quad \xi_2(t) = \begin{bmatrix} 1 \\ -1 \end{bmatrix} e^{-t}$$

both satisfy $\dot{\xi} = A\xi$, and the general solution is

$$\xi(t) = c_1 \begin{bmatrix} 2 \\ 1 \end{bmatrix} e^{2t} + c_2 \begin{bmatrix} 1 \\ -1 \end{bmatrix} e^{-t}.$$

Substituting $t = 0$ and the initial condition gives

$$\xi(0) = c_1 \begin{bmatrix} 2 \\ 1 \end{bmatrix} + c_2 \begin{bmatrix} 1 \\ -1 \end{bmatrix} = \begin{bmatrix} 1 \\ 0 \end{bmatrix},$$

which may be rearranged as

$$\begin{bmatrix} 2 & 1 \\ 1 & -1 \end{bmatrix} \begin{bmatrix} c_1 \\ c_2 \end{bmatrix} = \begin{bmatrix} 1 \\ 0 \end{bmatrix} \implies c_1 = \frac{1}{3}, \quad c_2 = \frac{1}{3},$$

so

$$\xi(t) = \frac{1}{3} \begin{bmatrix} 2 \\ 1 \end{bmatrix} e^{2t} + \frac{1}{3} \begin{bmatrix} 1 \\ -1 \end{bmatrix} e^{-t}$$

is the solution to the initial value problem.

Example 6.4. Find the solution to the initial value problem

$$\frac{d}{dt} \begin{bmatrix} \xi_1 \\ \xi_2 \\ \xi_3 \end{bmatrix} = \begin{bmatrix} -4 & 1 & 0 \\ 0 & -6 & 0 \\ 0 & 0 & -2 \end{bmatrix} \begin{bmatrix} \xi_1 \\ \xi_2 \\ \xi_3 \end{bmatrix}, \quad \xi(0) = \begin{bmatrix} 3 \\ 4 \\ 3 \end{bmatrix}.$$

The eigenvalues of A are computed by

$$\begin{vmatrix} -4-\lambda & 1 & 0 \\ 0 & -6-\lambda & 0 \\ 0 & 0 & -2-\lambda \end{vmatrix} = 0.$$

Computing the determinant by an expansion about the third row and column gives

$$(-2-\lambda)[(-4-\lambda)(-6-\lambda)] = -(\lambda+2)(\lambda^2+10\lambda+24) = 0$$

so $\lambda_1 = -2$, $\lambda_2 = -4$, and $\lambda_3 = -6$, which are distinct, so from Theorem 6.1 we are guaranteed that the set of eigenvectors is linearly independent.

The eigenvector corresponding to $\lambda_1 = -2$ is computed from

$$\begin{bmatrix} -4-(-2) & 1 & 0 & | & 0 \\ 0 & -6-(-2) & 0 & | & 0 \\ 0 & 0 & -2-(-2) & | & 0 \end{bmatrix} \iff \begin{bmatrix} -2 & 1 & 0 & | & 0 \\ 0 & -4 & 0 & | & 0 \\ 0 & 0 & 0 & | & 0 \end{bmatrix} \implies \hat{\xi}^1 = \begin{bmatrix} 0 \\ 0 \\ 1 \end{bmatrix}.$$

The eigenvector corresponding to $\lambda_1 = -4$ is computed from

$$\begin{bmatrix} -4-(-4) & 1 & 0 & | & 0 \\ 0 & -6-(-4) & 0 & | & 0 \\ 0 & 0 & -2-(-4) & | & 0 \end{bmatrix} \iff \begin{bmatrix} 0 & 1 & 0 & | & 0 \\ 0 & -2 & 0 & | & 0 \\ 0 & 0 & 2 & | & 0 \end{bmatrix} \implies \hat{\xi}^2 = \begin{bmatrix} 1 \\ 0 \\ 0 \end{bmatrix}.$$

The eigenvector corresponding to $\lambda_1 = -6$ is computed from

$$\begin{bmatrix} -4-(-6) & 1 & 0 & | & 0 \\ 0 & -6-(-6) & 0 & | & 0 \\ 0 & 0 & -2-(-6) & | & 0 \end{bmatrix} \iff \begin{bmatrix} 2 & 1 & 0 & | & 0 \\ 0 & 0 & 0 & | & 0 \\ 0 & 0 & 4 & | & 0 \end{bmatrix} \iff \begin{bmatrix} 2 & 1 & 0 & | & 0 \\ 0 & 0 & 4 & | & 0 \\ 0 & 0 & 0 & | & 0 \end{bmatrix}.$$

The second row requires that the third component be zero; that is, $\hat{\xi}_3^3 = 0$. Setting the first component to be one gives that the second component must be two, from the first equation. Hence

$$\hat{\xi}^3 = \begin{bmatrix} -1 \\ 2 \\ 0 \end{bmatrix}.$$

The eigenvectors are guaranteed to be linearly independent, therefore the general solution is

$$\xi(t) = c_1 \hat{\xi}^1 e^{\lambda_1 t} + c_2 \hat{\xi}^2 e^{\lambda_2 t} + c_3 \hat{\xi}^3 e^{\lambda_3 t}$$

$$= c_1 \begin{bmatrix} 0 \\ 0 \\ 1 \end{bmatrix} e^{-2t} + c_2 \begin{bmatrix} 1 \\ 0 \\ 0 \end{bmatrix} e^{-4t} + c_3 \begin{bmatrix} -1 \\ 2 \\ 0 \end{bmatrix} e^{-6t}.$$

To satisfy the initial conditions, substitute $t = 0$ into the general solution, which gives the system of equations and associated augmented matrix

$$c_1 \begin{bmatrix} 0 \\ 0 \\ 1 \end{bmatrix} + c_2 \begin{bmatrix} 1 \\ 0 \\ 0 \end{bmatrix} + c_3 \begin{bmatrix} -1 \\ 2 \\ 0 \end{bmatrix} = \begin{bmatrix} 3 \\ 4 \\ 3 \end{bmatrix} \implies \begin{bmatrix} 0 & 1 & -1 & 3 \\ 0 & 0 & 2 & 4 \\ 1 & 0 & 0 & 3 \end{bmatrix} \iff \begin{bmatrix} 1 & 0 & 0 & 3 \\ 0 & 1 & -1 & 3 \\ 0 & 0 & 2 & 4 \end{bmatrix}$$

from which we get $c_3 = 2$, $c_2 = 5$, and $c_1 = 3$.

Hence, the solution is

$$\xi(t) = 3 \begin{bmatrix} 0 \\ 0 \\ 1 \end{bmatrix} e^{-2t} + 5 \begin{bmatrix} 1 \\ 0 \\ 0 \end{bmatrix} e^{-4t} + 2 \begin{bmatrix} -1 \\ 2 \\ 0 \end{bmatrix} e^{-6t} = \begin{bmatrix} 5e^{-4t} - 2e^{-6t} \\ 4e^{-6t} \\ 3e^{-2t} \end{bmatrix}.$$

The final example is with a symmetric matrix with repeated eigenvalues.

Example 6.5. Find the general solution to $\dot{\xi} = A\xi$ where

$$A = \begin{bmatrix} 5 & 0 & -1 \\ 0 & 4 & 0 \\ -1 & 0 & 5 \end{bmatrix}.$$

Because $A = A^T$, by Theorem 6.2, the set of three eigenvalues will be linearly independent (and orthogonal too). Computing the characteristic equation

$$\begin{vmatrix} (5 - \lambda) & 0 & -1 \\ 0 & (4 - \lambda) & 0 \\ -1 & 0 & (5 - \lambda) \end{vmatrix} = (4 - \lambda) \left[(5 - \lambda)^2 - 1 \right]$$

$$= (4 - \lambda) \left[\lambda^2 - 10\lambda + 24 \right]$$

$$= (4 - \lambda) \left[(\lambda - 4)(\lambda - 6) \right] = 0,$$

so the eigenvalues are $\lambda_1 = 4$, $\lambda_2 = 4$, and $\lambda_3 = 6$. These are not distinct, so Theorem 6.1 does not apply; however, Theorem 6.2 does apply, so we know that A still has a set of three linearly independent eigenvectors.

For $\lambda = 4$,

$$\begin{bmatrix} 1 & 0 & -1 \\ 0 & 0 & 0 \\ -1 & 0 & 1 \end{bmatrix} \begin{bmatrix} \hat{\xi}_1 \\ \hat{\xi}_2 \\ \hat{\xi}_3 \end{bmatrix} = \begin{bmatrix} 0 \\ 0 \\ 0 \end{bmatrix} \implies \hat{\xi}^1 = \begin{bmatrix} 1 \\ 0 \\ 1 \end{bmatrix}, \quad \hat{\xi}^2 = \begin{bmatrix} 0 \\ 1 \\ 0 \end{bmatrix}.$$

and skipping the computational details,

$$\hat{\xi}^3 = \begin{bmatrix} -1 \\ 0 \\ 1 \end{bmatrix}.$$

Hence, the general solution is

$$\xi(t) = c_1 \begin{bmatrix} 0 \\ 1 \\ 0 \end{bmatrix} e^{4t} + c_2 \begin{bmatrix} 1 \\ 0 \\ 1 \end{bmatrix} e^{4t} + c_3 \begin{bmatrix} -1 \\ 0 \\ 1 \end{bmatrix} e^{6t}.$$

6.4 Complex Eigenvectors

The eigenvalues are computed from a polynomial, thus complex values are possible. In a manner similar to complex roots in Chapter 3, it is desirable to convert the complex exponentials into trigonometric functions.

Example 6.6. Again consider the mass–spring–damper system illustrated in Figure 6.1. Let

$$m_1 = 1, \quad k_1 = 10, \quad b_1 = 0.1, \quad m_2 = 1, \quad k_2 = 1, \quad b_2 = 0.1.$$

As a starting observation, because the damping is relatively light, oscillatory solutions should be expected. Substituting these values into the A matrix in Equation (6.2) gives

$$A = \begin{bmatrix} 0 & 1 & 0 & 0 \\ -11 & -0.2 & 1 & 0.1 \\ 0 & 0 & 0 & 1 \\ 1 & 0.1 & -1 & -0.1 \end{bmatrix}$$

which has eigenvalues

$$\lambda_1 = -0.1093 + 3.3285i, \quad \lambda_2 = -0.1093 - 3.3285i,$$
$$\lambda_3 = -0.0407 + 0.9487i, \quad \lambda_4 = -0.0407 - 0.9487i,$$

and corresponding eigenvectors

$$\hat{\xi}^1 = \begin{bmatrix} -0.0094 - 0.2859i \\ 0.9527 \\ -0.0074 + 0.0287i \\ -0.0946 - 0.0278i \end{bmatrix}, \quad \hat{\xi}^2 = \begin{bmatrix} -0.0094 + 0.2859i \\ 0.9527 \\ -0.0074 - 0.0287i \\ -0.0946 + 0.0278i \end{bmatrix}$$

$$\hat{\xi}^3 = \begin{bmatrix} 0.0713 + 0.0060i \\ -0.0086 + 0.0674i \\ 0.7216 \\ -0.0294 + 0.6846i \end{bmatrix}, \quad \hat{\xi}^4 = \begin{bmatrix} 0.0713 - 0.0060i \\ -0.0086 - 0.0674i \\ 0.7216 \\ -0.0294 - 0.6846i \end{bmatrix}.$$

Observe that the eigenvalues occur in complex conjugate pairs. This should be obviously expected inasmuch as eigenvalues are the roots of a polynomial. Less obvious, but probably not surprising, is that the eigenvectors also occur in complex conjugate pairs. The reason this is true is given by the proof of the following.

Proposition 6.1. *If $A \in \mathbb{R}^{n \times n}$ and two eigenvalues of A are such that $\lambda_i = \overline{\lambda}_j$, then if $\hat{\xi}^i$ is the eigenvector corresponding to λ_i, $\overline{\hat{\xi}^i}$ is an eigenvector corresponding to λ_j.*

Proof. Eigenvector $\hat{\xi}^i$ satisfies

$$(A - \lambda_i I)\hat{\xi}^i = 0.$$

Taking the complex conjugate of both sides gives

$$\overline{(A - \lambda_i I)\hat{\xi}^i} = \overline{0} \implies \left(A - \overline{\lambda}_i I\right)\overline{\hat{\xi}^i} = 0 \implies (A - \lambda_j I)\overline{\hat{\xi}^i} = 0.$$

Thus, $\hat{\xi}^j = \overline{\hat{\xi}^i}$. $\qquad\qquad\qquad\qquad\qquad\qquad\qquad\qquad\qquad\qquad\square$

To solve the initial value problem

$$\dot{\xi} = A\xi, \quad \xi(0) = \xi_0$$

we may proceed as before and simply write the general solution

$$\xi(t) = c_1\hat{\xi}^1 e^{\lambda_1 t} + \cdots c_1\hat{\xi}^n e^{\lambda_n t},$$

substitute $t = 0$

$$\xi(0) = c_1\hat{\xi}^1 + \cdots c_1\hat{\xi}^n,$$

and solve for the unknown coefficients c_i. The following example illustrates that fact.

Example 6.7. Solve

$$\dot{\xi} = A\xi \qquad \xi(0) = \begin{bmatrix} 1 \\ 1 \end{bmatrix},$$

where

$$A = \begin{bmatrix} 1 & -2 \\ 2 & 1 \end{bmatrix}.$$

Computing the eigenvalues gives

$$\det(A - \lambda I) = (1 - \lambda)^2 + 4 = 0 \qquad \Longrightarrow \qquad \lambda = 1 \pm 2i.$$

For $\lambda_1 \doteq 1 + 2i$

$$\begin{bmatrix} -2i & -2 \\ 2 & -2i \end{bmatrix} \begin{bmatrix} x_1 \\ x_2 \end{bmatrix} = \begin{bmatrix} 0 \\ 0 \end{bmatrix} \qquad \Longrightarrow \qquad \hat{\xi}^1 = \begin{bmatrix} \xi_1^1 \\ \xi_2^1 \end{bmatrix} = \begin{bmatrix} 1 \\ -i \end{bmatrix}.$$

We do not need to compute the second eigenvector because it is guaranteed to be the complex conjugate, but just as a check we compute it this time. For $\lambda_2 = 1 - 2i$,

$$\begin{bmatrix} 2i & -2 \\ 2 & 2i \end{bmatrix} \begin{bmatrix} x_1 \\ x_2 \end{bmatrix} = \begin{bmatrix} 0 \\ 0 \end{bmatrix} \qquad \Longrightarrow \qquad \hat{\xi}^2 = \begin{bmatrix} \xi_1^2 \\ \xi_2^2 \end{bmatrix} = \begin{bmatrix} 1 \\ i \end{bmatrix}.$$

So the general solution is

$$\xi(t) = c_1 \begin{bmatrix} 1 \\ -i \end{bmatrix} e^{(1+2i)t} + c_2 \begin{bmatrix} 1 \\ i \end{bmatrix} e^{(1-2i)t}$$

and at $t = 0$,

$$\xi(0) = c_1 \begin{bmatrix} 1 \\ -i \end{bmatrix} + c_2 \begin{bmatrix} 1 \\ i \end{bmatrix} = \begin{bmatrix} 1 & 1 \\ -i & i \end{bmatrix} \begin{bmatrix} c_1 \\ c_2 \end{bmatrix} = \begin{bmatrix} 1 \\ 1 \end{bmatrix}.$$

Either solving for c_1 and c_2 by inverting the matrix or by eliminating one coefficient from one equation and substituting into the other gives

$$c_1 = \frac{1}{2} + \frac{1}{2}i, \quad c_2 = \frac{1}{2} - \frac{1}{2}i.$$

Finally, substituting c_1 and c_2 into the general solution gives

$$\xi(t) = \begin{bmatrix} \frac{1}{2} + \frac{1}{2}i \\ \frac{1}{2} - \frac{1}{2}i \end{bmatrix} e^{(1+2i)t} + \begin{bmatrix} \frac{1}{2} - \frac{1}{2}i \\ \frac{1}{2} + \frac{1}{2}i \end{bmatrix} e^{(1-2i)t}. \tag{6.11}$$

This is the correct answer, however, it is somewhat dissatisfying in that it is complex whereas the matrix A and the initial conditions were all real. If the complex exponentials are expanded using Euler's formula, then

$$\xi(t) = \begin{bmatrix} \cos 2t - \sin 2t \\ \cos 2t + \sin 2t \end{bmatrix} e^t \tag{6.12}$$

is obtained. Interestingly, the imaginary components of the terms in Equation (6.11) are identically zero, although, it is certainly difficult to see that without all the work to convert from Equation (6.11) to Equation (6.12).

The preceding example illustrates that the general solution may still be correctly expressed as a linear combination of the eigenvalues times the exponential of the corresponding eigenvectors. However, there are several reasons to reformulate the answer.

1. The solution may not "naturally" result in a purely real expression for ξ, which is what is expected.
2. Furthermore, and perhaps arduous, manipulation may be necessary to determine the form of the solution that is purely real.
3. Many computations involving complex numbers, requiring four operations for multiplication and two operations for addition, are involved in computing the solution.
4. The fact that the eigenvalues and eigenvectors occur in complex conjugate pairs was not exploited at all.

In order to make the computations less burdensome, an alternative approach which is analogous to the approach in the case of second-order systems with a characteristic equation with complex roots is utilized. Fundamentally, the "shortcut" to this approach is based upon the complex conjugate nature of the eigenvalues and eigenvectors.

Consider a matrix with n eigenvalues, and assume that the first two are a pair of complex conjugate eigenvalues. Denote them by by

$$\lambda_1 = \mu + i\omega, \quad \lambda_2 = \mu - i\omega \tag{6.13}$$

and the eigenvectors by

$$\hat{\xi}^1 = \hat{\xi}_r^1 + i\hat{\xi}_i^1, \quad \hat{\xi}^2 = \hat{\xi}_r^1 - i\hat{\xi}_i^1, \tag{6.14}$$

where $\hat{\xi}_r^1$ and $\hat{\xi}_i^1$ are the real and imaginary parts of the eigenvector $\hat{\xi}^1$, respectively.

The general solution is

$$\xi(t) = c_1 \hat{\xi}^1 e^{\lambda_1 t} + c_2 \hat{\xi}^2 e^{\lambda_2 t} + \cdots,$$

where only the complex terms are written. Substituting for the components of λ_1, λ_2, $\hat{\xi}^1$ and $\hat{\xi}^2$ and using Euler's formula gives

$$
\begin{aligned}
\xi(t) &= c_1 \hat{\xi}^1 e^{\lambda_1 t} + c_2 \hat{\xi}^2 e^{\lambda_2 t} + \cdots \\
&= c_1 \left(\hat{\xi}_r^1 + i\hat{\xi}_i^1 \right) e^{(\mu+i\omega)t} + c_2 \left(\hat{\xi}_r^1 - i\hat{\xi}_i^1 \right) e^{(\mu-i\omega)t} + \cdots \\
&= c_1 \left(\hat{\xi}_r^1 + i\hat{\xi}_i^1 \right) e^{\mu t} (\cos \omega t + i \sin \omega t) + c_2 \left(\hat{\xi}_r^1 - i\hat{\xi}_i^1 \right) e^{\mu t} (\cos \omega t - i \sin \omega t) + \cdots \\
&= e^{\mu t} \left[c_1 \hat{\xi}_r^1 \cos \omega t - c_1 \hat{\xi}_i^1 \sin \omega t + i c_1 \hat{\xi}_r^1 \sin \omega t + i c_1 \hat{\xi}_i^1 \cos \omega t \right. \\
&\qquad \left. + c_2 \hat{\xi}_r^1 \cos \omega t - c_2 \hat{\xi}_i^1 \sin \omega t - i c_2 \hat{\xi}_i^1 \cos \omega t - i c_2 \hat{\xi}_r^1 \sin \omega t \right] + \cdots \\
&= e^{\mu t} \left[(c_1 + c_2) \hat{\xi}_r^1 \cos \omega t - (c_1 + c_2) \hat{\xi}_i^1 \sin \omega t \right] \\
&\qquad + e^{\mu t} i \left[(c_1 - c_2) \hat{\xi}_r^1 \sin \omega t + (c_1 - c_2) \hat{\xi}_i^1 \cos \omega t \right] + \cdots.
\end{aligned}
$$

Let

$$k_1 = c_1 + c_2, \quad k_2 = i(c_1 - c_2)$$

and substitute into $\xi(t)$ to give

$$\boxed{\xi(t) = k_1 e^{\mu t}\left(\hat{\xi}_r^1 \cos \omega t - \hat{\xi}_i^1 \sin \omega t\right) + k_2 e^{\mu t}\left(\hat{\xi}_r^1 \sin \omega t + \hat{\xi}_i^1 \cos \omega t\right) + \cdots,}$$
(6.15)

where the terms in the equations are defined in Equations (6.13) and (6.14).

Example 6.8. Returning to the mass–spring–damper system in Example 6.6, observe that we have

$$\mu_1 = -0.1093 \quad \omega_1 = 3.3285$$
$$\mu_2 = -0.0407 \quad \omega_2 = 0.9487$$

and

$$\hat{\xi}_r^1 = \begin{bmatrix} -0.0094 \\ 0.9527 \\ -0.0074 \\ -0.0946 \end{bmatrix} \quad \hat{\xi}_i^1 = \begin{bmatrix} -0.2859 \\ 0 \\ 0.0287 \\ -0.0278 \end{bmatrix}$$

$$\hat{\xi}_r^2 = \begin{bmatrix} 0.0713 \\ -0.0086 \\ 0.7216 \\ -0.0294 \end{bmatrix} \quad \hat{\xi}_r^2 = \begin{bmatrix} 0.0060 \\ 0.0674 \\ 0 \\ 0.6846 \end{bmatrix}.$$

From Equation (6.15), the general solution is of the form

$$\xi(t) = k_1 e^{\mu_1 t}\left(\hat{\xi}_r^1 \cos \omega_1 t - \hat{\xi}_i^1 \sin \omega_1 t\right) + k_2 e^{\mu_1 t}\left(\hat{\xi}_r^1 \sin \omega_1 t + \hat{\xi}_i^1 \cos \omega_1 t\right)$$
$$+ k_3 e^{\mu_2 t}\left(\hat{\xi}_r^2 \cos \omega_2 t - \hat{\xi}_i^2 \sin \omega_2 t\right) + k_4 e^{\mu_2 t}\left(\hat{\xi}_r^2 \sin \omega_2 t + \hat{\xi}_i^2 \cos \omega_2 t\right),$$

or substituting all the numerical values

$$\xi(t) = k_1 e^{-0.1093t} \left(\begin{bmatrix} -0.0094 \\ 0.9527 \\ -0.0074 \\ -0.0946 \end{bmatrix} \cos 3.3285t - \begin{bmatrix} -0.2859 \\ 0 \\ 0.0287 \\ -0.0278 \end{bmatrix} \sin 3.3285t \right)$$

$$+ k_2 e^{-0.1093t} \left(\begin{bmatrix} -0.0094 \\ 0.9527 \\ -0.0074 \\ -0.0946 \end{bmatrix} \sin 3.3285t + \begin{bmatrix} -0.2859 \\ 0 \\ 0.0287 \\ -0.0278 \end{bmatrix} \cos 3.3285t \right)$$

$$+ k_3 e^{-0.0407t} \left(\begin{bmatrix} 0.0713 \\ -0.0086 \\ 0.7216 \\ -0.0294 \end{bmatrix} \cos 0.9487t - \begin{bmatrix} 0.0060 \\ 0.0674 \\ 0 \\ 0.6846 \end{bmatrix} \sin 0.9487t \right)$$

$$+ k_4 e^{-0.0407t} \left(\begin{bmatrix} 0.0713 \\ -0.0086 \\ 0.7216 \\ -0.0294 \end{bmatrix} \sin 0.9487t + \begin{bmatrix} 0.0060 \\ 0.0674 \\ 0 \\ 0.6846 \end{bmatrix} \cos 0.9487t \right).$$

Example 6.9. Returning to Example 6.7,

$$\lambda_1 = 1 + 2i$$

and

$$\hat{\xi}^1 = \begin{bmatrix} 1 \\ -i \end{bmatrix}.$$

Hence $\mu = 1$ and $\omega = 1$, and

$$\hat{\xi}_r^1 = \begin{bmatrix} 1 \\ 0 \end{bmatrix}, \quad \hat{\xi}_i^1 = \begin{bmatrix} 0 \\ -1 \end{bmatrix}.$$

Substituting into Equation (6.15) gives

$$\begin{bmatrix} \xi_1(t) \\ \xi_2(t) \end{bmatrix} = k_1 e^t \left(\begin{bmatrix} 1 \\ 0 \end{bmatrix} \cos 2t - \begin{bmatrix} 0 \\ -1 \end{bmatrix} \sin 2t \right) + k_2 \left(\begin{bmatrix} 1 \\ 0 \end{bmatrix} \sin 2t + \begin{bmatrix} 0 \\ -1 \end{bmatrix} \cos 2t \right).$$

The initial condition is

$$\xi(0) = \begin{bmatrix} 1 \\ 1 \end{bmatrix}.$$

Substituting $t = 0$ into the solution and equating it to the initial condition gives

$$\begin{bmatrix} 1 \\ 1 \end{bmatrix} = k_1 \begin{bmatrix} 1 \\ 0 \end{bmatrix} + k_2 \begin{bmatrix} 0 \\ -1 \end{bmatrix}$$

which gives

$$k_1 = 1, \quad k_2 = -1.$$

Hence,

$$\begin{bmatrix} \xi_1(t) \\ \xi_2(t) \end{bmatrix} = e^t \left(\begin{bmatrix} 1 \\ 0 \end{bmatrix} \cos 2t - \begin{bmatrix} 0 \\ -1 \end{bmatrix} \sin 2t \right) - \left(\begin{bmatrix} 1 \\ 0 \end{bmatrix} \sin 2t + \begin{bmatrix} 0 \\ -1 \end{bmatrix} \cos 2t \right)$$

$$= e^t \left(\begin{bmatrix} 1 \\ 1 \end{bmatrix} \cos 2t + \begin{bmatrix} -1 \\ 1 \end{bmatrix} \sin 2t \right)$$

which is the same as Equation (6.12).

This next example contains one real eigenvalue and one complex conjugate pair of eigenvalues.

Example 6.10. Determine the general solution to

$$\dot{\xi} = A\xi,$$

where

$$A = \begin{bmatrix} -7 & 0 & 8 \\ 0 & -2 & 0 \\ -4 & 0 & 1 \end{bmatrix}.$$

Computing

$$\det(A - \lambda I) = \begin{vmatrix} -7 - \lambda & 0 & 8 \\ 0 & -2 - \lambda & 0 \\ -4 & 0 & 1 - \lambda \end{vmatrix}$$

by a cofactor expansion across the second row gives

$$-1(-2-\lambda)[(-7-\lambda)(1-\lambda)+32] = (2+\lambda)(\lambda^2 + 6\lambda + 25) = 0.$$

Hence, $\lambda_1 = -2$ and

$$\lambda_{2,3} = \frac{-6 \pm \sqrt{36 - 100}}{2} = -3 \pm 4i.$$

For $\lambda_1 = -2$, $(A + 2I)\hat{\xi} = 0$ is computed by

$$\begin{bmatrix} -5 & 0 & 8 & | & 0 \\ 0 & 0 & 0 & | & 0 \\ -4 & 0 & 3 & | & 0 \end{bmatrix} \iff \begin{bmatrix} -5 & 0 & 8 & | & 0 \\ 0 & 0 & -\frac{17}{5} & | & 0 \\ 0 & 0 & 0 & | & 0 \end{bmatrix}$$

which gives

$$\hat{\xi}^1 = \begin{bmatrix} 0 \\ 1 \\ 0 \end{bmatrix}.$$

For $\lambda_2 = -3 + 4i$,

$$(A + (3 - 4i)I)\hat{\xi} = 0$$

is computed by

$$
\begin{bmatrix} -4-4i & 0 & 8 & 0 \\ 0 & 1-4i & 0 & 0 \\ -4 & 0 & 4-4i & 0 \end{bmatrix} \iff \begin{bmatrix} -4-4i & 0 & 8 & 0 \\ 0 & 1-4i & 0 & 0 \\ 0 & 0 & 0 & 0 \end{bmatrix},
$$

which was obtained by dividing the first row by $1+i$ and subtracting the result from the third row. If we let $\hat{\xi}_3^2 = 1$, then

$$
\hat{\xi}^2 = \begin{bmatrix} 1-i \\ 0 \\ 1 \end{bmatrix}.
$$

Both the eigenvalues and eigenvectors must occur in complex conjugate pairs, therefore for $\lambda_3 = -3 - 4i$,

$$
\hat{\xi}^3 = \begin{bmatrix} 1+i \\ 0 \\ 1 \end{bmatrix}.
$$

Using the second eigenvalue $\mu = -3$, $\omega = 4$,

$$
\hat{\xi}_r^1 = \begin{bmatrix} 1 \\ 0 \\ 1 \end{bmatrix}, \quad \hat{\xi}_i^1 = \begin{bmatrix} -1 \\ 0 \\ 0 \end{bmatrix}.
$$

Hence,

$$
\xi(t) = c_1 e^{-2t} \begin{bmatrix} 0 \\ 1 \\ 0 \end{bmatrix} + c_2 e^{-3t} \left(\begin{bmatrix} 1 \\ 0 \\ 1 \end{bmatrix} \cos 4t - \begin{bmatrix} -1 \\ 0 \\ 0 \end{bmatrix} \sin 4t \right)
$$

$$
+ c_3 e^{-3t} \left(\begin{bmatrix} 1 \\ 0 \\ 1 \end{bmatrix} \sin 4t + \begin{bmatrix} -1 \\ 0 \\ 0 \end{bmatrix} \cos 4t \right).
$$

6.5 Generalized Eigenvectors

The case where some of the eigenvalues are repeated is the most complicated. This is because when there are repeated eigenvalues, there may or may not be a complete set of linearly independent eigenvectors associated with the repeated eigenvalue. The next set of examples illustrates this fact.

Example 6.11. Consider $\dot{\xi} = A\xi$ where

$$
A = \begin{bmatrix} 2 & 1 \\ 0 & 2 \end{bmatrix}.
$$

Computing the eigenvalues gives

$$(2 - \lambda)^2 = 0 \quad \Longrightarrow \quad \lambda = 2.$$

Computing the eigenvectors,

$$\begin{bmatrix} 0 & 1 \\ 0 & 0 \end{bmatrix} \begin{bmatrix} x_1 \\ x_2 \end{bmatrix} = \begin{bmatrix} 0 \\ 0 \end{bmatrix} \quad \Longrightarrow \quad \hat{\xi} = \begin{bmatrix} 1 \\ 0 \end{bmatrix}.$$

In the preceding example, the eigenvalue $\lambda = 2$ was repeated. It may not be surprising that there also is only one eigenvector $\hat{\xi}$ as well. However, things are not so simple. Consider the following example.

Example 6.12. Consider $\dot{\xi} = A\xi$ where

$$A = \begin{bmatrix} 2 & 0 \\ 0 & 2 \end{bmatrix}.$$

Computing the eigenvalues gives

$$(2 - \lambda)^2 = 0 \quad \Longrightarrow \quad \lambda = 2,$$

which is exactly the same as before. Now computing the eigenvectors,

$$\begin{bmatrix} 0 & 0 \\ 0 & 0 \end{bmatrix} \begin{bmatrix} x_1 \\ x_2 \end{bmatrix} = \begin{bmatrix} 0 \\ 0 \end{bmatrix}.$$

In this case, however, we have that

$$\hat{\xi}^1 = \begin{bmatrix} 1 \\ 0 \end{bmatrix}, \quad \hat{\xi}^2 = \begin{bmatrix} 0 \\ 1 \end{bmatrix}$$

both satisfy the eigenvector equation and are linearly independent.

These two examples illustrate the fact that when an eigenvalue is repeated m times, there may or may not be a set of m linearly independent eigenvectors associated with it. This is problematic in that to use the approach utilized so far to solve $\dot{\xi} = A\xi$ we need a full set of linearly independent eigenvectors in order to obtain a general solution.

First we address the practical computational matter of determining how many linearly independent eigenvectors are associated with a repeated eigenvalue. Then we delineate the solution techniques for each case.

6.5.1 Geometric and Algebraic Multiplicities

The number of times that an eigenvalue is repeated is called its *algebraic multiplicity*. Similarly, the number of linearly independent eigenvectors associated with an eigenvalue is called its *geometric multiplicity*.

Definition 6.1. Let $A \in \mathbb{R}^{n \times n}$ and let

$$\det(A - \lambda I) = \sum_{i=1}^{m} (\lambda - \lambda_i)^{k_i},$$

where each λ_i is distinct. Note that $\sum_{i=1}^{m} k_i = n$. The number k_i is the *algebraic multiplicity* of eigenvalue λ_i.

Example 6.13. Consider

$$A = \begin{bmatrix} 1 & 0 & 0 & 0 \\ -1 & 2 & 0 & 0 \\ -1 & 0 & 1 & 1 \\ -1 & 0 & -1 & 3 \end{bmatrix}.$$

The characteristic equation is

$$\det(A - 2I) = (1 - \lambda)(2 - \lambda)^3.$$

Hence the algebraic multiplicity of $\lambda = 1$ is one and the algebraic multiplicity of $\lambda = 2$ is three.

Definition 6.2. Let $A \in \mathbb{R}^{n \times n}$. The dimension of the null space of $(A - \lambda_i I)$ is the *geometric multiplicity* of eigenvalue λ_i.

The definition of geometric multiplicity should make sense. Because the definition of an eigenvector is a nonzero vector, $\hat{\xi}$ satisfying

$$(A - \lambda I)\hat{\xi} = 0,$$

and the null space of a matrix is the set of vectors that, when multiplied into the matrix produce the zero vector, the number of linearly independent vectors that produce the zero vector is the dimension of the null space.

First we consider a matrix with distinct eigenvalues to illustrate the concept of the dimension of the null space of $(A - \lambda I)$ being the number of linearly independent eigenvectors associated with an eigenvalue as well as the simple procedural aspect of computing it.

Example 6.14. Determine all the linearly independent eigenvectors of

$$A = \begin{bmatrix} 1 & 0 & 1 \\ 0 & 1 & 1 \\ 0 & -2 & 4 \end{bmatrix}.$$

From an expansion down the first column, the characteristic equation is

$$\begin{vmatrix} (1 - \lambda) & 0 & 1 \\ 0 & (1 - \lambda) & 1 \\ 0 & -2 & (4 - \lambda) \end{vmatrix} = (1 - \lambda)[(1 - \lambda)(4 - \lambda) + 2]$$

$$= (1 - \lambda)[(\lambda - 2)(\lambda - 3)] = 0$$

so the eigenvalues are $\lambda_1 = 1$, $\lambda_2 = 2$ and $\lambda_3 = 3$. The eigenvalues are distinct, by Theorem 6.1, therefore each should have one linearly independent eigenvector associated with it and $\dim\left(\mathcal{N}\left(A - \lambda_i I\right)\right) = 1$ for each λ_i.

In detail, for $\lambda_1 = 1$ the associated eigenvalue satisfies

$$(A - \lambda_1 I)\,\hat{\xi}^1 = (A - I)\,\hat{\xi}^1 = 0.$$

The augmented matrix is

$$\begin{bmatrix} 1-\lambda & 0 & 1 & \big| & 0 \\ 0 & 1-\lambda & 1 & \big| & 0 \\ 0 & -2 & 4-\lambda & \big| & 0 \end{bmatrix}. \tag{6.16}$$

Substituting $\lambda_1 = 1$ and making a couple of elementary row manipulations yields

$$\begin{bmatrix} 0 & 0 & 1 & \big| & 0 \\ 0 & 0 & 1 & \big| & 0 \\ 0 & -2 & 3 & \big| & 0 \end{bmatrix} \iff \begin{bmatrix} 0 & -2 & 3 & \big| & 0 \\ 0 & 0 & 1 & \big| & 0 \\ 0 & 0 & 1 & \big| & 0 \end{bmatrix} \iff \begin{bmatrix} 0 & -2 & 3 & \big| & 0 \\ 0 & 0 & 1 & \big| & 0 \\ 0 & 0 & 0 & \big| & 0 \end{bmatrix}.$$

The last augmented matrix has one row of zeros, indicating that the dimension of its null space is one, so there is one linearly independent eigenvector associated with $\lambda_1 = 1$. From the second row, the third component $\hat{\xi}_3^1$ clearly must be zero. Using this fact and noting the first row indicates that the second component must also be zero. Finally, the first component of $\hat{\xi}_1^1$ is clearly arbitrary. Thus, the eigenvector must be

$$\hat{\xi}^1 = \begin{bmatrix} 1 \\ 0 \\ 0 \end{bmatrix}.$$

Similarly, substituting $\lambda_2 = 2$ into Equation (6.16) gives

$$\begin{bmatrix} -1 & 0 & 1 & \big| & 0 \\ 0 & -1 & 1 & \big| & 0 \\ 0 & -2 & 2 & \big| & 0 \end{bmatrix} \iff \begin{bmatrix} -1 & 0 & 1 & \big| & 0 \\ 0 & -1 & 1 & \big| & 0 \\ 0 & 0 & 0 & \big| & 0 \end{bmatrix}.$$

Picking the third component $\hat{\xi}_3^2$ to be one, we have

$$\hat{\xi}^2 = \begin{bmatrix} 1 \\ 1 \\ 1 \end{bmatrix}.$$

Finally, for $\lambda_3 = 3$

$$\begin{bmatrix} -2 & 0 & 1 & \big| & 0 \\ 0 & -2 & 1 & \big| & 0 \\ 0 & -2 & 1 & \big| & 0 \end{bmatrix} \iff \begin{bmatrix} -2 & 0 & 1 & \big| & 0 \\ 0 & -2 & 1 & \big| & 0 \\ 0 & 0 & 0 & \big| & 0 \end{bmatrix}.$$

This time picking the third component, $\hat{\xi}_3^3$, to be 2 gives

$$\hat{\xi}^3 = \begin{bmatrix} 1 \\ 1 \\ 2 \end{bmatrix}.$$

Now consider an example with repeated eigenvalues.

Example 6.15. Determine the eigenvalues and eigenvectors of

$$A = \begin{bmatrix} 0 & 1 & 1 \\ -4 & 5 & 1 \\ -5 & 1 & 5 \end{bmatrix}.$$

The characteristic equation is

$$\lambda^3 - 10\lambda^2 + 32\lambda - 32 = 0,$$

so (using a numerical root finder)[2] the eigenvalues are $\lambda_1 = 2$, $\lambda_2 = 4$ and $\lambda_3 = 4$. For $\lambda_1 = 2$,

$$\begin{bmatrix} -2 & 1 & 1 & | & 0 \\ -4 & 3 & 1 & | & 0 \\ -4 & 1 & 3 & | & 0 \end{bmatrix} \iff \begin{bmatrix} -2 & 1 & 1 & | & 0 \\ 0 & 1 & -1 & | & 0 \\ 0 & -1 & 1 & | & 0 \end{bmatrix} \iff \begin{bmatrix} -2 & 1 & 1 & | & 0 \\ 0 & 1 & -1 & | & 0 \\ 0 & 0 & 0 & | & 0 \end{bmatrix}.$$

Because there is one row of zeros, there is one linearly independent eigenvalue associated with $\lambda_1 = 2$, which is expected inasmuch as it is not repeated. Picking the third component of $\hat{\xi}^1$ to be one,

$$\hat{\xi}^1 = \begin{bmatrix} 1 \\ 1 \\ 1 \end{bmatrix}.$$

Now, for $\lambda_2 = 4$

$$\begin{bmatrix} -4 & 1 & 1 & | & 0 \\ -4 & 1 & 1 & | & 0 \\ -4 & 1 & 1 & | & 0 \end{bmatrix} \iff \begin{bmatrix} -4 & 1 & 1 & | & 0 \\ 0 & 0 & 0 & | & 0 \\ 0 & 0 & 0 & | & 0 \end{bmatrix}.$$

There are two rows of zeros, thus there are two linearly independent eigenvectors associated with $\lambda_2 = 4$. Picking the third component of $\hat{\xi}^2$ to be 4 and the second component to be zero, we have

$$\hat{\xi}^2 = \begin{bmatrix} 1 \\ 0 \\ 4 \end{bmatrix}.$$

---

[2] Both MATLAB and Octave have the `roots()` command to calculate the roots of polynomials.

There are two rows of zeros, thus we can find another solution to the equations. To determine one, we pick another combination of variables with the only restriction that it cannot be a scaled version of two of the components of $\hat{\xi}^2$. Picking the third component to be zero and the second component to be 4 gives

$$\hat{\xi}^3 = \begin{bmatrix} 1 \\ 4 \\ 0 \end{bmatrix}.$$

The fact that there were two rows of zeros in upper triangular form of the augmented matrix indicates that the dimension of the null space of $(A - 4I)$ was two. Thus, we were able to determine two linearly independent eigenvectors associated with the repeated eigenvalue.

Finally, just to complete the picture, the following is an example of an eigenvalue with algebraic multiplicity two but a geometric multiplicity of one.

Example 6.16. Returning to the matrix from Example 6.11 with

$$A = \begin{bmatrix} 2 & 1 \\ 0 & 2 \end{bmatrix},$$

we computed previously that $\lambda = 2$ was the only eigenvalue and that it had an algebraic multiplicity of two. Constructing the augmented matrix for $A - 2I$ gives

$$\begin{bmatrix} 0 & 1 & 1 \\ 0 & 0 & 0 \end{bmatrix}.$$

Because there is one row of zeros, the geometric multiplicity is one. Clearly the first component of the eigenvector is arbitrary and the second component must be zero. Thus, for example,

$$\hat{\xi}^1 = \begin{bmatrix} 1 \\ 0 \end{bmatrix}.$$

Finally, after this rather extensive detour into the realm of the nature of repeated eigenvalues and the computational details of computing the associated eigenvectors, we return to the main task at hand which is to solve $\dot{\xi} = A\xi$.

6.5.2 Homogeneous Solutions with Repeated Eigenvalues

Because there may or may not be a full set of linearly independent eigenvectors when there are repeated eigenvalues, we have to consider each case separately.

6.5.2.1 Equal Algebraic and Geometric Multiplicities

This is the case for which to hope because the solution technique is identical to the case of distinct eigenvalues. Even if there are repeated eigenvalues, the general solution is simply

$$\xi(t) = c_1 \hat{\xi}^1 e^{\lambda_1 t} + c_2 \hat{\xi}^2 e^{\lambda_2 t} + \cdots + c_n \hat{\xi}^n e^{\lambda_n t}.$$

This is, in fact, the general solution. The set of eigenvectors is linearly independent, therefore it will always be possible to solve for the coefficients for a specified initial condition regardless of the fact that some of the eigenvalues are repeated.

6.5.2.2 Algebraic Multiplicity Greater Than the Geometric Multiplicity

The case where the geometric multiplicity of an eigenvalue is less than its algebraic multiplicity is much more interesting, but unfortunately, requires a bit more work. In this case, if we simply compute eigenvectors, we will have a set of homogeneous solutions of the form

$$\xi(t) = \hat{\xi}^i e^{\lambda_i t},$$

but we will not have n linearly independent eigenvalues, so

$$\xi(t) = c_1 \hat{\xi}^1 e^{\lambda_1 t} + c_2 \hat{\xi}^2 e^{\lambda_2 t} + \cdots + c_m \hat{\xi}^m e^{\lambda_m t},$$

where $m < n$ is a solution, but not a general solution. In this case, it is not possible to compute coefficients c_i to satisfy any set of initial conditions because there is not a full set of linearly independent eigenvectors.

Recall from Chapter 3 that in the case of repeated roots, the approach was to multiply the one homogeneous solution by the independent variable t and add it to the first solution. The following two examples illustrate that fact, but also then go to make a connection to the matrix approach that is the subject of this chapter.

Example 6.17. Find the general solution to

$$\ddot{x} + 4\dot{x} + 4x = 0. \tag{6.17}$$

Assuming $x(t) = e^{\lambda t}$ and substituting gives

$$\lambda^2 + 4\lambda + 4 = 0 \quad \Longrightarrow \quad (\lambda + 2)^2 = 0. \tag{6.18}$$

So, $\lambda = 2$ is the solution. Hence, $x_h(t) = e^{-2t}$ is a homogeneous solution. Because there is no other root to the characteristic equation, the approach (which was fully detailed in Chapter 3) is to assume a second homogeneous solution of the form $x_h(t) = te^{-2t}$. The fact that this is a second homogeneous solution can be verified by substituting it into Equation (6.17) and the fact that it is linearly independent can be verified by computing the Wronskian. Thus the general solution to Equation (6.17) is

$$x(t) = c_1 e^{-2t} + c_2 t e^{-2t}. \tag{6.19}$$

Example 6.18. Consider the same equation as in Equation (6.17), but first convert it into a system of two first-order equations. The equivalent system is

$$\frac{d}{dt} \begin{bmatrix} x \\ \dot{x} \end{bmatrix} = \begin{bmatrix} 0 & 1 \\ -4 & -4 \end{bmatrix} \begin{bmatrix} x \\ \dot{x} \end{bmatrix}.$$

Computing the eigenvalues for the matrix in the preceding equation gives

$$\begin{vmatrix} -\lambda & 1 \\ -4 & -4-\lambda \end{vmatrix} = \lambda^2 + 4\lambda + 4 = (\lambda + 2)^2 = 0.$$

It is no coincidence that the characteristic equation for the eigenvalue problem is exactly the same as Equation (6.18). Thus, the only distinction is one of nomenclature: there are "repeated eigenvalues" instead of "repeated roots." Now computing the eigenvectors corresponding to $\lambda_1 = -2$ gives

$$\begin{bmatrix} 2 & 1 & | & 0 \\ -4 & -2 & | & 0 \end{bmatrix} \quad \Longleftrightarrow \quad \begin{bmatrix} 2 & 1 & | & 0 \\ 0 & 0 & | & 0 \end{bmatrix}.$$

Thus, there is one linearly independent eigenvector,

$$\hat{\xi}^1 = \begin{bmatrix} 1 \\ -2 \end{bmatrix}.$$

The goal is obviously to construct a solution that is equivalent to the general solution in Equation (6.19). Differentiating Equation (6.19) gives

$$\dot{x}(t) = -2c_1 e^{-2t} + c_2 e^{-2t} - 2c_2 t e^{-2t},$$

or in vector form

$$\frac{d}{dt} \begin{bmatrix} x \\ \dot{x} \end{bmatrix} = \frac{d}{dt} \begin{bmatrix} \xi_1 \\ \xi_2 \end{bmatrix}$$

$$= c_1 \begin{bmatrix} 1 \\ -2 \end{bmatrix} e^{-2t} + c_2 \left(\begin{bmatrix} 0 \\ 1 \end{bmatrix} e^{-2t} + t \begin{bmatrix} 1 \\ -2 \end{bmatrix} e^{-2t} \right)$$

$$= c_1 \hat{\xi}^1 e^{\lambda_1 t} + c_2 \left(\hat{\xi}^2 e^{\lambda_1 t} + t \hat{\xi}^1 e^{\lambda_1 t} \right).$$

Clearly, in the notation of the last line in the above equation, $\hat{\xi}^1$ is simply the eigenvector that we already computed. The question is how to compute $\hat{\xi}^2$, the other vector.[3]

    Recall that the whole business regarding eigenvalues and eigenvectors came about by simply assuming solutions of the form $\xi(t) = \hat{\xi} e^{\lambda t}$. Substituting this into

---

[3] Note that the superscripts for the $\hat{\xi}$s are indices, not powers.

$\dot{\xi} = A\xi$ then indicated that $\hat{\xi}$ had to be an eigenvector and λ had to be an eigenvalue. The approach now is pretty obvious: substitute the assumed form of the second homogeneous solution

$$\xi_h(t) = \left(\hat{\xi}^2 + t\hat{\xi}^1 \right) e^{\lambda t}$$

to verify first that $\hat{\xi}^1$ indeed satisfies the eigenvector equation (so that the fact that they are the same in this example is not a coincidence) and second, to determine what sort of equation $\hat{\xi}^2$ must satisfy. Differentiating and substituting gives

$$\lambda \left(\hat{\xi}^2 + t\hat{\xi}^1 \right) e^{\lambda t} + \hat{\xi}^1 e^{\lambda t} = A \left(\hat{\xi}^2 + t\hat{\xi}^1 \right) e^{\lambda t}.$$

Because this must hold for all t, the coefficients of the different powers of t must be equal. Therefore, collecting terms multiplying the same powers of t gives

$$t^0 : \qquad\qquad \lambda \left(\hat{\xi}^2 + \hat{\xi}^1 \right) e^{\lambda t} = A\hat{\xi}^2 e^{\lambda t}$$

$$t^1 : \qquad\qquad \lambda \hat{\xi}^1 e^{\lambda t} = A\hat{\xi}^1 e^{\lambda t}.$$

Inasmuch as $e^{\lambda t}$ is never zero we have the following two equations

$$(A - \lambda I)\hat{\xi}^1 = 0$$
$$(A - \lambda I)\hat{\xi}^2 = \hat{\xi}^1.$$

The first equation has already been solved, so

$$\hat{\xi}^1 = \begin{bmatrix} 1 \\ -2 \end{bmatrix}.$$

For the second equation we have

$$\begin{bmatrix} 2 & 1 & 1 \\ -4 & -2 & -2 \end{bmatrix} \quad \Longleftrightarrow \quad \begin{bmatrix} 2 & 1 & 1 \\ 0 & 0 & 0 \end{bmatrix}.$$

As with eigenvectors, the solution is determined only up to an arbitrary scaling constant. In this case, clearly, the vector

$$\hat{\xi}^2 = \begin{bmatrix} 0 \\ 1 \end{bmatrix}$$

satisfies the equation for $\hat{\xi}^2$.

The task now is to generalize the approach to systems of n equations where the multiplicity of a repeated eigenvalue may be greater than two. So consider the general case of

$$\dot{\xi} = A\xi, \quad A \in \mathbb{R}^{n \times n},$$

and assume that the algebraic multiplicity of eigenvalue λ_i is m but that the geometric multiplicity is less than m. Motivated by the above example, clearly the approach is to multiply exponential solutions by t to obtain additional linearly independent solutions. In the example, because the system was second-order, the highest power of t in the general solution was one; however, in the case where the algebraic multiplicity is greater than two, additional powers of t may be necessary. Therefore, let us propose the following homogeneous solution corresponding to eigenvalue λ_i with algebraic multiplicity m,

$$\xi(t) = \left(\hat{\xi}^m + t\hat{\xi}^{m-1} + \frac{t^2}{2!}\hat{\xi}^{m-2} + \cdots + \frac{t^{m-1}}{(m-1)!}\hat{\xi}^1 \right) e^{\lambda_i t}. \tag{6.20}$$

Differentiating this proposed solution gives

$$\begin{aligned}\dot{\xi}(t) = \lambda_i &\left(\hat{\xi}^m + t\hat{\xi}^{m-1} + \frac{t^2}{2!}\hat{\xi}^{m-2} + \cdots + \frac{t^{m-1}}{(m-1)!}\hat{\xi}^1 \right) e^{\lambda_i t} \\ &+ \left(\hat{\xi}^{m-1} + t\hat{\xi}^{m-2} + \cdots + \frac{t^{m-2}}{(m-2)!}\hat{\xi}^1 \right) e^{\lambda_i t}.\end{aligned} \tag{6.21}$$

Also,

$$A\xi(t) = A\left(\hat{\xi}^m + t\hat{\xi}^{m-1} + t^2\hat{\xi}^{m-2} + \cdots + t^{m-1}\hat{\xi}^1 \right) e^{\lambda_i t}. \tag{6.22}$$

Because $e^{\lambda_i t}$ is never zero it can be canceled from both equations and because $\dot{\xi} = At$ must hold for all t, equating the coefficients of each power of t in Equations (6.21) and (6.22), gives

$$\begin{aligned} t^0: \qquad & \lambda_i\hat{\xi}^m + \hat{\xi}^{m-1} = A\hat{\xi}^m \\ t^1: \qquad & \lambda_i\hat{\xi}^{m-1} + \hat{\xi}^{m-2} = A\hat{\xi}^{m-1} \\ t^2: \qquad & \lambda_i\hat{\xi}^{m-2} + \hat{\xi}^{m-2} = A\hat{\xi}^{m-2} \\ \vdots \qquad & \qquad \vdots \;\; \vdots \\ t^{m-1}: \qquad & \lambda_i\hat{\xi}^1 = A\hat{\xi}^1. \end{aligned}$$

Thus, the following sequence is obtained

$$\begin{aligned} (A - \lambda_i I)\,\hat{\xi}^1 &= 0 \\ (A - \lambda_i I)\,\hat{\xi}^2 &= \hat{\xi}^1 \\ (A - \lambda_i I)\,\hat{\xi}^3 &= \hat{\xi}^2 \\ \vdots &= \vdots \\ (A - \lambda_i I)\,\hat{\xi}^m &= \hat{\xi}^{m-1}. \end{aligned} \tag{6.23}$$

The first equation is simply the equation for a regular eigenvalue. The vectors $\hat{\xi}^2$ through $\hat{\xi}^m$ are called *generalized eigenvectors* and are determined by sequentially solving the second through mth equations.

Note that if the second line of Equation (6.23) is multiplied on the left by $(A - \lambda_i I)$ then

$$(A - \lambda_i I)(A - \lambda_i I)\hat{\xi}^2 = (A - \lambda_i I)\hat{\xi}^1,$$

but because

$$(A - \lambda_i I)\hat{\xi}^1 = 0$$

then

$$(A - \lambda_i I)^2 \hat{\xi}^2 = 0.$$

Similarly, multiplying the jth line in Equation (6.23) by $(A - \lambda_i I)^j$, where $1 < j < m$, gives

$$(A - \lambda_i I)^j \hat{\xi}^j = 0.$$

Further note that

$$(A - \lambda_i I)^m \hat{\xi}^j = (A - \lambda_i I)^{m-j}(A - \lambda_i I)^j \hat{\xi}^j = 0.$$

Hence, all the eigenvectors and generalized eigenvectors associated with λ_i are in the null space of $(A - \lambda_i I)^m$, which motivates the following definition.

Definition 6.3. The null space of $(A - \lambda_i I)^m$ is the *generalized eigenspace of A associated with* λ_i.

The following theorem assures us that the dimension of the generalized eigenspace associated with λ_i is the same as the algebraic multiplicity of λ_i. This fact is necessary in order to ensure that enough generalized eigenvectors exist to generate a full set of linearly independent homogeneous solutions to construct a general solution.

Theorem 6.3. *The dimension of the generalized eigenspace of A associated with* λ_i *is equal to the algebraic multiplicity of the eigenvalue* λ_i; *that is, if the algebraic multiplicity of the eigenvalue* λ_i *is m, then*

$$\dim(\mathcal{N}(A - \lambda_i I)^m) = m.$$

Proof. The reader is referred to [18] and [22].

This book refers to any $\xi \in \mathcal{N}(A - \lambda I)^m$ when $m > 1$ as a generalized eigenvector. So the set of generalized eigenvectors also contains some "regular" eigenvectors that satisfy $\xi \in \mathcal{N}(A - \lambda I)$. The following theorem gives the form of the homogeneous solution for any vector in generalized eigenspace of λ_i.

Theorem 6.4. *For* $A \in \mathbb{R}^{n \times n}$ *and* λ_i *an eigenvector of A with algebraic multiplicity m, if*

$$(A - \lambda_i I)^m \hat{\xi} = 0,$$

then

$$\xi(t) = \left(\hat{\xi} + t \left(A - \lambda_i I \right) \hat{\xi} + \frac{t^2}{2!} \left(A - \lambda_i I \right)^2 \hat{\xi} + \cdots + \frac{t^{m-1}}{(m-1)!} \left(A - \lambda_i I \right)^{m-1} \hat{\xi} \right) e^{\lambda_i t}$$

$$(6.24)$$

satisfies

$$\dot{\xi} = A\xi.$$

Proof. This is by direct computation. Simply differentiate $\xi(t)$ and substitute into $\dot{\xi} = A\xi$. □

So, finally we have the following solution technique for $\dot{\xi} = A\xi$, for $A \in \mathbb{R}^{n \times n}$ where λ_i has an algebraic multiplicity of m.

1. For the nonrepeated eigenvalues λ_j the corresponding homogeneous solution is $\xi(t) = \hat{\xi}^j e^{\lambda_j t}$. If two of these eigenvalues are a complex conjugate pair, then converting the homogeneous solution to sines and cosines as outlined in Section 6.4 is preferable.
2. For each repeated λ_i,

   a. Find a basis for the generalized eigenspace of λ_i, which is m generalized eigenvectors $\hat{\xi}^i$; that is,

   $$\left(A - \lambda_i I \right)^m \hat{\xi} = 0.$$

   These $\hat{\xi}$ may be regular eigenvectors, generalized eigenvectors, or linear combinations thereof.

   b. The homogeneous solution corresponding to each $\hat{\xi}^i$ is

   $$\xi(t) = \left(\hat{\xi} + t \left(A - \lambda_i I \right) \hat{\xi} + \frac{t^2}{2!} \left(A - \lambda_i I \right)^2 \hat{\xi} + \cdots + \frac{t^{m-1}}{(m-1)!} \left(A - \lambda_i I \right)^{m-1} \hat{\xi} \right) e^{\lambda_i t}.$$

Remark 6.1. Many, if not most, texts use the sequence in Equation (6.23) for the definition and computation of generalized eigenvectors. Such an approach suffers from the drawback that if there is more than one solution to the first equation in the sequence, the solution to the second may depend on choosing a specific linear combination of those solutions for the right-hand side of the second equation, which is difficult. The approach presented here, which is to compute a basis for the generalized eigenspace of the matrix, will always work.

A few examples help illustrate the approach.

Example 6.19. Determine the general solution to $\dot{\xi} = A\xi$ where

$$A = \begin{bmatrix} 1 & 0 & 0 & 0 \\ 0 & 2 & 1 & 0 \\ 0 & 0 & 2 & 1 \\ 0 & 0 & 0 & 2 \end{bmatrix}.$$

The matrix is triangular, therefore the eigenvalues are the values along the diagonal. Thus

$$\lambda_1 = 1, \quad \lambda_2 = 2, \quad \lambda_3 = 2, \quad \lambda_4 = 2.$$

Thus, $\lambda = 2$ is an eigenvalue with algebraic multiplicity of three. For $\lambda_1 = 1$, the eigenvector is

$$(A - \lambda_1 I)\,\hat{\xi}^1 = 0 \quad \Longleftrightarrow \quad \begin{bmatrix} 0 & 0 & 0 & 0 & | & 0 \\ 0 & 2 & 1 & 0 & | & 0 \\ 0 & 0 & 2 & 1 & | & 0 \\ 0 & 0 & 0 & 2 & | & 0 \end{bmatrix} \quad \Longleftrightarrow \quad \hat{\xi}^1 = \begin{bmatrix} 1 \\ 0 \\ 0 \\ 0 \end{bmatrix}.$$

For $\lambda_2 = \lambda_3 = \lambda_4 = 2$ we need to find all three vectors that satisfy $(A - 2I)^3\,\hat{\xi} = 0$, so we must compute

$$(A - 2I)^3 = \begin{bmatrix} -1 & 0 & 0 & 0 \\ 0 & 0 & 0 & 0 \\ 0 & 0 & 0 & 0 \\ 0 & 0 & 0 & 0 \end{bmatrix}.$$

Hence we need to find three solutions to

$$\begin{bmatrix} -1 & 0 & 0 & 0 & | & 0 \\ 0 & 0 & 0 & 0 & | & 0 \\ 0 & 0 & 0 & 0 & | & 0 \\ 0 & 0 & 0 & 0 & | & 0 \end{bmatrix}.$$

The free components are obviously the second, third, and fourth components. Hence, one choice is

$$\hat{\xi}^2 = \begin{bmatrix} 0 \\ 1 \\ 0 \\ 0 \end{bmatrix}, \quad \hat{\xi}^3 = \begin{bmatrix} 0 \\ 0 \\ 1 \\ 0 \end{bmatrix}, \quad \hat{\xi}^4 = \begin{bmatrix} 0 \\ 0 \\ 0 \\ 1 \end{bmatrix}.$$

Thus, the general solution is

$$\xi(t) = c_1 \hat{\xi}_1 e^{\lambda_1 t}$$
$$+ c_2 \left(\hat{\xi}^2 + t(A - 2I)\hat{\xi}^2 + \frac{t^2}{2!}(A - 2I)^2 \hat{\xi}^2 \right) e^{\lambda_2 t}$$
$$+ c_3 \left(\hat{\xi}^3 + t(A - 2I)\hat{\xi}^3 + \frac{t^2}{2!}(A - 2I)^2 \hat{\xi}^3 \right) e^{\lambda_2 t}$$
$$+ c_4 \left(\hat{\xi}^4 + t(A - 2I)\hat{\xi}^4 + \frac{t^2}{2!}(A - 2I)^2 \hat{\xi}^4 \right) e^{\lambda_2 t}.$$

Because we need them in the answer, observe that

$$(A - 2I) = \begin{bmatrix} 1 & 0 & 0 & 0 \\ 0 & 0 & 1 & 0 \\ 0 & 0 & 0 & 1 \\ 0 & 0 & 0 & 0 \end{bmatrix}, \quad (A - 2I)^2 = \begin{bmatrix} 1 & 0 & 0 & 0 \\ 0 & 0 & 0 & 1 \\ 0 & 0 & 0 & 0 \\ 0 & 0 & 0 & 0 \end{bmatrix}$$

and hence

$$(A - 2I)\,\hat{\xi}^2 = \begin{bmatrix} 0 \\ 0 \\ 0 \\ 0 \end{bmatrix}, \quad (A - 2I)^2\,\hat{\xi}^2 = \begin{bmatrix} 0 \\ 0 \\ 0 \\ 0 \end{bmatrix},$$

$$(A - 2I)\,\hat{\xi}^3 = \begin{bmatrix} 0 \\ 1 \\ 0 \\ 0 \end{bmatrix}, \quad (A - 2I)^2\,\hat{\xi}^3 = \begin{bmatrix} 0 \\ 0 \\ 0 \\ 0 \end{bmatrix},$$

$$(A - 2I)\,\hat{\xi}^4 = \begin{bmatrix} 0 \\ 0 \\ 1 \\ 0 \end{bmatrix}, \quad (A - 2I)^2\,\hat{\xi}^4 = \begin{bmatrix} 0 \\ 1 \\ 0 \\ 0 \end{bmatrix}.$$

So, finally, the general solution is

$$\xi(t) = c_1 \begin{bmatrix} 1 \\ 0 \\ 0 \\ 0 \end{bmatrix} e^t + c_2 \begin{bmatrix} 0 \\ 1 \\ 0 \\ 0 \end{bmatrix} e^{2t} + c_3 \left(\begin{bmatrix} 0 \\ 0 \\ 1 \\ 0 \end{bmatrix} + t \begin{bmatrix} 0 \\ 1 \\ 0 \\ 0 \end{bmatrix} \right) e^{2t}$$

$$+ c_4 \left(\begin{bmatrix} 0 \\ 0 \\ 0 \\ 1 \end{bmatrix} + t \begin{bmatrix} 0 \\ 0 \\ 1 \\ 0 \end{bmatrix} + \frac{t^2}{2} \begin{bmatrix} 0 \\ 1 \\ 0 \\ 0 \end{bmatrix} \right) e^{2t}.$$

That this is a solution may be verified by directly substituting this into the original differential equation.

Example 6.20. Determine the general solution to

$$\dot{\xi} = A\xi,$$

where

$$A = \begin{bmatrix} 3 & -1 & 0 \\ 1 & 1 & 0 \\ 0 & 0 & 2 \end{bmatrix}.$$

Computing

$$\begin{vmatrix} 3 - \lambda & -1 & 0 \\ 1 & 1 - \lambda & 0 \\ 0 & 0 & 2 - \lambda \end{vmatrix} = 0$$

using a cofactor expansion across the third row gives

$$(2-\lambda)\left[(3-\lambda)(1-\lambda)+1\right] =$$
$$(2-\lambda)\left[\lambda^2-4\lambda+4\right] =$$
$$(2-\lambda)\left[(2-\lambda)^2\right] = 0.$$

Hence, $\lambda = 2$ has an algebraic multiplicity of three.

Next we must determine the vectors that span the null space of $(A-\lambda I)^3$. Substituting $\lambda = 2$ gives

$$(A-2I) = \begin{bmatrix} 1 & -1 & 0 \\ 1 & -1 & 0 \\ 0 & 0 & 0 \end{bmatrix}$$

and a simple calculation shows that

$$(A-2I)^2 = (A-2I)^3 = \begin{bmatrix} 0 & 0 & 0 \\ 0 & 0 & 0 \\ 0 & 0 & 0 \end{bmatrix},$$

so we may choose any three vectors that span $\mathbb{R}^3$. Just for fun, we choose

$$\hat{\xi}^1 = \begin{bmatrix} 1 \\ 0 \\ 0 \end{bmatrix}, \quad \hat{\xi}^2 = \begin{bmatrix} 1 \\ 1 \\ 0 \end{bmatrix}, \quad \hat{\xi}^3 = \begin{bmatrix} 1 \\ 1 \\ 1 \end{bmatrix}.$$

All that is left to do is to substitute into Equation (6.24), which gives

$$\xi(t) = c_1 \left(\begin{bmatrix} 1 \\ 0 \\ 0 \end{bmatrix} + t \begin{bmatrix} 1 & -1 & 0 \\ 1 & -1 & 0 \\ 0 & 0 & 0 \end{bmatrix} \begin{bmatrix} 1 \\ 0 \\ 0 \end{bmatrix} \right) e^{2t}$$

$$+ c_2 \left(\begin{bmatrix} 1 \\ 1 \\ 0 \end{bmatrix} + t \begin{bmatrix} 1 & -1 & 0 \\ 1 & -1 & 0 \\ 0 & 0 & 0 \end{bmatrix} \begin{bmatrix} 1 \\ 1 \\ 0 \end{bmatrix} \right) e^{2t}$$

$$+ c_3 \left(\begin{bmatrix} 1 \\ 1 \\ 1 \end{bmatrix} + t \begin{bmatrix} 1 & -1 & 0 \\ 1 & -1 & 0 \\ 0 & 0 & 0 \end{bmatrix} \begin{bmatrix} 1 \\ 1 \\ 1 \end{bmatrix} \right) e^{2t}$$

$$= c_1 \left(\begin{bmatrix} 1 \\ 0 \\ 0 \end{bmatrix} + t \begin{bmatrix} 1 \\ 1 \\ 0 \end{bmatrix} \right) e^{2t} + c_2 \begin{bmatrix} 1 \\ 1 \\ 0 \end{bmatrix} e^{2t} + c_3 \begin{bmatrix} 1 \\ 1 \\ 1 \end{bmatrix} e^{2t}.$$

Just to complete the picture, let us repeat the previous example, but choose the usual basis for $\mathbb{R}^3$ instead.

Example 6.21. Returning to Example 6.20, choose

$$\hat{\xi}^1 = \begin{bmatrix} 1 \\ 0 \\ 0 \end{bmatrix}, \quad \hat{\xi}^2 = \begin{bmatrix} 0 \\ 1 \\ 0 \end{bmatrix}, \quad \hat{\xi}^3 = \begin{bmatrix} 0 \\ 0 \\ 1 \end{bmatrix}.$$

Substituting into Equation (6.24) gives

$$\xi(t) = c_1 \left(\begin{bmatrix} 1 \\ 0 \\ 0 \end{bmatrix} + t \begin{bmatrix} 1 & -1 & 0 \\ 1 & -1 & 0 \\ 0 & 0 & 0 \end{bmatrix} \begin{bmatrix} 1 \\ 0 \\ 0 \end{bmatrix} \right) e^{2t}$$

$$+ c_2 \left(\begin{bmatrix} 0 \\ 1 \\ 0 \end{bmatrix} + t \begin{bmatrix} 1 & -1 & 0 \\ 1 & -1 & 0 \\ 0 & 0 & 0 \end{bmatrix} \begin{bmatrix} 0 \\ 1 \\ 0 \end{bmatrix} \right) e^{2t}$$

$$+ c_3 \left(\begin{bmatrix} 0 \\ 0 \\ 1 \end{bmatrix} + t \begin{bmatrix} 1 & -1 & 0 \\ 1 & -1 & 0 \\ 0 & 0 & 0 \end{bmatrix} \begin{bmatrix} 0 \\ 0 \\ 1 \end{bmatrix} \right) e^{2t}$$

$$= c_1 \left(\begin{bmatrix} 1 \\ 0 \\ 0 \end{bmatrix} + t \begin{bmatrix} 1 \\ 1 \\ 0 \end{bmatrix} \right) e^{2t} + c_1 \left(\begin{bmatrix} 0 \\ 1 \\ 0 \end{bmatrix} + t \begin{bmatrix} -1 \\ -1 \\ 0 \end{bmatrix} \right) e^{2t} + c_3 \begin{bmatrix} 0 \\ 0 \\ 1 \end{bmatrix} e^{2t}.$$

This answer may appear to be different from the answer in Example 6.20, however, if we let

$$k_1 = c_1, \quad k_2 = c_1 + c_2, \quad k_3 = c_1 + c_2 + c_3$$

the answer is

$$\xi(t) = k_1 e^{2t} \left(\begin{bmatrix} 1 \\ 0 \\ 0 \end{bmatrix} + t \begin{bmatrix} 1 \\ 1 \\ 0 \end{bmatrix} \right) + k_2 e^{2t} \begin{bmatrix} 1 \\ 1 \\ 0 \end{bmatrix} + k_3 e^{2t} \begin{bmatrix} 1 \\ 1 \\ 1 \end{bmatrix},$$

which is the same.

6.6 Diagonalization

The fundamental idea underlying this approach is to convert the system of coupled first-order equations into decoupled equations. What this means mathematically is apparent shortly, but the consequence of each equation (or row) can be solved individually, or one at a time. First we need to investigate the concept of converting a matrix to diagonal form. The correct way to think of this conceptually is that the equations are transformed to a new set of variables, and in those variables the equations are decoupled.

For a homogeneous system of the form

$$\dot{\xi} = A\xi$$

we consider the easier case where A has a full set of n linearly independent eigen-vectors, $\hat{\xi}^1, \ldots, \hat{\xi}^n$, and define the matrix T as the matrix with the eigenvectors of A as its columns; that is,

$$T = \left[\hat{\xi}^1 \ \hat{\xi}^2 \cdots \ \hat{\xi}^n \right].$$

The definition of an eigenvector is

$$A\hat{\xi}^i = \lambda_i \hat{\xi}^i,$$

therefore

$$AT = \left[\lambda_1 \hat{\xi}^1 \ \lambda_2 \hat{\xi}^2 \ \cdots \ \lambda_n \hat{\xi}^n \right].$$

Because λ_i is a scalar, each term $\lambda_i \hat{\xi}^i$ is a vector and constitutes one column of the matrix AT.

Now, because we assumed that $\hat{\xi}^1, \hat{\xi}^2, \ldots, \hat{\xi}^n$ were linearly independent, then T is invertible. Note that by definition

$$T^{-1}T = \begin{bmatrix} 1 & 0 & \cdots & 0 \\ 0 & 1 & \cdots & 0 \\ \vdots & \vdots & \ddots & \vdots \\ 0 & 0 & \cdots & 1 \end{bmatrix}.$$

Considering this equation column-by-column, we have

$$T^{-1}\hat{\xi}^1 = \begin{bmatrix} 1 \\ 0 \\ \vdots \\ 0 \end{bmatrix}, \quad T^{-1}\hat{\xi}^2 = \begin{bmatrix} 0 \\ 1 \\ \vdots \\ 0 \end{bmatrix}, \quad \cdots, \quad T^{-1}\hat{\xi}^n = \begin{bmatrix} 0 \\ 0 \\ \vdots \\ 1 \end{bmatrix}.$$

Also, because $A\hat{\xi}^i = \lambda_i \hat{\xi}^i$

$$T^{-1}A\hat{\xi}^1 = T^{-1}\lambda_1\hat{\xi}^1 = \lambda_i T^{-1}\hat{\xi}^1 = \lambda_1 \begin{bmatrix} 1 \\ 0 \\ \vdots \\ 0 \end{bmatrix} = \begin{bmatrix} \lambda_1 \\ 0 \\ \vdots \\ 0 \end{bmatrix},$$

$$T^{-1}A\hat{\xi}^2 = T^{-1}\lambda_2\hat{\xi}^2 = \lambda_i T^{-1}\hat{\xi}^2 = \lambda_2 \begin{bmatrix} 0 \\ 1 \\ \vdots \\ 0 \end{bmatrix} = \begin{bmatrix} 0 \\ \lambda_2 \\ \vdots \\ 0 \end{bmatrix},$$

and so forth until

$$T^{-1}A\hat{\xi}^n = T^{-1}\lambda_n\hat{\xi}^n = \lambda_i T^{-1}\hat{\xi}^n = \lambda_n \begin{bmatrix} 0 \\ 0 \\ \vdots \\ 1 \end{bmatrix} = \begin{bmatrix} 0 \\ 0 \\ \vdots \\ \lambda_n \end{bmatrix}.$$

Finally, putting it all together gives the important relation

$$T^{-1}AT = \begin{bmatrix} \lambda_1 & 0 & 0 & \cdots & 0 \\ 0 & \lambda_2 & 0 & \cdots & 0 \\ 0 & 0 & \lambda_3 & \cdots & 0 \\ \vdots & \vdots & \vdots & \ddots & \vdots \\ 0 & 0 & 0 & \cdots & \lambda_n \end{bmatrix}.$$

A common notation for a diagonal matrix is

$$\Lambda = T^{-1}AT.$$

The way to exploit this means to transform A into a diagonal matrix with the eigenvalues of A on the diagonal and to consider the change of coordinates

$$\xi = \begin{bmatrix} \xi_1 \\ \vdots \\ \xi_n \end{bmatrix} \iff \psi = \begin{bmatrix} \psi_1 \\ \vdots \\ \psi_n \end{bmatrix},$$

given by

$$\xi = T\psi \iff \psi = T^{-1}\xi.$$

We are assuming that A has n linearly independent eigenvalues, thus T is invertible. Thus, using this change of coordinates on

$$\dot{\xi} = A\xi$$

gives

$$T\dot{\psi} = AT\psi.$$

Note that because the columns of T are eigenvectors, it is a constant matrix and thus $\dot{T} = 0$. Multiplying each equation on the left by T^{-1}

$$\dot{\psi} = T^{-1}AT\psi = \Lambda\psi.$$

Inasmuch as Λ is diagonal, each equation is decoupled and it is possible to solve them individually. After finding $\psi(t)$, the answer in the original coordinates is given by

$$\xi(t) = T\psi(t).$$

The following example illustrates the means to diagonalize a matrix to solve a system of equations. Observe that for homogeneous solutions it does not really save any effort because the eigenvalues and eigenvectors are needed to diagonalize A, which are exactly what are needed in the solution methods outlined above. However, as is apparent subsequently, it is useful for nonhomogeneous systems.

Example 6.22. Solve the system from Example 6.4 where

$$A = \begin{bmatrix} -4 & 1 & 0 \\ 0 & -6 & 0 \\ 0 & 0 & -2 \end{bmatrix}.$$

In that example, we computed the eigenvalues and eigenvectors as $\lambda_1 = -2$, $\lambda_2 = -4$, and $\lambda_3 = -6$ and the corresponding eigenvectors are

$$\hat{\xi}^1 = \begin{bmatrix} 0 \\ 0 \\ 1 \end{bmatrix}, \quad \hat{\xi}^2 = \begin{bmatrix} 1 \\ 0 \\ 0 \end{bmatrix}, \quad \hat{\xi}^3 = \begin{bmatrix} -1 \\ 2 \\ 0 \end{bmatrix}, \quad \Longrightarrow \quad T = \begin{bmatrix} 0 & 1 & -1 \\ 0 & 0 & 2 \\ 1 & 0 & 0 \end{bmatrix}.$$

Because $\det T = 2$ and the cofactor matrix for T is

$$C = \begin{bmatrix} 0 & 2 & 0 \\ 0 & 1 & 1 \\ 2 & 0 & 0 \end{bmatrix},$$

then

$$T^{-1} = \frac{1}{2} \begin{bmatrix} 0 & 0 & 2 \\ 2 & 1 & 0 \\ 0 & 1 & 0 \end{bmatrix} = \begin{bmatrix} 0 & 0 & 1 \\ 1 & \frac{1}{2} & 0 \\ 0 & \frac{1}{2} & 0 \end{bmatrix},$$

and

$$\Lambda = T^{-1}AT = \begin{bmatrix} 0 & 0 & 1 \\ 1 & \frac{1}{2} & 0 \\ 0 & \frac{1}{2} & 0 \end{bmatrix} \begin{bmatrix} -4 & 1 & 0 \\ 0 & -6 & 0 \\ 0 & 0 & -2 \end{bmatrix} \begin{bmatrix} 0 & 1 & -1 \\ 0 & 0 & 2 \\ 1 & 0 & 0 \end{bmatrix} = \begin{bmatrix} -2 & 0 & 0 \\ 0 & -4 & 0 \\ 0 & 0 & -6 \end{bmatrix}.$$

Hence,

$$\frac{d}{dt} \begin{bmatrix} \psi_1 \\ \psi_2 \\ \psi_3 \end{bmatrix} = \begin{bmatrix} -2 & 0 & 0 \\ 0 & -4 & 0 \\ 0 & 0 & -6 \end{bmatrix} \begin{bmatrix} \psi_1 \\ \psi_2 \\ \psi_3 \end{bmatrix}$$

or

$$\begin{aligned} \dot{\psi}_1 &= -2\psi_1 &\Longrightarrow&\quad \psi_1 = c_1 e^{-2t} \\ \dot{\psi}_2 &= -4\psi_2 &\Longrightarrow&\quad \psi_2 = c_2 e^{-4t} \\ \dot{\psi}_3 &= -6\psi_3 &\Longrightarrow&\quad \psi_3 = c_3 e^{-6t}. \end{aligned}$$

Multiplying $\psi(t)$ by T gives $\xi(t)$,

$$\xi(t) = T\psi(t) = \begin{bmatrix} 0 & 1 & -1 \\ 0 & 0 & 2 \\ 1 & 0 & 0 \end{bmatrix} \begin{bmatrix} c_1 e^{-2t} \\ c_2 e^{-4t} \\ c_3 e^{-6t} \end{bmatrix} = \begin{bmatrix} c_2 e^{-4t} - c_3 e^{-6t} \\ 2c_3 e^{-6t} \\ c_1 e^{-2t} \end{bmatrix},$$

which is the same answer as the general solution in Example 6.4.

6.7 The Matrix Exponential

Chapter 2 showed that the solution to $\dot{x} = ax$ is

$$x(t) = e^{at} x(0)$$

and this section generalizes the definition of an exponential so that we may write the solution to $\dot{\xi} = A\xi$ as

$$\xi(t) = e^{At}\xi(0). \tag{6.25}$$

Note that A is a *matrix*, so we need to define what the exponential term in Equation (6.25) means. To do so, consider the Taylor series about $t = 0$ in the scalar case, which is

$$e^{at} = 1 + at + \frac{1}{2}(at)^2 + \frac{1}{3!}(at)^3 + \cdots = \sum_{n=0}^{\infty} \frac{(at)^n}{n!}.$$

Motivated by this, we can define the *matrix exponential* as follows.

Definition 6.4. For $A \in \mathbb{R}^{n \times n}$ define the *matrix exponential* as

$$e^{At} = I + At + \frac{1}{2}A^2 t^2 + \frac{1}{3!}A^3 t^3 + \cdots = \sum_{n=0}^{\infty} \frac{A^n t^n}{n!}, \tag{6.26}$$

where I is the $n \times n$ identify matrix.

If this series converges, then it has the following useful property. Differentiating Equation (6.26) term-by-term gives

$$\frac{d}{dt}e^{At} = \frac{d}{dt}\left(I + At + \frac{1}{2}A^2 t^2 + \frac{1}{3!}A^3 t^3 + \cdots\right)$$

$$= A + A^2 t + \frac{1}{2}A^3 t^2 + \frac{1}{3!}A^4 t^3 + \cdots$$

$$= A\left(I + At + \frac{1}{2}A^2 t^2 + \frac{1}{3!}A^3 t^3 + \cdots\right)$$

$$= Ae^{At},$$

or using the summation notation

$$\frac{d}{dt}e^{At} = \sum_{n=1}^{\infty} n\frac{A^n t^{n-1}}{n!} = A\sum_{n=1}^{\infty} \frac{A^{n-1} t^{n-1}}{(n-1)!} = A\sum_{n=0}^{\infty} \frac{A^n t^n}{n!}.$$

Hence, it follows the same differentiation rule as the scalar case.

In that case, if we write

$$\xi(t) = e^{At}\xi(0)$$

then substituting into $\dot{\xi} = A\xi$ gives

$$Ae^{At}\xi(0) = Ae^{At}\xi(0)$$

which shows it is a solution. Also substituting $t = 0$ into the solution gives

$$\xi(0) = e^{A0}\xi(0) = I\xi(0) = \xi(0)$$

so it also satisfies the initial condition.

Although this is a very nice expression for the solution and particularly convenient and intuitive given the obvious relationship with the solution in the scalar case, the obvious difficulty is in computing the exponential because it is defined as an infinite series. Using a computer, it would be relatively easy to get a decent approximation by computing a partial sum with many terms, however, to do it "by hand" would be very difficult.[4] Fortunately, some computations related to diagonalization are useful in this case.

Assuming A has a full set of linearly independent eigenvectors, then, as in Section 6.6, if the columns of T are the eigenvectors of A, then

$$\Lambda = T^{-1}AT$$

is diagonal with the eigenvalues along the diagonal. Solving this equation for A gives

$$A = T\Lambda T^{-1}$$

and substituting this into the definition of the exponential of A gives

$$e^{At} = I + \left(T\Lambda T^{-1}\right)t + \frac{1}{2}\left(T\Lambda T^{-1}\right)^2 t^2 + \frac{1}{3!}\left(T\Lambda T^{-1}\right)^3 t^3 + \cdots.$$

Observe that

$$\left(T\Lambda T^{-1}\right)^n = T\Lambda T^{-1}T\Lambda T^{-1}\cdots T\Lambda T^{-1}$$
$$= T\Lambda^n T^{-t}$$

and that

$$\Lambda = \begin{bmatrix} \lambda_1 & 0 & \cdots & 0 \\ 0 & \lambda_2 & \cdots & 0 \\ \vdots & & \ddots & \vdots \\ 0 & 0 & \cdots & \lambda_n \end{bmatrix} \implies \Lambda^n = \begin{bmatrix} \lambda_1^n & 0 & \cdots & 0 \\ 0 & \lambda_2^n & \cdots & 0 \\ \vdots & & \ddots & \vdots \\ 0 & 0 & \cdots & \lambda_n^n \end{bmatrix}.$$

This decouples the diagonal terms in the series and the exponential of Λ becomes

---

[4] Both MATLAB and Octave have the expm() function to compute matrix exponentials.

$$
e^{\Lambda t} = \begin{bmatrix} 1 & 0 & 0 & 0 \\ 0 & 1 & 0 & 0 \\ \vdots & & \ddots & \vdots \\ 0 & 0 & 0 & 1 \end{bmatrix} + \begin{bmatrix} \lambda_1 & 0 & \cdots & 0 \\ 0 & \lambda_2 & \cdots & 0 \\ \vdots & & \ddots & \vdots \\ 0 & 0 & \cdots & \lambda_n \end{bmatrix} t + \frac{1}{2} \begin{bmatrix} \lambda_1 & 0 & \cdots & 0 \\ 0 & \lambda_2 & \cdots & 0 \\ \vdots & & \ddots & \vdots \\ 0 & 0 & \cdots & \lambda_n \end{bmatrix}^2 t^2 + \cdots
$$

$$
= \begin{bmatrix} 1 & 0 & 0 & 0 \\ 0 & 1 & 0 & 0 \\ \vdots & & \ddots & \vdots \\ 0 & 0 & 0 & 1 \end{bmatrix} + \begin{bmatrix} \lambda_1 t & 0 & \cdots & 0 \\ 0 & \lambda_2 t & \cdots & 0 \\ \vdots & & \ddots & \vdots \\ 0 & 0 & \cdots & \lambda_n 5 \end{bmatrix} + \begin{bmatrix} \frac{1}{2}\lambda_1^2 t^2 & 0 & \cdots & 0 \\ 0 & \frac{1}{2}\lambda_2^2 t^2 & \cdots & 0 \\ \vdots & & \ddots & \vdots \\ 0 & 0 & \cdots & \frac{1}{2}\lambda_n^2 t^2 \end{bmatrix} + \cdots
$$

$$
= \begin{bmatrix} e^{\lambda_1 t} & 0 & \cdots & 0 \\ 0 & e^{\lambda_2 t} & \cdots & 0 \\ \vdots & & \ddots & \vdots \\ 0 & 0 & \cdots & e^{\lambda_n t} \end{bmatrix}. \tag{6.27}
$$

Using this fact, if A has a linearly independent set of n eigenvalues, then

$$
\begin{aligned}
e^{At} &= I + \left(T\Lambda T^{-1}\right)t + \frac{1}{2}\left(T\Lambda T^{-1}\right)^2 t^2 + \frac{1}{3!}\left(T\Lambda T^{-1}\right)^3 t^3 + \cdots \\
&= TT^{-1} + T\Lambda T^{-1}t + \frac{1}{2}T\Lambda^2 T^{-1}t^2 + \frac{1}{3!}T\Lambda^3 T^{-1}t^3 + \cdots \\
&= T\left(I + \Lambda t + \frac{1}{2}\Lambda^2 t^2 + \frac{1}{3!}\Lambda^3 t^3 + \cdots\right)T^{-1} \\
&= Te^{\Lambda t}T^{-1},
\end{aligned}
$$

which is easy to compute using Equation (6.27). Thus the solution to $\dot{\xi} = A\xi$, in a computable form, is

$$
\boxed{\xi(t) = Te^{\Lambda t}T^{-1}\xi(0).} \tag{6.28}
$$

Remark 6.2. Observe that other than including the definition of the matrix exponential, this is equivalent to, and involves many of the same steps, as diagonalizing the system. It just leaves out the explicit step of defining ψ, which is easily seen by multiplying Equation (6.28) on the left by T^{-1}.

Example 6.23. Solve $\dot{\xi} = A\xi$ where

$$
A = \begin{bmatrix} -4 & 1 & 0 \\ 0 & -6 & 0 \\ 0 & 0 & -2 \end{bmatrix}, \quad \xi(0) = \begin{bmatrix} 3 \\ 4 \\ 3 \end{bmatrix}.
$$

This is the same matrix considered previously in Examples 6.4 and 6.22, so we have from before that

$$
T = \begin{bmatrix} 0 & 1 & -1 \\ 0 & 0 & 2 \\ 1 & 0 & 0 \end{bmatrix}
$$

and

$$\Lambda = T^{-1}AT = \begin{bmatrix} -2 & 0 & 0 \\ 0 & -4 & 0 \\ 0 & 0 & -6 \end{bmatrix}.$$

Hence,

$$e^{\Lambda t} = \begin{bmatrix} e^{-2t} & 0 & 0 \\ 0 & e^{-4t} & 0 \\ 0 & 0 & e^{-6t} \end{bmatrix}$$

and

$$e^{At} = \begin{bmatrix} 0 & 1 & -1 \\ 0 & 0 & 2 \\ 1 & 0 & 0 \end{bmatrix} \begin{bmatrix} e^{-2t} & 0 & 0 \\ 0 & e^{-4t} & 0 \\ 0 & 0 & e^{-6t} \end{bmatrix} \begin{bmatrix} 0 & 0 & 1 \\ 1 & \frac{1}{2} & 0 \\ 0 & \frac{1}{2} & 0 \end{bmatrix} = \begin{bmatrix} e^{-4t} & \frac{1}{2}e^{-4t} - \frac{1}{2}e^{-6t} & 0 \\ 0 & e^{-6t} & 0 \\ 0 & 0 & e^{-2t} \end{bmatrix}.$$

Thus,

$$\xi(t) = e^{At}\xi(0) = \begin{bmatrix} 5e^{-4t} - 2e^{-6t} \\ 4e^{-6t} \\ 3e^{-3t} \end{bmatrix},$$

which is the same answer as was obtained in Example 6.4.

6.8 Nonhomogeneous Systems of First-Order Equations

Now we consider how to solve systems of the type

$$\dot{\xi} = A\xi + g(t), \tag{6.29}$$

where $A \in \mathbb{R}^{n \times n}$, $\xi \in \mathbb{R}^n$, and $g(t) \in \mathbb{R}^n$, or in detail

$$\frac{d}{dt} \begin{bmatrix} \xi_1 \\ \xi_2 \\ \vdots \\ \xi_n \end{bmatrix} = \begin{bmatrix} a_{11} & a_{12} & \cdots & a_{1n} \\ a_{21} & a_{22} & \cdots & a_{2n} \\ \vdots & \vdots & \ddots & \vdots \\ a_{n1} & a_{n2} & \cdots & a_{nn} \end{bmatrix} \begin{bmatrix} \xi_1 \\ \xi_2 \\ \vdots \\ \xi_n \end{bmatrix} + \begin{bmatrix} g_1(t) \\ g_2(t) \\ \vdots \\ g_n(t) \end{bmatrix}. \tag{6.30}$$

First consider a mechanical example that gives rise to equations of this nature.

Example 6.24. As an example of a type of system that is modeled by such a set of equations, consider again the system illustrated in Figure 6.1, but unlike before we do not assume that $F(t) = 0$. As before, if

$$\xi_1 = x_1, \quad \xi_2 = \dot{x}_1, \quad \xi_3 = x_2, \quad \xi_4 = \dot{x}_2$$

then the equations of motion given in Equation (6.1) are equivalent to

$$\frac{d}{dt}\begin{bmatrix}\xi_1\\\xi_2\\\xi_3\\\xi_4\end{bmatrix} = \begin{bmatrix} 0 & 1 & 0 & 0\\ -\frac{k_1+k_2}{m_1} & -\frac{b_1+b_2}{m_1} & \frac{k_2}{m_1} & \frac{b_2}{m_1}\\ 0 & 0 & 0 & 1\\ \frac{k_2}{m_2} & \frac{b_2}{m_2} & -\frac{k_2}{m_2} & -\frac{b_2}{m_2}\end{bmatrix}\begin{bmatrix}\xi_1\\\xi_2\\\xi_3\\\xi_4\end{bmatrix} + \begin{bmatrix}0\\0\\0\\\frac{F(t)}{m_2}\end{bmatrix}.$$

The following three methods are appropriate for solving nonhomogeneous systems of first-order linear ordinary differential equations.

6.8.1 Diagonalization

This method uses the coordinate transformation that diagonalizes the differential equations considered in Section 6.6, with the only complication that the effect of the transformation on the inhomogeneous term must be considered. As before, if the columns of T are the eigenvectors of A and we let

$$\xi = T\psi,$$

then substituting into Equation (6.29) gives

$$T\dot{\psi} = AT\psi + g(t),$$

or

$$\dot{\psi} = T^{-1}AT\psi + T^{-1}g(t).$$

This decouples the ψ equations which can be solved individually. In detail, this looks like

$$\frac{d}{dt}\begin{bmatrix}\psi_1\\\psi_2\\\psi_3\\\vdots\\\psi_n\end{bmatrix} = \begin{bmatrix}\lambda_1 & 0 & 0 & \cdots & 0\\ 0 & \lambda_2 & 0 & \cdots & 0\\ 0 & 0 & \lambda_3 & \cdots & 0\\ \vdots & \vdots & \vdots & \ddots & \vdots\\ 0 & 0 & 0 & \cdots & \lambda_n\end{bmatrix}\begin{bmatrix}\psi_1\\\psi_2\\\psi_3\\\vdots\\\psi_n\end{bmatrix} + T^{-1}\begin{bmatrix}g_1(t)\\g_2(t)\\g_3(t)\\\vdots\\g_n(t)\end{bmatrix}$$

$$= \begin{bmatrix}\lambda_1\psi_1\\\lambda_2\psi_2\\\lambda_3\psi_3\\\vdots\\\lambda_n\psi_n\end{bmatrix} + T^{-1}\begin{bmatrix}g_1(t)\\g_2(t)\\g_3(t)\\\vdots\\g_n(t)\end{bmatrix} = \begin{bmatrix}\lambda_1\psi_1\\\lambda_2\psi_2\\\lambda_3\psi_3\\\vdots\\\lambda_n\psi_n\end{bmatrix} + \begin{bmatrix}h_1(t)\\h_2(t)\\h_3(t)\\\vdots\\h_n(t)\end{bmatrix},$$

(6.31)

where $h(t) = T^{-1}g(t)$.

The significance of Equation (6.31) is that each of the ψ_i equations is *decoupled* and in the form of

$$\dot{\psi}_i = \lambda_i\psi_i + h_i(t).$$

Hence, each can be solved independently using the appropriate method from Chapter 2. For example, using an integrating factor

$$\frac{d}{dt}\psi_i - \lambda_i\psi_i = h_i(t)$$

$$e^{-\lambda_i t}\left(\frac{d}{dt}\psi_i - \lambda_i\psi_i\right) = e^{-\lambda_i t}h_i(t)$$

$$\frac{d}{dt}\left(e^{-\lambda_i t}\psi_i\right) = e^{-\lambda_i t}h_i(t).$$

Hence, integrating both sides gives

$$\int_0^t \frac{d}{d\tau}\left(e^{-\lambda_i \tau}\psi_i(\tau)\right)d\tau = \int_0^t e^{-\lambda_i \tau}h_i(\tau)d\tau$$

$$e^{-\lambda_i t}\psi_i(t) - \psi_i(0) = \int_0^t e^{-\lambda_i \tau}h_i(\tau)d\tau.$$

Hence

$$\psi_i(t) = e^{\lambda_i t}\int_0^t e^{-\lambda_i \tau}h_i(\tau)d\tau + \psi_i(0)e^{\lambda_i t},$$

if the initial condition is specified or

$$\psi_i(t) = e^{\lambda_i t}\int_0^t e^{-\lambda_i \tau}h_i(\tau)d\tau + c_i e^{\lambda_i t},$$

if the general solution is desired. After solving all the $\psi_i(t)$ equations, the solution for the ξ variables is computed using the original equation $\xi = T\psi$.

Example 6.25. Determine the general solution to

$$\frac{d}{dt}\begin{bmatrix} \xi_1 \\ \xi_2 \\ \xi_3 \end{bmatrix} = \begin{bmatrix} 1 & 1 & 1 \\ 2 & 1 & -1 \\ -8 & -5 & -3 \end{bmatrix}\begin{bmatrix} \xi_1 \\ \xi_2 \\ \xi_3 \end{bmatrix} + \begin{bmatrix} 0 \\ 0 \\ \cos t \end{bmatrix}.$$

Computing the eigenvalues and eigenvectors gives $\lambda_1 = -2$, $\lambda_2 = -1$, and $\lambda_3 = 2$ with

$$\hat{\xi}^1 = \begin{bmatrix} -4 \\ 5 \\ 7 \end{bmatrix}, \quad \hat{\xi}^2 = \begin{bmatrix} -3 \\ 4 \\ 2 \end{bmatrix}, \quad \hat{\xi}^3 = \begin{bmatrix} 0 \\ -1 \\ 1 \end{bmatrix}.$$

Thus

$$T = \begin{bmatrix} -4 & -3 & 0 \\ 5 & 4 & -1 \\ 7 & 2 & 1 \end{bmatrix} \iff T^{-1} = \begin{bmatrix} \frac{1}{2} & \frac{1}{4} & \frac{1}{4} \\ -1 & -\frac{1}{3} & -\frac{1}{3} \\ -\frac{3}{2} & -\frac{13}{12} & -\frac{1}{12} \end{bmatrix}.$$

Computing $T^{-1}AT$ and $T^{-1}g(t)$ gives the following equations for ψ

$$\frac{d}{dt}\begin{bmatrix} \psi_1 \\ \psi_2 \\ \psi_3 \end{bmatrix} = \begin{bmatrix} -2 & 0 & 0 \\ 0 & -1 & 0 \\ 0 & 0 & 2 \end{bmatrix} \begin{bmatrix} \psi_1 \\ \psi_2 \\ \psi_3 \end{bmatrix} + \begin{bmatrix} \frac{1}{4}\cos t \\ -\frac{1}{3}\cos t \\ -\frac{1}{12}\cos t \end{bmatrix},$$

or as individual equations

$$\dot{\psi}_1 = -2\psi_1 + \frac{1}{4}\cos t$$

$$\dot{\psi}_2 = -\psi_2 - \frac{1}{3}\cos t$$

$$\dot{\psi}_3 = 2\psi_3 - \frac{1}{12}\cos t.$$

The solutions to these equations are

$$\psi_1 = e^{-2t}\int_0^t e^{2t}\frac{1}{4}\cos\tau d\tau + \psi_1(0)e^{-2t}$$

$$\psi_2 = -e^{-t}\int_0^t e^t\frac{1}{3}\cos\tau d\tau + \psi_2(0)e^{-t}$$

$$\psi_3 = -e^{2t}\int_0^t e^{-2t}\frac{1}{12}\cos\tau d\tau + \psi_3(0)e^{2t},$$

or

$$\psi_1(t) = c_1 e^{-2t} + \frac{1}{10}\cos t + \frac{1}{20}\sin t$$

$$\psi_2(t) = c_2 e^{-t} - \frac{1}{6}\cos t - \frac{1}{6}\sin t$$

$$\psi_3(t) = c_3 e^{2t} + \frac{1}{30}\cos t - \frac{1}{60}\sin t.$$

The final solution is computed by determining $\xi = T\psi$, which is

$$\begin{bmatrix} \xi_1 \\ \xi_2 \\ \xi_3 \end{bmatrix} = \begin{bmatrix} -4 & -3 & 0 \\ 5 & 4 & -1 \\ 7 & 2 & 1 \end{bmatrix} \begin{bmatrix} c_1 e^{-2t} + \frac{1}{10}\cos t + \frac{1}{20}\sin t \\ c_2 e^{-t} - \frac{1}{6}\cos t - \frac{1}{6}\sin t \\ c_3 e^{2t} + \frac{1}{30}\cos t - \frac{1}{60}\sin t \end{bmatrix} = \begin{bmatrix} -4\psi_1(t) - 3\psi_2(t) \\ 5\psi_1(t) + 4\psi_2(t) - \psi_3(t) \\ 7\psi_1(t) + 2\psi_2(t) + \psi_3(t) \end{bmatrix},$$

which gives

$$\xi_1(t) = -4c_1 e^{-2t} - \frac{2}{5}\cos t - \frac{1}{5}\sin t - 3c_2 e^{-t} + \frac{1}{2}\cos t + \frac{1}{2}\sin t$$

$$\xi_2(t) = 5c_1 e^{-2t} + \frac{1}{2}\cos t + \frac{1}{4}\sin t + 4c_2 e^{-t} - \frac{2}{3}\cos t - \frac{2}{3}\sin t$$

$$- c_3 e^{2t} - \frac{1}{30}\cos t + \frac{1}{60}\sin t$$

$$\xi_3(t) = 7c_1 e^{-2t} + \frac{7}{10}\cos t + \frac{7}{20}\sin t + 2c_2 e^{-t} - \frac{1}{3}\cos t - \frac{1}{3}\sin t$$

$$+ c_3 e^{2t} + \frac{1}{30}\cos t - \frac{1}{60}\sin t.$$

6.8.2 Undetermined Coefficients

Recall that the method of undetermined coefficients from Section 3.4.1 was based upon the fact that derivatives of linear combinations of functions of the form

1. $\sin \omega t$ and $\cos \omega t$
2. $e^{\alpha t}$
3. $\alpha_0 t^n + \alpha_1 t^{n-1} + \alpha_2 t^{n-2} + \cdots + \alpha_{n-1} t + \alpha_n$
4. Products or sums of them

are linear combinations of the same functions. Thus when the nonhomogeneous term contains functions of this type, the particular solution of an ordinary differential equation will be a combination of the same type of function. There are two slight complications that are necessary to distinguish the approach for systems of first-order equations from one scalar second-order system.

6.8.2.1 General Form of Particular Solution

The first complication is that even though the nonhomogeneous term may appear in one component of the differential equation, the form of the solution must have undetermined coefficients for all of the components, as is illustrated by the following example.

Example 6.26. Find the general solution to

$$\frac{d}{dt}\begin{bmatrix} \xi_1 \\ \xi_2 \end{bmatrix} = \begin{bmatrix} 2 & 1 \\ 0 & 3 \end{bmatrix}\begin{bmatrix} \xi_1 \\ \xi_2 \end{bmatrix} + \begin{bmatrix} 0 \\ \cos 4t \end{bmatrix}.$$

In the scalar case, the assumed form of the solution would simply be $x_p(t) = a\cos 4t + b\sin 4t$, so for this problem we assume

$$\xi_p(t) = a\cos 4t + b\sin 4t = \begin{bmatrix} a_1 \\ a_2 \end{bmatrix}\cos 4t + \begin{bmatrix} b_1 \\ b_2 \end{bmatrix}\sin 4t.$$

Note that we are looking for $\cos 4t$ and $\sin 4t$ terms for *both* rows of the equation. As we show, a_1 and b_1 are not zero, so it is incorrect only to assume the corresponding terms for undetermined coefficients in the rows where they appear in the inhomogeneous term.

The rest of the procedure is exactly as before. Substitute the assumed form of the particular solution into the differential equations and equate the coefficients of different functions of t. Thus,

$$\dot{\xi}_p(t) = -4a\sin 4t + 4b\cos 4t,$$

and substituting gives

$$\begin{bmatrix} -4a_1\sin 4t + 4b_1\cos 4t \\ -4a_2\sin 4t + 4b_2\cos 4t \end{bmatrix} = \begin{bmatrix} 2 & 1 \\ 0 & 3 \end{bmatrix} \begin{bmatrix} a_1\cos 4t + b_1\sin 4t \\ a_2\cos 4t + b_2\sin 4t \end{bmatrix} + \begin{bmatrix} 0 \\ \cos 4t \end{bmatrix}.$$

This must be true for all time, therefore the coefficients of the sine and cosine terms in each equation must be equal. Thus, the coefficients are determined by the following four equations,

$$\begin{aligned}
\text{sine term, first equation} &\Longrightarrow & -4a_1 &= 2b_1 + b_2 \\
\text{cosine term, first equation} &\Longrightarrow & 4b_1 &= 2a_1 + a_2 \\
\text{sine term, second equation} &\Longrightarrow & -4a_2 &= 3b_2 \\
\text{cosine term, second equation} &\Longrightarrow & 4b_2 &= 3a_2 + 1.
\end{aligned}$$

Solving these gives

$$a_1 = -\frac{1}{50}, \quad a_2 = -\frac{3}{25}, \quad b_1 = -\frac{1}{25}, \quad b_2 = \frac{4}{25}.$$

Thus the particular solution is

$$\xi_p(t) = \begin{bmatrix} -\frac{1}{50} \\ -\frac{3}{25} \end{bmatrix} \cos 4t + \begin{bmatrix} -\frac{1}{25} \\ -\frac{4}{25} \end{bmatrix} \sin 4t.$$

To compute the general solution, the homogeneous solution, that is, the solution to $\dot{\xi} = A\xi$, is needed. A simple computation shows that the eigenvalues and eigenvectors of A are

$$\lambda_1 = 3, \quad \hat{\xi}^1 = \begin{bmatrix} 1 \\ 1 \end{bmatrix}, \quad \lambda_2 = 2, \quad \hat{\xi}^2 = \begin{bmatrix} 1 \\ 0 \end{bmatrix}.$$

Thus, the general solution is

$$\xi(t) = c_1 \begin{bmatrix} 1 \\ 1 \end{bmatrix} e^{3t} + c_2 \begin{bmatrix} 1 \\ 0 \end{bmatrix} e^{2t} + \begin{bmatrix} -\frac{1}{50} \\ -\frac{3}{25} \end{bmatrix} \cos 4t + \begin{bmatrix} -\frac{1}{25} \\ -\frac{4}{25} \end{bmatrix} \sin 4t.$$

In the previous example, note that the sine and cosine terms appear in both components of the solution even though the nonhomogeneous term contains $\cos 4t$ only in the second term. This is due to the fact that the equations are coupled, and the effect of the inhomogeneity is not limited to the line in which it appears.

6.8.2.2 Equivalent Homogeneous Solution and Nonhomogeneous Term

The second complication is when the nonhomogeneous term is the exponential of an eigenvalue of the matrix A. When confronted with this problem in Chapter 3, the approach was to multiply the assumed form of the particular solution by the dependent variable.

Remark 6.3. An important distinction with systems of equations is that it is necessary to include both the "original" form (not multiplied by t) as well as the form multiplied by t in the assumed form for ξ_p. This is because the equations are coupled and the form not multiplied by t may be needed by some of the equations and the form multiplied by t may be needed for the others. Furthermore, because part of the assumed form will be a homogeneous solution for some of the equations, it may be that there is not a unique solution for the coefficients.

Exercise 6.11 is included to illustrate that both the original form and the form multiplied by t must be included in x_p. The following example illustrates the correct approach.

Example 6.27. Consider

$$\frac{d}{dt}\begin{bmatrix} \xi_1 \\ \xi_2 \end{bmatrix} = \begin{bmatrix} 2 & 1 \\ 0 & 3 \end{bmatrix}\begin{bmatrix} \xi_1 \\ \xi_2 \end{bmatrix} + \begin{bmatrix} 0 \\ e^{3t} \end{bmatrix}.$$

Observing that $\lambda = 3$ is an eigenvalue of A we assume

$$\xi_p(t) = ate^{3t} + be^{3t} = \begin{bmatrix} a_1 \\ a_2 \end{bmatrix} te^{3t} + \begin{bmatrix} b_1 \\ b_2 \end{bmatrix} e^{3t}.$$

Thus

$$\dot{\xi}_p(t) = 3ate^{3t} + ae^{3t} + 3be^{3t}$$

and substituting into the differential equation gives

$$\left(3t\begin{bmatrix} a_1 \\ a_2 \end{bmatrix} + \begin{bmatrix} a_1 \\ a_2 \end{bmatrix} + 3\begin{bmatrix} b_1 \\ b_2 \end{bmatrix}\right)e^{3t} = \begin{bmatrix} 2 & 1 \\ 0 & 3 \end{bmatrix}\left(\begin{bmatrix} a_1 \\ a_2 \end{bmatrix} t + \begin{bmatrix} b_1 \\ b_2 \end{bmatrix}\right)e^{3t} + \begin{bmatrix} 0 \\ e^{3t} \end{bmatrix}.$$

Equating coefficients of e^{3t} and te^{3t} in each equation gives

$$a_1 + 3b_1 = 2b_1 + b_2$$
$$3a_1 = 2a_1 + a_2$$
$$a_2 + 3b_2 = 3b_2 + 1$$
$$3a_2 = 3a_2.$$

Observe that the last equation is satisfied for any value of a_2 so there really are only three equations. The reason there are fewer than four equations, and hence this is not a unique solution, is because the vector b in the assumed form of the solution must be an eigenvector of A and hence can be combined in any linear way with one of the homogeneous solutions. One solution to the above three equations is

$$a_1 = 1, \quad a_2 = 1, \quad b_1 = 0, \quad b_2 = 1,$$

and hence

$$\xi_p(t) = \begin{bmatrix} 1 \\ 1 \end{bmatrix} te^{3t} + \begin{bmatrix} 0 \\ 1 \end{bmatrix} e^{3t}. \tag{6.32}$$

This particular solution is not unique. Indeed,

$$a_1 = 1, \quad a_2 = 1, \quad b_1 = -1, \quad b_2 = 0,$$

also work giving

$$\xi_p(t) = \begin{bmatrix} 1 \\ 1 \end{bmatrix} te^{3t} + \begin{bmatrix} -1 \\ 0 \end{bmatrix} e^{3t}. \tag{6.33}$$

The reason both particular solutions work is that when they are combined with the homogeneous solution, they yield the same solution. In particular, the homogeneous solution is

$$\xi_h(t) = c_1 \begin{bmatrix} 1 \\ 0 \end{bmatrix} e^{2t} + c_2 \begin{bmatrix} 1 \\ 1 \end{bmatrix} e^{3t}.$$

Then the general solution using the particular solution from Equation (6.32) gives

$$\xi(t) = c_1 \begin{bmatrix} 1 \\ 0 \end{bmatrix} e^{2t} + c_2 \begin{bmatrix} 1 \\ 1 \end{bmatrix} e^{3t} + \begin{bmatrix} 1 \\ 1 \end{bmatrix} te^{3t} + \begin{bmatrix} 0 \\ 1 \end{bmatrix} e^{3t},$$

and the general solution using the particular solution from Equation (6.33) gives

$$\xi(t) = \hat{c}_1 \begin{bmatrix} 1 \\ 0 \end{bmatrix} e^{2t} + \hat{c}_2 \begin{bmatrix} 1 \\ 1 \end{bmatrix} e^{3t} + \begin{bmatrix} 1 \\ 1 \end{bmatrix} te^{3t} + \begin{bmatrix} -1 \\ 0 \end{bmatrix} e^{3t}.$$

For c_2 from the first equation and $\hat{c}_2$ from the second equation, if $\hat{c}_2 = c_2 + 1$ the equations are identical.

6.8.3 Variation of Parameters

With all the complications involved in the method of undetermined coefficients, one may be hesitant to even venture into the realm of variation of parameters because, at least in Chapter 3 the derivation was rather complicated. Thankfully, in the case of nonhomogeneous systems of first-order equations, variation of parameters is even more straightforward than in the scalar second-order case.

Given

$$\dot{\xi} = A\xi + g(t), \tag{6.34}$$

where

$$A \in \mathbb{R}^{n \times n}, \quad \xi \in \mathbb{R}^n, \quad g(t) \in \mathbb{R}^n,$$

assume that $\xi_{1_h}, \xi_{2_h}, \ldots, \xi_{n_h}$ are n linearly independent homogeneous solutions to Equation (6.34); that is, they satisfy

$$\dot{\xi}_{i_h} = A\xi_{i_h}.$$

Because it is useful subsequently, we first construct and define a matrix $\Xi(t)$ where the columns of $\Xi(t)$ are the homogeneous solutions $\xi_{i_h}(t)$.

Definition 6.5. Let $\xi_{1_h}, \xi_{2_h}, \ldots, \xi_{n_h}$ satisfy

$$\dot{\xi}_{i_h} = A\xi_{i_h}.$$

The *fundamental matrix solution* is the matrix

$$\Xi(t) = \begin{bmatrix} \xi_{1_h}(t) & \xi_{2_h}(t) & \cdots & \xi_{n_h}(t) \end{bmatrix};$$

that is, the columns of $\Xi(t)$ are the homogeneous solutions.

Example 6.28. Consider the general solution to $\dot{\xi} = A\xi$ where

$$A = \begin{bmatrix} 2 & 0 & 0 & 0 \\ 0 & 2 & 0 & 0 \\ 0 & 0 & 3 & 1 \\ 0 & 0 & 0 & 3 \end{bmatrix}.$$

Skipping the details the general solution is

$$\xi(t) = c_1 \hat{\xi}^1 e^{\lambda_1 t} + c_2 \hat{\xi}^2 e^{\lambda_1 t} + c_3 \hat{\xi}^3 e^{\lambda_3 t} + c_4 \left(\hat{\xi}^4 + t \hat{\xi}^3 \right) e^{\lambda_3 t}$$

$$= c_1 \begin{bmatrix} 1 \\ 0 \\ 0 \\ 0 \end{bmatrix} e^{2t} + c_2 \begin{bmatrix} 0 \\ 1 \\ 0 \\ 0 \end{bmatrix} e^{2t} + c_3 \begin{bmatrix} 0 \\ 0 \\ 1 \\ 0 \end{bmatrix} e^{3t} + c_4 \left(\begin{bmatrix} 0 \\ 0 \\ 0 \\ 1 \end{bmatrix} + t \begin{bmatrix} 0 \\ 0 \\ 1 \\ 0 \end{bmatrix} \right) e^{3t}.$$

Each term that is multiplied by a constant c_i is a homogeneous solution, thus we can construct a matrix with each one as a column to construct the fundamental matrix solution

$$\Xi(t) = \left[\hat{\xi}^1 e^{\lambda_1 t} \; \hat{\xi}^2 e^{\lambda_2 t} \; \hat{\xi}^3 e^{\lambda_3 t} \; \left(\hat{\xi}^4 + t \hat{\xi}^3 \right) e^{\lambda_3 t} \right]$$

$$= \left[\begin{bmatrix} 1 \\ 0 \\ 0 \\ 0 \end{bmatrix} e^{2t} \; \begin{bmatrix} 0 \\ 1 \\ 0 \\ 0 \end{bmatrix} e^{2t} \; \begin{bmatrix} 0 \\ 0 \\ 1 \\ 0 \end{bmatrix} e^{3t} \; \left(\begin{bmatrix} 0 \\ 0 \\ 0 \\ 1 \end{bmatrix} + t \begin{bmatrix} 0 \\ 0 \\ 1 \\ 0 \end{bmatrix} \right) e^{3t} \right]$$

$$= \begin{bmatrix} e^{2t} & 0 & 0 & 0 \\ 0 & e^{2t} & 0 & 0 \\ 0 & 0 & e^{3t} & te^{3t} \\ 0 & 0 & 0 & e^{3t} \end{bmatrix}.$$

The fundamental matrix solution has one important property that is used in the derivation of the variation of parameters solution; namely, the whole matrix satisfies the homogeneous equation. In other words, if $\Xi(t)$ is the fundamental matrix solution to $\dot{\xi} = A\xi$ then

$$\dot{\Xi} = A\Xi.$$

This is true because each column of $\Xi(t)$ is a homogeneous solution and is illustrated by the following example.

Example 6.29. From Example 6.28 we have

$$\Xi(t) = \begin{bmatrix} e^{2t} & 0 & 0 & 0 \\ 0 & e^{2t} & 0 & 0 \\ 0 & 0 & e^{3t} & te^{3t} \\ 0 & 0 & 0 & e^{3t} \end{bmatrix}$$

so

$$\dot{\Xi}(t) = \begin{bmatrix} 2e^{2t} & 0 & 0 & 0 \\ 0 & 2e^{2t} & 0 & 0 \\ 0 & 0 & 3e^{3t} & 3te^{3t}+e^{3t} \\ 0 & 0 & 0 & 3e^{3t} \end{bmatrix} = \begin{bmatrix} 2 & 0 & 0 & 0 \\ 0 & 2 & 0 & 0 \\ 0 & 0 & 3 & 1 \\ 0 & 0 & 0 & 3 \end{bmatrix} \begin{bmatrix} e^{2t} & 0 & 0 & 0 \\ 0 & e^{2t} & 0 & 0 \\ 0 & 0 & e^{3t} & te^{3t} \\ 0 & 0 & 0 & e^{3t} \end{bmatrix}.$$

Thus $\dot{\Xi} = A\Xi$.

Similar to the approach for second-order equations, the approach to find the particular solution for a nonhomogeneous system of first-order equations is to assume that the particular solution is of the form of

$$\xi_p(t) = \Xi(t)u(t),$$

where $u(t)$ is a vector of unknown functions. To determine $u(t)$, substitute into Equation (6.34). First note that (dropping the explicit dependence on t)

$$\dot{\xi}_p = \dot{\Xi}u + \Xi\dot{u}.$$

Substituting into Equation (6.34) gives

$$\dot{\Xi}u + \Xi\dot{u} = A\Xi u + g.$$

Inasmuch as

$$\dot{\Xi} = A\Xi \quad \Longrightarrow \quad \dot{\Xi}u = A\Xi u,$$

so

$$\Xi \dot{u} = g.$$

Because Ξ contains n linearly independent solutions, it is invertible and hence

$$\dot{u} = \Xi^{-1}g \quad \Longrightarrow \quad u(t) = \int_{t_0}^{t} \Xi^{-1}(\tau)g(\tau)d\tau.$$

Substituting into the assumed form of the particular solution gives a complete expression for the particular solution as

$$\xi_p(t) = \Xi \int_{t_0}^{t} \Xi^{-1}(\tau)g(\tau)d\tau.$$

Note that to even compute the particular solution we need the fundamental matrix which contains a full set of homogeneous solutions. Because any linear combination of the homogeneous solutions can be expressed as

$$c_1\xi_{1_h} + c_2\xi_{2_h} + \cdots + c_n\xi_{n_h} = \Xi(t)c,$$

where

$$c = \begin{bmatrix} c_1 \\ c_2 \\ \vdots \\ c_n \end{bmatrix},$$

the general solution to Equation (6.34) is

$$\boxed{\xi(t) = \Xi(t)c + \Xi(t)\int_{t_0}^{t} \Xi^{-1}(\tau)g(\tau)d\tau.} \tag{6.35}$$

Finally, if the initial conditions $\xi(t_0)$ are specified, then

$$\xi(t_0) = \Xi(t_0)c$$

because the integral with the same upper and lower limits is zero. Hence

$$c = \Xi^{-1}(t_0)\xi(t_0)$$

and substituting into the general solution gives the entire answer as

$$\boxed{\xi(t) = \Xi(t)\Xi^{-1}(t_0)\xi(t_0) + \Xi(t)\int_{t_0}^{t} \Xi^{-1}(\tau)g(\tau)d\tau.} \tag{6.36}$$

An example illustrates the straightforward application of this method.

Example 6.30. Solve

$$\frac{d}{dt}\begin{bmatrix}\xi_1\\\xi_2\end{bmatrix} = \begin{bmatrix}-3 & 1\\1 & -3\end{bmatrix}\begin{bmatrix}\xi_1\\\xi_2\end{bmatrix} + \begin{bmatrix}e^{-4t}\\0\end{bmatrix}.$$

The eigenvalues and eigenvectors for the matrix are

$$\lambda_1 = -4, \quad \hat{\xi}^1 = \begin{bmatrix}-1\\1\end{bmatrix}, \quad \lambda_2 = -2 \quad \hat{\xi}^2 = \begin{bmatrix}1\\1\end{bmatrix},$$

and thus

$$\Xi(t) = \begin{bmatrix}-e^{-4t} & e^{-2t}\\e^{-4t} & e^{-2t}\end{bmatrix}.$$

A simple computation determines that

$$\Xi^{-1}(t) = \frac{1}{2}\begin{bmatrix}-e^{4t} & e^{4t}\\e^{2t} & e^{2t}\end{bmatrix},$$

and

$$\Xi^{-1}(t)g(t) = \begin{bmatrix}-\frac{1}{2}\\\frac{1}{2}e^{-2t}\end{bmatrix}.$$

Assuming that $t_0 = 0$,

$$\int_0^t \Xi^{-1}(\tau)g(\tau)d\tau = \int_0^t \begin{bmatrix}-\frac{1}{2}\\\frac{1}{2}e^{-2\tau}\end{bmatrix}d\tau = \begin{bmatrix}-\frac{1}{2}\tau\\\frac{1}{4}\left(1-e^{-2t}\right)\end{bmatrix}.$$

Then

$$\Xi(t)\int_0^t \Xi^{-1}(\tau)g(\tau)d\tau = \begin{bmatrix}\frac{1}{4}\left(e^{-2t}+2te^{-4t}-e^{-4t}\right)\\\frac{1}{4}\left(e^{-2t}-2te^{-4t}-e^{-4t}\right)\end{bmatrix}.$$

So finally we have

$$\xi(t) = \Xi(t)c + \Xi(t)\int_{t_0}^t \Xi^{-1}(\tau)g(\tau)d\tau$$

$$= c_1\begin{bmatrix}-e^{-4t}\\e^{-4t}\end{bmatrix} + c_2\begin{bmatrix}e^{-2t}\\e^{-2t}\end{bmatrix} + \begin{bmatrix}\frac{1}{4}\left(e^{-2t}+2te^{-4t}-e^{-4t}\right)\\\frac{1}{4}\left(e^{-2t}-2te^{-4t}-e^{-4t}\right).\end{bmatrix}.$$

6.9 Exercises

It is possible to complete all of these exercises by hand.

6.1. Each of the matrices in this problem has a full set of linearly independent eigenvectors. For each one, indicate whether Theorem 6.1 or 6.2 applies and find the general solution to $\dot{\xi} = A\xi$ for:

$$A_1 = \begin{bmatrix} 12 & -3 \\ 8 & 2 \end{bmatrix}, \qquad A_2 = \begin{bmatrix} 12 & -3 \\ 8 & 1 \end{bmatrix}, \qquad A_3 = \begin{bmatrix} 13 & -6 \\ 16 & -7 \end{bmatrix},$$

$$A_4 = \begin{bmatrix} 7 & -5 \\ 2 & 0 \end{bmatrix}, \qquad A_5 = \begin{bmatrix} -7 & 10 \\ -4 & 7 \end{bmatrix}, \qquad A_6 = \begin{bmatrix} 14 & -5 \\ 40 & -16 \end{bmatrix}.$$

6.2. Each of the matrices in this problem has a full set of linearly independent eigenvectors. For each one, indicate whether Theorem 6.1 or 6.2 applies and find the general solution to $\dot{\xi} = A\xi$ for:

$$A_1 = \begin{bmatrix} 6 & -4 \\ 0 & 2 \end{bmatrix}, \qquad A_2 = \begin{bmatrix} -3 & 0 & 0 \\ 0 & -3 & 1 \\ 0 & 1 & -3 \end{bmatrix}, \qquad A_3 = \begin{bmatrix} -3 & 0 & 0 \\ -1 & -3 & 1 \\ -1 & 1 & -3 \end{bmatrix},$$

$$A_4 = \begin{bmatrix} -8 & 7 & 1 \\ 0 & -1 & 1 \\ 0 & 0 & 0 \end{bmatrix}, \qquad A_5 = \begin{bmatrix} 3 & 2 & 0 & 0 \\ 2 & 3 & 0 & 0 \\ 0 & 0 & 1 & 4 \\ 0 & 0 & 4 & 1 \end{bmatrix}, \qquad A_6 = \begin{bmatrix} -2 & 0 & 0 & 0 \\ 0 & -2 & 0 & 0 \\ 0 & 0 & 0 & 2 \\ 0 & 0 & 2 & 0 \end{bmatrix},$$

$$A_7 = \begin{bmatrix} 2 & 0 & 0 & 0 \\ 0 & 2 & 0 & 9 \\ 0 & 2 & 1 & 4 \\ 0 & -4 & 0 & 14 \end{bmatrix}, \qquad A_8 = \begin{bmatrix} -3 & 1 & 0 \\ 0 & -2 & 0 \\ 1 & 1 & -4 \end{bmatrix}, \qquad A_9 = \begin{bmatrix} 2 & 0 & 3 \\ 0 & -5 & 0 \\ 3 & 0 & 2 \end{bmatrix}.$$

6.3. For A_2, A_3, A_4, A_8, and A_9 in Exercise 6.2, determine the solution if $\xi_1(0) = 1$, $\xi_2(0) = 2$, and $\xi_3(0) = 4$.

6.4. Each of the matrices in this problem has some complex eigenvalues. Determine the general solution to $\dot{\xi} = A\xi$ for:

$$A_1 = \begin{bmatrix} -1 & 2 \\ -2 & -1 \end{bmatrix}, \qquad A_2 = \begin{bmatrix} -1 & 1 \\ -10 & 5 \end{bmatrix}, \qquad A_3 = \begin{bmatrix} -12 & 10 \\ -20 & 16 \end{bmatrix},$$

$$A_4 = \begin{bmatrix} 8 & -10 \\ 4 & -4 \end{bmatrix}, \qquad A_5 = \begin{bmatrix} 4 & 4 \\ -2 & 0 \end{bmatrix}, \qquad A_6 = \begin{bmatrix} -2 & 2 \\ -1 & -4 \end{bmatrix}.$$

6.5. Each of the matrices in this problem has some complex eigenvalues. Determine the general solution to $\dot{\xi} = A\xi$ for:

$$A_1 = \begin{bmatrix} 0 & 1 & 0 & 0 \\ -4 & 4 & 0 & 0 \\ 0 & 0 & 3 & 2 \\ 0 & 0 & -2 & 3 \end{bmatrix}, \quad A_2 = \begin{bmatrix} -\frac{7}{2} & \frac{15}{2} & -3 \\ -\frac{3}{2} & -\frac{1}{2} & 3 \\ 0 & 0 & 1 \end{bmatrix}, \quad A_3 = \begin{bmatrix} -1 & -4 \\ 4 & -1 \end{bmatrix},$$

$$A_4 = \begin{bmatrix} 11 & 0 & 17 \\ 0 & -6 & 0 \\ -2 & 0 & 1 \end{bmatrix}, \quad A_5 = \begin{bmatrix} -5 & 1 & 0 & 0 \\ -1 & -3 & 0 & 0 \\ 0 & 0 & -1 & -4 \\ 0 & 0 & 2 & -5 \end{bmatrix}, \quad A_6 = \begin{bmatrix} -5 & 0 & 0 & 0 & 0 \\ 0 & -3 & 2 & 0 & 0 \\ 0 & -4 & 1 & 0 & 0 \\ 0 & 0 & 0 & -5 & 1 \\ 0 & 0 & 0 & -1 & -7 \end{bmatrix},$$

$$A_7 = \begin{bmatrix} -3 & 0 & 0 \\ 0 & -5 & 6 \\ 0 & -3 & 1 \end{bmatrix}, \quad A_8 = \begin{bmatrix} 2 & 0 & -3 \\ 0 & -5 & 0 \\ 3 & 0 & 2 \end{bmatrix}, \quad A_9 = \begin{bmatrix} -1 & 2 & 0 & 0 \\ -2 & -1 & 0 & 0 \\ 0 & 0 & -1 & 2 \\ 0 & 0 & -2 & -1 \end{bmatrix}.$$

6.6. For A_2, A_4, A_7, and A_8 in Exercise 6.5, determine the solution if $\xi_1(0) = 1$, $\xi_2(0) = 1$, and $\xi_3(0) = 0$.

6.7. Each of the matrices in this problem has some repeated eigenvalues. Determine the general solution to $\dot{\xi} = A\xi$ for:

$$A_1 = \begin{bmatrix} -2 & 1 \\ 0 & -2 \end{bmatrix}, \quad A_2 = \begin{bmatrix} -3 & 1 \\ -1 & -5 \end{bmatrix}, \quad A_3 = \begin{bmatrix} -8 & 1 \\ -4 & -4 \end{bmatrix},$$

$$A_4 = \begin{bmatrix} -11 & 1 \\ -9 & -5 \end{bmatrix}, \quad A_5 = \begin{bmatrix} -11 & 1 \\ -4 & -7 \end{bmatrix}, \quad A_6 = \begin{bmatrix} -7 & 2 \\ -8 & 1 \end{bmatrix}.$$

6.8. Each of the matrices in this problem has some repeated eigenvalues. Determine the general solution to $\dot{\xi} = A\xi$ for:

$$A_1 = \begin{bmatrix} -1 & 0 & 0 \\ 0 & -2 & 0 \\ 0 & 0 & -2 \end{bmatrix}, \quad A_2 = \begin{bmatrix} -1 & 0 & 0 \\ 0 & -2 & 1 \\ 0 & 0 & -2 \end{bmatrix}, \quad A_3 = \begin{bmatrix} -2 & 1 & 0 \\ 0 & -2 & 0 \\ 0 & 1 & -2 \end{bmatrix},$$

$$A_4 = \begin{bmatrix} 6 & 0 & 0 \\ 1 & 5 & 1 \\ 1 & -1 & 7 \end{bmatrix}, \quad A_5 = \begin{bmatrix} -4 & 1 & 0 & 0 & 0 \\ 0 & -4 & 0 & 0 & 0 \\ 0 & 0 & -4 & 0 & 0 \\ 0 & 0 & 0 & -1 & 1 \\ 0 & 0 & 0 & 0 & -1 \end{bmatrix}, \quad A_6 = \begin{bmatrix} -4 & 1 & 0 & 0 & 0 \\ 0 & -4 & 1 & 0 & 0 \\ 0 & 0 & -4 & 0 & 0 \\ 0 & 0 & 0 & -1 & 0 \\ 0 & 0 & 0 & 0 & -1 \end{bmatrix}.$$

6.9. Prove Theorem 6.4 by substituting Equation (6.24) into $\dot{\xi} = A\xi$ and making use of the properties of generalized eigenvectors.

6.10. Find the general solution to

$$\frac{d}{dt} \begin{bmatrix} \xi_1 \\ \xi_2 \end{bmatrix} = \begin{bmatrix} 2 & 1 \\ 0 & 3 \end{bmatrix} \begin{bmatrix} \xi_1 \\ \xi_2 \end{bmatrix} + \begin{bmatrix} 0 \\ e^{-t} \end{bmatrix}.$$

6.11. Consider

$$\frac{d}{dt}\begin{bmatrix} \xi_1 \\ \xi_2 \end{bmatrix} = \begin{bmatrix} 2 & 1 \\ 0 & 3 \end{bmatrix}\begin{bmatrix} \xi_1 \\ \xi_2 \end{bmatrix} + \begin{bmatrix} 0 \\ e^{3t} \end{bmatrix}.$$

- What happens when you assume

$$\xi_p(t) = ae^{3t} = \begin{bmatrix} a_1 \\ a_2 \end{bmatrix} e^{3t}?$$

Explain why it does not work.
- What happens when you assume

$$\xi_p(t) = ate^{3t} = t\begin{bmatrix} a_1 \\ a_2 \end{bmatrix} e^{3t}?$$

Explain why it does not work.

6.12. Determine the solution to $\dot{\xi} = A\xi + g(t)$ where

$$A = \begin{bmatrix} -3 & 1 & 0 \\ 0 & -2 & 0 \\ 1 & 1 & -4 \end{bmatrix}, \quad g(t) = \begin{bmatrix} 0 \\ 0 \\ \cos t \end{bmatrix}$$

- Using the method of undetermined coefficients
- By determining a coordinate transformation that diagonalizes A
- Using the method of variation of parameters

6.13. Determine the solution to $\dot{\xi} = A\xi + g(t)$ where

$$A = \begin{bmatrix} -3 & 0 & 1 \\ 0 & -2 & 0 \\ 1 & 0 & -3 \end{bmatrix}, \quad g(t) = \begin{bmatrix} e^{-4t} \\ 0 \\ 0 \end{bmatrix}$$

- Using the method of undetermined coefficients
- By determining a coordinate transformation that diagonalizes A
- Using the method of variation of parameters.

Because $A = A^T$, make use of the fact that $T^{-1} = T^T$ (as long as you normalize all the eigenvectors to have unit length). Verify this fact by showing that $T^T T = I$.

6.14. Compute the matrix exponential for A_1, A_2, and A_9 from Exercise 6.2. For the initial condition given in Exercise 6.3, A_3 and A_9 verify that

$$\xi(t) = e^{At}\xi(0)$$

is the same solution as was computed in Exercise 6.3.

6.15. When an $n \times n$ matrix does not have a linearly independent set of n eigenvectors, the matrix cannot be diagonalized. However, it is still possible to convert it into a simpler and useful form called *Jordan canonical form*. For the matrices A_3 and A_4 from Exercise 6.8, construct the matrix T from the generalized eigenvectors and compute $T^{-1}AT$. Explain the manner in which the resulting matrix would be useful to solve either a homogeneous or inhomogeneous set of differential equations containing A.

6.16. Determine the general solution to

$$\frac{d}{dt} \begin{bmatrix} \xi_1 \\ \xi_2 \\ \xi_3 \end{bmatrix} = \begin{bmatrix} -6 & 0 & 0 \\ 1 & -5 & 1 \\ 1 & -1 & -7 \end{bmatrix} \begin{bmatrix} \xi_1 \\ \xi_2 \\ \xi_3 \end{bmatrix} + \begin{bmatrix} 0 \\ t \\ 0 \end{bmatrix}.$$

Plot the solutions. Write a computer program to determine an approximate numerical solution and compare the answers.

6.17. Determine the solution to

$$\ddot{x} + \dot{x} + x - y = 0$$
$$\dot{y} - x + 2y = e^{-t},$$

where $x(0) = 1$, $\dot{x}(0) = -1$ and $y(0) = 2$.

6.18. Prove that the principle of superposition holds for systems of linear first-order ordinary differential equations; that is, if $\xi_1(t)$ and $\xi_2(t)$ both satisfy $\dot{\xi} = A\xi$, then $\xi(t) = c_1\xi_1(t) + c_2\xi_2(t)$ also satisfies it.

Chapter 7
Applications of Systems of First-Order Equations

There are many important engineering applications of systems of first-order, linear, constant-coefficient, ordinary differential equations. This chapter considers the study of linear vibrations for systems with more than one mass and then two topics from modern control theory are introduced, which are *pole placement* and the *linear quadratic regulator*.

7.1 Multidegree of Freedom Vibrations

This section considers systems of the type illustrated in Figure 7.1. We first analyze this system using the approach from classical vibrations theory and then relate it to the material covered in Chapter 6.

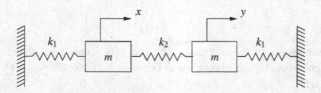

Fig. 7.1 Two degree of freedom mass–spring system.

A simple analysis of the free body diagrams for the two masses yields the following equations of motion

$$m\ddot{x} + (k_1 + k_2)x - k_2 y = 0$$
$$m\ddot{y} + (k_1 + k_2)y - k_2 x = 0. \tag{7.1}$$

Remark 7.1. In this book, all of the presentations concerning vibration systems with more than one mass are without damping. Damping greatly increases the complexity of the algebra necessary for the analysis, but has little to add pedagogically. Based on the results from the study of vibrations in Chapter 4, when light damping is added the resonance frequencies will change very little, and so the results for the undamped analysis should still apply. When the damping is significant, extra work will be required. In exercises, light damping is usually added so that in steady-state, it is possible to distinguish between the homogeneous and particular solutions.

7.1.1 Classical Approach

The general approach presented in this book to solve these types of problems is to convert them to systems of first-order equations. As presented in the next section, the eigenvalues and eigenvectors associated with the system correspond to the frequencies of oscillation and the relative motions of the masses, respectively, for the system. However, there is something at least somewhat dissatisfying converting some perfectly good second-order equations into first-order ones. Of course, they are mathematically equivalent, but the second-order equations are more transparently related to $F = ma$, and also there are no "extra" variables created for the velocities. So, this section presents some basic results that can be obtained by keeping the equations in the original, second-order, form, and is often the approach taken in vibrations texts such as [21].

These are linear, constant coefficient, homogeneous equations, so exponential solutions seem reasonable to assume. Because there are initial conditions corresponding to the position and velocity of each mass, we seek four linearly independent solutions to form a fundamental set of solutions. We need to be precise about what we mean by the term *solution* in this case. The masses are connected by a spring, therefore the solutions for the two masses are related to each other and hence, a solution is the pair,

$$z(t) = \begin{bmatrix} x(t) \\ y(t) \end{bmatrix}.$$

The two components of the solution, $x(t)$ and $y(t)$, cannot be changed independently of each other, which makes sense because if, for example, $y(t)$ changes, it will change $x(t)$.

If $z_1(t)$ and $z_2(t)$ are solutions, then a linear combination of solutions of the form

$$c_1 z_1(t) + c_2 z_2(t) = \begin{bmatrix} c_1 x_1(t) + c_2 x_2(t) \\ c_1 y_1(t) + c_2 y_2(t) \end{bmatrix}$$

is also a solution. Inasmuch as there is no damping, we know the solutions will be oscillatory, so we may as well just assume sine and cosine solutions from the beginning. In particular, let

$$x(t) = a_1 \cos \omega t$$
$$y(t) = a_2 \cos \omega t.$$

At this point this is nothing more than a guess. It may work out that we also need to add in some sine functions, but let us proceed to see what happens. Observe that there are three unknowns: the magnitudes a_1 and a_2 as well as the frequency ω, and we can probably only expect to solve for two out of the three because there are only two equations of motion.

Remark 7.2. Note that this is not the method of undetermined coefficients considered previously in this book because the equations in (7.1) are homogeneous. The reason there are coefficients in this equation is because we do not know the relative magnitudes of the motion of the two masses a priori. In the scalar case, the homogeneous solution can be multiplied by a constant and still satisfy the differential equation. In this case, we need to determine the relationship between the magnitudes of the motion of the two masses. Once that is established, then they both may be scaled by the same constant. The coefficients a_1 and a_2 play a role equivalent to that of components of an eigenvector, which is only defined up to a scale multiple.

Differentiating the proposed solution a couple of times and substituting gives

$$-m\omega^2 a_1 \cos \omega t + (k_1 + k_2) a_1 \cos \omega t - k_2 a_2 \cos \omega t = 0 \qquad (7.2)$$

for the left mass and

$$-m\omega^2 a_2 \cos \omega t + (k_1 + k_2) a_2 \cos \omega t - k_2 a_1 \cos \omega t = 0 \qquad (7.3)$$

for the right mass. These may be written as

$$\begin{bmatrix} k_1 + k_2 - m\omega^2 & -k_2 \\ -k_2 & k_1 + k_2 - m\omega^2 \end{bmatrix} \begin{bmatrix} a_1 \\ a_2 \end{bmatrix} = \begin{bmatrix} 0 \\ 0 \end{bmatrix}. \qquad (7.4)$$

This only has a nonzero solution for a_1 and a_2 when the determinant of the matrix is zero, so we need

$$\left(k_1 + k_2 - m\omega^2 \right)^2 - k_2^2 = 0 \qquad (7.5)$$

which, expanding everything gives

$$\omega^4 - 2\omega^2 \frac{k_1 + k_2}{m} + \frac{k_1 (k_1 + 2k_2)}{m^2} = 0.$$

Note this is a quartic equation in ω, which seems good because we seek four solutions. Due to the absence of the odd powers of ω it may be considered a quadratic equation in ω^2, so we can use the quadratic equation to find solutions for ω^2. This has roots

$$\omega^2 = \frac{k_1 + k_2}{m} \pm \sqrt{\left(\frac{k_1 + k_2}{m} \right)^2 - \frac{k_1 (k_1 + 2k_2)}{m^2}} = \frac{k_1 + k_2}{m} \pm \frac{k_2}{m},$$

so there are two natural frequencies,

$$\omega_{n_1}^2 = \frac{k_1}{m}, \quad \omega_{n_2}^2 = \frac{k_1 + 2k_2}{m}.$$

Substituting $\omega^2 = k_1/m$ into Equation (7.4) gives

$$\begin{bmatrix} k_2 & -k_2 \\ -k_2 & k_2 \end{bmatrix} \begin{bmatrix} a_1 \\ a_2 \end{bmatrix} = \begin{bmatrix} 0 \\ 0 \end{bmatrix},$$

so, $a_1 = a_2$. Substituting $\omega^2 = (k_1 + 2k_2)/m$ gives

$$\begin{bmatrix} -k_2 & -k_2 \\ -k_2 & -k_2 \end{bmatrix} \begin{bmatrix} a_1 \\ a_2 \end{bmatrix} = \begin{bmatrix} 0 \\ 0 \end{bmatrix},$$

so, in that case, $a_1 = -a_2$.

Hence, the solution when $a_1 = a_2$ and setting $a_1 = 1$ is

$$z_1(t) = \begin{bmatrix} \cos \sqrt{\frac{k_1}{m}} t \\ \cos \sqrt{\frac{k_1}{m}} t \end{bmatrix} = \begin{bmatrix} 1 \\ 1 \end{bmatrix} \cos \omega_{n_1} t \tag{7.6}$$

and the solution when $a_1 = -a_2$ and setting $a_1 = 1$ is

$$z_2(t) = \begin{bmatrix} \cos \sqrt{\frac{k_1 + 2k_2}{m}} t \\ -\cos \sqrt{\frac{k_1 + 2k_2}{m}} t \end{bmatrix} = \begin{bmatrix} 1 \\ -1 \end{bmatrix} \cos \omega_{n_2} t. \tag{7.7}$$

Going through the exact same exercise with $\sin \omega t$ gives two more solutions

$$z_3(t) = \begin{bmatrix} \sin \sqrt{\frac{k_1}{m}} t \\ \sin \sqrt{\frac{k_1}{m}} t \end{bmatrix} = \begin{bmatrix} 1 \\ 1 \end{bmatrix} \sin \omega_{n_1} t \tag{7.8}$$

$$z_4(t) = \begin{bmatrix} \sin \sqrt{\frac{k_1 + 2k_2}{m}} t \\ -\sin \sqrt{\frac{k_1 + 2k_2}{m}} t \end{bmatrix} = \begin{bmatrix} 1 \\ -1 \end{bmatrix} \sin \omega_{n_2} t. \tag{7.9}$$

If

$$z(t) = c_1 z_1(t) + c_2 z_2(t) + c_3 z_3(t) + c_4 z_4(t), \tag{7.10}$$

can satisfy any set of initial conditions, then we take it as the general solution. It is left as an exercise to show that this is the case.

The difference between $z_1(t)$ and $z_3(t)$ is simply a phase shift, as is the difference between $z_2(t)$ and $z_4(t)$. Hence, it suffices to study $z_1(t)$ and $z_2(t)$. The two components of $z_1(t)$ are identical. Hence, the two masses move with the same frequency, in the same direction with the same magnitude of oscillation, which we call *mode one*, as is schematically illustrated in Figure 7.2. The two components of $z_2(t)$ have the same frequency, but coefficients with opposite signs. Hence, they move at the same frequency, but in opposite directions, which we call *mode two*, as is illustrated in Figure 7.3. Considering any solution for the system is made up of these solutions, perhaps the better way to consider a solution is not as a combination of the motion of two masses, but rather the superposition of of the two modes, as the following example illustrates.

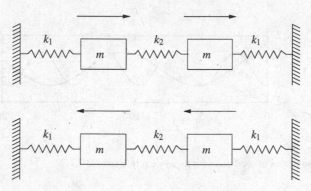

Fig. 7.2 Mode one oscillations.

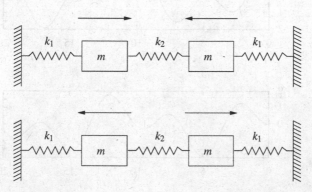

Fig. 7.3 Mode two oscillations.

Example 7.1. Let $k_1 = 1$, $k_2 = 2$, $m = 3$, $c_1 = c_3 = 1$, and $c_2 = c_3 = 0$ in Equation (7.10). Figure 7.4 illustrates the two modes of oscillation separately for each mass. Note that one pair of the curves is in phase and one pair is out of phase. The combined solution, which is the superposition of the two modes, is illustrated in Figure 7.5. The actual motion of the two masses and the relationship between them appears rather complicated from the plots for the actual solutions in Figure 7.5, despite the fact they arise from the superposition of two relatively simple modes.

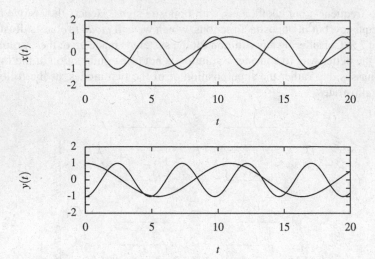

Fig. 7.4 The two modes of oscillation for the system in Example 7.1.

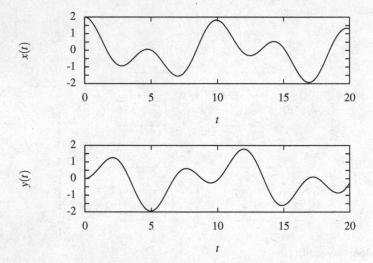

Fig. 7.5 Superposition of the two modes for the system in Example 7.1.

7.1.2 Eigenvalue and Eigenvector Approach

Considering the equations of motion for the system illustrated in Figure 7.1, given by Equation (7.1), if

$$\xi_1 = x_1, \quad \xi_2 = \dot{x}_1, \quad \xi_3 = x_2, \quad \xi_4 = \dot{x}_2,$$

then

$$\dot{\xi} = \frac{d}{dt}\begin{bmatrix}\xi_1 \\ \xi_2 \\ \xi_3 \\ \xi_4\end{bmatrix} = \begin{bmatrix} 0 & 1 & 0 & 0 \\ -\frac{k_1+k_2}{m} & 0 & \frac{k_2}{m} & 0 \\ 0 & 0 & 0 & 1 \\ \frac{k_2}{m} & 0 & -\frac{k_1+k_2}{m} & 0 \end{bmatrix}\begin{bmatrix}\xi_1 \\ \xi_2 \\ \xi_3 \\ \xi_4\end{bmatrix} = A\xi.$$

The eigenvalues of A are determined by the cofactor expansion

$$|A - \lambda I| = \begin{vmatrix} -\lambda & 1 & 0 & 0 \\ -\frac{k_1+k_2}{m} & -\lambda & \frac{k_2}{m} & 0 \\ 0 & 0 & -\lambda & 1 \\ \frac{k_2}{m} & 0 & -\frac{k_1+k_2}{m} & -\lambda \end{vmatrix}$$

$$= -\lambda\begin{vmatrix} -\lambda & \frac{k_2}{m} & 0 \\ 0 & -\lambda & 1 \\ 0 & -\frac{k_1+k_2}{m} & -\lambda \end{vmatrix} + (-1)\begin{vmatrix} -\frac{k_1+k_2}{m} & \frac{k_2}{m} & 0 \\ 0 & -\lambda & 1 \\ \frac{k_2}{m} & -\frac{k_1+k_2}{m} & -\lambda \end{vmatrix}$$

$$= \lambda^4 + 2\frac{k_1+k_2}{m}\lambda^2 + \left(\frac{k_1+k_2}{m}\right)^2 - \left(\frac{k_2}{m}\right)^2 = 0.$$

Hence

$$\lambda_1 = i\sqrt{\frac{k_1}{m}}, \quad \lambda_2 = -i\sqrt{\frac{k_1}{m}}, \quad \lambda_3 = i\sqrt{\frac{k_1+2k_2}{m}}, \quad \lambda_4 = -i\sqrt{\frac{k_1+2k_2}{m}}.$$

Now computing the eigenvectors gives

$$(A - \lambda_1 I)\hat{\xi}_1 = 0 \iff \left[\begin{array}{cccc|c} -i\sqrt{\frac{k_1}{m}} & 1 & 0 & 0 & 0 \\ -\frac{k_1+k_2}{m} & -i\sqrt{\frac{k_1}{m}} & \frac{k_2}{m} & 0 & 0 \\ 0 & 0 & -i\sqrt{\frac{k_1}{m}} & 1 & 0 \\ \frac{k_2}{m} & 0 & -\frac{k_1+k_2}{m} & -i\sqrt{\frac{k_1}{m}} & 0 \end{array}\right]$$

$$\iff \left[\begin{array}{cccc|c} -i\sqrt{\frac{k_1}{m}} & 1 & 0 & 0 & 0 \\ 0 & i\frac{k_2}{\sqrt{k_1 m}} & \frac{k_2}{m} & 0 & 0 \\ 0 & 0 & -i\sqrt{\frac{k_1}{m}} & 1 & 0 \\ \frac{k_2}{m} & 0 & -\frac{k_1+k_2}{m} & -i\sqrt{\frac{k_1}{m}} & 0 \end{array}\right] \iff \left[\begin{array}{cccc|c} -i\sqrt{\frac{k_1}{m}} & 1 & 0 & 0 & 0 \\ 0 & i\sqrt{\frac{k_2}{km}} & \frac{k_2}{m} & 0 & 0 \\ 0 & 0 & -i\sqrt{\frac{k_1}{m}} & 1 & 0 \\ 0 & -i\sqrt{\frac{k_2}{k_1 m}} & -\frac{k_1+k_2}{m} & -i\sqrt{\frac{k_1}{m}} & 0 \end{array}\right]$$

$$\iff \left[\begin{array}{cccc|c} -i\sqrt{\frac{k_1}{m}} & 1 & 0 & 0 & 0 \\ 0 & -i\sqrt{\frac{k_2}{k_1 m}} & \frac{k_2}{m} & 0 & 0 \\ 0 & 0 & -i\sqrt{\frac{k_1}{m}} & 1 & 0 \\ 0 & 0 & -\frac{k_1}{m} & -i\sqrt{\frac{k_1}{m}} & 0 \end{array}\right] \iff \left[\begin{array}{cccc|c} -i\sqrt{\frac{k_1}{m}} & 1 & 0 & 0 & 0 \\ 0 & -i\sqrt{\frac{k_2}{km}} & \frac{k_2}{m} & 0 & 0 \\ 0 & 0 & -i\sqrt{\frac{k_1}{m}} & 1 & 0 \\ 0 & 0 & 0 & 0 & 0 \end{array}\right].$$

Thus,

$$\hat{\xi}_1 = \begin{bmatrix} -i \\ \sqrt{\frac{k_1}{m}} \\ -i \\ \sqrt{\frac{k_1}{m}} \end{bmatrix}.$$

Similar computations show that

$$\hat{\xi}^2 = \begin{bmatrix} i \\ \sqrt{\frac{k_1}{m}} \\ i \\ \sqrt{\frac{k_1}{m}} \end{bmatrix}, \quad \hat{\xi}^3 = \begin{bmatrix} i \\ -\sqrt{\frac{k_1+2k_2}{m}} \\ -i \\ \sqrt{\frac{k_1+2k_2}{m}} \end{bmatrix}, \quad \hat{\xi}^4 = \begin{bmatrix} i \\ \sqrt{\frac{k_1+2k_2}{m}} \\ -i \\ -\sqrt{\frac{k_1+2k_2}{m}} \end{bmatrix}.$$

Using

$$\lambda_1 = i\omega_{n_1}, \quad \hat{\xi}^1 = \begin{bmatrix} 0 \\ \omega_{n_1} \\ 0 \\ \omega_{n_1} \end{bmatrix} + i \begin{bmatrix} -1 \\ 0 \\ -1 \\ 0 \end{bmatrix}, \quad \lambda_3 = i\omega_{n_2}, \quad \hat{\xi}^3 = \begin{bmatrix} 0 \\ -\omega_{n_2} \\ 0 \\ \omega_{n_2} \end{bmatrix} + i \begin{bmatrix} 1 \\ 0 \\ -1 \\ 0 \end{bmatrix},$$

and substituting into Equation (6.15), the solution is

$$\xi(t) = c_1 \left(\begin{bmatrix} 0 \\ \omega_{n_1} \\ 0 \\ \omega_{n_1} \end{bmatrix} \cos \omega_{n_1} t + \begin{bmatrix} 1 \\ 0 \\ 1 \\ 0 \end{bmatrix} \sin \omega_{n_1} t \right)$$

$$+ c_2 \left(\begin{bmatrix} 0 \\ \omega_{n_1} \\ 0 \\ \omega_{n_1} \end{bmatrix} \sin \omega_{n_1} t - \begin{bmatrix} 1 \\ 0 \\ 1 \\ 0 \end{bmatrix} \cos \omega_{n_1} t \right)$$

$$+ c_3 \left(\begin{bmatrix} 0 \\ -\omega_{n_2} \\ 0 \\ \omega_{n_2} \end{bmatrix} \cos \omega_{n_1} t + \begin{bmatrix} -1 \\ 0 \\ 1 \\ 0 \end{bmatrix} \sin \omega_{n_2} t \right)$$

$$+ c_4 \left(\begin{bmatrix} 0 \\ -\omega_{n_1} \\ 0 \\ \omega_{n_2} \end{bmatrix} \sin \omega_{n_1} t + \begin{bmatrix} 1 \\ 0 \\ -1 \\ 0 \end{bmatrix} \cos \omega_{n_1} t \right).$$

The first and third rows are equivalent to a linear combination of the solutions given in Equations (7.6) through (7.9), and the second and fourth rows are the derivatives of the first and third rows, respectively, that is, the velocities.

The important points of this example are twofold:

1. The magnitude of the eigenvalues are exactly the same as the frequencies computed using the classical method.
2. The eigenvectors represent the modes as well; that is,

   a. The first and third components of $\hat{\xi}_1$ and $\hat{\xi}_2$ are identical, which is a consequence of the fact that $a_1 = a_2$ in the case where the frequency is $\omega = \sqrt{k_1/m}$.

   b. The first and third components of $\hat{\xi}_3$ and $\hat{\xi}_4$ have the same magnitude but opposite sign, which is a consequence of the fact that $a_1 = -a_2$ in the case where the frequency is $\omega = \sqrt{(k_1 + 2k_2)/m}$.

7.1.3 Forced Undamped Multidegree of Freedom Systems

Now consider the case where one of the masses is subjected to a harmonic forcing function as is illustrated in Figure 7.6. The homogeneous solution was solved in the preceding two sections; this section focuses on the particular solution. The equations of motion are

$$m\ddot{x} + (k_1 + k_2)x - k_2 y = F\cos\omega t$$
$$m\ddot{y} + (k_1 + k_2)y - k_2 x = 0. \tag{7.11}$$

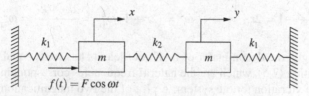

Fig. 7.6 Two degrees of freedom forced mass–spring system.

Because there is no damping, we may assume a solution of the form[1]

$$x_p(t) = c_1\cos\omega t, \quad y_p(t) = c_2\cos\omega t.$$

Note that in this case, ω is equal to the frequency of the forcing function. It is not an unknown natural frequency that must be determined, as was the case previously in this chapter. Differentiating, substituting, and equating the coefficients of the cosine terms gives

$$-m\omega^2 c_1 + (k_1 + k_2)c_1 - k_2 c_2 = F$$
$$-m\omega^2 c_2 + (k_1 + k_2)c_2 - k_2 c_1 = 0,$$

---

[1] If this does not work, we could always go back and add the sine term.

and solving for c_1 and c_2 gives

$$x_p(t) = -F \left(\frac{m\omega^2 - k_1 - k_2}{(m\omega^2 - k_1 - k_2)^2 - k_2^2} \right) \cos \omega t = -\delta \left(\frac{\frac{\omega^2}{\omega_{n_1}^2} - 1 - k_r}{\left(\frac{\omega^2}{\omega_{n_1}^2} - 1 - k_r \right)^2 - k_r^2} \right) \cos \omega t$$

$$= \delta M_1 \cos \omega t$$

and

$$y_p(t) = F \left(\frac{k_2}{(m\omega^2 - k_1 - k_2)^2 - k_2^2} \right) \cos \omega t = \delta \left(\frac{k_r}{\left(\frac{\omega^2}{\omega_{n_1}^2} - 1 - k_r \right)^2 - k_r^2} \right) \cos \omega t$$

$$= \delta M_2 \cos \omega t,$$

where $\omega_{n_1} = \sqrt{k_1/m}$, $\delta = F/k_1$ is the static deflection of the first mass subjected to a static force of magnitude F, $k_r = k_2/k_1$ is the ratio of spring constants, and M_1 and M_2 are the magnification factors.[2]

Note that the denominator is equal to zero when $\omega^2/\omega_{n_1}^1 = 1$ or

$$\frac{\omega^2}{\omega_{n_1}^2} - 1 = 2r \quad \Longleftrightarrow \quad \omega^2 = (2r+1)\,\omega_{n_1}^2 = \frac{2k_2 + k_1}{m} = \omega_{n_2}^2.$$

Not surprisingly, the denominator of both coefficients has roots equal to the solutions of Equation (7.5), which are the natural frequencies corresponding to the two modes of free vibration for the system. A plot of the two magnification factors versus frequency ratio for the case where $k_r = 2$ is illustrated in Figure 7.7.

Figure 7.7 contains a lot of information about the particular solution. For example, below the first resonance, the motion of the two masses is in phase with the forcing and the magnitude of the motion of $x_p(t)$ is greater than $y_p(t)$. Slightly above the first resonance, the motion of both masses is out of phase with the forcing, but the magnitude of $y_p(t)$ is greater than $x_p(t)$, and so on.

Remark 7.3. Note that Figure 7.7 only holds for the rather specific case illustrated in Figure 7.1 where the two outer springs are identical, the masses are the same, and there is no damping. If only light damping is added, then referring to the effect of damping in the scalar case, illustrated by Figure 4.16, it is reasonable to assume that the figure will only be slightly altered. If any properties of the system are otherwise changed, (e.g., unequal masses, etc.), there will still be two resonance frequencies, but otherwise the details for the magnification factors will not be the same.

---

[2] Note that we normalized the system using ω_{n_1}. It is left as an exercise to normalize with respect to ω_{n_2}.

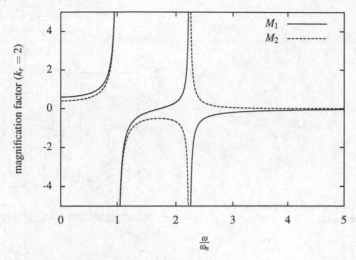

Fig. 7.7 Magnification factors for two-mass system.

7.1.4 Vibration Absorbers

If a system is subjected to a harmonic force that is near its resonance, large oscillations naturally result, which are often undesirable. One approach to reducing the magnitude of the oscillations may be to remove or eliminate the force, change the frequency of the force, or change the natural frequency of the system. Sometimes, such approaches are not feasible, however, in which case a *vibration absorber* may be the best solution.

Consider the system illustrated in Figure 7.8. The problem is to select the spring and mass for the absorber, m_a and k_a, respectively, so that the magnitude of the motion of the mass m is significantly reduced. Ideally, m_a would be small.

Assuming all the displacements are measured in an inertial coordinate system, the equations of motion for the system with the absorber are

$$m\ddot{x} = -kx + f(t) - k_a(x - x_a)$$
$$m_a\ddot{x}_a = k_a(x - x_a).$$

Assuming $f(t) = F\cos\omega t$ and rearranging gives

$$m\ddot{x} + (k + k_a)x - k_ax_a = F\cos\omega t$$
$$m_a\ddot{x}_a + k_ax_a - k_ax = 0.$$

Exercise 7.14 verifies that the particular solutions for x and x_a are

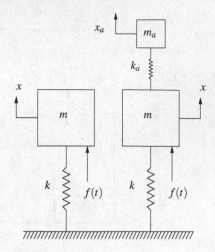

Fig. 7.8 Original system (left) and system
with absorber (right).

$$x_p = \delta \left[\frac{1 - \frac{\omega^2}{\omega_a^2}}{\left(1 - \frac{\omega^2}{\omega_a^2}\right)\left(1 + \frac{k_a}{k} - \frac{\omega^2}{\omega_n^2}\right) - \frac{k_a}{k}} \right] \cos \omega t = \delta M \cos \omega t$$

$$x_{a_p} = \delta \left[\frac{1}{\left(1 - \frac{\omega^2}{\omega_a^2}\right)\left(1 + \frac{k_a}{k} - \frac{\omega^2}{\omega_n^2}\right) - \frac{k_a}{k}} \right] \cos \omega t = \delta M_a \cos \omega t,$$

(7.12)

where δ is the static deflection for the original mass; that is, $\delta = F/k$, $\omega_n = \sqrt{k/m}$, $\omega_a = \sqrt{k_a/m_a}$, and M and M_a are the magnification factors for the original mass and absorber mass, respectively.

One immediate observation is that if the natural frequency of the absorber is designed to be equal to the forcing frequency, $\omega_a = \sqrt{k_a/m_a} = \omega$, the motion of the original mass is zero; that is, it is completely absorbed. It is left as an exercise to show that the intuitively appealing interpretation that the oscillating absorber exerts a force on the original mass that is equal and opposite to the applied force $f(t)$.

As mentioned, the main reason for adding a vibration absorber would be when the forcing frequency is near the resonant frequency of the original mass, so typically we design $\omega_a = \omega_n$, so

$$\frac{k_a}{m_a} = \frac{k}{m}.$$

Let the ratio of the spring constants, which is also the ratio of the masses, be denoted by

$$\frac{k_a}{k} = \frac{m_a}{m} = \rho.$$

Then resonance for the system with the absorber added is when the denominators of both particular solutions in Equation (7.12) are zero; that is,

$$\left(1-\frac{\omega^2}{\omega_a^2}\right)\left(1+\frac{k_a}{k}-\frac{\omega^2}{\omega_a^2}\right)-\frac{k_a}{k}=\left(\frac{\omega^2}{\omega_a^2}\right)^2-\left(\frac{\omega^2}{\omega_a^2}\right)(2+\rho)+1=0,$$

which has two solutions

$$\left(\frac{\omega^2}{\omega_a^2}\right)=\frac{2+\rho}{2}\pm\sqrt{\frac{\rho\,(4+\rho)}{4}},\qquad(7.13)$$

which correspond to the two resonance frequencies (keep in mind that this equation reflects the choice of $\omega_a=\omega_n$). A plot of the solutions to Equation (7.13) is illustrated in Figure 7.9, and it gives the two resonance frequencies for any value of the mass ratio of the absorber to the original mass. For example, if the absorber has a mass that is one-fifth of the original mass, the two resonance frequencies are approximately 80% and 120% of the original resonance.

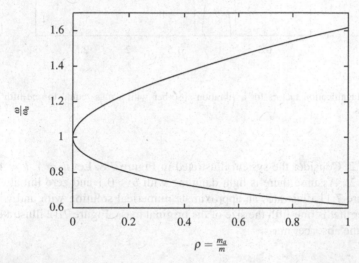

Fig. 7.9 The two resonance frequencies for a system with a vibration absorber as a function of mass ratio.

Note that $\omega/\omega_a=\omega/\omega_n=1$ corresponded to resonance for the original system without an absorber. What Equation (7.13) gives is the ratio of values of the two resonance frequencies for the system with the absorber to the original resonance frequency as a function of ρ, which is the ratio of the mass of the absorber to the original mass. An intuitive way to think of the operation of the absorber is that it splits the resonance of the original system into two resonances, one of which is at a frequency lower than the original resonance and the other of which is at a frequency higher than the original resonance.

Figure 7.10 illustrates the magnitudes of the motion of the original mass and the absorber mass. Note that when $\omega=\omega_n$ the original mass is motionless. Perhaps counter-intuitively, the magnitude of the absorber motion is near a local minimum as well. Of course the "cost" of the vibration absorber is that there are now two resonances instead of one.

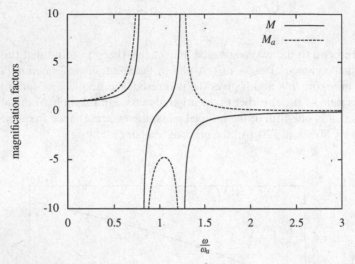

Fig. 7.10 Magnification factors for a vibration absorber with a mass equal to one fifth of the original mass.

Example 7.2. Consider the system illustrated in Figure 7.8. Let $m = 1$, $k = 4$, and $f(t) = \cos 2t$. Assume there is light damping with $b = 0.1$ and zero initial conditions. Figure 7.11 illustrates an approximate numerical solution with and without an absorber that is one-fifth the size of the original mass. Figure 7.12 illustrates the motion of the absorber mass.

7.2 Introduction to "Modern" Control

The subject of feedback control analysis and design is a vast one. This section briefly considers two subsets of the subject and gives a overview of state-space control, which is often referred to as *modern control*. This subject is outlined in this section, but a complete exposition requires an entire course (at least). An interested reader is referred to [3] for a complete treatment, or for the chapters on modern control in [13, 42] for a shorter discussion. Chapters 9 and 10 in this book cover the area of *classical control*.

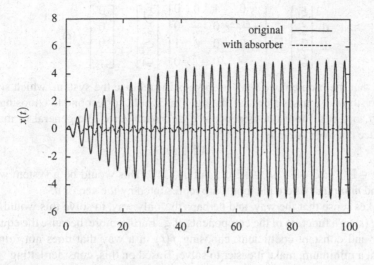

Fig. 7.11 Motion of mass with and without absorber for Example 7.2.

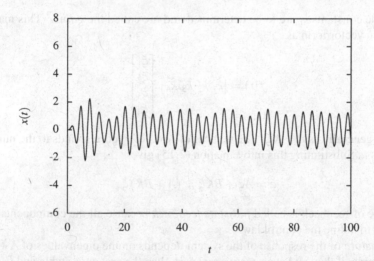

Fig. 7.12 Motion of absorber mass for Example 7.2.

7.2.1 State-Space Control Systems

Consider again the system from Example 6.1 illustrated in Figure 6.1. To make the problem more interesting, let us assume that there is no damping (i.e., $b_1 = b_2 = 0$) and that our goal is to stop the masses from oscillating by using the input force $f(t)$. As a system of first-order, linear, constant coefficient equations, this would be of the form

$$\frac{d}{dt} \begin{bmatrix} \xi_1 \\ \xi_2 \\ \xi_3 \\ \xi_4 \end{bmatrix} = \begin{bmatrix} 0 & 1 & 0 & 0 \\ -\frac{k_1+k_2}{m_1} & 0 & \frac{k_2}{m_1} & 0 \\ 0 & 0 & 0 & 1 \\ \frac{k_2}{m_2} & 0 & -\frac{k_2}{m_2} & 0 \end{bmatrix} \begin{bmatrix} \xi_1 \\ \xi_2 \\ \xi_3 \\ \xi_4 \end{bmatrix} + \begin{bmatrix} 0 \\ 0 \\ 0 \\ \frac{1}{m_2} \end{bmatrix} f(t). \tag{7.14}$$

Assume that the problem we want to solve is to stabilize the system, which simply means to stop the masses from oscillating. This would correspond to choosing $f(t)$ in a way so that all four components of ξ approach zero. The general form for a state-space control system is

$$\dot{\xi} = A\xi + Bu, \tag{7.15}$$

where $A \in \mathbb{R}^{n \times n}$, $\xi \in \mathbb{R}^n$, $B \in \mathbb{R}^{n \times m}$, and $u \in \mathbb{R}^m$. This would be a system with n states and m inputs, where the inputs are represented by the vector u.

It makes sense that the way, and perhaps the only way, to solve this would be to choose $f(t)$ as a function of the components of ξ. Furthermore, because the equation is linear and constant coefficient, choosing $f(t)$ in a way that does not ruin that would, at a minimum, make it easier to solve. Based on this, consider letting

$$f(t) = \hat{k}_1 \xi_1 + \hat{k}_2 \xi_2 + \hat{k}_3 \xi_3 + \hat{k}_4 \xi_4,$$

where the constants $\hat{k}_i$ are to be determined and are called the *gains*.[3] This may be written in vector form as

$$f(t) = \begin{bmatrix} \hat{k}_1 & \hat{k}_2 & \hat{k}_3 & \hat{k}_4 \end{bmatrix} \begin{bmatrix} \xi_1 \\ \xi_2 \\ \xi_3 \\ \xi_4 \end{bmatrix}$$

or more generally as $u = K\xi$. The number of rows of K corresponds to the number of inputs u. Substituting this into Equation (7.15) gives

$$\dot{\xi} = A\xi + BK\xi = (A + BK)\xi.$$

This type of feedback is called *full-state feedback* because all the components of ξ are used to define the control law.

The nature of the response of the system depends on the eigenvalues of $A + BK$. For example, if they all have negative real part, then the system is stable and $\xi(t) \to 0$ as $t \to \infty$. If any eigenvalues are complex conjugate pairs, then the solutions will oscillate, and so on. What makes this problem difficult is that we must choose the elements of the matrix K to make this happen.

---

[3] In controls, gains are almost universally designated with the variable k. Unfortunately, in mechanical systems, spring constants are almost universally designated with the variable k. For this example only, they are distinguished with hats on the gains, but if the problem does not also involve springs, then the hats are dropped for variables representing the gains.

7.2.2 Pole Placement

The *pole placement problem* is to pick the elements of K so that $A + BK$ has eigenvalues at specified locations. Another more descriptive name is the *eigenvalue assignment problem*. The most basic approach to this problem is to compute the characteristic equation for the system with the desired eigenvalues and the characteristic equation for the system with the unspecified gains, and then equate the coefficients of the powers of λ in the two characteristic polynomials. The following example illustrates this with the problem to stabilize the two masses.

Example 7.3. Returning to Example 6.1, let $b_1 = b_2 = 0$ and $m_1 = m_2 = k_1 = k_2 = 1$. The equations of motion are then

$$
\frac{d}{dt}\begin{bmatrix} \xi_1 \\ \xi_2 \\ \xi_3 \\ \xi_4 \end{bmatrix} = \begin{bmatrix} 0 & 1 & 0 & 0 \\ -2 & 0 & 1 & 0 \\ 0 & 0 & 0 & 1 \\ 1 & 0 & -1 & 0 \end{bmatrix} + \begin{bmatrix} 0 \\ 0 \\ 0 \\ 1 \end{bmatrix} f(t)
$$

and if $f(t) = K\xi$, then

$$
\frac{d}{dt}\begin{bmatrix} \xi_1 \\ \xi_2 \\ \xi_3 \\ \xi_4 \end{bmatrix} = \begin{bmatrix} 0 & 1 & 0 & 0 \\ -2 & 0 & 1 & 0 \\ 0 & 0 & 0 & 1 \\ 1 & 0 & -1 & 0a \end{bmatrix}\begin{bmatrix} \xi_1 \\ \xi_2 \\ \xi_3 \\ \xi_4 \end{bmatrix} + \begin{bmatrix} 0 \\ 0 \\ 0 \\ 1 \end{bmatrix}\begin{bmatrix} k_1 & k_2 & k_3 & k_4 \end{bmatrix}\begin{bmatrix} \xi_1 \\ \xi_2 \\ \xi_3 \\ \xi_4 \end{bmatrix}
$$

$$
= \begin{bmatrix} 0 & 1 & 0 & 0 \\ -2 & 0 & 1 & 0 \\ 0 & 0 & 0 & 1 \\ k_1+1 & k_2 & k_3-1 & k_4 \end{bmatrix}\begin{bmatrix} \xi_1 \\ \xi_2 \\ \xi_3 \\ \xi_4 \end{bmatrix}.
$$

Computing the characteristic polynomial gives

$$
\begin{vmatrix} -\lambda & 1 & 0 & 0 \\ -2 & -\lambda & 1 & 0 \\ 0 & 0 & -\lambda & 1 \\ k_1+1 & k_2 & k_3-1 & k_4-\lambda \end{vmatrix}
$$

$$
= -\lambda\left[-\lambda\left(\lambda^2 - k_4\lambda - k_3 + 1 + k_2\right)\right] - \left[\left(-2\left(\lambda^2 - k_4\lambda - k_3 + 1 + k_2\right) - k_1 + 1\right)\right]
$$

$$
= \lambda^4 - k_4\lambda^3 + (3 - k_3)\lambda^2 - (2k_4 + k_2)\lambda + (1 - k_2 - 2k_3) = 0.
$$

(7.16)

Fortunately, we do not have to factor this because we can decide where we want the eigenvalues to be and then determine the characteristic polynomial corresponding to that and equate the coefficients. If we want the eigenvalues at, for example,

$$
\lambda_1 = -1, \quad \lambda_2 = -2, \quad \lambda_3 = -3, \quad \lambda_4 = -4,
$$

then

$$(\lambda + 1)(\lambda + 2)(\lambda + 3)(\lambda + 4) = \lambda^4 + 10\lambda_3 + 35\lambda_2 + 50\lambda + 24, \qquad (7.17)$$

so we need to equate the coefficients of the powers of λ in Equations (7.16) and (7.17) and solve

$$-k_4 = 10, \quad 3 - k_3 = 35, \quad -k_2 - 2k_4 = 50, \quad 1 - k_1 - 2k_3 = 24,$$

which gives
$$k_1 = 41, \quad k_2 = -30, \quad k_3 = -32, \quad k_4 = -10.$$

A plot of the position of the two masses versus time with initial conditions

$$x_1(0) = 1, \quad \dot{x}_1(0) = 0, \quad x_2(0) = -1, \quad \dot{x}_2(0) = 0,$$

is illustrated in Figure 7.13.

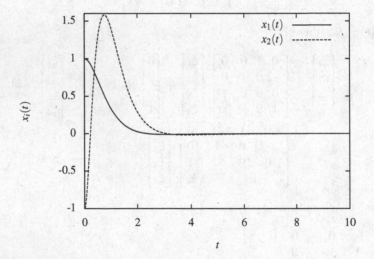

Fig. 7.13 Motion of the masses in Example 7.3.

Keep in mind what was accomplished in the previous example: using one force on one of the masses, the motion of both masses was stopped. The system itself had no damping, yet the masses were stopped in an exponential manner. This was because we designed the motion by specifying where we wanted the eigenvalues. This may seem great, but a drawback to the design is that the magnitude of the necessary force is very high, which should make some sense because part of the solution was specified to be of the order of $\exp(-4t)$. Exercise 7.10 investigates the magnitude of the force and the effect of poles in different locations, that is, specifying other eigenvalues.

This is the manner in which the simplest pole placement problems are solved. However, if there is more than one input, or if the dimension of the problem is much higher, then solving this problem "by hand" by equating coefficients in the characteristic equation with the desired characteristic equation may become difficult. One common method for more complicated problems is a coordinate transformation that is similar, but not identical, to diagonalization. The references mentioned previously give a good overview of these more general approaches. In practice, of course, numerical methods are used.[4]

A final remark is that not any system can have its eigenvalues arbitrarily assigned by appropriate selections of the values for the components of the matrix K. For example, in the previous example, if A were zero, then regardless of K, the only component of ξ that would change would be the fourth component. The property that A and B must have to have sufficient coupling in the dynamics of the system to be able to arbitrarily assign the eigenvalues is called *controllability*.

7.2.3 The Linear Quadratic Regulator

The *linear quadratic regulator* (LQR) problem chooses the gains for the controller to minimize a performance specification for the *regulation* problem, which is to stabilize the system to the origin. Typically, what is minimized is

$$J = \int_0^\infty \left(\xi^T Q \xi + u^T R u \right) dt,$$

where Q and R are symmetric, positive definite matrices. The matrices Q and R provide the relative weighting of the speed at which the system is stabilized and the amount of control effort expended, respectively. A positive definite matrix Q has the property that

$$\xi^T Q \xi > 0$$

if $\xi \neq 0$. A result we use in the following examples is that a matrix is positive definite if all of its eigenvalues are positive. This should make some intuitive sense in the case where Q or R is diagonalized.

For a system of the form

$$\dot{\xi} = A\xi + Bu,$$

the solution to this problem is given by

$$\boxed{u = -R^{-1}B^T P \xi,}$$

where P is the symmetric positive definite solution to the algebraic Riccati equation

---

[4] Both MATLAB and Octave have the function place() which solve the pole placement problem.

$$\boxed{A^T P + PA - PBR^{-1}B^T P + Q = 0.}$$

Although most subjects in this book are fully developed, this is one of the few exceptions. It is important for a student to be aware of the LQR problem, but, unfortunately, the proof of the result is based on subjects that are different from the focus of this book. The approach is illustrated with one example and the reader is referred to [2, 30] for a more complete exposition.

Example 7.4. Consider the system illustrated in Figure 4.1 with $m = k = 1$ and $b = 0$. The equations of motion are

$$\frac{\mathrm{d}}{\mathrm{d}t} \begin{bmatrix} x \\ \dot{x} \end{bmatrix} = \begin{bmatrix} 0 & 1 \\ -1 & 0 \end{bmatrix} + \begin{bmatrix} 0 \\ 1 \end{bmatrix} u,$$

where $u(t) = f(t)$. We work out this problem for several different cases where

$$Q = \begin{bmatrix} \alpha & 0 \\ 0 & 1 \end{bmatrix}, \quad R = 1.$$

To find the solution we need the matrix P. The Riccati equation is

$$\begin{bmatrix} 0 & -1 \\ 1 & 0 \end{bmatrix} \begin{bmatrix} p_{11} & p_{12} \\ p_{12} & p_{22} \end{bmatrix} + \begin{bmatrix} p_{11} & p_{12} \\ p_{12} & p_{22} \end{bmatrix} \begin{bmatrix} 0 & 1 \\ -1 & 0 \end{bmatrix} - \begin{bmatrix} p_{11} & p_{12} \\ p_{12} & p_{22} \end{bmatrix} \begin{bmatrix} 0 \\ 1 \end{bmatrix} \begin{bmatrix} 0 & 1 \end{bmatrix} \begin{bmatrix} p_{11} & p_{12} \\ p_{12} & p_{22} \end{bmatrix}$$

$$+ \begin{bmatrix} \alpha & 0 \\ 0 & 1 \end{bmatrix} = \begin{bmatrix} 0 & 0 \\ 0 & 0 \end{bmatrix}.$$

Because P must be symmetric, the same term is placed on the off-diagonal components. Expanding everything results in the system of equations

$$p_{12}^2 + 2p_{12} - \alpha = 0$$
$$p_{22}^2 - 2p_{12} - 1 = 0$$
$$p_{11} - p_{22} - p_{12}p_{22} = 0.$$

The first equation has two solutions for p_{12}, which, when substituted into the second equation gives two solutions for p_{22} for each value of p_{12}, and the third equation gives one solution for p_{11} for each. Hence, there are four solutions. However, P must be positive definite. Checking the eigenvalues of P for each case gives only one solution for each α.

• For $\alpha = 1$,

$$P = \begin{bmatrix} 1.912 & 0.414 \\ 0.414 & 1.35 \end{bmatrix}$$

is the only one with positive eigenvalues. Hence,

$$f(t) = -R^{-1}B^T P \xi = -\begin{bmatrix} 0.414 & 1.35 \end{bmatrix} \xi.$$

- For $\alpha = 10$,

$$P = \begin{bmatrix} 7.87 & 2.32 \\ 2.32 & 2.37 \end{bmatrix}$$

is the only one with positive eigenvalues. Hence,

$$f(t) = -R^{-1}B^T P \xi = -[2.32 \ 2.37] \, \xi.$$

Plots of the responses are illustrated in Figure 7.14 and plots of the forces are illustrated in Figure 7.15. With $\alpha = 10$ the input force has a significantly larger magnitude and $x(t)$ is driven to zero more quickly.[5]

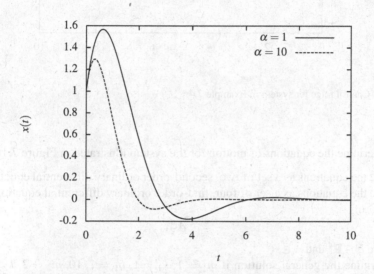

Fig. 7.14 LQR response for system in Example 7.4.

7.3 Exercises

7.1. Show that

$$z(t) = c_1 z_1(t) + c_2 z_2(t) + c_3 z_3(t) + c_4 z_4(t),$$

where $z_1(t)$ through $z_4(t)$ are given by Equations (7.6) through (7.9) can satisfy any initial conditions by an appropriate choice of constants c_1 through c_4. To simplify the problem, assume that the initial conditions are specified at $t = 0$.

---

[5] As was the case with pole placement the system must be controllable for LQR to work. In practice, most most problems are solved numerically using the MATLAB or Octave lqr() command.

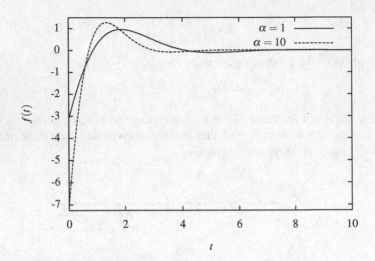

Fig. 7.15 Control force for system in Example 7.4.

7.2. Determine the equations of motion for the system illustrated in Figure 7.16.

1. Write the equations as a set of two, second-order ordinary differential equations.
2. Write the equations as a set of four, first-order ordinary differential equations of the form

$$\dot{\xi} = A\xi,$$

   where $\xi \in \mathbb{R}^4$ and $A \in \mathbb{R}^{4\times4}$.
3. Determine the general solution if $m_1 = 1$, $k_1 = 1$, $b_1 = 1/10$, $m_2 = 2$, $k_2 = 2$, $b_2 = 1/20$, $k_3 = 3$, and $b_3 = 1/5$. You may use a numerical computation package to compute the eigenvalues and eigenvectors.
4. Using the results from previous part, on the same plot, graph the motion of the two masses if they start from rest and $x_1(0) = 1$ and $x_2(0) = -1/2$.

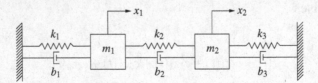

Fig. 7.16 Two-mass system for Exercise 7.2.

7.3. Determine the equations of motion for the system illustrated in Figure 7.17.

1. Write the equations as a set of three, second-order ordinary differential equations.

2. Write the equations as a set of six, first-order ordinary differential equations of the form

$$\dot{\xi} = A\xi,$$

where $\xi \in \mathbb{R}^6$ and $A \in \mathbb{R}^{6 \times 6}$.

3. Write a computer program or use a numerical computation package for the following.

   a. Determine the eigenvalues and eigenvectors of the system when you assign some numerical values to the system parameters.

   b. Compute an approximate numerical solution for the system when it starts with some nonzero initial conditions.

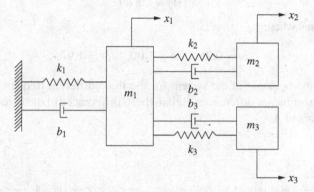

Fig. 7.17 Three mass system for Problem 7.3.

7.4. Determine the equations of motion for the system illustrated in Figure 7.18.

1. Write the second-order differential equation which is the equation of motion for masses 1, 2, i, and n.

2. Write the equations in the form

$$\dot{\xi} = A\xi$$

where $\xi \in \mathbb{R}^{2n}$ and $A \in \mathbb{R}^{2n \times 2n}$. Because n is not specified, it is acceptable for the matrix A to contain ellipses.

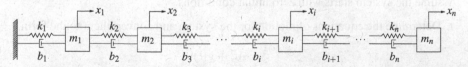

Fig. 7.18 System with n masses for Problem 7.4.

7.5. Consider the system with ten masses illustrated in Figure 7.19. Assume that all the masses have a mass of one and all the spring have a spring constant of one except the spring between the second to last and last mass which has a spring constant of five. Assume the system starts with zero initial conditions.

1. Determine the equations of motion for the system and convert them to the form

$$\dot{\xi} = A\xi + g(t).$$

Compute the eigenvalues and eigenvectors of the matrix A if $i = 10$. You may use a computer program to do this computation.

2. Write a computer program to determine an approximate numerical solution for the system when

$$f(t) = \sin \omega t$$

for the cases where

$$\omega = 0.25, \quad \omega = 1.00, \quad \omega = 1.97.$$

Compare the response of the system for the three different frequencies and explain any significant differences. Relate these differences to the eigenvalues and eigenvectors of A.

Fig. 7.19 Ten-mass system for Exercise 7.5.

7.6. Consider the system illustrated on the left in Figure 7.20.

1. Determine the equations of motion if x_1 and x_2 are measured from the unstretched position of the springs.
2. Determine the equations of motion if x_1 and x_2 are measured from the equilibrium position of the masses.

7.7. Consider the structure illustrated on the right in Figure 7.20. Assume that all the masses have a mass of one and all the springs have a spring constant of one. Assume the system starts with zero initial conditions.

1. Determine the equations of motion for the system and convert them to the form

$$\dot{\xi} = A\xi + g(t).$$

Compute the eigenvalues and eigenvectors of the matrix A. You may use a computer program to do this computation.

2. Write a computer program to determine an approximate numerical solution for the system when

$$f(t) = \sin \omega t$$

for the cases where

$$\omega = 0.25, \quad \omega = 1.00, \quad \omega = 1.97.$$

Compare the response of the system for the three different frequencies and explain any significant differences. Relate these differences to the eigenvalues and eigenvectors of A.

7.8. Consider the mass–spring system illustrated in Figure 7.21 where $m = k = 1$.

1. Determine the equations of motion and convert them to the form

$$\dot{\xi} = A\xi + g(t).$$

2. Let the force $f(t)$ be of the form

$$f(t) = k_1 x + k_2 \dot{x}.$$

Determine the values for k_1 and k_2 such that the eigenvalues for the resulting system are $\lambda_1 = -2$ and $\lambda_2 = -4$.
3. Solve the resulting system if $x(0) = 1$ and $\dot{x}(0) = 0$. On the same plot sketch x and $\dot{x}$ versus t.

7.9. Plot $f(t)$ versus time for the solution in Example 7.3 and comment on its feasibility. For each of the following sets of eigenvalues, plot $x_1(t)$ and $x_2(t)$ versus time and explain why the nature of the solutions makes sense:

$$\lambda = -.1, -.2, -.3, -.4$$
$$\lambda = -1 \pm i, -1, -2$$
$$\lambda = 1, -2, -3, -4.$$

The point of this controls problem would be to stop the system, thus it typically would be the case that the initial conditions are not known. Hence, specify your own initial conditions. Also plot $f(t)$ versus time for those cases. Of these, which solution seems the best and why?

7.10. Plot the magnitude of $f(t)$ for Example 7.3. Do the problem again, but double the magnitude of the eigenvalues; that is, place the eigenvalues at $\lambda_1 = -2$, $\lambda_2 = -4$, $\lambda_3 = -6$ and $\lambda_4 = -8$. What happens to the magnitude of $f(t)$? What happens to the response of the system? Repeat it again, but reduce the magnitude of the eigenvalues by a factor of two. What happens to the magnitude of $f(t)$? What happens to the response? What can you say in general about the trade-off between the magnitude of $f(t)$ and the speed of the response?

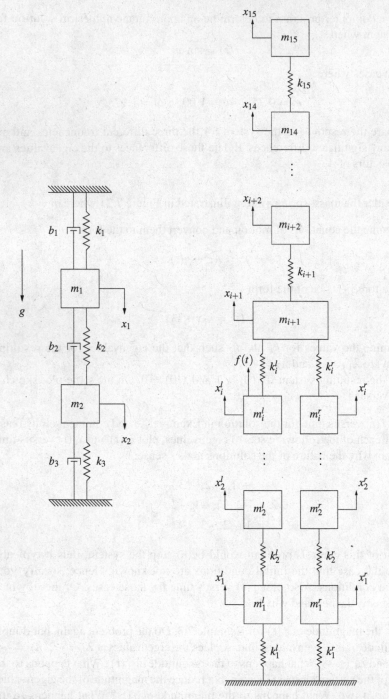

Fig. 7.20 Two-mass system for Exercise 7.6 (left) and structure for Exercise 7.7 (right).

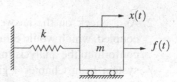

Fig. 7.21 System for Exercise 7.8.

7.11. Do Exercise 7.9 but have the force acting on the other mass.

7.12. Consider the system illustrated in Figure 7.19. Assume there are five masses and that all the masses have a mass of one and all the spring constants are one.

1. Use the `place()` command in MATLAB or Octave to find a full state feedback control law to place all the eigenvalues of the system with full state feedback system for each of the following cases.

   a. Place the eigenvalues at $\lambda = -1, -1.1, -1.2, -1.3$ and -1.4.
   b. Place all the eigenvalues at $\lambda = -.1, -.2, \ldots, -.5$.
   c. Place four of the eigenvalues at the same place as the first four eigenvalues in part 1a and one at $\lambda = 0.1$.
   d. Place three of the eigenvalues at the same place as the first three eigenvalues in part 1a and two at $-.1 \pm 2i$.

   Specify your own initial conditions. Plot the response of each of the five masses and, on a separate plot, the control force. By referring to the plots, explain the features of the solutions that are a result of the differences in the eigenvalue assignments you selected.

2. Use the `lqr()` command in MATLAB or Octave to find an LQR controller for the following cases.

   a. Q is the 5×5 identity matrix and $R = 1$.
   b. The element in the first row and first column of Q is changed to be 100, and the rest of Q is the same as before and $R = 1$.
   c. The element in the first row and first column of Q is changed back to 1, the element in the fourth row and fourth column of Q is changed to be 100, and the rest of Q is the same as before and $R = 1$.
   d. Q is the 10×10 identity matrix and $R = 100$.

   Specify your own initial conditions. In each case, plot the response of the five masses and a separate plot of the control force. By referring to the plots, explain the features of the solutions that are a result of the differences in the Q and R matrices you specified.

7.13. If, in Example 7.4, R is dimensionless, what are the units for the four elements of Q?

7.14. Consider the particular solution for the vibration absorber given by Equation (7.12).

1. Use the method of undetermined coefficients to derive it.

2. As a check on the answer, if $k_a \to \infty$, then the two masses would be rigidly coupled, which would effectively make it a system with one mass. Does this then conform to the analysis for the particular solution for a single degree of freedom system from Chapter 4 given as part of the solution in Equation (4.8)? What about when $m_a \to 0$ or $k_a \to 0$?

3. When the forcing frequency ω is equal to the natural frequency of the absorber, $\omega_a = \sqrt{k_a/m_a}$, what is the magnitude of the motion of the original mass and the absorber mass?

4. When the forcing frequency is equal to the natural frequency of the absorber, show that the force that the absorber exerts on the mass is equal and opposite to the applied force.

7.15. Consider the two-mass system illustrated in Figure 7.1. Which of the two modes illustrated in Figures 7.2 and 7.3 has a higher frequency? Using the physical properties of the system, explain why this makes sense.

7.16. In the derivation of the magnification factors leading to the plot in Figure 7.7, the magnification factors were a function of ω/ω_{n_1} and $k_r k_2/k_1$. Derive the magnification factors in terms of ω/ω_{n_2} and $k_r = k_2/(k_1 + 2k_2)$. Choose a ratio for the spring constants and construct a plot similar to Figure 7.7 by plotting the two magnification factors versus frequency ratio.

7.17. Consider the two-mass system illustrated in Figure 7.6.

- Let $F = 5$, $m = 1$, $k_1 = 2$, and $k_2 = 3$. For these values, make a plot of the magnification factors versus the frequency of the forcing.
- Add dampers to the system in parallel with the springs with $b_1 = 0.01$ and $b_2 = 0.01$. What are the equations of motion for the system? With very light damping such as this, your plot of the magnification factors should be fairly accurate.
- Verify your plot by writing a computer program to determine an approximate numerical solution for this system. Choose the forcing frequency to be well below the first resonance, slightly below the first resonance, slightly above the first resonance, slightly below the second resonance, slightly above the second resonance, and well above the second resonance. Plot $x(t)$, $y(t)$ and $F \cos \omega t$ for an interval of time after the transient response has decayed. Verify each case by checking whether $x(t)$ and $y(t)$ are in or out of phase with each other and in or out of phase with the forcing function and how that is indicated in your plot of the magnification factors. Also, check the relative magnitudes of $x(t)$, $y(t)$, and A.

7.18. Write a computer program that determines an approximate numerical solution for the equations given by (7.1) with the parameter values given by Example 7.1. Pick initial conditions so that the solution is composed of only one of the two modes of oscillation. Pick a different set of initial conditions so that the solution is composed only of the other mode.

7.19. Consider the system illustrated in Figure 7.22. What are the natural frequencies for the system?

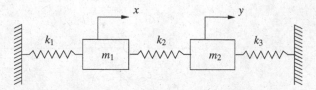

Fig. 7.22 System for Exercise 7.19.

7.20. Consider the system illustrated in Figure 7.23 where $k_1 = 1$, $k_2 = 2$, $k_3 = 3$, $m_1 = 4$, and $m_2 = 5$. Determine the particular solutions and write them in the form $x_p(t) = M_x \cos \omega t$ and $y_p(t) = M_y \cos \omega t$. Plot M_x and M_y versus ω. If you did Exercise 7.19, do the peaks of M_x and M_y correspond to the natural frequencies?

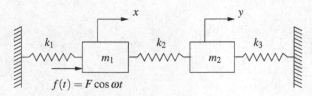

Fig. 7.23 System for Exercise 7.20.

7.21. Using the solutions to the vibration absorber problem given in Equation (7.12), show that if $\omega = \omega_a$, then the force exerted by the absorber on the original mass is exactly equal and opposite to the applied force.

7.22. Consider a mass–spring–damper system with $m = 2$, $k = 32$, $b = 0.05$, and a forcing function $f(t) = \cos 4t$. What should k_a and m_a be if you want the absorber mass to be one-eighth of the original mass? Write a computer program to compute an approximate numerical solution for this system.

1. Compare the vibration of the mass with and without the absorber.
2. Plot the motion of the absorber mass.
3. Use Figure 7.9 to predict the two resonance frequencies and verify it by changing the forcing frequency to those values.

Chapter 8
The Laplace Transform

The Laplace transform is an integral transformation that converts solving ordinary differential equations into solving a system of algebraic equations. Various types of integral transform methods exist, but due to its central role in control theory, this text focuses on Laplace transforms.

8.1 Motivational Example

Integral transform methods are sufficiently abstract that it may be useful to demonstrate their utility up front. The steps involved with the following example are not the obvious ones to the uninitiated, but nonetheless are intended to illustrate the following.

1. They may be used to solve linear, constant-coefficient ordinary differential equations.
2. If one can tolerate the "overhead" of computing the transforms, it converts solving a differential equation into algebra.

Example 8.1. Consider

$$\dot{x} + 2x = 6e^{4t},$$

where $x(0) = 2$. Let us start the exercise by presenting two integrals, both of which have some unstated assumptions that are addressed subsequently:

$$\int_0^\infty e^{-at} e^{-st} \, dt = \frac{1}{s+a}, \tag{8.1}$$

and

$$\int_0^\infty \frac{dx}{dt}(t) e^{-st} \, dt = s \int_0^\infty x(t) e^{-st} \, dt - x(0). \tag{8.2}$$

Both of these facts can be verified by simply evaluating the integrals.

B. Goodwine, *Engineering Differential Equations: Theory and Applications*,
DOI 10.1007/978-1-4419-7919-3_8, © Springer Science+Business Media, LLC 2011

Returning to the problem at hand, multiply each side of the differential equation by e^{-st} and integrate from 0 to ∞ with respect to t,

$$\int_0^\infty e^{-st} \left(\frac{dx}{dt}(t) + 2x(t) \right) dt = \int_0^\infty \frac{dx}{dt}(t) e^{-st} dt + 2 \int_0^\infty x(t) e^{-st} dt$$
$$= 6 \int_0^\infty e^{4t} e^{-st} dt.$$

Clearly, the whole point of the exercise is to find $x(t)$, so there is not too much that can be done with the right-hand side of the first equation except to get rid of the derivative of $x(t)$ in the first integral by making use of Equation (8.2). Also, because we do not know what $x(t)$ is, for the time being, let

$$X(s) = \int_0^\infty x(t) e^{-st} dt.$$

Note that the second equation can be evaluated using Equation (8.1), so

$$6 \int_0^\infty e^{4t} e^{-st} dt = \frac{6}{s-4}.$$

Substituting these into the original differential equation gives

$$sX(s) - x(0) + 2X(s) = \frac{6}{s-4} \tag{8.3}$$

and substituting for $x(0)$ and solving for $X(s)$ gives

$$X(s) = \frac{1}{s+2} \left(\frac{6}{s-4} + 2 \right) = \frac{2s-2}{(s-4)(s+2)} = \frac{1}{s+2} + \frac{1}{s-4}. \tag{8.4}$$

Referring to Equation (8.1) it is clear that the right-hand side of this equation is simply the same transform (multiply by e^{-st} and integrate) that was originally used on the differential equation of the sum of two exponentials. Hence, it is reasonable to assume that

$$x(t) = e^{-2t} + e^{4t}$$

is the solution to the differential equation. A quick substitution shows that indeed it satisfies the differential equation as well as the initial condition.

There are a few important details added subsequently, but for purposes of this example take note that the integral

$$F(s) = \int_0^\infty f(t) e^{-st} dt$$

is called the Laplace transform of $f(t)$.
    Observe the following.

1. Much as with Equation (8.1), for a given function $f(t)$, the Laplace transform only needs to be computed once. Hence, tables of Laplace transforms may be compiled that essentially eliminate the need for actually evaluating the integrals most of the time.
2. Once the equation was fully transformed, which is represented in Equation (8.3), solving for $X(s)$ is simply algebra!
3. Converting from $X(s)$ back to $x(t)$ required determining which functions transformed to $x(t)$, so this step can also usually be handled by tables.
4. The initial condition was handled automatically.
5. The answer is only valid for $t \geq 0$ because the integrals used in the transform have 0 for the lower limit. Any information for $t < 0$ is lost.

So, it is clearly justified to conclude that as long as the work involved in appropriately transforming the differential equation and then inverting the transform at the end is not too great, this is a handy way to solve some types of differential equations. The general manner in which to do this is outlined subsequently. However, before that a short review of a related concept, Fourier transforms, is in order.

8.2 Fourier Transforms

This section presents a brief description of the Fourier transform. This is not strictly necessary for the Laplace transform material that follows, but inasmuch as many students may already be familiar with it, and it is a bit easier to understand than the Laplace transform, it is included here.

First, recall the definition of an improper integral

$$\int_a^\infty f(t)\mathrm{d}t = \lim_{b \to \infty} \int_a^b f(t)\mathrm{d}t.$$

Similarly for the lower limit of integration

$$\int_{-\infty}^b f(t)\mathrm{d}t = \lim_{a \to -\infty} \int_a^b f(t)\mathrm{d}t,$$

and

$$\int_{-\infty}^\infty f(t)\mathrm{d}t = \lim_{a \to -\infty} \lim_{b \to \infty} \int_a^b f(t)\mathrm{d}t.$$

Definition 8.1. For a function $f(t)$, the *Fourier transform* is given by

$$\mathcal{F}(\omega) = \int_{-\infty}^\infty f(t)\mathrm{e}^{\mathrm{i}\omega t}\,\mathrm{d}t$$

if the integral converges.

Note, by Euler's formula, the Fourier transform may also be written as

$$\mathcal{F}(\omega) = \int_{-\infty}^{\infty} f(t)(\cos \omega t + i \sin \omega t)\, dt.$$

Using this expression, the usual interpretation of the Fourier transform as providing the "frequency content" of the signal $f(t)$ is obvious. For a given ω, the cosine and sine functions are in phase with the components of the signal of $f(t)$ that have the same frequency and thus integrate to some nonzero value. For a given ω if there is no component of the signal $f(t)$ with that frequency, the integral is zero. The relative contribution of the real and imaginary components of the transform gives the phase of a given frequency in the signal $f(t)$.

Just for completeness, the inverse Fourier transform is given by

$$f(t) = \frac{1}{2\pi} \int_{-\infty}^{\infty} \mathcal{F}(\omega) e^{i\omega t}\, d\omega.$$

8.3 Laplace Transforms

This section defines the Laplace transform and considers some of its properties.

Definition 8.2. Define the *Laplace transform* of a function $f(t)$ to be

$$F(s) = \int_{0^-}^{\infty} f(t) e^{-st}\, dt,$$

where $s \in \mathbb{C}$; that is, s is a complex number.

First, we clarify some notation. With respect to the limits of integration of the Laplace transform, define an integral of a function with lower limit 0^- and upper limit ∞ to be

$$\int_{0^-}^{\infty} f(t) dt = \lim_{\varepsilon \uparrow 0} \int_{\varepsilon}^{\infty} f(t) dt,$$

where the notation $\lim_{\varepsilon \uparrow 0}$ means that the limit approaches 0 from below. The reason for having the lower limit be 0^- instead of simply 0 is because sometimes something that can affect the value of the integral, such as an impulse, occurs exactly at $t = 0$, having the lower limit equal to 0 is ambiguous as to whether that effect is included in the integral.

Second, with respect to the variable s, because it is a complex number it has a real and imaginary part. If it is denoted by $s = \sigma + i\omega$, then the Laplace transform becomes

$$F(s) = \int_{0^-}^{\infty} f(t) e^{-\sigma t}(\cos \omega t + i \sin \omega t)\, dt.$$

So, one way to interpret the Laplace transform is that it is similar to the Fourier transform in that it provides some information about the frequency content of $f(t)$, but has, for positive values of σ, a multiplicative decaying exponential term.

The Laplace transform is a transform, thus we frequently use an operator notation to represent it. If we are considering the function $f(t)$, the Laplace transform is denoted by $\mathcal{L}$; that is,

$$F(s) = \mathcal{L}(f(t)) = \int_{0^-}^{\infty} e^{-st} f(t) dt.$$

The fundamental concept to keep in mind regarding the transformation is that it transforms the function from the *time domain t* to the *frequency domain s*.

The Laplace transform has an inverse. This is important because it guarantees that there is one and only one $F(s)$ corresponding to $\mathcal{L}(f(t))$, so if we use the Laplace transform of a function to solve a differential equation, it will correspond to the unique solution.[1]

Definition 8.3. The inverse Laplace transform is given by

$$f(t) = \mathcal{L}^{-1}(F(s)) = \frac{1}{2\pi i} \int_{\sigma-i\infty}^{\sigma+i\infty} F(s) e^{st} ds,$$

where σ is a real number such that $F(s)$ converges. Typically this requires that σ be larger than the real part of all values of s for which the denominator of $F(s)$ is equal to zero.

As is made clear subsequently, the values of s for which the denominator and numerator of $F(s)$ are zero provide almost all the essential information we need regarding the properties of the time domain function $f(t) = \mathcal{L}^{-1}(F(s))$. For example, referring back to Example 8.1, observe that the values for which the denominator of $X(s)$ in Equation (8.4) is equal to zero are $s = -2$ and $s = 4$. It is no coincidence that these are exactly the values of the coefficients of t in the exponents of the time domain answer

$$x(t) = e^{-2t} + e^{4t}.$$

Because we refer to these values frequently, they are given names.

Definition 8.4. The values of s for which the denominator of $F(s)$ is equal to zero are called the *poles of $F(s)$*.

Definition 8.5. The values of s for which the numerator of $F(s)$ is equal to zero are called the *zeros of $F(s)$*.

Example 8.2. The frequency domain function

$$F(s) = \frac{s+2}{(s-3)(s+10)}$$

has one zero at $s = -2$ and two poles, one at $s = 3$ and one at $s = -10$.

---

[1] This is not completely correct. For example, two functions that are the same but are different at only isolated points will have the same Laplace transfer because the isolated differences will not change the integral. If, however, we look for continuous functions when computing inverse Laplace transforms, the inverse will be unique. See [9] for a discussion.

Of course, the poles and zeros may occur at values where s is complex, as is illustrated by the following example.

Example 8.3. The function

$$F(s) = \frac{s+a}{(s+a)^2 + b^2}$$

has a zero at $s = -a$ and two poles that comprise a complex-conjugate pair at $s = -a \pm ib$.

8.3.1 The Laplace Transform of Some Common Functions

This section computes the Laplace transform of some functions common in engineering. Unless otherwise stated, we make the following assumption for all computations regarding Laplace transforms.

Assumption 8.1. *In this book, whenever a Laplace transform or inverse Laplace transform is computed, the values for s are assumed to be such that all the required integrals converge.*

Example 8.4. The Laplace transform of $f(t) = e^{at}$ is

$$\mathcal{L}\left(e^{at}\right) = \int_{0^-}^{\infty} e^{-st} e^{at} dt = \int_{0^-}^{\infty} e^{(a-s)t} dt = \frac{1}{a-s} e^{(a-s)t} \Big|_0^{\infty} = \frac{1}{a-s}(0-1) = \frac{1}{s-a}.$$

Hence

$$\mathcal{L}\left(e^{at}\right) = \frac{1}{s-a}.$$

With regard to Assumption 8.1, note that the upper limit of integration only converges if the real part of s is greater than a. Although this is a mathematical necessity, it is one that fortunately rarely concerns us in application and, consistent with the assumption, in this example we assume that the values of s are appropriately restricted. Henceforth, we always implicitly assume whatever restriction is necessary for convergence of any integral involved in computations associated with Laplace transforms.

Example 8.5. Compute the Laplace transform of $f(t) = \sin \omega t$. We want to evaluate

$$\mathcal{L}\left(\sin \omega t\right) = \int_{0^-}^{\infty} e^{-st} \sin(\omega t) dt.$$

Integrating once by parts gives

$$\int_{0^-}^{\infty} e^{-st}\sin(\omega t)\,dt = \left(-\frac{1}{\omega}e^{-st}\cos(\omega t)\right)\bigg|_{0^-}^{\infty} - \frac{s}{\omega}\int_{0^-}^{\infty} e^{-st}\cos(\omega t)\,dt$$

$$= \frac{1}{\omega} - \frac{s}{\omega}\int_{0^-}^{\infty} e^{-st}\cos(\omega t)\,dt. \qquad (8.5)$$

Integrating the last term by parts gives

$$\int_{0^-}^{\infty} e^{-st}\cos\omega t\,dt = \left(\frac{1}{\omega}e^{-st}\sin(\omega t)\right)\bigg|_{0^-}^{\infty} + \frac{s}{\omega}\int_{0^-}^{\infty} e^{-st}\sin(\omega t)\,dt$$

$$= \frac{s}{\omega}\int_{0^-}^{\infty} e^{-st}\sin(\omega t)\,dt.$$

Substituting this into Equation (8.5) gives

$$\int_{0^-}^{\infty} e^{-st}\sin(\omega t)\,dt = \frac{1}{\omega} - \frac{s^2}{\omega^2}\int_{0^-}^{\infty} e^{-st}\sin(\omega t)\,dt,$$

and solving for the original integral, $\mathcal{L}(\sin\omega t)$ gives

$$\int_{0^-}^{\infty} e^{-st}\sin(\omega t)\,dt = \frac{\frac{1}{\omega}}{1 + \frac{s^2}{\omega^2}} = \frac{\omega}{\omega^2 + s^2}.$$

One set of functions that may appear in differential equations for which the Laplace transform is particularly useful are those with discontinuities. So next we consider *step functions* and *impulses*.

Definition 8.6. The function

$$\mathbb{1}(t) = \begin{cases} 0, & t < 0 \\ 1, & t \geq 0, \end{cases}$$

is called the *step function*.

The step function is illustrated in Figure 8.1. It is useful in two ways. First, it is very common in controls because it represents the situation when some control command is activated; that is, at $t = 0$ the control command switches from "off" to "on." Second, it allows us to easily piece together some discontinuous functions.

Example 8.6. Compute the Laplace transform of the step function. Evaluating the transform gives

$$\int_{0^-}^{\infty} e^{-st}\mathbb{1}(t)\,dt = \int_{0^-}^{\infty} e^{-st}\,dt = \frac{1}{-s}\left(e^{-st}\right)\bigg|_{0^-}^{\infty} = \frac{1}{-s}(0-1) = \frac{1}{s}.$$

We occasionally need step functions where the discontinuity does not occur at zero. Note that the function $\mathbb{1}(t - \tau)$ has the discontinuity occur at time $t = \tau$. A plot of $\mathbb{1}(t - 1.5)$ is also illustrated in Figure 8.1. Note that the proper interpretation

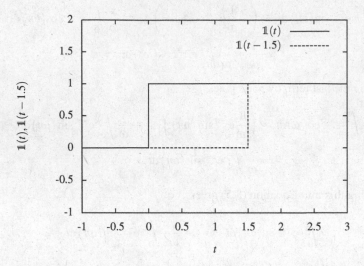

Fig. 8.1 The step functions $\mathbb{1}(t)$ and $\mathbb{1}(t-1.5)$.

of $\mathbb{1}(t-\tau)$ is that the function $1(t)$ is shifted by an amount τ. We consider time shifts of arbitrary functions in Section 8.3.2.

Example 8.7. Compute the Laplace transform of $f(t) = \mathbb{1}(t-\tau)$. Assuming $\tau \geq 0$ substituting into the definition of the Laplace transform gives

$$\int_{0^-}^{\infty} e^{-st}\mathbb{1}(t-\tau)\,dt = \int_{0^-}^{\tau} 0e^{-st}\,dt + \int_{\tau}^{\infty} 1e^{-st}\,dt = \frac{1}{-s}\left(e^{-st}\right)\big|_{\tau}^{\infty}$$

$$= \frac{1}{-s}\left(0 - e^{s\tau}\right) = e^{-s\tau}\frac{1}{s}.$$

Another object that is elegantly handled by the Laplace transform is the impulse. It provides a manner to model, for example, extremely large forces that occur over a very short period of time. An example of an impulse would be the force exerted by a bat on a ball.

Definition 8.7. Consider the function

$$\delta_\varepsilon(t) = \begin{cases} \frac{1}{2\varepsilon}, & |t| \leq \varepsilon, \\ 0, & |t| > \varepsilon, \end{cases}$$

which is illustrated in Figure 8.2 for various values of ε. Define

$$\delta(t) = \lim_{\varepsilon \to 0} \delta_\varepsilon(t).$$

Note that although $\delta(t)^2$ is zero everywhere except at the origin, it still satisfies

$$\int_{-\infty}^{\infty} \delta(t)dt = 1,$$

and then furthermore for any function $f(t)$,

$$\int_{-\infty}^{\infty} \delta(t)f(t)dt = f(0),$$

and similarly if shifted

$$\int_{-\infty}^{\infty} \delta(t-\tau)f(t)dt = f(\tau).$$

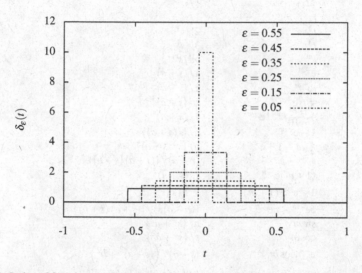

Fig. 8.2 Series of functions leading to the definition of an impulse.

Example 8.8. Compute the Laplace transform of $f(t) = \delta(t)$. Substituting into the definition of the Laplace transform gives

$$\int_{0^-}^{\infty} e^{-st}\delta(t)dt = e^0 = 1.$$

The reason that the lower limit of the integral in the definition of the Laplace transform is 0^- is so that it is clear whether to include impulses that occur at $t = 0$.

---

[2] Although $\delta(t)$ is commonly called the *delta function* or *Dirac delta function*, it is not a function. This is because it is zero everywhere except precisely at the point where we care about it, which is $t = 0$. It is actually a *distribution* and a reader interested in pursuing the matter further is referred to [48].

Because the impulse has zero width, if the lower limit were simply 0, then whether the impulse is included in the integral would be ambiguous.

Example 8.9. Compute the Laplace transform of $f(t) = \delta(t - \tau)$. Substituting into the definition of the Laplace transform gives

$$\int_{0^-}^{\infty} e^{-st} \delta(t - \tau) \mathrm{d}t = e^{-s\tau}.$$

Table 8.1 summarizes the Laplace transform of some common functions in engineering.

$f(t), t \geq 0$	$F(s)$
$\delta(t)$	1
$\mathbb{1}(t)$	$1/s$
t	$1/s^2$
t^2	$2!/s^3$
t^3	$3!/s^4$
t^m	$m!/s^{m+1}$
e^{-at}	$1/(s+a)$
te^{-at}	$1/(s+a)^2$
$\frac{1}{2!}t^2 e^{-at}$	$1/(s+a)^3$
$\frac{1}{(m-1)!}t^{m-1}e^{-at}$	$1/(s+a)^m$
$1 - e^{-at}$	$a/(s(s+a))$
$\frac{1}{a}(at - 1 + e^{-at})$	$a/(s^2(s+a))$
$e^{-at} - e^{-bt}$	$(b-a)/((s+a)(s+b))$
$(1 - at)e^{-at}$	$s/(s+a)^2$
$1 - e^{-at}(1 + at)$	$a^2/(s(s+a)^2)$
$be^{-bt} - ae^{-at}$	$((b-a)s)/((s+a)(s+b))$
$\sin at$	$a/(s^2 + a^2)$
$\cos at$	$s/(s^2 + a^2)$
$e^{-at}\cos bt$	$(s+a)/((s+a)^2 + b^2)$
$e^{-at}\sin bt$	$b/((s+a)^2 + b^2)$
$t \sin at$	$2as/(s^2 + a^2)^2$
$t \cos at$	$(s^2 - a^2)/(s^2 + a^2)^2$
$1 - e^{at}(\cos bt + \frac{a}{b}\sin bt)$	$(a^2 + b^2)/(s[(s+a)^2 + b^2])$

Table 8.1 Table of Laplace transform pairs for functions common in engineering

8.3.2 *Properties of the Laplace Transform*

It is useful to study the definition of the Laplace transform to determine some of its generic properties that we may exploit when using it. The first property we consider is how the derivative of a function acts under a Laplace transform. It turns out that it is very simple and extremely useful. It is simple in that the Laplace transform of a derivative of a function is algebraically related to the Laplace transform of the function itself. In particular, it is simply multiplication of $F(s)$ by s. So, in the frequency domain, differentiation by t is replaced by multiplication by s. This is also its utility in that the Laplace transform then transforms differential equations into algebraic equations.

Theorem 8.1. *If the Laplace transform of a function $f(t)$ is $\mathcal{L}(f(t)) = F(s)$, then*

$$\mathcal{L}\left(\frac{df}{dt}(t)\right) = sF(s) - f(0).$$

Proof. The proof is simply evaluating the integral by the following,

$$\int_{0-}^{\infty} \frac{df}{dt}(t)e^{-st}dt = \left(e^{-st}f(t)\right)\big|_{0-}^{\infty} + s\int_{0-}^{\infty} f(t)e^{-st}dt$$
$$= \left(e^{-st}f(t)\right)\big|_{0-}^{\infty} + sF(s)$$
$$= (0 - f(0)) + sF(s)$$
$$= sF(s) - f(0).$$

$\square$

The second property we consider is called the *final value theorem*. It is useful because it allows us to determine the steady-state values of a solution to a differential equation without having to determine the inverse Laplace transform.

Theorem 8.2. *If all the poles of $sF(s)$ are in the left half of the complex plane, then*

$$\lim_{t \to \infty} f(t) = \lim_{s \to 0} sF(s). \tag{8.6}$$

Proof. Consider

$$\lim_{s \to 0} \int_{0-}^{\infty} e^{-st} \frac{df}{dt}(t)dt = \int_{0-}^{\infty} \left(\lim_{s \to 0} e^{-st} \frac{df}{dt}(t)\right)dt = \int_{0-}^{\infty} \frac{df}{dt}(t)dt = \lim_{t \to \infty} f(t) - f(0).$$
$$\tag{8.7}$$

Also, by Theorem 8.1

$$\lim_{s \to 0} \int_{0-}^{\infty} e^{-st} \frac{df}{dt}(t)dt = \lim_{s \to 0} (sF(s) - f(0)). \tag{8.8}$$

Setting Equation (8.7) equal to Equation (8.8) gives

$$\lim_{t \to \infty} f(t) - f(0) = \lim_{s \to 0} (sF(s) - f(0)) \quad \Longrightarrow \quad \lim_{t \to \infty} f(t) = \lim_{s \to 0} (sF(s)).$$

$\square$

Another very useful property of Laplace transforms is that shifts in time have a very simple form.

Theorem 8.3. *If* $\mathcal{L}(f(t)) = F(s)$, *then the Laplace transform of a function shifted in time satisfies*

$$\boxed{\mathcal{L}(f(t-\tau)\mathbb{1}(t-\tau)) = e^{-s\tau}F(s)}$$

for $\tau \geq 0$.

Proof. The proof is based upon a change of variables. If we let $\hat{t} = t - \tau$, then

$$\mathcal{L}\left(f\left(t-\tau\right)\mathbb{1}(t-\tau)\right) = \int_{0^-}^{\infty} e^{-st} f\left(t-\tau\right)\mathbb{1}(t-\tau)\mathrm{d}t = \int_{-\tau^-}^{\infty} e^{-s(\hat{t}+\tau)} f\left(\hat{t}\right)\mathbb{1}\left(\hat{t}\right)\mathrm{d}\hat{t}$$

$$= e^{-s\tau}\left(\int_{-\tau^-}^{0^-} e^{-s\hat{t}} f\left(\hat{t}\right)\mathbb{1}\left(\hat{t}\right)\mathrm{d}\hat{t} + \int_{0^-}^{\infty} e^{-s\hat{t}} f\left(\hat{t}\right)\mathbb{1}\left(\hat{t}\right)\mathrm{d}\hat{t}\right)$$

$$= e^{-s\tau}\int_{0^-}^{\infty} e^{-s\hat{t}} f\left(\hat{t}\right)\mathbb{1}\left(\hat{t}\right)\mathrm{d}\hat{t} = e^{-s\tau}F(s).$$

$\square$

The proper interpretation of Theorem 8.3 takes some care, especially with respect to the step function appearing in it. Figure 8.3 illustrates a function as well as that function shifted by an amount $\tau = 3/4$. Because the lower limit of the Laplace transform is $t = 0^-$, the values for $f(t)$ for $t < 0$ do not affect the Laplace transform. Mathematically, $\mathcal{L}(f(t)\mathbb{1}(t)) = \mathcal{L}(f(t))$.

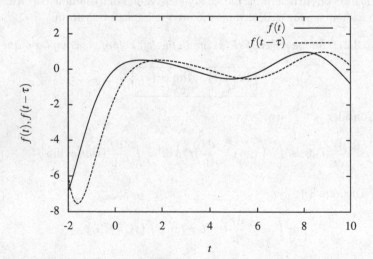

Fig. 8.3 A function $f(t)$ compared to $f(t-\tau)$.

When $f(t)$ is shifted by a positive τ, then we need to either account for the part of $f(t)$ that originally corresponded to $t < 0$ that was shifted into positive times, or exclude it. If we want to include it, then we must reevaluate the integral in the transform, because $F(s)$ only contains information about $f(t)$ for positive time. If we do want to use $F(s)$ and not evaluate the integral, then we must exclude the part of $f(t)$ shifted into positive time. This is accomplished by multiplying $f(t - \tau)$ by $\mathbb{1}(t - \tau)$ because the step function will be zero for $t < \tau$, which corresponds exactly to the part of $f(t)$ that $F(s)$ does not represent.

So, the functions to which Theorem 8.3 applies are illustrated in Figure 8.4. The portion of $f(t)$ for $t \geq 0$ is shifted by an amount τ, but for $t < \tau$, the shifted function must be zero. This fact appears in the proof of Theorem 8.3 in the line where the integral with lower limit τ^- and upper limit 0^- is evaluated to zero.

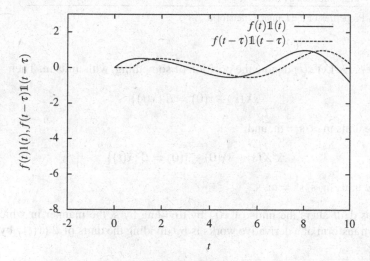

Fig. 8.4 A function $f(t)\mathbb{1}(t)$ compared to $f(t - \tau)\mathbb{1}(t - \tau)$ for which $\mathcal{L}(f(t)) = \mathcal{L}(f(t - \tau)\mathbb{1}(t - \tau))$ and Theorem 8.3 properly applies.

Finally we consider units. From the definition of the Laplace transform of a function $f(t)$,

$$F(s) = \int_{0^-}^{\infty} f(t)e^{-st}\,dt,$$

inasmuch as t has units of seconds, $F(s)$ will have the units of $f(t)$ times seconds. Inasmuch as the exponent of e must be dimensionless, s must have units of one divided by seconds, hence the term *frequency domain*.

Example 8.10. Let $x(t)$ denote the position of something, with units m. Then

$$X(s) = \mathcal{L}\{x(t)\}$$

will have units $m \cdot s$.

The derivative works as expected as an operator.

Name	Time Function	Laplace Transform
Transform pair	$f(t)$	$F(s)$
Superposition	$\alpha f_1(t) + \beta f_2(t)$	$\alpha F_1(s) + \beta F_2(s)$
Differentiation	$\frac{d^m}{dt^m} f(t)$	$s^m F(s) - s^{m-1} f(0) - s^{m-2}\dot{f}(0) -$
		$\cdots - s \frac{d^{m-2}}{dt^{m-2}} f(0) - \frac{d^{m-1}}{dt^{m-1}} f(0)$
Time delay $(\tau \geq 0)$	$f(t-\tau)\mathbb{1}(t-\tau)$	$F(s)e^{-s\tau}$
Time scaling	$f(at)$	$F(s/a)/|a|$
Frequency shift	$e^{-at}f(t)$	$F(s+a)$
Integration	$\int f(\xi)d\xi$	$F(s)/s$
Convolution	$f_1(t) * f_2(t)$	$F_1(s)F_2(s)$
Initial value theorem	$f(0^+)$	$\lim_{s \to \infty} sF(s)$
Final value theorem	$\lim_{t \to \infty} f(t)$	$\lim_{s \to 0} sF(s)$
Time product	$f_1(t)f_2(t)$	$1/(2\pi i) \int_{c-i\infty}^{c+i\infty} F_1(\xi)F_s(s-\xi)d\xi$
Multiplication by time	$t \cdot f(t)$	$-\frac{d}{ds}F(s)$

Table 8.2 Properties of the Laplace transform

Example 8.11. Let $x(t)$ denote the position of something, with units m. Then

$$sX(s) - x(0) = \mathcal{L}\{\dot{x}(t)\}$$

will have units $m \cdot s/s = m$, and

$$s^2 X(s) - sx(0) - \dot{x}(0) = \mathcal{L}\{\ddot{x}(t)\}$$

will have units $m \cdot s/s^2 = m/s$.

Just as d/dt alters the units of $x(t)$ by dividing by s, the manner in which the Laplace transform of a derivative works is by dividing the units of $\mathcal{L}\{x(t)\}$ by s.

8.4 Initial Value Problems and Discontinuous Forcing

Laplace transforms may be used to solve initial value problems for linear, constant-coefficient ordinary differential equations. There are two attributes worth noting. First, there is no need to separate the solution method into homogeneous and particular solutions. Second, the method works particularly well for system where the inhomogeneous term is discontinuous. In such a case the methods from Chapters 2 and 3 would require that we "piece together" solutions, which potentially could be arduous.

We illustrate the means to use Laplace transforms to solve initial value problems with a few examples. The procedure is the same as in Example 8.1, which is to take the Laplace transform of each side of the equation, algebraically solve for the dependent variable, and then determine the inverse Laplace transform to find the time domain function for the dependent variable.

Example 8.12. Find the solution to

$$\ddot{x} + 4\dot{x} + 13x = 20\cos 5t - 12\sin 5t,$$

where $x(0) = 1$ and $\dot{x}(0) = 15$. Taking the Laplace transform gives

$$\left(s^2 X(s) - sx(0) - \dot{x}(0)\right) + 4\left(sX(s) - x(0)\right) + 13X(s) = 20\frac{s}{s^2 + 25} - 12\frac{5}{s^2 + 25}.$$

Substituting the initial conditions gives

$$\left(s^2 X(s) - s - 15\right) + 4\left(sX(s) - 1\right) + 13X(s) = 20\frac{s}{s^2 + 25} - 12\frac{5}{s^2 + 25}.$$

Rearranging some gives

$$X(s)\left(s^2 + 4s + 13\right) = \frac{20s - 60}{s^2 + 25} + s + 19,$$

or

$$X(s) = \frac{20s - 60}{(s^2 + 25)(s^2 + 4s + 13)} + \frac{s + 19}{s^2 + 4s + 13}.$$

Now we want to convert the right-hand side into a combination of terms that appear in Table 8.1. Attempting to factor the denominator $s^2 + 4s + 13$ will show that it has the complex roots, $s = -2 \pm 3i$, and is, by completing the square, equivalent to $(s + 2)^2 + 9$, which is of the form of a denominator in the table. So

$$X(s) = \frac{20s - 60}{(s^2 + 25)\left((s + 2)^2 + 9\right)} + \frac{s + 19}{(s + 2)^2 + 9}.$$

A partial fraction expansion[3] of the first term gives

$$X(s) = \frac{as + b}{s^2 + 25} + \frac{cs + d}{(s + 2)^2 + 9} + \frac{s + 19}{(s + 2)^2 + 9}$$

$$= \frac{(as + b)\left(s^2 + 4s + 13\right) + (cs + d)\left(s^2 + 25\right)}{(s^2 + 25)\left((s + 2)^2 + 9\right)} + \frac{s + 19}{(s + 2)^2 + 9}.$$

Equating numerators in the first term gives

$$(a + c)s^3 + (4a + b + d)s^2 + (13a + 4b + 25c)s + (13b + 25d) = 20s - 60,$$

and some tedious algebra gives $a = 0$, $b = 5$, $c = 0$, and $d = -5$. So,

---

[3] Readers not familiar with partial fractions are referred to Appendix A.3.

$$X(s) = \frac{5}{s^2 + 25} - \frac{5}{(s+2)^2 + 9} + \frac{s+19}{(s+2)^2 + 9}$$

$$= \frac{5}{s^2 + 25} + \frac{s+14}{(s+2)^2 + 9}.$$

Referring to the table, we want either $s+2$ or 3 in the numerator of the second term, so we split the second term into two terms as follows,

$$X(s) = \frac{5}{s^2 + 25} + \frac{s+2}{(s+2)^2 + 9} + \frac{12}{(s+2)^2 + 9}$$

$$= \frac{5}{s^2 + 25} + \frac{s+2}{(s+2)^2 + 9} + 4\frac{3}{(s+2)^2 + 9}.$$

Now all the terms are entries in Table 8.1 and the solution is

$$x(t) = \sin 5t + e^{-2t}\cos 3t + 4e^{-2t}\sin 3t, \quad t \geq 0.$$

Remark 8.1. The Laplace transform only accounts for events for $t \geq 0$, therefore any solution to a differential equation using them is only valid for $t \geq 0$ also. Some texts denote this by multiplying the solutions by $\mathbb{1}(t)$. Rather than adopt the extra notation for this, this text simply remarks on it this one time and leaves it clear from the context that if a Laplace transform was used, the solutions are only valid for $t \geq 0$.

Step functions and time shifts may be combined in useful ways to easily evaluate differential equations that have inhomogeneous terms with discontinuities.

Example 8.13. Determine the solution to

$$\dot{x} + x = f(t), \tag{8.9}$$

where $x(0) = 0$ and

$$f(t) = \begin{cases} 1, & 2 \leq t < 3, \\ 0, & \text{otherwise.} \end{cases}$$

The function $f(t)$ is illustrated in Figure 8.5.

For purposes of using the tools at our disposal to solve this differential equation, the critical observation is that we may write

$$f(t) = \mathbb{1}(t-2) - \mathbb{1}(t-3),$$

which is illustrated in Figure 8.6.

So, now we take the Laplace transform of

$$\dot{x} + x = \mathbb{1}(t-2) - \mathbb{1}(t-3)$$

with $x(0) = 0$ to get

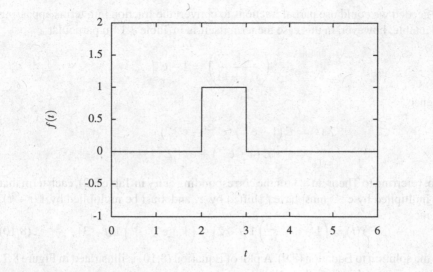

Fig. 8.5 Function for Example 8.13.

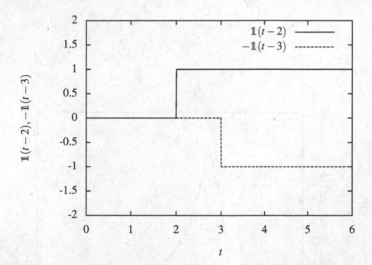

Fig. 8.6 Two-step function combined to give $f(t)$ in Figure 8.5 from Example 8.13.

$$sX(s) + X(s) = \frac{e^{-2s}}{s} - \frac{e^{-3s}}{s},$$

and solving for $X(s)$ gives

$$X(s) = \frac{1}{s(s+1)} \left(e^{-2s} - e^{-3s} \right).$$

If needed we could use partial fractions to convert the fraction into terms appearing in a table; however, in this case the term itself is in Table 8.1. In particular

$$\mathcal{L}^{-1}\left(\frac{1}{s(s+1)}\right) = 1 - e^{-t}.$$

Hence,

$$X(s) = \mathcal{L}\left(1 - e^{-t}\right)\left(e^{-2s} - e^{-3s}\right)$$
$$= e^{-2s}\mathcal{L}\left(1 - e^{-t}\right) - e^{-3s}\mathcal{L}\left(1 - e^{-t}\right).$$

So, referring to Theorem 8.3 (or the corresponding entry in Table 8.2), each term that is multiplied by $e^{-\tau s}$ must have t shifted by τ, and must be multiplied by $\mathbb{1}(t - \tau)$. Hence

$$x(t) = \left(1 - e^{-(t-2)}\right)\mathbb{1}(t-2) - \left(1 - e^{-(t-3)}\right)\mathbb{1}(t-3), \qquad (8.10)$$

is the solution to Equation (8.9). A plot of Equation (8.10) is illustrated in Figure 8.7. Written in another form this solution is

$$x(t) = \begin{cases} 0, & t < 2, \\ 1 - e^{-(t-2)}, & 2 \leq t < 3, \\ e^{-(t-3)} - e^{-(t-2)}, & t \geq 3. \end{cases}$$

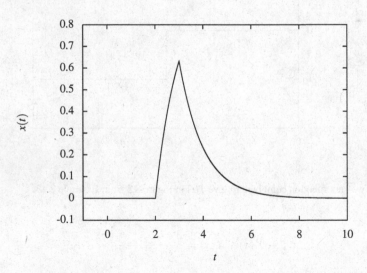

Fig. 8.7 Solution for Example 8.13.

At this point we can recognize that if we are able to piece together step functions to be either one (or negative one) for specific ranges in time, then we can use such a structure to multiply other functions to have them appear for only a limited period of time. The next example illustrates that fact.

Example 8.14. Find the solution to

$$\dot{x} + x = \begin{cases} 0, & t < 1, \\ 3t^2, & 1 \le t < 2, \\ 0, & t \ge 2, \end{cases}$$

with $x(0) = 0$. We can write the inhomogeneous term as a combination of step functions as

$$\dot{x} + x = 3t^2 \left[\mathbb{1}(t-1) - \mathbb{1}(t-2) \right] = 3t^2 \mathbb{1}(t-1) - 3t^2 \mathbb{1}(t-2).$$

If we denote $f(t) = t^2$, neither of the two terms on the right-hand side is in the appropriate form to use Theorem 8.3. For the first one, we need

$$f(t-1) = (t-1)^2 = t^2 - 2t + 1$$

and for the second one we need

$$f(t-2) = (t-2)^2 = t^2 - 4t + 4.$$

So, to make the equation amenable for use by Theorem 8.3, write

$$\begin{aligned} \dot{x} + x &= 3 \left[t^2 \mathbb{1}(t-1) - t^2 \mathbb{1}(t-2) \right] \\ &= 3 \left[\left((t-1)^2 + 2t - 1 \right) \mathbb{1}(t-1) - \left((t-2)^2 + 4t - 4 \right) \mathbb{1}(t-2) \right] \\ &= 3 \left[(t-1)^2 \mathbb{1}(t-1) - (t-2)^2 \mathbb{1}(t-2) \right. \\ &\quad \left. + (2t-1) \mathbb{1}(t-1) - (4t-4) \mathbb{1}(t-2) \right]. \end{aligned}$$

The first two terms may make use of Theorem 8.3, but now we need to take care of the terms that were added, that is, the $2t - 1$ and $4t - 4$ terms. So, write

$$2t - 1 = 2(t-1) + 1$$

and

$$4t - 4 = 4(t-2) + 4$$

and substituting gives

$$\begin{aligned} \dot{x} + x &= \left[(t-1)^2 \mathbb{1}(t-1) - (t-2)^2 \mathbb{1}(t-2) \right. \\ &\quad \left. + (2(t-1)+1) \mathbb{1}(t-1) - (4(t-2)+4) \mathbb{1}(t-2) \right]. \end{aligned}$$

Let us consider this term by term using the relationship

$$\mathcal{L}(f(t-\tau)\mathbb{1}(t-\tau)) = e^{-\tau s} \mathcal{L}(f(t)).$$

1. For the first term

$$\mathcal{L}\left((t-1)^2 \mathbb{1}(t-1)\right) = e^{-s}\mathcal{L}\left(t^2\right) = e^{-s}\frac{2}{s^3}.$$

2. For the second term

$$\mathcal{L}\left((t-2)^2 \mathbb{1}(t-2)\right) = e^{-2s}\mathcal{L}\left(t^2\right) = e^{-2s}\frac{2}{s^3}.$$

3. For the third term

$$\mathcal{L}((2(t-1)+1)\mathbb{1}(t-1)) = e^{-s}\mathcal{L}(2t+1) = e^{-s}\left(\frac{2}{s^2}+\frac{1}{s}\right).$$

4. For the last term

$$\mathcal{L}((4(t-2)+4)\mathbb{1}(t-2)) = e^{-2s}\mathcal{L}(4t+4) = e^{-2s}\left(\frac{4}{s^2}+\frac{4}{s}\right).$$

Taking the Laplace transform of the entire equation gives

$$sX(s)+X(s) = e^{-s}\frac{2}{s^3} - e^{-2s}\frac{2}{s^3} + e^{-s}\left(\frac{2}{s^2}+\frac{1}{s}\right) - e^{-2s}\left(\frac{4}{s^2}+\frac{4}{s}\right).$$

So

$$X(s) = e^{-s}\left(\frac{2}{s^3(s+1)}+\frac{2}{s^2(s+1)}+\frac{1}{s(s+1)}\right)$$
$$- e^{-2s}\left(\frac{2}{s^3(s+1)}+\frac{4}{s^2(s+1)}+\frac{4}{s(s+1)}\right).$$

From Table 8.1 we can find the inverse Laplace transform of the second two terms

$$\mathcal{L}^{-1}\left(\frac{1}{s(s+1)}\right) = 1-e^{-t}, \quad \mathcal{L}^{-1}\left(\frac{1}{s^2(s+1)}\right) = t-1+e^{-t}.$$

It is left as an exercise to show that

$$\mathcal{L}^{-1}\left(\frac{1}{s^3(s+1)}\right) = \frac{1}{2}t^2 - t + 1 - e^{-t}.$$

Finally, remembering to replace t by $t-1$ or $t-2$ depending on whether the Laplace transform is multiplied by e^{-s} or e^{-2s}, respectively,

$$x(t) = 2\left(\frac{1}{2}(t-1)^2 - (t-1) + 1 - e^{-(t-1)}\right)\mathbb{1}(t-1)$$

$$+ 2\left((t-1) - 1 + e^{-(t-1)}\right)\mathbb{1}(t-1) + \left(1 - e^{-(t-1)}\right)\mathbb{1}(t-1)$$

$$- 2\left(\frac{1}{2}(t-2)^2 - (t-2) + 1 - e^{-(t-2)}\right)\mathbb{1}(t-2)$$

$$- 4\left((t-2) - 1 + e^{-(t-2)}\right)\mathbb{1}(t-2) - 4\left(1 - e^{-(t-2)}\right)\mathbb{1}(t-2),$$

which simplifies to

$$x(t) = \left[(t-1)^2 + 1 - e^{-(t-1)}\right]\mathbb{1}(t-1)$$

$$- \left[(t-2)^2 + 2(t-2) + 2 - 2e^{-(t-2)}\right]\mathbb{1}(t-2).$$

Finally, another example involves some trigonometric functions.

Example 8.15. Find the solution to

$$\dot{x} + 2x = f(t), \tag{8.11}$$

where $x(0) = 1$ and

$$f(t) = \begin{cases} 1, & t < \pi, \\ \cos 2t, & \pi \le t < \frac{7\pi}{2}, \\ -e^{-(t-7\pi/2)}, & t > \frac{7\pi}{2}. \end{cases}$$

This function is illustrated in Figure 8.8.

To express $f(t)$ in a manner that is convenient to compute the Laplace transform, we may write $f(t)$ as the sum of three functions, $f(t) = f_1(t) + f_2(t) + f_3(t)$, where

$$f_1(t) = \begin{cases} 1, & 0 \le t < \pi, \\ 0, & \text{otherwise,} \end{cases}$$

$$f_2(t) = \begin{cases} \cos 2t, & \pi \le t < \frac{7\pi}{2}, \\ 0, & \text{otherwise,} \end{cases}$$

$$f_3(t) = \begin{cases} -e^{-(t-7\pi/2)}, & t \ge \frac{7\pi}{2}, \\ 0, & \text{otherwise.} \end{cases}$$

Each of these functions may be written as a single expression using step functions as

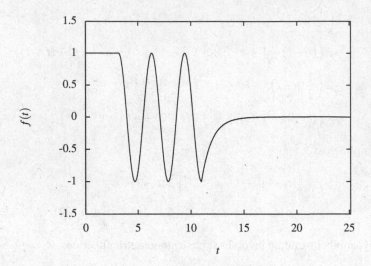

Fig. 8.8 Inhomogeneous term for Equation (8.11) in Example 8.15.

$$f_1(t) = \mathbb{1}(t) - \mathbb{1}(t - \pi)$$

$$f_2(t) = \mathbb{1}(t - \pi)\cos 2t - \mathbb{1}\left(t - \frac{7\pi}{2}\right)\cos 2t$$

$$f_3(t) = -\mathbb{1}\left(t - \frac{7\pi}{2}\right)e^{-(t - 7\pi/2)}.$$

The second function $f_2(t)$ is not in a form that allows us to use Theorem 8.3 because the arguments to the step functions and the cosine function do not match. What we need is to convert $\cos 2t$ to a function of $t - \pi$ and $t - 7\pi/2$ for each of the step functions. Observing that

$$\cos(2(t - \pi)) = \cos 2t$$

$$\cos\left(2\left(t - \frac{7\pi}{2}\right)\right) = -\cos 2t$$

we then have

$$f_2(t) = \mathbb{1}(t - \pi)\cos 2(t - \pi) + \mathbb{1}\left(t - \frac{7\pi}{2}\right)\cos 2\left(t - \frac{7\pi}{2}\right).$$

So,

$$\mathcal{L}(f(t)) = \left(1 - e^{-\pi s}\right)\frac{1}{s} + \left(e^{-\pi s} + e^{-(7\pi/2)s}\right)\frac{s}{s^2 + 4} - e^{-(7\pi/2)s}\frac{1}{s + 1}$$

Computing the Laplace transform of both sides of Equation (8.11) gives

$$(sX(s) - 1) + 2X(s) = \mathcal{L}(f(t))$$

$$= (1 - e^{-\pi s})\frac{1}{s} + \left(e^{-\pi s} + e^{-(7\pi/2)s}\right)\frac{s}{s^2 + 4} - e^{-(7\pi/2)s}\frac{1}{s+1}$$

and solving for $X(s)$ gives

$$X(s) = \left((1 - e^{-\pi s})\frac{1}{s} + \left(e^{-\pi s} + e^{-(7\pi/2)s}\right)\frac{s}{s^2 + 4} - e^{-(7\pi/2)s}\frac{1}{s+1} + 1\right)\frac{1}{s+2}.$$

Considering the inverse Laplace transform term-by-term gives

1. Rearranging the first term

$$(1 - e^{-\pi s})\frac{1}{s(s+2)} = \frac{1}{2}(1 - e^{-\pi s})\frac{2}{s(s+2)}$$

so

$$\mathcal{L}^{-1}\left(\frac{1}{2}(1 - e^{-\pi s})\frac{1}{s(s+2)}\right)$$

$$= \frac{1}{2}\left[\left(1 - e^{-2t}\right)\mathbb{1}(t) - \left(1 - e^{-2(t-\pi)}\right)\mathbb{1}(t - \pi)\right].$$

2. The product in the second term needs to be expanded as

$$\frac{s}{s^2 + 4}\frac{1}{s+2} = \frac{as + b}{s^2 + 4} + \frac{c}{s+2} = \frac{(a+c)s^2 + (2a+b)s + (2b+4c)}{(s^2+4)(s+2)},$$

Equating numerators gives

$$(a+c)s^2 + (2a+b)s + (2b+4c) = s.$$

Because this must be true for arbitrary s, the coefficients of different powers of s must be equal, so

$$a + c = 0, \quad 2a + b = 1, \quad 2b + 4c = 0,$$

and solving for a, b, and c and substituting gives

$$\frac{s}{s^2+4}\frac{1}{s+2} = \frac{\frac{1}{4}s + \frac{1}{2}}{s^2+4} + \frac{-\frac{1}{4}}{s+2} = \frac{1}{4}\left(\frac{s+2}{s^2+4} - \frac{1}{s+2}\right)$$

$$= \frac{1}{4}\left(\frac{s}{s^2+4} + \frac{2}{s^2+4} - \frac{1}{s+2}\right),$$

where each term appears in Table 8.1. Hence

$$\mathcal{L}^{-1}\left(\left(e^{-\pi s}+e^{-\frac{7\pi}{2}s}\right)\frac{s}{s^2+4}\frac{1}{s+2}\right)$$

$$=\mathcal{L}^{-1}\left(\frac{1}{4}\left(e^{-\pi s}+e^{-\frac{7\pi}{2}s}\right)\left(\frac{s}{s^2+4}+\frac{2}{s^2+4}-\frac{1}{s+2}\right)\right)$$

$$=\frac{1}{4}\left[\mathbb{1}(t-\pi)\left(\cos2(t-\pi)+\sin2(t-\pi)-e^{-2(t-\pi)}\right)\right.$$

$$\left.+\mathbb{1}\left(t-\frac{7\pi}{2}\right)\left(\cos2\left(t-\frac{7\pi}{2}\right)+\sin2\left(t-\frac{7\pi}{2}\right)-e^{-2\left(t-\frac{7\pi}{s}\right)}\right)\right].$$

3. The product in the next term can be expanded as

$$\frac{1}{s+1}\frac{1}{s+2}=\frac{a}{s+1}+\frac{b}{s+2}=\frac{a(s+2)+b(s+1)}{(s+1)(s+2)}=\frac{(a+b)s+(2a+b)}{(s+1)(s+2)}.$$

Equating powers of s in the numerator gives

$$\frac{1}{s+1}\frac{1}{s+2}=\frac{1}{s+1}-\frac{1}{s+2}$$

both of which are in Table 8.1. Hence

$$\mathcal{L}^{-1}\left(-e^{-(7\pi/2)s}\frac{1}{s+1}\frac{1}{s+2}\right)=\mathbb{1}\left(t-\frac{7\pi}{s}\right)\left(e^{-(t-7\pi/2)}-e^{-2(t-7\pi/2)}\right).$$

4. Finally, the last term gives

$$\mathcal{L}^{-1}\left(\frac{1}{s+2}\right)=e^{-2t}.$$

The entire solution is, of course, the sum of these four terms and is

$$x(t)=\frac{1}{2}\left[\left(1-e^{-2t}\right)\mathbb{1}(t)-\left(1-e^{-2(t-\pi)}\right)\mathbb{1}(t-\pi)\right]$$

$$+\frac{1}{4}\left[\mathbb{1}(t-\pi)\left(\cos2(t-\pi)+\sin2(t-\pi)-e^{-2(t-\pi)}\right)\right.$$

$$\left.+\mathbb{1}\left(t-\frac{7\pi}{2}\right)\left(\cos2\left(t-\frac{7\pi}{2}\right)+\sin2\left(t-\frac{7\pi}{2}\right)-e^{-2(t-7\pi/2)}\right)\right]$$

$$-\mathbb{1}\left(t-\frac{7\pi}{s}\right)\left(e^{-(t-7\pi/2)}-e^{-2(t-7\pi/2)}\right)+e^{-2t}.$$

We can check the solution by evaluating it in each of the regions in which $f(t)$ has a different form. In particular

1. For $0\le t<\pi$,

$$x(t)=\frac{1}{2}\left(1-e^{-2t}\right)+e^{-2t} \tag{8.12}$$

and

$$\dot{x}(t) = -e^{-2t}. \tag{8.13}$$

Hence, substituting into Equation (8.11) gives

$$\dot{x} + 2x = -e^{-2t} + 2\left(\frac{1}{2}\left(1 - e^{-2t}\right) + e^{-2t}\right) = 1.$$

Also checking the initial condition gives

$$x(0) = \frac{1}{2}\left(1 - e^0\right) + e^0 = 1.$$

2. For $\pi \leq t < 7\pi/2$, $x(t)$ is the same as in Equation (8.12) with the addition of the terms multiplied by $\mathbb{1}(t - \pi)$,

$$x(t) = \frac{1}{2}\left(1 - e^{-2t}\right) + e^{-2t} - \frac{1}{2}\left(1 - e^{-2(t-\pi)}\right)$$
$$+ \frac{1}{4}\left(\cos 2(t - \pi) + \sin 2(t - \pi) - e^{-2(t-\pi)}\right) \tag{8.14}$$

and

$$\dot{x}(t) = -\frac{1}{2}e^{-2t} - e^{-2(t-\pi)} + \frac{1}{2}\left(-\sin(2(t - \pi)) + \cos(2(t - \pi)) - e^{-2(t-\pi)}\right).$$

Substituting into Equation (8.11) gives

$$\dot{x} + 2x = \cos(2(t - \pi)) = \cos 2t.$$

Also, the solutions in Equations (8.12) and (8.14) must match at $t = \pi$. Substituting $t = \pi$ into Equation (8.12) gives

$$x(\pi) = \frac{1}{2}\left(1 - e^{-2\pi}\right) + e^{-2\pi} = \frac{1}{2}\left(1 + e^{-2\pi}\right).$$

Substituting $t = \pi$ into Equation (8.14) gives

$$x(\pi) = \frac{1}{2}\left(1 - e^{-2\pi}\right) + e^{-2\pi} - \frac{1}{2}\left(1 + e^{-2(\pi-\pi)}\right)$$
$$+ \frac{1}{4}\left(\cos 2(\pi - \pi) + \sin 2(\pi - \pi) - e^{-2(\pi-\pi)}\right)$$
$$= \frac{1}{2}\left(1 - e^{-2\pi}\right) + \frac{1}{4}(1 + 0 - 1)$$
$$= \frac{1}{2}\left(1 - e^{-2\pi}\right),$$

so the two solutions match at $t = \pi$.

3. Verifying for $t \geq 7\pi/2$ is left as an exercise.

8.5 Transfer Functions

The notion of a transfer function is particularly useful in engineering because it is a concise representation of the relationship between the input and output of a system. In order to determine transfer functions in engineering a student must have basic abilities to model engineering components. If that is not something that comes naturally, perhaps a review of the material from Section 1.5 would be useful before proceeding. A simple example helps illustrate the concept of a transfer function.

Example 8.16. Consider the task of controlling the system illustrated in Figure 8.9. What is desired is to control the position of mass two with the input force $f(t)$. Exactly how to control it is addressed subsequently. Now we consider the task of determining a convenient way to express its behavior mathematically.

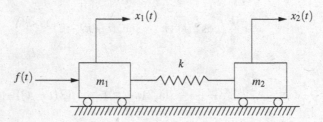

Fig. 8.9 System to control for Example 8.16.

The equations of motion are simple to determine:

$$m_1 \ddot{x}_1 = k(x_2 - x_1) + f(t) \tag{8.15}$$
$$m_2 \ddot{x}_2 = k(x_1 - x_2). \tag{8.16}$$

Clearly these are coupled and, in the present form it is impossible to determine $x_2(t)$ without simultaneously solving for $x_1(t)$. The same is true if we represent it as a system of first-order equations by setting

$$\xi_1 = x_1, \quad \xi_2 = \dot{x}_1, \quad \xi_3 = x_2, \quad \xi_4 = \dot{x}_2$$

which gives

$$\frac{d}{dt}\begin{bmatrix} \xi_1 \\ \xi_2 \\ \xi_3 \\ \xi_4 \end{bmatrix} = \begin{bmatrix} 0 & 1 & 0 & 0 \\ -\frac{k}{m_1} & 0 & \frac{k}{m_1} & 0 \\ 0 & 0 & 0 & 1 \\ \frac{k}{m_2} & 0 & -\frac{k}{m_2} & 0 \end{bmatrix} + \begin{bmatrix} 0 \\ \frac{1}{m_1} \\ 0 \\ 0 \end{bmatrix} f(t).$$

Solving these equations is no problem, however, it would be especially convenient if we could have a more concise representation of the relationship between the input $f(t)$ and the output $x_2(t)$. Recalling that a main feature of Laplace transforms is that, once transformed, solving the differential equations is reduced to algebra, if we compute the Laplace transform of the equations of motion, it may be possible to algebraically eliminate the intermediate variable(s).

Assuming that the initial conditions are all zero, that is,

$$x_1(0) = 0, \quad \dot{x}_1(0) = 0, \quad x_2(0) = 0, \quad \dot{x}_2(0) = 0,$$

and computing the Laplace transform of Equations (8.15) and (8.16) gives

$$m_1 s^2 X_1(s) = k(X_2(s) - X_1(s)) + F(s) \tag{8.17}$$

$$m_2 s^2 X_2(s) = k(X_1(s) - X_2(s)). \tag{8.18}$$

These are two equations that are linear in three functions, $X_1(s)$, $X_2(s)$ and $F(s)$. Hence, we may use one of the equations to eliminate one of the variables. Since we are interested in the relationship between the input force, $f(t)$ and the position of mass two, $x_2(t)$, it makes sense to solve one equation for $X_1(s)$ and substitute into the other equation. Solving Equation (8.18) for $X_1(s)$ gives

$$X_1(s) = \frac{m_2 s^2 + k}{k} X_2(s).$$

Substituting this into Equation (8.17) and rearranging gives

$$X_2(s) = \frac{k}{s^2(m_1 m_2 s^2 + k(m_1 + m_2))} F(s). \tag{8.19}$$

Because it directly relates the effect of the input force on the position of the output mass, call the function

$$\frac{X_2(s)}{F(s)} = \frac{k}{s^2(m_1 m_2 s^2 + k(m_1 + m_2))}$$

the *transfer function* from the input $F(s)$ to the output $X_2(s)$.

Observe the following about Equation (8.19).

1. This is a concise relationship between the input force and the position of mass two. In fact, the variable representing the position of mass one does not explicitly appear in the equation at all.
2. Given an input force $f(t)$, we could compute its Laplace transform $F(s) = \mathcal{L}(f(t))$, substitute $F(s)$ into Equation (8.19) and compute the inverse Laplace transform of $X_2(s)$ to find the motion of $x_2(t)$.
3. Although the variable $X_1(s)$ does not explicitly appear in the equation, it is implicitly in the equation in the terms in the denominator of the transfer function. In fact, it should be obvious that it cannot be eliminated in some complete sense. After all, the only way mass two moves is by the force accelerating mass one, and the motion of mass one affecting mass two through the spring.

In light of the usefulness of the formulation of the relationship between the force and position of the mass represented by Equation (8.19), we may define a transfer function in the following manner.

Definition 8.8. A *transfer function* is the ratio of the Laplace transform of the output to the input of some system assuming all the initial conditions are zero.

What exactly the *input* and *output* of a system are depends on the problem and either must be stated or should be clear from the context of the problem. Subsequently it is apparent that the output of one system may be the input to another. For example, the output of a motor, which may be the torque or position of the motor shaft, is the input to whatever it is driving.

As is clear subsequently, the denominator of the transfer function is of particular importance.

Definition 8.9. Let

$$G(s) = \frac{N(s)}{D(s)}$$

be a transfer function.

- The equation

$$D(s) = 0,$$

that is, setting the denominator equal to zero, is called the *characteristic equation*.
- The lowest power of s in the polynomial in the denominator is called the *system type*.
- If the order of the polynomial in the denominator is greater than the order of the polynomial in the numerator, the transfer function is called *proper*.

Unless otherwise specified, all transfer functions in this text are proper. Also, for this chapter system type is just something to define and observe. It is of importance in Chapter 10 with respect to the steady-state error of the response of a control system.

Now, we make the problem in Example 8.16 more complicated by replacing the general forcing function $f(t)$ with something more realistic.

Example 8.17. Consider the same system as in Example 8.16 but where the force is generated by a belt attached to a pulley attached to a dc motor which is driven by an electric circuit, as illustrated in Figures 8.10 and 8.11. In Figure 8.10, the first mass is attached to a belt that driven by a pulley. The pulley on the left is attached to a dc motor that is driven by the circuit illustrated in Figure 8.11. The pulley on the right is identical to the pulley on the left except it is not driven and is free to rotate. Each pulley has a radius r and moment of inertia J about its center. Assume that the belt is light so that its mass may be ignored and that it does not slip on the pulleys. The motor circuit is comprised of an ideal current source, a resistor, and a dc motor attached to the output. The dc motor has a torque constant of k_τ and a back emf

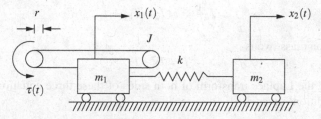

Fig. 8.10 System to control for example 8.17.

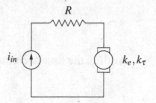

Fig. 8.11 Motor driving circuit for Example 8.17.

constant of k_e. We wish to determine the transfer function from the input current to the circuit to the position of the mass on the right.

The Laplace transform of the differential equations for the two masses are given in Example 8.16 in Equations (8.17) and (8.18). So what is left is to model the belt and pulley system as well as the circuit. Free body diagrams of the two pulleys are illustrated in Figure 8.12. The bottom portion of the belt is attached to the mass, thus if the mass is accelerating the tension on each side of the mass must be different. Because there is no mechanical component between the pulleys on the top, the tension in the top belt is constant along its length. Denote the tension in the top portion of the belt by $T_1(t)$, and the tension in the bottom of the belt to the left and right of the mass by $T_2(t)$ and $T_3(t)$, respectively.

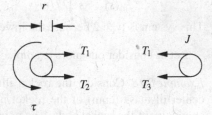

Fig. 8.12 Free body diagrams of the pulleys from Example 8.17.

If we denote the angular position of both pulleys by θ, because the belt does not slip, θ is related to the position of the mass by $r\theta = x_1$. Newton's law on the right pulley gives

$$J\frac{\ddot{x}_1}{r} = r(T_1 - T_3)$$

and on the left pulley gives

$$J\frac{\ddot{x}_1}{r} = r(T_2 - T_1) + \tau.$$

The force on mass two is

$$f = T_2 - T_3.$$

Computing the Laplace transform of both sides of these three equations with zero initial conditions gives

$$Js^2X_1(s) = r^2(T_1(s) - T_3(s))$$
$$Js^2X_1(s) = r^2(T_2(s) - T_1(s)) + rT(s)$$
$$F(s) = T_2(s) - T_3(s).$$

Adding the first two equations gives

$$2Js^2X_1(s) = r^2(T_2(s) - T_3(s)) + rT(s)$$

and using the last equation

$$2Js^2X_1(s) = r^2F(s) + rT(s). \tag{8.20}$$

Inasmuch as the circuit has a current source, the torque produced by the motor is

$$\tau = k_\tau i \quad \Longrightarrow \quad T(s) = k_\tau I(s).$$

Substituting into Equation (8.20) gives

$$2Js^2X_1(s) = r^2F(s) + rk_\tau I(s),$$

and eliminating $X_1(s)$ and $F(s)$ from this equation and Equations (8.17) and (8.18) gives

$$\frac{X_2(s)}{I_{in}(s)} = \frac{k_\tau kr}{s^2[2J(m_2s^2 + k) - r^2(k(m_1 + m_2) + m_1m_2s^2)]}.$$

This system is type 2 because the lowest power of s in the denominator is two.

Let us consider one more example which probably qualifies as rocket science.

Example 8.18. Consider the rocket illustrated in Figure 8.13. The velocity of the center of mass (com) of the rocket is at an angle θ_r with respect to the axis of symmetry of the rocket body. The point through which all aerodynamic forces may be resolved is called the *center of pressure* (cop). The component of the aerodynamic force along the axis of symmetry of the rocket body is called the *drag* and the component orthogonal to the drag is called the *lift*. The lift force is denoted by f_l. The mass moment of inertia of the rocket about its center of mass is denoted by J_r. Assume the distance between the center of mass and center of pressure is l_1 and the distance between the center of mass and the location in the rocket nozzle where the thrust force acts is l_2.

The rocket is controlled by *thrust vectoring*, which means that the nozzle of the rocket engine is gimballed and can pivot. The thrust of the rocket engine is denoted by f_t and the angle of the nozzle with respect to the center-line of the rocket body is denoted by θ_n.

This is rocket is *unstable* because the center of pressure is above the center of mass. This would typically be considered a poor design; however, if we want the rocket to be highly maneuverable, then perhaps it is a good feature. The problem is to find the transfer function from the nozzle angle to the angle of attack, θ_r of the rocket. To simplify the analysis, we assume that the velocity of the rocket is constant.

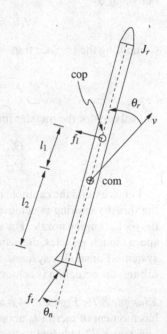

Fig. 8.13 Rocket for Example 8.18.

Basic aerodynamics provides a formula for the lift force, which is

$$f_l = C_l \frac{\rho \|v\|^2 A}{2},$$

where C_l is the coefficient of lift, ρ is the density of the air, $\|v\|$ is the magnitude of the velocity of the rocket, and A is the reference area, which is a function of the lateral area of the rocket exposed to the sideways flow due to a nonzero angle of attack. Because A is proportional to the angle of attack θ_r, then

$$A = A_{\text{ref}} \sin \theta_r$$

and then

$$f_l = \frac{1}{2} \rho C_l \|v\|^2 A_{\text{ref}} \sin \theta_r$$

and for $\theta_r \ll 1$, then

$$f_l \approx \frac{1}{2}\rho C_l \|v\|^2 A_{\mathrm{ref}}\theta_r.$$

Assuming $\theta_r \ll 1$ is reasonable; otherwise the rocket would essentially be flying "sideways."

Newton's law for the rotation of the rocket body gives

$$J_r \ddot{\theta}_r = f_t l_2 \sin\theta_n + \frac{1}{2}\rho C_l \|v\|^2 A_{\mathrm{ref}} l_1 \theta_r.$$

For small θ_n,

$$J_r \ddot{\theta}_r = f_t l_2 \theta_n + \frac{1}{2}\rho C_l \|v\|^2 A_{\mathrm{ref}} l_1 \theta_r.$$

Computing the Laplace transform and assuming zero initial conditions gives

$$J_r s^2 \Theta_r(s) = f_t l_2 \Theta_n(s) + \frac{1}{2}\rho C_l \|v\|^2 A_{\mathrm{ref}} l_1 \Theta_r(s) \tag{8.21}$$

and solving for the transfer function gives

$$\frac{\Theta_r(s)}{\Theta_n(s)} = \frac{f_t l_2}{J_r s^2 - \frac{1}{2}\rho C_l \|v\|^2 A_{\mathrm{ref}} l_1}.$$

Let us extend the example now to include some sort of actuation. We assume that the thrust vectoring is achieved by attaching a dc motor to the axis of rotation of the rocket engine nozzle. For very large rocket engines, such as the main engines on space launch vehicles, the actuation for the thrust vectoring is achieved by hydraulic systems. For smaller systems, such as the maneuvering thrusters for the space shuttle orbiter, the actuation is achieved by dc servo motors.[4]

Example 8.19. Figure 8.14 is a schematic of the nozzle actuation system. The nozzle has moment of inertia J_n about its pivot point and there are two springs with spring constant $k_n/2$ attached to the nozzle a length l_3 from the pivot point. A dc motor with torque constant K_τ and back emf constant k_e is attached to the pivot point that rotates the nozzle and provides a torque τ. The circuit driving the motor is illustrated in Figure 8.15. Find the transfer function from the input voltage to the circuit to the angle of the nozzle, and then find the transfer function from the input voltage to the angle of attack of the rocket. Assume that the overall rotation of the nozzle is small.

If $\theta_n \ll 1$, then the restoring torque about the pivot point due to the displacement of the springs is approximately

$$\tau_s = l_3^2 k \sin\theta_n \approx l_3^2 k_n \theta_n.$$

---

[4] A servo motor is a unit where the angle of the motor is controlled. A signal to the servo motor, typically a *pulse width modulated* signal indicates what the angle of the shaft of the motor should be, and internal feedback control circuitry controls the output angle of the shaft so that it is accomplished. The means to do this is covered when we consider feedback in the following sections.

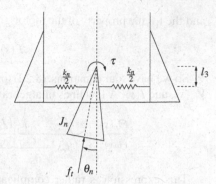

Fig. 8.14 Vectored thrust rocket nozzle
for Example 8.19.

The only torques about the pivot point are τ_s from the springs and τ from the dc
motor. Hence, Newton's law about the pivot point is

$$J_n\ddot{\theta}_n = \tau - \tau_s = \tau - l_3^2 k_n \theta_n.$$

Computing the Laplace transform with zero initial conditions gives

$$s^2 J_n \Theta_n(s) = T(s) - l_3^2 k_n \Theta_n(s),$$

where $T(s) = \mathcal{L}(\tau(t))$. Thus the transfer function from the motor torque to the nozzle angle is

$$\frac{\Theta_n(s)}{T(s)} = \frac{1}{J_n s^2 + l_3^2 k_n}. \tag{8.22}$$

Returning to Equation (8.21), the torque required to pivot the nozzle has an equal
and opposite effect on the rocket body. In particular, Equation (8.21) is now

$$J_r s^2 \Theta_r(s) = f_t l_2 \Theta_n(s) + \frac{C_l \|v\|^2 A_{\text{ref}}}{2} l_1 \Theta_r(s) + T(s). \tag{8.23}$$

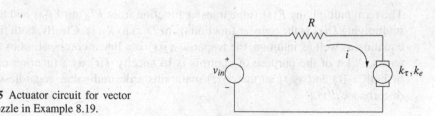

Fig. 8.15 Actuator circuit for vector
thrust nozzle in Example 8.19.

Now considering the circuit, Kirchhoff's voltage law around the circuit gives

$$v_{in} = iR + k_e \dot{\theta}_n$$

or

$$V_{in}(s) = I(s)R + sk_e \Theta_n(s) \tag{8.24}$$

and the torque property of the motor gives $\tau = ik_\tau$ or

$$T(s) = k_\tau I(s). \tag{8.25}$$

So we have four equations, (8.22) to (8.25) and five variables, $\Theta_r(s)$, $\Theta_n(s)$, $T(s)$, $V_{in}(s)$, and $I(s)$. A few lines of algebra gives

$$\frac{\Theta_r(s)}{V_{in}(s)} = \frac{k_\tau \left(J_n s^2 + \left(k_n l_3^2 + f_t l_2\right)\right)}{\left(J_r s^2 - \frac{C_l \|v\|^2 A_{ref}}{2} l_1\right)\left(J_n R s^2 + k_e k_\tau s + k_n l_3^2 R\right)}.$$

This expression is rather complicated, but it is not surprising: the effect of a voltage through a circuit with a motor attached to a nozzle that directs the angle of attack of a rocket is not necessarily very simple.

Many systems have more than one input and more than one output. Even for control systems where we want to control a single variable with one input, there will often be external disturbances. The following example illustrates this fact.

Example 8.20. Consider the mechanical system illustrated in Figure 8.16, which is the same as the system in Example 8.16 except now an external disturbance force $d(t)$ is acting on the second mass. The equations of motion for each mass are

$$m_1 \ddot{x}_1(t) = k(x_2(t) - x_1(t)) + f(t), \quad m_2 \ddot{x}_2(t) = k(x_1(t) - x_2(t)) - d(t),$$

so

$$\left(m_1 s^2 + k\right) X_1(s) = k X_2(s) + F(s), \quad \left(m_2 s^2 + k\right) X_2(s) = k X_1(s) + D(s).$$

Eliminating $X_1(s)$ gives

$$X_2(s) = \frac{k}{(m_1 s^2 + k)(m_2 s^2 + k) - k^2} F(s) + \frac{m_1 s^2 + k}{(m_1 s^2 + k)(m_2 s^2 + k) - k^2} D(s).$$

The term multiplying $F(s)$ is the transfer function from $F(s)$ to $X_2(s)$ and the term multiplying $D(s)$ is the transfer function from $D(s)$ to $X_2(s)$. Clearly, both from the equation as well as intuition, the response $x_2(t)$ is a linear combination of the two terms. A lot of the purpose of controls is to specify $f(t)$ as a function of either or both $x_1(t)$ and $x_2(t)$ so that $x_2(t)$ maintains a desired value, regardless of the disturbance $d(t)$.

8.6 Block Diagram Representation and Algebra

Block diagrams are a graphical means to represent transfer functions and feedback control systems. They are particularly convenient because they represent feedback

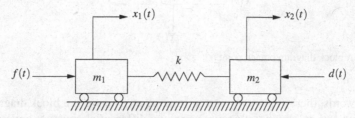

Fig. 8.16 System to control for Example 8.20.

in a visually intuitive manner, the various components are often isolated and the overall representation is simpler. The salient point to keep in mind is that they are simply an alternative representation, and that this alternative representation is as rigorous as the algebraic representation.

Block diagrams are comprised of four types of components.

1. A *block* represents a transfer function describing the relationship between some input and output. It is usually graphically represented by a rectangle. The output is equal to the input times the transfer function inside the block.
2. *Arrows* represent signals, which are the Laplace transform of some time domain function. Arrows directed into blocks represent input signals and arrows directed out of blocks represent output signals from that transfer function. A block with an input and output arrow is illustrated in Figure 8.17.

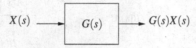

Fig. 8.17 A block with an input and output arrow.

3. *Comparators* add or subtract multiple signals, as illustrated in Figure 8.18. The sign associated with any signal is indicated near the corresponding arrow where it enters the comparator.

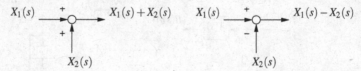

Fig. 8.18 A block diagram comparator.

4. *Branch points* distribute a signal concurrently to multiple arrows, as illustrated in Figure 8.19. They do not "split" or "divide" the signal.

Because the elements of a block diagram are defined with mathematical precision it is important to keep in mind that they are an exact representation of a system.

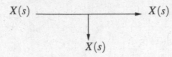

Fig. 8.19 A block diagram branch point.

In other words, there is a one-to-one correspondence between a block diagram representation and an equation that represents the differential equation governing the system.

Components of a block diagram have explicit algebraic meaning, thus just as it is possible to algebraically manipulate an equation, it is possible to algebraically manipulate a block diagram. All of these are relatively straightforward and a few examples should help elucidate the concept.

Example 8.21. A branch point carries a signal concurrently along multiple arrows. If a signal is multiplied by a transfer function in a block before a branch point, the arrows out of the branch point are both multiplied by the transfer function inside the block. In order to move a branch point from the output side of a block to the input side, both arrows must then have the transfer function inside a block so that they carry the same signal. This is represented in Figure 8.20.

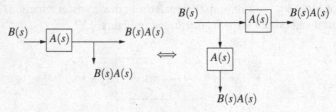

Fig. 8.20 Moving a branch point to the input side of a block.

Similarly, the manner in which to move a branch point from the input side of a block to the output side of the block is illustrated in Figure 8.21.

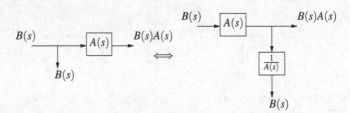

Fig. 8.21 Moving a branch point to the output side of a block.

Example 8.22. The previous example illustrated how to move a branch point to another side of a block. Mathematically it represents the algebraic property of distribution. The algebraic property that multiplication distributes over is represented by the equality

$$D(s) = \big(B(s) + C(s)\big)A(s) = B(s)A(s) + C(s)A(s).$$

In a block diagram, it is represented by the fact that the two block diagrams in Figure 8.22 are equivalent.

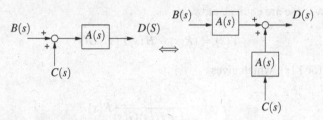

Fig. 8.22 Equivalent block diagrams representing the fact that multiplication distributes over addition.

Similarly, the relationship

$$D(s) = B(s)A(s) + C(s) = \left(B(s) + \frac{C(s)}{A(s)}\right)A(s)$$

is represented in Figure 8.23.

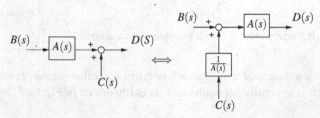

Fig. 8.23 Equivalent block diagrams based on factoring out a transfer function $A(s)$.

So, we now have a rule to move a comparator to either side of a block. If a comparator is moved to the output side of a block, each arrow entering the comparator must multiply the block. If a comparator is moved to the input side of the block, the arrow that originally did not multiply the block must have a block that inverts the multiplication of the block.

The next example illustrates what is perhaps the most important block diagram manipulation that we commonly utilize.

Example 8.23. Consider the feedback system illustrated on the left in Figure 8.24. We show that it is equivalent to the block diagram on the right.

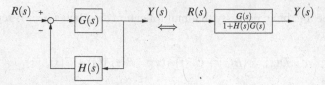

Fig. 8.24 Feedback transfer function.

To show these are equivalent, write

$$Y(s) = \big(R(s) - H(s)Y(s)\big)G(s)$$

and solve for $Y(s)$, which gives

$$Y(s) = \frac{G(s)}{1 + H(s)G(s)}R(s).$$

As the next example shows, the order of branch points may be switched as long as there is no component between them. However, in general switching the order of a comparator and branch point will require some care.

Example 8.24. The two block diagrams in Figure 8.25 are equivalent.

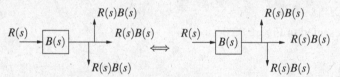

Fig. 8.25 Switching the order of branch points in a block diagram.

Switching a comparator and branch point in a similar manner results in a block diagram that is generally not equivalent, as is illustrated in Figure 8.26.

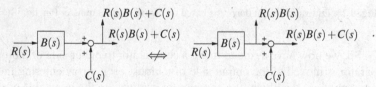

Fig. 8.26 Switching the order of a branch point and comparator in a block diagram.

These and a few other manipulations are summarized in Table 8.3.

The canonical form for a feedback block diagram is the form on the left in Figure 8.24, where there is one *feedforward* block leading from the input to output and one *feedback* block. This form is convenient because it is natural minimal representation for a feedback system, and many analysis and design methods in controls start

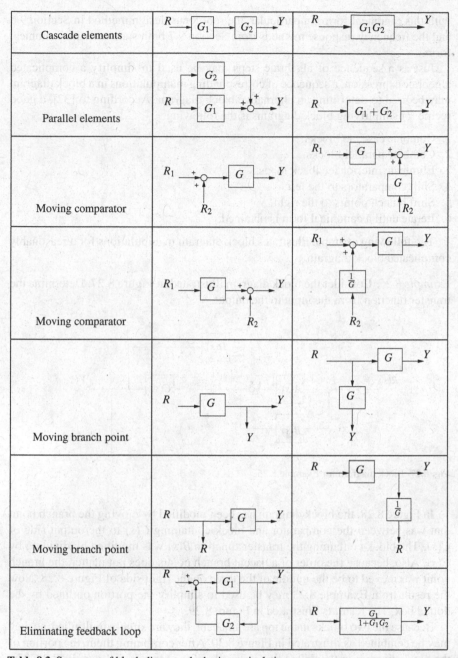

Table 8.3 Summary of block diagram algebraic manipulations

with this canonical form. In particular, the root locus design method in Section 9.6 and the frequency response methods from Section 9.7 both start with this canonical form.

Just as a sequence of algebraic steps may be used to simplify a complicated algebraic expression, a sequence of corresponding manipulations in a block diagram may be used to determine an alternative block diagram. According to [32], a good recipe for simplifying block diagrams is the following.

1. Combine cascade blocks.
2. Combine parallel blocks.
3. Eliminate interior feedback loops.
4. Shift comparators to the left.
5. Shift branch points to the right.
6. Iterate until a canonical form is obtained.

The following example illustrates block diagram manipulations for a reasonably complicated block diagram.

Example 8.25. Consider the block diagram illustrated in Figure 8.27. Determine the transfer function from the input to the output.

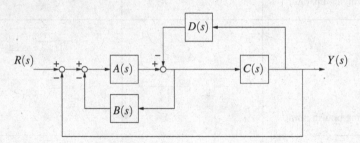

Fig. 8.27 Block diagram for Example 8.25.

In Figure 8.28, the block diagram has been modified by moving the branch point that was between the comparator and block containing $C(s)$ to the output side of $C(s)$. The block containing the transfer function $B(s)$ was modified by dividing by $C(s)$. Also, because the order of adjacent branch points does not matter, the branch point was moved to be the middle of the three on the right side of Figure 8.28. Now the result from Example 8.23 may be used to simplify the portion outlined by the dotted box. The result is illustrated in Figure 8.29.

Because the two blocks in the top are adjacent, they are simply multiplied, so they may be combined as illustrated in Figure 8.30. After combining them, the portion of the block diagram in the dotted line is exactly of the form from Example 8.23. The simplified result is illustrated in Figure 8.31 after the simplification of

$$\frac{\frac{A(s)C(s)}{1+C(s)D(s)}}{1+\frac{B(s)}{C(s)}\frac{A(s)C(s)}{1+C(s)D(s)}} = \frac{A(s)C(s)}{1+C(s)D(s)+A(s)B(s)}.$$

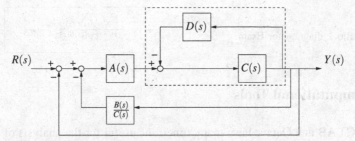

Fig. 8.28 Block diagram for Example 8.25.

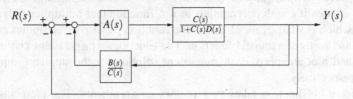

Fig. 8.29 Block diagram for Example 8.25.

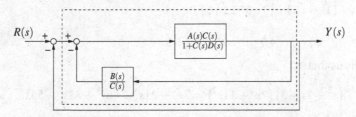

Fig. 8.30 Block diagram for Example 8.25.

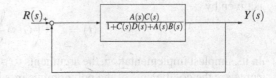

Fig. 8.31 Block diagram for Example 8.25.

Finally, all of Figure 8.31 is of the form of the feedback loop from Example 8.23, so this would be the usual stopping point for this problem. Just for completeness we take it one step further and reduce it to one block with one transfer function which, after some simplification, is illustrated in Figure 8.32.

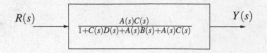

Fig. 8.32 Block diagram for Example 8.25.

8.7 Computational Tools

Both MATLAB and Octave have many functions useful for the analysis of transfer functions. This section presents an overview of commonly used functions.

Four MATLAB functions are particularly useful and are highlighted here. Because we are dealing with polynomials in s, a function that multiplies polynomials is handy, which is what `conv()` does. The function `pzmap()` computes and plots the poles and zeros of a transfer function. The functions `step()` and `impulse()` compute and plot an approximate numerical solution for the step and impulse responses, respectively.

The `conv()` function takes two vectors as arguments. The elements of the vectors are the coefficients of the powers of s in a polynomial. For example, $\left(s^2 + 3s + 5\right)\left(7s^4 + 11s^2\right)$ is computed as follows.

```
>> conv([1 3 5],[7 0 11 0])
ans =
        7      21      46      33      55       0
```

which tells us that

$$\left(s^2 + 3s + 5\right)\left(7s^4 + 11s^2\right) = 7s^5 + 21s^4 + 46s^3 + 33s^2 + 55s.$$

Note that the 0s are necessary, both in the vectors entered into `conv()` as well as in the answer, to determine to what power of s the coefficient belongs.

The `step()` function computes an approximate numerical solution to the step response of a transfer function. If $G(s)$ is a transfer function, then the step response is given by

$$y(t) = \mathcal{L}^{-1}\left(G(s)\frac{1}{s}\right).$$

In its simplest implementation, the arguments to `step()` are vectors whose components are the coefficients of the polynomials in s in the numerator and denominator of $G(s)$, respectively. For example, to compute and plot the step response for

$$G(s) = \frac{s+2}{s^2 + 5s + 10}$$

which is

$$y(t) = \mathcal{L}^{-1}\left(G(s)\frac{1}{s}\right)$$

enter

```
>> step([1 2],[1 5 10])
```

at the command prompt. If there is a need to record the response, enter

```
>> [y,t] = step([1 2],[1 5 10])
```

and then the vector y would contain the step response, and each element of y would correspond to the time contained in the corresponding element of t.

The impulse() function is the same as step() except it determines a numerical solution for the impulse response. The pzmap() function takes the input in the same format as step() and impulse(), but it plots the location of the poles and zeros of the transfer function. This is useful for transfer functions with polynomials that are of higher order than can be factored by hand.

The syntax for Octave is very similar to MATLAB, with the only exception that the step() and impulse() functions require that the transfer function be designated as such with the tf() function. The conv() are the same as for MATLAB.

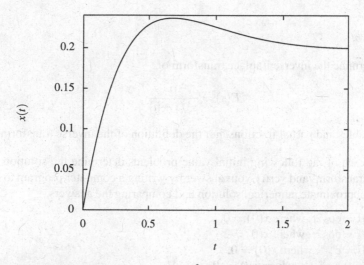

Fig. 8.33 The step response of $G(s) = (s+2)/(s^2+5s+10)$ produced by the Octave command step(tf([1 2],[1 5 10])).

The step() function computes an approximate numerical solution to the step response of a transfer function. For the same system as was illustrated in the preceding section for MATLAB, enter

```
octave:> step(tf([1 2],[1 5 10]))
```

at the command prompt. The output to this function is illustrated in Figure 8.33. If there is a need to record the response, enter

```
octave:> [y,t] = step(tf([1 2],[1 5 10]))
```

and then the vector y would contain the step response, and each element of y would correspond to the time contained in the corresponding element of t.

As would be expected the impulse() function is the same as step() except it determines a numerical solution for the impulse response. Similar to step() in Octave, it needs the transfer function to be expressed inside the tf() function.

8.8 Exercises

8.1. Compute the Laplace transform of the following functions using the definition of the Laplace transform; that is, evaluate the integral in Definition 8.2.

1. $\cos \omega t$
2. $t^2 + 2t + 2$
3. t^n
4. te^{2t}
5. $e^{at} \cos bt$

8.2. Determine the inverse Laplace transform of

$$F(s) = \frac{a}{s^3(s+a)}.$$

Use the tables and partial fractions, not the definition of the inverse transform.

8.3. For each of the following initial value problems determine the solution using Laplace transforms and verify your answer by writing a computer program to determine an approximate numerical solution and comparing the answers.

1. $2\dot{x} + 3x = \cos 2t$, where $x(0) = 0$.
2. $2\dot{x} + 3x = \cos 2t$, where $x(0) = 2$.
3. $2\dot{x} + 3x = e^{-3t}$, where $x(0) = 0$.
4. $2\dot{x} + 3x = e^{-3t} + t$, where $x(0) = 0$.
5. $2\dot{x} - 3x = 0$, where $x(0) = -1$.
6. $2\dot{x} - 3x = 1$, where $x(0) = -1$.
7. $\ddot{x} + 9x = \cos 2t$, where $x(0) = 1$ and $\dot{x}(0) = 1$.
8. $\ddot{x} + 4x = \cos 5t$, where $x(0) = 1$ and $\dot{x}(0) = 1$.
9. $\ddot{x} + 16x = 0$, where $x(0) = 1$ and $\dot{x}(0) = 0$.
10. $\ddot{x} + 16x = 0$, where $x(0) = 1$ and $\dot{x}(0) = 1$.
11. $\ddot{x} + 16x = \cos 4t$, where $x(0) = 1$ and $\dot{x}(0) = 1$.
12. $\ddot{x} + 6\dot{x} + 9x = e^{-3t}$, where $x(0) = 0$ and $\dot{x}(0) = 0$.
13. $\ddot{x} + x = \sin t$, where $x(0) = 1$ and $\dot{x}(0) = 0$.
14. $\ddot{x} + 5\dot{x} + 6x = 0$, where $x(0) = 2$ and $\dot{x}(0) = -5$.
15. $\ddot{x} - 5\dot{x} + 4x = e^{3t}$, where $x(0) = 1$ and $\dot{x}(0) = -1$.
16. $2\ddot{x} + \dot{x} - x = e^{4t}$, where $x(0) = 3$ and $\dot{x}(0) = 0$.
17. $\ddot{x} = 0$, where $x(0) = 0$ and $\dot{x}(0) = 1$.

18. $\ddot{x} + 2\dot{x} + 5x = 6\cos 3t - 4\sin 3t$, where $x(0) = 0$ and $\dot{x}(0) = 5$.

8.4. Consider number 11 of Exercise 8.3. Explain any complications that would arise if this problem were solved using the method of undetermined coefficients. Using Laplace transforms, does one need to be careful if the inhomogeneous terms contain a homogeneous solution?

8.5. For each of the following initial value problems determine the solution using Laplace transforms and verify your answer by writing a computer program to determine an approximate numerical solution and comparing the answers.

1. $\ddot{x} + 16x = \delta(t)$, where $x(0) = 0$ and $\dot{x}(0) = 0$.
2. $\ddot{x} + 16x = \delta(t)$, where $x(0) = 1$ and $\dot{x}(0) = 0$.
3. $\ddot{x} + 16x = \delta(t)$, where $x(0) = 0$ and $\dot{x}(0) = 1$.
4. $\ddot{x} + 16x = \delta(t-2)$, where $x(0) = 0$ and $\dot{x}(0) = 0$.
5. $\ddot{x} + 9x = 1(t)$, where $x(0) = 0$ and $\dot{x}(0) = 0$.
6. $\ddot{x} + 9x = 1(t-3)$, where $x(0) = 1$ and $\dot{x}(0) = 0$.

8.6. Determine the solution to each of the following differential equations.

1. $\dot{x} - 5x = f(t)$ where $x(0) = 0$ and

$$f(t) = \begin{cases} 0, & t < 3, \\ t, & 3 \le t < 4, \\ 0, & 4 \le t. \end{cases}$$

2. $\ddot{x} + 4x = f(t)$ where $x(0) = 0$, $\dot{x}(0) = 0$ and

$$f(t) = \begin{cases} \cos t, & 0 \le t < \pi, \\ -1, & \pi \le t. \end{cases}$$

3. $\ddot{x} + 4x = f(t)$ where $x(0) = 1$, $\dot{x}(0) = 0$ and

$$f(t) = \begin{cases} \sin 2t, & 0 \le t < 4\pi, \\ 0, & 4\pi \le t. \end{cases}$$

4. $\dot{x} + 2x = f(t)$ where $x(0) = 1$ and

$$f(t) = \begin{cases} 0, & 0 < t \le 1, \\ 2, & 1 < t \le 2, \\ 0, & 2 \le t. \end{cases}$$

5. $\dot{x} + x = f(t)$, where $x(0) = 1$ and

$$f(t) = \begin{cases} 0, & 0 < t \le 2, \\ t, & t > 2. \end{cases}$$

In each case, plot your answer. Compare your answer with an approximate numerical solution for the differential equation obtained by writing a computer program or using a computer package.

8.7. Determine the solution to using Laplace transforms.

8.8. Determine the solution to where $x(0) = 0$ and $\dot{x}(0) = 0$ using Laplace transforms. Plot your answer. Compare your answer with an approximate numerical solution for the differential equation obtained by writing a computer program or using a computer package.

8.9. Determine the solution to

$$\ddot{x} + 25x = \begin{cases} t, & 0 \le t < 1, \\ \cos(t-1), & 1 \le t, \end{cases}$$

where $x(0) = 0$ and $\dot{x}(0) = 0$ using Laplace transforms. Plot your answer. Compare your answer with an approximate numerical solution for the differential equation obtained by writing a computer program or using a computer package.

8.10. Determine the solution to

$$\ddot{x} + \dot{x} + x = \begin{cases} \sin t, & 0 \le t < \frac{\pi}{2}, \\ 1, & \frac{\pi}{2} \le t, \end{cases}$$

where $x(0) = 0$ and $\dot{x}(0) = 0$ using Laplace transforms. Plot your answer. Compare your answer with an approximate numerical solution for the differential equation obtained by writing a computer program or using a computer package.

8.11. Because differentiation in the time domain corresponds to multiplication by s in the frequency domain, and

$$\mathcal{L}\{\sin 2t\} = \frac{2}{s^2+4}$$

why does

$$\mathcal{L}\{\cos 2t\} \neq \frac{2s}{s^2+4}?$$

8.12. We have glossed over some technical issues with respect to Laplace transforms. Specifically, not all functions have a Laplace transform because the integral may not converge. What is required for a function $f(t)$ to have a Laplace transform is:

1. It must be piece-wise continuous.
2. It must be of exponential order, which means there exist constants M and c such that
$$|f(t)| \le Me^{ct}, \quad 0 \le t < \infty.$$

Show that the function t^n is of exponential order, and hence has a Laplace transform, but that e^{t^2} is not of exponential order, and hence does not have a Laplace transform. See [9] for a more complete discussion.

8.13. Laplace transforms have "built into them" the fact that initial conditions occur at $t = 0$. If the initial conditions are not specified at $t = 0$, a simple change of variables with a time shift may be used so employ the method. For example, to solve

$$\ddot{x} + 3\dot{x} + 2x = e^{-t},$$

where $x(3) = 1$ and $\dot{x}(3) = 0$, let $y(t) = x(t-3)$. Solve this equation using Laplace transforms with that change of variables, and compare the answer to a solution obtained using a different method from Chapter 3.

8.14. Prove that if $\mathcal{L}\{f(t)\} = F(s)$, then

$$\mathcal{L}\{-tf(t)\} = \frac{d}{ds}F(s).$$

Hint: The right hand side of the equality is the derivative of $F(s)$ with respect to s, thus differentiate the definition of the Laplace transform with respect to s.

8.15. This problem is going to find the transfer function for a loudspeaker. From physics, if a wire of length l carries a current of i amperes and is arranged at a right angle to a magnetic field of strength B tesla, then the force (in newtons) on the wire is at a right angle to the plane of the wire and magnetic field and has a magnitude

$$f = Bli. \tag{8.26}$$

In a speaker, the wire is usually coiled to fit a longer length in a small space.

This is illustrated schematically in Figure 8.34. A current i through the coil c causes a force f on the mass (which, in this exercise, is the magnet) in the direction shown with a magnitude given by Equation (8.26).

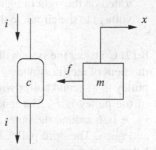

Fig. 8.34 Speaker model for Exercise 8.15.

1. Find the transfer function from the current through the speaker coil i to the location of the speaker mass x.

2. Now we attach a highpass filter to the speaker. The circuit is illustrated in Figure 8.35. An analysis of the properties of highpass filters is presented subsequently in Section 10.4.

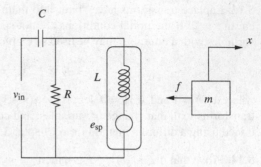

Fig. 8.35 Speaker model for Exercise 8.15.

Everything in the circuit should be obvious except the circle labeled e_{sp}. Just as in a dc motor, there is a voltage drop across the speaker due to the speaker moving. It is given by $e_{sp} = Bl\dot{x}$. Find the transfer function from v_{in} to x.

8.16. Consider the inverted pendulum illustrated on the left in Figure 8.36.

1. Determine the equation of motion for the system. Is it linear or nonlinear?
2. Assume that $\theta \ll 1$, for which $\sin\theta \approx \theta$ and $\cos\theta \approx 1$. If you substitute these approximations, is the equation linear or nonlinear?
3. Using the approximation, determine the transfer function from the input torque, τ to the angle of the pendulum θ.
4. Determine the transfer function from the input torque to the angular velocity of the pendulum.
5. Assume the torque is produced by a dc motor that is driven by the circuit illustrated on the right in Figure 8.36. Determine the transfer function from the input voltage to the circuit to the pendulum angle θ.
6. Assume the torque is produced by a dc motor that is driven by the circuit illustrated on the right in Figure 8.36. Determine the transfer function from the input voltage to the circuit to the pendulum angle angular velocity.

8.17. Consider the system illustrated on the right in Figure 8.8. A pulley with a mass moment of inertia J_1 and r_1 is subjected to a torque τ. A light belt connects the first pulley to a second pulley with an inner and outer spool. The mass moment of inertia of the pulley is J_2. The inner spool has radius r_1 and the outer spool has a radius r_2. A belt around the outer spool of the second pulley is attached to a third pulley and mass. The third pulley has radius r_2 and mass moment of inertia J_3. The mass has a mass m and is attached to a linear spring with spring constant k. The variable $x(t)$ represents the displacement of the mass and the variables $\theta_1(t)$, $\theta_2(t)$, and $\theta_3(t)$ represent the angular displacements of the pulleys.

1. Determine the transfer function from $\tau(t)$ to $x(t)$.

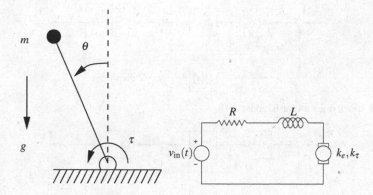

Fig. 8.36 System for Problem 8.16.

2. Assume the torque τ is imposed on the first pulley by a dc motor driven by the circuit illustrated on the left in Figure 8.8. Find the transfer function from the input voltage of the circuit $v_{in}(t)$ to $x(t)$.

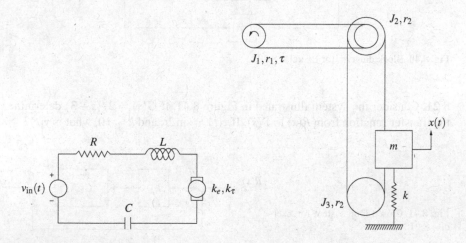

Fig. 8.37 System for Exercise 8.17.

8.18. Consider the block diagram illustrated in Figure 8.38. Determine the transfer function $G(s) = Y(s)/R(s)$.

8.19. Consider the block diagram illustrated in Figure 8.39. Determine the transfer function $G(s) = Y(s)/R(s)$.

8.20. Consider the block diagram illustrated in Figure 8.40. Determine the transfer function $G(s) = Y(s)/R(s)$.

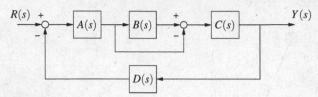

Fig. 8.38 Block diagram for Exercise 8.18.

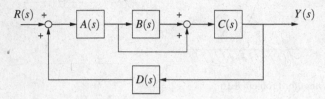

Fig. 8.39 Block diagram for Exercise 8.19.

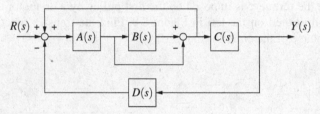

Fig. 8.40 Block diagram for Exercise 8.20.

8.21. Consider the system illustrated in Figure 8.41. If $G(s) = 2/(s+3)$ determine the transfer function from $R(s)$ to $Y(s)$. If $r(t) = \sin 2t$, and $k = 10$, what is $y(t)$?

Fig. 8.41 Closed loop system for Exercise 8.21.

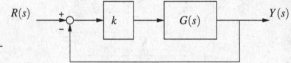

Chapter 9
Classical Control Theory: Analysis

The exploitation of *feedback* was fundamental to many engineering breakthroughs of the twentieth century. Although feedback was certainly manifested well before that, such as in Watt's steam engine governor, it was the need for and development of feedback amplifiers in the first half of the century that drove the development of the theory and analysis that made the use of feedback of general utility.

The utility of feedback has several aspects.

1. It may stabilize an otherwise unstable system.
2. It may improve the performance of a system.
3. It may make a system operate similarly regardless of variability in the components or operating conditions.
4. It may increase the bandwidth of the response of a system.

This chapter presents the basic analysis tools common to *classical control theory*, which is mainly concerned with problems formulated in the frequency domain with only one input and one output. The subject of feedback control is certainly much broader than that, extending to linear multi-input, multi-output systems ("modern control"), nonlinear systems, and so on.

In order to develop some intuition regarding feedback and because it is ubiquitous, Section 9.1 presents an introduction to proportional–derivative–integral control. It is intended as an introduction to this very common control methodology and as an introduction to the concept of feedback. Section 9.2 provides the definition of various quantities that are commonly used to specify desired control system behavior. The most critical section is Section 9.3 which discusses how system response is a function of the location in the complex plane of the poles of a transfer function. Section 9.4 presents a computational tool that is useful for determining the number of roots of a polynomial that are in the right-half of the complex plane. Right-half plane poles correspond to unstable responses, thus this is essentially a stability test. Section 9.5 considers the special case of second-order control systems. These are useful because many aspects of the system response for a second-order system can be exactly quantified based on the pole locations of the transfer function. Higher-order systems can also be approximated by second-order systems, so these results

B. Goodwine, *Engineering Differential Equations: Theory and Applications*,
DOI 10.1007/978-1-4419-7919-3_9, © Springer Science+Business Media, LLC 2011

may extend to those cases as well. The root locus method is the subject of Section 9.6, which is a systematic way to plot how the pole locations of a transfer function change as a gain is changed, which is useful as the most basic controller design tool. Finally, Section 9.7 presents the frequency response analysis methods.

9.1 PID Control

It is accurate to say that the vast majority of feedback control, particularly of mechanical systems, in industry is the so-called *proportional plus integral plus derivative (PID)* control. Designing PID controllers is usually somewhat ad hoc, however, this section will be devoted to the analysis of the features of these controllers as well as presenting a few "rules of thumb" with respect to designing them. The approach is by way of an exhaustive example.

Example 9.1. Consider the simple "robot arm" illustrated in Figure 9.1. The arm is a rigid link constrained to rotate about the fixed point A. The arm has a moment of inertia J and a center of mass located at a length l from the point A. The arm has a mass m and is subjected to gravity. The robot is fitted with a sensor that is able to determine the angle θ, which is measured from the horizontal position as indicated. Finally, a motor provides a torque τ about the point A.

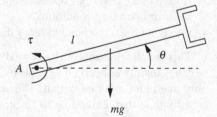

Fig. 9.1 Robot arm mechanism.

The purpose of *feedback control* is to determine a *control law* that makes the arm move to a desired angle, say θ_d, and either stay there if it is constant, or track it if it varies with time, despite any variable forces that may be applied to the arm (say by manipulating different objects of different masses). The idea of *feedback* is that the sensor measures θ which is then used (fed back) to determine a good value for the torque τ.

Using Newton's law, the equation of motion for the system is

$$J\ddot{\theta} = \tau - mgl\cos\theta.$$

This is an ordinary, second-order, nonlinear differential equation. In order to make it much more amenable to analysis, we assume that $\theta \ll 1$ so that $\cos\theta \approx 1$. In such a case, then the equation of motion is

$$J\ddot{\theta} = \tau - mgl. \tag{9.1}$$

9.1.1 Proportional Control

The idea of proportional control is simple and has an obvious intuitive appeal: have the control input be proportional to the error in the system.

Example 9.2. Returning to the system in Example 9.1, using proportional control would be to specify that

$$\tau(t) = k_p\left(\theta_d(t) - \theta(t)\right), \tag{9.2}$$

where, as stated previously, $\theta_d(t)$ is the desired position of the arm at time t. Thus, the torque τ is proportional to the error $\theta_d(t) - \theta(t)$. The proportionality constant k_p is called the *proportional gain*.

Depending upon the system, sometimes proportional control suffices. However, in the case at hand, it is straightforward to illustrate that the approach has several drawbacks. Substituting the control law from Equation (9.3) into Equation (9.1) gives

$$J\ddot{\theta} = k_p\left(\theta_d - \theta\right) - mgl$$

or

$$J\ddot{\theta} + k_p\theta = k_p\theta_d - mgl, \tag{9.3}$$

which is an ordinary, second-order, constant-coefficient, linear, inhomogeneous differential equation. The rest of this example analyzes this system using the tools and methods from Chapter 3. Obviously the homogeneous solution is

$$\theta_h(t) = c_1\cos\left(\sqrt{\frac{k_p}{J}}t\right) + c_2\sin\left(\sqrt{\frac{k_p}{J}}t\right)$$

and the particular solution depends upon the form of $\theta_d(t)$.

In order to proceed with the analysis, let θ_d be a specified constant. In that case,

$$\theta_p(t) = \theta_d - \frac{mgl}{k_p},$$

and the general solution is

$$\theta(t) = c_1\cos\left(\sqrt{\frac{k_p}{J}}t\right) + c_2\sin\left(\sqrt{\frac{k_p}{J}}t\right) + \theta_d - \frac{mgl}{k_p}.$$

In order to proceed further and plot some solutions, let us specify some numerical values for the initial conditions and all parameter values except for k_p; namely,

$$J = 1, \quad mgl = 1, \quad \theta(0) = 0, \quad \dot{\theta}(0) = 0, \quad \theta_d = 1,$$

in which case

$$\theta(t) = \left(\frac{1}{k_p} - 1\right) \cos\sqrt{k_p}t + 1 - \frac{1}{k_p}. \tag{9.4}$$

Although $\theta_d = 1$ violates the assumption that θ is small, because Equation (9.2) is linear, the nature of the solutions will be qualitatively the same as the case when the assumption is satisfied. In other words, due to linearity, the shape of the response will be the same regardless of whether the desired value is one or 0.01. The value of one is used simply to have the equations in a somewhat "normalized" form.

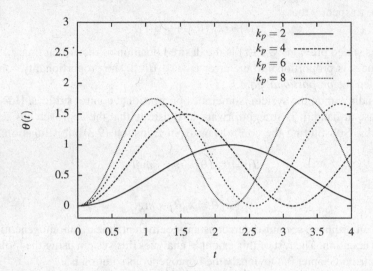

Fig. 9.2 Response of robot arm under proportional control.

A plot of the movement of the robot arm for various values of k_p is illustrated in Figure 9.2. Note the following features of proportional control.

1. The solutions are oscillatory and are not decaying.
2. As k_p increases the frequency of oscillation increases.
3. As k_p increases the average value of the oscillation approaches $\theta_d = 1$.
4. As k_p increases, the earliest time at which $\theta = \theta_d$ decreases.

Clearly, using proportional control for this example is not adequate if we desire that the robot arm approach θ_d and not oscillate about it.

Now, the same analysis is repeated using Laplace transforms and block diagrams.

Example 9.3. Referring back to Example 9.2, the equation of motion for proportional feedback is

$$J\ddot{\theta} + k_p\theta = k_p\theta_d - mgl.$$

Assuming zero initial conditions, the Laplace transform of the above equation is

$$Js^2\Theta(s) + k_p\Theta(s) = k_p\Theta_d(s) - \frac{mgl}{s}.$$

and

$$\Theta(s) = \frac{k_p\Theta_d s - mgl}{s(Js^2 + k_p)}. \tag{9.5}$$

Assuming, as before, $mgl = J = 1$ and $\theta_d = 1$ so $\Theta_d(s) = 1/s$, then

$$\Theta(s) = \frac{k_p - 1}{s(s^2 + k_p)}. \tag{9.6}$$

The inverse Laplace transform for Equation (9.6) is exactly the same as Equation (9.4), and for various k_p values must give the same response curves as are illustrated in Figure 9.2

Figure 9.3 illustrates a block diagram for proportional control. Using the rules for the block diagram representation of transfer functions we verify that this is the same representation as determined in Equation (9.6).

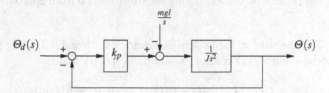

Fig. 9.3 Proportional control of a robot arm in Example 9.3.

The signal coming out of the first comparator is the error $E(s) = \Theta_d(s) - \Theta(s)$. Then it is multiplied by the proportional gain to give the torque $T(s) = k_p(\Theta_d(s) - \Theta(s))$. Figure 9.4 illustrates the same block diagram with these two signals labeled. Then it is added to the gravity term and finally multiplied by the robot dynamics to give $\Theta(s)$. Mathematically,

$$\Theta(s) = \left[k_p(\Theta_d(s) - \Theta(s)) - \frac{mgl}{s}\right]\frac{1}{Js^2}.$$

Solving for the arm angle gives

$$\Theta(s) = \frac{k_p\Theta_d s - mgl}{s(Js^2 + k_p)},$$

which is the same as Equation (9.5).

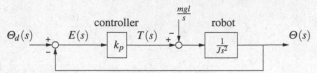

Fig. 9.4 Proportional control of a robot arm in Example 9.3.

9.1.2 Proportional plus Derivative Control

The idea of proportional plus derivative control is that, in contrast to proportional control, the control law should also reflect the derivative of the error. The intuition is that although the error may be positive or negative, how large the control input should be should also depend upon whether the error is increasing or decreasing.

Referring to Figure 9.2, the idea is that, for example, for the case of $k_p = 8$ and $0 < t < 0.6$ where the error, $\theta_d - \theta > 0$, because the error is decreasing, reducing τ relative to what it is for just proportional control should reduce the amount by which the response "overshoots" during the time interval from approximately $0.6 < t < 1.6$.

Example 9.4. Returning to the system in Examples 9.1 and 9.2, using proportional plus derivative control (PD control) would be to specify that

$$\tau(t) = k_p \left(\theta_d(t) - \theta(t)\right) + k_d \left(\dot{\theta}_d(t) - \dot{\theta}(t)\right),$$

where, as stated previously, $\theta_d(t)$ is the desired position of the arm at time t. Thus, the torque τ is not simply proportional to the error, but also includes a term proportional to the derivative of the error. The proportionality constant for the derivative term k_d is called the *derivative gain*.

Substituting this into the equation of motion and rearranging gives

$$J\ddot{\theta} + k_d\dot{\theta} + k_p\theta = k_p\theta_d + k_d\dot{\theta}_d - mgl. \tag{9.7}$$

In this case, the homogeneous solution is

$$\theta_h = e^{-k_d t/2J} \left(c_1 \cos\left(\frac{\sqrt{4k_p J - k_d^2}}{2J}t\right) + c_2 \sin\left(\frac{\sqrt{4k_p J - k_d^2}}{2J}t\right) \right).$$

Note that the oscillations due to the homogeneous solution decay with time as long as $k_p, k_d, J > 0$, so this potentially improves the performance over proportional control because the continued oscillations present in proportional control decay with derivative control. Thus, the steady-state solution depends only upon the form of the particular solution, which, of course, depends upon the exact nature of $\theta_d(t)$.

In order to continue the analysis as before, let us consider the case where θ_d is a constant. In that case, inasmuch as $\dot{\theta}_d = 0$, the particular solution is the same as in the proportional control case in Example 9.2 and hence

$$\theta(t) = e^{-k_d t/2J} \left[c_1 \cos \left(\frac{\sqrt{4k_p J - k_d^2}}{2J} t \right) + c_2 \sin \left(\frac{\sqrt{4k_p J - k_d^2}}{2J} t \right) \right] + \theta_d - \frac{mgl}{k_p}.$$

Clearly, because the homogeneous solution decays for positive k_d, k_p, and J, the steady-state solution is

$$\theta_{ss}(t) = \theta_d - \frac{mgl}{k_p}.$$

Before plotting some solutions with numerical values, note for the steady state response, a very large k_p is desirable because it makes $\theta_{ss} \to \theta_d$. Also, if k_d increases, any oscillations should decay more quickly.

To plot a solution, let $J = 1$, $mgl = 1$, $\theta_d = 1$, $\theta(0) = 0$, and $\dot{\theta}(0) = 0$ in which case

$$\theta(t) = e^{-k_d t/2} \left[\left(\frac{1}{k_p} - 1 \right) \cos \left(\frac{\sqrt{4k_p - k_d^2}}{2} t \right) \right.$$

$$\left. + \left(\frac{k_d (1 - k_p)}{k_p \sqrt{4k_p - k_d^2}} \right) \sin \left(\frac{\sqrt{4k_p - k_d^2}}{2} t \right) \right] + 1 - \frac{1}{k_p}. \tag{9.8}$$

Figure 9.5 illustrates the response for a fixed $k_p = 8.0$ and various k_d values. Note that as k_d is increased, the oscillations decay more quickly and the value of the first maximum (near $t = 1.0$) is decreased. Also note that changing k_d does not affect the steady-state value of $\theta(t)$ and that the steady-state error is nonzero.

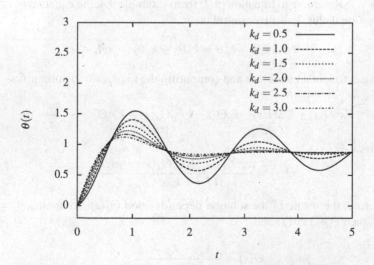

Fig. 9.5 Response of robot arm under PD control with fixed $k_p = 8$ and various k_d.

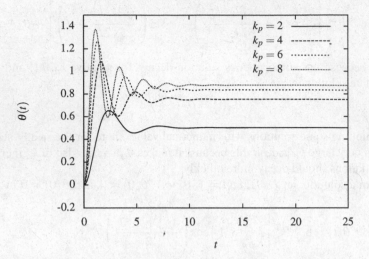

Fig. 9.6 Response of robot arm under PD control with fixed $k_d = 1$ and various k_p.

Figure 9.6 illustrates the response for a fixed $k_d = 1.0$ and various k_p values. Note that as k_p is increased, the final steady-state error decreases (recall $\theta_d = 1$), but that the initial overshoot is greatly increased and the frequency of oscillation increases.

Now, the same analysis is repeated using Laplace transforms.

Example 9.5. Returning to Equation (9.7) from Example 9.4, the equation of motion for proportional plus derivative control is

$$J\ddot{\theta} + k_d\dot{\theta} + k_p\theta = k_p\theta_d + k_d\dot{\theta}_d - mgl.$$

Assuming zero initial conditions and computing the Laplace transform gives

$$Js^2\Theta(s) + k_ds\Theta(s) + k_p\Theta(s) = k_p\Theta_d(s) + k_ds\Theta_d(s) - \frac{mgl}{s}$$

which gives

$$\Theta(s) = \frac{k_d\Theta_d(s)s^2 + k_p\Theta_d(s)s - mgl}{s\left(Js^2 + k_ds + k_p\right)}.$$

As before, the nature of the solution depends upon $\Theta_d(s)$. Assuming $\theta_d = J = mgl = 1$, then $\Theta_d(s) = 1/s$ and

$$\Theta(s) = \frac{k_ds + k_p - 1}{s\left(s^2 + k_ds + k_p\right)},$$

and the inverse Laplace transform is the same as the solution from Example 9.4 given by Equation (9.8) and plotted in Figures 9.5 and 9.6 for various gain values.

Also, using the final value theorem (Theorem 8.2),

$$\lim_{t \to \infty} \theta(t) = \lim_{s \to 0} s\Theta(s) = \lim_{s \to 0} s \frac{k_p - 1}{s\left(s^2 + k_d s + k_p\right)} = 1 - \frac{1}{k_p},$$

which shows that as k_p is increased, the steady-state error decreases.

9.1.3 Proportional plus Integral plus Derivative Control

With proportional plus derivative plus integral control, a third term is added to proportional plus derivative control that is, naturally, proportional to the integral of the error. In Examples 9.2–9.5 there was always a steady-state error, $\lim_{t \to \infty} \theta(t) \neq \theta_d$, and the effect of integral control is to reduce or, in the case of a second-order system, completely eliminate that error.

The idea behind integral control is that as time increases, if there is a consistent error, the input to the system will increase with time to compensate for the error. The need for integral control in many problems is obvious considering the robot arm from these examples because a robot is not very useful if it does not end up where we want it to be, and in the case of proportional and proportional plus derivative control considered previously, there was a steady-state error.

To see why this is the case, for both proportional and proportional plus derivative control, if there is no error, that is, $\theta = \theta_d$ then $\tau = 0$. If the torque is zero, then there is nothing to offset the torque caused by gravity and the arm cannot stay at the desired location. The steady-state value in the case of PD control is the angle at which the error is great enough to cause an error that will result in a torque that will offset the torque due to gravity.

The following example illustrates the the efficacy of integral control with respect to eliminating steady-state error.

Example 9.6. Returning yet again to the system from Example 9.1, adding integral control yields an expression for the torque of the form

$$\tau = k_p \left(\theta_d - \theta\right) + k_d \left(\dot{\theta}_d - \dot{\theta}\right) + k_i \int_0^t \theta_d(\hat{t}) - \theta(\hat{t}) \mathrm{d}\hat{t}$$

so the equation of motion for the robot arm becomes

$$J\ddot{\theta} + k_d\dot{\theta} + k_p\theta = k_p\theta_d + k_d\dot{\theta}_d + k_i \int_0^t \theta_d(\hat{t}) - \theta(\hat{t}) \mathrm{d}\hat{t} - mgl. \tag{9.9}$$

This is a second-order integral-differential equation and there are various ways to handle the integral term.

One approach to solving Equation (9.9) is to convert the system into a coupled set of ordinary differential equations. Note that because

$$\frac{d}{dt} \int_0^t \theta_d(\hat{t}) - \theta(\hat{t}) d\hat{t} = \theta_d(t) - \theta(t),$$

if we define a new variable, $\hat{I}$ (I for "integral"), then Equation (9.9) is equivalent to the two ordinary differential equations

$$J\ddot{\theta} + k_d \dot{\theta} + k_p \theta = k_p \theta_d + k_d \dot{\theta}_d + k_i \hat{I} - mgl$$

$$\dot{\hat{I}} = \theta_d - \theta.$$

If

$$x_1 = \theta, \quad x_2 = \dot{\theta}, \quad x_3 = \hat{I},$$

then

$$\frac{d}{dt} \begin{bmatrix} x_1 \\ x_2 \\ x_3 \end{bmatrix} = \begin{bmatrix} x_2 \\ \frac{k_p \theta_d + k_d \dot{\theta}_d + k_i x_3 - k_d x_2 - k_p x_1 - mgl}{J} \\ \theta_d - x_1 \end{bmatrix},$$

which can be solved using the methods from Chapter 6 or, perhaps more conveniently, can be solved numerically using the methods from Chapter 12.

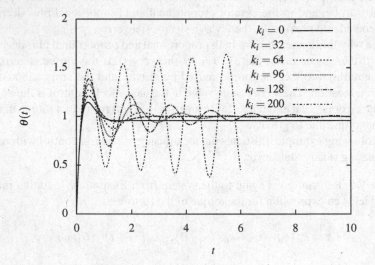

Fig. 9.7 Response of robot arm under PID control for fixed $k_p = 24$ and $k_d = 8$ and various k_i.

Figure 9.7 illustrates the response of the arm (computed numerically) for fixed values of k_p, k_d and various k_i with $J = mgl = 1$. Note that any nonzero value for k_i eliminates the steady-state error; however, increasing k_i increases the magnitude and duration of the transient oscillations, and, if large enough, destabilizes the system ($k_i = 200$).

Finally, for completeness, we determine the PID control equations using Laplace transforms.

Example 9.7. The equation of motion for the robot arm under PID control was given in Equation (9.9) and is

$$J\ddot{\theta} + k_d\dot{\theta} + k_p\theta = k_p\theta_d + k_d\dot{\theta}_d + k_i \int_0^t \theta_d(\hat{t}) - \theta(\hat{t})d\hat{t} - mgl.$$

Computing the Laplace transform with zero initial conditions gives

$$\left(Js^2 + k_ds + k_p + \frac{k_i}{s}\right)\Theta(s) = \left(k_p + k_ds + \frac{k_i}{s}\right)\Theta_d(s) - \frac{mgl}{s}$$

or

$$\Theta(s) = \frac{k_ds^2 + k_ps + k_i}{Js^3 + k_ds^2 + k_ps + k_i}\Theta_d(s) - \frac{mgl}{Js^3 + k_ds^2 + k_ps + k_i}. \qquad (9.10)$$

Using Theorem 8.2 and assuming $\Theta_d(s) = 1/s$, that is, θ_d is a unit step input

$$\lim_{t\to\infty}\theta(t) = \lim_{s\to 0} s\Theta(s) = \lim_{s\to 0} s\frac{k_ds^2 + k_ps + k_i}{Js^3 + k_ds^2 + k_ps + k_i}\frac{1}{s} - s\frac{mgl}{Js^3 + k_ds^2 + k_ps + k_i} = 1;$$

that is, there is no steady-state error.

9.2 Time Domain Specifications

The qualitative discussions regarding the effect of altering controller gains in Section 9.1 practically beg us to be more precise and quantitative about the nature of the response of a system. Consider a generic system response to a unit step input illustrated in Figure 9.8.

In the diagram, the following quantities are defined.

1. The *rise time* t_r is the time at which the response is first equal to the magnitude of the input. For a unit step input, it is the time at which the response is first equal to one. If the system is overdamped, then the response may only asymptotically approach the desired value. In that case the rise time may be defined to be the time it takes to achieve 90% of the steady-state value. Unless otherwise specified, in this book the rise time refers to the second definition.
2. The *peak time* t_p is the time at which the response reaches its maximum value.
3. The *settling time* t_s is the time after which the response always stays within a specified range of its steady-state value. In Figure 9.8, this is illustrated as 0.9 ± 0.05, but other ranges may be specified as a certain percentage, such as, "the 3% settling time."
4. The *maximum percentage overshoot* O is defined to be the percentage that the peak value y_p exceeds the desired value, y_d; that is,

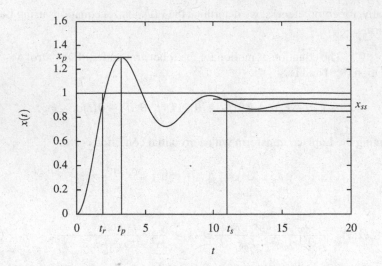

Fig. 9.8 Time domain specifications definitions for a unit step input.

$$O = \frac{x_p - x_{ss}}{x_{ss}}.$$

Collectively these terms are referred to as the *transient response* because they describe how the system transitions from the initial conditions to the steady-state behavior, but do not describe the steady-state behavior. With regard to the steady state, the *steady-state error* is the difference between the steady-state value of the response and the desired value; that is,

$$e_{ss} = x_d - x_{ss}.$$

The tools used to determine the nature of the transient response are discussed subsequently in Section 9.3. The usual tool used to compute the steady-state error is Theorem 8.2, the final value theorem.

These time domain specifications may be used to specify the manner in which a control system should respond. For example, it may be desired that a control surface on an airplane wing, say an aileron, respond with less than one second rise time, less than 1% overshoot, and a settling time less than three seconds. As is the case with many design problems, it may or may not be possible to meet all the specifications. Whether it is possible depends, among other things, upon the dynamics of the system and the nature of the actuation.

9.3 Response Versus Pole Location

This section considers the mathematical basis for the rest of this chapter. Understanding this section is critical for a fundamental understanding of what follows. The main concept is that the nature of the response of a system is governed by the location in the complex plane of the poles of the transfer function describing the system.

If we consider a generic transfer function $G(s)$ and the relationship between the reference signal for the system $R(s)$ and the output $Y(s)$, we have

$$Y(s) = G(s)R(s).$$

If we were to solve this for the time domain response of the system $y(t)$, we would need to know the input $R(s)$ and then would compute a partial fraction expansion of $G(s)R(s)$ to algebraically manipulate the expression to be a combination of terms that appear in a Laplace transform table.

Example 9.8. To solve

$$Y(s) = -\frac{6s+5}{(s+3)(s^2+4)} \tag{9.11}$$

for $y(t)$, we would convert

$$Y(s) = -\frac{6s+5}{(s+3)(s^2+4)} = \frac{c_1}{s+3} + \frac{c_2 s + c_3}{s^2+4} = \frac{1}{s+3} - \frac{s+3}{s^2+4}$$
$$= \frac{1}{s+3} - \frac{s}{s^2+4} - \frac{3}{2}\frac{2}{s^2+4},$$

which corresponds to

$$y(t) = e^{-3t} - \cos 2t - \frac{3}{2}\sin 2t. \tag{9.12}$$

Now, if we look at the original transfer function in Equation (9.11), the poles (the values of s for which the denominator is equal to zero; see Definition 8.4), are $s = -3$ and $s = \pm 2i$. It is no coincidence that the solution in Equation (9.12) is a linear combination of an exponential with a negative three in the exponent, corresponding to the pole at $s = -3$ and sine and cosine functions with a frequency of two, corresponding to the complex conjugate pair of poles at $s = \pm 2i$.

Because they are commonly used to characterize the nature of a transfer function, the time domain solution of the output for two specific inputs are given names.

Definition 9.1. For a transfer function $G(s)$, input $R(s)$, and output $Y(s)$, where

$$Y(s) = G(s)R(s),$$

the *unit impulse response*, or simply the *impulse response* is the inverse Laplace transform of the output when the input is an impulse. Because $\mathcal{L}(\delta(t)) = 1$, the impulse response is given by the inverse Laplace transform of the transfer function

$$y_\delta(t) = \mathcal{L}^{-1}(G(s)).$$

Definition 9.2. For a transfer function $G(s)$, input $R(s)$, and output $Y(s)$, where

$$Y(s) = G(s)R(s),$$

the *unit step response*, or simply the *step response* is the inverse Laplace transform of the output when the input is a unit step function. Because $\mathcal{L}(\mathbb{1}(t)) = 1/s$, the impulse response is given by

$$y_{\mathbb{1}}(t) = \mathcal{L}^{-1}\left(\frac{G(s)}{s}\right).$$

A detailed study of Table 8.1 makes it clear that what differentiates the fundamental nature of the response of a system is the location of the poles. We consider the various possible cases that depend upon whether the pole is real, zero, purely imaginary, or complex.

9.3.1 Real Poles

First we consider the case where a transfer function has a pole that is real. Consider

$$Y(s) = \frac{1}{s-p}R(s). \tag{9.13}$$

Note that $Y(s)$ has a pole at $s = p$. Regardless of the nature of $R(s)$, a partial fraction expansion of Equation (9.13) is of the form

$$Y(s) = \frac{c_1}{s-p} + \sum \hat{R}(s),$$

where $\sum \hat{R}(s)$ are the terms in the partial fraction expansion due to the input.

So, regardless of the input, if p is real, $y(t)$ will contain a term of the form e^{pt}; that is,

$$y(t) = c_1 e^{pt} + \text{other terms}.$$

Hence, we have the following proposition.

Proposition 9.1. *If a transfer function has a pole that is a real at $s = p$, it will have an exponential term in the time domain solution and that exponential term will decay to zero if $p < 0$, will grow unbounded if $p > 0$, and will be a constant if $p = 0$.*

Example 9.9. Predict the unit step response of the two transfer functions

$$G_1(s) = \frac{2}{s+2}, \quad G_2(s) = \frac{4}{s+4}$$

without actually computing the inverse Laplace transform.

We want to compare

$$Y_1(s) = G_1(s)R(s) = \frac{2}{s+2}\frac{1}{s}, \quad Y_2(s) = G_2(s)R(s) = \frac{4}{s+4}\frac{1}{s}.$$

Using Theorem 8.2,

$$\lim_{t\to\infty} y_1(t) = \lim_{s\to 0} s\frac{2}{s+2}\frac{1}{s} = 1,$$

and

$$\lim_{t\to\infty} y_2(t) = \lim_{s\to 0} s\frac{4}{s+4}\frac{1}{s} = 1.$$

The final value theorem may be applied to both of these because all the poles of both $sG_1(s)R(s)$ and $sG_2(s)R(s)$ are in the left half-plane if $R(s)$ is a step function. If we were to compute them, the partial fraction expansions would be of the form

$$Y_1(s) = \frac{c_1}{s+2} + \frac{c_2}{s},$$

and

$$Y_2(s) = \frac{c_1}{s+4} + \frac{c_2}{s}.$$

Observe that in both cases the first term gives an exponential solution with a negative coefficient in the exponent and the second term gives a constant value. Because $G_1(s)$ has a pole at $s = -2$ and $G_2(s)$ has a pole at $s = -4$, as is illustrated in Figure 9.9, the exponential part of the solution in $Y_2(s)$ decays more quickly than the exponential part in $Y_1(s)$. Hence we may conclude that $y_2(t)$ converges more quickly to the steady state value than $y_1(t)$. This is verified in Figure 9.10 which compares the two solutions.

The association between pole locations, Figure 9.9, and the nature of the response, Figure 9.10, cannot be emphasized enough. In this particular example, if another system were compared that had a pole farther to the left, then its response would be even faster, and if another system had a pole farther to the right (but still less than zero), its response would be slower. If any system has a pole to the right of the imaginary axis, the solution will blow up; that is, the system will be unstable.

For comparison, the impulse responses of two transfer functions with pole locations at $s = -2$ and $s = -4$ are illustrated in Figure 9.11. Again, the system with the pole farther to the left has a faster decaying transient response.

9.3.2 Poles at the Origin

Now we consider some poles at the origin. Consider

$$Y(s) = \frac{1}{s}R(s). \tag{9.14}$$

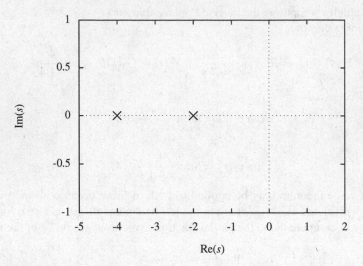

Fig. 9.9 Pole locations for $G_1(s) = 2/(s+2)$ and $G_2(s) = 4/(s+4)$.

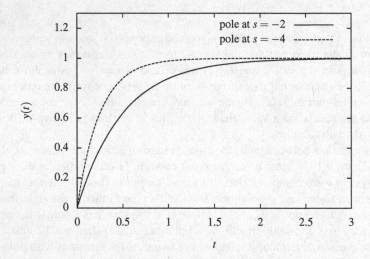

Fig. 9.10 Step response for $G_1(s) = 2/(s+2)$ and $G_2(s) = 4/(s+4)$.

Note that $Y(s)$ has a pole at $s = 0$. Regardless of the nature of $R(s)$, a partial fraction expansion of Equation (9.14) will be of the form

$$Y(s) = \frac{c_0}{s} + \sum \hat{R}(s),$$

where $\sum \hat{R}(s)$ are the terms in the partial fraction expansion due to the input.

So, regardless of the input, $y(t)$ will contain a term of the form c_0; that is,

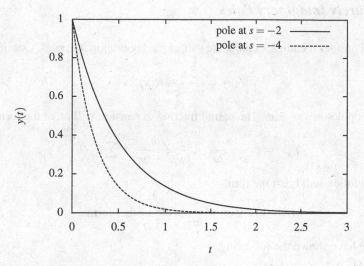

Fig. 9.11 Impulse response for $G_1(s) = 1/(s+2)$ and $G_2(s) = 1/(s+4)$.

$$y(t) = c_0 + \text{other terms}.$$

We may conclude from that, in general, if a transfer function has a pole at the origin it will have a constant term in the solution.

If the transfer function has multiple poles at the origin; that is,

$$Y(s) = \frac{1}{s^n} R(s),$$

then the partial fraction expansion will be of the form

$$Y(s) = \frac{c_0 s^{n-1} + c_1 s^{n-2} + \cdots + c_{n-1}}{s^n} + \sum \hat{R}(s) = \frac{c_0}{s} + \frac{c_1}{s^2} + \cdots + \frac{c_{n-1}}{s^n} + \sum \hat{R}(s).$$

So, *regardless of the input*, $y(t)$ will contain an $(n-1)$th-order polynomial in t; that is,

$$y(t) = c_0 + c_1 t + \frac{c_2}{2} t^2 + \cdots + \frac{c_{n-1}}{(n-1)!} t^{n-1} + \text{other terms}.$$

Hence, we have the following.

Proposition 9.2. *If a transfer function has multiple poles at the origin, it will have a polynomial term in the solution that has an order one less than the multiplicity of the pole at the origin.*

9.3.3 Purely Imaginary Poles

Now we consider a complex conjugate pair of purely imaginary poles. Consider

$$Y(s) = \frac{1}{s^2 + \omega^2} R(s),$$

which has poles at $s = \pm i\omega$. The partial fraction expansion will be of the form

$$Y(s) = c_1 \frac{s}{s^2 + \omega^2} + \frac{c_2}{\omega} \frac{\omega}{s^2 + \omega^2} + \sum \hat{R}(s)$$

and the solution will be of the form

$$y(t) = c_1 \cos \omega t + \frac{c_2}{\omega} \sin \omega t + \text{other terms.}$$

So, we have shown the following.

Proposition 9.3. *If a transfer function has a purely imaginary complex conjugate pair of poles, it will have sine and cosine terms in the solution. The distance from the real axis to the pole is equal to the frequency of oscillation.*

Example 9.10. The poles of

$$G_1(s) = \frac{2}{s^2 + 4}$$

and

$$G_2(s) = \frac{8}{s^2 + 16}$$

are plotted in Figure 9.12. The corresponding step responses are plotted in Figure 9.13. Note as the poles move farther from the real axis, the frequency of oscillation increases. For comparison, the impulse response, when $R(s) = 1$, for both cases is illustrated in Figure 9.14.

9.3.4 Complex Conjugate Poles

Finally, the last case to consider is when a transfer function contains a complex conjugate pair of poles with nonzero real and imaginary parts. Consider

$$Y(s) = \frac{1}{(s+a)^2 + b^2} R(s)$$

which has a complex conjugate pair of poles at $s = -a \pm bi$. The partial fraction expansion is of the form

$$Y(s) = c_1 \frac{s+a}{(s+a)^2 + b^2} + c_2 \frac{b}{(s+a)^2 + b^2} + \sum \hat{R}(s)$$

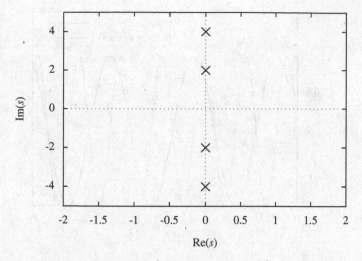

Fig. 9.12 Pole locations for $G_1(s) = 2/(s^2 + 4)$ and $G_2(s) = 8/(s^2 + 16)$.

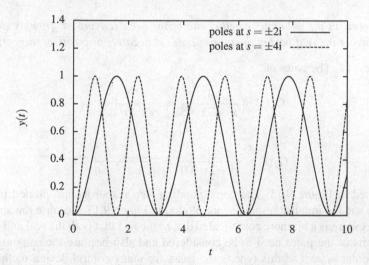

Fig. 9.13 Step response for $G_1(s) = 2/(s^2 + 4)$ and $G_2(s) = 8/(s^2 + 16)$.

and hence the solution is of the form

$$y(t) = c_1 e^{-at} \cos bt + c_2 e^{-at} \sin bt + \text{other terms}.$$

This shows the following.

Proposition 9.4. *If a transfer function contains a complex conjugate pair of poles with nonzero real part, it will have exponentially decaying or growing sine and*

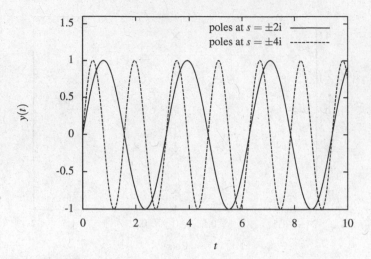

Fig. 9.14 Impulse response for $G_1(s) = 2/(s^2+4)$ and $G_2(s) = 4/(s^2+16)$.

cosine terms in the solution. Whether the terms are decaying or growing depend upon whether the real part of the pair of poles is negative or positive, respectively.

Example 9.11. The poles of

$$G_1(s) = \frac{1}{s^2+2s+5} = \frac{1}{(s+1)^2+4}$$

and

$$G_2(s) = \frac{1}{s^2+2s+10} = \frac{1}{(s+1)^2+9}$$

are plotted in Figure 9.15. The corresponding step responses are plotted in Figure 9.16 and the impulse responses are plotted in Figure 9.17. Because the analysis of the response is a bit more complicated due to the fact that both the real and imaginary parts of the poles need to be considered and also because the response of a second-order system of this type is the basis for many control design methods, a complete analysis of a system with complex conjugate poles is discussed subsequently, in Section 9.5.

A summary of these results is illustrated in Figure 9.18. Any poles in the right half-plane lead to instabilities. Complex conjugate purely imaginary poles contribute harmonic solutions. Poles at the origin contribute polynomial solutions. Negative real poles contribute to decaying exponential terms and complex conjugate poles with negative real part contribute decaying sine and cosine terms.

Also, because the real part of any pole corresponds exactly to the coefficient of time in an exponential, we may talk about "fast" and "slow" poles. In particular, for poles with negative real part, the farther the pole is to the left, the faster it decays.

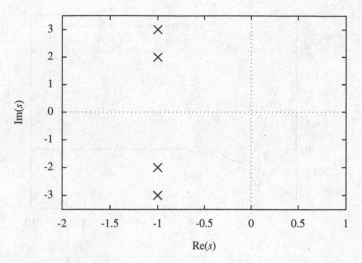

Fig. 9.15 Pole locations for $G_1(s) = 1/\left((s+1)^2+4\right)$ and $G_2(s) = 1/\left((s+1)^2+9\right)$.

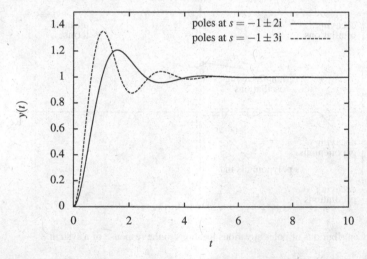

Fig. 9.16 Step response for $G_1(s) = 1/\left((s+1)^2+4\right)$ and $G_2(s) = 1\left((s+1)^2+9\right)$.

All poles with positive real part are unstable; however, the larger the magnitude of the real part of the positive pole, the faster the instability grows. This is qualitatively summarized in Figure 9.19.

Based upon what we know so far, we can state the following important result.

Proposition 9.5. *Given a transfer function $G(s)$, the impulse and step responses are stable if and only if all the poles of*

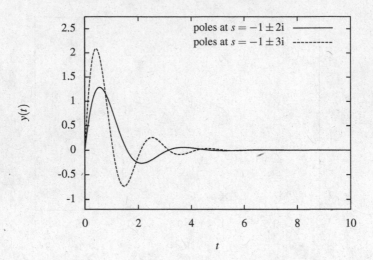

Fig. 9.17 Impulse response for $G_1(s) = 1/\left((s+1)^2 + 4\right)$ and $G_2(s) = 1/\left((s+1)^2 + 9\right)$.

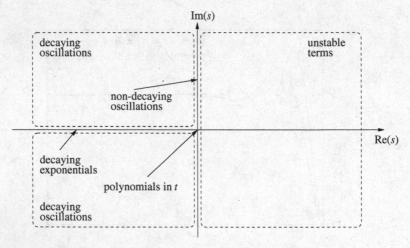

Fig. 9.18 Contributions of poles at various locations to the response of a system.

$$Y(s) = G(s)R(s)$$

are in the left half of the complex plane, where the input $R(s) = 1$ or $R(s) = 1/s$, in the case of the impulse and step responses, respectively.

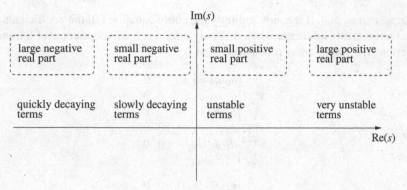

Fig. 9.19 The effect of the magnitude of the real part of a pole on the nature of its contribution to the solution.

9.4 Stability

From Section 9.3 it is clear that if a transfer function has any poles in the right half complex plane then corresponding components of the solution will grow unbounded. Hence, it is desirable to know if a transfer function has any poles in the right half of the complex plane. In such a case, regardless of the input, the effect of the initial conditions will persist and grow exponentially in magnitude, that is, "blow up." Poles with zero real part on the imaginary axis correspond to sinusoidal solutions that do not decay. However, for engineering systems it is arguably impossible to determine if the poles are exactly on the imaginary axis, and if we are not able to tell on which side of the axis a pole is, then caution would dictate treating it as unstable. Most of the focus in classical control theory is concerned with altering a system through feedback to be stable with only left half-plane poles.

Hence, what we want is a means to determine if all the poles of a transfer function are in the left half of the complex plane. For first- and second-order systems, this is easy to do because the polynomial can be easily factored, but for higher-order systems that are difficult to factor by hand, it is not. There are various methods to determine when a system has any right half-plane poles,[1] but the one typically covered in undergraduate controls courses is the so-called *Routh criterion*.

The method is based upon constructing an array and examining the signs of the elements in the first column of the array. First we define the array and then we present the stability test. Consider an nth-order polynomial of the form

$$D(s) = a_0 s^n + a_1 s^{n-1} + \cdots + a_{n-1} s + a_n,$$

and consider the nature of the solutions to

$$a_0 s^n + a_1 s^{n-1} + \cdots + a_{n-1} s + a_n = 0.$$

---

[1] The Hurwitz criterion, the Hermite criterion, the Liénard–Chipart criterion, and the Kharitonov test are all examples of such stability tests.

Assume that $a_0 > 0$. If it is not, multiply the entire equation (all the coefficients) by -1. Our interest is the case when this polynomial is the denominator of a transfer function. The *Routh array* is constructed as follows,

$$
\begin{array}{c|cccccc}
s^n & a_0 & a_2 & a_4 & a_6 & \cdots & 0 \\
s^{n-1} & a_1 & a_3 & a_5 & a_7 & \cdots & 0 \\
s^{n-2} & b_1 & b_2 & b_3 & \cdots & & 0 \\
s^{n-3} & c_1 & c_2 & c_3 & \cdots & 0 & 0 \\
s^{n-4} & d_1 & d_2 & d_3 & \cdots & 0 & 0 \\
\vdots & \vdots & & & & & \\
s^0 & e_1 & 0 & 0 & 0 & 0 & 0
\end{array}
$$

and where the a_i terms are the coefficients of the characteristic polynomial and the other terms are computed as

$$
b_1 = -\frac{\begin{vmatrix} a_0 & a_2 \\ a_1 & a_3 \end{vmatrix}}{a_1} \qquad c_1 = -\frac{\begin{vmatrix} a_1 & a_3 \\ b_1 & b_2 \end{vmatrix}}{b_1} \qquad d_1 = -\frac{\begin{vmatrix} b_1 & b_2 \\ c_1 & c_2 \end{vmatrix}}{c_1}
$$

$$
b_2 = -\frac{\begin{vmatrix} a_0 & a_4 \\ a_1 & a_5 \end{vmatrix}}{a_1} \qquad c_2 = -\frac{\begin{vmatrix} a_1 & a_5 \\ b_1 & b_3 \end{vmatrix}}{b_1} \qquad d_2 = -\frac{\begin{vmatrix} b_1 & b_3 \\ c_1 & c_3 \end{vmatrix}}{c_1}
$$

$$
b_3 = -\frac{\begin{vmatrix} a_0 & a_6 \\ a_1 & a_7 \end{vmatrix}}{a_1} \qquad c_3 = -\frac{\begin{vmatrix} a_1 & a_7 \\ b_1 & b_4 \end{vmatrix}}{b_1} \qquad d_3 = -\frac{\begin{vmatrix} b_1 & b_4 \\ c_1 & c_4 \end{vmatrix}}{c_1}
$$

$$\vdots \qquad\qquad\qquad \vdots \qquad\qquad\qquad \vdots$$

where any term that is not defined is zero.

Finally, the point of all of this is the following theorem that provides necessary and sufficient conditions for stability of a transfer function.

Theorem 9.1. *All the solutions to*

$$
a_0 s^n + a_1 s^{n-1} + \cdots + a_{n-1} s + a_n = 0
$$

are in the left half of the complex plane if and only if all the elements of the first column of the corresponding Routh array are positive.[2]

A proof is not presented inasmuch as it depends on results that are generally outside the main focus of this text. An interested reader is referred to some of the original work of Routh [46].

Before we present any examples, a necessary condition for stability is presented.

---

[2] In fact, the Routh array gives even more information. The number of sign changes in the first column is equal to the number of roots in the right half-plane.

Proposition 9.6. *If $a_0 > 0$, and if any of the coefficients in*

$$D(s) = a_0 s^n + a_1 s^{n-1} + \cdots + a_{n-1} s + a_n.$$

are not positive, then $D(s)$ will have at least one root that is not in the left half of the complex plane.

The proof is left as an exercise.[3] Note this is a necessary condition, meaning that if any of the coefficients are zero or negative, then we know there is at least one root that is not in the left half-plane; however, if they are all positive we cannot conclude anything.

Example 9.12. Determine if all the poles of

$$G(s) = \frac{s+6}{s^4 + 7s^3 + 18s^2 + 22s + 12}$$

are in the left half of the complex plane. This is, of course, equivalent to determining if all the solutions to

$$s^4 + 7s^3 + 18s^2 + 22s + 12 = 0 \tag{9.15}$$

have a negative real part.

So, the start of the Routh array is

$$\begin{array}{c|cccc} s^4 & 1 & 18 & 12 & 0 \\ s^3 & 7 & 22 & 0 & 0 \end{array}$$

Computing the next row,

$$b_1 = \frac{(7)(18) - 22}{7} = \frac{104}{7}$$

$$b_2 = \frac{(7)(12) - 0}{7} = 12$$

so the array is

$$\begin{array}{c|cccc} s^4 & 1 & 18 & 12 & 0 \\ s^3 & 7 & 22 & 0 & 0 \\ s^2 & \frac{104}{7} & 12 & 0 & 0 \end{array}$$

Computing the next row,

---

[3] Basically all that is necessary is to show that if factors with negative real parts are multiplied together to construct the polynomial, all the coefficients are positive.

$$c_1 = \frac{22\frac{104}{7} - (7)(12)}{\frac{104}{7}} = \frac{425}{26}$$

$$c_2 = 0,$$

so the array is

$$
\begin{array}{c|cccc}
s^4 & 1 & 18 & 12 & 0 \\
s^3 & 7 & 22 & 0 & 0 \\
s^2 & 104 & 12 & 0 & 0 \\
s^1 & \frac{425}{26} & 0 & 0 & 0
\end{array}
$$

Finally,

$$d_1 = 12,$$

so the complete array is

$$
\begin{array}{c|cccc}
s^4 & 1 & 18 & 12 & 0 \\
s^3 & 7 & 22 & 0 & 0 \\
s^2 & 104 & 12 & 0 & 0 \\
s^1 & \frac{425}{26} & 0 & 0 & 0 \\
s^0 & 12
\end{array}
$$

Because there are no sign changes in the first row, all the solutions to Equation (9.15) are in the left half-plane.

We can use this to accomplish a bit more than simply determine whether poles are in the right half-plane. After all, with a modern computer, it is easy to numerically factor polynomials, such as with the `roots()` command in MATLAB or Octave. In particular, the Routh array allows us to determine ranges of parameter values for which a transfer function is stable.

Example 9.13. Determine the values of k for which

$$G(s) = \frac{k\frac{s+2}{s^2-2s+2}}{1 + k\frac{s+2}{s^2-2s+2}}$$

is stable.

Simplifying the denominator gives

$$D(s) = s^2 + (k-2)s + (2+2k).$$

Constructing the Routh array gives

$$
\begin{array}{c|ccc}
s^2 & 1 & 2+2k & 0 \\
s^1 & k-2 & 0 & 0
\end{array}
$$

Computing the last row gives the complete array

$$
\begin{array}{c|ccc}
s^2 & 1 & 2+2k & 0 \\
s^1 & k-2 & 0 & 0 \\
s^0 & 2k+2 & 0 & 0
\end{array}.
$$

In order for there to be no sign change from the s^2 row to the s^1 row, we need that $k > 2$. In order for the first element s^0 to be greater than zero we need $k > -1$. In order to satisfy both, we need $k > 2$.

A final example illustrates the obvious fact that there will not necessarily be any values for k to make some transfer functions stable.

Example 9.14. Determine the values of k for which

$$
G(s) = \frac{k\frac{s-2}{s^2-2s+2}}{1+k\frac{s-2}{s^2-2s+2}}
$$

is stable.

In this case the characteristic polynomial is

$$
D(s) = s^2 + (k-2)s + (2-2k),
$$

and the Routh array is

$$
\begin{array}{c|ccc}
s^2 & 1 & 2-2k & 0 \\
s^1 & k-2 & 0 & 0 \\
s^0 & 2-2k & 0 & 0
\end{array}.
$$

In order for the first element in the s^1 row to be positive we need that $k > 2$. In order for the first element in the s^0 row to be positive we need $k < 1$. There are no values of k that can satisfy both.

9.5 Response of a Second-Order System

Because second-order systems are common and easy to solve, it is easy to develop some rules regarding the relationship between the time domain specifications for the response of a system and the pole locations for a second-order system. Even when a system is not second-order, it is sometimes possible to approximate it by a second-order system by ensuring that the response is dominated by a pair of complex conjugate poles that are closer to the imaginary axis than the other poles. Finally, in the case where there is only one more pole or zero added to a second-order system, it is possible to determine some rules of thumb regarding the nature of the response.

Recall that the canonical form for a generic second-order system from Section 3.3.0.1 is

$$
m\ddot{x} + b\dot{x} + kx = f(t) \quad \Longleftrightarrow \quad \ddot{x} + 2\zeta\omega_n\dot{x} + \omega_n x = \frac{f(t)}{m},
$$

where

$$\omega_n = \sqrt{\frac{k}{m}}, \quad \zeta = \frac{b}{2\sqrt{mk}}.$$

Computing the Laplace transform with zero initial conditions gives

$$X(s) = \frac{1}{s^2 + 2\zeta\omega_n s + \omega_n^2}\frac{F(s)}{m} = G(s)R(s).$$

If $0 \le \zeta < 1$, the poles of $G(s)$ are

$$s = -\zeta\omega_n \pm i\omega_n\sqrt{1-\zeta^2} = \omega_n\left(-\zeta \pm i\sqrt{1-\zeta^2}\right) = -\zeta\omega_n \pm i\omega_d,$$

where

$$\omega_d = \omega_n\sqrt{1-\zeta^2}.$$

Using the notation from Table 8.1, the poles are located at

$$s = -a \pm ib = \omega_n\left(-\zeta \pm i\sqrt{1-\zeta^2}\right).$$

As is illustrated in Figure 9.20, the relationship between the pole location and the parameters in the canonical second-order system are as follows.

1. The length of the vector from the origin to the pole is ω_n.
2. If the angle from the imaginary axis to the vector from the origin to the pole is denoted by θ, then

$$\zeta = \sin\theta.$$

3. The damped natural frequency ω_d is the imaginary component of the pole,

$$\omega_d = b = \omega_n\sqrt{1-\zeta^2}.$$

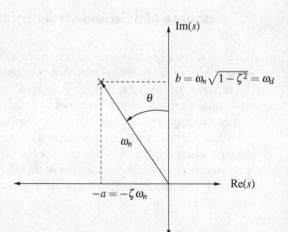

Fig. 9.20 Relationship between the location of complex conjugate poles and ω_n, ζ, and ω_d.

From the discussion in the previous section regarding the response when a transfer function contains a complex conjugate pole, we can deduce that the solution will have terms of the form $e^{-\zeta\omega_n t}\sin\omega_d t$ and $e^{-\zeta\omega_n t}\cos\omega_d t$. Hence, the effects of moving the location of a complex conjugate pole with negative real part are as follows.

1. If the imaginary part of the pole is increased and the real part is held constant, then the frequency of the response will increase and the damping ratio will decrease.
2. If the real part of the pole is decreased and the imaginary part is held constant, then the damping ratio is increased and the frequency of the response will be constant.
3. If the angle between the imaginary axis and the vector from the origin to the pole is held constant and the magnitude of the vector is increased, then the damping remains constant and the frequency of the response increases.

All three of these cases are illustrated in Figure 9.21.

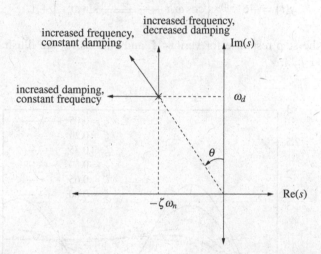

Fig. 9.21 Effect of moving the location of a complex conjugate pole.

9.5.1 Second-Order System Step Response

Now let us relate the location of the poles for a second-order system to the time domain specifications defined in Section 9.2 for a unit step input. Consider

$$G(s) = \frac{\omega_n^2}{s^2 + 2\zeta\omega_n s + \omega_n^2}.$$

For a step input to this transfer function, we have

$$Y(s) = \frac{\omega_n^2}{s^2 + 2\zeta\omega_n s + \omega_n^2} \frac{1}{s}$$

$$= -\frac{s + 2\zeta\omega_n}{s^2 + 2\zeta\omega_n s + \omega_n^2} + \frac{1}{s}$$

$$= -\frac{s + 2\zeta\omega_n}{(s + \zeta\omega_n)^2 + \omega_d^2} + \frac{1}{s}$$

$$= -\frac{s + \zeta\omega_n}{(s + \zeta\omega_n)^2 + \omega_d^2} - \frac{\zeta\omega_n}{\omega_d}\frac{\omega_d}{(s + \zeta\omega_n)^2 + \omega_d^2} + \frac{1}{s}$$

$$= -\frac{s + \zeta\omega_n}{(s + \zeta\omega_n)^2 + \omega_d^2} - \frac{\zeta}{\sqrt{1 - \zeta^2}}\frac{\omega_d}{(s + \zeta\omega_n)^2 + \omega_d^2} + \frac{1}{s}$$

so

$$y(t) = -e^{-\zeta\omega_n t}\left(\cos\omega_d t + \frac{\zeta}{\sqrt{1 - \zeta^2}}\sin\omega_d t\right) + 1. \qquad (9.16)$$

Plots of the step response for various ζ and various ω_n are illustrated in Figures 9.22 and 9.23.

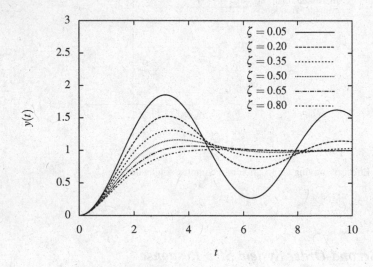

Fig. 9.22 Step response of second-order system with $\omega_n = 1$ and for various ζ.

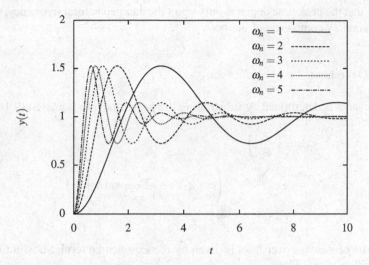

Fig. 9.23 Step response of second-order system with $\zeta = 0.2$ and for various ω_n.

9.5.1.1 Peak Time

The peak time is determined by finding the time when the derivative of Equation (9.16) is zero. Hence

$$
\begin{aligned}
\frac{\mathrm{d}}{\mathrm{d}t} y(t) &= \zeta \omega_n e^{-\zeta \omega_n t} \left(\cos \omega_d t + \frac{\zeta}{\sqrt{1-\zeta^2}} \sin \omega_d t \right) \\
&\quad - e^{-\zeta \omega_n t} \left(-\sin \omega_d t + \frac{\zeta \omega_d}{\sqrt{1-\zeta^2}} \cos \omega_d t \right) \\
&= \zeta \omega_n e^{-\zeta \omega_n t} \left(\cos \omega_d t + \frac{\zeta}{\sqrt{1-\zeta^2}} \sin \omega_d t \right) \\
&\quad - e^{-\zeta \omega_n t} \left(-\sin \omega_d t + \zeta \omega_n \cos \omega_d t \right) \\
&= \left(\frac{\zeta^2 \omega_n}{\sqrt{1-\zeta^2}} + 1 \right) \sin \omega_d t \\
&= 0.
\end{aligned}
$$

The first positive time for which this is zero is the peak time and is

$$
\boxed{t_p = \frac{\pi}{\omega_d}.}
$$

Observe that the peak time depends only upon the damped natural frequency, which is the imaginary component of the pole.

9.5.1.2 Overshoot

The overshoot is determined by substituting the peak time into Equation (9.16):

$$y_p = y\left(\frac{\pi}{\omega_d}\right)$$

$$= -e^{-\zeta\omega_n\pi/\omega_d}\left(\cos\pi + \frac{\zeta}{\sqrt{1-\zeta^2}}\sin\pi\right) + 1$$

$$= 1 + e^{-\zeta\omega_n\pi/\omega_d}.$$

Hence, the *percentage* overshoot is given by the exponential term. Substituting for the definition of ω_d gives

$$\boxed{O = \exp\left(-\frac{\pi\zeta}{\sqrt{1-\zeta^2}}\right).} \qquad (9.17)$$

Observe that the percentage overshoot depends only upon the damping ratio. A plot of O versus ζ is given in Figure 9.24. The maximum overshoot is a function of only the damping ratio, thus there is a simple geometric interpretation for second-order poles that will meet an overshoot specification, as is illustrated by the following example.

Example 9.15. Determine the region in the complex plane where the poles should be located in order for a second-order system to have a maximum overshoot of less than 10%. Either referring to Figure 9.24 or solving Equation (9.17) gives

$$O < 0.1 \quad \Longleftrightarrow \quad \zeta > 0.6.$$

Because $\sin\theta = \zeta$ if θ is measured from the imaginary axis, we need $\theta > 36.9°$. So, the region in the complex plane where a pair of second-order poles must be located to satisfy this specification is illustrated in Figure 9.25.

9.5.1.3 Settling Time

To determine the settling time, note that the rate at which the transient response decays is governed by the exponential term $e^{-\zeta\omega_n t}$. Hence, for the 5% settling time,

$$0.05 = e^{-\zeta\omega_n t_s} \quad \Longleftrightarrow \quad t_s = -\frac{\ln(0.05)}{\zeta\omega_n} \approx \frac{3}{\zeta\omega_n}.$$

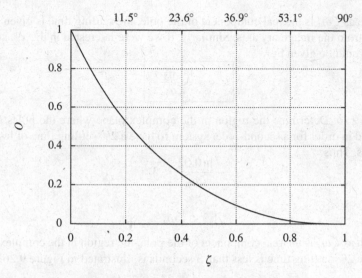

Fig. 9.24 Percentage overshoot O versus damping ratio ζ and angle between imaginary axis and the pole θ.

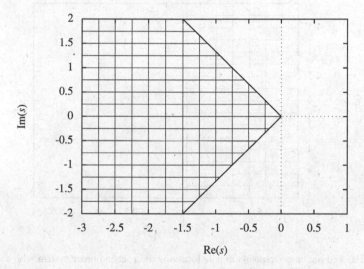

Fig. 9.25 Hatched region corresponds to pole locations for a second-order system with less than 10% overshoot.

Inasmuch as $\zeta \omega_n$ is the real component of the pole, the settling time is given by the distance from the imaginary axis. Similarly, if we were interested in the $x\%$ settling time, it would be given by

$$t_s = -\frac{\ln\left(\frac{x}{100}\right)}{\zeta \omega_n}.$$

Example 9.16. Determine the region in the complex plane where the poles should be located in order for a second-order system to have a 2% settling time of less than 3 seconds. Thus

$$-\frac{\ln 0.02}{\zeta \omega_n} < 3$$

which gives

$$1.3 < \zeta \omega_n.$$

Inasmuch as $\zeta \omega_n$ is the real component of the pole, the region in the complex plane where the 2% settling time is less than 3 seconds is illustrated in Figure 9.26.

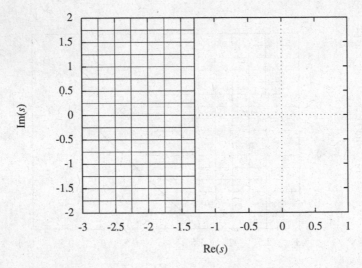

Fig. 9.26 Hatched region corresponds to pole locations for a second-order system with a settling time of less than 3 seconds.

9.5.1.4 Rise Time

Unlike the overshoot and settling time, only an approximation for the rise time is available with a simple geometric interpretation. Different approximations for the rise time appear in different texts. For example, [17] gives

$$t_r \approx \frac{1.8}{\omega_n},$$ (9.18)

and [28] gives

$$t_r \approx \frac{0.8 + 2.5\zeta}{\omega_n} \quad \text{or} \quad t_r \approx \frac{1 - 0.4167\zeta + 2.917\zeta^2}{\omega_n}.$$

One reason for the different formulae is because of different definitions of the rise time. For example, in [28] it is defined as the time to go from 10% to 90% of the final value. We use the approximation in Equation (9.18), because it only depends on ω_n, which is easily related to pole locations because it is the distance from the origin to the pole. For example, the region in the complex plane corresponding to a rise time of less than 1.8 seconds is illustrated in Figure 9.27, because

$$\frac{1.8}{\omega_n} < 1.8 \quad \Longleftrightarrow \quad 1 < \omega_n,$$

and ω_n is the distance from the origin.

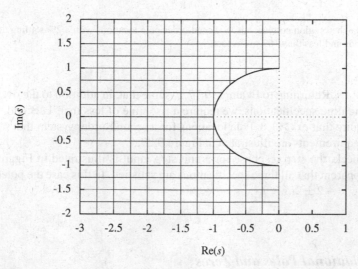

Fig. 9.27 Hatched region corresponds to pole locations for a second-order system with an approximate rise time of less than 1.8 seconds.

We can now combine these specifications to determine pole locations that satisfy more than one specification.

Example 9.17. To determine the region in the complex plane where the poles should be located in order for a second-order system to have a 2% settling time of less than

3 seconds and a maximum overshoot less than 10% we can take the intersection of
the regions in Figures 9.25 and 9.26. This region is illustrated in Figure 9.28.

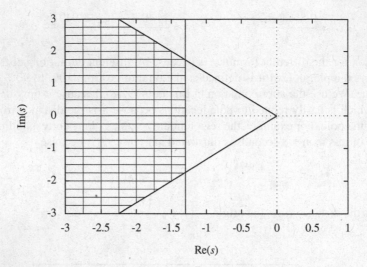

Fig. 9.28 Hatched region corresponds to second-order pole locations with a 2% settling time less
than 3 seconds and less than 10% overshoot.

Example 9.18. Returning to Example 9.17, assume that, in addition to the overshoot
and settling time specifications, we require a rise time of less than 1 second, which
would require that $\omega_n \geq 1.8$. Pole locations for a second-order system that satisfies
all three requirements are illustrated in Figure 9.29.

As a check, the step response for a unit step input is illustrated in Figure 9.30,
and it is apparent that all three specifications are satisfied. In this case the poles were
located at $s = -2 \pm 2i$.

9.5.2 Additional Poles and Zeros

At this point we have the correspondence between the time domain specifications
and pole locations for a second-order system. Unfortunately, the world is not composed entirely of second-order systems, so it is useful to relate, when possible, the
response of a system that is not second order to the second-order response we know
so well. We consider several ways in which a system may deviate from a canonical
second-order system.

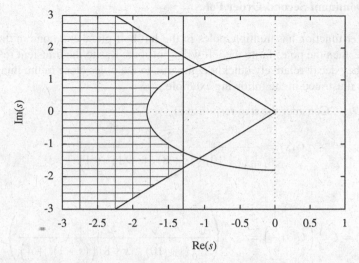

Fig. 9.29 Hatched region corresponds to second-order pole locations with a 2% settling time less than 3 seconds, less than 10% overshoot, and less than a 1 second rise time.

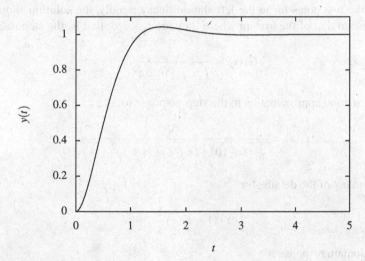

Fig. 9.30 Step response of $G(s) = 8/\left((s+2)^2 + 2^2\right)$, which has poles within the region satisfying all three specifications in Example 9.18.

9.5.2.1 Dominant Second-Order Poles

If a transfer function has multiple poles in the left half-plane and none in the right half-plane, then the poles far to the left will contribute less to the transient response because they decay relatively quickly compared to the poles close to the imaginary axis, as is illustrated in the following example.

Example 9.19. Consider the step response of

$$G(s) = \frac{5}{\frac{1}{10}(s+10)\frac{1}{8}(s+8)\left((s+1)^2+4\right)};$$

that is,

$$y(t) = \mathcal{L}^{-1}\left(G(s)\frac{1}{s}\right) = \mathcal{L}^{-1}\left(\frac{5}{s\frac{1}{10}(s+10)\frac{1}{8}(s+8)\left((s+1)^2+4\right)}\right).$$

Before we solve this, we observe that two poles are pretty far to the left and two are relatively close to the imaginary axis as illustrated in Figure 9.31. Because the effect of the two poles far to the left should decay rapidly, the solution should be rather close to that of the system where they are ignored; that is, the step response of

$$G_1(s) = \frac{5}{\left((s+1)^2+4\right)}$$

should be a good approximation to the step response to

$$G(s) = \frac{5}{\frac{1}{10}(s+10)\frac{1}{8}(s+8)\left((s+1)^2+4\right)}.$$

Skipping many of the details, for

$$Y(s) = G(s)\frac{1}{s}$$

the time domain response is

$$y(t) = 1 + \frac{4}{17}e^{-10t} - \frac{25}{53}e^{-8t} - \frac{688}{901}e^{-t}\cos 2t - \frac{984}{901}e^{-t}\sin 2t$$

and for

$$Y_1(s) = G_1(s)\frac{1}{s} \tag{9.19}$$

the time domain response is

$$y_1(t) = 1 - e^{-t}\left(\cos 2t + \frac{1}{2}\sin 2t\right). \tag{9.20}$$

The two step responses are plotted in Figure 9.32. Clearly, the step response of $G_1(s)$ is a fairly good approximation of the step response of $G(s)$. The reason for this is because the e^{-10t} and e^{-8t} terms decay so rapidly. It is not at all clear, however, from a casual observation of the time domain solutions given in Equations (9.19) and (9.20) that the solutions are approximately the same. In fact, even ignoring the decaying exponentials corresponding to the fast poles, the coefficients of the sine and cosine terms are not approximately equal. However, with a conceptual understanding of the fact that the poles at $s = -10$ and $s = -8$ decay very quickly, it is obvious from the location of the poles illustrated in Figure 9.31 that the complex conjugate pair will dominate the response.

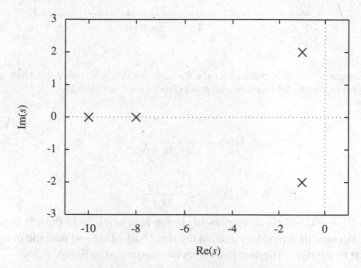

Fig. 9.31 Pole locations for transfer function in Example 9.19. The two poles near the imaginary axis should dominate the response.

Because it may not always be the case that there is a dominant pair of poles for a system that is higher than second-order, we next determine the effect of one additional pole or zero, regardless of whether it is far to the left.

9.5.2.2 Additional Real Zero

If a transfer function has a complex conjugate pair of poles and one real zero, the effect of the zero on the response will depend on the location of the zero. This section draws some conclusions based upon inference from an example. The proof of the conclusions is left to Exercise 9.10.

Example 9.20. Consider

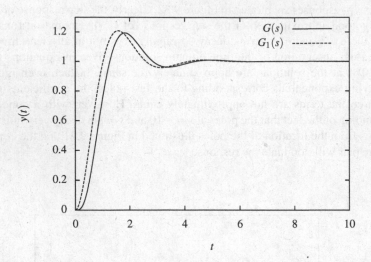

Fig. 9.32 Comparison of step responses of the transfer function with four poles $G(s)$, and the transfer function with only the two dominant poles $G_1(s)$ from Example 9.19.

$$G(s) = \frac{5}{s^2 + 2s + 5}$$

and

$$G_1(s) = \frac{\frac{5}{r}(s+r)}{s^2 + 2s + 5},$$

where $r = 10, 1, -1,$ and 10, corresponding to the zero being far to the left, in the left half-plane but near the imaginary axis, in the right half-plane and near the imaginary axis and far to the right. The step responses are illustrated in Figure 9.33.

From this example we may infer the following general rules.

- If the zero is far from the imaginary axis, then it has little effect on the step response.
- If the zero is in the left half plane and close to the imaginary axis, it will decrease the rise time and increase the overshoot.
- If the zero is in the right half plane and close to the imaginary axis, it will increase the rise time, perhaps increase the overshoot, and perhaps the system will initially move in the "wrong direction."

Systems with left half-plane zeros are called *nonminimum phase* and are discussed subsequently in Section 9.7. One common characteristic of nonminimum phase systems is that they may initially start in the "wrong" direction, as illustrated in the example.

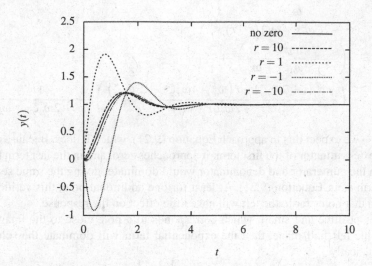

Fig. 9.33 The effect of an additional real zero on the second-order step response: the step response of $G_1(s) = \left(5\,(s+r)\right)/\left(r\,(s^2+2s+5)\right)$ for various zero locations (various r).

9.5.2.3 Additional Real Pole

Consider

$$G(s) = \frac{\omega_n^2}{s^2 + 2\zeta\,\omega_n s + \omega_n^2}$$

and

$$G_1(s) = \frac{\omega_n^2}{(s^2 + 2\zeta\,\omega_n s + \omega_n^2)\,\frac{1}{r}\,(s+r)}.$$

From Equation (9.16), the partial fraction expansion for $Y(s) = G(s)/s$ is

$$Y(s) = -\frac{s + 2\zeta\,\omega_n}{(s + \zeta\,\omega_n)^2 + \omega_d^2} + \frac{1}{s} \tag{9.21}$$

and the step response is

$$y(t) = -e^{-\zeta\,\omega_n t}\left(\cos\omega_d t + \frac{\zeta}{\sqrt{1-\zeta^2}}\sin\omega_d t\right) + 1.$$

Computing the partial fraction expansion for $Y(s) = G(s)/s$ gives

$$Y(s) = \frac{\omega_n^2 r}{(s^2 + 2\zeta\omega_n s + \omega_n^2)(s+r)} \frac{1}{s}$$

$$= -\left(\frac{\omega_n^2}{r^2 - 2\omega_n\zeta r + \omega_n^2}\right)\frac{1}{s+r} + \frac{1}{s} \tag{9.22}$$

$$+ \left(\frac{-r^2(s + 2\omega_n\zeta) - r(\omega_n^2 - 4\omega_n^2\zeta^2 - 2s\omega_n\zeta)}{(r^2 - 2\omega_n\zeta r + \omega_n^2)}\right)\frac{1}{(s^2 + 2\omega_n\zeta s + \omega_n^2)}.$$

As $r \to \infty$ we expect this to approach Equation (9.21), which it does. Because of the r^2 in the denominator of the first term, it approaches zero, and in the last term the r^2 terms in the numerator and denominator would dominate, giving the same second-order term as in Equation (9.21). At least for one additional pole, this verifies our intuition that poles far to the left will have little effect on the response.

If r is positive and small, which corresponds to a pole close to the imaginary axis in the left half-plane, then the exponential term will dominate the solution. Mathematically

$$\lim_{r\downarrow 0} Y(s) = -\frac{1}{s+r} + \frac{1}{s},$$

so for small r,

$$y(t) \approx 1 - e^{-rt},$$

which has no overshoot and a large rise time. Inasmuch as r is small, the exponential terms decay very slowly and hence $y(t)$ approaches the steady-state value very slowly.

Conceptually interpolating between these two extremes we can conclude that if a second-order system has an additional pole in the left half-plane then is is characterized by the following features.

- If the additional pole is far to the left, it will have little effect on the response.
- If the additional pole is very close to the imaginary axis compared to the second-order poles, it will dominate the response and the solution will slowly asymptotically approach the steady-state value.
- If the additional pole is of the same order as the second-order poles it should increase the rise time and decrease the overshoot.
- If the additional pole is anywhere in the right half plane, then the solution will be unstable.

Example 9.21. Consider

$$G(s) = \frac{5}{s^2 + 2s + 5}$$

and

$$G_1(s) = \frac{5r}{(s+r)(s^2 + 2s + 5)},$$

where $r = 10, 1, 0.1$, and -1, corresponding to the pole being far to the left, in the left half-plane of the same order of magnitude as the complex conjugate pair of poles, very near the imaginary axis and in the right half-plane. The step responses are illustrated in Figure 9.34.

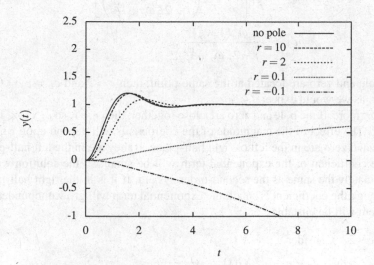

Fig. 9.34 The effect of an additional real pole on the second-order step response: the step response of $G_1(s) = (5r)/\left((s+r)\left(s^2+2s+5\right)\right)$ for various pole locations (various r).

9.5.2.4 Poles and Zeros Close Together

If a pole and zero are close together, algebraically they nearly cancel. It is natural to expect, then, that their effect in the solution would nearly cancel as well. In fact, this is true which we demonstrate with one zero and one pole located near each other.

Consider

$$G(s) = \frac{\omega_n^2}{(s^2+2\zeta\omega_n s+\omega_n^2)}\frac{s+\hat{r}}{s+r}.$$

A partial fraction expansion gives

$$G(s) = \frac{c_1 s+c_2}{s^2+2\zeta\omega_n s+\omega_n^2} + \frac{c_3}{s+r},$$

where

$$c_1 = \frac{\omega_n^2 \left(1 - \frac{\hat{r}}{r}\right)}{r - 2\zeta\omega_n + \frac{\omega_n^2}{r}}$$

$$c_2 = \frac{\omega_n^2}{r} \left[\hat{r} + \frac{\omega_n^2}{r} \left(\frac{1 - \frac{\hat{r}}{r}}{r - 2\zeta\omega_n + \frac{\omega_n^2}{r}} \right) \right]$$

$$c_3 = -\frac{\omega_n^2 \left(1 - \frac{\hat{r}}{r}\right)}{r - 2\zeta\omega_n + \frac{\omega_n^2}{r}}.$$

If the pole and zero are located at the same point, then $r = \hat{r}$ and $c_1 = c_3 = 0$ and $c_2 = \omega_n^2$, as we would expect.

Furthermore, if the pole and zero are close together, then $r \approx \hat{r}$, so $c_1, c_3 \ll 1$, and $c_2 \approx \omega_n^2$. The effect of the magnitude of the coefficients will depend upon whether the pole and zero are in the left or right half-plane. If they are in the left half-plane, then the coefficient of the exponential term will be small, so the solution will be approximately the same as the second-order system. If it is in the right half-plane, even though the coefficient is small, the exponential term will grow unbounded and the system will be unstable.

Example 9.22. Consider

$$G(s) = \frac{5}{s^2 + 2s + 5},$$

$$G_1(s) = \frac{5}{s^2 + 2s + 5} \left(\frac{s + 1}{\frac{1}{.95}(s + 0.95)} \right),$$

$$G_2(s) = \frac{5}{s^2 + 2s + 5} \left(\frac{s + 1}{\frac{1}{2}(s + 2)} \right),$$

and

$$G_3(s) = \frac{5}{s^2 + 2s + 5} \left(\frac{s - 1}{\frac{1}{.95}(s - 0.95)} \right).$$

The first transfer function $G(s)$ only has a complex conjugate pair of poles. The second $G_1(s)$ has an additional pole at $s = -.95$ and an additional zero at $s = -1$. The third has a pole and zero that are not close together and finally, the fourth $G_3(s)$ has a pole and zero that are close, but in the right half-plane.

The step responses are illustrated in Figure 9.35. Observe that if the pole and zero are in the left half-plane and are close together, then they almost cancel and the step response is much like that of $G(s)$. If they are not close together then they have a substantial effect on the response. If they are in the right half-plane, then even if they are close together the system is unstable.

Remark 9.1. It is true that mathematically if there are a pole and zero in the right half plane that exactly cancel, then they will have no effect on the response of the

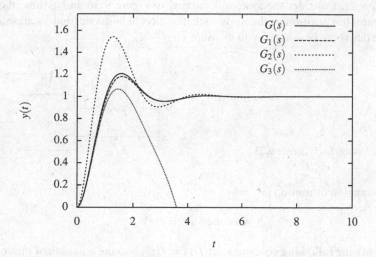

Fig. 9.35 Effect of additional poles and zeros that are close together from Example 9.22.

system. However, for a real engineering system, if there is a pole in the right half-plane attempting to cancel it with a zero will not work because it is impossible to characterize any real system exactly.

9.6 Root Locus Analysis

The root locus design method is probably the most basic feedback control design methodology. This section develops the rules for constructing root locus plots and presents examples illustrating the utility of the method for control design.

9.6.1 Motivational Example

From Section 9.3 it should be clear that the nature of the response of a system is dictated by the pole locations of the transfer function that describes it. A *root locus plot* is a plot of how the poles of a transfer function change as some parameter in the system is varied. This is useful because it may give us a means to determine the value of such a parameter that gives the system some desired response such as a specified rise time, maximum overshoot, settling time, and so on. We motivate this by a particular example and then in the following sections develop the rules that allow us to sketch a root locus plot by hand.

Example 9.23. Consider the system illustrated in Figure 9.36 and assume the task is to control the position of the mass so that it stays at some desired location $x_d(t)$. Assume that there is some way to measure $x(t)$.

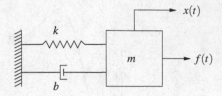

Fig. 9.36 System for Example 9.23.

The equation of motion is

$$m\ddot{x} + b\dot{x} + kx = f(t).$$

To simplify the following equations, let $\hat{f}(t) = f(t)/k$ so the equation of motion can be written

$$\ddot{x} + \frac{b}{m}\dot{x} + \frac{k}{m}x = \frac{k}{m}\hat{f}(t)$$

and the transfer function from the input force to the position of the mass is given by

$$\frac{X(s)}{\hat{F}(s)} = \frac{\frac{k}{m}}{s^2 + \frac{b}{m}s + \frac{k}{m}} = \frac{\omega_n^2}{s^2 + 2\zeta\omega_n s + \omega_n^2}.$$

We use proportional control so that

$$\hat{f}(t) = k_p(x_d - x(t))$$

or

$$\hat{F}(s) = k_p(X_d(s) - X(s)).$$

A block diagram representation of the system with proportional control is illustrated in Figure 9.37. Included in the figure are labels for the error signal $E(s)$ and the force $\hat{F}(s)$. The transfer function from the desired position of the mass to the actual position is given by

$$\frac{X(s)}{X_d(s)} = \frac{k_p\omega_n^2}{s^2 + 2\zeta\omega_n s + \omega_n^2(k_p + 1)}. \tag{9.23}$$

The nature of the transient response is easy to determine from the poles of Equation (9.23), which are simply given by the quadratic equation

$$p = -\zeta\omega_n \pm \omega_n\sqrt{\zeta^2 - (k_p + 1)}.$$

To proceed, let us assume some numerical values for the system parameters. If $\omega_n = 1$ and $\zeta = 2$, then

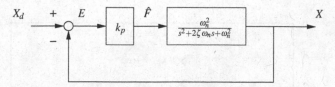

Fig. 9.37 Block diagram for proportional control for Example 9.23.

$$p = -2 \pm \sqrt{4 - (k_p + 1)},$$

so if $k_p < 3$ the solutions are exponentials and if $k_p > 3$ the solutions are damped oscillations. A plot of the pole locations for various k_p is illustrated in Figure 9.38. Observe that for very small k_p there will be two real poles. One will be near the origin and the other will be near $s = -4$. As k_p increases the poles move toward each other along the real axis and at $k_p = 3$ they will both be at $s = -2$. Further increasing k_p will result in complex conjugate poles. The real part of the poles for $k_p \geq 3$ is fixed at $s = -2$ and the imaginary part increases as k_p increases.

Before we solve for the step responses we can observe the following regarding the nature of the step response.

1. For $k_p \geq 3$ the settling time will not be changed by altering k_p. For $k_p < 3$ the settling time will be larger than for $k_p \geq 3$ because one of the poles will have a real part to the right of $s = -2$. Thus, the best we can do for settling time is at $k_p \geq 3$.
2. There will be no overshoot for $k_p \leq 3$ because the solutions will be exponentials. For $k_p > 3$ increasing k_p will increase the overshoot because it will decrease the angle between the pole and the imaginary axis.
3. For $k_p > 3$ the rise time will decrease as k_p increases. Note that it may be the case that it will be possible to satisfy either a rise time or an overshoot specification, but not both, because one gets worse with increasing k_p and the other gets better.

To verify our analysis, the corresponding step responses are illustrated in Figure 9.39. Note that for each value of k there are two solutions. This makes sense because the equation is second order. Subsequently, we construct continuous curves for the infinite number of possible k values, and the number of branches is equal to the order of the characteristic polynomial.

At least in one respect our attempt to control the location of the mass in Example 9.23 is deficient because the steady-state value of x depends on k_p and the steady-state error is not zero. The way to remedy this should be obvious from Section 9.1, which is to add integral control. In the following example we do just that. As expected, this eliminates the steady-state error. The larger point of the example, though, is that once integral control is added, plotting the pole locations will not be easy because the denominator of the transfer function will be a third-order polynomial, so a method to plot how the poles move as a parameter is varied that works for higher-order polynomials is needed.

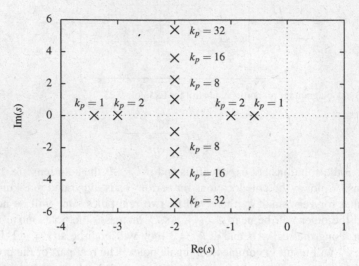

Fig. 9.38 Pole locations for various k_p ($k_p = 4$ not labeled) for Example 9.23.

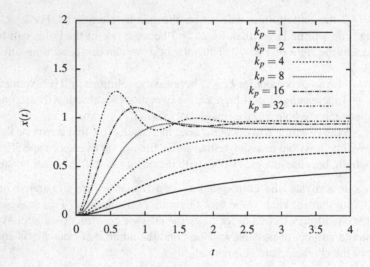

Fig. 9.39 Step response for various k_p for Example 9.23.

Example 9.24. In this example we add integral control to try to control the location of the mass in Figure 9.36. Let

$$\hat{f}(t) = k_p (x_d - x(t)) + k_i \int_0^t x_d(\tau) - x(\tau) d\tau.$$

In general, it is necessary to consider how altering both k_p and k_i affect the nature of the solution. Because we are considering how the system responds when one parameter is varied, we fix the relationship between k_p and k_i and if a satisfactory result is not obtained, change the relationship between them and start over.

Somewhat arbitrarily let $k_i = k_p/2$, so

$$\hat{f}(t) = k \left((x_d(t) - x(t)) + \frac{1}{2} \int_0^t x_d(\tau) - x(\tau) d\tau \right),$$

where $k_p = k$ and $k_i = k/2$. The Laplace transform of the control law is

$$\hat{F}(s) = k \left((X_d(s) - X(s)) + \frac{1}{2s} \right) (X_d(s) - X(s)) = k \left(1 + \frac{1}{2s} \right) (X_d(s) - X(s))$$

and the block diagram representation of this system is illustrated in Figure 9.40.

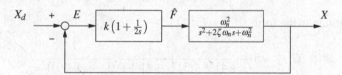

Fig. 9.40 Block diagram for proportional plus integral control for Example 9.24.

Using this control law, the transfer function, after a bit of algebra, is

$$\frac{X(s)}{X_d(s)} = \frac{k\omega_n^2 \left(s + \frac{1}{2} \right)}{s^3 + 2\zeta \omega_n s^2 + \omega_n^2 (1 + k) s + \frac{k\omega_n^2}{2}}$$

and if $\omega_n = 1$ and $\zeta = 2$

$$\frac{X(s)}{X_d(s)} = \frac{k \left(s + \frac{1}{2} \right)}{s^3 + 4s^2 + (1 + k) s + \frac{k}{2}} = \frac{2ks + k}{2s^3 + 8s^2 + 2(1 + k)s + k}.$$

A critical point regarding the preceding equation is that, in contrast to the system in Example 9.23, a tool as simple as the quadratic equation is not available to check how the poles of the transfer function vary as the parameter k is varied. Of course it may be done numerically, but having methods available to do the analysis by hand is extremely important because it allows us to gain insight into such systems. We return to this example subsequently after we develop a method for doing exactly that.

9.6.2 A Quick Review of Functions of a Complex Variable

A more detailed review of complex variable theory is contained in Appendix A. This section highlights the just results necessary for sketching root locus plots.

Consider a transfer function of the form

$$G(s) = \frac{N(s)}{D(s)}.$$

Regardless of whether we can do it by hand, we may write the numerator and denominator in factored form. In particular, we may write

$$G(s) = \frac{N(s)}{D(s)} = \hat{k} \frac{\prod_{i=1}^{n_z} (s - z_i)}{\prod_{i=1}^{n_p} (s - p_i)}, \tag{9.24}$$

where $N(s)$ and $D(s)$ are polynomials and the z_i are the zeros, the p_i are the poles, n_z is the number of zeros, and n_p is the number of poles of $G(s)$.

Example 9.25. The transfer function

$$G(s) = \frac{5(s+3)}{s^4 + 11s^3 + 40s^2 + 58s + 40}$$

may be written in the form

$$G(s) = 5 \frac{s+3}{(s+4)(s+5)(s+(1+i))(s+(1-i))}.$$

A fundamental property of complex numbers is that they may be represented in a Cartesian manner, which is typically of the form $s = a + ib$ where a is the real component and b is the imaginary component of s. An alternative representation is in polar coordinates where s is represented by a magnitude and phase which are the usual Euclidean norm and phase if the number is plotted in its Cartesian coordinates.[4] Referring to Figure 9.41, if $s = a + ib$, then

$$r = \sqrt{a^2 + b^2} = |s|$$

and

$$\theta = \tan^{-1} \left(\frac{b}{a} \right) = \angle s.$$

The Cartesian form is easy for addition and subtraction because if $s_1 = a_1 + ib_1$ and $s_2 = a_2 + ib_2$, then

$$s_1 + s_2 = (a_1 + a_2) + i(b_1 + b_2).$$

---

[4] The usual name for the angle, or phase, of a complex number in the mathematics literature, is the *argument*. We typically use the more colloquial terms *angle* or *phase*.

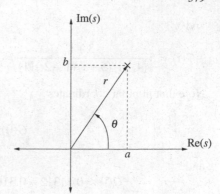

Fig. 9.41 Cartesian, $s = a + ib$, and polar, $s = (r, \theta)$, forms of a complex number s.

However, multiplication is easier in polar form. In particular, if $s_1 = (r_1, \theta_1)$ and $s_2 = (r_2, \theta_2)$, then the product is

$$s_1 s_2 = (r_1 r_2, \theta_1 + \theta_2)$$

and the quotient is

$$\frac{s_1}{s_2} = \left(\frac{r_1}{r_2}, \theta_1 - \theta_2 \right).$$

Proving these two results is simple and is left as an exercise.

The critical concept in this section relates how to evaluate a transfer function in terms of the location in the complex plane of s and the location of its poles and zeros. Returning to Equation (9.24),

$$G(s) = \hat{k} \frac{\prod_{i=1}^{n_z} (s - z_i)}{\prod_{i=1}^{n_p} (s - p_i)} = \hat{k} \frac{(s - z_1)(s - z_2) \cdots (s - z_{n_z})}{(s - p_1)(s - p_2) \cdots (s - p_{n_p})},$$

note that the numerator is simply the product of the difference between s and all of the zeros of $G(s)$. Similarly, the denominator is the product of the difference between s and each of the poles of $G(s)$.

This concept is critical to understanding the development that follows. If it is still not clear after the following example, the reader is strongly encouraged to reread it before proceeding to the next section.

Example 9.26. Returning to the transfer function from Example 9.25 let

$$G(s) = 5 \frac{s + 3}{(s + 4)(s + 5)(s + (1 + i))(s + (1 - i))}.$$

If we wish to determine $G(s)$ at a particular value for s the easiest thing would be just to substitute it into G. For example,

$$G(0) = 5 \frac{3}{(4)(5)(1 + i)(1 - i)} = \frac{3}{8}.$$

and

$$G(i) = 5\frac{3+i}{(4+i)(5+i)(1+2i)(1)} = 5\frac{3+i}{1+47i} = 5\frac{3+i}{1+47i}\frac{1-47i}{1-47i} = \frac{50-140i}{442}.$$

Note that in polar coordinates

$$G(0) = \left(\frac{3}{8}, 0\right)$$

and

$$G(i) \approx 0.11312 - 0.31674i = (0.33634, -70.346°).$$

"Plugging and chugging" may be best to evaluate the Cartesian form of $G(s)$. In polar form there is a geometric interpretation as well. Figure 9.42 plots the poles and zeros of $G(s)$ and marks $s = i$ with a $+$. Now consider each term in the numerator and denominator of $G(s)$. Each is of the form $s - z$ or $s - p$ and one way to interpret $s - z$ or $s - p$ is that it is the vector from z or p, respectively, to the point s, as illustrated in Figure 9.43.

So, an alternative way to evaluate $G(s)$ is to consider the vectors from all the zeros and poles of $G(s)$ to the point s. The magnitude of $G(s)$ will be $\hat{k}$ times the product of the magnitudes of all the vectors from the zeros of $G(s)$ to s divided by the product of the magnitudes of all the vectors from the poles of $G(s)$ to s. Mathematically,

$$|G(s)| = \hat{k}\frac{\prod_{i=1}^{n_z}|s-z_i|}{\prod_{i=1}^{n_p}|s-p_i|} = \hat{k}\frac{|s-z_1||s-z_2|\cdots|s-z_{n_z}|}{|s-p_1||s-p_2|\cdots|s-p_{n_p}|}.$$

In words, we may graphically measure the length of all the arrows from the poles and zeros of $G(s)$ to s and multiply them for zeros and divide by them for poles to determine the magnitude of $G(s)$. In all cases, the phase is measured counter-clockwise from the horizontal, as is illustrated in Figure 9.41.

Because the angles of complex numbers add when they are multiplied, then the angle of $G(s)$ is determined by summing the angles from all the zeros to s and by subtracting the angle from all the poles to s. Mathematically,

$$\angle G(s) = \sum_{i=1}^{n_z}\angle(s-z_i) - \sum_{i=1}^{n_p}\angle(s-p_i)$$

$$= \angle(s-z_1) + \angle(s-z_2) + \cdots + \angle(s-z_{n_z})$$

$$- \angle(s-p_1) - \angle(s-p_2) - \cdots - \angle(s-p_{n_p}).$$

In words, we can measure the angle from all the zeros and poles to s and sum the angles from the zeros and subtract the angles from the poles.

Now evaluating $G(s)$ using this graphical interpretation

$$|G(\mathrm{i})| = 5\frac{\left|\sqrt{10}\right|}{|1|\left|\sqrt{5}\right|\left|\sqrt{17}\right|\left|\sqrt{26}\right|} \approx 0.33634,$$

and

$$\angle G(\mathrm{i}) = \tan^{-1}\left(\frac{1}{3}\right) - \tan^{-1}\left(\frac{0}{1}\right) - \tan^{-1}\left(\frac{2}{1}\right) - \tan^{-1}\left(\frac{1}{4}\right) - \tan^{-1}\left(\frac{1}{5}\right)$$
$$\approx -70.346°.$$

Observe that with an even scale on the graph, a ruler, and protractor, one could evaluate $G(s)$ with pretty decent accuracy.

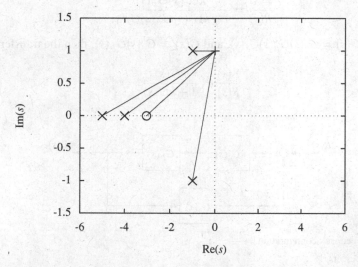

Fig. 9.42 The poles and zeros of $G(s) = \left(5\,(s+3)\right)/\left((s+4)\,(s+5)\,(s+(1+\mathrm{i}))\,(s+(1-\mathrm{i}))\right)$ and the point $s = \mathrm{i}$ (+) for Example 9.26.

9.6.3 Root Locus Plotting Rules

Consider the transfer function described by the block diagram illustrated in Figure 9.44. This is a typical *cascade compensation* configuration where $G_p(s)$ is the plant transfer function, $G_c(s)$ is the compensator, k is a gain, and $G_s(s)$ represents the transfer function for the sensor that is measuring the output. Often sensor dynamics are negligible compared to the rest of the system, in which case $G_s(s) = 1$ and the system is referred to as *unity feedback*. The transfer function for this system is

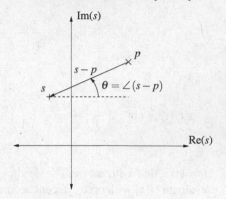

Fig. 9.43 The vector $(s - p)$ from p to s.

$$\frac{Y(s)}{R(s)} = \frac{kG_c(s)G_p(s)}{1 + kG_c(s)G_p(s)G_s(s)}.$$

If we let $G(s) = G_c(s)G_p(s)G_s(s)$, and $\hat{G}(s) = G_c(s)G_p(s)$, then the transfer function is

$$\frac{Y(s)}{R(s)} = \frac{k\hat{G}(s)}{1 + kG(s)}. \tag{9.25}$$

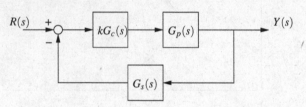

Fig. 9.44 Feedback configuration.

We wish to study how the poles of $k\hat{G}(s)/(1 + kG(s))$ change as k is varied, so we need to know the poles change with k. A pole is a value of s that satisfied

$$1 + kG(s) = 0.$$

Hence, s is a pole if

$$G(s) = -\frac{1}{k}.$$

We limit our attention to k values that are zero or real and positive. As k goes from 0 to $+\infty$, $-1/k$ will go from $-\infty$ to the origin along the negative real axis as illustrated in Figure 9.45. We assume that the transfer function $G(s)$ is of the form

$$G(s) = \hat{k}\frac{(s - z_1)(s - z_2)\cdots(s - z_{n_z})}{(s - p_1)(s - p_2)\cdots(s - p_{n_p})}.$$

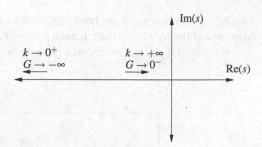

Fig. 9.45 The relationship on the complex plane of k and $G(s)$ if $1 + kG(s) = 0$.

Inasmuch as $-1/k$ is a negative real number that has an angle of $\pm 180°$, if we can determine all s values for which $\angle G(s) = 180°$, we will have plotted all the solutions to $1 + kG(s) = 0$ for positive k values. In the process we can determine how the poles of the transfer function in Equation (9.25) change with k, and hence will be able to determine properties of the response of Equation (9.25) such as the percent overshoot, settling time, rise time, and so on. This is the *root locus plot* for the transfer function $Y(s)/R(s)$.

First let us consider the two limiting cases where $k = 0$ and $k \to +\infty$. In the case where $k = 0$, the only way for $1 + kG(s)$ to be zero is if $G(s)$ is unbounded. So we can state the following rule.

Rule 9.1. *If the denominator of $G(s)$ is nth order, then at $k = 0$, one of each of the n branches of the root locus will start at one of the poles of $G(s)$.*

As $k \to +\infty$, the only way for $1 + kG(s)$ to equal zero is for $G(s) \to 0$. We are considering only proper transfer functions where the order of the denominator is greater than the numerator, therefore the only way for $|G(s)| \to \infty$ is for s to approach a pole. In contrast, there are two ways that $|G(s)|$ may approach 0. The first, obviously, is if s approaches a zero. Also, if s grows unbounded in any direction, $G(s)$ will approach 0 because the transfer function is proper and the order of the denominator is greater than the order of the numerator. So, we can state the second rule.

Rule 9.2. *As $k \to +\infty$, the root locus either approaches a zero of $G(s)$ or grows unbounded.*

An example may be useful at this point. Because we only have two rules so far, many of the features of the root locus unrelated to these two rules are not obvious.

Example 9.27. The solid lines in Figure 9.46 illustrate the solutions of

$$1 + k\frac{s+3}{(s+4)(s+5)(s^2+2s+3)} = 0$$

as k goes from 0 to $+\infty$ and the poles and zeros are marked as usual. Because, if the denominator is cleared, this is a fourth-order polynomial, for any k value, there will be four solutions to the equation. In this particular case, there is one solution on each of the four separate branches in the plot. As is clear from the figure, the branches of the root locus start at each pole of $G(s)$ and one of them approaches the one zero of $G(s)$ and the other three grow unbounded as $k \to +\infty$.

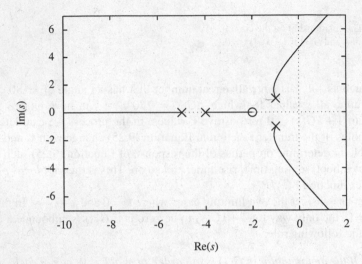

Fig. 9.46 The root locus plot for Example 9.27. The poles of $G(s)$ are marked with a $\times$ and the zeros of $G(s)$ are marked with a $\circ$.

Before we determine exactly how the solutions to $1 + kG(s) = 0$ grow unbounded as $k \to +\infty$, observe that the root locus is comprised of *branches*. For a transfer function with a characteristic equation of order n, the fundamental theorem of algebra requires that there be n solutions. Indeed, in the previous example, it appears that for any given value of k, there are four solutions and as k varies, these solutions move along continuous lines in the complex plane. The fact that these lines are continuous should make sense. If k is only slightly altered, then the n solutions will only be slightly altered as well. Hence, as k varies continuously from 0 to $+\infty$, the solutions to $1 + kG(s) = 0$ will vary continuously as well. The root locus starts at $k = 0$ at the poles of $G(s)$, thus each branch that corresponds to one of the solutions will start at one of the poles. We need to refer back to it, therefore we restate this argument as a proposition.

Proposition 9.7. *The solutions to $1 + kG(s) = 0$ depend continuously on k.*

Referring to Figure 9.46 with the root locus for Example 9.27, it is clear that the three branches that grow unbounded do so along specific asymptotes. Specifically, the branch of the locus that leaves pole $p_1 = -1 + \sqrt{2}i$ grows unbounded in a manner where both the real and imaginary parts of s approach $+\infty$. The branch that leaves $p_2 = -1 - \sqrt{2}i$ has an imaginary part that grows to $-\infty$ whereas the real part approaches $+\infty$. This branch also appears symmetric to the first branch about the real axis. In fact this must be so because if a complex number is a root of a polynomial, its complex conjugate must also be a root. Finally, the third branch grows unbounded with a zero imaginary part and a real part that approaches $-\infty$.

In order to determine these asymptotes, consider a map of the poles and zeros of $G(s)$ that has a very large scale, such as is illustrated in Figure 9.47 for the transfer function in the previous examples. If we desire to determine whether a point, indicated by a cross in the figure is on one of the branches of the root locus, we use the fact that it will be on the root locus only if $\angle G(s) = \pm 180°$.

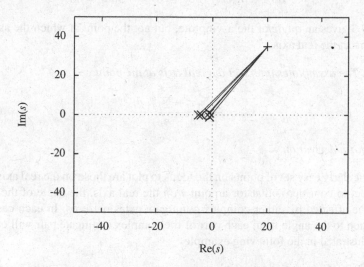

Fig. 9.47 The poles and zeros of a transfer function with a large scale appear in a small cluster.

Recall that

$$\angle G(s) = \sum_{i=1}^{n_z} \angle (s - z_i) - \sum_{i=1}^{n_p} \angle (s - p_i).$$

For very large s, the angle from all the poles and zeros is approximately the same. Hence if θ is this angle then

$$\angle G(s) = \theta (n_z - n_p)$$

and s will be on the root locus if

$$\theta (n_z - n_p) = \pm 180°.$$

We can always add or subtract 360° from the angle. Doing so and solving for θ gives

$$\theta = \frac{(180° + n360°)}{(n_z - n_p)} \quad n = 0, 1, \ldots (n_p - n_z).$$

Rule 9.3. *The branches of the root locus that grow unbounded do so along asymptotes with angles*

$$\theta_n = \frac{(180° + n360°)}{(n_z - n_p)}.$$

Let us verify this rule on the previous example.

Example 9.28. Because $G(s)$ from Example 9.27 had four poles and one zero, these asymptotes will have angles

$$\theta_0 = -60°, \quad \theta_1 = -180°, \quad \theta_2 = -300° = 60°.$$

Rule 9.3 gives an *angle* of the asymptote, but not the point at which the asymptotes intersect the real axis.

Rule 9.4. *The asymptotes intersect the real axis at the point*

$$s_{int} = \frac{\sum_{i=1}^{n_z} z_i - \sum_{i=1}^{n_p} p_i}{n_z - n_p}. \tag{9.26}$$

See [17] *for a derivation.*

A particularly easy set of points on the locus to plot are those on the real axis. The critical fact to consider is that for a point s on the real axis, the angle of the point will not be affected by either complex conjugate poles or zeros. In each case the contribution to the angle from each part of the complex conjugate pair will cancel. This is illustrated in the following example.

Example 9.29. Consider once again

$$1 + k\frac{s+3}{(s+5)(s+5)(s^2+2s+2)} = 0$$

and let $s = -2.5$. The poles and zeros of $G(s)$ are plotted in Figure 9.48 and the point s is illustrated by the cross. Observe that the contribution to the angle of $G(s)$ by the complex conjugate pair of poles in this example is 360°, which is equivalent to 0°. Hence, when evaluating

$$\angle G(s) = \sum_{i=1}^{n_z} \angle(s - z_i) - \sum_{i=1}^{n_p} \angle(s - p_i)$$

they do not matter.

From the preceding example it is, one hopes, obvious that all complex conjugate pairs of poles or zeros will contribute nothing to $\angle G(s)$. Hence, for real s only the poles and zeros on the real axis affect $\angle G(s)$. Furthermore, if s is real and the only poles and zeros that matter are real, all of the angles in the sum

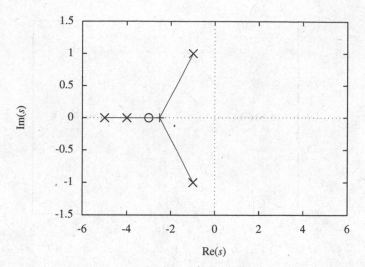

Fig. 9.48 Evaluating $\angle G(s) = 0°$ by considering the geometry of s relative to the poles and zeros of $G(s)$ on the complex plane.

$$\angle G(s) = \sum_{i=1}^{n_z} \angle (s - z_i) - \sum_{i=1}^{n_p} \angle (s - p_i)$$

will either be $0°$ or $180°$ depending upon whether the point s is to the right or left, respectively, of the pole or zero in question.

In fact, if s is to the right of all the real poles and zeros of $G(s)$, then $\angle G(s) = 0$ because the angle from each of them is zero. If s is decreased and crosses to the left of the first pole or zero, then $\angle G(s) = \pm 180°$ where the sign of the angle depends upon whether it was a pole or a zero.

Example 9.30. Returning to the previous series of examples, if $s = -3.5$, as illustrated in Figure 9.49, then $\angle G(s) = 180°$. This should be clear from the figure because s is to the left of z_1, so $\angle (s - z_1) = 180°$. Because s is to the right of p_2 and p_3, $\angle (s - p_2) = \angle (s - p_3) = 0$.

If we continue to decrease s so that it passes another pole or zero, then the angle of $G(s)$ will increase or decrease by $180°$. Regardless, $\angle G(s) = 0°$ because it will either algebraically sum to zero or it will be $360°$, which is equivalent to zero. Once it passes another one, $\angle G(s) = \pm 180°$ and then when it passes the next, $\angle G(s) = 0$, and so on. Hence, we have the following rule.

Rule 9.5. *On the real axis, the root locus is to the left of an odd number of zeros and poles.*

This rule certainly holds in the example case we have been considering if we refer back to Figure 9.46.

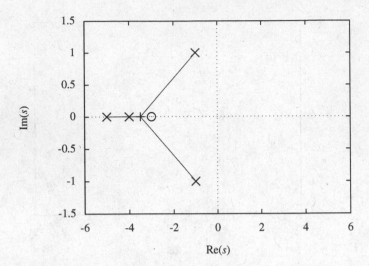

Fig. 9.49 Evaluating $\angle G(s) = 180°$ by considering the geometry of s relative to the poles and zeros of $G(s)$ on the complex plane.

At this point we have considered every feature of the root locus in Figure 9.46 except one, which is the angle where the locus departs the complex conjugate pair of poles. The loci appear to depart p_1 at approximately $90°$ and depart p_2 at approximately $-90°$. Instead of zooming out to consider a very large s as we did to find the asymptote angles, we zoom in and consider a point s very close to a pole. We should be able to determine, for example, that a point s very close to p_1 must be at an angle approximately equal to $90°$ from p_1.

In fact, in order to determine the departure angle from a pole or zero, all we must do is consider a point very close to it. If s is very close to a pole p_i, or zero z_i, the angle from all the other poles and zeros to it is approximately the same as the angle from the other poles and zeros to the pole or zero to which s is close, and all we must do is solve

$$\angle G(s) = \sum_{i=1}^{n_z} \angle (s - z_i) - \sum_{i=1}^{n_p} \angle (s - p_i)$$

for the term $(s - p_i)$ or $(s - z_i)$ and substitute p_i or z_i for s in all the terms except the one to which s is adjacent.

Rule 9.6. *The angle at which a branch of the root locus leaves a pole p_j is given by*

$$\angle (s - p_j) = \sum_{i=1}^{n_z} \angle (p_j - z_i) - \sum_{i=1,i \neq j}^{n_p} \angle (p_j - p_i) - 180°$$

and the angle at which it approaches a zero z_j is given by

$$\angle(s - z_j) = 180° - \sum_{i=1, i \neq j}^{n_z} \angle(z_j - z_i) + \sum_{i=1, i \neq j}^{n_p} \angle(z_j - p_i).$$

Rule 9.6 works for real as well as complex poles and zeros. However, there is no point in doing the computation for the real poles and zeros because Rule 9.5 gives the appropriate angle.

The final rule addresses a feature not present in Figure 9.46, so we present another example that reviews a couple of the rules we know so far and introduces the need for the final rule.

Example 9.31. Consider

$$G(s) = \frac{s + 6}{(s + 1)(s + 3)}.$$

The poles and zeros of $G(s)$ are illustrated in Figure 9.50. The first rule we apply is Rule 9.5, so we know the root locus will be between the two poles and then to the left of the zero. If we compute the asymptote angles, we get only one asymptote at $\theta_0 = 180°$, which coincides with the part to the left of the zero already completed by Rule 9.5. This part of the root locus is illustrated in Figure 9.51.

Now, consider Rules 9.1 and 9.2, which require that the branches of the root locus start at the poles of $G(s)$ and end at either zeros of $G(s)$ or grow unbounded. So far the root locus does have branches that start at the poles and it does end at the zero and does grow unbounded.

However, recall Proposition 9.7 which states that the root locus must be continuous. Therefore there must be a way that the branches which start from the poles are connected to the branches that go to the zero or infinity. They cannot connect along the real axis because between the middle pole and the zero $\angle G(s) = 0°$. Hence, the only way it may happen is that they "break away" from the real axis between the poles and "break in" to the real axis to the left of the zero. The root locus, computed numerically, is illustrated in Figure 9.52.

This example is actually quite interesting. For small k, the poles are both real, then as k is increased they are a complex conjugate pair, and as k is even further increased they become real again, which corresponds to starting as exponential solutions for small k, sine and cosine solutions for intermediate k values, and exponential solutions again for large k values.

First we determine the rule to compute exactly where the root locus will break in and away from the real axis. Then we present an argument as to why the curve between the break-in and away points is a rather nice near-circle, as opposed to being, for example, very wavy between the break in and away points.

The root locus starts at the poles, therefore the point at which the locus will break away corresponds to the maximum value that k attains on the real axis. Correspondingly, the branches will break in at a minimum value for k. Hence the break-away and break-in points can be determined by solving $1 + kG(s) = 0$ for k and determining the points on the part of the locus on the real axis for which the derivative of $k = -1/G(s)$ with respect to s is zero. Inasmuch as we are searching for where the derivative is zero, we can drop the minus sign for simplicity.

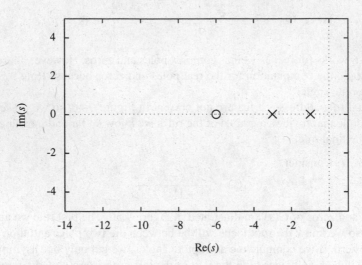

Fig. 9.50 Poles and zeros of $G(s) = (s+6)/\big((s+1)(s+3)\big)$ for Example 9.31.

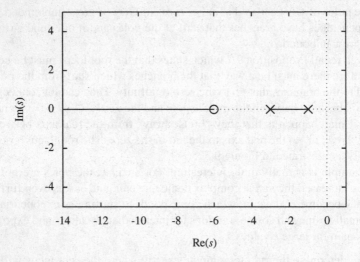

Fig. 9.51 A partial root locus plot for Example 9.31.

Rule 9.7. *For the part of the root locus on the real axis (determined by Rule 9.5), the locus breaks in or breaks away at points where*

$$\frac{\mathrm{d}}{\mathrm{d}s}\left(\frac{1}{G(s)}\right) = 0. \tag{9.27}$$

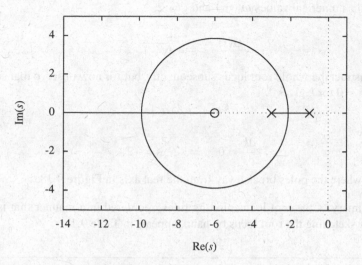

Fig. 9.52 The root locus plot for $G(s) = (s+6)/((s+1)(s+3))$.

Note that there may be other points at which the derivative in Equation (9.27) are zero, but if they are not on the real axis to the left of an odd number of poles and zeros they are not relevant. The reason these may occur includes, for example, an extremum for k that corresponds to a negative value for k.

Example 9.32. Returning to Example 9.31, solving $1 + kG(s) = 0$ for k gives

$$k = \frac{(s+1)(s+3)}{s+6}.$$

Differentiating with respect to s gives

$$\frac{dk}{ds} = \frac{(2s+4)(s+6) - (1)(s^2+4s+3)}{(s+6)^2} = \frac{s^2+12s+21}{(s+6)^2}.$$

Hence,

$$\frac{dk}{ds} = 0 \implies s = -6 \pm \sqrt{15}$$

or

$$s \approx -2.1270, \quad s \approx -9.8730,$$

which conforms to Figure 9.52.

Let us revisit the very first motivational example.

Example 9.33. The transfer function from Example 9.23 was

$$G(s) = \frac{\omega_n^2}{s^2 + 2\zeta\omega_n s + \omega_n^2},$$

or using the numerical values $\omega_n = 1$ and $\zeta = 2$,

$$G(s) = \frac{1}{s^2 + 4s + 1}.$$

We reconstruct the whole root locus subsequently, but for now observe that solving $1 + kG(s) = 0$ for k gives

$$k = s^2 + 4s + 1$$

and hence

$$\frac{dk}{ds} = 0 \quad \Longrightarrow \quad s = -2,$$

which is where the poles break away from the real axis in Figure 9.38.

A summary of the root locus plotting rules, reordered in a manner that is most useful for sketching the root locus by hand, appears in Table 9.1.

Rules to plot the solutions of $1 + kG(s) = 0$ for $k \in [0, \infty)$.

---

1. Plot the poles and zeros of $G(s)$. There will be $n_p - n_z$ branches and each branch of the root locus starts at one of the poles. If $G(s)$ has n_p poles and n_z zeros, n_z of the branches will end at the zeros (Rules 9.1 and 9.2)
2. Draw the root locus on the real axis to the left of an odd number of poles plus zeros. (Rule 9.5)
3. Compute the asymptote angles using

$$\theta_n = \frac{(180° + n360°)}{(n_z - n_p)}.$$

Sketch the asymptotes, which intersect the real axis at

$$s_{int} = \frac{\sum_{i=1}^{n_z} z_i - \sum_{i=1}^{n_p} p_i}{n_z - n_p}.$$

(Rules 9.3 and 9.4)
4. If $G(s)$ has any complex conjugate pairs of poles or zeros, compute the departure or arrival angles, respectively, by taking a point very close to one of them and computing the angle from the pole or zero that would be necessary to ensure $\angle G(s) = 180°$. (Rule 9.6)
5. Compute the break-away or break-in points from the real axis, if any, by computing the values for which

$$\frac{d}{ds}\left(\frac{1}{G(s)}\right) = 0.$$

(Rule 9.7)
6. Complete the root locus keeping in mind that the branch connecting two sections cannot be too complicated if the order of the numerator and denominator of $G(s)$ is not too large.

Table 9.1 Root locus plotting rules

9.6.4 Examples

This section presents a few examples that illustrate the application of the root locus plotting rules in Table 9.1.

Example 9.34. Let us return to the PI control problem from Example 9.24. In that problem the transfer function was expressed in the block diagram in Figure 9.40. If $\omega_n = 1$ and $\zeta = 2$, the transfer function is

$$\frac{X(s)}{X_d(s)} = \frac{k\frac{1+\frac{1}{2s}}{s^2+4s+1}}{1+k\frac{1+\frac{1}{2s}}{s^2+4s+1}} \tag{9.28}$$

Hence, the transfer function to use in all the plotting rules is

$$G(s) = \frac{1+\frac{1}{2s}}{s^2+4s+1} = \frac{s+\frac{1}{2}}{s(s^2+4s+1)}.$$

Recall that when we added integral control in Example 9.24 we could not proceed any further than determining the transfer function because the denominator was third order. Now, after all the work in the preceding section, we can accomplish what we wanted, which was to see how the poles of the transfer function in Equation (9.28) vary as the gain k is varied from 0 to $+\infty$.

Let us follow the steps exactly as they appear in Table 9.1.

1. $G(s)$ has a zero at $s = -1/2$ and three poles at $s = 0$, $s \approx -3.73205$ and $s \approx -0.26795$, all of which are easy to determine by hand. A plot of the poles and zeros for $G(s)$ appears in Figure 9.53.
2. Now filling in to the left of an odd number of zeros plus poles results in the partial root locus plot illustrated in Figure 9.54.
3. There are three poles and one zero, so $n_z - n_p = -2$. Hence the two asymptote angles are

$$\theta_0 = -90°, \quad \theta_1 = 90°,$$

and the intersection with the real axis is at

$$s_{int} = \frac{(-1/2) - (0 - 3.73205 - 0.26795)}{1 - 3} = -1.75.$$

The asymptotes are sketched on the root locus diagram by the vertical line in Figure 9.55.
4. Step 4 does not apply because there are no complex conjugate poles and zeros.
5. Differentiating $k = 1/G(s)$ with respect to s gives

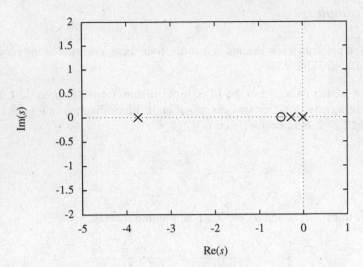

Fig. 9.53 Partial root locus plot for $G(s) = (s + 1/2) / (s (s^2 + 4s + 1))$ after step 1.

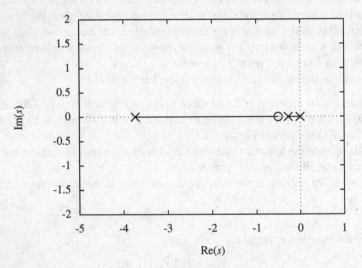

Fig. 9.54 Partial root locus plot for $G(s) = (s + 1/2) / (s (s^2 + 4s + 1))$ after step 2.

$$
\begin{aligned}
\frac{\mathrm{d}}{\mathrm{d}s} \left(\frac{1}{G(s)} \right) &= \frac{\mathrm{d}}{\mathrm{d}s} \frac{s (s^2 + 4s + 1)}{s + \frac{1}{2}} \\
&= \frac{\left(3s^2 + 8s + 1\right) \left(s + \frac{1}{2}\right) - \left(s^3 + 4s^2 + s\right)}{\left(s + \frac{1}{s}\right)^2} \\
&= \frac{2s^3 + \frac{11}{2}s^2 + 4s + \frac{1}{2}}{\left(s + \frac{1}{2}\right)^2}.
\end{aligned}
$$

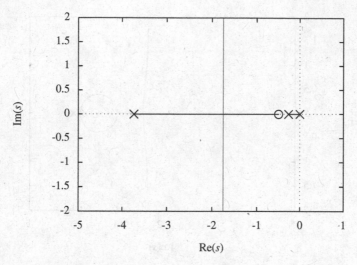

Fig. 9.55 Partial root locus plot for $G(s) = (s + 1/2) / \left(s \left(s^2 + 4s + 1\right)\right)$ after Step 3.

Finding the zeros of the cubic polynomial in the numerator unfortunately is a bit hard to do by hand. Hence, we just give the answer computed numerically. The derivative is zero, $dk/ds = 0$, at the values

$$s = -1.59, \quad s = -1.00, \quad s = -0.157.$$

The root locus must break out from the point $s = -0.157$ inasmuch as it is between two poles. The other two points are also on the root locus on the real line, so one must be a break-in point and one a break-away point for the loci to grow unbounded along the asymptotes.

6. The completed root locus is illustrated in Figure 9.56.

Example 9.35. Sketch the root locus plot for

$$G(s) = \frac{s + 3}{(s + 1)(s + 2)}.$$

1. The poles are at $s = -1$ and $s = -2$. There is one zero at $s = -3$.
2. The root locus on the real axis is to the left of an odd number of zeros plus poles, as is illustrated in Figure 9.57.
3. There are two poles and one zero, so the only asymptote is at $\theta = 180°$, which has already been plotted by the step dealing with the root locus on the real axis.
4. There are no complex conjugate poles or zeros of $G(s)$, so this step does not apply.
5. The break-in and break-away points are where

$$\frac{dk}{ds} = 0$$

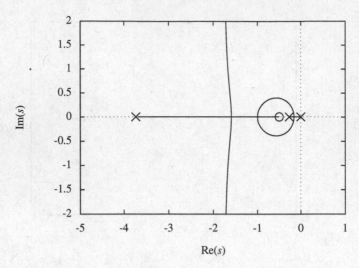

Fig. 9.56 The completed root locus plot for $G(s) = (s + 1/2) / (s(s^2 + 4s + 1))$.

or

$$\frac{dk}{ds} = \frac{d}{ds}\frac{1}{G(s)} = \frac{d}{ds}\left(\frac{(s+1)(s+2)}{s+3}\right) = \frac{d}{ds}\left(\frac{s^2+3s+2}{s+3}\right)$$

$$= \frac{(2s+3)(s+3) - (s^2+3s+2)}{(s+3)^2} = \frac{s^2+6s+7}{(s+3)^2},$$

which is equal to zero at $s \approx -4.4142$ and $s \approx -1.5858$. The first must be a break-in point and the latter a break-away point.

6. Because the root locus must break out between the poles and break in to the left of the zero, the root locus must be comprised of complex conjugate pairs between the two. Because we are sketching the roots of a relatively low-order polynomial, the path between the break-away and break-in points must be relatively low-order. The completed root locus is illustrated in Figure 9.58.

Example 9.36. Sketch the root locus for

$$G(s) = 7\frac{s-1}{s^2+2s+5}.$$

1. There is a zero at $s = 1$ and two poles at $s = -1 \pm 2i$.
2. On the real axis, the root locus will be to the left of the zero, as is illustrated in Figure 9.60.
3. There are two poles and one zero, so the only asymptote is at $180°$, which has already been completed by the rule for the locus on the real axis.

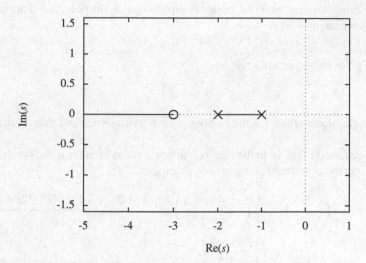

Fig. 9.57 Partial root locus for Example 9.35.

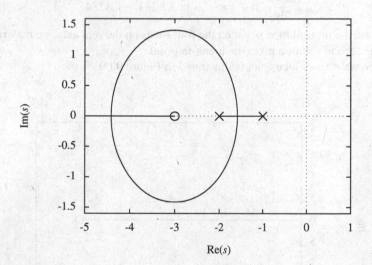

Fig. 9.58 Root locus for Example 9.35.

4. Considering a point near the upper complex conjugate pole and determining $\angle G(s)$, we have

$$135° - 90° - \theta = \pm 180°$$

which gives the departure angle as

$$\theta = 225°.$$

The angle of departure for the bottom pole is symmetric, and thus is equal to 135°.

5. The locus must break in to the real axis at some point because it ends at the zero and along the asymptote going to $-\infty$. Compute

$$\frac{dk}{ds} = \frac{d}{ds}\left(\frac{1}{G(s)}\right) = \frac{d}{ds}\left(\frac{s^2+2s+5}{s-1}\right) = \frac{(2s+2)(s-1)-(s^2+2s+5)}{(s-1)^2}$$
$$= \frac{s^2-2s-7}{(s-1)^2},$$

which gives

$$\frac{dk}{ds} = 0 \quad \Longrightarrow \quad s \approx 3.8284, -1.8284.$$

Because the first solution is not on the root locus on the real axis, we may ignore it. The second solution gives the break-in point.

6. The complete root locus plot is illustrated in Figure 9.60.

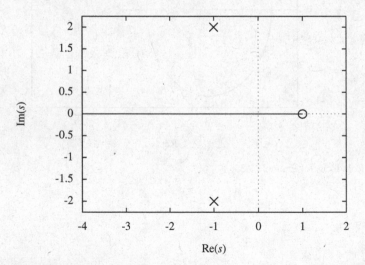

Fig. 9.59 Partial root locus for Example 9.36.

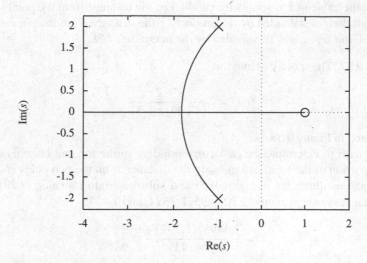

Fig. 9.60 Root locus for Example 9.36.

9.6.5 Determining the Gain

Although they make for some pretty pictures, the real reason to sketch a root locus plot is that it serves as a tool to determine a good value, if any, for k. Once the root locus plot has been sketched, if it appears that it passes through a region where it will have the desired response characteristics, then it will be necessary to determine the gain k value that corresponds to a point on the locus in that region. Fortunately, this is relatively easy.

Points on the root locus satisfy

$$1 + kG(s) = 0$$

thus

$$|k| = \left| -\frac{1}{G(s)} \right| = \frac{1}{|G(s)|}.$$

Also, because

$$|G(s)| = \left| \hat{k} \frac{\prod_{i=1}^{n_z} (s - z_i)}{\prod_{i=1}^{n_p} (s - p_i)} \right| = |\hat{k}| \frac{|s - z_1| \, |s - z_2| \cdots |s - z_{n_z}|}{|s - p_1| \, |s - p_2| \cdots |s - p_{n_p}|},$$

we have

$$k = \frac{|s - p_1| \, |s - p_2| \cdots |s - p_{n_p}|}{|\hat{k}| \, |s - z_1| \, |s - z_2| \cdots |s - z_{n_z}|}. \tag{9.29}$$

In words, the value of k is simply the product of the distance from the point on the locus to all the poles divided by the product of the distance from the point on the locus to all the zeros, and also divided by the magnitude of $\hat{k}$.

Example 9.37. The root locus plot for

$$G(s) = 2\frac{s+2}{(s+1)(s+3)}$$

is illustrated in Figure 9.58.

If we wish to determine the gain corresponding to the top and bottoms of the "circle" portion of the locus, we measure the distance from the two poles and zero to the point, as illustrated in Figure 9.61 and substitute into Equation (9.29). The three distances are approximately, 2.4495, 1.7321, and 1.4142

$$k \approx \frac{(2.4495)(1.7321)}{2(1.4142)} \approx \frac{3}{2}.$$

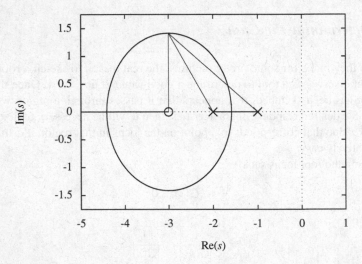

Fig. 9.61 Measuring the distance from the poles and zeros of $G(s)$ to the point of interest to determine the gain for Example 9.37.

9.6.6 Computational Tools

Both MATLAB and Octave have functions to graph root locus plots.

The MATLAB command to plot a root locus is `rlocus()`. As the reader probably expects, it takes the numerator and denominator of $G(s)$ as arguments and then plots the solutions to $1 + kG(s) = 0$ for $k \in [0, +\infty)$

Example 9.38. To use MATLAB to plot the root locus for the system in Example 9.31 where

$$G(s) = \frac{s+6}{(s+1)(s+3)},$$

enter

```
>> rlocus([1 6],conv([1 1],[1 3]))
```

or equivalently

```
>> rlocus([1 6],[1 4 3])
```

to create the plot.

The Octave command to plot a root locus is also `rlocus()`, but the transfer function must be specified with the `tf()` function.

Example 9.39. To use Octave to plot the root locus for the system in Example 9.31 where

$$G(s) = \frac{s+6}{(s+1)(s+3)},$$

enter

```
octave:> rlocus(tf([1 6],conv([1 1],[1 3])))
```

or equivalently

```
octave:> rlocus(tf([1 6],[1 4 3]))
```

to create the plot.

9.7 Frequency Response Analysis

Frequency response analysis of a system focuses upon analyzing the relationship between the input and output of a transfer function when the input is a purely sinusoidal signal. As shown subsequently, if an input to a transfer function is a pure sinusoid, $r(t) = \sin \omega t$, the output will be a sinusoid of the same frequency, but with a magnitude and phase shift that depend on the frequency of the input.

An advantage of frequency response methods is that the data necessary for the analyses based on them may be determined experimentally. For a very complicated system, it may be very difficult to determine the equations governing the system based on first principles, so these approaches that are amenable to experimental system identification are very useful for real-world control problems.

Example 9.40. Consider

$$\frac{R(s)}{Y(s)} = \frac{2}{s+2}$$

and two input signals

$$r_1(t) = \sin t, \quad r_2(t) = \sin 3t,$$

or

$$R_1(s) = \frac{1}{s^2+1}, \quad R_2(s) = \frac{3}{s^2+9}.$$

Solving either

$$y_1(t) = \mathcal{L}^{-1}\left(\frac{2}{s+2}\frac{1}{s^2+1}\right), \quad y_2(t) = \mathcal{L}^{-1}\left(\frac{2}{s+2}\frac{3}{s^2+9}\right)$$

or

$$\dot{y}_1 + 2y = 2\sin t, \quad \dot{y}_2 + 2y = 2\sin 3t$$

(the latter with zero initial conditions) gives

$$y_1(t) = \frac{2}{5}\left(e^{-2t} + 2\sin t - \cos t\right), \quad y_2(t) = \frac{2}{13}\left(3e^{-2t} + 2\sin 3t - 3\cos 3t\right).$$

Hence, for large t, the steady state solutions are

$$y_{1,ss}(t) = \frac{2}{5}\left(2\sin t - \cos t\right), \quad y_{2,ss}(t) = \frac{2}{13}\left(2\sin 3t - 3\cos 3t\right).$$

Using the relationship

$$\sin(\omega t + \phi) = \cos\phi\sin\omega t + \sin\phi\cos\omega t$$

these may be written as

$$y_{1,ss}(t) = \frac{2}{5}\sqrt{5}\sin(t + \phi_1), \qquad \phi_1 = \tan^{-1}\left(\frac{-1}{2}\right)$$

$$y_{2,ss}(t) = \frac{2}{13}\sqrt{13}\sin(3t + \phi_2), \qquad \phi_2 = \tan^{-1}\left(\frac{-3}{2}\right).$$

Observe that the steady-state solution is a sinusoid of the same frequency as the input, but the magnitude is scaled and there may be a phase shift, both of which may change as the forcing frequency ω changes. Foreshadowing what is to come, note that if we substitute $s = i\omega$ into $G(s)$, we get for each case, respectively,

$$G(i) = \frac{2}{i+2} = \frac{4-2i}{5}$$

$$G(3i) = \frac{2}{3i+2} = \frac{4-6i}{13}.$$

The magnitude of these is

$$|G(i)| = \frac{1}{5}\sqrt{20} = \frac{2}{\sqrt{5}}$$

$$|G(3i)| = \frac{1}{13}\sqrt{52} = \frac{2}{\sqrt{13}}.$$

So, it appears that if we simply substitute $s = i\omega$ into $G(s)$ and determine its magnitude, it gives the magnitude of the response.

Similarly, if we compute

$$\angle G(i) = \tan^{-1}\left(\frac{-2}{4}\right) = \tan^{-1}\left(\frac{-1}{2}\right)$$

$$\angle G(3i) = \tan^{-1}\left(\frac{-6}{4}\right) = \tan^{-1}\left(\frac{-3}{2}\right),$$

it appears that the phase shift in the steady-state response is given by the angle of $G(i\omega)$.

The next proposition shows that if a transfer function is stable, then if the input is $A \sin \omega t$ the steady-state solution will be scaled by $|G(i\omega)|$ and have a phase shift of $\phi = \tan^{-1}(\mathrm{Im}(i\omega)/\mathrm{Re}(i\omega))$.

Proposition 9.8. *If all the poles of $G(s)$ are in the left half-plane, and if*

$$y(t) = \mathcal{L}^{-1}\left(G(s)\frac{A\omega}{s^2 + \omega^2}\right)$$

as t becomes large, the steady-state solution is given by

$$y_{ss}(t) = A|G(i\omega)|\sin(\omega t + \phi),$$

where

$$\phi = \tan^{-1}\left(\frac{\mathrm{Im}(i\omega)}{\mathrm{Re}(i\omega)}\right).$$

Proof. Let

$$G(s) = \frac{N(s)}{D(s)};$$

then one form of a partial fraction expansion will be

$$G(s)\frac{A\omega}{s^2 + \omega^2} = \frac{N(s)}{D(s)}\frac{A\omega}{s^2 + \omega^2} = \frac{C_1(s)}{D(s)} + \frac{A\omega(c_1 s + c_2)}{s^2 + \omega^2}.$$

Using the method from Appendix A.3, to determine $C_2(s)$, multiply both sides of this equation by $(s^2 + \omega^2)$ and take the limit as $s \to i\omega$; that is,

$$\lim_{s \to i\omega} \left(G(s) \frac{A\omega}{s^2 + \omega^2} \left(s^2 + \omega^2\right) \right)$$

$$= \lim_{s \to i\omega} \left(\frac{AC_1(s)}{D(s)} \left(s^2 + \omega^2\right) + \frac{A\omega (c_1 s + c_2)}{s^2 + \omega^2} \left(s^2 + \omega^2\right) \right)$$

which gives

$$c_1 i\omega + c_2 = G(i\omega),$$

so

$$c_1 = \frac{1}{\omega} \operatorname{Im}(G(i\omega)), \quad c_2 = \operatorname{Re}(G(i\omega)).$$

Referring to Table 8.1, the c_1 term corresponds to the cosine component in the steady state solution and the c_2 term corresponds to the sine component.

Hence,

$$y_{ss}(t) = A \left(\operatorname{Re}(G(i\omega)) \sin\omega t + \operatorname{Im}(G(i\omega)) \cos\omega t \right)$$

$$= \sqrt{[\operatorname{Re}(G(i\omega))]^2 + [\operatorname{Im}(G(i\omega))]^2} \sin(\omega t + \phi)$$

$$= |G(i\omega)| \sin(\omega t + \phi),$$

where

$$\phi = \tan^{-1} \left(\frac{\operatorname{Im}(G(i\omega))}{\operatorname{Re}(G(i\omega))} \right).$$

$\square$

It turns out that it is relatively easy to sketch $|G(i\omega)|$ and $\phi = \tan^{-1}(\operatorname{Im}(G(i\omega)) / \operatorname{Re}(G(i\omega)))$ by hand, so it is not too difficult to obtain information about the steady-state response of the system to sinusoidal inputs. It is more important that very useful information regarding the stability of a system under unity feedback and information on designing feedback controllers may be obtained by graphs of the magnitude and phase of the steady-state response to a sinusoidal input. This type of analysis is referred to as a *frequency response analysis* and is common in control theory, particularly in electrical engineering.

9.7.1 Bode Plots

A *Bode plot* is a log–log plot of the magnitude and phase of $G(i\omega)$ versus ω. It is conventional to plot the magnitude on a log scale and it is conventional to do so in *decibels*. For our purposes the definition of a decibel is

$$|G(i\omega)|_{dB} = 20 \log_{10} |G(i\omega)|.$$

Because the plot is on a logarithmic scale, we may construct the plot for individual terms in the product of a transfer function and then add (or subtract) them to construct the overall plot. In order to do this, we need to have handy the plot for the typical components of a transfer function.

Example 9.41. Consider

$$G(s) = \frac{10}{s+10}.$$

A computer-generated Bode plot of $G(s)$ is illustrated in Figure 9.62.

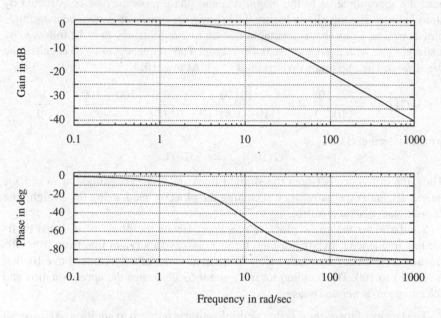

Fig. 9.62 Bode plot for $G(s) = 10/(s+10)$.

Let us first rewrite $G(i\omega)$ in a way that subsequently helps with our analysis,

$$G(i\omega) = \frac{10}{i\omega + 10} = \frac{10}{i\omega + 10}\frac{10 - i\omega}{10 - i\omega} = \frac{100 - 10i\omega}{100 + \omega^2}.$$

For $\omega \ll 10$,

$$G(i\omega) \approx 1, \quad \Longrightarrow \quad |G(i\omega)| \approx 1 = 0\,\text{dB}$$

and

$$\angle G(i\omega) \approx 0°.$$

For small frequencies, the Bode plot in Figure 9.62 corresponds to this.
For $\omega \gg 10$

$$G(i\omega) \approx -\frac{10i}{\omega}, \quad \Longrightarrow \quad |G(i\omega)| \approx \frac{10}{\omega}.$$

As ω increases, $|G(i\omega)|$ decreases, and in particular whenever ω increases by a factor of 10, $|G(i\omega)|$ decreases by a factor of 10. A decrease by a factor of 10 corresponds to a decrease of 20 dB, thus the slope of the magnitude curve at high ω should be -20 dB/decade (-20 dB for every increase in order of magnitude of ω). Also, for $\omega \gg 10$, $G(i\omega)$ is almost a purely negative imaginary number, so

$$\angle G(i\omega) \approx -90°.$$

For large frequencies, the Bode plot in Figure 9.62 corresponds to this.

Because of the manner in which the logarithmic scale presents the data, a fairly accurate approximation to the magnitude and phase plots may be constructed by straight lines. The top plot in Figure 9.63 shows the computer-generated magnitude plot as well as a plot that is a straight line up to the frequency $\omega = 10$ followed by a straight line with a slope of -20 dB/decade. The frequency, $\omega = 10$ is called the *breakpoint*. The maximum error is at $\omega = 10$. At $\omega = 10$,

$$|G(10i)| = \left| \frac{10}{10 + 10i} \right| = \left| \frac{10(10 - 10i)}{(10 + 10i)(10 - 10i)} \right| = \left| \frac{100 - 100i}{200} \right| = \frac{1}{\sqrt{2}}$$

which in decibels is

$$|G(10i)|_{dB} = -3.0103.$$

Thus, the difference between the straight lines and exact magnitude at $\omega = 10$ is only 3 dB, and, especially when sketching the plots by hand, using the straight line approximation should suffice.

Similarly, for the phase plot, we can approximate the phase of $|G(i\omega)|$ by $0°$ for low frequencies, by $90°$ for high frequencies, and a straight line from $0°$ to $90°$ starting an order of magnitude below 10 up to an order of magnitude above 10, that is, from 1 to 100. The bottom plot in Figure 9.63 illustrates the approximation and the computer-generated phase.

The beauty of logarithms is that multiplication is reduced to addition. Also, recall that when complex numbers are multiplied, the phases add. We may exploit this fact when sketching Bode plots by sketching each term in a transfer function individually and then adding them, as the following example illustrates.

Example 9.42. Sketch the Bode plot for

$$G(s) = \frac{100}{s(10s + 1)}.$$

Explicitly writing all three terms we have

$$G(s) = 100 \cdot \frac{1}{s} \cdot \frac{1}{10s + 1}$$

so

$$|G(i\omega)| = \left| 100 \cdot \frac{1}{i\omega} \cdot \frac{1}{10\omega i + 1} \right| = |100| \cdot \left| \frac{1}{i\omega} \right| \cdot \left| \frac{1}{10\omega i + 1} \right|$$

or, in decibels and making use of the fact that the logarithm of a product is the sum of the logarithms

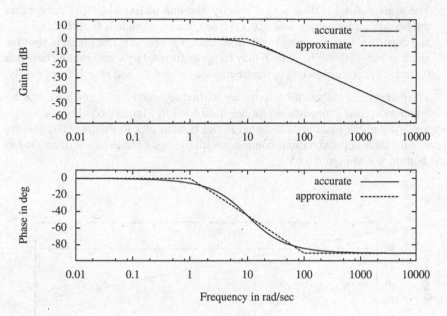

Fig. 9.63 Approximations in a Bode plot for a real pole.

$$|G(\mathrm{i}\omega)|_{\mathrm{dB}} = |100|_{\mathrm{dB}} + \left|\frac{1}{\mathrm{i}\omega}\right|_{\mathrm{dB}} + \left|\frac{1}{10\omega\mathrm{i}+1}\right|_{\mathrm{dB}}. \qquad (9.30)$$

Similarly, because in polar coordinates the phase of complex numbers add when complex numbers are multiplied we have

$$\angle G(\mathrm{i}\omega) = \angle\left(100 \cdot \frac{1}{\mathrm{i}\omega} \cdot \frac{1}{10\omega\mathrm{i}+10}\right)$$

$$= \angle 100 + \angle\frac{1}{\mathrm{i}\omega} + \angle\frac{1}{10\omega\mathrm{i}+1}. \qquad (9.31)$$

Consider each term in Equation (9.30).

1. $|100|_{\mathrm{dB}} = 40$ and does not depend on ω. Because it is a positive real number, $\angle 100 = 0°$.

2. $|1/(\mathrm{i}\omega)|_{\mathrm{dB}}$ has a slope of -20 dB/decade and has a value of 0 dB at $\omega = 1$. Because

$$\frac{1}{\mathrm{i}\omega} = -\frac{\mathrm{i}}{\omega}$$

it is a negative imaginary number, and hence

$$\angle\frac{1}{\mathrm{i}\omega} = -90°.$$

3. The analysis for $1/(10i\omega + 1)$ is exactly like that of Example 9.41 except that the breakpoint frequency is $\omega = 0.1$. Hence, the magnitude is 0 dB for $\omega < 0.1$, and decreases with a slope of -20 dB/decade for $\omega > 20$. The phase is zero for small ω and $-90°$ for large ω. It may be approximated by a straight line between $0°$ and $-90°$ corresponding to the frequencies $\omega = 0.01$ and $\omega = 1$, respectively.

The magnitudes of these three terms are plotted separately in Figure 9.64, and the sum and an accurate magnitude plot are illustrated in Figure 9.65. The individual approximate phase plots are illustrated in the bottom plot in Figure 9.64, and the sum of the three approximations compared with the exact phase plot is illustrated in the bottom plot in Figure 9.65.

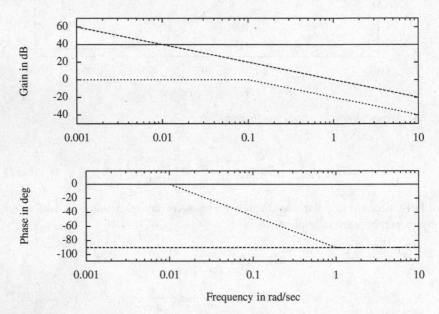

Fig. 9.64 Straight-line approximations for the three terms in the Bode magnitude plot for Example 9.42: 100 is the solid line, $1/s$ is the dashed line, and $1/(10s + 1)$ is the dotted line.

9.7.1.1 Sketching Bode Plots

Bode plots are relatively straightforward to sketch by hand. This section develops a procedure to do so for the types of terms in transfer functions; namely, a constant gain, poles, zeros, and complex conjugate pairs of poles and zeros.

For a constant transfer function

$$G(s) = \hat{k},$$

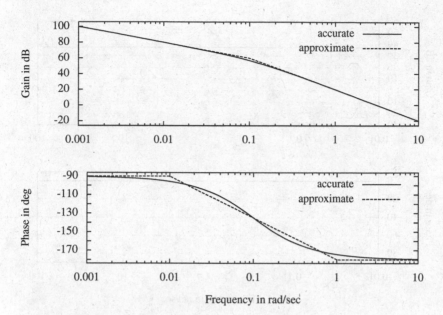

Fig. 9.65 Sum of the straight-line approximations for the three terms in the Bode phase plot for Example 9.42 compared to the accurate plot.

the Bode plot is simple because the constant does not depend on ω. In particular, the magnitude in decibels is simply

$$|G(i\omega)|_{dB} = 20 \log_{10} \hat{k}.$$

If $\hat{k} > 0$, then $\angle \hat{k} = 0°$. If $\hat{k} < 0$, then $\angle \hat{k} = \pm 180°$.

Example 9.43. The Bode plot for

$$G(s) = 75$$

is illustrated in Figure 9.66.

For a real zero,

$$G(s) = \frac{s}{\omega^z} + 1$$

then

$$G(i\omega) = 1 + i\frac{\omega}{\omega^z}.$$

For the case where $\omega \ll \omega^z$, then

$$G(i\omega) \approx 1$$

and hence for $\omega \ll \omega^z$,

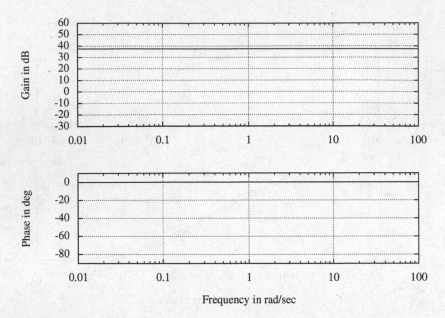

Fig. 9.66 Bode plot for $G(s) = 75$.

$$|G(i\omega)|_{dB} \approx 0$$

and

$$\angle G(i\omega) \approx 0°.$$

For the case where $\omega \gg \omega^z$, then

$$G(i\omega) \approx i\frac{\omega}{\omega^z}$$

and hence for $\omega \gg \omega^z$, if ω increases by a factor of 10, then $|G(i\omega)|$ will increase by a factor of 10 and $20\log_{10}|G(i\omega)|_{dB}$ will increase by 20 dB. Hence, for $\omega \gg \omega^z$, the magnitude plot will have a slope of 20 dB/decade. Also, for $\omega \gg \omega^z$,

$$\angle G(i\omega) \approx 90°.$$

These are illustrated in Figure 9.67.

Example 9.44. The Bode plot for

$$G(s) = s + 100$$

is illustrated in Figure 9.68. It may be sketched simply by considering small and large ω relative to 100, or by rewriting the transfer function as

$$G(s) = 100\left(\frac{s}{100} + 1\right).$$

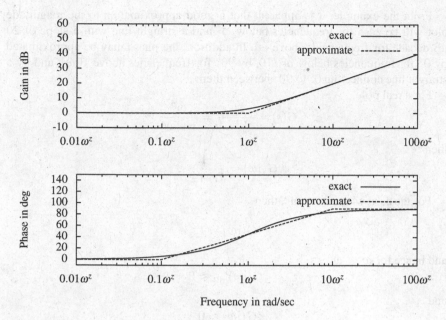

Fig. 9.67 Bode plot for $G(s) = s/\omega^z + 1$.

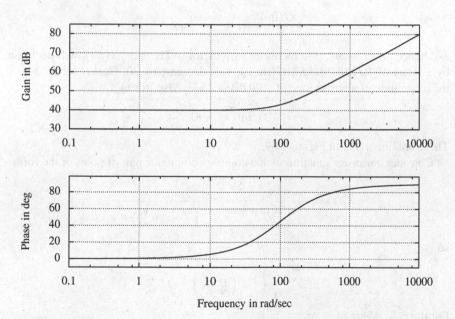

Fig. 9.68 Bode plot for $G(s) = s/100 + 1$.

From the example, it is apparent that a good approximation to the magnitude plot will be zero for frequencies below ω^z and a straight line with a slope of 20 dB/decade for frequencies above ω^z. In addition, the phase may be approximated by $0°$ for frequencies below $\omega^z/10$, by $90°$ for frequencies above $10\omega^z$, and by a straight line connecting $0°$ to $90°$ between them.

For a real pole,

$$G(s) = \frac{1}{\frac{s}{\omega^p}+1} = \frac{\omega^p}{s+\omega^p}$$

then

$$G(i\omega) = \frac{1}{1+i\frac{\omega}{\omega^p}}.$$

For the case where $\omega \ll \omega^p$, then

$$G(i\omega) \approx 1$$

and hence for $\omega \ll \omega^p$,

$$|G(i\omega)|_{dB} \approx 0$$

and

$$\angle G(i\omega) \approx 0°.$$

For the case where $\omega \gg \omega^p$, then

$$G(i\omega) \approx \frac{1}{i\frac{\omega}{\omega^p}} = -i\frac{\omega^p}{\omega},$$

and hence for $\omega \gg \omega^p$, if ω increases by a factor of 10, then $|G(i\omega)|$ will decrease by a factor of 10 and $20\log_{10}|G(i\omega)|_{dB}$ will decrease by 20. Hence, for $\omega \gg \omega^p$, the magnitude plot has a slope of -20 dB/decade. Also, for $\omega \gg \omega^p$,

$$\angle G(i\omega) \approx -90°.$$

These are illustrated in Figure 9.69.

Consider a transfer function with a complex conjugate pair of poles of the form

$$G(s) = \frac{\omega_n^2}{s^2 + 2\zeta\omega_n s + \omega_n^2} = \frac{1}{\left(\frac{s}{\omega_n}\right)^2 + \frac{2\zeta s}{\omega_n} + 1},$$

so

$$G(i\omega) = \frac{1}{\left(1-\left(\frac{\omega}{\omega_n}\right)^2\right) + i2\zeta\frac{\omega}{\omega_n}}.$$

For the case where $\omega \ll \omega_n$,

$$G(i\omega) \approx 1$$

so if $\omega \ll \omega_n$,

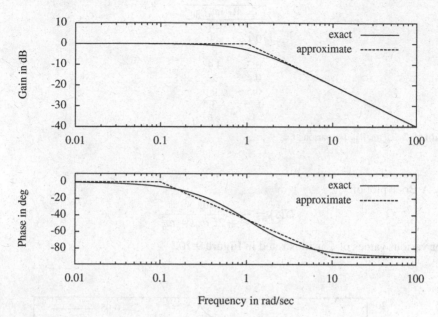

Fig. 9.69 Bode plot for $G(s) = 1/(s/\omega^p + 1)$.

$$|G(i\omega)|_{dB} \approx 0$$

and

$$\angle G(i\omega) \approx 0°.$$

At the other extreme where $\omega \gg \omega_n$

$$G(i\omega) \approx -\left(\frac{\omega_n}{\omega}\right)^2.$$

Hence, when ω increases by a factor of 10, the magnitude of $G(i\omega)$ decreases by a factor of 100. Hence, in decibels, the slope of the magnitude plot for $\omega \gg \omega_n$ will have a slope of -40 dB/decade. Also, for $\omega \gg \omega_n$

$$\angle G(i\omega) = -180°.$$

In the complex conjugate case, we also need to consider the case where $\omega \approx \omega_n$. In that case

$$G(i\omega) \approx \frac{1}{i2\zeta} = -\frac{i}{2\zeta}.$$

Hence, when $\omega \approx \omega_n$

$$|G(i\omega)|_{dB} \approx 20\log_{10}\left(\frac{1}{2\zeta}\right).$$

Table 9.2 presents values of the magnitude of $G(i\omega_n)$ for various values of ζ.

ζ	$|G(i\omega_n)|_{dB}$
0.05	20
0.1	14
0.2	8
0.3	4.4
0.4	2
0.5	0
0.75	−3.5
1	−6

Table 9.2 $|G(i\omega_n)|_{dB}$ for various ζ

A Bode plot of

$$G(s) = \frac{\omega_n^2}{s^2 + 2\zeta\,\omega_n s + \omega_n^2}$$

for various values of ζ is illustrated in Figure 9.70.

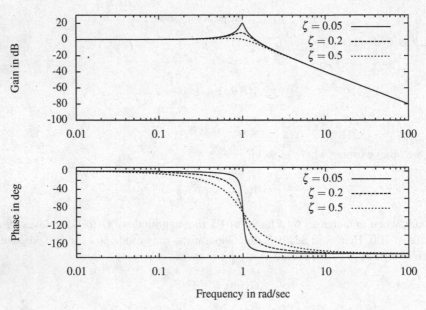

Fig. 9.70 Bode plot for $G(s) = \omega_n^2 / (s^2 + 2\zeta\omega_n s + \omega_n^2)$ for various ζ.

Similarly, for a transfer function with a complex conjugate pair of zeros, the Bode plot of

$$G(s) = \left(\frac{s}{\omega_n}\right)^2 + \frac{2\zeta s}{\omega_n} + 1$$

for various values of ζ is illustrated in Figure 9.71.

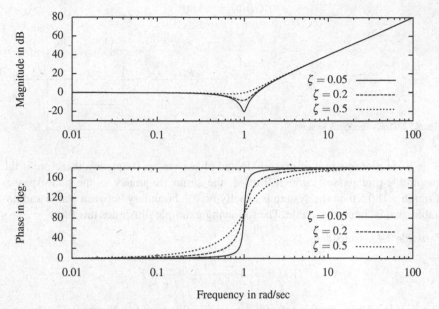

Fig. 9.71 Bode plot for $G(s) = (s/\omega_n)^2 + 2\zeta s/\omega_n + 1$ for various ζ.

9.7.1.2 Gain and Phase Margins

It is possible to determine stability of a system under unity feedback, such as the one illustrated in Figure 9.72, using Bode plots. A system is on the boundary between stable and unstable when it has a pole (or poles) that are exactly on the imaginary axis on the complex plane. Because the root locus plot is a plot of all the points that satisfy

$$1 + kG(s) = 0$$

when a branch crosses the imaginary axis, s is a purely imaginary number, that is, $s = i\omega$, and hence for the gain value corresponding to that point $G(s)$ satisfies

$$1 + kG(i\omega) = 0$$

or

$$kG(i\omega) = -1.$$

In terms of magnitude and phase, then at the point between stable and unstable $G(s)$ satisfies

$$|kG(i\omega)| = 1 \quad \Longleftrightarrow \quad |kG(i\omega)|_{dB} = 0$$

and

$$\angle kG(i\omega) = \pm 180°,$$

or because k only affects the magnitude, but not the phase

$$\angle G(i\omega) = \pm 180°.$$

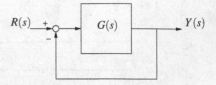

Fig. 9.72 Unity feedback system.

A Bode plot is a plot of magnitude and phase versus frequency, therefore if the magnitude plot passes through 0 dB at the same frequency as the phase passes through $\pm 180°$, then the system is exactly on the boundary between stable and unstable, that is, neutrally stable. The following example illustrates this fact.

Example 9.45. Consider

$$G(s) = \frac{1}{(s+1)(s+2)(s+3)}.$$

The root locus plot for this transfer function is illustrated in Figure 9.73. The gain value where the branches cross the imaginary axis may be determined geometrically from the root locus plot or by using the Routh criterion from Section 9.4. In either case, $k = 60$ corresponds to the point where the branches cross the imaginary axis. For $k > 60$ the system is unstable and for $k < 60$ the system is stable.

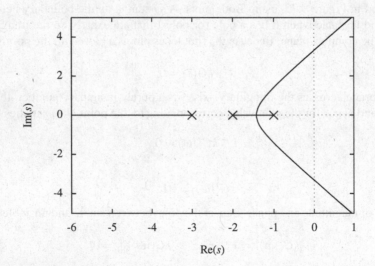

Fig. 9.73 Root locus plot for system in Example 9.45.

Next consider the Bode plot for

$$kG(s) = 60\frac{1}{(s+1)(s+2)(s+3)},$$

which is illustrated in Figure 9.74. In the figure, the magnitude plot crosses 0 dB at the same frequency where the phase passes through $-180°$, which must be the case from our analysis above.

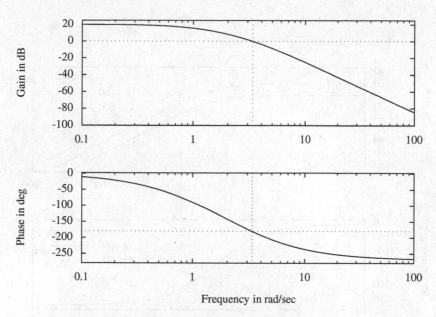

Fig. 9.74 Bode plot for neutrally stable system in Example 9.45.

Because the magnitude plot is a plot of the magnitude of a transfer function, the effect of changing k on the Bode plot of $kG(s)$ will be to shift the magnitude curve up or down, as the k is increased or decreased, respectively, because if k is increased, $|kG(s)|$ increases. Also, because

$$\angle kG(s) = \tan^{-1}\left(\frac{\operatorname{Re} kG(s)}{\operatorname{Im} kG(s)}\right) = \tan^{-1}\left(\frac{k\operatorname{Re} G(s)}{k\operatorname{Im} G(s)}\right) = \tan^{-1}\left(\frac{\operatorname{Re} G(s)}{\operatorname{Im} G(s)}\right),$$

the phase plot is unaffected by changes in k.

Consider the system in Example 9.45: if k is increased the system will become unstable because, from the root locus plot, the poles will move into the right half of the complex plane. The effect of increasing k on the Bode plot will be to shift the magnitude curve up, which will make the magnitude cross through 0 dB after the phase crosses through $-180°$. Conversely, if k is decreased, the system will be stable. On the Bode plot, if k is decreased, then the magnitude curve will pass through 0 dB before the phase goes through $-180°$.

Example 9.46. Considering again

$$G(s) = \frac{1}{(s+1)(s+2)(s+3)}$$

from Example 9.45, Figure 9.75 illustrates the Bode plot for the cases where $k = 6$, $k = 60$, and $k = 600$. For $k = 6$, the magnitude is always below 0 dB. For $k = 600$, the magnitude plot crosses through 0 dB at a higher frequency than the phase passes through $-180°$.

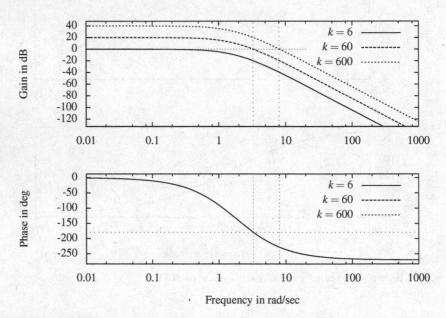

Fig. 9.75 Bode plot for Example 9.46.

Hence, we have the situation that if $G(s)$ is of the nature that increasing k in Figure 9.72 destabilizes the system, we may use the Bode plot to check for stability. One type of system that can go unstable only as k is increased is when all the poles and zeros of $G(s)$ are in the left half-plane. Such a system is described as *minimum phase*.

Definition 9.3. A transfer function $G(s)$ is *minimum phase* if all the poles and zeros of $G(s)$ are in the left half of the complex plane.[5]

So, for a minimum phase system, we have the following proposition.

---

[5] Note that some texts define minimum phase to have no right half-plane zeros, but have no restriction on the poles, such as [13]; others, such as [5] require that there are no right half-plane poles *or* zeros.

Proposition 9.9. *For a minimum phase proper transfer function, $G(s)$, the unity feedback closed-loop transfer function*

$$\frac{Y(s)}{R(s)} = \frac{kG(s)}{1+kG(s)}$$

is stable if $|G(i\omega)|_{dB} = 0$ at a lower frequency than $\angle G(i\omega) = \pm 180°$. Also, if $\angle G(i\omega)$ never passes through $\pm 180°$, the system is stable and if $|G(i\omega)|_{dB}$ is always less than zero the system is stable.

An observant reader will recognize that up until this point in this book a system has been either unstable, neutrally stable, or stable. It would be desirable to have a measure of stability, which are the concepts of *gain margin* and *phase margin*.

Definition 9.4. For a minimum phase system, the *gain margin*, denoted by g_m, is the difference between 0 dB and $|kG(i\omega)|$ where ω is the frequency at which $\angle G(i\omega) = -180°$.

Definition 9.5. For a minimum phase system, the *phase margin*, denoted by ϕ_m, is the difference between $\angle G(i\omega)$ and $-180°$ where ω is the frequency at which $|G(i\omega)|_{dB} = 0$.

When the gain and phase margins are positive, they provide a measure of stability. If the phase never passes through $-180°$ then the gain margin is infinite, which means that the gain can be increased to an arbitrarily large value and still remain stable. The following example illustrates these concepts.

Example 9.47. Figure 9.76 illustrates the gain and phase margins for

$$G(s) = \frac{3000}{(s+10)(s^2+20s+200)}.$$

Because the gain margin is 10.45 dB, that means if $G(s)$ is placed in the unity feedback loop illustrated in Figure 9.72, it will be stable as long as the gain k satisfies

$$k < 10.45 \text{ dB} = 10^{10.45/20} \approx 3.35.$$

In the next chapter, we often use the phase margin as a measure of stability. In such cases, the design objective would normally be to increase the phase margin.

9.7.1.3 Bode Plots and System Type

Recall that the system type is the lowest power of s in the denominator of a transfer function. The system type is straightforward to determine from the low frequency portion of the magnitude plot of Bode plot, which is illustrated with a couple of examples.

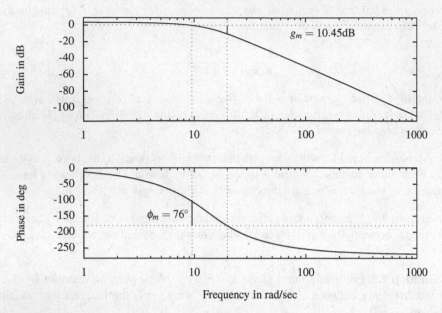

Fig. 9.76 Gain and phase margins for system in Example 9.47.

Example 9.48. Consider the type 1 system

$$G(s) = 10000\frac{s+10}{s\,(s+100)\,(s+1000)}$$

$$= 10000\left(\frac{1}{s}\right)(s+10)\left(\frac{1}{s+100}\right)\left(\frac{1}{s+1000}\right).$$

The last three terms in the product are a zero and two poles, respectively. Considering the term that is a pole at the origin, $1/s$, substituting $s = i\omega$ gives

$$\frac{1}{i\omega} = -i\frac{1}{\omega},$$

which has a constant phase of $-90°$. On the Bode plot, the magnitude curve is a straight line with slope of $-20\,\text{dB/decade}$, which can be seen because as ω increases by a factor of 10, the magnitude decreases by a factor of 10 which corresponds to a change of $-20\,\text{dB}$. Unlike poles and zeros away from the origin, this magnitude curve is a straight line for all frequencies, and persists at low frequencies. Thus, if the slope of the magnitude curve is $-20\,\text{dB/decade}$ at frequencies below all other poles and zeros, the system is type 1. The Bode plot for this system is illustrated in Figure 9.77.

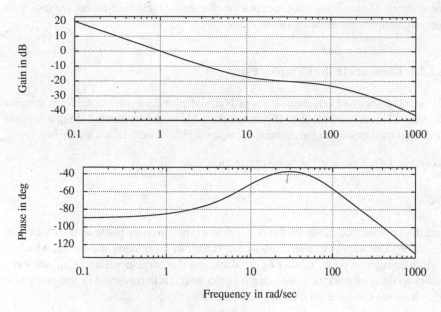

Fig. 9.77 Bode plot of the type 1 system from Example 9.48.

A similar analysis shows that type 0 systems have a slope of zero and type 2 systems have a slope of -40 dB/decade on the low frequency portion of the magnitude curves on Bode plots, as indicated in Table 9.3.

System Type	Low Frequency Slope
type 0	0 dB/decade
type 1	-20 dB/decade
type 2	-40 dB/decade

Table 9.3 Relationship between system type and magnitude curves for low frequencies on Bode magnitude plots

9.7.2 Nyquist Plots

A *Nyquist plot* basically combines both plots in a Bode plot into one plot. As is outlined subsequently, unlike Bode plots, it may be used to determine the stability of a system regardless of whether it is minimum phase. The concept behind a Nyquist plot is that if a transfer function is evaluated along a closed contour in the complex plane, it is possible to determine information about the poles and zeros inside the contour. Hence if the contour encloses the entire right half-plane, we can infer something about the stability of the system.

9.7.2.1 Contours in the Complex Plane

To develop some understanding about Nyquist plots, we first consider a simpler situation. This is meant both to illustrate how a function is evaluated along a contour and what happens when the contour encircles a pole or zero of a transfer function.

Example 9.49. Consider the transfer function

$$G(s) = \frac{1}{s+2}$$

and the contour that traverses the unit circle in the complex plane in the clockwise direction. The idea is to plot $G(s)$ as the values of s go along the circle, which is simple enough to do by tabulating s values and the corresponding $G(s)$, and even easier to do on a computer. Because it is the unit circle traversed in the clockwise direction, we can let θ go from 0 to 2π and let

$$s = \cos\theta - \mathrm{i}\sin\theta.$$

Incrementing θ by $\pi/4$ produces the data in Table 9.4 and a plot of the s-contour and the $G(s)$-contour is illustrated in Figure 9.78. The points from the table are indicated by markers on the plot. Observe from the table although the s contour is clockwise, the resulting contour for $G(s)$ is counterclockwise.

θ	s	$G(s)$ (Exact)	$G(s)$ (Decimal)
0	1	$\frac{1}{3}$	0.33
$\frac{\pi}{4}$	$\frac{1}{\sqrt{2}}(1-\mathrm{i})$	$\frac{1}{34}\left[\left(16-3\sqrt{2}\right)+\mathrm{i}\left(5\sqrt{2}-4\right)\right]$	0.35 + 0.09i
$\frac{\pi}{2}$	$-\mathrm{i}$	$\frac{1}{5}(2+\mathrm{i})$	0.40 + 0.20i
$\frac{3\pi}{4}$	$-\frac{1}{\sqrt{2}}(1+\mathrm{i})$	$\frac{1}{34}\left[\left(16+3\sqrt{2}\right)+\mathrm{i}\left(4+5\sqrt{2}\right)\right]$	0.60 + 0.33i
π	-1	1	1.00
$\frac{5\pi}{4}$	$-\frac{1}{\sqrt{2}}(1-\mathrm{i})$	$\frac{1}{34}\left[\left(16+3\sqrt{2}\right)-\mathrm{i}\left(4+5\sqrt{2}\right)\right]$	0.60 - 0.33i
$\frac{3\pi}{2}$	i	$\frac{1}{5}(2-\mathrm{i})$	0.40 - 0.20i
$\frac{7\pi}{4}$	$\frac{1}{\sqrt{2}}(1+\mathrm{i})$	$\frac{1}{34}\left[\left(16-3\sqrt{2}\right)+\mathrm{i}\left(4-5\sqrt{2}\right)\right]$	0.35 - 0.09i
2π	1	$\frac{1}{3}$	0.33

Table 9.4 Evaluating $G(s) = 1/(s+2)$ for discrete points as s traverses the unit circle in the clockwise direction.

Now we increase the radius of the contour of s so that the pole of $G(s)$ is inside it. The plot of the two contours is illustrated in Figure 9.79. The important consequence

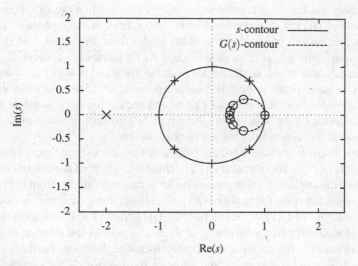

Fig. 9.78 A plot of s as it traverses the unit circle and $G(s) = 1/(s+2)$ evaluated along the contour of s. The values for s from Table 9.4 are indicated with a $+$ and the values of $G(s)$ from the table are indicated by a $\circ$.

is that when the pole is in the interior of the s-contour, the resulting $G(s)$-contour encircles the origin.

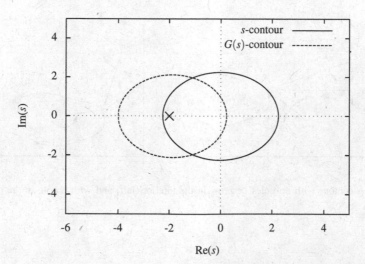

Fig. 9.79 A plot of s as it traverses a circle centered at the origin that encircles the pole of $G(s) = 1/(s+2)$. The resulting plot of $G(s)$ then encircles the origin.

The reason that the $G(s)$-contour encircles the origin when the s-contour encircles the pole of $G(s)$ is illustrated in Figure 9.80. Recall that the phase of $G(s)$ as a complex number is equal to the sum of the angles from all the zeros of $G(s)$ to s minus the sum of the angles from all the poles. As s traverses the contour, if it starts and ends at the same point, all the angles from the zeros and poles of $G(s)$ will start and end at the same point as well. If there is a pole or zero inside the contour, then, although the angle from that point to s starts and ends at the same angle, it will increase or decrease by 2π. Conversely, for a pole or zero outside the contour, it will start and end at the same value, but not increase by 2π.

If there is a zero in the interior of the s-contour, because its angle increases by 2π, then the angle of $G(s)$ also increases by 2π. If there is a pole in the interior of the s-contour, then the angle of $G(s)$ decreases by 2π because the angles from the poles of $G(s)$ are subtracted from the angle of $G(s)$. Therefore, if the s-contour is clockwise and there is a zero of $G(s)$ in the interior, the $G(s)$-contour will encircle the origin in the clockwise direction; conversely, if there is a pole in the interior, the $G(s)$-contour will encircle the origin in the counterclockwise direction. Furthermore, the rule generalizes to multiple poles and zeros in the interior of the contour as we would expect, as is presented in the following proposition, which these arguments have proved.

Proposition 9.10. *If C is a closed contour in the complex plane, and if s traverses C in the clockwise direction, the number of times that the contour of $G(s)$ encircles the origin in the clockwise direction is equal to the number of zeros of $G(s)$ minus the number of poles of $G(s)$ in the interior of C.*

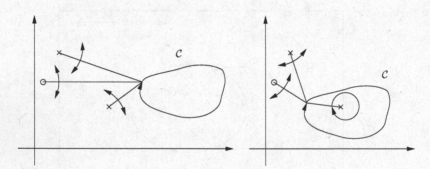

Fig. 9.80 A contour with no poles or zeros in the interior (left) and with a pole in the interior (right).

9.7.2.2 The Nyquist Stability Criterion

The Nyquist criterion is to encircle the entire right half complex plane to check if there are any poles of a transfer function there, that would correspond to the transfer function being unstable. We are primarily concerned with feedback systems, therefore we consider the common case where the denominator of the closed-loop transfer function can be put in the form, $D(s) = 1 + kG(s)$. This is the denominator, thus we look for zeros of $1 + kG(s)$ that are in the right half-plane. Finally, observe that the effect of the one that is added to $kG(s)$ is to shift the contour to the right by one; hence, we can drop the one and instead of focusing on encirclements of the origin, check for encirclements of the point $s = -1$.

Thus, to determine stability, if we evaluate $kG(s)$ as s encircles the entire right half of the complex plane, the number of zeros of $kG(s)$ in the right half-plane minus the number of poles of $G(s)$ in the right half-plane will be equal to the number of clockwise encirclements of the point $s = -1$. The contour that is commonly used starts at the origin, goes "up" along the positive imaginary axis to $s = +i\infty$, then proceeds clockwise at $s = \infty$ around the right half-plane to the negative imaginary axis and finally returns to the origin from $s = -i\infty$, as is illustrated schematically in Figure 9.81. If $G(s)$ is proper, then the part of the contour at infinity does not need to be considered, because at all of those points $G(s)$ will be zero. The result of all of this is the famous Nyquist stability criterion, that is restated as follows.

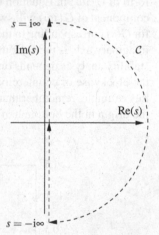

Fig. 9.81 Contour for the Nyquist stability criterion.

Proposition 9.11. *If $kG(s)$ is proper, has no poles or zeros on the imaginary axis and if $kG(s)$ has p poles in the right half-plane, then the number of zeros of $1 + kG(s)$ in the right half-plane is equal to the number of times the $kG(s)$-contour encircles the point $s = -1$ in the clockwise direction minus p.*

So, for stability then, we have the following.

Corollary 9.1. *If $kG(s)$ is proper, has no poles or zeros on the imaginary axis, and if $kG(s)$ has p poles in the right half-plane, then there will be no zeros of $1 + kG(s)$ in the right half-plane if the $kG(s)$-contour encircles the point $s = -1$ p times in the counterclockwise direction.*

Plotting Nyquist plots is relatively straightforward and is illustrated with a few examples.

Example 9.50. Plot the Nyquist plot for

$$G(s) = \frac{1}{s+1}.$$

Inasmuch as

$$G(i\omega) = \frac{1}{i\omega + 1} = \frac{1 - i\omega}{1 + \omega^2}, \qquad (9.32)$$

we can see that for very small ω, the plot starts at the point $s = 1$ and initially will have a negative imaginary component. Because of the ω^2 term in the denominator, as ω becomes large, the plot will approach the origin. Furthermore, for very large ω,

$$\omega \gg 1 \quad \Longrightarrow \quad G(i\omega) \approx -\frac{i}{\omega},$$

the plot will approach the origin tangent to the negative imaginary axis. From the form of $G(i\omega)$ in Equation (9.32), it is clear that neither the real nor the imaginary component of $G(i\omega)$ will switch signs for $\omega > 0$. Hence, the segment of the contour for $G(s)$ corresponding to the segment of the s-contour corresponding to the positive imaginary axis is the bottom half of the contour plotted in Figure 9.82. Because the stability analysis depends on whether the plot for $G(s)$ encircles the point $s = -1$ in the clockwise or counterclockwise direction, we must keep track of orientation. In this example, remember that the plot for $G(s)$ started at the point $s = 1$ and went *to* the origin in the bottom half of the complex plane.

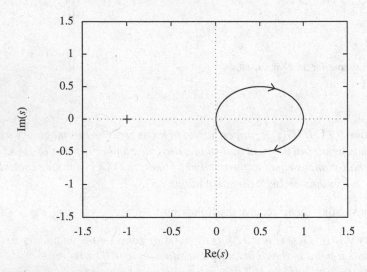

Fig. 9.82 Nyquist plot for Example 9.50.

The transfer function is proper, therefore the whole $G(s)$-contour corresponding to the s-contour that is at infinity is at the origin. Referring to Equation (9.32), the plot corresponding to the s-contour on the negative imaginary axis will be a reflection about the real axis of the first part of the plot because the only difference is that ω changed sign. Hence, the segment of the contour for $G(s)$ corresponding to the segment of the s-contour corresponding to the negative imaginary axis is the top half of the contour plotted in Figure 9.82.

Remark 9.2. For a proper transfer function, it is always the case that the Nyquist plot is symmetric about the real axis. This is because, for $s = i\omega$ the real part of $G(s)$ is determined by the even powers of s and the imaginary part by the odd powers. Hence, the difference between the segments of the s-contour corresponding to the positive and negative imaginary axes will only affect the imaginary part of the contour for $G(s)$. Hence, as a practical matter, we only need to plot the part of the Nyquist plot corresponding to $s = +i\omega$.

Example 9.51. Plot the Nyquist plot for

$$G(s) = \frac{\omega_n^2}{s^2 + 2\zeta\omega_n s + \omega_n^2}.$$

Substituting $s = i\omega$ gives

$$G(i\omega) = \frac{\omega_n^2}{(\omega_n^2 - \omega^2) + i(2\zeta\omega_n\omega)}.$$

Note the following.

- For $\omega = 0$, the contour for $G(s)$ starts at $s = 1$.
- The plot for $G(i\omega)$ will have a negative imaginary component initially. If it is necessary to see this, multiply $G(i\omega)$ by the complex conjugate of the denominator.
- The imaginary component will always be negative for $\omega > 0$.
- As $\omega \to +\infty$, the real part dominates, and hence the plot approaches the origin tangent to the real axis.
- The real part of $G(i\omega)$ will switch signs by crossing the negative imaginary axis when ω passes the value of ω_n.

The Nyquist plot for

$$G(s) = \frac{1}{s^2 + s + 1}$$

is illustrated in Figure 9.83, showing those features.

The following example demonstrates that the Nyquist analysis works for a system with left half-plane poles.

Example 9.52. The Nyquist plot for

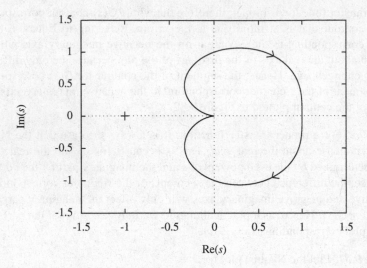

Fig. 9.83 Nyquist plot for a second-order system in Example 9.51.

$$G(s) = k\frac{1}{s-2}$$

with $k = 1$ and $k = 3$ is illustrated in Figure 9.84. Because k scales $G(s)$, the larger contour corresponds to the larger k.

Because

$$G(i\omega) = \frac{k}{i\omega - 2} = k\frac{-2 - i\omega}{4 + \omega^2}$$

the plot starts at $s = k/2$ and initially has a negative imaginary component. For very large ω it approaches the origin. Hence the contour has a counterclockwise orientation. The number of right half plane zeros of $1 + kG(s)$ is equal to the number of counterclockwise encirclements of $s = -1$ minus the number of right half-plane poles of $G(s)$, of which there is one, therefore there is one right half-plane zero of $1 + kG(s)$ for $k = 1$ and zero for $k = 3$. Hence the system is unstable for the smaller k and stable for the larger k.

A third-order system is one where the system may actually go unstable even if the open loop transfer function has all left half-plane poles.

Example 9.53. Plot the Nyquist plot for

$$G(s) = \frac{k}{s^3 + 6s^2 + 11s + 6} \tag{9.33}$$

and determine the value for k at which the system goes unstable under unity feedback.

Setting $k = 1$ and substituting $s = i\omega$ gives

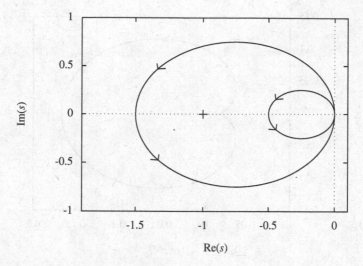

Fig. 9.84 Nyquist plot for system in Example 9.52.

$$G(i\omega) = \frac{1}{6(1 - \omega^2) + i\omega(11 - \omega^2)}.$$

- Substituting $s = 0$, shows that the plot starts at $G(0) = 1/6$.
- For $\omega \ll 1$, $G(i\omega)$ has a negative imaginary component, so the contour of $G(s)$ starts downward.
- The real part will switch signs at $\omega = 1$, and hence the contour of $G(s)$ will cross the imaginary axis.
- The imaginary part will switch signs at $\omega = \sqrt{11}$ and hence cross the real axis.
- For $\omega \gg 1$, the contour of $G(i\omega)$ will approach the origin.
- For $s = -i\omega$ the contour of $G(s)$ will be the same as the contour when $s = i\omega$, except the imaginary part will have the opposite sign. Hence, it is a reflection about the real axis of the contour when $s = i\omega$.

The plot is illustrated in Figure 9.85. Because the contour crosses the real axis when $\omega = \sqrt{11}$, this occurs at

$$G\left(i\sqrt{11}\right) = \frac{1}{6(1 - 11)} = \frac{1}{60}.$$

Hence, the magnitude of $G(s)$ can be increased by a factor of 60 before it encircles $s = -1$. Because $G(s)$ has no right half-plane poles, this corresponds to the value of k where the closed-loop system is neutrally stable.

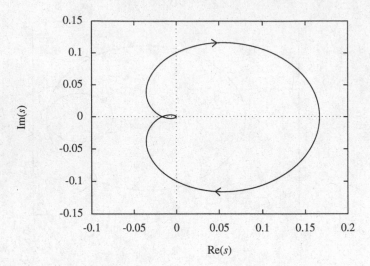

Fig. 9.85 Nyquist plot for system in Example 9.53.

9.7.2.3 Stability Margins and Nyquist Plots

The same stability margins as were defined with Bode plots may be obtained from a Nyquist plot. The gain margin is the amount that the open-loop gain can be increased before the closed-loop system becomes unstable, thus it is related to the point where the Nyquist plot crosses the real axis between the points $s = 0$ and $s = -1$. In particular, the point at which it crosses is the reciprocal of the gain margin. Also, the angle measured from the negative imaginary axis to the point at which the contour for $G(s)$ has a magnitude of one is the phase margin. Both concepts are illustrated in Figure 9.86, which shows the portion of the Nyquist plot for

$$G(s) = \frac{10}{s^3 + 4s^2 + 6s + 6}$$

corresponding to the portion of the s-contour along the positive imaginary axis, that is, $+i\omega$.

A single stability measure that has features superior to those of the gain and phase margins is the minimum distance in the complex plane that the Nyquist contour passes to the critical point $s = -1$. This is superior in the sense that it is one metric instead of two. Furthermore, although perhaps unusual, it is possible for a system to have large gain and phase margins and still respond in a manner that characterizes neutral stability. Such a system is explored in Exercise 9.38. For the transfer function in Equation (9.33), Figure 9.87 also illustrates a circle of radius $r = 0.3$ centered at $s = -1$. Because the circle is tangent to the Nyquist plot, that is the closest the plot passes to the point $s = -1$, and hence the stability margin is $s_m = 0.3$.

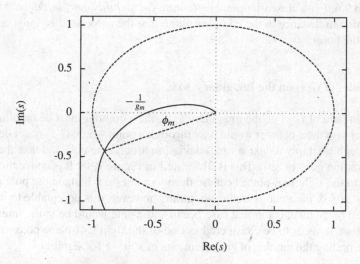

Fig. 9.86 Gain and phase margins from a Nyquist plot.

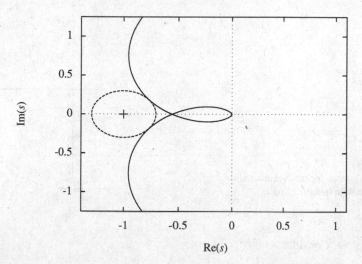

Fig. 9.87 The circle centered at $s = -1$ has a radius of 0.3; hence, the minimum distance the Nyquist contour passes to the point $s = -1$ is $s_m = 0.3$.

Definition 9.6. For a closed-loop stable system, the *stability margin*, denoted by s_m, is the minimum distance in the complex plane that the associated Nyquist contour passes to the point $s = -1$.

9.7.2.4 Poles of $G(s)$ on the Imaginary Axis

If there is a pole of $G(s)$ on the imaginary axis, the s-contour must be modified because otherwise the s-contour would pass through a point where $G(s)$ is not defined. The approach is simply to take a small deviation around the pole and take the limit as the deviation goes to zero. This is illustrated in Figure 9.88. It is conventional to deviate into the right half-plane because doing so does not include the pole on the imaginary axis in the interior of the s-contour; however, it is acceptable to deviate into the left half-plane, but in that case because the pole would be in the interior of the s-contour, it needs to be considered as one of the right half-plane poles of $G(s)$ when interpreting the number of encirclements of $s = -1$ for stability.

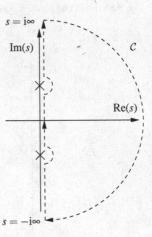

Fig. 9.88 Contour for the Nyquist stability criterion with imaginary poles.

Example 9.54. Consider

$$G(s) = \frac{1}{s(s+2)}. \tag{9.34}$$

Because $G(s)$ is not defined at the origin the contour must deviate around the pole at the origin. Inasmuch as we normally started the plot by beginning with $\omega = 0$ and proceeding up the positive imaginary axis, we still simply start at a small positive value and deal with the deviation about the origin at the end.

Substituting $s = i\omega$ gives

$$G(i\omega) = \frac{1}{-\omega^2 + 2\omega i} = \frac{-\omega^2 - 2\omega i}{\omega^2(\omega^2 + 4)} = -\frac{1}{\omega^2 + 4} - i\frac{2}{\omega(\omega^2 + 4)}.$$

Thus for $0 < \omega \ll 1$, $G(i\omega) \approx -1/4 - i/(2\omega)$. So, in words, it is coming from minus imaginary infinity along a real value of $-1/4$. Neither the real nor imaginary parts of the $G(s)$-contour will change sign as the s-contour proceeds up the positive imaginary axis, and it will approach the origin as $\omega \to \infty$. The segment corresponding to the s-contour at infinity is all at the origin, and the segment for the portion of the s-contour up the negative imaginary axis from negative infinity up close to, but not including the origin, will be a reflection across the real axis of the portion of the $G(s)$-contour corresponding to the positive $i\omega$ segment. We have not yet dealt with the deviation about the origin, but the Nyquist contour constructed so far is illustrated in Figure 9.89.

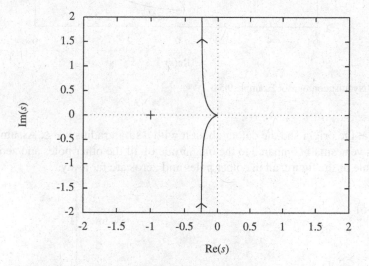

Fig. 9.89 Partial Nyquist contour for Example 9.54.

The very small deviation about the origin has large consequences. Because in that detour, s is very small, the magnitude of $G(s)$ will be very large. This portion of the $G(s)$ contour will connect the segment of the $G(s)$-contour that is going up to positive $i\infty$ to the one that is coming up from $-i\infty$. Numerically computing a semi-circular contour that deviates about the origin in a circle with radius $1/10$ results in the contour illustrated in Figure 9.90. Note that the point $s = -1$ is not encircled, but it would have been if the $G(s)$-contour corresponding to the deviation about the origin had "wrapped around" to the left instead of to the right, as can happen. The way to interpret Figure 9.90 is that, as the deviation about the origin approaches zero, the semicircular part of the $G(s)$-contour will increase in magnitude to be a contour at infinity.

The final issue to resolve is to verify theoretically, as opposed to simply numerically for a single example, that the semicircular portion of the contour in Figure 9.90 is guaranteed to be a semi-circle (180°) connecting the other two branches of the $G(s)$-contour, as opposed to, say, a circle and a half (540°). Figure 9.91 illustrates

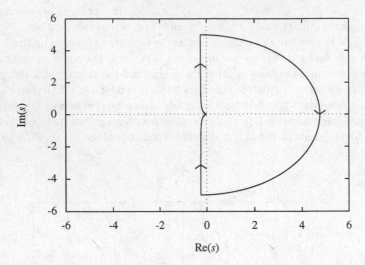

Fig. 9.90 Nyquist contour for Example 9.54.

the pole at the origin and the detour about it with a small radius $r = \varepsilon$. Assume that detour is very small compared to the magnitude of all the other poles and zeros, so in the scale of the figure, all the other poles and zeros are far away.

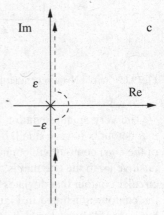

Fig. 9.91 Small deviation of the contour about a pole at the origin.

Observe that as the values for s traverse the detour, the angle from the pole at the origin to s changes by 180°. If all the other poles and zeros are far away, the angle from them to s changes negligibly. Hence, $G(s)$ must undergo a net change in the angle of 180°, meaning that it can only be a semicircle.

Remark 9.3. Based on the analysis in the last paragraph of the preceding example, if there are two poles at the origin, then $G(s)$ undergoes a net change in angle of approximately $360°$ as the s-contour detours the origin. Similarly, if there are three poles, the net change is $540°$, and so on. Hence, it is easy to determine the system type from a Nyquist plot.

9.8 Exercises

9.1. Write a computer program that determines an approximate numerical solution to the equations of motion for the robot arm from Example 9.1. Assuming zero initial conditions and a small desired angle, use your program to verify the following "rules of thumb" for PID control for a step input.

Your program should be for the original nonlinear model, not the linearized one where we assumed $|\theta| \ll 1$. The idea is to verify that what we determined using the linearized version works for the nonlinear case as well, as long as the desired angle of the robot arm is small.

1. For proportional control, that is, $k_p > 0$, $k_d = 0$, and $k_i = 0$, the solutions are oscillatory, and increasing k_p increases the frequency of oscillation (which decreases the rise time and peak time) but decreases the mean steady-state error. The settling time is infinite. *Hint:* Pick a starting value of $k_p = 5$.
2. Add derivative control the proportional controller (i.e., $k_p > 0$, $k_d > 0$, and $k_i = 0$) and verify the following.

   a. For small k_d the solutions are decaying oscillations.
   b. Increasing k_d decreases the settling time.
   c. Increasing k_d to a sufficiently large value eliminates the oscillatory behavior completely, resulting in an solution that exponentially decays to the final steady-state value.
   d. Increasing k_p decreases the final steady state error.
   e. Increasing k_p decreases the rise time.

   Hint: Pick a starting value of about $k_d = 0.5$.
3. Add integral control (PID control) and verify the following.

   a. PID control eliminates the steady-state error, even for small values of k_p.
   b. Increasing k_i generally increases the overshoot and settling time.
   c. Increasing k_p decreases rise time, but may increase overshoot.
   d. Increasing k_d increases damping and stability.

   Hint: Pick a starting value of about $k_i = 0.5$.
4. Choose a set of gain values from the above simulations that seems to work well. Use those for an attempt to have the desired angle be large. Does it still work well?

9.2. Consider the robot arm from Example 9.1 where the torque is from a dc motor, as is illustrated in Figure 9.92. Let $m = 1$, $J = 2$, $l = 2$, $g = 9.81$, $k_e = 4$, $k_\tau = 5$, $R = 10$, and the initial conditions be zero. Write a computer program to determine an approximate numerical solution for the system. Use the nonlinear equations and do not assume that $\theta \ll 1$. Use proportional control to specify the voltage supplied to the motor to investigate its efficacy. If necessary add derivative and integral control. Experiment to find good values for the gains such that the transient response has what you consider a good response (low rise time, low overshoot, and fast settling time) and very small or zero steady-state error. Consider each of the following cases.

1. θ_d is small and constant.
2. $\theta_d \approx 1$ and constant.
3. $\theta_d \approx -1$ and constant.
4. $\pi/2 < \theta_d < \pi$ and constant.
5. $-\pi < \theta_d < -\pi/2$ and constant.
6. $\theta_d(t) = \sin t$.

When $\theta_d(t)$ varies with time, the problem is usually called *tracking*.

9.3. This problem investigates the relationship between pole locations and the stability of the response for the robot arm example with PID control. Using the transfer function from $\Theta_d(s)$ to $\Theta(s)$ given by Equation (9.10), let $J = mgl = 1$, $k_p = 24$, and $k_d = 8$. The denominator is a third-order polynomial, thus it is difficult to factor by hand, so use the `pzmap()` function in MATLAB or Octave, or something equivalent, to determine the pole locations for the cases illustrated in Figure 9.7. What is the difference between the pole locations for the stable responses and the unstable response?

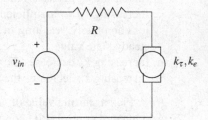

Fig. 9.92 DC motor for Exercises 9.2 and 9.12.

9.4. Verify the results in Figure 9.21 by using a computer to compute a numerical solution to the step response to

$$G(s) = \frac{\omega_n^2}{s^2 + 2\zeta\omega_n s + \omega_n^2}$$

and by appropriately choosing and varying ω_n and ζ so that the poles move in the three directions indicated in the figure. Create plots illustrating the pole locations and corresponding step responses and whether the change in the step response when the pole is moved in one of the three directions indicated is as predicted.

9.5. If $s_1 = a_1 + ib_1$ and $s_2 = a_2 + ib_2$, and $s_1 s_2 = (a_1 a_2 - b_1 b_2) + i(a_1 b_2 + a_2 b_1)$ use the fact that $r = \sqrt{a^2 + b^2}$ and

$$\theta = \tan^{-1}\left(\frac{b}{a}\right)$$

to show that in polar coordinates

$$s_1 s_2 = (r_1 r_2, \theta_1 + \theta_2), \quad \frac{s_1}{s_2} = \left(\frac{r_1}{r_2}, \theta_1 - \theta_2\right).$$

9.6. Show that if the transfer function has multiple poles at the origin; that is

$$Y(s) = \frac{1}{s^n} R(s)$$

then regardless of the input, $y(t)$ will contain an $(n-1)$th-order polynomial in t; that is,

$$y(t) = c_0 + c_1 t + \frac{c_2}{2} t^2 + \cdots + \frac{c_{n-1}}{(n-1)!} t^{n-1} + \text{other terms}.$$

9.7. Use a partial fraction expansion to compute $x(t)$ when

$$X(s) = \frac{4}{s^2 + 2s + 4}\left(\frac{1}{s}\right).$$

Use a partial fraction expansion to compute x(t) when

$$X(s) = \frac{4}{(s^2 + 2s + 4)(s + 20)}\left(\frac{1}{s}\right).$$

Are the responses similar? Explain whether this was expected or unexpected.

9.8. Consider

$$G(s) = \frac{\omega_n^2}{s^2 + 2\zeta\omega_n s + \omega_n^2}.$$

Referring to Figure 9.28, choose ζ and ω_n such that the poles are

1. In the hatched region where both the settling time and the overshoot specifications are satisfied
2. Where the settling time specification is satisfied but the overshoot specification is not
3. Where the overshoot specification is satisfied but the settling time specification is not
4. Where neither specification is satisfied

Use the MATLAB or Octave step() command, or something equivalent, to verify the relationship between the pole location and step response characteristics.

9.9. Figure 9.93 contains six plots of poles and zeros for different transfer functions. Figure 9.94 contains step responses. Match the pole–zero maps with the corresponding step responses. Sketch them next to each other and indicate on the figures the attributes of the pole and zero locations that correspond to attributes in the step responses.

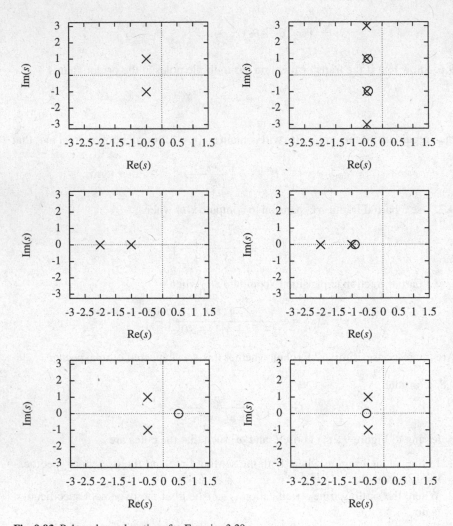

Fig. 9.93 Pole and zero locations for Exercise 3.29.

9.10. The step response of

$$G(s) = \frac{\omega_n^2}{s^2 + 2\zeta\omega_n s + \omega_n^2}$$

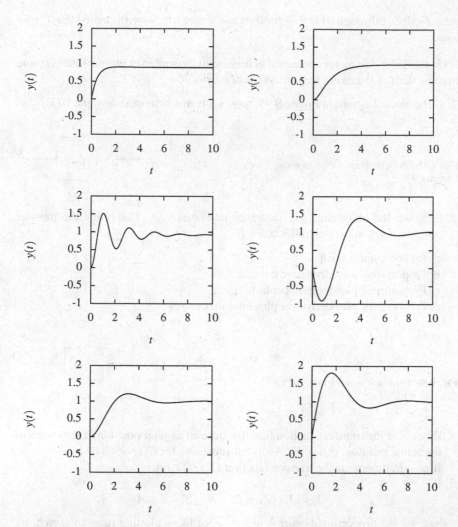

Fig. 9.94 Step responses for Exercise 3.29.

is given by Equation (9.16). Now consider

$$G(s) = \frac{\frac{\omega_n^2}{r}(s+r)}{s^2 + 2\zeta\omega_n s + \omega_n^2}.$$

Compute the partial fraction expansion of the step response of $G(s)$, and using the resulting time function, explain why the rules for an additional real zero added to a second-order system are true.

9.11. The root locus plots we considered in Section 9.6 considered only the case where $k \in [0, +\infty)$ and is often called the 180° root locus. Determine each of the rules (Rule 9.1 through Rule 9.7) for the case where $k \in (-\infty, 0]$, called the $0°$ root locus.

9.12. Consider a dc motor connected to the circuit illustrated in Figure 9.92. Assume that the shaft of the motor has a moment of inertia J.

1. If the block diagram in Figure 9.95, represents this system, determine $G(s)$.

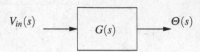

Fig. 9.95 Block diagram for dc motor in
Exercise 9.12.

2. Consider the block diagram illustrated in Figure 9.96. Determine the transfer function, $C(s)$ in the controller block for

   a. Proportional control
   b. Proportional plus derivative control
   c. Proportional plus integral control
   d. Proportional plus derivative plus integral control

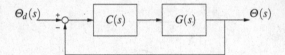

Fig. 9.96 Feedback control loop for
Exercise 9.12.

3. Determine the transfer function from the desired angular position of the motor to the actual position, $\Theta_d(s)/\Theta(s)$ (do not substitute for $C(s)$ or $G(s)$).
4. If $\omega = \dot{\theta}$, determine the transfer function $\Omega(s)/\Omega_d(s)$.
5. If

$$k_e = 1, \quad k_\tau = 2, \quad R = 3, \quad J = 4$$

   and we use proportional control, use the root locus plotting rules to sketch, by hand, how the poles of $\Theta(s)/\Theta_d(s)$ vary as the proportional gain is varied from 0 to $+\infty$. Determine the approximate gain value, if any, that gives a damping ratio of approximately $1/2$.
6. For the same parameter values as above, use PD control and fix the ratio between the proportional gain and the derivative gain to be $1/2$; that is,

$$v_{in} = k_p(\theta_d - \theta) + k_d(\dot{\theta}_d - \dot{\theta})$$
$$= k\left[(\theta_d - \theta) + \frac{1}{2}(\dot{\theta}_d - \dot{\theta})\right],$$

   sketch the root locus plot for the system. Discuss qualitatively what will happen to the rise time, the percentage overshoot, and the settling time as k increases.

9.13. In this problem we design a servo motor. Consider the circuit illustrated in Figure 9.97 which is comprised of a voltage source, a resistor, and a dc motor.

The mass moment of inertia of the load on the motor is J, the torque and back emf constants of the motor are k_τ and k_e, respectively, the angular position of the motor is θ, and the other parameters are as indicated in the figure.

All the physical parameters have values greater than zero.

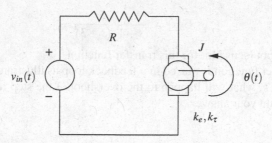

Fig. 9.97 Servo motor for Exercise 9.13.

1. Determine the transfer function from the input voltage to the angular position of the motor shaft.
2. Draw a block diagram for the system if proportional control is used where the input to the block diagram is the desired position of the servo and the output is the actual position. (You don't need this for the problem, but proportional control is easy to implement for this circuit by attaching a potentiometer to the output of the motor.)
3. If proportional control is used, what will be the steady-state error to a unit step input. Does the steady-state error depend on the gain used in proportional control or the physical parameters of the motor and circuit?
4. Without using any numerical values, sketch what the root locus for this system would look like.
5. If a load is increased so that J increases, sketch how the root locus would change. If k is such that $0 < \zeta < 1$ for the first case before the load was increased and k is not changed when J increases, what will be the effect on the response characteristics; specifically,

   a. rise time (use $t_r \approx 1.8/\omega_n$)
   b. overshoot
   c. settling time
   d. steady-state error

Justify your answer by referring to the root locus plots.

9.14. Consider

$$G(s) = \frac{4}{(s+1)(s+3)}.$$

1. Sketch the root locus plot for this transfer function.
2. If this transfer function is placed in a feedback loop as illustrated in Figure 9.96, with $C(s) = k$, what will happen to the overshoot of the step response as k gets large? Explain your answer.
3. Determine the maximum value for k so that the percentage overshoot remains under 20%.

9.15. Consider

$$G(s) = \frac{s+5}{(s+1)(s+3)}.$$

1. Sketch the root locus plot for this transfer function.
2. If this transfer function is placed in a feedback loop as illustrated in Figure 9.96, with $C(s) = k$, what will happen to the overshoot of the step response as k gets large? Explain your answer.

9.16. Consider

$$G(s) = \frac{1}{(s+1)(s+3)(s+5)}.$$

1. Sketch the root locus plot for this transfer function.
2. What can you say about the stability of the response of the system under unity feedback as k gets large?

9.17. Consider

$$G(s) = \frac{4}{(s+2)(s^2+2s+2)}.$$

1. Sketch the root locus plot for this transfer function.
2. From the root locus plot, determine the range of positive gain values for which the transfer function under unity feedback will be stable.
3. Verify your answer in the previous part by using the Routh array.
4. Sketch the Bode plot for this transfer function.
5. By referring to the Bode plot, is it possible to place a lead compensator of the form

$$G_c(s) = \frac{s-z}{s-p}$$

in series with the plant so that it is stable for all positive gains? Explain your answer.

9.18. Consider

$$G_p(s) = \frac{1}{(s-1)(s+1)}, \quad G_c(s) = \frac{s+2}{s+3}, \quad G_s(s) = 1$$

in the block diagram illustrated in Figure 9.44. Sketch the root locus plot for the system and by referring to the plot, determine the range of k values for which the system is stable.

9.19. A minor complication occurs if a transfer function has two more poles or zeros at the same location. The root locus plot for

$$G(s) = \frac{s+3}{s^2(s+2)},$$

which has a double pole at the origin is illustrated in Figure 9.98 and the root locus for

$$G(s) = \frac{s+3}{s(s+2)^2},$$

which has a double pole at $s = -2$ is illustrated in Figure 9.99.

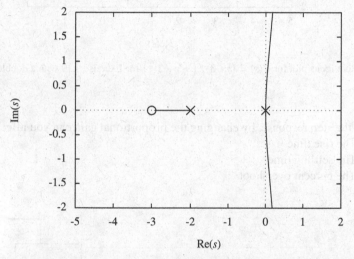

Fig. 9.98 Root locus plot for $G(s) = (s+3) / (s^2(s+2))$ for Exercise 9.19 with a double pole at the origin.

Do all the rules summarized in Table 9.1 still apply? Explain your answer for each of the rules by specifically referring to the features of Figures 9.98 and 9.99.

9.20. In order to do this problem, you must understand how to deal with multiple poles in the same location, which was considered in Exercise 9.19. Consider the system illustrated in Figure 9.100.

1. Determine the transfer function from the applied force $f(t)$ to the position of the mass $x(t)$.
2. Sketch the root locus plot for this transfer functions if $m = 1$.
3. What does the root locus plot tell you about using proportional control to control the position of the mass? Specifically,

   a. Will it be stable, unstable, or on the margin

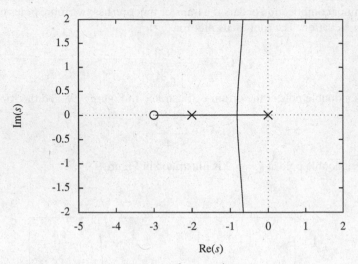

Fig. 9.99 Root locus plot for $G(s) = (s+3)/\left(s(s+2)^2\right)$ for Exercise 9.19 with a double pole at $s = -2$.

  b. For the step response, by changing the proportional gain can you affect
   i. The rise time
   ii. The settling time
   iii. The percent overshoot

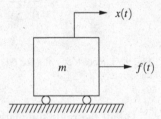

Fig. 9.100 System for Exercise 9.20.

9.21. Consider again the system illustrated in Figure 9.100.

1. Determine the transfer function from the applied force $f(t)$ to the velocity of the mass $\dot{x}(t)$.
2. Sketch the root locus plot for this transfer function.
3. Discuss the use of proportional control for this system. What characteristics of the response of the system can you affect by altering the proportional gain?

9.22. Sketch the root locus plot for

$$\frac{Y(s)}{R(s)} = \frac{kG(s)}{1+kG(s)},$$

where

$$G(s) = \frac{1}{s^2 + 4s + 5}.$$

1. Indicate on your root locus plot the region on the complex plan where the the maximum percent overshoot for the step response for a complex conjugate pair of poles is less than 16%. Label any angles that you use in this determination.
2. Compute and then indicate on the root locus plot the region on the complex plan where the the rise time for the step response for a complex conjugate pair of poles is less than .65 seconds. Use the approximation $t_r \approx 1.8/\omega_n$. Label any angles or distances that you used in this determination.
3. Use the root locus plot to determine the approximate range of values for the parameter k that satisfy both the rise time and overshoot specifications.

9.23. Sketch the root locus plot for

$$\frac{Y(s)}{R(s)} = \frac{kG(s)}{1 + kG(s)},$$

where

$$G(s) = \frac{1}{(s+3)(s^2 + 4s + 5)}.$$

Be sure to include the details of all your computations.

1. Use your sketch on the previous page to determine the *approximate* value for k at which the root locus crosses the imaginary axis.
2. Use the Routh array to determine the *exact* value of k at which the root locus crosses the imaginary axis.

9.24. Sketch the root locus plot for

$$G(s) = \frac{20}{s^2 + s + 10}.$$

A *phase lead compensator*, which is considered in the next chapter, is of the form

$$C(s) = \frac{s + 10}{s + 20}.$$

Sketch the root locus plot for $C(s)G(s)$ and explain why this phase lead compensator increases the stability of the system under unity feedback.

9.25. Consider

$$G(s) = \frac{1}{s(s+2)(s+4)}.$$

1. Sketch the root locus plot for this transfer function.
2. Determine the closed-loop transfer function (i.e., $Y(s)/R(s)$) if $G(s)$ is in the block diagram illustrated in Figure 9.101. Use the Routh array to determine the values for k for which the closed-loop transfer function is stable.

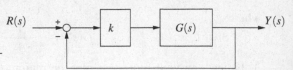

Fig. 9.101 Closed loop system for Exercise 9.25.

3. Verify your computation from the previous step by using the root locus plot to determine the values for k for which the closed-loop transfer function is stable.
4. Sketch the Bode diagram for gain values much smaller, equal to, and much larger than the gain values you determined in the previous steps and determine the gain and phase margins in each case.

9.26. Sketch the Bode plot for

$$G(s) = \frac{100}{(s+10)(s+100)}.$$

9.27. Sketch the Bode plot for

$$G(s) = \frac{100}{(s+10)(s+100)(s+1000)}.$$

9.28. Sketch the Bode plot for

$$G(s) = \frac{s}{(s+10)(s+100)(s+1000)}.$$

9.29. Sketch the Bode plot for

$$G(s) = \frac{s+100}{(s+10)(s+10000)}.$$

9.30. A Bode plot that was determined experimentally for a system that has all of its poles and zeros in the left half of the complex plane is illustrated in Figure 9.102.

1. If this system were placed under unity feedback sketch what the unit step response would look like for $k = 1$, $k = 10$, and $k = 20$. Explain your answer.
2. For the $k = 1$ case, what is the steady-state value to a unit step response under unity feedback? Explain your answer.

9.31. Consider the lowpass filter illustrated in Figure 9.103. Determine the transfer function from the input voltage to the output voltage. Sketch the Bode plot for this circuit if $R_{HP} = 100$ and $C_{HP} = 100$ and explain why it is called a lowpass filter.

9.32. Consider the highpass filter illustrated in Figure 9.104. Determine the transfer function from the input voltage to the output voltage. Sketch the Bode plot for this circuit if $R_{HP} = 10$ and $C_{HP} = 10$ and explain why it is called a highpass filter.

9.33. Sketch the Bode plot for

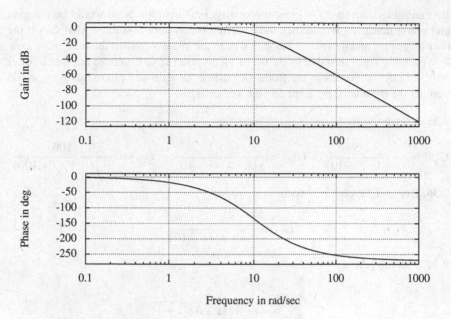

Fig. 9.102 Bode plot for Exercise 9.30.

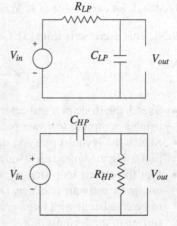

Fig. 9.103 Lowpass filter for Exercise 9.31.

Fig. 9.104 Highpass filter for Exercise 9.32.

$$G(s) = \frac{20000}{(s+10)(s+1000)}.$$

9.34. Sketch the Bode plot for

$$G(s) = \frac{1}{(s+10)(s-1)}.$$

Be careful to treat the $(s-1)$ term properly. For this plot, what would be the gain and phase margins? Would they indicate that the system is stable? Also sketch the root locus plot and determine the gain where the system crosses the imaginary axis. Is it crossing from unstable to stable, or vice versa? If you were to attempt to infer stability information from the Bode plot using the gain and phase margins, would you get the right answer? Explain your reasoning.

9.35. Sketch the Bode and Nyquist plots for

$$G(s) = \frac{s+5}{(s+1)(s+10)}, \quad G(s) = \frac{1}{s(s+1)(s+10)}, \quad G(s) = \frac{100}{s^2(s+1)(s+100)}.$$

9.36. Repeat Exercise 9.34 with

$$G(s) = \frac{s-2}{(s+1)(s+10)}.$$

9.37. Plot the Nyquist plot for

$$G(s) = \frac{k}{(s-1)(s+3)(s+4)}$$

when $k = 1$, $k = 15$, and $k = 60$. Determine the stability of the system under unity feedback for each value of k. Verify your Nyquist analysis using a root locus plot.

9.38. This exercise is from [5]. Consider

$$G(s) = \frac{0.38\left(s^2+0.1s+0.55\right)}{s(s+1)(s^2+0.06s+0.5)}.$$

- Sketch the Bode plot and determine the gain and phase margins. Verify your plot using a computer software package.
- Sketch the Nyquist plot and determine the gain and phase margins. Verify your plot using a computer software package.
- Plot the closed-loop step response of the system under unity feedback using a computer software package. Describe whether the response of the system seems to be consistent with large gain and phase margins. Explain your answer by referring to the Nyquist plot.

9.39. Plot the Nyquist contour and determine the range of gains for stability under unity feedback for the open-loop transfer function

$$G(s) = \frac{1}{s-2}.$$

Chapter 10
Classical Control Theory: Design

This chapter considers some standard methods to design controllers. As is the case with design in engineering in general, there are many different ways to do this, but this book is limited to two types. This chapter focuses on lead–lag control. A lead compensator generally enhances the stability of a system and is designed to meet the transient response specifications for the system. A lag compensator generally decreases the steady-state error without significantly altering the transient response characteristics of the system. The previous chapter, by way of introduction to controls, developed the usual rules for design of PID controllers.

Various configurations are possible for feedback compensation. This book particularly focuses on compensation added in a feedback loop in the configuration illustrated in Figure 10.1 which is commonly called *cascade compensation*. In Figure 10.1, the block with $G_p(s)$ represents the plant dynamics, the output of which we desire to control. The block $G_c(s)$ is the compensator block, which must be designed based upon the performance specifications for the system. The block with $G_s(s)$ represents the sensor dynamics. In this text, this is often idealized as the identity; however, in most applications the dynamics (or, at a minimum, the gain) of the sensors must be considered.

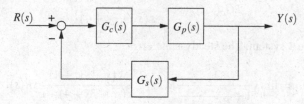

Fig. 10.1 Cascade compensation configuration.

B. Goodwine, *Engineering Differential Equations: Theory and Applications*,
DOI 10.1007/978-1-4419-7919-3_10, © Springer Science+Business Media, LLC 2011

10.1 System Type and Steady-State Error

There is a very simple relationship between system type and steady-state error for different types of inputs. Consider the unity feedback system in Figure 10.2. If the error is defined as the signal out of the comparator, then

$$E(s) = R(s) - Y(s) = \left(1 - \frac{Y(s)}{R(s)}\right)R(s) = \left(1 - \frac{kG_c(s)G_p(s)}{1 + kG_c(s)G_p(s)}\right)R(s)$$

$$= \frac{1}{1 + kG_c(s)G_p(s)}R(s).$$

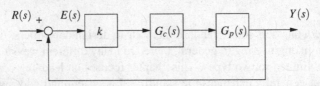

Fig. 10.2 Unity feedback system.

Recall the system type is the lowest power of s in the polynomial in the denominator of the system. We focus on the open-loop compensated system, so this depends on the denominator of $G_c(s)G_p(s)$. The types of inputs considered are *step inputs*, *ramp inputs*, and *parabolic inputs*. The details are simple, so a couple of examples should suffice to elucidate the concept. For an input $R(s)$, the steady-state error is given by the final value theorem

$$e_{ss}(t) = \lim_{s \to 0} s \frac{1}{1 + kG_c(s)G_p(s)}R(s).$$

Example 10.1. Consider

$$kG_c(s)G_p(s) = \frac{k}{s(s^2 + 3s + 4)},$$

which is a type 1 system. The steady-state error is

$$e_{ss} = \lim_{s \to 0} s \frac{1}{1 + \frac{k}{s(s^2+3s+4)}}R(s) = \frac{s(s^2 + 3s + 4)}{s(s^2 + 3s + 4) + k}sR(s).$$

Observe the following.

1. If the input is a step (i.e., $R(s) = 1/s$), then $e_{ss} = 0$.
2. If the input is a ramp (i.e., $R(s) = 1/s^2$), then $e_{ss} = 4/k$.
3. If the input is a parabola (i.e., $R(s) = 1/s^3$), then $e_{ss} \to \infty$.

These all follow from the fact that the system is type 1, which results in one being the lowest power of s in the numerator of $E(s)$. Combining this with the s in the final value theorem, and the various powers of s in the denominator of $R(s)$ for a step, ramp, and so on, the result follows. Finally, observe that in the case where the steady-state error is finite and nonzero, the error is decreased when the gain k is increased.

A plot illustrating the closed-loop response of the system to a ramp input with two different values of k appears in Figure 10.3. Note that because the system is type 1, the closed-loop steady-state error is nonzero and finite, and when k is increased, the error is decreased.

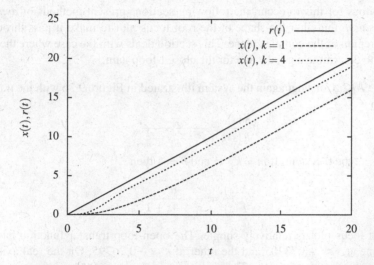

Fig. 10.3 The response of the type 1 system from Example 10.1 to a ramp input illustrating that the finite steady-state error decreases as the gain is increased.

Following exactly the same analysis as Example 10.1 gives the results summarized in Table 10.1. If there is also a block in the feedback loop, $G_s(s)$ will appear in these formulae (Exercise 10.9).

	$R(s) = 1/s$	$R(s) = 1/s^2$	$R(s) = 1/s^3$
Type 0:	$1/(1+kG_c(0)G_p(0))$	∞	∞
Type 1:	0	$1/(kG_c(0)G_p(0))$	∞
Type 2:	0	0	$1/(kG_c(0)G_p(0))$

Table 10.1 Steady-state errors versus system type.

10.2 Controller Design Using a Root Locus Plot

Root locus plots are very useful for control design because they present pole locations for every positive value of the gain.

10.2.1 Proportional Control

Root locus plots may be used to determine a "good" value for the feedback gain if it happens that the root locus plot goes through the region in the complex plane corresponding to the desired response characteristics. Because typically it would only be fortuitous for this to occur, the following sections present methods of *dynamic compensation* that alter the shape of the root locus plot to make it pass through a desired region of the complex plane. This section deals with the case where the only issue is to determine a good value for the closed-loop gain.

Example 10.2. Consider again the system illustrated in Figure 9.36 with the transfer function

$$\frac{X(s)}{\hat{F}(s)} = \frac{\frac{k}{m}}{s^2 + \frac{b}{m}s + \frac{k}{m}},$$

which is a type 0 system. If $m = k = 1$ and $b = 4$ then

$$\frac{X(s)}{\hat{F}(s)} = \frac{1}{s^2 + 4s + 1}.$$

The root locus plot is relatively simple. The open-loop transfer function has two poles, one at $s \approx -3.73205$ and the other at $s \approx -0.26795$. On the real axis, the locus is between the two poles. The asymptote angles are $\pm 90°$ and the asymptotes intersect the real axis at $s = -2$. The break-away point is at $s = -2$. The complete root locus plot is illustrated in Figure 10.4.

Focusing on the transient response for the moment, assume it is desired that the percentage overshoot be less than 10%. From Figure 9.24, the damping ratio must be greater than 0.6. Because $\sin^{-1} 0.6 \approx 37°$, we need that k be in the region between the lines of constant damping illustrated in Figure 10.5. Picking the point $s = -2 \pm 2.5i$ to locate the poles of the closed-loop transfer function, we may determine k from the distance from the two poles of the open-loop transfer function. By Equation (9.29), we need to know the distance from all of the open-loop poles and zeros to the desired pole location of the closed-loop transfer function. Figure 10.6 indicates the two relevant distances, both of which are equal to $\sqrt{1.7^2 + 2.5^2} \approx 3.0414$. Hence we use $k = 9.25$.

To verify the answer, we compute the step response of the closed-loop system using the computed gain. The closed-loop transfer function is

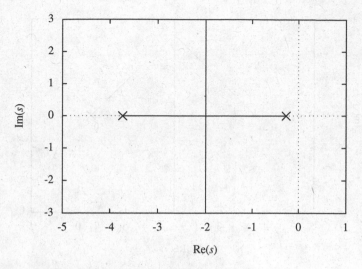

Fig. 10.4 Root locus plot for the system from Example 10.2.

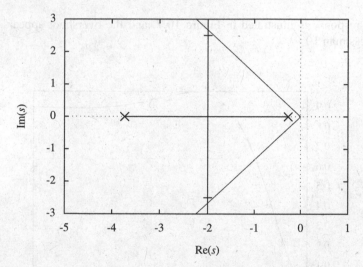

Fig. 10.5 Pole locations that result in less than a 10% overshoot for Example 10.2.

$$\frac{Y(s)}{R(s)} = \frac{k\frac{1}{s^2+4s+1}}{1+k\frac{1}{s^2+4s+1}} = \frac{k}{s^2+4s+1+k},$$

and for $k = 4$,

$$\frac{Y(s)}{R(s)} = \frac{4}{s^2+4s+5}.$$

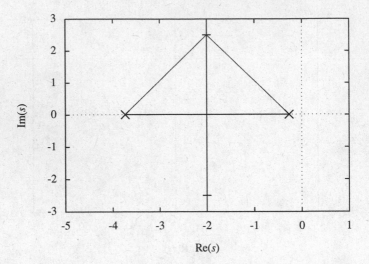

Fig. 10.6 Distances to determine the gain for Example 10.2.

The step response is illustrated in Figure 10.7, and the overshoot appears to be slightly less than 10%.

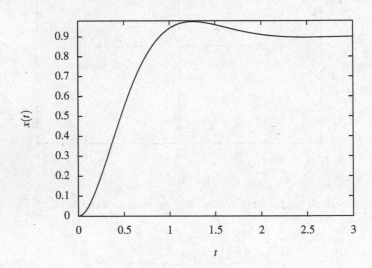

Fig. 10.7 Step response demonstrating the desired overshoot for Example 10.2.

Let us make the preceding example more difficult by adding a rise time specification to the problem as well.

Example 10.3. For the system in Example 10.2, in addition to requiring the closed-loop step response to have less than a 10% overshoot, assume also that we desire the rise time to be less than 0.5 seconds. Use the approximation $t_r \approx 1.8/\omega_n$. If we require

$$t_r \leq 0.5$$

then

$$\frac{1.8}{\omega_n} \leq 0.5 \implies \omega_n \geq 3.6.$$

The region in the complex plane where the closed-loop system will satisfy this requirement is outside the semicircle illustrated in Figure 10.8. Also plotted are the lines corresponding to the damping ratio that satisfy the overshoot requirement.

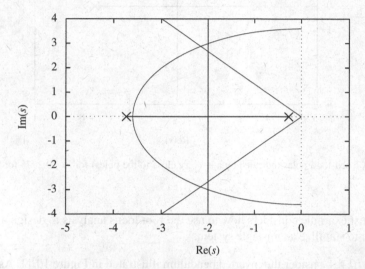

Fig. 10.8 Complex plane regions satisfying the overshoot and rise time requirements for Example 10.3.

Observe that it is impossible to choose a gain value that corresponds to a point on the root locus that is between the lines that indicate the overshoot specification and outside the semicircle that indicates the rise time specification. This is true even for the part of the root locus on the real axis. If we choose a point that is outside the semicircle on the root locus on the real axis there is another pole on the branch that is coming from the other pole, which does not satisfy the specifications. For example, if we place a closed-loop pole at $s = -3.65$, which seemingly satisfies the specification, this corresponds to a k value determined by

$$k \approx 0.2775.$$

For $k = 0.2775$ the closed-loop poles are located at $s = -3.36$ and $s = -0.35$, as is illustrated in Figure 10.9, and the latter does not satisfy the rise time specification. Hoping it will anyway, we can compute the step response, which is illustrated in Figure 10.10. Clearly, it does not work.

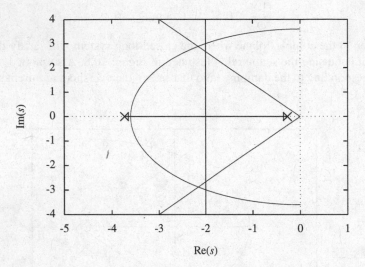

Fig. 10.9 Closed-loop poles indicated by a $+$ (very close to the poles) for $k = 0.2775$ for Example 10.3.

The next example considers how to use the root locus analysis to design a good controller to stabilize an unstable system.

Example 10.4. Consider the inverted pendulum illustrated in Figure 10.11. Assume the bar has a length l, is light with negligible inertia, and that the mass moves under the influence of gravity and a torque τ that is applied about the point of rotation of the pendulum. Assume that we require an overshoot less than 25% and a rise time less than 0.6 seconds.

Using Newton's second law for a planar system rotating about a point, the equation of motion is

$$ml^2 \ddot{\theta} = mgl \sin \theta + \tau.$$

This is a nonlinear equation due to the $\sin \theta$ term. For small θ, $\sin \theta \approx \theta$, and making this approximation we have

$$ml^2 \ddot{\theta} - mgl\theta = \tau.$$

For computational purposes, let $mgl = ml^2 = 1$. Using proportional feedback for τ, the transfer function from a specified desired angle $\Theta_d(s)$ to the actual angle $\Theta(s)$ is illustrated in Figure 10.12 where the controller $G_c(s) = k_p$.

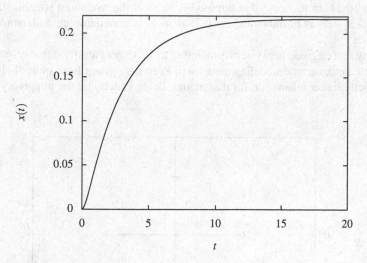

Fig. 10.10 Closed-loop step response for $k = 0.2775$ for Example 10.3.

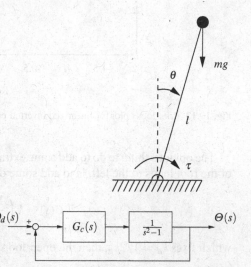

Fig. 10.11 Inverted pendulum system for Example 10.4.

Fig. 10.12 Block diagram for feedback control of the inverted pendulum in Example 10.4.

The root locus plot for the open-loop transfer function

$$G(s) = \frac{1}{s^2 - 1}$$

is illustrated in Figure 10.13. From the root locus plot we can conclude that for $k_p \leq 1$, the system will be unstable and for $k_p > 1$ the system will be neutrally stable because $k_p = 1$ corresponds to the break-away point at the origin. In other words, the linearized equation will have nondecaying sinusoidal solutions. The step responses of the linearized system with $k_p = 0.1$, $k_p = 1$, and $k_p = 2$ are illustrated in Figure 10.14. In this case it is impossible to meet the overshoot specification. If $k_p \leq 1$ the system is unstable and for $k_p > 1$ there is zero damping, independent of k_p.

For any real engineering system, predicting an exactly neutrally stable response is impossible because any modeling errors will keep the system from either behaving in an exactly linear manner or, for that matter, being exactly on the imaginary axis.

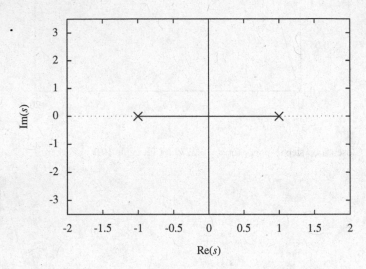

Fig. 10.13 Root locus plot for linearized inverted pendulum for Example 10.4.

The obvious thing to do to add some extra stability, and hence to pull the branches of the root locus to the left, is to add some derivative control. If we specify

$$C(s) = k\left(\frac{1}{2}s + 1\right)$$

which fixes $k_d = 1/2k_p$, then the open-loop transfer function is

$$G(s) = \frac{\frac{1}{2}s + 1}{s^2 - 1},$$

which is illustrated in Figure 10.15.

The regions in the complex plane where the overshoot and rise time specifications are met are illustrated in Figure 10.16. From Figure 9.24, an overshoot of less than 25% corresponds to a damping ratio of greater than 0.4, which corresponds to a pole location at an angle of $\sin^{-1}(0.4) \approx 25°$. Using the relationship $t_r \approx 1.8/\omega_n$ a rise time less than 0.6 seconds requires a natural frequency greater than 3. Inasmuch as the closed-loop poles start at the open-loop poles, an analysis of the root locus plot shows that any gain that meets the rise time specification will also meet the

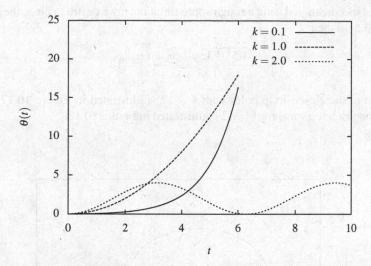

Fig. 10.14 Step responses for various proportional gains for unity feedback for the linearized inverted pendulum in Example 10.4.

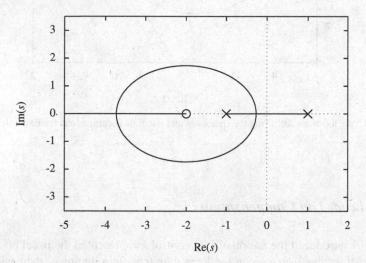

Fig. 10.15 Root locus plot for linearized inverted pendulum for Example 10.4 with PD control.

overshoot specification. Using a rough approximation, if we desire to place the poles at $s \approx -3.5 \pm i$ then

$$k \approx \frac{\sqrt{4.5^2 + 1^2}\sqrt{2.5^2 + 1^2}}{\sqrt{1.5^2 + 1^2}} = 12.$$

A plot of the closed-loop poles with $k = 12$ is illustrated in Figure 10.17. The closed-loop step response for $k = 12$ is illustrated in Figure 10.18.

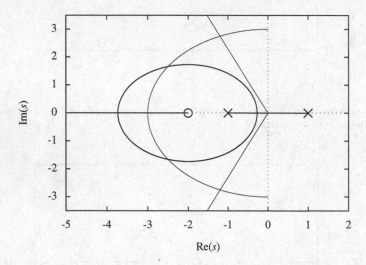

Fig. 10.16 Pole locations satisfying the overshoot and rise time specifications for Example 10.4.

10.2.2 Lead–Lag Compensation

Section 9.1 introduced the notion of PID control and presented the usual effects of each type of feedback on a second-order system (e.g., introducing or increasing the gain for derivative control increases damping). The mathematical analysis is useful and the proper point to initially consider the tool, however, what was missing was the means by which one could actually implement it in a real engineering system.

Lead and lag filters are easy to implement with analog circuits and are hence economical and effective means for control. As demonstrated subsequently, a lead compensator is a means to approximate PD control and a lag compensator is a means to approximate PI control. Combining them, obviously, results in an approximate manner to implement PID control. As a practical matter, lead compensation has the advantage over derivative control because high-frequency noise may have a very

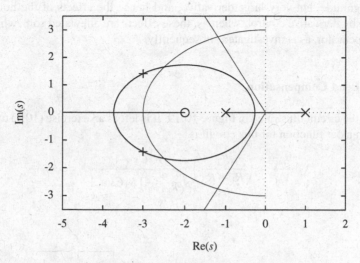

Fig. 10.17 Closed-loop pole locations for PD control with $k = 12$ for Example 10.4.

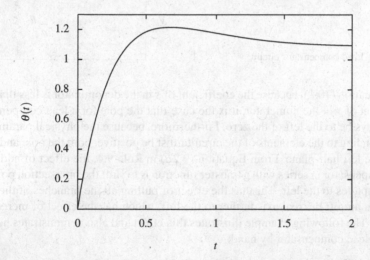

Fig. 10.18 Closed-loop step response with $k = 12$ and PD control for the linearized inverted pendulum in Example 10.4.

small magnitude, but very large derivative, and hence the effects of the noise are amplified by derivative control, whereas, these effects are alleviated somewhat in a lead compensator, as is investigated subsequently.

10.2.2.1 Lead Compensation

Consider the circuit illustrated in Figure 10.19. It is left as an exercise (10.1) to show that the transfer function for this circuit is

$$\frac{V_{out}}{V_{in}} = \left(\frac{R_2}{R_1 + R_2}\right)\frac{R_1 Cs + 1}{\left(\frac{R_2}{R_1 + R_2}\right)R_1 Cs + 1}. \tag{10.1}$$

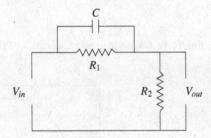

Fig. 10.19 Lead compensator circuit.

In Equation (10.1), because the coefficient of s in the denominator is less than the coefficient of s in the numerator, it is the case that the pole for a lead compensator will always be to the left of the zero. Furthermore, because the physical parameters corresponding to the elements of the circuit must be positive, both the pole and zero are in the left half-plane. From Equation (9.26) in Rule 9.4, the effect of adding a lead compensator in series with a transfer function is to shift the intersection point of the asymptotes to the left. This has the effect of pulling all the branches of the root locus that are off the real axis farther to the left, which has the effect of increasing stability. The following example illustrates this effect and also demonstrates how to design a lead compensator by hand.

Example 10.5. Consider the system illustrated in Figure 10.1 and assume

$$G_p(s) = \frac{1}{s(s+2)}, \quad G_s(s) = 1, \quad G_c(s) = k,$$

which is just proportional control for the system. The root locus plot for this system is illustrated in Figure 10.20.

Assume the approximate desired specifications are as follows.

- A maximum percent overshoot of 15% or less
- A rise time of 0.5 seconds or less
- A 5% settling time of 1 second or less

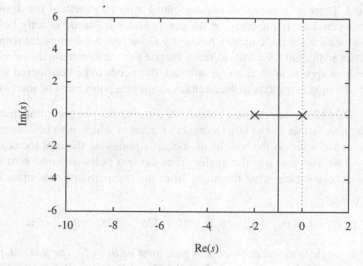

Fig. 10.20 Root locus plot for $G(s)$ in Example 10.5.

For a second-order system, Figure 9.24 indicates that the damping ratio should be approximately 0.55 or more, which corresponds to an angle measured from the imaginary axis of 33.4°. Referring to the Figure 10.20, this fixes a maximum value for k that corresponds to the point on the root locus that passes through $s = -1 \pm 2i$. Because the distance from each of the poles is $\sqrt{5}$, we have that $k = 5$ corresponds to this point on the root locus. At this point, the rise time is approximately $1.8/\sqrt{5} \approx 0.8$ seconds, so this criterion can be met too. However, the 5% settling time is approximately three seconds, so that criterion cannot be met.

First we use a lead compensation of the form

$$G_c(s) = \frac{s+z}{s+p}$$

and determine the pole and zero locations to meet the performance specifications. After that we determine values for the components of the circuit.

Observe that if we hold the angle from the imaginary axis fixed, but move farther from the origin, we simultaneously decrease the rise and settling times without altering the percentage overshoot. If we can add a lead compensator so that the root locus passes through the points $s = -3 \pm 3i$, then both specifications are met. To attempt to offset the increased overshoot due to the zero of the compensator, we attempt to make the root locus pass through $s = -3 \pm 2i$. To attempt to do this, we place the zero of the compensator under the desired point (i.e., $z = -3$) and then use the $\angle G(s) = 180°$ rule to compute the location of the pole for the compensator. The only reason to place the zero at $= -3$ is to make the computations easier, and of course it is possible to do the converse; fix the location of the pole and compute the necessary location of the zero.

Remark 10.1. There is a complication associated with this method for designing the lead compensator. If the zero for the compensator is placed exactly between the desired location for the complex poles, it will influence the transient response. Recall that an additional zero will decrease the rise time and increase the overshoot. If the overshoot specification must be satisfied, this needs to be considered and the angle from the imaginary axis to the complex-conjugate poles must be increased.

Figure 10.21 illustrates the two poles of $G_p(s)$, the zero of the compensator at $s = -3$, the pole farther to the left, the exact location of which is to be determined, and the point we want on the root locus. Because points on the root locus satisfy $\angle G(s) = -180°$, we can use the angles from the two poles and one zero to the desired point to compute what the angle from the compensator pole must be. In particular,

$$\angle G(s) = \theta_z - \theta_{p_1} - \theta_{p_2} - \theta_{p_3} = 90° - 135° - 108° - \theta_p = -180°$$

gives that the angle from the compensator pole must be $\theta_p = 27°$, or $p \approx -9$. Referring to Equation (10.1), we need that $R_1 C = 1/3$ and $R_2/(R_1 + R_2) = 1/3$. Picking $R_2 = 1$, then $R_1 = 2$ and $C = 1/6$. Substituting those values into the transfer function for the circuit gives

$$\frac{V_{out}}{V_{in}} = \frac{1}{3} \frac{\frac{s}{3}+1}{\frac{1}{3}\frac{1}{3}+1} = \frac{s+3}{s+9}.$$

The root locus plot for

$$G(s) = G_c(s)G_p(s) = \frac{s+3}{s+9}\frac{1}{s(s+2)}$$

is illustrated in Figure 10.22. Measuring the distance from all the poles and the zero to the desired point, gives

$$k = \frac{\sqrt{3^3 + 3^3}\sqrt{1 + 3^2}\sqrt{5.9^2 + 3^2}}{3} \approx 30.$$

The step response is illustrated in Figure 10.23, illustrating that the three specifications are approximately met for the compensated system with $k = 30$ and is compared to the system with proportional control and $k = 2$. The rise time and settling time have decreased significantly. The overshoot has actually increased somewhat, due to the presence of the zero from the compensator. If this were problematic, then we would iterate on the design, probably by moving the pole to the left.

10.2.2.2 Lag Compensation

A lag compensator may be used to approximate integral control to reduce the steady-state error of the system response. The idea is that a pole and zero can be placed close to each other, so that their effect on the transient response cancels. However,

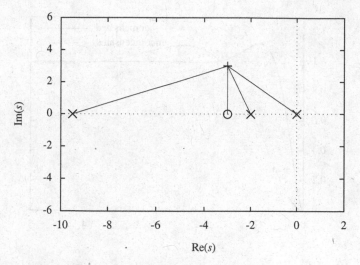

Fig. 10.21 Computations for lead compensator in Example 10.5.

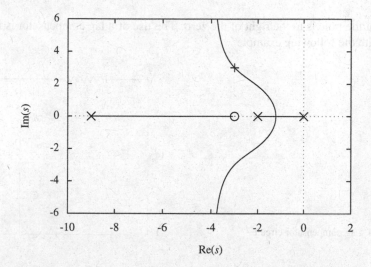

Fig. 10.22 Root locus plot for $G_c(s)G_p(s)$ in Example 10.5.

if they are placed near the origin, then the ratio of the magnitude of the zero to the pole can be large, which may significantly reduce the steady-state error.

Consider the circuit illustrated in Figure 10.24. It is left as an exercise (10.7) to show that the transfer function for this circuit is

$$\frac{V_{out}}{V_{in}} = \frac{CR_2s + 1}{(R_1 + R_2)Cs + 1}. \tag{10.2}$$

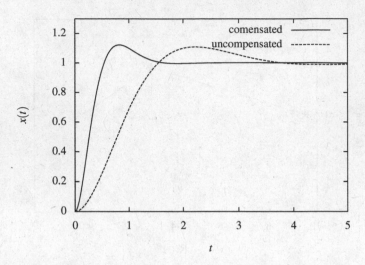

Fig. 10.23 Step response for compensated system in Example 10.5.

Note that the pole is to the right of the zero. The use of a lag compensator is illustrated with the following example.

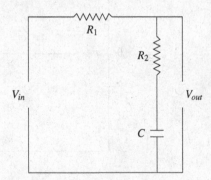

Fig. 10.24 Lag compensator circuit.

Example 10.6. Consider the system illustrated in Figure 10.1 and assume

$$G_p(s) = \frac{1}{(s+1)(s+3)}, \quad G_s(s) = 1, \quad G_c(s) = k,$$

which is just proportional control for the system. The closed-loop transfer function is given by

$$\frac{Y(s)}{R(s)} = \frac{k\frac{1}{(s+1)(s+3)}}{1+k\frac{1}{(s+1)(s+3)}} = \frac{k}{s^2+4s+3+k},$$

and the steady state value for $y(t)$ for a unit step input is given by the final value theorem

$$y_{ss} = \lim_{s \to 0} s \frac{k \frac{1}{(s+1)(s+3)}}{1 + k \frac{1}{(s+1)(s+3)}} = \frac{k}{s^2 + 4s + 3 + k} \frac{1}{s} = \frac{k}{3+k},$$

and hence the steady-state error is $e_{ss} = 3/(3+k)$. For example, for $k = 5$, the steady-state error is $e_{ss} = 3/8$.

Without yet worrying about computing the circuit parameter values, consider

$$G_c = \frac{s + \frac{1}{10}}{s + \frac{1}{100}}.$$

Using this compensator, the closed-loop transfer function is

$$\frac{Y(s)}{R(s)} = \frac{k \frac{s + \frac{1}{10}}{s + \frac{1}{100}} \frac{1}{(s+1)(s+3)}}{1 + k \frac{s + \frac{1}{10}}{s + \frac{1}{100}} \frac{1}{(s+1)(s+3)}}$$

and the steady-state error is

$$e_{ss} = 1 - \frac{k \frac{\frac{1}{10}}{\frac{1}{100}} \frac{1}{3}}{1 + k \frac{\frac{1}{10}}{\frac{1}{100}} \frac{1}{3}} = 1 - \frac{10k}{3 + 10k} = \frac{3}{3 + 10k}.$$

For $k = 5$, $e_{ss} = 3/53$ which is significantly reduced.

The next example uses the robot arm considered in Section 9.1 and uses both a lead and lag compensator to meet both transient and steady-state time domain specifications.

Example 10.7. If the robot arm is near the vertical position and $J = mgl = 1$, then

$$G_p(s) = \frac{\Theta(s)}{T(s)} = \frac{1}{s^2 - 1}, \quad G_s(s) = 1, \quad G_c(s) = k,$$

where $T(s)$ is the Laplace transform of the torque signal. Design a compensator so that for a unit step input the system satisfies the following.

- The percentage overshoot is less than 60%.
- The rise time is less than 1 second.
- The settling time is less than 4 seconds.
- The steady-state error is less than 5%.

The root locus plot for $G_p(s)$ is illustrated in Figure 10.25. The root locus is entirely on the imaginary axis, so regardless of the proportional feedback gain the system always oscillates. In order to meet the transient response requirements, consider making the root locus pass through the point $s = -1 + 3i$. At that point the damping ratio will be 0.32 which corresponds to a percentage overshoot slightly less than 40%. The rise time will be approximately $t_r \approx 1.8/\sqrt{10} \approx 0.56$ and the settling time will be three seconds.

Placing the zero of the lead compensator at $s = -2$, it is easy to compute and verify that the pole should be located at approximately $s = -6$, as is illustrated in Figure 10.26 and the gain corresponding to the point $s = -1 + 3i$ is $k = 17$. The step response of the closed-loop system with this compensator and gain is illustrated in Figure 10.28. All the transient response specifications are met; however, the steady-state error is too large.

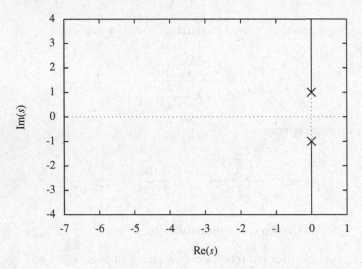

Fig. 10.25 Root locus for proportional control of the robot arm from Example 10.7.

The root locus plot for the system with a lag compensator with a zero at $s = -0.3$ and a pole at $s = -0.03$ in addition to the lead compensator designed above is in Figure 10.27. Because the pole and zero for the lag compensator are close together, the shape of the rest of the root locus plot is not significantly altered. The step response is illustrated in Figure 10.28. Compared with the response with lead compensation only, the steady-state error is significantly reduced.

10.3 Frequency Response Design

Compensator design can also be carried out using frequency response tools. These are particularly useful when dealing with a very complex system in which the transfer function is only known through experimental data in the form of a Nyquist or Bode plot. The frequency response characteristics of lead and lag compensators are

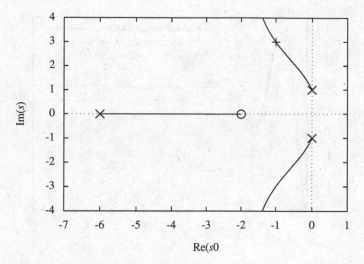

Fig. 10.26 Root locus for control of the robot arm with lead compensation from Example 10.7.

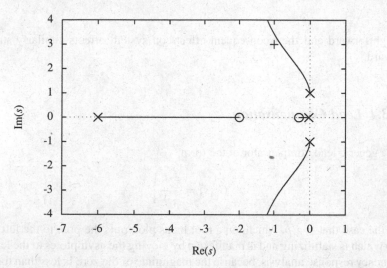

Fig. 10.27 Root locus plot for control of the robot arm with lead–lag compensation from Example 10.7.

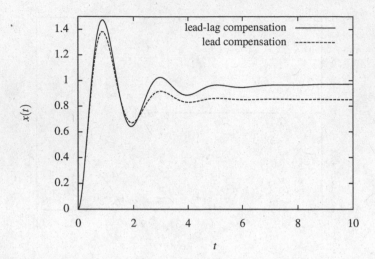

Fig. 10.28 Step response for closed-loop control of the robot arm with lead–lag compensation from Example 10.7.

straightforward and their consequent effects on system effects similarly straightforward.

10.3.1 Lead Compensation

For a generic lead compensator of the form

$$G_c(s) = \frac{\frac{s}{z} + 1}{\frac{s}{p} + 1}$$

it is the case that $z < p$, which, on a root locus plot, puts the pole to the left of the zero which is stabilizing and is manifested by moving the asymptotes to the left. In a frequency response analysis, because the magnitude of the zero is less than the pole, its effect is manifested at lower frequencies on a Bode plot. At frequencies between the zero and the pole, closer to the zero the effect of the zero will dominate and closer to the pole the effect of the pole will more nearly cancel the effect of the zero. A zero has a frequency response phase curve that has positive phase, therefore between the zero and the pole a lead compensator will have positive phase. When placed in series, then, with another transfer function, a lead compensator will increase the phase between the zero and pole. Thus with the zero and pole placed properly, the phase margin may be increased.

Example 10.8. The Bode plot for

$$G_c(s) = \frac{\frac{s}{10}+1}{\frac{s}{100}+1}$$

is illustrated in Figure 10.29. Note that the maximum value of the phase is half way between the 10 and 100 on the log scale, and hence

$$\log \omega_m = \frac{1}{2}\left(\log 10 + \log 100\right) = \log \sqrt{1000} \approx \log 31.5,$$

where the notation ω_m indicates for a lead compensator the frequency at which the maximum phase value occurs.

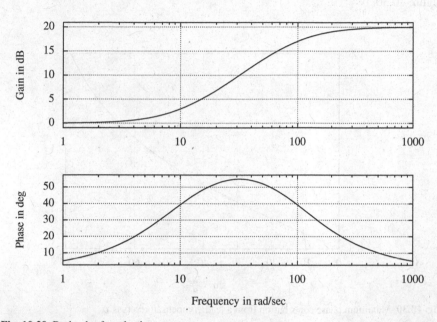

Fig. 10.29 Bode plot for a lead compensator.

In general, the maximum phase contribution of the lead compensator will occur at the geometric mean of the zero and pole frequencies; that is,

$$\boxed{\omega_m = \sqrt{zp}.}$$

Substituting this frequency into the transfer function for the lead compensator gives

$$|G_c(i\sqrt{zp})| = \frac{\sqrt{\frac{p}{z}}+1}{\sqrt{\frac{z}{p}}+1} \qquad (10.3)$$

and (Exercise 10.11)

$$\angle G_c\left(i\sqrt{zp}\right) = \sin^{-1}\frac{p-z}{p+z}.$$

In order to facilitate the steps used to design the compensator developed subsequently, let

$$\alpha = \frac{p}{z}.$$

Then

$$\boxed{\angle G_c\left(i\sqrt{zp}\right) = \sin^{-1}\frac{\alpha-1}{\alpha+1}.}\qquad(10.4)$$

A plot of the phase of the lead compensator at $\omega_m = \sqrt{pz}$ versus α is illustrated in Figure 10.30.

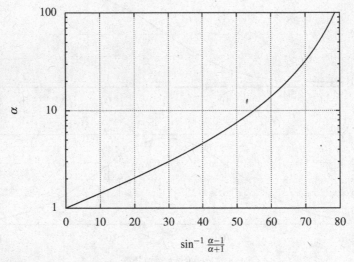

Fig. 10.30 Maximum phase contribution from a lead compensator versus α.

The value of α clearly establishes the ratio of the pole and zero values. The other step to design the compensator is to fix the value of ω_m such that the maximum phase contributed by the compensator occurs at a desirable value. The steps in designing a lead compensator used in a cascade configuration using a Bode plot thus are as follows.

1. Determine α based on how much additional phase margin is desired. As shown subsequently, the value for α must be increased beyond the exact amount by which the phase margin should be increased. If the desired amount is substantial, more than one lead compensator in series may be necessary.

2. Place ω_m so that the maximum phase added by the compensator is at the frequency where the compensated magnitude plot crosses through 0 dB, which will increase the phase margin by the maximum possible amount.

As shown in the examples, the second step is not as trivial as it may initially seem because the lead compensator also modifies the magnitude plot, so the crossover point for the compensated system will not be the same as for the uncompensated system. Also, because of this fact, the amount of phase that must be added in the first step is not exactly the amount by which the phase margin must be increased because the phase of the original system at the compensated crossover point will be, in general, less than at the original crossover point, which necessitates increasing α.

To determine ω_m note that Equation (10.3) gives the gain of the compensator at $\omega_m = \sqrt{zp}$, which in terms of α is

$$|G_c\left(i\sqrt{zp}\right)|_{dB} = 20\log\left(\frac{\sqrt{\alpha}+1}{\frac{1}{\sqrt{\alpha}}+1}\right).$$

Hence, if we select $\omega_m = \sqrt{zp}$ to be at the frequency where the *uncompensated* gain is

$$|G\left(i\sqrt{zp}\right)|_{dB} = -20\log\left(\frac{\sqrt{\alpha}+1}{\frac{1}{\sqrt{\alpha}}+1}\right) \tag{10.5}$$

(note the minus sign) then the compensated gain will pass through 0 dB right at the point where the maximum phase is contributed by the lead compensator. At this point it is possible to check if the value for α is sufficient because the phase of the uncompensated system at ω_n is then known. If α is not great enough, then it must be increased, which will then change ω_n, thus possibly necessitating iteration.

Example 10.9. Consider a system that is unknown[1] other than the plant having the frequency response characteristics illustrated in Figure 10.31. Assume that the steady-state error to a step input must be less than $1/4$ and the phase margin must be greater than $20°$. For simplicity, it would be desirable to meet these specifications with only a lead compensator and op-amp.

From the Bode plot, because the slope of the magnitude curve for low frequencies is zero, we may infer it is a type 0 system and also $\lim_{s\to 0} G_p(s) \approx -20_{dB} \approx 1/10$. Inasmuch as, for a type 0 system

$$e_{ss} = \frac{1}{1+kG_p(0)},$$

---

[1] If the reader wants to reproduce this example, the transfer function used was

$$G_p(s) = \frac{s+1}{(s^2+2s+2)(2s^2+5s+4)},$$

but we do not use any specific information from the plant transfer function $G_p(s)$ other than its frequency response.

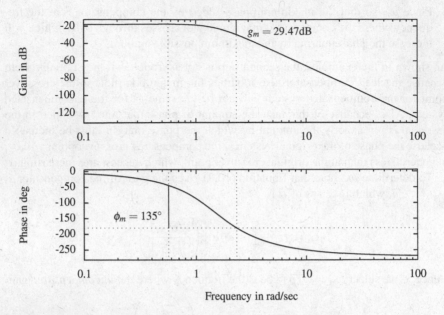

Fig. 10.31 Frequency response of the plant for Example 10.9.

a gain of $k = 30$ would meet the steady-state error specification.

Referring back to Figure 10.31, because in $k = 30$, converting to dB gives

$$k = 10 \log 30 \approx 29.5 \text{ dB},$$

it is very close to the gain margin for the system and hence, is unacceptable. We use a lead compensator to increase the stability margin without having to alter the gain which provides satisfactory steady-state performance.

Figure 10.30 may be used to determine α, to which, noting the logarithmic scale, $\alpha = 3$ corresponds approximately to an maximum phase for the lead compensator of $25°$. To account for the additional phase needed as described above, we select $\alpha = 5$. Substituting $\alpha = 5$ into the right-hand side of Equation (10.5) gives

$$G\left(i\sqrt{zp}\right) \approx -7.$$

We are using $k = 30$ and $20 \log 30 = 29.5$, therefore we need the frequency in Figure 10.31 where the uncompensated gain is approximately -37 dB, which is $\sqrt{zp} = 3$. From the phase plot at $\omega = 3$, the phase for the uncompensated system is approximately $-210°$, which is $30°$ below the crossover phase of $-180°$, so the phase added by the compensator must be at least $50°$. Referring to Figure 10.30, $\alpha = 5$ is not enough. A value of $\alpha = 9$ seems necessary.

Iterating, then, and substituting $\alpha = 9$ into the right-hand side of Equation (10.5)

$$G\left(i\sqrt{zp}\right) \approx -9.5.$$

Referring to Figure 10.31, where the uncompensated gain is approximately -40 dB, is closer to $\sqrt{zp} = 4$.

Solving $\alpha = p/z = 9$ and $\sqrt{zp} = 4$ for p and z gives $z = 4/3$ and $p = 12$. Hence

$$G_c(s) = \frac{\frac{4s}{3}+1}{\frac{s}{12}+1}.$$

Figure 10.32 contains three Bode plots. In the magnitude plot, the lowest curve is the original uncompensated system. The middle curve is the system multiplied by a gain of $k = 30$ to reduce the steady-state error. The top curve is with the lead compensator. The gain and phase margins indicated are for the fully compensated system. In the phase plot, the system with the lead compensator is the top curve and the other two systems are the same and are the bottom curve. The step response of the compensated system is illustrated in Figure 10.33.

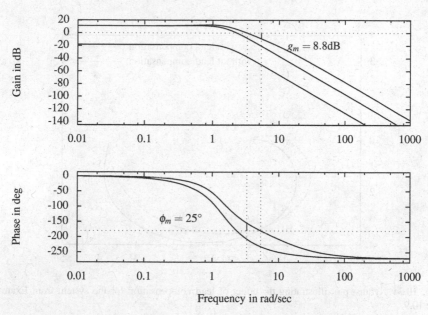

Fig. 10.32 Comparison of compensated and uncompensated Bode plots for the system from Example 10.9. The two small vertical lines indicate the gain and phase margins.

The Nyquist plot corresponding to the positive $i\omega$ segment of the s-contour for the system multiplied by the gain of $k = 30$ and the system with the gain of $k = 30$ and the lead compensator are illustrated in Figure 10.34. The effect of the lead compensator is to add phase, which rotates the middle portion of the Nyquist plot counterclockwise, which increases the stability margin.

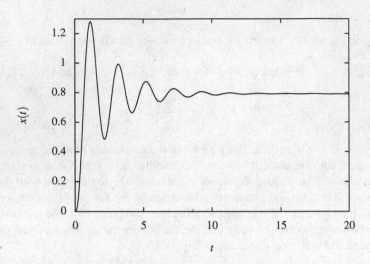

Fig. 10.33 Step response of compensated system from Example 10.9.

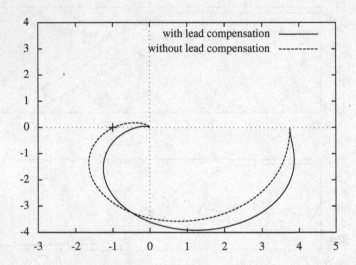

Fig. 10.34 Nyquist plot illustrating the effect of lead compensation for the system from Example 10.9.

10.3.2 Lag Compensation

A lag compensator works by increasing the magnitude of the transfer function for low frequencies. If the frequencies of the pole and zero of the lag compensator are significantly lower than the crossover frequency for the system, the stability properties will not be altered. This is demonstrated by a simple example.

Example 10.10. Modify the compensated system from Example 10.9 to further decrease the steady-state error to be 1/13. Using a lag compensator with the zero at $s = 1/10$ and the pole at $s = 1/100$ in series with the lead compensator from Example 10.9 gives

$$G_c(s) = \frac{s + \frac{1}{10}}{s + \frac{1}{100}} \frac{\frac{3s}{4} + 1}{\frac{s}{12} + 1}.$$

Inasmuch as, for a type 0 system, the steady-state error to a step is given by

$$e_{ss}(t) = \frac{1}{1 + kG_c(0)G_p(0)},$$

and adding the lag term increases $G_c(0)$ by an order of magnitude, then

$$e_{ss} = \frac{1}{1 + 30} = \frac{1}{31}.$$

The Bode plot for the system with this lag compensator is illustrated in Figures 10.35. Note that the very low frequency portion of the magnitude plot is increased by 20 dB, which correspondingly decreases the steady-state error by an order of magnitude.

10.4 Filters

This section considers the characteristics of low- and highpass filters. There are many types of filters, some of which are explored in the exercises, and this section only provides the most basic overview. In control engineering, filters are often used to eliminate frequency components of inputs that are undesirable. For example, the resonant frequency of a system should not be excited by a control signal, so it may be desirable to filter the input to reduce the effect of that frequency.

10.4.1 Lowpass Filters

Consider the circuit illustrated in Figure 10.36. For reasons that are addressed subsequently, this is called a *lowpass filter*. Kirchhoff's voltage law around the circuit

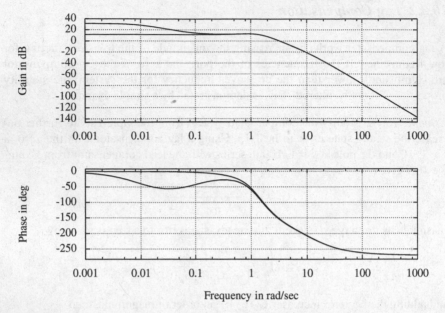

Fig. 10.35 Bode plot for system from Example 10.10 illustrating the increase in the low frequency gain.

gives

$$v_{in} = v_R + v_C = v_r + v_{out},$$

where v_R and v_C are the voltage drops across the resistor and capacitor, respectively. Because

$$v_R = iR, \quad i = C\frac{dv_C}{dt}$$

in the frequency domain we have

$$V_R(s) = I(s)R, \quad I(s) = CsV_C(s) = CsV_{out}(s)$$

and substituting into the voltage equation gives

$$V_{in}(s) = (CRs + 1)V_{out}(s).$$

So the transfer function is

$$\frac{V_{out}}{V_{in}} = \frac{1}{CRs + 1}.$$

Clearly, this circuit has a pole at $s = -1/(CR)$. The frequency $\omega = 1/(CR)$ is called the *cutoff frequency*. For the case where $C = R = 10$, the Bode plot is illustrated in Figure 10.37. Frequencies below $\omega \approx 0.01$ are passed through the filter without any amplification or attenuation; in contrast, frequencies above $\omega \approx 0.01$ are attenuated.

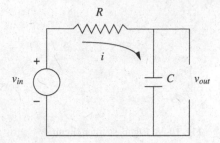

Fig. 10.36 Lowpass filter circuit.

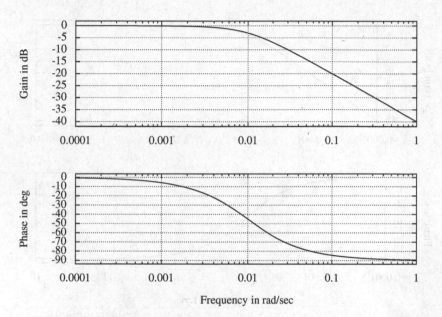

Fig. 10.37 Frequency response of a lowpass filter with $C = 10$ and $R = 10$.

10.4.2 Highpass Filters

If the output voltage is measured across the resistor instead of the capacitor, the circuit is a highpass filter, which is illustrated in Figure 10.38. An easy circuit analysis gives the transfer function as

$$\frac{V_{out}}{V_{in}} = \frac{CRs}{CRs + 1}$$

and the frequency response is illustrated in Figure 10.39. Frequencies above $\omega \approx 0.01$ are passed through the filter without any amplification or attenuation; in contrast, frequencies below $\omega \approx 0.01$ are attenuated.

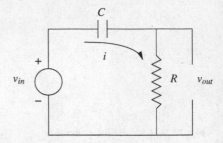

Fig. 10.38 Highpass filter circuit.

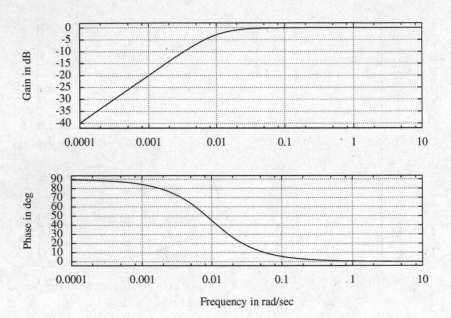

Fig. 10.39 Frequency response of a highpass filter with $C = 10$ and $R = 10$.

10.5 Exercises

10.1. Determine the transfer function from the input voltage to the output voltage for the circuit illustrated in Figure 10.19. Show that for this circuit the pole and zero are in the left half-plane and that the pole is always to the left of the zero.

10.2. Write a computer program to simulate proportional control on the nonlinear dynamics of robot arm in Example 9.1 about the vertical position. Is the result stable or unstable?

10.3. Write a computer program to simulate PD control for the inverted pendulum system from Example 10.4. Remember that the root locus plot is based on the linear equation. Compare the response of the linear model to the nonlinear one with increasingly large step inputs. At what point, if any, does the linear solution deviate significantly from the nonlinear one?

10.4. Verify the magnitude and phase plots in Figure 9.71 for

$$G(s) = \frac{s^2 + 2\zeta\omega_n s + \omega_n^2}{\omega_n^2}$$

for the cases where $\omega \ll \omega_n$ and $\omega \gg \omega_n$. Verify the magnitude plot for when $\omega = \omega_n$.

10.5. Show that the transfer function for the notch filter illustrated in Figure 10.40 is given by

$$\frac{V_{out}}{V_{in}} = \frac{(C_L R_H + 1)(C_H R_L + 1)}{C_L C_H R_L R_H s^2 + (C_H R_L + C_L R_H + C_L R_L)s + 1}.$$

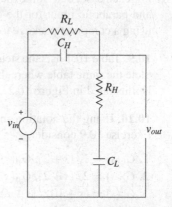

Fig. 10.40 Notch filter circuit for Exercise 10.4.

10.6. Consider connecting a lowpass filter and highpass filter together in series, as is illustrated in Figure 10.41.

1. Determine the transfer function from the input voltage to the output voltage. Is the transfer for the circuits in series equal to the product of the individual transfer functions? Explain the answer.
2. If $R_{HP} = 10$, $R_{LP} = 100$, $C_{HP} = 10$, and $C_{LP} = 100$ sketch the Bode plot for this transfer function. Explain why this may be called a *bandpass filter*.
3. How would you modify the circuit to make the bandpass region either narrower or wider? Do so and either sketch or use a computer package to generate the Bode diagram.
4. How would you make the transitions in the bandpass filter sharper, that is, steeper transitions? Do so and either sketch or use a computer package to generate the Bode diagram.

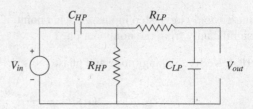

Fig. 10.41 Bandpass filter for Exercise 10.6.

10.7. Show that the circuit in Figure 10.24 has the transfer function given in Equation (10.2).

10.8. Using the notation and configuration from Figure 10.2, consider

1. $G_c(s) = 1/(s+2)$, $G_p(s) = 10/(s^2+5s+10)$, $G_s(s) = 1$.
2. $G_c(s) = 1/(s+2)$, $G_p(s) = 10/(s^2+5s)$, $G_s(s) = 1$.
3. $G_c(s) = 1/(s^2+2s)$, $G_p(s) = 10/(s^3+5s^2+10s)$, $G_s(s) = 1$.

For each case, determine the system type and the steady-state error for a step, ramp, and parabolic input for the cases where $k = 1$ and $k = 10$. Verify your answer by using a computer package to plot the responses for each case.

10.9. Table 10.1 lists the steady-state errors in the case of unity feedback only. Complete the same table where there is a transfer function $G_s(s)$ in the feedback loop, as is illustrated in Figure 10.2.

10.10. Using the notation and configuration from Figure 10.2, and your result from Exercise 10.9 consider

1. $G_c(s) = 1/(s+2)$, $G_p(s) = 10/(s^2+5s+10)$, $G_s(s) = 10000/(s+10000)$.
2. $G_c(s) = 2/(s+2)$, $G_p(s) = 10/(s^2+5s)$, $G_s(s) = 1000/(s+10000)$.
3. $G_c(s) = (s+10)/(s+2)$, $G_p(s) = 10/(s^3+5s^2+10s)$, $G_s(s) = 1/s$.

For each case, determine the system type and the steady-state error for a step, ramp and parabolic input for the cases where $k = 1$ and $k = 10$. Verify your answer by using a computer package to plot the responses for each case.

10.11. Evaluate the phase of

$$G_c(s) = \frac{\frac{s}{p}+1}{\frac{s}{p}+1}$$

at the frequency where the phase is maximum to verify Equation (10.4).

10.12. Referring to Figure 10.16, what will happen to the rise time if k is increased significantly beyond $k = 12$?

10.13. Sketch the Nyquist plot for each of the following transfer functions and determine the gain and phase margins.

1. $G(s) = 1/((s+1)(s+5))$.

2. $G(s) = 1/((s+1)(s+2)(s+3))$.
3. $G(s) = 1/(s^2 + 3s + 5)$.
4. $G(s) = 1/(s^3 + s^2 + 3s + 1)$.

10.14. For each of the following transfer functions design a lead compensator so that the root locus for the compensated system passes through the indicated point.

1. $G(s) = 1/(s(s+2))$, $s = -3 + 3i$.
2. $G(s) = 1/(s^2 + s + 2)$, $s = -2 + 3i$.
3. $G(s) = (s+1)/((s+2)(s^2 + 2s + 2))$, $s = -3 + 3i$.
4. $G(s) = 1/(s^2 + 4)$, $s = -1 + 3i$.

For each problem,

- Plot both the uncompensated and compensated root locus plot.
- Determine the gain that places the poles of the closed-loop system at the indicated pole location.
- Plot the closed-loop step response with the gain determined in the previous step and indicate whether or not the transient response characteristics are what you expect based on the pole and zero locations for the closed-loop system.

10.15. Design a lead compensator to satisfy the stated specifications for each of the following transfer functions under unity feedback.

1. $G(s) = 1/(s(s+1))$, $O < 5\%$, $t_r < 0.7$ seconds.
2. $G(s) = 1/(s(s+1))$, $O < 25\%$, 5% settling time satisfying $t_s < 1$ second.
3. $G(s) = 1/(s(s+1)(s+2))$, $t_r < 1.8$ seconds and $O < 50\%$.
4. $G(s) = 1/(s^2 + 2s + 2)$, $t_r < 0.6$, $O < 10\%$,
5. $G(s) = 1/((s+1)(s^2 + 2s + 2))$, $\zeta \approx 0.3$ and 5% settling time satisfying $t_s < 5.5$ second.
6. $G(s) = 1/(s^2 + 2s + 2)$, no overshoot and $t_r < 1.8/8$.

In each case,

- Sketch the root locus plot for the uncompensated system.
- Choose a point that the root locus should pass through to meet the specifications and design a lead compensator to meet the indicated specifications and plot the compensated root locus plot.
- Determine the gain needed to satisfy the specifications.
- Verify the specifications are met by plotting the step response.

Note that because the rise time relationship to the natural frequency is an approximation and also because the compensator adds a zero, if the specifications include rise time or overshoot specifications, it may be necessary to iterate on the design. Also, as with any set of design specifications, sometimes it just is not possible to meet them all, in which case determine a suitable compromise.

10.16. For each of the systems in Exercise 10.15, determine the steady-state error to a unit step input using the final value theorem and verify your computation using the MATLAB or Octave step() command or something similar. If the steady-state error is not zero, design a lag compensator to be placed in series with the lead compensator that reduces the steady-state error by an order of magnitude but that does not significantly alter the transient response that was the result of the lead compensator. Verify the compensator works by plotting the step response for the system without and with the lag compensator using the MATLAB or Octave step() command.

10.17. Design a lead compensator for $G(s) = 1/(s(s+1))$ so that $O < 10\%$ and $t_r < 1$ seconds. Verify your design using the MATLAB or Octave step() command. What is the system type? What will be the steady-state error to a step input and a ramp input? Design a lag compensator that reduces the steady-state error to a ramp input by a factor of five. Determine how to use the MATLAB or Octave step() command to plot the response to a unit *ramp* input, and verify your lag compensator reduces the steady-state error by the desired amount.

10.18. For each of the following transfer functions and gains, design a lead compensator that increases the phase margin by $10°$.

1. $G(s) = 1/(s(s+1))$, $k = 1$.
2. $G(s) = 1/(s(s+1))$, $k = 10$.
3. $G(s) = 1/((s+2)(s+1))$, $k = 100$.
4. $G(s) = 10000/((s+1)(s+10)(s+100))$, $k = 1$.
5. $G(s) = 10000/((s+1)(s+10)(s+100))$, $k = 10$.

For each one

- Plot the Bode plot for the uncompensated and uncompensated systems.
- Compare the unit step response for the uncompensated and compensated systems.
- Plot the Nyquist plots for the uncompensated and compensated systems and explain the manner in which the compensator changes the Nyquist plot and the reason why it has that effect.

10.19. Predict the difference in the output of a lead compensator and a compensator with derivative control when the signal into the compensator has a noise component given by

$$n(t) = \frac{1}{100} \sin 200t.$$

Plot the output signal for each with $n(t)$ as the input.

Chapter 11
Partial Differential Equations

This chapter considers techniques for solving some types of partial differential equations. The solution method that is considered in this book is the *separation of variables* method. Linear partial differential equations can be categorized by type, similar to categorizing ordinary differential equations. However, in contrast to ordinary differential equations, for the limited number of partial differential equations considered in this book, the categorization does not affect the solution method, but rather is a reflection of the properties of the resulting solution, which itself is a result of the underlying physics.

The outline of this chapter is first to present three common engineering problems that lead to different types of partial differential equations. As is apparent, there are some broad commonalities with respect to the solution technique. An extension of the theory developed by the engineering problems is investigated later in the chapter in Section 11.6.

11.1 The One-Dimensional Wave Equation

The so-called *wave equation* describes many different physical wavelike phenomena. It is motivated and initially solved using the example of a vibrating string.

11.1.1 Derivation of the Wave Equation

Consider the elastic string illustrated in Figure 11.1. Let x denote the location along a straight line between the endpoints, u denote the displacement of the string, and L denote the length between the endpoints. The function u is a function of the position along the string x as well as time, that is, $u(x,t)$. Solving the wave equation amounts to determining the function $u(x,t)$ that gives the displacement of the string at time t and location x. Let the tension in the string be denoted by τ and the mass per unit

B. Goodwine, *Engineering Differential Equations: Theory and Applications*, DOI 10.1007/978-1-4419-7919-3_11, © Springer Science+Business Media, LLC 2011

length be denoted by ρ. The string is assumed to be elastic, which means it may have an internal tension τ but no moment is needed to bend it.

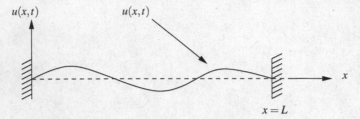

Fig. 11.1 Vibrating string.

The derivation of the wave equation is simply using Newton's law on a infinitesimal segment of the string. Consider the small section illustrated in Figure 11.2. Newton's law on the element in the vertical direction gives

$$\rho \Delta x \frac{\partial^2 u}{\partial t^2}\left(x+\frac{\Delta x}{2},t\right) = \tau(x+\Delta x,t)\sin(\theta(x+\Delta x,t)) - \tau(x,t)\sin(\theta(x,t)).$$

(11.1)

Expanding each of the terms in a Taylor series individually gives

$$\frac{\partial^2 u}{\partial t^2}\left(x+\frac{\Delta x}{2},t\right) = \frac{\partial^2 u}{\partial t^2}(x,t) + \frac{\partial^3 u}{\partial t^2 \partial x}(x,t)\frac{\Delta x}{2}+\cdots$$

$$\tau(x+\Delta x,t) = \tau(x,t) + \frac{\partial \tau}{\partial x}(x,t)\Delta x+\cdots$$

$$\sin(\theta(x+\Delta x,t)) = \sin(\theta(x,t)) + \frac{\mathrm{d}}{\mathrm{d}\theta}\sin(\theta)\frac{\partial \theta}{\partial x}(x,t)\Delta x+\cdots$$

$$= \sin(\theta(x,t)) + \cos(\theta(x,t))\frac{\partial \theta}{\partial x}(x,t)\Delta x+\cdots.$$

Substituting into Equation (11.1) and keeping terms only up to Δx, that is, assuming $\Delta x \ll 1$, gives

$$\rho \Delta x \frac{\partial^2 u}{\partial t^2}(x,t) = \tau(x,t)\cos(\theta(x,t))\frac{\partial \theta}{\partial x}(x,t)\Delta x + \sin(\theta(x,t))\frac{\partial \tau}{\partial x}(x,t)\Delta x,$$

or

$$\rho\frac{\partial^2 u}{\partial t^2} = \tau\cos\theta\frac{\partial \theta}{\partial x} + \sin\theta\frac{\partial \tau}{\partial x}$$

(11.2)

where all the terms are evaluated at (x,t).

To proceed any further, we need some assumptions. Assume that the string only undergoes small displacements; that is, $u(x,t) \ll 1$ and furthermore that the slope of the string is small; that is, $\partial u/\partial x \ll 1$. This would imply immediately that

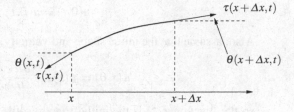

Fig. 11.2 Infinitesimal element of the string.

$$\sin(\theta(x,t)) \approx \theta(x,t), \quad \cos(\theta(x,t)) \approx 1, \quad \theta(x,t) \approx \tan(\theta(x,t)) = \frac{\partial u}{\partial x}(x,t).$$

Also, express the tension in the string as

$$\tau(x,t) = \tau + \hat{\tau}(x,t),$$

where τ is a constant and is the tension in the string when it is still ($u(x,t) = 0$). For small motions, it is the case that $\hat{\tau}(x,t) \ll 1$ and $\partial \tau / \partial x \ll 1$.

Both terms in the second term of the sum on the right-hand side of Equation (11.2) are small, thus

$$\rho \frac{\partial^2 u}{\partial t^2}(x,t) = \tau \frac{\partial \theta}{\partial x}(x,t), \tag{11.3}$$

and so

$$\boxed{\rho \frac{\partial^2 u}{\partial t^2}(x,t) = \tau \frac{\partial^2 u}{\partial x^2}(x,t)} \tag{11.4}$$

is the usual expression for the *one-dimensional wave equation*. This has the interpretation that the net force is proportional to the curvature in the string (right-hand side of the equation). The left-hand side is mass times acceleration.

11.1.2 Boundary Conditions

In general, the wave equation is of the form

$$\frac{\partial^2 u}{\partial t^2} = \alpha^2 \frac{\partial^2 u}{\partial x^2}.$$

Analogous to ordinary differential equations, in order to solve this equation conditions on $u(x,t)$ at the initial time for the problem, usually $t = 0$ *as well as* conditions on $u(x,t)$ on the physical boundaries of the problem must be specified. The latter are normally called *boundary conditions* and play a fundamental role in the solution of the problem.

To proceed, assume the ends of the string are fixed, that is,

$$u(0,t) = u(L,t) = 0.$$

Also, assume that the initial shape and velocity of the string are known, that is,

$$u(x,0) = f(x), \quad \frac{\partial u}{\partial t}(x,0) = g(x),$$

so the function $f(x)$ is the initial shape profile of the string and $g(x)$ is the initial velocity profile.

11.1.3 Separation of Variables

The basic idea behind the method of *separation of variables* is that the solution to the wave equation can be expressed in the form

$$u(x,t) = X(x)T(t);$$

that is, the solution is the product of two functions, where one of the functions only depends on the spatial variable x, and the other function only depends on the temporal variable t. Note that due to the assumed form of $u(x,t)$

$$\frac{\partial^2 u}{\partial t^2}(x,t) = X(x)\frac{d^2 T}{dt^2}(t) = X(x)T''(t)$$

$$\frac{\partial^2 u}{\partial x^2}(x,t) = \frac{d^2 X}{dx^2}(x)T(t) = X''(x)T(t).$$

So, substituting into the wave equation gives

$$X(x)T''(t) = \alpha^2 X''(x)T(t).$$

or

$$\frac{X''(x)}{X(x)} = \frac{1}{\alpha^2}\frac{T''(t)}{T(t)}. \tag{11.5}$$

The critical feature of Equation (11.5) is that the left-hand side depends only upon x, the right-hand side depends only upon t, and they are equal. The only way a function of x can equal a function of t for arbitrary x and t is for both sides to be equal to a constant. One way to think of this is that because x and t are the independent variables, we can choose them arbitrarily. Hence, for example, we can hold x fixed, which holds $X(x)$ fixed and hence the ratio $X'(x)/X(x)$ fixed, but we could choose to vary t. Hence, $T'(t)/T(t)$ must be fixed, or a constant, even if t varies because it is equal to $X'(x)/X(x)$. Note, this does not mean $X(x)$ is a constant and $T(t)$ is a constant; rather, the ratios $X''(x)/X(x)$ and $T''(t)/T(t)$ must be constant. That constant is denoted by $-\lambda$, and is called an *eigenvalue*.[1] Thus

$$\frac{X''(x)}{X(x)} = \frac{1}{\alpha^2}\frac{T''(t)}{T(t)} = -\lambda$$

which actually represents *two* equations

$$\frac{d^2X}{dx^2}(x) + \lambda X(x) = 0$$

$$\frac{d^2T}{dt^2}(t) + \alpha^2 \lambda T(t) = 0.$$

Solutions to these two equations are, it is hoped, obvious from inspection:

$$X(x) = c_1 \sin\sqrt{\lambda}x + c_2\cos\sqrt{\lambda}x \tag{11.6}$$

$$T(t) = c_3 \sin\alpha\sqrt{\lambda}t + c_4\cos\alpha\sqrt{\lambda}t, \tag{11.7}$$

if $\lambda > 0$,

$$X(x) = c_1 e^{\sqrt{-\lambda}x} + c_2 e^{-\sqrt{-\lambda}x} \tag{11.8}$$

$$T(t) = c_3 e^{\alpha\sqrt{-\lambda}t} + c_4 e^{-\alpha\sqrt{-\lambda}t}, \tag{11.9}$$

if $\lambda < 0$, and

$$X(x) = c_1 + c_2 x \tag{11.10}$$

$$T(t) = c_3 + c_4 t,$$

if $\lambda = 0$. At this point we have to consider all three possibilities for λ inasmuch as we do not know anything about it other than it is a constant. However, only one of the three will be able to satisfy the boundary conditions.

If we consider the boundary conditions

$$u(0,t) = 0, \quad u(L,t) = 0.$$

Substituting $u(x,t) = X(x)T(t)$ into the left boundary condition gives

$$X(0)T(t) = X(L)T(t) = 0,$$

which gives

$$X(0) = X(L) = 0,$$

because otherwise $T(t)$ would have to be zero for all time, which would then make $u(x,t)$ zero for all time.

---

[1] The negative sign appears in front of the eigenvalue only because of some foresight that it will make some of the notation easier. It could be positive too with no complications other than some messier equations subsequently.

Substituting $x = 0$ into Equation (11.6) gives that $c_2 = 0$. Then substituting $x = L$ gives

$$c_1 \sin \sqrt{\lambda} L = 0,$$

which requires either $c_1 = 0$ or

$$\sqrt{\lambda} L = n\pi, \quad n = 1, 2, \ldots.$$

Note that $c_1 = 0$ leads to the trivial solution $(u(x,t) = X(x)T(t) = 0T(t) = 0)$ which is not able to satisfy any initial shape and velocity profiles (unless they are also zero, which is a pretty boring vibrating string). Hence, the constant λ is determined by the boundary conditions and must be

$$\lambda = \left(\frac{n\pi}{L}\right)^2, \quad n = 1, 2, \ldots. \tag{11.11}$$

Note also that there are an infinite number of solutions, one for each $n = 1, 2, \ldots$. The role of λ is to scale the frequency of the sine function so that it passes through zero at $x = L$. This is illustrated subsequently in Figure 11.4.

Substituting $x = 0$ into Equation (11.8) gives that $c_1 = -c_2$. Then substituting $x = L$ gives

$$c_1 \left(e^{-\sqrt{-\lambda} L} - e^{\sqrt{-\lambda} L} \right) = 0$$

which has no solutions for $\lambda < 0$ and $L \neq 0$.

Substituting $x = 0$ into Equation (11.10) gives that $c_1 = 0$. Then substituting $x = L$ gives $c_2 = 0$, which results in the trivial solution, which, as mentioned above, is not able to satisfy nonzero initial conditions.

Hence, by considering the boundary conditions, it is clear that the constant λ must be positive and furthermore, must have one of the values given in Equation (11.11). Substituting those values for λ into $u(x,t) = X(x)T(t)$ gives an infinite number of solutions

$$u_n(x,t) = c_1 \sin \frac{n\pi x}{L} \left(c_{3,n} \sin \frac{\alpha n \pi t}{L} + c_{4,n} \cos \frac{\alpha n \pi t}{L} \right), \quad n = 1, 2, \ldots$$

or

$$u_n(x,t) = \sin \frac{n\pi x}{L} \left(a_n \sin \frac{\alpha n \pi t}{L} + b_n \cos \frac{\alpha n \pi t}{L} \right), \quad n = 1, 2, \ldots \tag{11.12}$$

where the constants were combined into a_n and b_n.

Observe the following very important point. Any of the $u_n(x,t)$ satisfies the wave equation as well as the two boundary conditions, as does any linear combination of the $u_n(x,t)$ because the wave equation is linear and homogeneous.

The last task is to satisfy the initial conditions, which were

$$u(x,0) = f(x), \quad \frac{\partial u}{\partial t}(x,0) = g(x).$$

This may seem like an impossible task at first, but perhaps the availability of an infinite number of solutions will be of some help. In fact, let us just go for it and try to combine all the infinite number of solutions together in the form

$$u(x,t) = \sum_{n=1}^{\infty} \sin\frac{n\pi x}{L}\left(a_n\sin\frac{\alpha n\pi t}{L} + b_n\cos\frac{\alpha n\pi t}{L}\right). \tag{11.13}$$

Note that in this form, the initial conditions are

$$u(x,0) = \sum_{n=1}^{\infty} b_n\sin\frac{n\pi x}{L} = f(x) \tag{11.14}$$

$$\frac{\partial u}{\partial t}(x,0) = \sum_{n=1}^{\infty} a_n\frac{\alpha n\pi}{L}\sin\frac{n\pi x}{L} = g(x). \tag{11.15}$$

Finally, what at first probably seems like a trick, but is ultimately shown to be pretty general and useful, is to multiply Equation (11.14) by $\sin(m\pi x/L)$ and integrate from 0 to L

$$\int_0^L \sin\frac{m\pi x}{L}\left(\sum_{n=1}^{\infty} b_n\sin\frac{n\pi x}{L}\right)dx = \sum_{n=1}^{\infty} b_n\int_0^L \sin\frac{m\pi x}{L}\sin\frac{n\pi x}{L}dx$$
$$= \int_0^L \sin\frac{m\pi x}{L}f(x)dx. \tag{11.16}$$

An amazing fact, the details of which are in Appendix C.1, is that if m and n are integers, then

$$\int_0^L \sin\frac{m\pi x}{L}\sin\frac{n\pi x}{L}dx = \begin{cases} 0, & m \neq n \\ \frac{L}{2}, & m = n; \end{cases}$$

that is, every single term in the series is zero except for $m = n$, which nicely kills off all but one of the infinite number of terms in the series.[2] Hence

$$b_n\int_0^L \left(\sin\frac{n\pi x}{L}\right)^2 dx = \int_0^L \sin\frac{n\pi x}{L}f(x)dx,$$

or

$$b_n = \frac{2}{L}\int_0^L f(x)\sin\frac{n\pi x}{L}dx. \tag{11.17}$$

Using these values for b_n, Equation (11.14) is called the *Fourier sine series* for $f(x)$. The following example illustrates the computations involved in computing the Fourier sine series as well as gives an indication of the convergence properties of such a series.

Example 11.1. Let $L = 3$ and

---

[2] Note that switching the order if the integration and summation is not necessarily valid of the series does not converge. In this case, it is allowable. The reader is referred to [38] for a more complete justification of the theory underlying this topic.

$$f(x) = \begin{cases} x & x < 1 \\ \frac{3-x}{2} & 1 \le x \le 3 \end{cases} \tag{11.18}$$

which is illustrated in Figure 11.3. Keep in mind that the task is to combine an infinite number of sine functions to be equal to this triangle. Computing the Fourier coefficients,

$$
\begin{aligned}
b_n &= \frac{2}{3} \int_0^3 f(x) \sin \frac{n\pi x}{3} dx \\
&= \frac{2}{3} \left[\int_0^1 x \sin \frac{n\pi x}{3} dx + \int_1^3 \frac{3-x}{2} \sin \frac{n\pi x}{3} dx \right] \\
&= \frac{2}{3} \left[-\frac{3x}{n\pi} \cos \frac{n\pi x}{3} \Big|_0^1 + \int_0^1 \frac{3}{n\pi} \cos \frac{n\pi x}{3} dx + \frac{3}{2} \int_1^3 \sin \frac{n\pi x}{3} dx \right. \\
&\quad \left. + \frac{3x}{2n\pi} \cos \frac{n\pi x}{3} \Big|_1^3 - \frac{1}{2} \int_1^3 \frac{3}{n\pi} \cos \frac{n\pi x}{3} dx \right] \\
&= \frac{2}{3} \left[-\frac{3x}{n\pi} \cos \frac{n\pi x}{3} \Big|_0^1 + \frac{9}{n^2\pi^2} \sin \frac{n\pi x}{3} \Big|_0^1 - \frac{9}{2n\pi} \cos \frac{n\pi x}{3} \Big|_1^3 \right. \\
&\quad \left. + \frac{3x}{2n\pi} \cos \frac{n\pi x}{3} \Big|_1^3 - \frac{9}{2n^2\pi^2} \sin \frac{n\pi x}{3} \Big|_1^3 \right] \\
&= \frac{2}{3} \left[-\frac{3}{n\pi} \cos \frac{n\pi}{3} + \frac{9}{n^2\pi^2} \sin \frac{n\pi}{3} - \frac{9}{2n\pi} \left(\cos n\pi - \cos \frac{n\pi}{3} \right) \right. \\
&\quad \left. + \left(\frac{9}{2n\pi} \cos n\pi - \frac{3}{2n\pi} \cos \frac{n\pi}{3} \right) + \frac{9}{2n^2\pi^2} \sin \frac{n\pi}{3} \right] \\
&= \frac{9}{n^2\pi^2} \sin \frac{n\pi}{3}.
\end{aligned}
$$

Figure 11.4 is a plot of the first four terms in the Fourier series; namely,

$$
\begin{aligned}
f_1(x) &= \frac{9}{\pi^2} \sin \frac{\pi}{3} \sin \frac{\pi x}{3} = \frac{9\sqrt{3}}{2\pi^2} \sin \frac{\pi x}{3} \\
f_2(x) &= \frac{9}{2^2\pi^2} \sin \frac{2\pi}{3} \sin \frac{2\pi x}{3} = \frac{9\sqrt{3}}{8\pi^2} \sin \frac{2\pi x}{3} \\
f_3(x) &= \frac{9}{3^2\pi^2} \sin \frac{3\pi}{3} \sin \frac{3\pi x}{3} = 0 \\
f_4(x) &= \frac{9}{4^2\pi^2} \sin \frac{4\pi}{3} \sin \frac{4\pi x}{3} = -\frac{9\sqrt{3}}{32\pi^2} \sin \frac{4\pi x}{3}.
\end{aligned}
$$

The series is given by

$$f(x) = \sum_{n=1}^{\infty} \left(\frac{9}{n^2\pi^2} \sin \frac{n\pi}{3} \right) \sin \frac{n\pi x}{L}$$

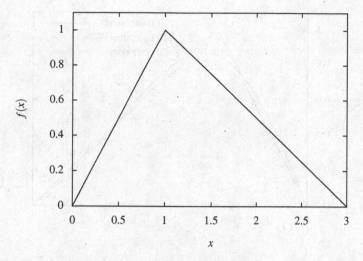

Fig. 11.3 Function for the Fourier series for Example 11.1.

and Figure 11.5 illustrates the sum of the first 3, 10, and 20 components. Note that the curve converges to $f(x)$ as the number of components increases.

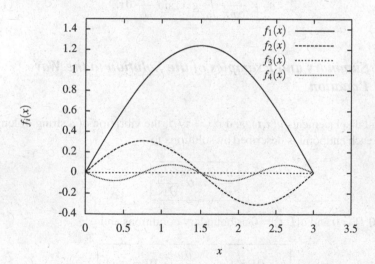

Fig. 11.4 First four Fourier sine components of Equation (11.18).

Now return to the solution for the vibrating string in Equation (11.13). The b_n coefficients have already been determined by Equation (11.17) and the a_n coeffi-

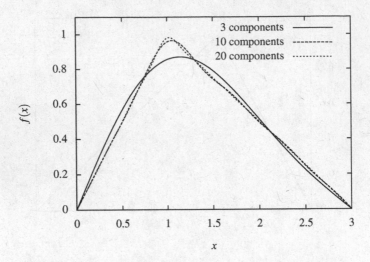

Fig. 11.5 Truncated Fourier sine series converging to $f(x)$ from Equation (11.18).

cients are computed similarly; that is, multiply each side of Equation (11.15) by $\sin(m\pi x/L)$ and integrate from 0 to L which gives

$$a_n = \frac{2}{\alpha n \pi} \int_0^L g(x) \sin \frac{n\pi x}{L} dx. \tag{11.19}$$

11.1.4 Summary and Examples of the Solution to the Wave Equation

For small displacements $u(x,t)$, and $\alpha^2 = \tau/\rho$, the vibration of a string of length L fixed at each endpoint is described by solutions to

$$\boxed{\frac{\partial^2 u}{\partial t^2} = \alpha^2 \frac{\partial^2 u}{\partial x^2}}$$

with $u(0,t) = 0$ and $u(L,t) = 0$ as boundary conditions and

$$\boxed{u(x,0) = f(x), \quad \frac{\partial u}{\partial t}(x,0) = g(x),}$$

as initial conditions, where $f(x)$ and $g(x)$ are the initial shape of the string and initial velocity profile, respectively.

From the preceding analysis, the solution for the vibrating string problem is

$$u(x,t) = \sum_{n=1}^{\infty} \left[\sin \frac{n\pi x}{L} \left(a_n \sin \frac{\alpha n\pi t}{L} + b_n \cos \frac{\alpha n\pi t}{L} \right) \right] \qquad (11.20)$$

where

$$a_n = \frac{2}{\alpha n\pi} \int_0^L g(x) \sin \frac{n\pi x}{L} dx \qquad (11.21)$$

and

$$b_n = \frac{2}{L} \int_0^L f(x) \sin \frac{n\pi x}{L} dx. \qquad (11.22)$$

In physical terms this answer makes sense for the following reasons.

1. If the tension τ is increased, α increases, which increases the frequency of oscillation.
2. If the mass per unit length ρ is increased, α decreases, which decreases the frequency of oscillation.
3. If the length L is increased, the frequency decreases.

Example 11.2. Solve

$$\frac{\partial^2 u}{\partial t^2} = \alpha^2 \frac{\partial^2 u}{\partial x^2},$$

where $L = 3$ and $\alpha = 2$ subjected to the boundary conditions $u(0,t) = 0$, $u(L,t) = 0$, and initial conditions

$$u(x,0) = \begin{cases} x, & x < 1, \\ \frac{3-x}{2}, & 1 \le x \le 3, \end{cases}$$

$$\frac{\partial u}{\partial t}(x,0) = 0.$$

This represents a string plucked $1/3$ of the way along its length with zero initial velocity.

All the computations for this problem have already been carried out. Substituting for b_n from Example 11.1 and $a_n = 0$ (because the initial velocity is zero) into Equation (11.13) gives

$$u(x,t) = \sum_{n=1}^{\infty} b_n \sin \frac{n\pi x}{3} \cos \frac{2n\pi t}{3} = \sum_{n=1}^{\infty} \left(\frac{9}{n^2 \pi^2} \sin \frac{n\pi}{3} \right) \sin \frac{n\pi x}{3} \cos \frac{2n\pi t}{3}.$$

A plot of the motion of the string for various t values including the first 20 terms in the Fourier series is illustrated in Figure 11.6. A plot of the magnitude of the coefficient b_n versus frequency $\alpha n\pi/L$ is illustrated in Figure 11.7. This is called the *spectrum* of the response and is an illustration of the contribution of each mode to the overall response of the system. Typically the lowest frequency is called the *fundamental frequency* and the higher modes are called *harmonics*. Note that not only do the higher modes have a different shape, as was illustrated in Figure 11.4, the higher modes have higher frequencies as well.

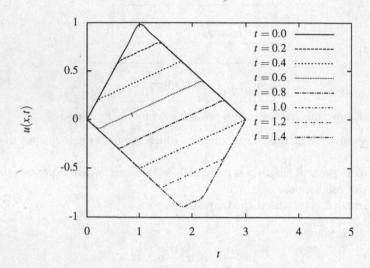

Fig. 11.6 Response of a plucked string from Example 11.2.

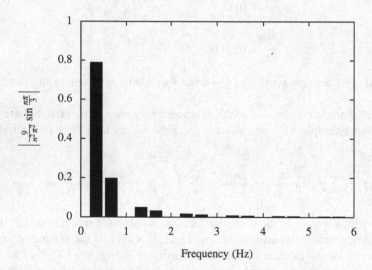

Fig. 11.7 Spectrum for plucked string in Example 11.2.

Example 11.3. Consider the same string as in Example 11.2 but instead of having the string plucked, like a guitar or banjo, consider it being impacted by a small hammer over a small segment of its length, like a piano. Thus, solve

$$\frac{\partial^2 u}{\partial t^2} = \alpha^2 \frac{\partial^2 u}{\partial x^2}$$

where $L = 3$ and $\alpha = 2$ subjected to the boundary conditions $u(0,t) = 0$, $u(L,t) = 0$ and initial conditions

$$u(x,0) = 0$$

$$\frac{\partial u}{\partial t}(x,0) = \begin{cases} 0, & 0 < x \le \frac{3}{4}, \\ 1, & \frac{3}{4} < x \le 1, \\ 0, & 1 < x \le 3. \end{cases}$$

Substituting into Equation (11.21) gives

$$a_n = \frac{1}{n\pi} \int_0^3 g(x) \sin\frac{n\pi x}{3} dx = \frac{1}{n\pi} \int_{\frac{3}{4}}^1 \sin\frac{n\pi x}{3} dx = -\left. \frac{3}{n^2\pi^2} \cos\frac{n\pi x}{3} \right|_{\frac{3}{4}}^1$$

$$= \frac{3}{n^2\pi^2} \left(\cos\frac{n\pi}{4} - \cos\frac{n\pi}{3} \right)$$

and Equation (11.22) gives that $b_n = 0$. Hence, from Equation (11.20)

$$u(x,t) = \sum_{n=1}^{\infty} \frac{3}{n^2\pi^2} \left(\cos\frac{n\pi}{4} - \cos\frac{n\pi}{3} \right) \sin\frac{n\pi x}{3} \sin\frac{2n\pi t}{3}$$

A plot of the spectrum, that is, the magnitude of the coefficient b_n versus frequency $\alpha n\pi/L$, is illustrated in Figure 11.8. Note that the relative contributions of the harmonics in this example are different from the plucked example (Example 11.2). This explains why a plucked and struck string do not sound the same, even if they are the same note.

11.2 Fourier Series

Motivated by our apparent ability to use an infinite series of sine and cosine functions to match any initial conditions for the wave equation defined on the length of the string (Equations 911.17) and (11.19)), we now consider the general problem of representing an arbitrary periodic function as a trigonometric series.

Motivated by the form of the solution to the wave equation, consider the series

$$f(x) = \sum_{n=0}^{\infty} \left[a_n \sin\frac{n\pi x}{L} + b_n \cos\frac{n\pi x}{L} \right]. \tag{11.23}$$

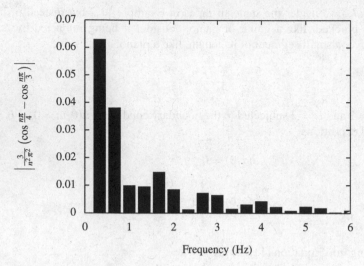

Fig. 11.8 Spectrum for hammered string in Example 11.3.

The question to consider is under what conditions will we be able to compute the infinite number of coefficients a_n and b_n so that this series converges to a specified function? There are a variety of reasons to pursue this, not the least of which are the following.

- We may be forced to represent a function in this manner, as was the case for satisfying the initial conditions for the wave equation.
- Even though it is an infinite series, sine and cosine functions are generally pretty easy to deal with, so, in the right context, it may be worth the effort to represent some given function as a trigonometric series of this nature because it may be more expedient elsewhere in a problem.

An example of the second case is considered in the exercises.

11.2.1 Periodic Functions

As an initial observation, it is worth noting that because of the periodic nature of the trigonometric functions, it will probably not be possible to represent any function by a series of the form of Equation (11.23). In particular, observe that

$$f(x+2L) = \sum_{n=0}^{\infty} \left[a_n \sin \frac{n\pi\,(x+2L)}{L} + b_n \cos \frac{n\pi\,(x+2L)}{L} \right]$$

$$= \sum_{n=0}^{\infty} \left[a_n \sin \left(\frac{n\pi x}{L} + 2n\pi \right) + b_n \cos \left(\frac{n\pi x}{L} + 2n\pi \right) \right]$$

$$= \sum_{n=0}^{\infty} \left(a_n \sin \frac{n\pi x}{L} + b_n \cos \frac{n\pi x}{L} \right)$$

$$= f(x).$$

Of course, mathematically what this represents is that the series repeats itself over every interval of $2L$. Observe that similarly

$$f(x) = f(x+2L) = f(x+4L) = f(x+6L) = \cdots = f(x+2mL)$$

where m is a natural number (positive integer). Motivated by this we define a periodic function as follows.

Definition 11.1. A function $f(x)$ is *periodic with period T*, if T is the smallest number such that $f(x) = f(x+T)$.

Having defined a periodic function and observed that the series we are considering is periodic with period $2L$, it is obvious to conclude that the class of functions for which the series converges *must be periodic*. In the case of the wave equation and other partial differential equations considered subsequently, the initial shape of the string was not periodic; however, we were only interested in its shape over the length of the string. If we had plotted the Fourier series for the initial condition outside the domain of $x = 0$ to $x = L$ we would have observed that, in fact, the function was periodic, but we were only interested in it over the length of one half of its period.

If we wish to consider the properties of the series for general periodic functions, because the length L was one half of the period, we could substitute $L = T/2$ in the sine and cosine functions in the series to put it in the form

$$f(x) = \sum_{n=0}^{\infty} \left[a_n \sin \frac{2n\pi x}{T} + b_n \cos \frac{2n\pi x}{T} \right], \tag{11.24}$$

that is, in terms of the period T rather than the length L.

11.2.2 Inner Products

Not surprisingly, the "trick" (Equation (11.16)) that allowed us to compute the infinite number of coefficients in the Fourier series is used in a similar manner here. However, instead of simply considering it to be a trick whose only redeeming feature is one of mathematical manipulation, we investigate things a bit further to see that, in fact, it is nothing more than using the usual dot product to project one vector onto another. In the rest of this section we consider the generic properties of the dot product and its geometric interpretation which include the important concept of orthogonality.

Recall from vector algebra that the *dot product* between two vectors is defined as

$$\mathbf{x} \cdot \mathbf{y} = \begin{bmatrix} x_1 \\ x_2 \\ x_3 \\ \vdots \\ x_n \end{bmatrix} \cdot \begin{bmatrix} y_1 \\ y_2 \\ y_3 \\ \vdots \\ y_n \end{bmatrix} = x_1 y_1 + x_2 y_2 + \cdots + x_n y_n = \sum_{i=1}^{n} x_i y_i.$$

So, in words, the dot product is simply the sum of the product of all of the corresponding components of the vectors $\mathbf{x}$ and $\mathbf{y}$.

To generalize this idea to functions, first note that, loosely speaking, one may think of a function as a vector by "sampling" its values at various points (perhaps an infinite number of points) along its domain; that is,

$$f(x) = \begin{bmatrix} f(x_0) \\ f(x_1) \\ f(x_2) \\ \vdots \end{bmatrix}.$$

Now, considering the dot product between two functions $f(x)$ and $g(x)$ over an interval of $-L < x < L$ and taking the values of each at many points we may write

$$\begin{bmatrix} f(-L) \\ f(-L+dx) \\ f(-L+2dx) \\ \vdots \\ f(L) \end{bmatrix} \cdot \begin{bmatrix} g(-L) \\ g(-L+dx) \\ g(-L+2dx) \\ \vdots \\ g(L) \end{bmatrix} = \sum_{n=0}^{2L/dx} f(-L+ndx)g(-L+ndx).$$

Now clearly our goal is going to be to take the limit as $dx \to 0$; however in this limit this sum will typically not converge for nonzero $f(x)$ and $g(x)$ because it will be the infinite sum of finite values. However, if we modify it slightly by multiplying the product of f and g by dx, and taking the limit as $dx \to 0$ we have

$$\lim_{dx \to 0} \sum_{n=0}^{2L/dx} f(-L+ndx)g(-L+ndx)dx = \int_{-L}^{L} f(x)g(x)dx.$$

Motivated by this we define the inner product between two periodic functions with period $2L$.

Definition 11.2. Let $f(x)$ and $g(x)$ be periodic functions with period $2L$. The *inner product of f and g*, denoted by $\langle f, g \rangle$ is

$$\langle f,g\rangle = \int_{-L}^{L} f(x)g(x)\mathrm{d}x.$$

With this definition, it is clear that all the usual properties of the dot product generalize to this inner product.

1. $\langle f_1 + f_2, g\rangle = \langle f_1, g\rangle + \langle f_2, g\rangle$.
2. $\langle \alpha f, g\rangle = \alpha \langle f, g\rangle$.
3. $\langle f, g\rangle = \langle g, f\rangle$ (for *real* f and g).
4. $\langle f, f\rangle \neq 0$ unless $f = 0$.

Also observe that the integral may be evaluated over any interval of length $2L$ because the functions are periodic with period $2L$. In addition to the usual properties of a dot product holding for the generalization of the inner product to functions, the main intuitive idea also holds: *the inner product gives a measure of the degree of "alignment" of the functions.*

Example 11.4. Consider the three functions

$$f_1(x) = \sin x$$
$$f_2(x) = \sin 2x$$
$$f_3(x) = \begin{cases} x, & 0 \leq x \leq \frac{\pi}{2}, \\ \pi - x, & \frac{\pi}{2} < x \leq \frac{3\pi}{2}, \\ x - 2\pi & \frac{3\pi}{2} < x \leq 2\pi. \end{cases}$$

These three functions are plotted in Figure 11.9. Observe that $f_1(x)$ and $f_3(s)$ are well aligned over the interval, whereas $f_2(x)$ is not aligned with $f_1(x)$ or $f_3(x)$. In fact, careful inspection of Figure 11.9 makes it clear that for every value for x where $f_2(x)$ and the other two functions have the same sign, there is a point where they have the same magnitudes, but opposite signs. Thus, if the interpretation of the inner product is that it is a measure of alignment of the functions, we would expect that $\langle f_1(x), f_3(x)\rangle$ would be positive and that both $\langle f_2(x), f_1(x)\rangle$ and $\langle f_2(x), f_3(x)\rangle$ would be zero.

Computing the three inner products on the interval $[0, 2\pi]$ gives

$$\langle f_1(x), f_2(x)\rangle = \int_0^{2\pi} (\sin x)(\sin 2x)\,\mathrm{d}x = 0$$

by Proposition C.1. Computing

$$\langle f_1(x), f_3(x)\rangle = \int_0^{2\pi} (\sin x) f_3(x)\mathrm{d}x$$
$$= \int_0^{\frac{\pi}{2}} x\sin x\mathrm{d}x + \int_{\pi/2}^{3\pi/2} (\pi - x)(\sin x)\,\mathrm{d}x$$
$$+ \int_{3\pi/2}^{2\pi} (x - 2\pi)(\sin x)\,\mathrm{d}x = 4,$$

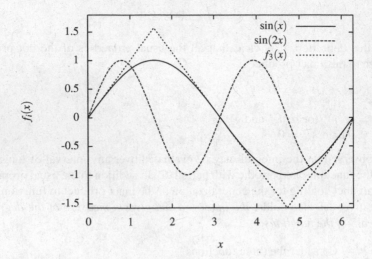

Fig. 11.9 Three functions from Example 11.4.

which makes sense that it is nonzero inasmuch as, by Figure 11.9, the functions are somewhat aligned. Also,

$$\langle f_2(x), f_3(x) \rangle = \int_0^{2\pi} f_3(x) \sin 2x \, dx$$

$$= \int_0^{\frac{\pi}{2}} x \sin 2x \, dx + \int_{\pi/2}^{3\pi/2} (\pi - x)(\sin 2x) \, dx$$

$$+ \int_{3\pi/2}^{2\pi} (x - 2\pi)(\sin 2x) \, dx = 0.$$

11.2.3 Orthogonality

Recall from vector algebra that two vectors are orthogonal if their dot product is zero. In Euclidean space this corresponds to the angle between the vectors being 90°. Given two vectors with varying orientation, the magnitude of the dot product will be maximum when they are perfectly aligned (colinear) and zero when they are perfectly "unaligned," that is, orthogonal. Using a similar notion, we define two functions to be orthogonal when their inner product is zero; that is, the functions f and g are orthogonal if $\langle f, g \rangle = 0$.

For the present case, the most important class of functions that are orthogonal are trigonometric and have already been used. In particular

$$\langle \sin \frac{n\pi x}{L}, \sin \frac{m\pi x}{L} \rangle = \int_{-L}^{L} \sin \frac{n\pi x}{L} \sin \frac{m\pi x}{L} dx = \begin{cases} 0, & m \neq n, \\ L, & m = n, \end{cases}$$

$$\langle \cos \frac{n\pi x}{L}, \cos \frac{m\pi x}{L} \rangle = \int_{-L}^{L} \cos \frac{n\pi x}{L} \cos \frac{m\pi x}{L} dx = \begin{cases} 0, & m \neq n, \\ L, & m = n, \end{cases}$$

$$\langle \sin \frac{n\pi x}{L}, \cos \frac{m\pi x}{L} \rangle = \int_{-L}^{L} \sin \frac{n\pi x}{L} \cos \frac{m\pi x}{L} dx = 0 \qquad \forall m, n.$$

Now we consider a fundamental role of the inner product, which is in the projection of one vector (or function) onto another. Consider the vector $\mathbf{z}$ illustrated in Figure 11.10. Its projection onto the other two vectors $\mathbf{x}$ and $\mathbf{y}$ are given by the dot product, that is, $\mathbf{x} \cdot \mathbf{y}$ and $\mathbf{x} \cdot \mathbf{z}$, or using the inner product notation, $\langle \mathbf{x}, \mathbf{z} \rangle$ and $\langle \mathbf{x}, \mathbf{z} \rangle$. The important point to observe is that it is only in the case of an orthogonal basis that the sum of the projections of a vector onto the basis elements is equal to the vector. In the right figure, with orthogonal basis vectors $\mathbf{x}$ and $\mathbf{y}$, if $\mathbf{z}$ is projected onto each one, then $\mathbf{z}$ is equal to the sum of the projections. In the left figure, where the basis elements are not orthogonal, the sum of the projections is not equal to the original vector.

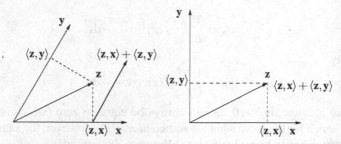

Fig. 11.10 Projection of a vector onto other vectors.

Now, we apply these concepts to periodic functions. The set of functions composed of $\sin(n\pi x/L)$ and $\cos(n\pi x/L)$ for $n = 0, 1, 2, \ldots$ is a basis for the set of periodic functions.[3] By defining the inner product as in Definition 11.2, the basis is orthogonal. That fact allows us to project any function onto the basis elements using the inner product to determine the components of that function in terms of that basis.

---

[3] There are subtleties associated with infinite-dimensional vector spaces that are not at all obvious to a reader with a background in basic linear algebra limited to finite-dimensional vector spaces. All of the computations and manipulations herein are valid, but one must exercise caution when applying all intuition based on finite-dimensional vector spaces to the infinite-dimensional case.

11.2.4 The General Fourier Series

Given a function $f(x)$, with period $T = 2L$, we now have all the tools to be able to express it as a Fourier series of the form

$$f(x) = \sum_{n=0}^{\infty} a_n \sin \frac{n\pi x}{L} + b_n \cos \frac{n\pi x}{L}.$$

In order to find the coefficients, multiply by $\sin(m\pi x/L)$ for the a_n and multiply by $\cos(m\pi x/L)$ for the b_n and integrate from $-L$ to L with respect to x. Due to the orthogonality of the sine and cosine functions, all the terms in the series vanish except for one of them, which allows us to solve for the coefficient. In particular,

$$\int_{-L}^{L} \sin \frac{m\pi x}{L} \left(\sum_{n=0}^{\infty} a_n \sin \frac{n\pi x}{L} + b_n \cos \frac{n\pi x}{L} \right) dx = \int_{-L}^{L} f(x) \sin \frac{m\pi x}{L} dx,$$

which gives

$$La_m = \int_{-L}^{L} f(x) \sin \frac{m\pi x}{L} dx$$

or

$$a_m = \frac{1}{L} \int_{-L}^{L} f(x) \sin \frac{m\pi x}{L} dx.$$

Similarly,

$$b_m = \frac{1}{L} \int_{-L}^{L} f(x) \cos \frac{m\pi x}{L} dx.$$

Note that because $\sin 0 = 0$, a_0 will always be equal to zero (which is why all the Fourier series for the wave equation started at $n = 1$). However, the same is not true for b_0, which has to be evaluated for each series. In particular,

$$\int_{-L}^{L} \cos \frac{0\pi x}{L} \left(\sum_{n=0}^{\infty} a_n \sin \frac{n\pi x}{L} + b_n \cos \frac{n\pi x}{L} \right) dx = \int_{-L}^{L} b_0 dx = 2Lb_0 = \int_{-L}^{L} f(x) dx,$$

which gives

$$b_0 = \frac{1}{2L} \int_{-L}^{L} f(x) dx.$$

Note that this expression has a factor of one-half that is not present in the equation for b_n for $n > 0$. Hence, it is conventional to write

$$\boxed{f(x) = \frac{b_0}{2} + \sum_{n=1}^{\infty} a_n \sin \frac{n\pi x}{L} + b_n \cos \frac{n\pi x}{L},} \tag{11.25}$$

where

$$a_n = \frac{1}{L}\int_{-L}^{L} f(x)\sin\frac{n\pi x}{L}dx, \quad n = 1,2,3,\ldots \qquad (11.26)$$

and

$$b_n = \frac{1}{L}\int_{-L}^{L} f(x)\cos\frac{\cdot n\pi x}{L}dx, \quad n = 0,1,2,3,\ldots \qquad (11.27)$$

which allows us to use the same formula for b_0 as the rest of the cosine coefficients.

Remark 11.1. Because all the functions involved are periodic, the integrals in Equations (11.26) and (11.27) may have any limits as long as the interval over which the integrals are evaluated has a length of $T = 2L$.

11.2.5 Examples of Fourier Series

Now we consider a few examples.

Example 11.5. Determine the Fourier series representation for the square wave function, given by

$$f(x) = \begin{cases} 1, & 0 < x \le 1, \\ -1, & 1 < x \le 2, \end{cases}$$

for $x \in (0,2]$ and with $f(x+2) = f(x)$ which is illustrated in Figure 11.11.

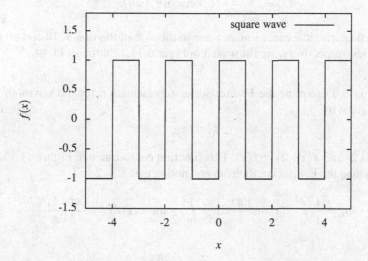

Fig. 11.11 Square wave function for Example 11.5.

Computing the Fourier coefficients,

$$a_n = \int_0^2 f(x)\sin\frac{2n\pi x}{2}dx$$

$$= \int_0^1 (1)\sin(n\pi x)\,dx + \int_1^2 (-1)\sin(n\pi x)\,dx$$

$$= \frac{1}{n\pi}\left[-\cos(n\pi x)|_0^1 - -\cos(n\pi x)|_1^2\right]$$

$$= \frac{1}{n\pi}\left[-\cos(n\pi) + 1 + \cos(2n\pi) - \cos(n\pi)\right]$$

$$= \frac{2}{n\pi}\left[1 - \cos(n\pi)\right].$$

and

$$b_n = \int_0^2 f(x)\cos\frac{2n\pi x}{2}dx$$

$$= \int_0^1 (1)\cos(n\pi x)\,dx + \int_1^2 (-1)\cos(n\pi x)\,dx$$

$$= \frac{1}{n\pi}\left[\sin(n\pi x)|_0^1 - \sin(n\pi x)|_1^2\right]$$

$$= 0.$$

Also,

$$b_0 = \int_0^2 f(x)dx = \int_0^1 1dx - \int_1^2 dx = x|_0^1 - x|_1^2 = 0.$$

Hence,

$$f(x) = \sum_{n=1}^{\infty} \frac{2}{n\pi}\left(1 - \cos(n\pi)\right)\sin(n\pi x).$$

Plots comparing the exact square wave to the sum of the first 5, 10 and 50 terms in the series, respectively, are illustrated in Figures 11.12 through 11.14.

Example 11.6. Determine the Fourier series representation for the sawtooth wave function given by

$$f(x) = \frac{x}{2}$$

for $x \in (0,2]$ and $f(x+2) = f(x)$. This function is illustrated in Figure 11.15.

Computing the Fourier coefficients and noting that $T = 2$ so $L = 1$

$$a_n = \frac{1}{1}\int_0^2 f(x)\sin\frac{n\pi x}{1}dx = \int_0^2 \frac{x}{2}\sin(n\pi x)\,dx = -\frac{\cos 2n\pi}{n\pi}$$

and

$$b_n = \frac{1}{1}\int_0^2 f(x)\cos\frac{n\pi x}{1}dx = 0$$

for $n \neq 0$. For $n = 0$,

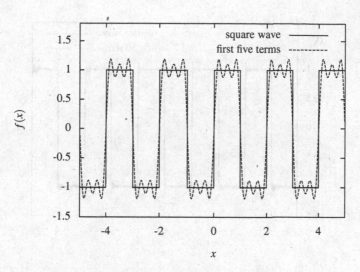

Fig. 11.12 The first 5 terms in the Fourier series for the square wave in Example 11.5.

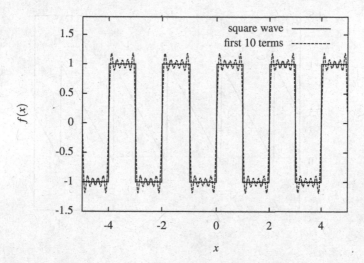

Fig. 11.13 The first 10 terms in the Fourier series for the square wave in Example 11.5.

$$b_0 = \frac{1}{1} \int_0^2 \frac{x}{2} dx = 1.$$

Hence

$$f(x) = \frac{1}{2} + \sum_{n=1}^{\infty} -\frac{\cos(2n\pi)}{n\pi} \sin(n\pi x).$$

A plot of the first 5 and 10 terms of the series is illustrated in Figure 11.16.

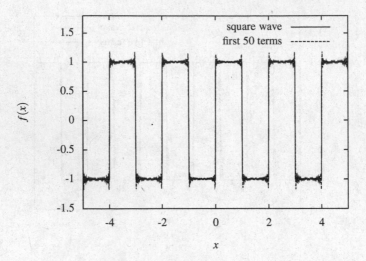

Fig. 11.14 The first 50 terms in the Fourier series for the square wave in Example 11.5.

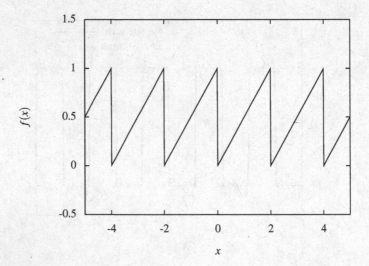

Fig. 11.15 Sawtooth wave for Example 11.6.

Example 11.7. Compute the Fourier series for the function

$$f(x) = \begin{cases} x, & 0 < x \leq 1, \\ 1, & 1 < x \leq 2, \end{cases}$$

where $f(x+2) = f(x)$.

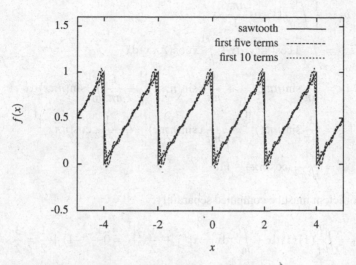

Fig. 11.16 First 5 and 10 terms in the Fourier series for the sawtooth function in Example 11.6.

The function is periodic with period $T = 2$; hence, $L = 1$. The coefficients are given by

$$
\begin{aligned}
a_n &= \frac{1}{1} \int_{-1}^{1} f(x) \sin \frac{n\pi x}{1} dx \\
&= \int_{-1}^{0} 1 \sin(n\pi x) dx + \int_{0}^{1} x \sin(n\pi x) dx \\
&= -\frac{1}{n\pi} \cos(n\pi x) \Big|_{-1}^{0} - \left(\frac{1}{n\pi} x \cos(n\pi x) \right) \Big|_{0}^{1} + \frac{1}{n\pi} \int_{0}^{1} \cos(n\pi x) dx \\
&= -\frac{1}{n\pi} \cos(n\pi x) \Big|_{-1}^{0} - \left(\frac{1}{n\pi} x \cos(n\pi x) \right) \Big|_{0}^{1} + \frac{1}{n^2 \pi^2} \sin(n\pi x) \Big|_{0}^{1} \\
&= -\frac{1}{n\pi} (1 - \cos(-n\pi)) - \frac{1}{n\pi} (\cos(n\pi) - 0) + \frac{1}{n^2 \pi^2} (0 - 0) \\
&= -\frac{1}{n\pi}
\end{aligned}
$$

and

$$b_n = \frac{1}{1} \int_{-1}^{1} f(x) \cos \frac{n\pi x}{1} dx$$

$$= \int_{-1}^{0} 1 \cos(n\pi x) dx + \int_{0}^{1} x \cos(n\pi x) dx$$

$$= \frac{1}{n\pi} \sin(n\pi x) \Big|_{-1}^{0} + \frac{1}{n\pi} x \sin(n\pi x) \Big|_{0}^{1} - \frac{1}{n\pi} \int_{0}^{1} \sin(n\pi x) dx$$

$$= \frac{1}{n\pi} \sin(n\pi x) \Big|_{-1}^{0} + \frac{1}{n\pi} x \sin(n\pi x) \Big|_{0}^{1} + \frac{1}{n^2\pi^2} \cos(n\pi x) \Big|_{0}^{1}$$

$$= \frac{1}{n^2\pi^2} (\cos(n\pi) - 1).$$

The b_0 coefficient must be computed separately,

$$b_0 = \int_{-1}^{0} (1)(1) dx + \int_{0}^{1} x dx = x \Big|_{-1}^{0} + \frac{1}{2} x^2 \Big|_{0}^{1} = 0 - (-1) + \frac{1}{2} = \frac{3}{2}.$$

Hence,

$$f(x) = \frac{3}{4} + \sum_{n=1}^{\infty} \left(-\frac{1}{n\pi} \sin(n\pi x) + \frac{\cos(n\pi) - 1}{n^2\pi^2} \cos(n\pi x) \right).$$

A plot of $f(x)$ as well as partial sums of the series including the first 10 and 20 terms of the series is illustrated in Figure 11.17.

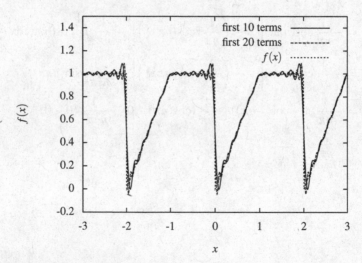

Fig. 11.17 Fourier series for Example 11.7.

11.2.6 Summary Remarks and Theory

Fourier series have been developed in the context of using a series of sine and cosine functions to represent a periodic function, however, recall in Section 11.1 that it originally arose in the context of representing a function in a finite interval, which in the case of the string, was along its length. A reader may initially be confused about the fact that for satisfying the initial conditions for the wave equation, the integrals were on the interval from 0 to L and for Fourier series in this section they were from $-L$ to L. Nothing is fundamentally different between the two problems. If we were to plot the solution to the wave equation outside the interval from 0 to L, it would be apparent that the solution over the larger domain is, in fact, periodic along the spatial dimension x.

We end this section with a theorem that provides some conditions on a function $f(x)$ to guarantee that the Fourier series will converge to $f(x)$. It is from [9] and the reader is referred to that text for a proof. The theorem states that the Fourier series converges to $f(x)$ where it is continuous and, where there is a discontinuity, it converges to the average of the values before and after the jumps.

Theorem 11.1. *Let f and f' be piecewise continuous on the interval $-L \le x \le L$. The series given by Equation (11.25) with coefficients give by Equations (11.26) and (11.27) converges to $f(x)$ if f is continuous at x and converges to $(f(x^+) - f(x^-))/2$, the average of the values of the limits of f as x is approached from above and below, respectively, if f is discontinuous at x.*

The observant reader may question the convergence result in this theorem. Figures 11.11 and 11.17 show clear "overshoots" in the Fourier series at points of discontinuity of the function $f(x)$. These cannot be eliminated and their existence is called *Gibb's phenomenon*. The nature of the convergence of the series to $f(x)$ is one of pointwise versus uniform convergence and is beyond the scope of this text. An interested reader is referred to [38] for a more complete exposition. In the case of the Gibbs phenomenon, pointwise convergence may be considered by observing that for any point x, except for the value of the actual discontinuity, if enough terms are included in the series, then the series converges to $f(x)$. If x is moved closer to the discontinuity, then the overshoot similarly shifts closer to the discontinuity as well. In contrast, if the Fourier series converged uniformly, which it does not at a discontinuity, then there would be no Gibbs phenomenon.

11.3 The One-Dimensional Heat Equation

Heat conduction is also a common phenomenon described by a partial differential equation. Consider the temperature distribution along a slender beam that is insulated everywhere except for the two ends, which have specified temperatures, as is illustrated on the left in Figure 11.18. Because the sides are insulated, the only mechanism by which the temperature is changed at any point along the beam is by heat *conduction* through the beam itself.

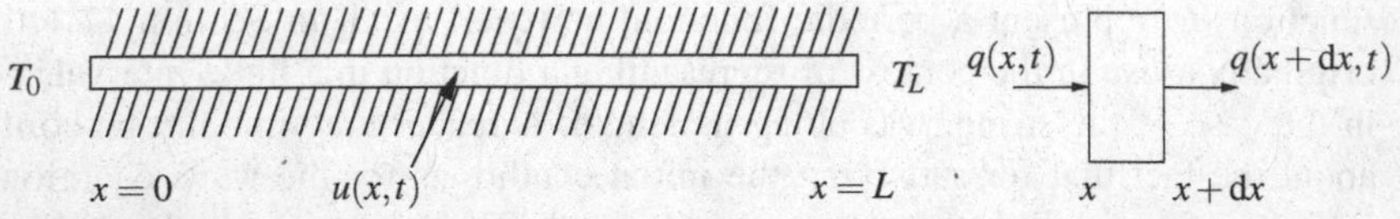

Fig. 11.18 One-dimensional heat conduction along the slender beam (left) and a small segment of the beam (right).

The physical principle upon which heat conduction is based is *Fourier's law*, which states that the rate of thermal energy transfer in the form of heat in the x-direction is proportional to the temperature gradient at that point; that is,

$$q(x,t) = -kA\frac{\partial u}{\partial x}(x,t), \tag{11.28}$$

where the heat rate q has units of W.[4] The proportionality constant k is called the *thermal conductivity* which has units $W/(m \cdot K)$ and is a property of the material. This law makes perfect sense in that energy in the form of heat is conducted from hot to cold sections of the material.

Considering a small segment of the beam with length dx, as is illustrated on the right in Figure 11.18, the rate at which energy will accumulate in a small segment of length dx is the difference between the heat rate in and the heat rate out, which, for small dx is given by

$$q(x,t) - q(x+dx,t) = q(x,t) - \left(q(x,t) + \frac{\partial q}{\partial x}(x,t)dx + \cdots\right) \approx -\frac{\partial q}{\partial x}(x,t)dx. \tag{11.29}$$

A basic property of any material is that the rate at which the temperature changes at a point is related to the rate at which thermal energy is added at that point. Using the Equation (11.29), this is expressed as

$$-\frac{\partial q}{\partial x}dx = \rho c_p A\frac{\partial u}{\partial t}dx.$$

Inasmuch as the heat rate is proportional to the temperature gradient, using Equation (11.28), we have

$$kA\frac{\partial^2 u}{\partial x^2}dx = \rho c_p\frac{\partial u}{\partial t}Adx$$

which can be expressed as

---

[4] Typically, Fourier's law is formulated in terms of the *heat flux*, which is the rate of energy transfer per unit area. In this development, because the temperature is assumed to be uniform on a cross-section, the heat rate is a more convenient representation.

$$\alpha^2 \frac{\partial^2 u}{\partial x^2}(x,t) = \frac{\partial u}{\partial t}(x,t). \tag{11.30}$$

The coefficient $\alpha = k/\rho c_p$ is called the *thermal diffusivity*. Note that if α increases, then the rate at which the temperature changes increases, and vice-versa. Elaborating on this by relating it to the material properties is left as an exercise.

Fourier's law is apparent in an interpretation of the heat equation that the temperature at a given point x will change if the curvature of the temperature profile is nonzero. Figure 11.19 contains two temperature profiles. In the one with no curvature, the rate of heat conduction along the entire bar will be constant. In the case where the temperature profile has a nonzero, there will be a higher rate of heat conduction where the gradient is steep and a lower rate where it is less steep. Thus, for the curved temperature profile, there will be more heat conduction from the right to the center than there will be from the center to the left boundary, the consequence of which will be that the temperature in the center of the bar will increase. In fact, as is shown subsequently, the steady-state solution is the solution with no curvature, that is, a straight line. Hence, the curved solution will approach the straight solution as $t \to \infty$.

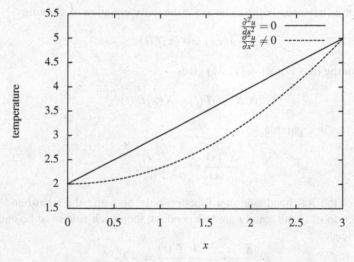

Fig. 11.19 Heat conduction with zero and nonzero temperature curvature.

11.3.1 *Solution to the Heat Conduction Equation with Homogeneous Boundary Conditions*

The usual approach to solve the heat equation is to solve it with *homogeneous boundary conditions* and then solve it with nonhomogeneous boundary conditions. Homogeneous boundary conditions are where boundary conditions are

$$u(0,t) = u(L,t) = 0;$$

that is, the temperature at both ends is zero. This is not a very realistic situation, but we use the solution for the homogeneous boundary conditions as part of the solution to the more realistic nonhomogeneous case.

The complete problem statement includes the differential equation and the boundary conditions as well as an initial temperature profile:

$$\alpha^2 \frac{\partial^2 u}{\partial x^2} = \frac{\partial u}{\partial t} \tag{11.31}$$

and

$$u(0,t) = 0, \quad u(L,t) = 0, \quad u(x,0) = f(x).$$

The approach is exactly the same as for the wave equation. Assuming

$$u(x,t) = X(x)T(t)$$

and substituting into Equation (11.31) gives

$$\alpha^2 X''(x)T(t) = X(x)T'(t).$$

As before, this is separable, so

$$\frac{X''(x)}{X(x)} = \frac{1}{\alpha^2} \frac{T'(t)}{T(t)},$$

and because the left-hand side is a function only of x and the right-hand side is only a function of t, and x and t are independent, then each side must be equal to a constant. Hence,

$$\frac{X''(x)}{X(x)} = \frac{1}{\alpha^2} \frac{T'(t)}{T(t)} = -\lambda$$

or

$$X''(x) + \lambda X(x) = 0, \quad T'(t) + \alpha^2 \lambda T(t) = 0.$$

This is similar to the wave equation except that the equation for $T(t)$ is a first-order equation instead of second-order. This should make sense inasmuch as we would not expect that the temperature profile in a bar would exhibit solutions that are oscillatory, which may be the case for a second-order equation.

We proceed as before by applying the boundary conditions to determine λ, which gives an infinite number of solutions for $X(x)$. We may then use the infinite number of solutions to satisfy the initial temperature profile by using a Fourier series. In fact, the homogeneous boundary conditions give rise to exactly the same case as for the wave equation. In particular, the general solution for $X(x)$ is

$$X(x) = c_1 \sin \sqrt{\lambda}x + c_2 \cos \sqrt{\lambda}x$$

and the boundary conditions require

$$u(0,t) = 0 \implies X(0) = 0 \implies c_2 = 0,$$

and

$$u(L,t) = 0 \implies X(L) = 0 \implies \lambda = \frac{n^2\pi^2}{L^2}, \quad n = 1,2,3,\ldots.$$

So, we have

$$X_n(x) = c_n \sin \frac{n\pi x}{L}.$$

Because the general solution for $T(t)$ is then

$$T(t) = e^{-\alpha^2 \lambda t} = e^{-\alpha^2 n^2 \pi^2 t / L^2}$$

the general solution to Equation (11.31) is

$$u(x,t) = \sum_{n=1}^{\infty} c_n \sin \frac{n\pi x}{L} e^{-(\alpha n\pi/L)^2 t}. \tag{11.32}$$

Given some initial temperature profile, at $t = 0$ the exponential term is one and the initial profile may be satisfied by a Fourier series; that is,

$$u(x,0) = \sum_{n=1}^{\infty} c_n \sin \frac{n\pi x}{L} e^{-(\alpha^2 n\pi/L)^2 0} = \sum_{n=1}^{\infty} c_n \sin \frac{n\pi x}{L} = f(x).$$

The coefficients are determined by exploiting orthogonality as before. In particular, multiplying by $\sin(m\pi x/L)$ and integrating from 0 to L with respect to x gives

$$\int_0^L \sin \frac{m\pi x}{L} \sum_{n=1}^{\infty} c_n \sin \frac{n\pi x}{L} dx = \int_0^L \sin \frac{m\pi x}{L} f(x) dx.$$

The sine functions are orthogonal except for the case where $n = m$, therefore the infinite series reduces to one term, which gives

$$c_n \int_0^L \sin \frac{n\pi x}{L} \sin \frac{n\pi x}{L} dx = \int_0^L \sin \frac{n\pi x}{L} f(x) dx.$$

Evaluating the integral on the left-hand side gives what is exactly the same answer as before for the wave equation; namely,

$$c_n = \frac{2}{L} \int_0^L \sin\frac{n\pi x}{L} f(x) \mathrm{d}x. \tag{11.33}$$

Hence, for homogeneous boundary conditions

$$u(0,t) = u(L,t) = 0$$

and initial condition

$$u(x,0) = f(x)$$

we have the general solution

$$u(x,t) = \sum_{n=1}^{\infty} c_n \sin\left(\frac{n\pi x}{L}\right) \exp\left(-\left(\frac{\alpha n\pi}{L}\right)^2 t\right), \tag{11.34}$$

where the coefficients c_n are given by Equation (11.33).

Example 11.8. Determine the solution to

$$4\frac{\partial^2 u}{\partial x^2} = \frac{\partial u}{\partial t}$$

with

$$u(0,t) = u(10,t) = 0$$

and

$$u(x,0) = \begin{cases} x, & 0 < x \le 5, \\ 10 - x, & 5 < x \le 10. \end{cases}$$

This is the case where $\alpha = 2$ and $L = 10$ and $u(x,0)$ is as illustrated in Figure 11.20, and the solution is simply given by substituting into Equation (11.34).

The only work is to determine the coefficients in the Fourier series to satisfy the initial condition,

$$\begin{aligned} c_n &= \frac{2}{L} \int_0^L \sin\frac{n\pi x}{L} f(x) \mathrm{d}x \\ &= \frac{2}{10}\left[\int_0^5 x\sin\frac{n\pi x}{10}\mathrm{d}x + \int_5^{10}(10-x)\sin\frac{n\pi x}{L}\mathrm{d}x\right]. \end{aligned}$$

Using the fact that

$$\int_a^b x\sin cx\mathrm{d}x = -\frac{1}{c}x\cos cx\Big|_a^b + \frac{1}{c^2}\sin cx\Big|_a^b,$$

which can easily be verified by integrating by parts, the coefficients are

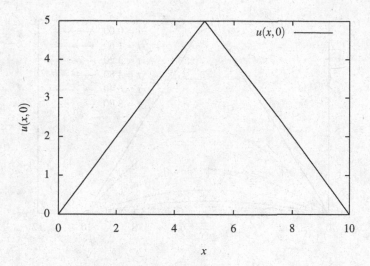

Fig. 11.20 Initial temperature profile for Example 11.8.

$$c_n = \frac{1}{5}\left[-\frac{10}{n\pi}x\cos\frac{n\pi x}{10}\Big|_0^5 + \left(\frac{10}{n\pi}\right)^2 \sin\frac{n\pi x}{10}\Big|_0^5 \right.$$

$$\left. -10\frac{10}{n\pi}\cos\frac{n\pi x}{10}\Big|_5^{10} \frac{10}{n\pi}x\cos\frac{n\pi x}{10}\Big|_5^{10} - \left(\frac{10}{n\pi}\right)^2 \sin\frac{n\pi x}{10}\Big|_5^{10}\right]$$

$$= \frac{2}{n\pi}\left[-5\cos\frac{n\pi}{2} - 0 + \frac{10}{n\pi}\left(\sin\frac{n\pi}{2} - 0\right) \right.$$

$$\left. -10\left(\cos n\pi - \cos\frac{n\pi}{2}\right) + 10\cos n\pi - 5\cos\frac{n\pi}{2} - \frac{10}{n\pi}\left(0 - \sin\frac{n\pi}{2}\right)\right]$$

$$= \frac{40}{n^2\pi^2}\sin\frac{n\pi}{2}.$$

Hence, by Equation (11.34), the solution is given by

$$u(x,t) = \sum_{n=1}^{\infty}\left(\frac{40}{n^2\pi^2}\sin\frac{n\pi}{2}\right)\sin\left(\frac{n\pi x}{10}\right)\exp\left(-\left(\frac{2n\pi}{10}\right)^2 t\right).$$

A plot of the partial sum of the solution through the tenth mode ($n = 10$) for various times is illustrated in Figure 11.21.

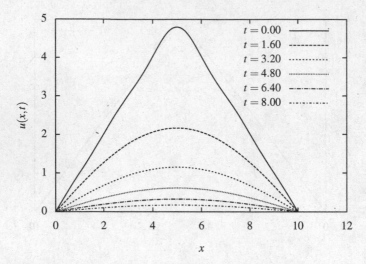

Fig. 11.21 Solution for the heat equation including the first 10 terms in the partial sum of the series solution in Example 11.8 for various times.

11.3.2 Solution to the Heat Equation with Inhomogeneous Boundary Conditions

Of course, the boundary conditions for the heat conduction equation are seldom both zero. Let us now consider the case where

$$u(0,t) = T_L, \quad u(L,t) = T_R.$$

From Section 11.3, the steady-state solution will be

$$\lim_{t \to \infty} u(x,t) = \frac{T_R - T_L}{L} x + T_L = u_{ss}(x) \tag{11.35}$$

which is simply a straight line from $u(0,t) = T_L$ at $x = 0$ to $u(L,t) = T_R$ at $x = L$. Because there is no curvature, this solution satisfies the heat equation inasmuch as it is constant in time. It also satisfies the boundary conditions.

The complement to the steady-state solution is the transient solution which, in fact, is related to the solution determined in Section 11.3.1 given by

$$u_{tr}(x,t) = \sum_{n=1}^{\infty} c_n \sin\left(\frac{n\pi x}{L}\right) e^{-(\alpha n\pi/L)^2 t}$$

and the steady-state solution from Equation (11.35). It makes sense to try to add them to find the complete solution. Along these lines, let

$$u(x,t) = u_{tr}(x,t) + u_{ss}(x)$$
$$= \frac{T_R - T_L}{L}x + T_L + \sum_{n=1}^{\infty} c_n \sin \frac{n\pi x}{L} e^{-(\alpha n\pi/L)^2 t}.$$

To finish this problem we need to do the following.

1. Check that it satisfies the heat equation.
2. Check that it satisfies the boundary conditions.
3. Find equations for the c_n so that it satisfies the initial conditions.

For the coefficients, substituting $t = 0$ into the general solution gives

$$u(x,0) = \sum_{n=1}^{\infty} c_n \sin \frac{n\pi x}{L} e^{-(\alpha n\pi/L)^2 0} + \frac{T_R - T_L}{L}x + T_L$$
$$= \sum_{n=1}^{\infty} c_n \sin \frac{n\pi x}{L} + \frac{T_R - T_L}{L}x + T_L$$
$$= f(x).$$

Thus, at $t = 0$ we may write

$$\sum_{n=1}^{\infty} c_n \sin \frac{n\pi x}{L} = f(x) - \frac{T_R - T_L}{L}x - T_L.$$

If we let

$$\hat{f}(x) = f(x) - \frac{T_R - T_L}{L}x - T_L$$

we may compute the Fourier coefficients in the usual manner:

$$c_n = \frac{2}{L} \int_0^L \hat{f}(x) \sin \frac{n\pi x}{L} dx. \tag{11.36}$$

Example 11.9. Solve the heat equation with $L = \alpha = 1$, $T_L = 0$, $T_R = 1$, and

$$u(x,0) = 0.$$

Substituting into Equation (11.36) gives

$$c_n = 2 \int_0^1 x \sin n\pi x \, dx = -\frac{2}{n\pi} \left(-x\cos n\pi x \right)\big|_0^1 = \frac{2}{n\pi} \cos n\pi.$$

Plots of the solution for various times are illustrated in Figure 11.22.

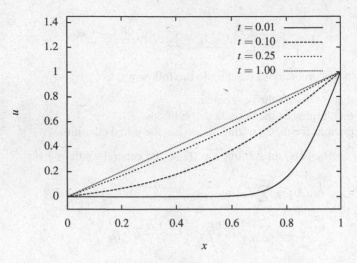

Fig. 11.22 Solution from Example 11.9 for various t.

11.3.3 Solution to the Heat Equation with an Insulated End

Because the rate of heat transfer is proportional to the derivative of the temperature, if an end is insulated the boundary condition at that end will be that the *derivative* must be zero. It is left as an exercise to show that the solution to

$$\alpha^2 \frac{\partial^2 u}{\partial x^2} = \frac{\partial u}{\partial t} \tag{11.37}$$

with boundary conditions

$$u(0,t) = 0, \quad \frac{\partial u}{\partial x}(L,t) = 0 \tag{11.38}$$

with

$$u(x,0) = f(x)$$

is given by

$$u(x,t) = \sum_{n=0}^{\infty} c_n \left(\sin \frac{(2n+1)\pi x}{2L} \right) \left(\exp \frac{-\alpha^2 (2n+1)^2 \pi^2 t}{4L^2} \right), \tag{11.39}$$

where

$$c_n = \frac{2}{L} \int_0^L f(x) \sin \frac{(2n+1)\pi x}{2L} dx. \tag{11.40}$$

Example 11.10. The solution to Equation (11.37) with boundary conditions given by Equation (11.38), $\alpha = L = 1$ and

$$u(x,0) = 1$$

is obtained by substituting into the equation for the coefficients. In particular,

$$
\begin{aligned}
c_n &= 2 \int_0^1 \sin \frac{(2n+1)\pi x}{2} \, dx \\
&= \frac{4}{(2n+1)\pi} \left(-\cos \frac{(2n+1)\pi x}{2} \right) \Bigg|_0^1 \\
&= \frac{4}{(2n+1)\pi} \left(1 - \cos \frac{(2n+1)\pi}{2} \right).
\end{aligned}
$$

Hence, the solution is

$$
u(x,t) = \sum_{n=0}^{\infty} \frac{4}{(2n+1)\pi} \left(1 - \cos \frac{(2n+1)\pi}{2} \right) \left(\sin \frac{(2n+1)\pi x}{2} \right) \\
\cdot \left(\exp \frac{-(2n+1)^2 \pi^2 t}{4} \right).
$$

Plots of the solution for various times are illustrated in Figure 11.23.

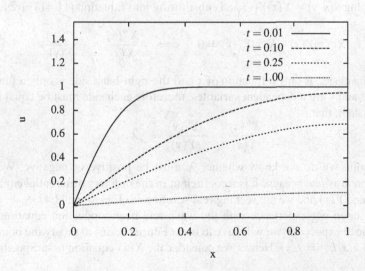

Fig. 11.23 Solution from Example 11.10 for various t.

11.4 Heat Conduction in Two Dimensions

The heat conduction equation has a relatively simple extension to higher dimensions. It is left as an exercise to show that the heat conduction equation in Cartesian coordinates in three dimensions is

$$\frac{\partial^2 u}{\partial x^2}(x,y,z,t) + \frac{\partial^2 u}{\partial y^2}(x,y,z,t) + \frac{\partial^2 u}{\partial z^2}(x,y,z,t) = \frac{\rho c_p}{k}\frac{\partial u}{\partial t}(x,y,z,t).$$

11.4.1 Laplace's Equation: Steady-State Temperature Distribution

In steady state, the temperature is constant, so the term in the heat conduction equation that is the derivative with respect to t is zero. This is called *Laplace's equation*, and in two dimensions is

$$\frac{\partial^2 u}{\partial x^2} + \frac{\partial^2 u}{\partial y^2} = 0. \tag{11.41}$$

Various combinations of boundary conditions are possible and are fully explored in the exercises. In this section we consider

$$u(0,y) = 0, \quad u(L_x,y) = 0, \quad u(x,0) = 0, \quad u(x,L_y) = f(x).$$

The exercises explore the solution method when the nonzero boundary condition is on one of the other three boundaries.

Assuming $u(x,y) = X(x)Y(y)$ and substituting into Equation (11.41) gives

$$X''(x)Y(y) + X(x)Y''(y) = 0 \quad \Longleftrightarrow \quad \frac{X''(x)}{X(x)} = -\frac{Y''(y)}{Y(y)}.$$

The left-hand side is only a function of x and the right-hand side is only a function of y and x and y are independent variables, therefore each side must be equal to the same constant; that is,

$$\frac{X''(x)}{X(x)} = -\frac{Y''(y)}{Y(y)} = -\lambda.$$

At this point we do not know whether λ must be positive or negative. We will assume that it is real because it is a coefficient in the ordinary differential equations for $X(x)$ and $Y(y)$ and we are seeking real solutions to Equation (11.41).

Based upon our experience with the wave and heat conduction equation, it is reasonable to expect that we will have to use a Fourier series to satisfy the boundary condition $u(x,L_y) = f(x)$. Hence, we consider the $X(x)$ equation first. Specifically, we have

$$X'' + \lambda X(x) = 0 \tag{11.42}$$

with

$$X(0) = 0, \quad X(L_x) = 0.$$

Regardless of the value of λ, the general solution to Equation (11.42) is

$$X(x) = c_1 e^{\sqrt{-\lambda}x} + c_2 e^{-\sqrt{-\lambda}x}.$$

The boundary condition at $x = 0$ gives

$$X(0) = c_1 + c_2 = 0 \quad \Longleftrightarrow \quad c_1 = -c_2.$$

Hence,

$$X(x) = c_1 \left(e^{\sqrt{-\lambda}x} - e^{-\sqrt{-\lambda}x} \right).$$

The boundary condition at $x = L_x$ requires that

$$X(a) = c_1 \left(e^{\sqrt{-\lambda}L_x} - e^{-\sqrt{-\lambda}L_x} \right) = 0.$$

If $\lambda < 0$, then $\sqrt{-\lambda}$ is real. Hence, either $c_1 = 0$ or $e^{\sqrt{-\lambda}L_x} = e^{-\sqrt{-\lambda}L_x}$. If $c_1 = 0$, then $u(x,y) = 0$ and the solution cannot satisfy the boundary condition at $y = L_y$ unless it happens that $f(x) = 0$. Furthermore, it is not possible for $e^{\sqrt{-\lambda}L_x} = e^{-\sqrt{-\lambda}L_x}$ if $\lambda < 0$. So, then either $\lambda = 0$ or $\lambda > 0$.

In the case where $\lambda = 0$, Equation (11.42) is of the form

$$X''(x) = 0$$

which has a general solution

$$X(x) = c_1 x + c_2.$$

Using the boundary conditions gives

$$X(0) = 0 \quad \Longrightarrow \quad c_1 = 0$$
$$X(L_x) = 0 \quad \Longrightarrow \quad c_2 = 0.$$

Again, unless $f(x) = 0$, this will not work.

Rapidly running out of options, consider the case where $\lambda > 0$. In that case

$$X(x) = c_1 e^{\sqrt{-\lambda}x} + c_2 e^{-\sqrt{-\lambda}x}$$
$$= c_1 e^{i\sqrt{\lambda}x} + c_2 e^{-i\sqrt{\lambda}x}$$
$$= (c_1 + c_2)\cos \sqrt{\lambda}x + i(c_1 - c_2)\sin \sqrt{\lambda}x$$
$$= \hat{c}_1 \cos \sqrt{\lambda}x + \hat{c}_2 \sin \sqrt{\lambda}x.$$

Applying the boundary condition at $x = 0$ gives

$$X(0) = 0 \quad \Longrightarrow \quad \hat{c}_1 = 0.$$

At $x = L_x$ the boundary condition requires that either $\hat{c}_2 = 0$ or

$$\sin \sqrt{\lambda} L_x = n\pi \quad n = 1, 2, 3, \ldots.$$

As before if $\hat{c}_2 = 0$, then $u(x, y) = 0$ which can not satisfy the boundary condition at $y = L_y$ unless $f(x) = 0$. So, finally we have that

$$\lambda = \left(\frac{n\pi}{L_x}\right)^2, \quad n = 1, 2, 3, \cdots.$$

Now, substituting the value of λ into the equation for $Y(y)$ gives

$$Y''(y) - \left(\frac{n\pi}{L_x}\right)^2 Y(y) = 0,$$

which has a general solution

$$Y(y) = k_1 e^{n\pi y/L_x} + k_2 e^{-n\pi y/L_x}.$$

Applying the boundary condition at $y = 0$ gives

$$Y(0) = k_1 + k_2 = 0 \quad \Longrightarrow \quad k_1 = -k_2.$$

Hence,

$$Y(y) = k_1 \left(e^{n\pi y/L_x} - e^{-n\pi y/L_x}\right).$$

We have an infinite number of general solutions for $X(x)$ and one solution for $Y(y)$. Combining them gives

$$u(x, y) = \sum_{n=1}^{\infty} c_n \sin \frac{n\pi x}{L_x} \left(e^{n\pi y/L_x} - e^{-n\pi y/L_x}\right).$$

To satisfy the boundary condition at $y = L_y$ we need that

$$u(x, L_y) = \sum_{n=1}^{\infty} c_n \sin \frac{n\pi x}{L_x} \left(e^{n\pi L_y/L_x} - e^{-n\pi L_y/L_x}\right) = f(x).$$

At this point, one hopes, it is obvious what must be done: multiply by $\sin(m\pi x/L_x)$ and integrate from 0 to L_x with respect to x. Doing so gives

$$\int_0^{L_x} \left(\sum_{n=1}^{\infty} c_n \sin \frac{n\pi x}{L_x} \left(e^{n\pi L_y/L_x} - e^{-n\pi L_y/L_x}\right)\right) \sin \frac{m\pi x}{L_x} dx = \int_0^{L_x} f(x) \sin \frac{m\pi x}{L_x} dx.$$

Rearranging the left hand side gives

$$\sum_{n=1}^{\infty} c_n \left(e^{n\pi L_y/L_x} - e^{-n\pi L_y/L_x}\right) \int_0^{L_x} \sin \frac{n\pi x}{L_x} \sin \frac{m\pi x}{L_x} dx = \int_0^{L_x} f(x) \sin \frac{m\pi x}{L_x} dx,$$

and due to the orthogonality of the sine functions the only nonzero term in the infinite series is the case where $n = m$, so

$$c_m \left(e^{m\pi L_y/L_x} - e^{-m\pi L_y/L_x} \right) \frac{L_x}{2} = \int_0^{L_x} f(x)\sin\frac{m\pi x}{L_x}dx$$

or, finally,

$$c_n = \frac{2}{L_x} \frac{1}{e^{n\pi L_y/L_x} - e^{-n\pi L_y/L_x}} \int_0^{L_x} f(x)\sin\frac{n\pi x}{L_x}dx.$$

So, in summary, the solution to

$$\frac{\partial^2 u}{\partial x^2} + \frac{\partial^2 u}{\partial y^2} = 0$$

with boundary conditions

$$u(0,y) = 0, \quad u(L_x,y) = 0, \quad u(x,0) = 0, \quad u(x,L_y) = f(x).$$

is

$$\boxed{u(x,y) = \sum_{n=1}^{\infty} c_n \sin\frac{n\pi x}{L_x} \left(e^{n\pi y/L_y} - e^{-n\pi y/L_x} \right).} \tag{11.43}$$

where

$$\boxed{c_n = \frac{2}{L\left(e^{n\pi L_y/L_x} - e^{-n\pi L_y/L_x} \right)} \int_0^{L_x} f(x)\sin\frac{n\pi x}{L_x}dx.} \tag{11.44}$$

Example 11.11. Find the solution to Laplace's equation in a rectangular domain where $L_x = 4$ and $L_y = 2$ and

$$u(x,2) = \begin{cases} x, & 0 < x \le 2 \\ 4-x, & 2 < x \le 4. \end{cases}$$

Evaluating the integrals for the Fourier coefficients gives

$$c_n = \frac{64e^{n\pi/2}\cos\left(\frac{n\pi}{4}\right)\sin^3\left(\frac{n\pi}{4}\right)}{(-1+e^{n\pi})n^2\pi^2}$$

A plot of the solution containing the first ten terms of the series is illustrated in Figure 11.24.

11.4.2 Unsteady Two-Dimensional Heat Conduction with Homogeneous Boundary Conditions

The two-dimensional heat conduction equation is given by

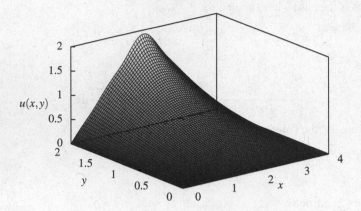

Fig. 11.24 Approximate solution to Laplace's equation from Example 11.11 using a partial sum containing the first ten terms.

$$\frac{\partial^2 u}{\partial x^2} + \frac{\partial^2 u}{\partial y^2} = \frac{1}{\alpha^2}\frac{\partial u}{\partial t}.$$

(11.45)

Homogeneous boundary conditions are given by

$$u(0,y,t) = 0, \quad u(L_x,y,t) = 0, \quad u(x,0,t) = 0, \quad u(x,L_y,t) = 0.$$

The initial condition is given by

$$u(x,y,0) = f(x,y),$$

(11.46)

where $f(x,y)$ is the initial temperature profile of the plate.

In a manner similar to the one-dimensional problem, due to the fact that the heat conduction equation is linear, we are able to combine the solution to Laplace's equation to the unsteady solution with homogeneous boundary conditions to determine the solution to the unsteady problem with inhomogeneous boundary conditions.

To solve this problem, assume a solution of the form

$$u(x,y,t) = X(x)Y(y)T(t).$$

Substituting into the differential equation gives

$$X''(x)Y(y)T(t) + X(x)Y''(y)T(t) = \frac{1}{\alpha^2}X(x)Y(y)T'(t)$$

and dividing by $X(x)Y(y)T(t)$ and dropping the arguments gives

$$\frac{X''}{X} + \frac{Y''}{Y} = \frac{1}{\alpha^2}\frac{T'}{T}. \tag{11.47}$$

The right-hand side of Equation (11.47) depends only on t and the left-hand side only depends on x and y and these variables are independent, thus each side must be equal to a constant. Hence,

$$\frac{X''}{X} + \frac{Y''}{Y} = \frac{1}{\alpha^2}\frac{T'}{T} = -\lambda$$

which gives the two equations

$$\frac{X''}{X} + \frac{Y''}{Y} = -\lambda \tag{11.48}$$

and

$$\frac{1}{\alpha^2}\frac{T'}{T} = -\lambda.$$

Until now, the systems considered had only two independent variables and we were able to find the general solution for each one from the corresponding ordinary differential equation, use the boundary conditions from one to determine the values for λ, and from there construct the general solution. That is not possible for this problem at this point because the equation with X and Y is not an ordinary differential equation because it has two independent variables. The equation for T is an ordinary differential equation, but the boundary conditions cannot be expressed in terms of $T(t)$.

This is resolved by taking the procedure one step further. Rearranging Equation (11.48) gives

$$\frac{X''}{X} + \lambda = -\frac{Y''}{Y}$$

and in this equation the left-hand side only depends on x and the right-hand side only depends on y, and hence both ratios must be a constant. Hence

$$\frac{X''}{X} + \lambda = -\frac{Y''}{Y} = \gamma,$$

which gives the two ordinary differential equations

$$Y'' + \gamma Y = 0, \quad X'' + (\lambda - \gamma)X = 0.$$

As the reader might suspect, even though there are two constants to determine, λ and γ, due to the fact that there are boundary conditions in two dimensions, x and y, it will be possible to use the boundary conditions to determine the constants.

In terms of the functions $X(x)$ and $Y(y)$ the boundary conditions are

$$u(0,y,t) = X(0)Y(y)T(t) = 0 \qquad \Longrightarrow \qquad X(0) = 0$$
$$u(L_x,y,t) = X(L_x)Y(y)T(t) = 0 \qquad \Longrightarrow \qquad X(L_x) = 0$$
$$u(x,0,t) = X(x)Y(0)T(t) = 0 \qquad \Longrightarrow \qquad Y(0) = 0$$
$$u(x,L_y,t) = X(x)Y(L_y)T(t) = 0 \qquad \Longrightarrow \qquad Y(L_y) = 0.$$

A similar argument as before requires that both γ and $\lambda - \gamma$ must be positive in order to possibly satisfy the boundary condition and hence the general solutions are

$$Y(y) = c_1 \cos \sqrt{\gamma} y + c_2 \sin \sqrt{\gamma} y$$
$$X(x) = d_1 \cos \sqrt{\lambda - \gamma} x + d_2 \sqrt{\lambda - \gamma} x.$$

The boundary conditions on both X and Y at zero require that both c_1 and d_1 be zero. The boundary condition on $Y(L_y)$ requires that

$$\gamma = \frac{m^2 \pi^2}{L_y^2}, \quad m = 1,2,3,\ldots.$$

The boundary condition on $X(L_x)$ requires that

$$\sqrt{\lambda - \gamma} L_x = n\pi, \quad \Longrightarrow \quad \lambda = \frac{n^2 \pi^2}{L_x^2} + \frac{m^2 \pi^2}{L_y^2}, \quad n = 1,2,3,\ldots.$$

Both ordinary differential equations for $X(x)$ and $Y(y)$ are linear, thus the linear combination of solutions for all the possible values for m and n are also solutions. Hence,

$$X(x) = \sum_{n=1}^{\infty} d_n \sin \frac{n\pi x}{L_x}, \quad n = 1,2,3,\ldots$$
$$Y(y) = \sum_{n=1}^{\infty} c_m \sin \frac{m\pi y}{L_y}, \quad m = 1,2,3,\ldots.$$

Finally, using λ, a solution to the $T(t)$ equation is

$$T(t) = e^{-\alpha \lambda t} = e^{-\alpha^2 \left((n^2 \pi^2 / L_x^2) + (m^2 \pi^2 / L_y^2) \right) t}.$$

Observe that the the solution for $T(t)$ is different for each combination of values of n and m. Substituting the solutions for $X(x)$, $Y(y)$, and $T(t)$ into $u(x,y,t) = X(x)Y(y)T(t)$ gives the general solution

$$\boxed{u(x,y,t) = \sum_{n=1}^{\infty} \sum_{m=1}^{\infty} \left[a_{n,m} \left(\sin \frac{n\pi x}{L_x} \right) \left(\sin \frac{m\pi y}{L_y} \right) e^{-\alpha^2 \left((n^2 \pi^2 / L_x^2) + (m^2 \pi^2 / L_y^2) \right) t} \right],}$$

$$(11.49)$$

where the coefficient $a_{n,m}$ is the combination of coefficients c_m and d_n when the series for X and Y are multiplied.

Equation (11.49) is the general solution to Equation (11.45) with homogeneous boundary conditions. To satisfy the initial condition, the orthogonality of the sine functions is exploited. Substituting $t = 0$ into the solution and equating it with the initial condition gives

$$\sum_{n=1}^{\infty} \sum_{m=1}^{\infty} a_{n,m} \sin \frac{n\pi x}{L_x} \sin \frac{m\pi y}{L_y} = f(x,y).$$

Multiplying by $\sin(\hat{m}\pi y/L_y)$ and integrating from 0 to L_y with respect to y gives

$$\int_0^{L_y} \sum_{n=1}^{\infty} \sum_{m=1}^{\infty} a_{n,m} \sin \frac{n\pi x}{L_x} \sin \frac{m\pi y}{L_y} \sin \frac{\hat{m}\pi y}{L_y} dy = \int_0^{L_y} f(x,y) \sin \frac{\hat{m}\pi y}{L_y} dy.$$

Because the sine functions are orthogonal, every term in the series indexed by m is zero except for when $\hat{m} = m$. Hence

$$\frac{L_y}{2} \sum_{n=1}^{\infty} a_{n,m} \sin \frac{n\pi x}{L_x} = \int_0^{L_y} f(x,y) \sin \frac{m\pi y}{L_y} dy.$$

Similarly multiplying by $\sin(\hat{n}\pi x/L_x)$ and integrating from 0 to L_x with respect to x eliminates all but one term in the remaining series. Doing so and solving for the coefficient gives

$$a_{n,m} = \frac{4}{L_x L_y} \int_0^{L_x} \int_0^{L_y} f(x,y) \sin \frac{n\pi x}{L_x} \sin \frac{m\pi y}{L_y} dy dx. \tag{11.50}$$

Together, Equations (11.49) and (11.50) are the solution to Equation (11.45) with homogeneous boundary conditions and initial temperature distribution given by Equation (11.46).

Example 11.12. Solve the two-dimensional unsteady heat conduction equation with homogeneous boundary conditions with $L_x = 2$, $L_y = 1$, $\alpha = 2$, and

$$f(x,y) = \begin{cases} 1, & 0.9 < x < 1.1, \quad 0.4 < y < 0.6 \\ 0, & \text{otherwise.} \end{cases}$$

The solution is obtained by substituting everything into Equation (11.50) to determine the coefficients, and then substituting the coefficients into the solution given by Equation (11.49). Doing so gives

$$a_{n,m} = \frac{4}{2} \int_{0.9}^{1.1} \int_{0.4}^{0.6} \sin\frac{n\pi x}{2} \sin(m\pi y)\,dy\,dx$$

$$= 2 \left(-\frac{2}{n\pi}\cos\frac{n\pi x}{2}\right)\Bigg|_{0.9}^{1.1} \left(-\frac{1}{m\pi}\cos m\pi y\right)\Bigg|_{0.4}^{0.6}$$

$$= \frac{4}{n\pi}\left(\cos(0.45n\pi) - \cos(0.55n\pi)\right)\frac{1}{m\pi}\left(\cos(0.4m\pi) - \cos(0.6m\pi)\right).$$

Plots of a partial sum solution including up to 50 terms in both series are illustrated in Figures 11.25 through 11.27.

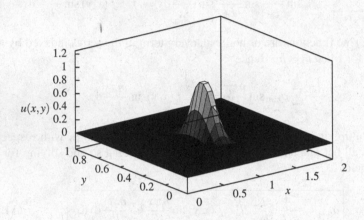

Fig. 11.25 Solution from Example 11.12 at $t = 0.05$.

11.4.3 Unsteady Two-Dimensional Heat Conduction with Inhomogeneous Boundary Conditions

The approach to solving the two-dimensional conduction equation with inhomogeneous boundary conditions is the same as in the one-dimensional case. If the solution is expressed as a transient solution plus a steady-state solution, the transient solution will satisfy the same differential equation and have homogeneous boundary conditions. The approach is illustrated by an example that basically combines Example 11.12 with Example 11.11.

Example 11.13. Find the solution to

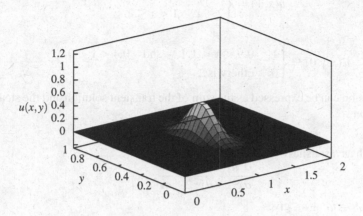

Fig. 11.26 Solution from Example 11.12 at $t = 0.15$.

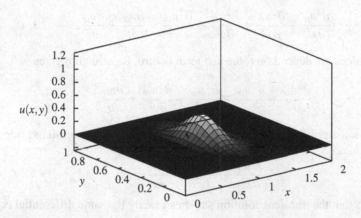

Fig. 11.27 Solution from Example 11.12 at $t = 0.25$.

$$\frac{\partial^2 u}{\partial x^2} + \frac{\partial^2 u}{\partial y^2} = \frac{\partial u}{\partial t}$$

subject to

$$u(0, y, t) = 0, \quad u(L_x, y, t) = 0, \quad u(x, 0, t) = 0, \quad u(x, L_y, t) = f(x),$$

where

$$u(x,1) = \begin{cases} x, & 0 < x \le 1, \\ 2 - x, & 1 < x \le 2. \end{cases}$$

and

$$u(x,y,0) = \begin{cases} 1, & 0.9 < x < 1.1, \quad \text{and} \quad 0.4 < y < 0.6 \\ 0, & \text{otherwise.} \end{cases}$$

The solution can be expressed as the sum of the transient solution and the steady-state solution

$$u(x,y,t) = u_{tr}(x,y,t) + u_{ss}(x,y,t).$$

Inasmuch as u satisfies

$$\frac{\partial^2 u}{\partial x^2} + \frac{\partial^2 u}{\partial y^2} = \frac{\partial u}{\partial t},$$

substituting $u = u_{tr} + u_{ss}$ gives

$$\frac{\partial^2}{\partial x^2}(u_{tr} + u_{ss}) + \frac{\partial^2}{\partial y^2}(u_{tr} + u_{ss}) = \frac{\partial}{\partial t}(u_{tr} + u_{ss}).$$

This may be written as

$$\frac{\partial^2 u_{tr}}{\partial x^2} + \frac{\partial^2 u_{ss}}{\partial x^2} + \frac{\partial^2 u_{tr}}{\partial y^2} + \frac{\partial^2 u_{ss}}{\partial y^2} = \frac{\partial u_{tr}}{\partial t} + \frac{\partial u_{tr}}{\partial t}.$$

Because u_{ss} does not depend on t, the last term is zero. Rearranging gives

$$\frac{\partial^2 u_{tr}}{\partial x^2} + \frac{\partial^2 u_{tr}}{\partial y^2} + \frac{\partial^2 u_{ss}}{\partial x^2} + \frac{\partial^2 u_{ss}}{\partial y^2} = \frac{\partial u_{tr}}{\partial t}.$$

Because u_{ss} satisfies Laplace's equation, the last two terms on the left-hand side add to zero, and hence

$$\frac{\partial^2 u_{tr}}{\partial x^2} + \frac{\partial^2 u_{tr}}{\partial y^2} = \frac{\partial u_{tr}}{\partial t},$$

which shows that the transient solution satisfies exactly the same differential equation as $u(x,y,t)$.

Expressing the boundary conditions in terms of the transient and steady-state solutions gives

$$u(0,y,t) = u_{tr}(0,y,t) + u_{ss}(0,y,t) = 0$$
$$u(L_x,y,t) = u_{tr}(L_x,y,t) + u_{ss}(L_x,y,t) = 0$$
$$u(x,0,t) = u_{tr}(x,0,t) + u_{ss}(x,0,t) = 0$$
$$u(x,L_y,t) = u_{tr}(x,L_y,t) + u_{ss}(x,L_y,t) = f(x).$$

The steady state solution must satisfy the boundary conditions, therefore it must be the case that the boundary conditions for the transient solution are zero, that is, homogeneous.

Finally, for the initial condition

$$u(x,y,0) = u_{tr}(x,y,0) + u_{ss}(x,y,0) = f(x,y)$$

so the initial condition for the transient solution is

$$u_{tr}(x,y,0) = f(x,y) - u_{ss}(x,y,0).$$

Because u_{ss} is the steady-state solution, it is the solution to Laplace's equation. Evaluating the coefficient given by Equation (11.44) gives

$$c_n = \frac{32e^{n\pi/2}\cos\frac{n\pi}{4}\sin^3\frac{n\pi}{4}}{(e^{n\pi}-1)n^2\pi^2},$$

so, substituting into Equation (11.43) gives

$$u_{ss}(x,y,t) = \sum_{n=1}^{\infty}\left(\frac{32e^{n\pi/2}\cos\frac{n\pi}{4}\sin^3\frac{n\pi}{4}}{(e^{n\pi}-1)n^2\pi^2}\right)\sin\frac{n\pi x}{L_x}\left(e^{n\pi y/L_y}-e^{-(n\pi y/L_x)}\right).$$

Evaluating the coefficients for the transient solution gives

$$
\begin{aligned}
a_{n,m} &= \int_0^{L_x}\int_0^{L_y} u_{tr}(x,y,0)\sin\frac{n\pi x}{L_x}\sin\frac{m\pi y}{L_y}dxdy \\
&= \int_0^{L_x}\int_0^{L_y} f(x,y)\,dxdy - \int_0^{L_x}\int_0^{L_y} u_{ss}(x,y,t)\,dxdy \\
&= \int_{0.9}^{1.1}\int_{0.4}^{0.6}\sin\frac{n\pi x}{2}\sin(m\pi y)\,dxdy \\
&\quad - \int_0^2\int_0^1\sum_{\hat{n}=1}^{\infty}\left(\frac{32e^{\hat{n}\pi/2}\cos\frac{\hat{n}\pi}{4}\sin^3\frac{\hat{n}\pi}{4}}{(e^{\hat{n}\pi}-1)\hat{n}^2\pi^2}\right)\sin\frac{\hat{n}\pi x}{L_x}\left(e^{(\hat{n}\pi y/L_y)}-e^{-(\hat{n}\pi y/L_x)}\right) \\
&\quad \cdot \sin\frac{n\pi x}{2}\sin(m\pi y)\,dxdy.
\end{aligned}
$$

The first integral has already been evaluated in Example 11.12. In the second integral, due to the orthogonality of the sine functions, the only term that survives from the infinite series is the one when $\hat{n} = n$, and hence

$$
\begin{aligned}
a_{n,m} &= \frac{4}{n\pi}(\cos(0.45n\pi)-\cos(0.55n\pi))\frac{1}{m\pi}(\cos(0.4m\pi)-\cos(0.6m\pi)) \\
&\quad - \int_0^2\int_0^1\left(\frac{32e^{n\pi/2}\cos\frac{n\pi}{4}\sin^3\frac{n\pi}{4}}{(e^{n\pi}-1)n^2\pi^2}\right)\sin\frac{n\pi x}{2}\left(e^{n\pi y}-e^{-(n\pi y/2)}\right) \\
&\quad \cdot \sin\frac{n\pi x}{2}\sin(m\pi y)\,dxdy \\
&= \frac{32n\pi e^{n\pi/2}\left(e^{n\pi/2}-1\right)\cos\frac{n\pi}{4}\sin^2\frac{n\pi}{4}}{n^4\pi^4\left(e^{n\pi/2}+1\right)}
\end{aligned}
$$

11.5 Vibrating Membranes

This section considers vibrating membranes, both with rectangular geometries as well as circular geometries. The latter are more commonly considered as drum heads.

11.5.1 The Two-Dimensional Wave Equation in Rectangular Coordinates

The two-dimensional wave equation is given by

$$\frac{\partial^2 u}{\partial y^2} + \frac{\partial^2 u}{\partial y^2} = \frac{1}{\alpha^2}\frac{\partial^2 u}{\partial t^2}, \tag{11.51}$$

with boundary conditions

$$u(0,y,t) = 0, \quad u(L_x,y,t) = 0, \quad u(x,0,t) = 0, \quad u(x,L_y,t) = 0$$

and initial conditions

$$u(x,y,0) = f(x,y), \quad \frac{\partial u}{\partial t}(x,y,0) = g(x,y).$$

Equation (11.51) describes small vibrations of a thin membrane on a rectangular domain with a length L_x in the x-direction and L_y in the y-direction. Recall that for the one-dimensional wave equation the net force on a small segment of the string was proportional to the curvature of the string. It makes intuitive sense that for the two-dimensional wave equation, the net force is proportional to the sum of the curvatures in the x- and y-directions.

It is left as an exercise to show that the general solution to Equation (11.51) with homogeneous boundary conditions is given by

$$u(x,y,t) = \sum_{n=1}^{\infty}\sum_{m=1}^{\infty} \sin\left(\frac{n\pi x}{L_x}\right)\sin\left(\frac{m\pi y}{L_y}\right)\left(a_{n,m}\sin\omega_{n,m} + b_{n,m}\cos\omega_{n,m}\right),$$

$$\tag{11.52}$$

where

$$\omega_{n,m} = \alpha\pi\sqrt{\frac{n^2}{L_x^2} + \frac{m^2}{L_y^2}}$$

and

$$a_{n,m} = \frac{4}{L_x L_y} \int_0^{L_y} \int_o^{L_x} f(x,y) \sin\left(\frac{n\pi x}{L_x}\right) \sin\left(\frac{m\pi y}{L_y}\right) dxdy$$

$$b_{n,m} = \frac{4}{L_x L_y \omega_{n,m}} \int_0^{L_y} \int_0^{L_x} g(x,y) \sin\left(\frac{n\pi x}{L_x}\right) \sin\left(\frac{m\pi y}{L_y}\right) dxdy.$$

The modes for this solution are exactly the same as for the two-dimensional heat equation. The only difference in the solution is that they oscillate in time instead of decaying.

11.5.2 The Two-Dimensional Wave Equation in Polar Coordinates

For a circular membrane, like a drum, because the boundary condition will hold at a fixed radius, as opposed to fixed values of x and y, it is much more convenient to solve it in polar coordinates.

The relationship between polar and Cartesian coordinates is given, as usual, by

$$x = r\cos\theta, \quad y = r\sin\theta$$

and the inverse transformation is given by

$$r = \sqrt{x^2 + y^2}, \quad \theta = \tan^{-1}\left(\frac{y}{x}\right).$$

We need to relate derivatives with respect to the variables x and y to derivatives with respect to the variables r and θ. Because we know the expressions for the change of coordinates, we can write the function as $u(r,\theta,t) = u(r(x,y), \theta(x,y), t)$. By the chain rule

$$\frac{\partial u}{\partial x} = \frac{\partial u}{\partial r}\frac{\partial r}{\partial x} + \frac{\partial u}{\partial \theta}\frac{\partial \theta}{\partial x} = \frac{\partial u}{\partial r}\frac{x}{\sqrt{x^2+y^2}} + \frac{\partial u}{\partial \theta}\left(-\frac{y}{x^2+y^2}\right) = \cos\theta\frac{\partial u}{\partial r} - \frac{\sin\theta}{r}\frac{\partial u}{\partial \theta}.$$

Evaluating the derivative again gives

$$
\begin{aligned}
\frac{\partial^2 u}{\partial x^2} &= \frac{\partial}{\partial x}\left(\frac{\partial u}{\partial x}\right) \\
&= \frac{\partial}{\partial r}\left(\frac{\partial u}{\partial x}\right)\frac{\partial r}{\partial x} + \frac{\partial}{\partial \theta}\left(\frac{\partial u}{\partial x}\right)\frac{\partial \theta}{\partial x} \\
&= \frac{\partial}{\partial r}\left(\cos\theta\,\frac{\partial u}{\partial r} - \frac{\sin\theta}{r}\frac{\partial u}{\partial \theta}\right)\frac{\partial r}{\partial x} + \frac{\partial}{\partial \theta}\left(\cos\theta\,\frac{\partial u}{\partial r} - \frac{\sin\theta}{r}\frac{\partial u}{\partial \theta}\right)\frac{\partial \theta}{\partial x} \\
&= \frac{\partial}{\partial r}\left(\cos\theta\,\frac{\partial u}{\partial r} - \frac{\sin\theta}{r}\frac{\partial u}{\partial \theta}\right)\cos\theta + \frac{\partial}{\partial \theta}\left(\cos\theta\,\frac{\partial u}{\partial r} - \frac{\sin\theta}{r}\frac{\partial u}{\partial \theta}\right)\left(-\frac{\sin\theta}{r}\right) \\
&= \left(\cos\theta\,\frac{\partial^2 u}{\partial r^2} + \frac{\sin\theta}{r^2}\frac{\partial u}{\partial \theta} - \frac{\sin\theta}{r}\frac{\partial^2 u}{\partial r\partial\theta}\right)\cos\theta \\
&\quad + \left(-\sin\theta\,\frac{\partial u}{\partial r} + \cos\theta\,\frac{\partial^2 u}{\partial\theta\partial r} - \frac{\cos\theta}{r}\frac{\partial u}{\partial \theta} - \frac{\sin\theta}{r}\frac{\partial^2 u}{\partial \theta^2}\right)\left(-\frac{\sin\theta}{r}\right)
\end{aligned}
$$

A similar computation gives

$$
\begin{aligned}
\frac{\partial^2 u}{\partial y^2} &= \left(\sin\theta\,\frac{\partial^2 u}{\partial r^2} - \frac{\cos\theta}{r^2}\frac{\partial u}{\partial \theta} + \frac{\cos\theta}{r}\frac{\partial^2 u}{\partial r\partial\theta}\right)\sin\theta \\
&\quad + \left(\cos\theta\,\frac{\partial u}{\partial r} + \sin\theta\,\frac{\partial^2 u}{\partial r\partial\theta} - \frac{\sin\theta}{r}\frac{\partial u}{\partial \theta} + \frac{\cos\theta}{r}\frac{\partial^2 u}{\partial \theta^2}\right)\frac{\cos\theta}{r}.
\end{aligned}
$$

Substituting these expressions into the right-hand side of Equation (11.51) gives the wave equation in polar coordinates.

$$
\boxed{\frac{\partial^2 u}{\partial r^2} + \frac{1}{r}\frac{\partial u}{\partial r} + \frac{1}{r^2}\frac{\partial^2 u}{\partial \theta^2} = \frac{1}{\alpha^2}\frac{\partial^2 u}{\partial t^2}.}
\tag{11.53}
$$

For a circular drum, the boundary condition is

$$
\boxed{u(\hat{r},\theta,t) = 0}
\tag{11.54}
$$

and the initial conditions are

$$
\boxed{u(r,\theta,0) = f(r,\theta), \quad \frac{\partial u}{\partial t}(r,\theta,0) = g(r,\theta).}
\tag{11.55}
$$

Assuming a solution of the form

$$
u(r,\theta,t) = R(r)\Theta(\theta)T(t)
$$

and substituting into Equation (11.53) gives

$$
R''(r)\Theta(\theta)T(t) + \frac{1}{r}R'(r)\Theta(\theta)T(t) + \frac{1}{r^2}R(r)\Theta''(\theta)T(t) = \frac{1}{\alpha^2}R(r)\Theta(\theta)T''(t)
$$

and dividing by $R(r)\Theta(\theta)T(t)$ gives

$$\frac{R''(r)}{R(r)} + \frac{1}{r}\frac{R'(r)}{R(r)} + \frac{1}{r^2}\frac{\Theta''(\theta)}{\Theta(\theta)} = \frac{1}{\alpha^2}\frac{T''(t)}{T(t)}.$$

The right side of the equation only depends on t and the left side depends only on r and θ, and all three variables are independent, therefore both sides must be constant; hence,

$$\frac{R''(r)}{R(r)} + \frac{1}{r}\frac{R'(r)}{R(r)} + \frac{1}{r^2}\frac{\Theta''(\theta)}{\Theta(\theta)} = \frac{1}{\alpha^2}\frac{T''(t)}{T(t)} = -\lambda,$$

where λ is a yet to be determined constant. Hence,

$$T''(t) + \alpha^2\lambda T(t) = 0 \tag{11.56}$$

and

$$\frac{R''(r)}{R(r)} + \frac{1}{r}\frac{R'(r)}{R(r)} + \frac{1}{r^2}\frac{\Theta''(\theta)}{\Theta(\theta)} = -\lambda.$$

Multiplying by r^2 and rearranging gives

$$r^2\frac{R''(r)}{R(r)} + r\frac{R'(r)}{R(r)} + r^2\lambda = -\frac{\Theta''(\theta)}{\Theta(\theta)}.$$

The left side of this equation only depends on r and the right side only depends on θ and the variables are independent, therefore these also must be equal to a constant, which is not necessarily the same as λ. Calling this constant γ, we have

$$r^2\frac{R''(r)}{R(r)} + r\frac{R'(r)}{R(r)} + r^2\lambda = -\frac{\Theta''(\theta)}{\Theta(\theta)} = \gamma.$$

Hence,

$$\Theta''(\theta) + \gamma\Theta(\theta) = 0 \tag{11.57}$$

and

$$r^2R''(r) + rR'(r) + (r^2\lambda - \gamma)R(r) = 0. \tag{11.58}$$

If we determine the solutions to Equations (11.56)–(11.58), we have a solution to Equation (11.53).

We proceed as we have done before by finding the general solutions to the ordinary differential equations for $R(r)$, $\Theta(\theta)$, and $T(t)$ and applying the boundary conditions. Although is appears that we only have one boundary condition given by Equation (11.54), there is also the fact that the solution for $\Theta(\theta)$ must be periodic, that is, $\Theta(\theta) = \Theta(\theta + 2\pi)$. Thus, γ must be positive and

$$\Theta(\theta) = c_1\sin\sqrt{\gamma}\theta + c_2\cos\sqrt{\gamma}\theta.$$

In order for $\Theta(\theta + 2\pi) = \Theta(\theta)$, $\sqrt{\gamma}$ must be an integer, or

$$\sqrt{\gamma} = m = 0, 1, 2, \ldots,$$

so

$$\Theta_m(\theta) = c_1 \sin m\theta + c_2 \cos m\theta, \qquad m = 1, 2, 3, \ldots.$$

Because $\sqrt{\gamma} = m$, and letting $x = \sqrt{\lambda}\, r$, Equation (11.58) becomes

$$x^2 R''(x) + x R'(x) + \left(x^2 - m^2\right) R(x) = 0,$$

which is the standard form for the Bessel equation. In the special case where m is an integer, which is the present case, the general solution may be written as

$$R(x) = c_1 J_m(x) + c_2 Y_m(x) \tag{11.59}$$

or

$$R(r) = c_1 J_m\left(\sqrt{\lambda}\, r\right) + c_2 Y_m\left(\sqrt{\lambda}\, r\right), \tag{11.60}$$

where

$$J_m\left(\sqrt{\lambda}\, r\right) = \sum_{n=0}^{\infty} \frac{(-1)^n}{n!\,(n-m)!} \left(\frac{\sqrt{\lambda}\, r}{2}\right)^{2n-k} \tag{11.61}$$

and

$$Y_m\left(\sqrt{\lambda}\, r\right) = \ln\left(\sqrt{\lambda}\, r\right) J_m\left(\sqrt{\lambda}\, r\right) - \frac{1}{2}\sum_{n=0}^{m-1} \frac{(m-n-1)!}{n!} \left(\frac{\sqrt{\lambda}\, r}{2}\right)^{2n-m}$$

$$- \frac{1}{2}\sum_{n=0}^{\infty} \frac{(-1)^n \left(\left(1 + \frac{1}{2} + \frac{1}{3} + \cdots + \frac{1}{n}\right) + \left(1 + \frac{1}{2} + \cdots + \frac{1}{n+m}\right)\right)}{n!\,(n+m)!} \left(\frac{\sqrt{\lambda}\, r}{2}\right)^{2n+m}.$$

The functions J_m and Y_m are the Bessel functions of the first and second kind, respectively, or order m. Note that as $r \to 0$, $Y_m(r) \to -\infty$, and hence we assume the motion of the center of the drum is bounded; then we must have $c_2 = 0$.

Hence,

$$R(r) = c_1 J_m\left(\sqrt{\lambda}\, r\right),$$

and because the boundary condition $u(\hat{r}, \theta, t) = 0$, requires that $R(\hat{r}) = 0$, either $c_1 = 0$, which would give the trivial solution or

$$J_m\left(\sqrt{\lambda}\, \hat{r}\right) = 0.$$

Referring to Figure 5.8, it is apparent that Bessel functions of the first kind of various orders are equal to zero for multiple values of r. In fact, they are tabulated in Table 5.2. Let $z_{m,n}$ denote the nth zero of the Bessel function of the first kind of order m. Hence,

$$\sqrt{\lambda}\, \hat{r} = z_{m,n} \quad \Longrightarrow \quad \lambda = \left(\frac{z_{m,n}}{\hat{r}}\right)^2, \quad n = 1, 2, 3, \ldots,$$

and

$$R(r) = c_1 J_m \left(\frac{z_{m,n} r}{\hat{r}} \right), \quad n = 1, 2, 3, \ldots.$$

The way to intuitively think of the role of $z_{m,n}$ is that it scales r in such a way that it will go through zero at the radius of the drum.[5] Figures 11.28 and 11.29 illustrate the Bessel functions of the first kind of order zero and one, respectively, with $\hat{r} = 5$ and $z_{m,n}$ equal to the first three zeros for each one. The feature to observe is that scaling the argument to be $z_{m,n} r / \hat{r}$ makes all the functions go through zero at $\hat{r} = 5$, which is what is necessary to match the boundary condition at the radius of the drum.

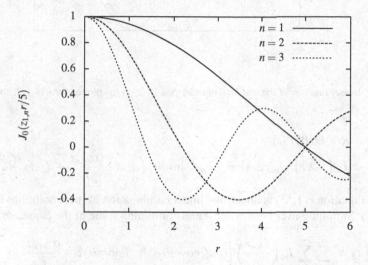

Fig. 11.28 Bessel function of the first kind of order zero, $J_0(z_{0,n} r/5)$ plotted for $n = 1, 2$, and 3.

At this point, both γ and λ have been determined. Returning to Equation (11.56), and substituting for λ gives

$$T''(t) + \left(\frac{\alpha z_{m,n}}{\hat{r}} \right)^2 T(t) = 0,$$

and hence

$$T(t) = d_1 \cos \frac{\alpha z_{m,n} t}{\hat{r}} + d_2 \sin \frac{\alpha z_{m,n} t}{\hat{r}}.$$

For fixed integers, m and n,

---

[5] This is exactly the same as before in the one-dimensional wave equation where we determined that $\lambda = n^2 \pi^2 / L^2$, which scaled the argument to the sine function so that it passed through zero at $x - L$.

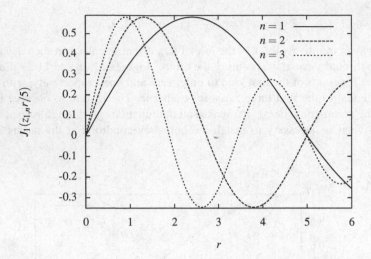

Fig. 11.29 Bessel function of the first kind of order one, $J_1(z_{1,n}r/5)$ plotted for $n = 1, 2$, and 3.

$$u(r,\theta,t) = R(r)\Theta(\theta)T(t)$$
$$= J_m\left(\sqrt{\lambda}\,r\right)(a_{m,n}\cos m\theta + b_{m,n}\sin m\theta)\left(d_1\cos\frac{\alpha z_{m,n}t}{\hat{r}} + d_2\sin\frac{\alpha z_{m,n}t}{\hat{r}}\right).$$

Because Equation (11.53) is linear, any linear combination of these solutions is also a solution. Summing over both m and n and combining some of the constants gives

$$u(r,\theta,t) = \sum_{m=0}^{\infty}\sum_{n=1}^{\infty} J_m\left(\frac{z_{m,n}r}{\hat{r}}\right)\left[(a_{m,n}\cos m\theta + b_{m,n}\sin m\theta)\cos\frac{\alpha z_{m,n}t}{\hat{r}}\right. \tag{11.62}$$
$$\left. + (c_{m,n}\cos m\theta + d_{m,n}\sin m\theta)\sin\frac{\alpha z_{m,n}t}{\hat{r}}\right].$$

This solution satisfies both Equation (11.53), the wave equation in polar coordinates, as well as Equation (11.54), the boundary condition. The initial conditions, given in Equation (11.55), still need to be satisfied. Substituting $t = 0$ into the solution in Equation (11.62) gives

$$u(r,\theta,0) = \sum_{m=0}^{\infty}\sum_{n=1}^{\infty} J_m\left(\frac{z_{m,n}r}{\hat{r}}\right)(a_{m,n}\cos m\theta + b_{m,n}\sin m\theta) = f(r,\theta).$$

To determine the coefficients, we make use of orthogonality of the sine and cosine functions, as well as the following fact,

$$\int_0^{\hat{r}} rJ_m\left(\frac{z_{m,n}r}{\hat{r}}\right)J_m\left(\frac{z_{m,\hat{n}}r}{\hat{r}}\right)dr = \begin{cases} 0, & n \neq \hat{n} \\ \frac{1}{2}\left(\frac{dJ_k}{dr}(z_{m,n})\right)^2, & n = \hat{n}. \end{cases} \tag{11.63}$$

This should seem quite remarkable. The fact that sine and cosine functions are orthogonal over the right interval makes some sense because the positive and negative parts will "cancel" in some way or another. At least a priori it may seem unexpected that Bessel functions have a similar property, but in fact they do.[6] Observe that the integral is weighted by r and also the Bessel function is the same kind and also the same order. The difference between the two Bessel functions in the integral is that a different $z_{m,n}$ appears in the argument to the function. The two terms in the integrand would be two of the curves in either Figure 11.28 or 11.29.

Hence, to determine the coefficients $a_{m,n}$ that satisfy

$$\sum_{m=0}^{\infty} \sum_{n=1}^{\infty} J_m\left(\frac{z_{m,n} r}{\hat{r}}\right) (a_{m,n} \cos m\theta + b_{m,n} \sin m\theta) = f(r, \theta),$$

multiply both sides of the equation by $\cos \hat{m}\theta$ and integrating from 0 to 2π gives

$$\sum_{m=0}^{\infty} \sum_{n=1}^{\infty} \int_0^{2\pi} J_m\left(\frac{z_{m,n} r}{\hat{r}}\right) (a_{m,n} \cos m\theta + b_{m,n} \sin m\theta) \cos \hat{m}\theta d\theta$$
$$= \int_0^{2\pi} \cos \hat{m}\theta f(r, \theta) d\theta.$$

Because of the orthogonality of the sine and cosine functions, every term in the series indexed by m is zero except for when $\hat{m} = m$, and hence

$$\sum_{n=1}^{\infty} J_m\left(\frac{z_{m,n} r}{\hat{r}}\right) \int_0^{2\pi} a_{m,n} \cos^2 m\theta d\theta = \int_0^{2\pi} f(r, \theta) \cos m\theta d\theta,$$

or

$$\sum_{n=1}^{\infty} a_{m,n} J_m\left(\frac{z_{m,n} r}{\hat{r}}\right) = \frac{\int_0^{2\pi} f(r, \theta) \cos m\theta d\theta}{\int_0^{2\pi} a_{m,n} \cos^2 m\theta d\theta}.$$

Now, multiplying by $r J(z_{m,\hat{n}} r / \hat{r})$ and integrating from 0 to $\hat{r}$ gives

$$\sum_{n=1}^{\infty} \int_0^{\hat{r}} a_{m,n} r J_m\left(\frac{z_{m,n} r}{\hat{r}}\right) J_m\left(\frac{z_{m,\hat{n}} r}{\hat{r}}\right) dr = \int_0^{\hat{r}} r J_m\left(\frac{z_{m,\hat{n}} r}{\hat{r}}\right) \frac{\int_0^{2\pi} f(r, \theta) \cos m\theta d\theta}{\int_0^{2\pi} a_{m,n} \cos^2 m\theta d\theta} dr,$$

or

$$a_{m,n} = \frac{\int_0^{\hat{r}} \int_0^{2\pi} r f(r, \theta) \cos m\theta J_m\left(\frac{z_{m,n} r}{\hat{r}}\right) d\theta dr}{\left(\int_0^{2\pi} \cos^2 m\theta d\theta\right) \left(\int_0^{\hat{r}} J_m^2\left(\frac{z_{m,n} r}{\hat{r}} dr\right)\right)}.$$

An analogous computation gives

$$b_{m,n} = \frac{\int_0^{\hat{r}} \int_0^{2\pi} r f(r, \theta) \sin m\theta J_m\left(\frac{z_{m,n} r}{\hat{r}}\right) d\theta dr}{\left(\int_0^{2\pi} \sin^2 m\theta d\theta\right) \left(\int_0^{\hat{r}} J_m^2\left(\frac{z_{m,n} r}{\hat{r}}\right) dr\right)}.$$

---

[6] In Section 11.6 we show they must be orthogonal.

To determine $c_{m,n}$ and $d_{m,n}$, differentiate Equation (11.62) with respect to time, substitute $t = 0$, and follow the same procedure, which gives

$$c_{m,n} = \frac{\hat{r}}{\alpha z_{m,n}} \frac{\int_0^{\hat{r}} \int_0^{2\pi} r f(r,\theta) \cos m\theta J_m\left(\frac{z_{m,n} r}{\hat{r}}\right) d\theta dr}{\left(\int_0^{2\pi} \cos^2 m\theta d\theta\right) \left(\int_0^{\hat{r}} J_m^2\left(\frac{z_{m,n} r}{\hat{r}}\right) dr\right)}$$

and

$$d_{m,n} = \frac{\hat{r}}{\alpha z_{m,n}} \frac{\int_0^{\hat{r}} \int_0^{2\pi} r f(r,\theta) \cos m\theta J_m\left(\frac{z_{m,n} r}{\hat{r}}\right) d\theta dr}{\left(\int_0^{2\pi} \cos^2 m\theta d\theta\right) \left(\int_0^{\hat{r}} J_m^2\left(\frac{z_{m,n} r}{\hat{r}}\right) dr\right)},$$

and with that, we have solved the two-dimensional wave equation in polar coordinates.

11.5.3 Summary of the Solution to the Wave Equation in Polar Coordinates

The solution to

$$\frac{\partial^2 u}{\partial r^2} + \frac{1}{r}\frac{\partial u}{\partial r} + \frac{1}{r^2}\frac{\partial^2 u}{\partial \theta^2} = \frac{1}{\alpha^2}\frac{\partial^2 u}{\partial t^2},$$

with

$$u(\hat{r},\theta,t) = 0, \quad u(r,\theta,0) = f(r,\theta), \quad \frac{\partial u}{\partial t}(r,\theta,0) = g(r,\theta)$$

is

$$u(r,\theta,t) = \sum_{m=0}^{\infty}\sum_{n=1}^{\infty} J_m\left(\frac{z_{m,n} r}{\hat{r}}\right)\left[(a_{m,n}\cos m\theta + b_{m,n}\sin m\theta)\cos\frac{\alpha z_{m,n} t}{\hat{r}} \right.$$
$$\left. + (c_{m,n}\cos m\theta + d_{m,n}\sin m\theta)\sin\frac{\alpha z_{m,n} t}{\hat{r}}\right] \tag{11.64}$$

where

$$a_{m,n} = \frac{\int_0^{\hat{r}}\int_0^{2\pi} r f(r,\theta)\cos m\theta J_m\left(\frac{z_{m,n} r}{\hat{r}}\right)d\theta dr}{\left(\int_0^{2\pi}\cos^2 m\theta d\theta\right)\left(\int_0^{\hat{r}} J_m^2\left(\frac{z_{m,n} r}{\hat{r}}\right)dr\right)}$$

$$b_{m,n} = \frac{\int_0^{\hat{r}}\int_0^{2\pi} r f(r,\theta)\sin m\theta J_m\left(\frac{z_{m,n} r}{\hat{r}}\right)d\theta dr}{\left(\int_0^{2\pi}\sin^2 m\theta d\theta\right)\left(\int_0^{\hat{r}} J_m^2\left(\frac{z_{m,n} r}{\hat{r}}\right)dr\right)}$$

$$c_{m,n} = \frac{\hat{r}}{\alpha z_{m,n}}\frac{\int_0^{\hat{r}}\int_0^{2\pi} r f(r,\theta)\cos m\theta J_m\left(\frac{z_{m,n} r}{\hat{r}}\right)d\theta dr}{\left(\int_0^{2\pi}\cos^2 m\theta d\theta\right)\left(\int_0^{\hat{r}} J_m^2\left(\frac{z_{m,n} r}{\hat{r}}\right)dr\right)}$$

$$d_{m,n} = \frac{\hat{r}}{\alpha z_{m,n}}\frac{\int_0^{\hat{r}}\int_0^{2\pi} r f(r,\theta)\cos m\theta J_m\left(\frac{z_{m,n} r}{\hat{r}}\right)d\theta dr}{\left(\int_0^{2\pi}\sin^2 m\theta d\theta\right)\left(\int_0^{\hat{r}} J_m^2\left(\frac{z_{m,n} r}{\hat{r}}\right)dr\right)}.$$

11.5.4 Modes of Vibration of a Circular Drum Head

By fixing m and n in Equation (11.64) the individual modes of vibration of the drum head can be isolated. Figure 11.30 illustrated several of the lower-order modes for the vibrating drum head. It is a plot of $J_m(z_{m,n}r)\cos m\theta$ for various m and n. Modes with the sine term would be rotated by 90 degrees relative to the modes illustrated. Any vibration of the drum head is a linear combination of these modes and the exact composition of the combination is determined by the initial conditions for the problem.

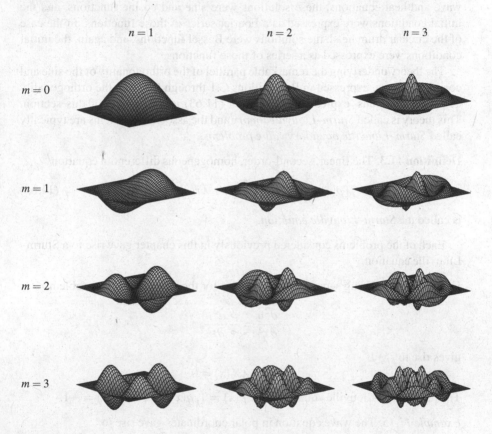

Fig. 11.30 Plots of $J_m(z_{m,n}r)\cos m\theta$ for various m and n, which are the modes of vibration for a circular drum head.

11.6 Sturm–Liouville Theory

In the previous sections of this chapter when using the method of separation of variables, for every problem we considered it was the case that there were an infinite
number of solutions to the ordinary differential equation with boundary value conditions that arose from the partial differential equation. Furthermore, in each case
it was possible to represent the functions that described the initial conditions in
the form of a series of these solutions, and even more remarkably, the coefficients
in this series were obtained by relatively simple computations. In the cases of the
wave and heat equations, these solutions were sine and cosine functions, and the
initial conditions were expressed as a Fourier series in those functions. In the case
of the circular drum head, the solutions were Bessel functions, and again, the initial
conditions were expressed as a series of those functions.

The theory underlying the remarkable parallel of the orthogonality of the sine and
cosine functions, expressed in Propositions C.1 through C.3, and the orthogonality
of Bessel functions, expressed in Equation (11.63) is the subject of this section.
This theory is called *Sturm–Liouville theory* and the associate problems are typically
called *Sturm–Liouville boundary value problems*.

Definition 11.3. The linear, second-order, homogeneous differential equation

$$\left(p(x)u'(x)\right)' - q(x)u(x) + \lambda r(x)u(x) = 0 \tag{11.65}$$

is called the *Sturm–Liouville equation*.

Each of the problems considered previously in this chapter gave rise to a Sturm–
Liouville equation.

Example 11.14. Using separation of variables for the vibrating string problem

$$\frac{\partial^2 u}{\partial x^2} = \frac{\rho}{\tau}\frac{\partial^2 u}{\partial t^2}$$

gives rise to

$$X''(x) + \lambda X(x) = 0.$$

This is a Sturm–Liouville equation with $p(x) = 1$, $q(x) = 0$, and $r(x) = -1$.

Example 11.15. The wave equation in polar coordinates gave rise to

$$r^2 R''(r) + rR'(r) + \left(r^2 - m^2\right)R(r) = 0,$$

which can be rewritten as

$$rR''(r) + R'(r) + rR(r) = \frac{m^2}{r}R(r),$$

or

$$\left(rR'(r)\right)' + rR(r) = \frac{m^2}{r}R(r).$$

This is a Sturm–Liouville equation with $\lambda = m^2$, $p(r) = r$, $q(r) = r$, and $r(r) = 1/r$.

The critical feature of the Sturm–Liouville equation is that it is a generalization of an eigenvalue problem. Recall (Definition B.4) for an $n \times n$ matrix A, λ is an eigenvalue, and $\hat{\xi}$ is an eigenvector if

$$A\hat{\xi} = \lambda\hat{\xi}.$$

Now, if we define the linear operator,

$$\mathrm{L}[u] = \frac{1}{r(x)}\left(-\frac{\mathrm{d}}{\mathrm{d}x}\left(p(x)\frac{\mathrm{d}u}{\mathrm{d}x}\right) + q(x)u\right), \tag{11.66}$$

then Equation (11.65) is of the form

$$\mathrm{L}[u] = \lambda u,$$

and hence determining solutions to this equation is clearly a generalization of determining the eigenvalues and eigenvectors of a matrix.

In Chapter 6, it was shown that Hermitian matrices had real eigenvalues and a full set of orthogonal eigenvectors. The simplest type of Hermitian matrix is a symmetric matrix, where $A = A^T$. To generalize this, consider the matrix product, inner product, and the dot product between two vectors. Specifically, for two vectors u and v, the dot product, matrix product, and inner product are all related by

$$u \cdot v = u^T v = \langle u, v \rangle.$$

If we multiply v by A, then

$$\langle u, Av \rangle = u^T Av.$$

Correspondingly,

$$\langle Au, v \rangle = u^T A^T v.$$

Recall that $(Au)^T = u^T A^T$. Hence, one way to define a matrix to be symmetric would be to require that, in an inner product,

$$\langle Au, v \rangle = \langle u, Av \rangle$$

for all vectors u and v in $\mathbb{R}^n$. The name for an operator that is a generalization of a symmetric matrix is a *self-adjoint operator.*

Definition 11.4. The linear operator L, is *self-adjoint* if

$$\langle \mathrm{L}[u], v \rangle = \langle u, \mathrm{L}[v] \rangle.$$

For the types of problems considered in this text, the inner product is defined as

$$\langle u, v \rangle = \int_0^L r(x)u(x)v(x)\mathrm{d}x.$$

As shown, a self-adjoint linear operator has properties similar to that of a symmetric matrix.

Definition 11.5. A Sturm–Liouville equation with boundary conditions

$$\alpha_1 u(0) + \alpha_2 u'(0) = 0 \tag{11.67}$$

$$\alpha_3 u(L) + \alpha_4 u'(L) = 0, \tag{11.68}$$

is called a *Sturm–Liouville boundary value problem.*

The most important fact at this point is the following.

Proposition 11.1. *If the operator* L *is defined by Equation (11.66) and the functions u and v satisfy homogeneous boundary conditions, then*

$$\langle L[u], v \rangle = \langle u, L[v] \rangle;$$

that is, L *is self-adjoint.*

Furthermore, if

1. $r(x) > 0$ and $p(x) > 0$ for $x \in [0, L]$
2. $p(x)$, $p'(x)$, $q(x)$ and $r(x)$ are continuous on $[0, L]$
3. The interval $[0, L]$ is finite

then the problem is called a *regular Sturm–Liouville problem.*

The foundation of the theory, and its main utility, is given by the following theorem. The reader is referred to [8, 9] for a slightly more comprehensive treatment, and the references therein for further study. The basic idea is a generalization of the fact that a symmetric matrix has real eigenvalues with orthogonal eigenvectors to the Sturm–Liouville operator.

Theorem 11.2. *For a regular Sturm–Liouville boundary value problem:*

1. *All of the eigenvalues are real.*
2. *All of the eigenvalues are simple, meaning that to each eigenvalue there is one and only one linearly independent eigenfunction, and there is a countable infinity of eigenvalues and corresponding eigenfunctions.*
3. *The eigenfunctions corresponding to different eigenvalues are orthogonal, that is, if u and v satisfy Equation (11.65) for different λ, then*

$$\langle u, v \rangle = \int_0^L r(x) u(x) v(x) \mathrm{d}x = 0.$$

4. *If $f(x)$ is continuously differentiable on the interval $[0, L]$ and $u_1(x), u_2(x), \ldots$ are the eigenfunctions, then $f(x)$ can be expressed as a convergent series of the eigenfunctions; that is,*

$$f(x) = \sum_{n=0}^{\infty} a_n u_n(x),$$

where

$$a_n = \frac{\int_0^L r(x)f(x)u_n(x)\mathrm{d}x}{\int_0^L r(x)u_n^2(x)\mathrm{d}x}.$$

The last item in the theorem is perhaps the most important. The set of solutions to regular Sturm–Liouville problems forms a basis for sets of functions. In the context of solving partial differential equations using separation of variables, it is the mechanism that allows us to satisfy the initial conditions for the problem in terms of the functions which for the solution. In the more general context, they provide different sets of functions that may be convenient series representations for different functions over different intervals.

Example 11.16. The problem arising from the the vibrating string with fixed endpoints

$$X''(x) + \lambda X(x) = 0$$

with $X(0) = 0$, and $X(L) = 0$, is a regular Sturm–Liouville boundary value problem with $p(x) = 1$, $q(x) = 0$, $r(x) = -1$, $\alpha_1 = \alpha_3 = 1$, and $\alpha_2 = \alpha_4 = 0$.

Example 11.17. The problem arising from the vibrating circular drum head with radius $r = 1$,

$$r^2 R''(r) + rR'(r) + \left(r^2\lambda - \gamma\right) = 0$$

with $R(1) = 0$ is not a regular Sturm–Liouville boundary value problem.

11.7 The Euler–Bernoulli Beam Equation

The *Euler–Bernoulli beam equation* is a partial differential equation describing small vibrations of beams. In contrast to strings, beams can support bending loads, which results in a higher-order partial differential equation describing its motion.

11.7.1 Derivation of the Beam Equation

Consider the cantilever beam illustrated in Figure 11.31. Assume that the beam is subjected to a distributed load that may vary in time $f(x,t)$, where the units of $f(x,t)$ are force per unit length. We make the following assumptions about the manner in which the beam deflects.

Assumption 11.1. *Assume the following.*

1. *The beam deflects in the vertical direction only and the deflection of the beam in the vertical direction is small.*
2. *The slope is also small.*
3. *Any planar cross-section of the beam remains planar when it is deflected.*

Consider the coordinate axes illustrated in Figure 11.31 with the y-axis directed into the page. Because the beam deflects in the z-direction only, all deflections remain within the plane of the page. Define the *neutral plane* to be the plane before deformation whose length is not changed when the beam is deformed. In Figure 11.32 the top of the beam is extended and the bottom of the beam is compressed. The neutral plane is illustrated by a dashed line. Let $u(x,t)$ represent the deflection of the beam's neutral plane in the negative z-direction at location x at time t from its unloaded equilibrium position, as is illustrated in Figure 11.32.

Now we may restate the assumption that the deformations are small with the expression $u(x,t) \ll 1$ and the assumption that the slope is small by $\partial u/\partial x \ll 1$. Having defined the neutral plane and coordinate axes we state another assumption.

Assumption 11.2.
Assume that a cross-section normal to the neutral plane does not change in height or width when the beam is deflected.

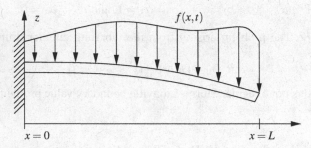

Fig. 11.31 Loaded beam.

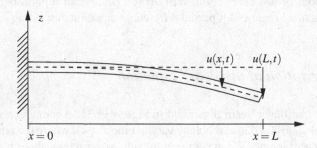

Fig. 11.32 Deflected beam.

To derive the equation of motion, consider a small segment of the beam, as is illustrated in Figure 11.33. Recall that u is defined to be positive in the downward direction and define the positive directions for the loading, shear, and moments to be as indicated in the figure. Let A be the cross-sectional area of the beam and ρ the density. Newton's law in the vertical direction for the segment gives

$$\rho A \frac{\partial^2 u}{\partial t^2}(x + \tfrac{1}{2}dx, t)dx = V(x + dx, t) - V(x, t) + \frac{1}{2}\left(f(x, t) + f(x + dx, t)\right)dx,$$

$$(11.69)$$

where the total applied load is computed as the average of $f(x, t)$ and $f(x + dx, t)$ times the length of the segment dx. Expanding $\partial^2 u / \partial t^2 \left(x + \tfrac{1}{2}dx, t\right)$, $V(x + dx, t)$, and $f(x + dx, t)$ in a Taylor series about x gives

$$\frac{\partial^2 u}{\partial t^2}\left(x + \frac{1}{2}dx, t\right) = \frac{\partial^2 u}{\partial t^2}(x, t) + \frac{\partial^3 u}{\partial t^2 \partial x}(x, t)\left(\frac{1}{2}dx\right) + \cdots$$

$$V(x + dx, t) = V(x, t) + \frac{\partial V}{\partial x}(x, t)dx + \cdots$$

$$f(x + dx, t) = f(x, t) + \frac{\partial f}{\partial x}(x, t)dx + \cdots.$$

Substituting into Equation (11.69) and ignoring the higher-order terms gives

$$\rho A \left(\frac{\partial^2 u}{\partial t^2}(x, t) + \frac{\partial^3 u}{\partial t^2 \partial x}(x, t)\left(\frac{1}{2}dx\right)\right)dx = \frac{\partial V}{\partial x}(x, t)dx$$

$$+ \frac{1}{2}\left(2f(x, t) + \frac{\partial f}{\partial x}(x, t)dx\right)dx,$$

or

$$\rho A \left(\frac{\partial^2 u}{\partial t^2}(x, t) + \frac{\partial^3 u}{\partial t^2 \partial x}(x, t)\left(\frac{1}{2}dx\right)\right) = \frac{\partial V}{\partial x}(x, t) + \frac{1}{2}\left(2f(x, t) + \frac{\partial f}{\partial x}(x, t)dx\right).$$

$$(11.70)$$

Taking the limit as $dx \to 0$ gives

$$\rho A \frac{\partial^2 u}{\partial t^2}(x, t) = \frac{\partial V}{\partial x}(x, t) + f(x, t). \qquad (11.71)$$

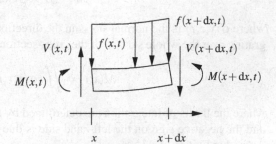

Fig. 11.33 Small segment of a beam.

Inasmuch as we are assuming the motion is only vertical, there is no angular acceleration, so the sum of the moments about any point must be zero. Computing the moments about the center of the right end of the beam segment in Figure 11.33 gives

$$-M(x,t) + M(x+dx,t) - V(x,t)dx + \frac{1}{2}(f(x,t)+f(x+dx,t))\frac{dx^2}{2} = 0,$$

where the moment due to the loading is approximated as the average load with an average moment arm of $dx/2$.

Using a Taylor series expansion for M and f,

$$M(x+dx) = M(x,t) + \frac{\partial M}{\partial x}(x,t)\,dx + \cdots$$

$$f(x+dx,t) = f(x,t) + \frac{\partial f}{\partial x}(x,t)dx + \cdots$$

and ignoring the higher-order terms gives

$$\frac{\partial M}{\partial x}(x,t)dx - V(x,t)dx + \frac{1}{2}\left(2f(x,t) + \frac{\partial f}{\partial x}(x,t)dx\right)\frac{dx^2}{2} = 0$$

or

$$\frac{\partial M}{\partial x}(x,t) - V(x,t) + \frac{1}{2}\left(2f(x,t) + \frac{\partial f}{\partial x}(x,t)dx\right)\frac{dx}{2} = 0.$$

Taking the limit as $dx \to 0$ gives

$$\frac{\partial M}{\partial x}(x,t) = V(x,t), \tag{11.72}$$

or, substituting into Equation (11.71) gives

$$\rho A \frac{\partial^2 u}{\partial t^2}(x,t) = \frac{\partial^2 M}{\partial x^2}(x,t) + f(x,t). \tag{11.73}$$

If we consider the normal stress over a small area of a cross-section, as illustrated in Figure 11.34, the moment due to the total force acting on that area is

$$-dM = z\sigma(x,z,t)dA,$$

where $\sigma(x,z,t)$ is the normal stress in the direction indicated in Figure 11.34. Integrating over the whole surface of the cross-section

$$-M(x,t) = \iint z\sigma(x,z,t)dzdy, \tag{11.74}$$

where the limits of integration are determined by the geometry of the cross-section and the negative sign on the left-hand side is due to the definition of a positive M, defined in Figure 11.33.

The basic constitutive law from solid mechanics is that normal stress and strain are related by

$$\sigma(x,z,t) = E\varepsilon(x,z,t), \tag{11.75}$$

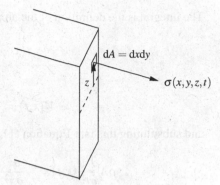

Fig. 11.34 Moment due to normal stress over a small area.

where E is the *modulus of elasticity* and has units of pascals, denoted by Pa where $1\text{Pa} = 1\text{N/m}^2$ and $\varepsilon(x,z,t)$ is the strain, which is dimensionless. Finally, to relate the strain to the deformation of the beam, consider the deflection of a small segment of the beam illustrated in Figure 11.35. Because the slope is small, $\theta \approx \partial u/\partial x$. Strain is defined as the displacement per unit length, therefore we have for the location z on the right face of the segment

$$
\begin{aligned}
\varepsilon(x,z,t) &= \frac{z\left(\sin\theta(x+dx,t) - \sin\theta(x,t)\right)}{dx} \\
&= \frac{z\left(\frac{\partial u}{\partial x}(x+dx,t) - \frac{\partial u}{\partial x}(x,t)\right)}{dx} \\
&= \frac{z\left(\left(\frac{\partial u}{\partial x}(x,t) + \left(\frac{\partial^2 u}{\partial x^2}(x,t)\right)dx\right) - \frac{\partial u}{\partial x}(x,t)\right)}{dx}(x,t) \\
&= z\frac{\partial^2 u}{\partial x^2}(x,t).
\end{aligned}
$$

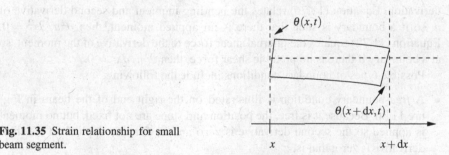

Fig. 11.35 Strain relationship for small beam segment.

Substituting this into Equation (11.75), and using that in Equation (11.74) gives

$$
-M(x,t) = \iint z^2 E\frac{\partial^2 u}{\partial x}(x,t)dydz = E\frac{\partial^2 u}{\partial x^2}(x,t)\iint z^2 dydz.
$$

The integral is the definition of the area moment of inertia, thus if we let

$$I(x) = \iint z^2 \mathrm{d}y\mathrm{d}z$$

we have

$$M(x,t) = -EI(x)\frac{\partial^2 u}{\partial x^2}(x,t) \tag{11.76}$$

and substituting this into Equation (11.73) gives

$$\rho A\frac{\partial^2 u}{\partial t^2}(x,t) = -\frac{\partial^2}{\partial x^2}\left(EI(x)\frac{\partial^2 u}{\partial x^2}(x,t)\right) + f(x,t).$$

Finally, if the cross-section of the beam is uniform along its length, then we have

$$\boxed{\rho A\frac{\partial^2 u}{\partial t^2}(x,t) = -EI\frac{\partial^4 u}{\partial x^4}(x,t) + f(x,t).} \tag{11.77}$$

If the beam is not loaded, then this may be written as

$$\boxed{\frac{\partial^2 u}{\partial t^2}(x,t) + \alpha^2\frac{\partial^4 u}{\partial x^4}(x,t) = 0} \tag{11.78}$$

and $\alpha^2 = EI/\rho A$.

11.7.2 Boundary Conditions for the Beam Equation

Because the beam is fourth-order in the spatial dimension four boundary conditions are needed to solve it. Possible boundary conditions include terms up to the third derivative in x, and the key to interpreting the higher derivative terms appear in the derivation. Equation (11.76) relates the bending moment and second derivative of u, so if a boundary is such that there is no applied moment, then $\partial^2 u/\partial x^2 = 0$. Equation (11.72) equates the internal shear force to the derivative of the moment, so at a boundary if there is no possible shear force, then $\partial^3 u/\partial x^3 = 0$.

Possible types of boundary conditions include the following.

- A *free* boundary condition is illustrated on the right end of the beam in Figure 11.36. Because it is free, the position and slope are not fixed, but no moment is applied so the second derivative is zero and no force is applied so the third derivative is zero; that is,

$$\frac{\partial^2 u}{\partial x^2}(L,t) = 0, \quad \frac{\partial^3 u}{\partial x^3}(L,t) = 0.$$

- A *cantilevered* boundary condition fixes both the position and the slope, such as a flagpole, and is illustrated on the left end of the beam in Figure 11.36. Because the position and slope are fixed geometrically, the boundary conditions are

$$u(0,t) = 0, \quad \frac{\partial u}{\partial x}(0,t) = 0.$$

- A *simply supported* boundary condition fixes the position of the support but allows it to freely rotate so that there is no applied moment at the attachment point. Conceptually, it is a pinned attachment as is illustrated on the right end of the beam in Figure 11.37 and is specified by

$$u(L,t) = 0, \quad \frac{\partial^2 u}{\partial x^2}(L,t) = 0.$$

- The left side of Figure 11.37 illustrates a type of boundary condition where the the slope and shear are zero, but the position and moment are unconstrained; that is,

$$\frac{\partial u}{\partial x}(0,t) = 0, \quad \frac{\partial^2 u}{\partial x^2}(0,t) = 0.$$

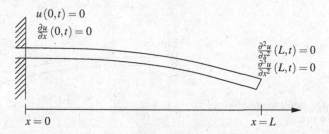

Fig. 11.36 Beam with cantilever boundary condition ($x = 0$) and free boundary condition ($x = L$).

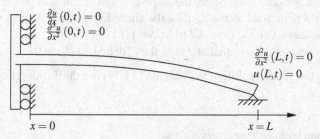

Fig. 11.37 Beam with a simply supported boundary condition at ($x = L$) and a boundary condition with fixed slope and shear ($x = 0$).

11.7.3 Solutions to the Beam Equation

Because the beam equation is fourth-order in the spatial dimension, even the static case is of interest. First a few static cases are solved, and then the time-varying case is addressed.

11.7.3.1 Static Deflection

Let us consider the case where

$$\frac{\partial^2 u}{\partial t^2}(x,t) = 0.$$

First, we consider the case where the beam is cantilever and subjected to a static force at the end, as illustrated in Figure 11.38. Because there is no acceleration and the solution does not depend upon time, the beam equation reduces to

$$EI\frac{d^4 u}{dx^4}(x) = 0. \tag{11.79}$$

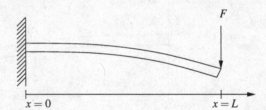

Fig. 11.38 Cantilever beam subjected to a force at the end.

The four boundary conditions are as follows.

1. The beam is fixed at zero, thus $u(0) = 0$.
2. The beam is a cantilever beam, the slope at zero is zero, thus $(du/dx)(0) = 0$.
3. There is a point load at $x = L$, thus the shear force at $x = L$ must equal F, and using Equation (11.72) gives $EI\left(d^3 u/dx^2\right)(L) = F$.
4. There is no moment applied at $x = L$, thus $\left(d^2 u/dx^2\right)(L) = 0$.

Clearly, the general solution to Equation (11.79) is a third-order polynomial in x,

$$u(x) = c_1 x^3 + c_2 x^2 + c_3 x + c_4.$$

Applying the boundary conditions gives

1. Fixed at zero:

$$u(0) = 0 \implies c_4 = 0;$$

2. Cantilever:

$$\frac{du}{dx}(0) = 0 \quad \Longrightarrow \quad c_3 = 0;$$

3. Shear at end:

$$6EIc_1 = F \quad \Longrightarrow \quad c_1 = \frac{F}{6EI};$$

and
4. No moment at end:

$$\frac{F}{6EI}6L + 2c_2 = 0 \quad \Longrightarrow \quad c_2 = -\frac{FL}{2EI}.$$

Hence

$$u(x) = \frac{F}{6EI}x^3 - \frac{FL}{2EI}x^2$$

and at $x = L$, an applied force of F produces a displacement of

$$u(L) = \frac{FL^3}{6EI} - \frac{FL}{2EI}L^2 = -\frac{L^3}{3EI}F.$$

Because the displacement is proportional to the applied force and the proportionality constant is $L^3/6EI$, we can conclude that a cantilever spring will have a spring constant of

$$k = \frac{3EI}{L^3}.$$

In the case of a rectangular beam with width w and height h,

$$I = \iint z^2 \mathrm{dzdy} = \frac{1}{12}wh^3$$

so

$$k = \frac{Ewh^3}{4L^3}.$$

11.7.3.2 Dynamic Solutions for the Cantilever–Free Beam

Only the simply supported beam has a straightforward solution procedure, so that one is left as an exercise. This section partially solves for the dynamic response of the cantilever beam illustrated in Figure 11.36 by determining the mode shapes and corresponding frequencies that make up the solution. Satisfying the initial conditions is beyond the scope of this book.

If the applied load is zero, then the beam equation can be expressed as

$$\frac{\partial^2 u}{\partial^2 t} + \alpha^2 \frac{\partial^4 u}{\partial x^4} = 0, \tag{11.80}$$

where

$$\alpha^2 = \frac{EI}{\rho A},$$

and the boundary conditions are

$$u(0,t) = 0, \quad \frac{\partial u}{\partial x}(0,t) = 0, \quad \frac{\partial^2 u}{\partial x^2}(L,t) = 0, \quad \frac{\partial^3 u}{\partial x^3}(L,t) = 0.$$

Using separation of variables, assuming

$$u(x,t) = X(x)T(t)$$

and substituting gives

$$X(x)T''(t) + \alpha^2 X''''(x)T(t) = 0 \quad \Longrightarrow \quad \frac{1}{\alpha^2}\frac{T''(t)}{T(t)} = -\frac{X''''(x)}{X(x)} = -\lambda^2 \quad (11.81)$$

which gives the two ordinary differential equations

$$X''''(x) - \lambda^2 X(x) = 0, \quad T''(t) + \alpha^2\lambda^2 T(t) = 0.$$

We used a little foresight in Equation (11.81) by setting the ratios equal to $-\lambda^2$. This makes some sense because the form of the ordinary differential equation for $T(t)$ will give oscillatory solutions, but at this point it is not necessarily guaranteed to work until we check what form the solutions to the $X(x)$ equations have and whether they can satisfy the boundary conditions.

The equation for $X(x)$ is linear, therefore it has solutions of the form

$$X(x) = e^{rx}.$$

Substituting and assuming $\lambda > 0$ gives

$$r^4 = \lambda^2 \quad \Longrightarrow \quad r^2 = \pm\lambda \quad \Longrightarrow \quad r = \pm\sqrt{\lambda}, \pm i\sqrt{\lambda}$$

or

$$X(x) = k_1 e^{\sqrt{\lambda}x} + k_2 e^{-\sqrt{\lambda}x} + k_3 \cos\left(\sqrt{\lambda}x\right) + k_4 \sin\left(\sqrt{\lambda}x\right).$$

It turns out that regrouping the solutions makes determining the coefficients that satisfy the boundary conditions easier. Recall the *hyperbolic functions*

$$\cosh x = \frac{1}{2}\left(e^x + e^{-x}\right), \quad \sinh x = \frac{1}{2}\left(e^x - e^{-x}\right).$$

Note with these definitions, $\cosh(0) = 1$, $\sinh(0) = 0$,

$$\frac{d}{dx}\cosh\sqrt{\lambda}x = \sqrt{\lambda}\sinh x, \quad \frac{d}{dx}\sinh\sqrt{\lambda}x = \sqrt{\lambda}\cosh x.$$

Using the hyperbolic trigonometric functions, $X(x)$ may be written as

$$X(x) = c_1 \left(\cos \sqrt{\lambda}x + \cosh \sqrt{\lambda}x\right) + c_2 \left(\cos \sqrt{\lambda}x - \cosh \sqrt{\lambda}x\right)$$
$$+ c_3 \left(\sin \sqrt{\lambda}x + \sinh \sqrt{\lambda}x\right) + c_4 \left(\sin \sqrt{\lambda}x - \sinh \sqrt{\lambda}x\right), \tag{11.82}$$

and also

$$X'(x) = \sqrt{\lambda} \left[c_1 \left(-\sin \sqrt{\lambda}x + \sinh \sqrt{\lambda}x\right) + c_2 \left(-\sin \sqrt{\lambda}x - \sinh \sqrt{\lambda}x\right) \right.$$
$$\left. + c_3 \left(\cos \sqrt{\lambda}x + \cosh \sqrt{\lambda}x\right) + c_4 \left(\cos \sqrt{\lambda}x - \cosh \sqrt{\lambda}x\right) \right].$$

Applying the boundary conditions at $x = 0$ gives the simple results

$$u(0,t) = 0 \implies c_1 = 0$$
$$\frac{\partial u}{\partial x}(0,t) = 0 \implies c_3 = 0.$$

For the boundary conditions at $x = L$ we need the second and third derivatives, which, when $c_1 = c_3 = 0$ are given by

$$X''(x) = \lambda \left[c_2 \left(-\cos \sqrt{\lambda}x - \cosh \sqrt{\lambda}x\right) + c_4 \left(-\sin \sqrt{\lambda}x - \sinh \sqrt{\lambda}x\right) \right]$$
$$X'''(x) = \lambda \sqrt{\lambda} \left[c_2 \left(\sin \sqrt{\lambda}x - \sinh \sqrt{\lambda}x\right) + c_4 \left(-\cos \sqrt{\lambda}x - \cosh \sqrt{\lambda}x\right) \right].$$

Applying the boundary conditions gives

$$\lambda \left[c_2 \left(-\cos \sqrt{\lambda}L - \cosh \sqrt{\lambda}L\right) + c_4 \left(-\sin \sqrt{\lambda}L - \sinh \sqrt{\lambda}L\right) \right] = 0 \tag{11.83}$$
$$\lambda \sqrt{\lambda} \left[c_2 \left(\sin \sqrt{\lambda}L - \sinh \sqrt{\lambda}L\right) + c_4 \left(-\cos \sqrt{\lambda}L - \cosh \sqrt{\lambda}L\right) \right] = 0. \tag{11.84}$$

Because $\lambda = 0$ would give a constant solution, $\lambda \neq 0$. Solving Equation (11.83) for c_2 gives

$$c_2 = -c_4 \frac{\sin \sqrt{\lambda}L + \sinh \sqrt{\lambda}L}{\cos \sqrt{\lambda}L + \cosh \sqrt{\lambda}L} \tag{11.85}$$

and substituting into Equation (11.84) gives

$$\left(\cos \sqrt{\lambda}L + \cosh \sqrt{\lambda}L\right) + \left(\sin \sqrt{\lambda}L - \sinh \sqrt{\lambda}L\right) \frac{\sin \sqrt{\lambda}L + \sinh \sqrt{\lambda}L}{\cos \sqrt{\lambda}L + \cosh \sqrt{\lambda}L} = 0. \tag{11.86}$$

A plot of the left-hand side of Equation (11.86) is illustrated in Figure 11.39. Each point where the curve passes through zero corresponds to a value for λ that satisfies Equation (11.86). Call this increasing sequence of values $\lambda_1, \lambda_2, \ldots, \lambda_n, \ldots$. Numerically solving for the first seven eigenvalues for $L = 1$ gives the frequencies listed in Table 11.1.

For a given λ_n value, the solution to

Eigenvalue	Frequency s^{-1}	Frequency Hz
λ_1	3.51	0.56
λ_2	22.03	3.51
λ_3	61.70	9.82
λ_4	120.90	19.24
λ_5	199.86	31.81
λ_6	298.56	47.52
λ_7	416.99	66.37

Table 11.1 Frequencies for cantilever beam

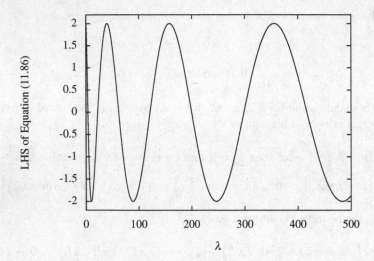

Fig. 11.39 Solutions to Equation (11.86) for $L = 1$ correspond to values of λ where the curve passes through zero.

$$X''''(x) - \lambda X(x) = 0$$

that satisfies all four boundary conditions is given substituting for c_2 from Equation (11.85) and for λ_n to give

$$X_n(x) = -\frac{\sin\sqrt{\lambda_n}L + \sinh\sqrt{\lambda_n}L}{\cos\sqrt{\lambda_n}L + \cosh\sqrt{\lambda_n}L}\left(\cos\sqrt{\lambda_n}x - \cosh\sqrt{\lambda_n}x\right) +$$

$$\left(\sin\sqrt{\lambda_n}x - \sinh\sqrt{\lambda_n}x\right). \quad (11.87)$$

A plot of the shapes of the first three modes is illustrated in Figure 11.40.

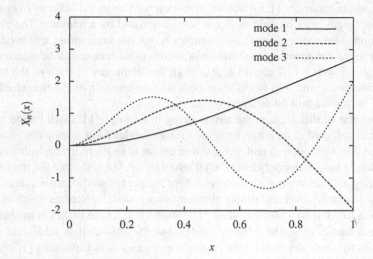

Fig. 11.40 First three mode shapes for an oscillating cantilever beam.

11.8 A Musical Interlude

Because the wave equation describes many phenomena related to music, this section gives a brief overview of some of the fundamentals of music. It does not address much of what should be considered *music theory*, meaning the structure of musical compositions, how they are composed, and the emotional impact on the listener, but rather the fundamental aspects of sound and its relationship to notes and music. Specifically, it outlines the structure of musical scales and then considers some of the mathematical details of why some notes sound good together and others do not. Interested readers are referred to [49] and [6] for a more detailed mathematical treatment and a historical treatment, respectively, of tuning and timbre, and [31] for a cognitive and psychological overview of music. Most of this section is an overview of material from [49].

As was illustrated by the solution to the wave equation, typically when a musical instrument plays a "note" it is actually made up of acoustic waves at many frequencies. This was manifested mathematically by the fact that when the string was set in motion, the motion of the string was the combination of an infinite number of modes. When the string was fixed at each end, each mode was sinusoidal in shape (dictated by the boundary conditions) with a different temporal frequency. In Equation (11.20), the shape of the mode was described by the $\sin(n\pi x/L)$ term and the motion in time was described by the $\sin(\alpha n\pi t/L)$ and $\cos(\alpha n\pi t/L)$ terms, where the frequency of motion is $(\alpha n)/(2L) = (\tau n)/(2\rho L)$ Hz, where τ is the tension in the string, ρ is the mass per unit length, and L is the overall length of the string.

From Examples 11.2 and 11.3, a string will have different solutions depending upon the manner in which it is set in motion. In Example 11.2, the string was plucked, and in Example 11.3 it was set in motion with an initial velocity over a portion of its length, which could model it being impacted by a hammer. The critical observation to take away from these examples is that the same string will produce a different sound because the combination of modes in each case will be different, as illustrated in Figures 11.7 and 11.8. Although the frequency of each of the modes is the same, the sound will be different because combination of the magnitudes of many of the modes will be different.

For the rest of this section, the struck string in Example 11.3 will be the prototypical example and is meant to serve as a rough model for a piano. The reader is cautioned that, obviously, a real musical instrument is much more complicated and the resulting sound is not only from the vibrating string, but also from the interaction of the string and acoustics with the structure of the piano, nonlinearities, and so on. However, as shown, even the simple vibrating string model describes much of what contributes to what we consider music. The sound that a note makes is made up of all the frequencies of all the different modes, and the amount that a different mode contributes to a note depends on the coefficients a_n and b_n in Equation (11.20). The "frequency" of a note is the *fundamental frequency*, which is the frequency of the lowest mode. Due to the fact that a_n and b_n decrease in magnitude as n gets large, which is necessary for the series to converge, the fundamental frequency usually has the largest magnitude, which means it is the loudest.

A basic feature of almost all music across all cultures is that if the frequency of a musical note is doubled or halved, the note sounds "the same," even to the point of being interchangeable. A manifestation of this fact is that when an adult and child sing a song together, although they are singing the same song with the same notes, the child sings the notes at a frequency that is, for example, double that of the adult. In music, the interval between two frequencies where the higher frequency is double that of the lower frequency is called an *octave*. A *scale* is a selection of which frequencies in an octave will be used as musical notes.

In modern western music, 12 frequencies (notes) are selected to make up an octave.[7] Other cultures have selected a different number of notes. After deciding the number of notes in an octave, the exact frequencies of the notes must be decided. If it were left to engineers, the frequency ratio between two adjacent notes would be $2^{1/12}$, so that whenever the 12 pitches in an octave are traversed the resulting frequency ratio would be

$$\left(2^{1/12}\right)^{12} = 2.$$

However, even if the number of pitches between octaves is settled at 12, it is not necessarily the case that all adjacent notes having the same frequency ratio results in music that sounds the best. To address this, we need to consider the concepts of the pair of antonyms *consonance* and *dissonance*. Roughly, two sounds are consonant when they qualitatively sound "good" together, and are dissonant when they sound "bad" together.[8] Although doing complete justice to neither concept, the following

---

[7] One would think there would be eight notes.

[8] Note that even at this basic level, not all music should be composed with consonant sounds because some music is supposed to sound dissonant. A famous example is the so-called "Tristan

two subsection are based on consonance and dissonance, respectively. First, using a very simple notion of consonance, a discussion of the complexity and history of determining which notes should be in an octave is presented. Then, a mathematical model of dissonance is presented and related to our simplified impacted string model. As shown, this captures much of the essence of why some chords (combinations of notes) sound good and others sound bad.

As a reference, a portion of the standard piano keyboard is illustrated in Figure 11.41 with each key labeled with the standard note name. The white keys correspond to notes named with the letters A through G. The black keys are "sharps" and "flats," indicated by $\sharp$ and $\flat$, respectively. Counting both the black and white keys, there are 12 pitches from any given note until it repeats. A repeating note corresponds to an octave and will have a frequency twice as large as the frequency corresponding to the note an octave lower.

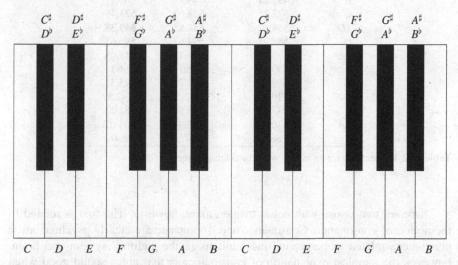

Fig. 11.41 Portion of a standard piano keyboard.

11.8.1 Defining Scales

Independent of the issue of frequency ratios, if more than one musical instrument is played at the same time, an absolute standard for some note must be established. A common standard is *Concert A*, which is the A note above middle C^9 which is set at

---

chord," which is the first chord in the opera *Tristan und Isolde*, by Richard Wagner, which sets the tone, literally, for the tragic plot of the opera.

^{9} On a piano, middle C is the fourth C, and in the usual western musical notation, middle C is located between the two staves in printed music.

440 Hz. As mentioned, an octave is a frequency interval where the higher frequency is twice that of the lower frequency, so the next A (an octave higher) will have a frequency of 880 Hz, the A above that will be at 1760 Hz, and so on.

If it is accepted that there should be 12 pitches in an octave, the next issue to resolve is what the frequency ratio between any of the pitches should be. In the mathematically appealing *equal temperament* the ratio of the pitches between adjacent notes is $2^{1/12}$. The names of the notes and the frequencies for equal temperament are listed in Table 11.2 in the first and second columns, respectively.

Note name	Frequency Hz (equal temperament)	Frequency Hz (perfect fifths)	Frequency Hz (wolf at $F^\sharp$)
A	440	440	440
$A^\sharp/B^\flat$	466.16	469.86	463.54
B	493.88	495	495
C	523.25	528.60	521.48
$C^\sharp/D^\flat$	554.37	556.88	549.38
D	587.33	594.67	586.67
$D^\sharp/E^\flat$	622.25	626.48	618.05
E	659.26	660	660
F	698.46	704.79	695.31
$F^\sharp/G^\flat$	739.99	742.5	742.5
G	783.99	792.89	782.22
$G^\sharp/A^\flat$	830.61	835.31	824.07
A	880	892	880

Table 11.2 Frequencies for an octave with equal temperament

There are two issues with equal temperament, however. The first is related to the notion of consonance. Consonance and dissonance are clearly psychoacoustic phenomena related to perceptual mechanisms in the auditory system and brain, however, the simplest explanation of consonance is that notes sound good when they are played together when the frequency intervals are ratios of small numbers. In other words, an octave (2/1), or a *perfect fifth*[10] (3/2) are consonant and sound good; whereas, the ratio of two adjacent notes ($2^{1/12} \approx 2119/2000$), in the words of Galileo, "keep the eardrums in perpetual torment" (quoted by [49]).

Referring to the second column in Table 11.2, the notes A and E are nearly, but not quite, in the 3/2 ratio. If E were tuned to 660 Hz instead of 659.26, it would presumably sound better with the A. Furthermore, it seems reasonable that before instrumentation existed to accurately measure frequency, tuning the E to be a perfect fifth above the A would be relatively easy to do simply by listening to the two notes together. Furthermore, because most instruments create notes that can be related to a geometric length, physically measuring a length that is 2/3 as long as the lower

---

[10] Note that a perfect fifth is the note that is *seven* notes above the fundamental.

note is also relatively easy to do.[11] Given that it is easy to hear an octave and perfect fifth, as well as to measure lengths that would produce such frequency ratios, it is not surprising that methods of tuning were developed that were based upon the fact that it is possible to work up and down successively by fifths to define the 12 notes in an octave. This is called the *Pythagorean scale*.

To build a scale using perfect fifths and octaves, consider starting with A at 440 Hz. The perfect fifth seven pitches above A is at E, which should be 660 Hz. The fifth above that will be the B above the octave, so dividing 660 by 2 to get 330 Hz and then determining the perfect fifth, gives the frequency for B in this octave which should be at $330(3/2) = 495$ Hz. The $F^\sharp$ is a fifth above the B, which should have a frequency of $495(3/2) = 742.5$ Hz, and so on. So, an algorithm to implement this to define the notes in an octave would be to start with A at 440 Hz and to do the following recursively.

- If seven pitches above the current note are within the octave of interest, then set that note to be $3/2$ of the frequency of the current note.
- If seven pitches above the current note are outside the octave, then divide the frequency by 2 and then multiply by $3/2$ to give the frequency of that note within the octave.

This results in the frequencies illustrated in the third column in Table 11.2. The ratio of any notes separated by a fifth will be exactly $3/2$. The obvious problem with this method is that the octave is no longer right! This should not be too surprising; if 12 equal pitches in an octave did not exactly give a perfect fifth, then building up the notes for perfect fifths should not be expected to give an exact octave. The amount by which the octave is off is called the *Pythagorean comma*.

One way to reconcile perfect fifths and the octave would be to make only one of the fifths not be exact, and to have it be off by an amount that will make the octave be the desired double frequency. This was commonly done and the note that did not have a perfect fifth above it was called the *wolf*. The fourth column in Table 11.2 is the Pythagorean scale with the wolf at $F^\sharp$. To see that this is indeed the wolf, take the frequency corresponding to any note in the fourth column and multiply it by $3/2$ and if the result is greater than 880, then divide the result by 2. The resulting frequency will correspond to the note seven pitches above (or five pitches below) the starting note, except for $F^\sharp$.

One way to compare the equal-tempered scale and the Pythagorean scale with a wolf, is that the former takes the Pythagorean comma, and instead of allocating all of it to the wolf, it spreads $1/12$ of the comma across all the notes. It seems that this would not be too bad, and indeed, the frequency for E is only 0.74 Hz lower than what would be a perfect fifth above the A. Unfortunately, some of the other intervals are not as close. For example, the *just minor third* has a ratio of $6/5$, and for a scale starting at 440 Hz, this would correspond to 528 Hz. In the equal-tempered scale in column two, the C is the closest note and this is off by a relatively large amount,

---

[11] According to [49], Pythagoras, of the $a^2 + b^2 = c^2$ fame, was the first to observe that the pitch of a string is related to its length. When the length is halved, the pitch increases by an octave. When the length is reduced by $2/3$, a fifth results; similarly, reducing the length by $3/4$ results in a fourth.

4.75 Hz. Similarly, the *just major third* (5/4), the *just minor sixth* (8/5), and the just major sixth (5/3) do not fare well either.

There are many other ways to define the 12 notes in an octave, and mathematically one way to consider them would be the means by which the wolf is allocated across the notes, and each method would have corresponding benefits and detriments. For example, *just intonation* is designed to make many of the thirds and fifths perfect, at the expense of the other thirds and fifths being worse. The drawback to this approach is that, depending on the composition, different just intonation tunings would be necessary to get the thirds and fifths that appear in the composition to be the ones that are perfectly tuned. *Meantone temperaments* can be considered ways to systematically spread the wolf through parts of the scale. For example, the wolf can be cut into four pieces and put in four different places, and so on. Even attempting a complete listing of ways that the 12 notes in an octave have been defined in western music is beyond the scope of this text and the interested reader is referred to the references listed at the beginning of this section.

Finally, regardless of the exact definition of the frequencies of the notes, the scale considered so far has 12 notes. This is called the *chromatic scale*. A reader that has studied Western music may be more familiar with scales that have seven notes ("do, re, me, fa, so, la, ti" and "do"). The common ones are called *diatonic scales*, which are made of of five steps that skip a note and two steps that do not. Because such scales are so common, the steps that skip a note are called whole steps, and the steps that do not are called half steps, and in the diatonic scales the half steps are placed as far apart as possible, meaning there are either two or three whole steps between each of the half steps.

The distinction between *major* and *minor* scales is the placement of the half steps. The *C major* is easy to read from the keyboard in Figure 11.41 because it starts with *C* and includes only the white keys up to the next *C*. So, the steps are whole, whole, half, whole, whole, whole and half. All 12 major scales have the same sequence of whole and half steps, but start with different notes. The *A natural minor* scale starts with an *A* and includes all the white keys to the next *A*, so the step are whole, half, whole, whole, half, whole and whole. The *natural minor* scales have the same sequence of steps.[12] Musical compositions are often composed primarily from the seven notes in a key, and because of the relationship between the notes in a key, music can have a qualitatively different sound in different keys.

11.8.2 A Simple Model for Dissonance

In our study of vibrations we have already encountered a basic phenomenon related to music and acoustics, which is commonly called *beating*. Figure 4.12 illustrated the solution to a mass–spring system forced near resonance from Example 4.6. The solution was the superposition of two cosines with slightly different frequencies

---

[12] There are also melodic and harmonic minor scales.

(approximately 0.16711 and 0.15915 Hz, respectively) and because the frequencies were close, the two solutions would slowly alternate between being in and out of phase, resulting in an obvious structure to the solution that is related to the difference between the two frequencies. In this example, $1/(0.16711 - 0.15915) = 125.63$, which is the time in seconds it takes the "beat" to repeat.

In acoustics, when two instruments play notes that differ by a small amount, say 1 Hz, this beating is easily perceptible. To model the unpleasantness of two tones, one experiment would be to play two sine waves at difference frequencies and ask listeners to indicate a level of unpleasantness. In [49], a mathematical model for dissonance was presented that is given by

$$d(f_1, f_2, l_1, l_2) = l_{12} \left(e^{-3.5s(f_2 - f_1)} - e^{-5.75s(f_2 - f_1)} \right) \qquad (11.88)$$

where f_1 is the lower frequency, f_2 is the higher frequency, l_1 and l_2 are the amplitudes of the two frequencies, and l_{12} is the minimum of the two amplitudes. The parameter s is given by

$$s = \frac{0.24}{0.021 f_1 + 19}. \qquad (11.89)$$

All the numerical values in Equations (11.88) and (11.89) are from least-squares optimization of the assumed form of the curves to experimental data based upon people listening to the two frequencies played at the same time and indicating the degree of pleasantness or unpleasantness of the sound. Using Concert A as a base frequency, $f_1 = 440$ and $l_1 = l_2 = 1$, a plot of the dissonance curve for frequencies from Concert A over an octave is illustrated in Figure 11.42. A sinusoid at 440 Hz sounds good with itself, so the measure of dissonance starts at zero. The peak seems to be near 470 Hz, and then tapers back to zero. This represents the phenomenon that beating is not necessarily a completely unpleasant sound (unless your goal is to tune two instruments, presumably), but somewhere between two notes being nearly the same and very distinct, there is a region of maximum dissonance.

Figure 11.42 measures the dissonance of pure sinusoids. However, any note from the simple piano model is made up of an infinite number of modes (but only a finite number are within the auditory range for humans), so for real instruments any measure of dissonance must take into account all the modes and the relationship between them. For example, on a piano two notes may be played where the fundamental frequencies are separated enough that the dissonance function for them would be relatively small. However, if some of the harmonics are related in a way where there is high dissonance, then as notes on the piano they will clash. To extend the dissonance model to the vibrating string example, we simply add the measure of dissonance for all the modes for each note.

Example 11.18. This example constructs the dissonance function for one octave starting at Concert A for our model of a piano. The frequency for the nth mode of the wave equation is $\alpha n \pi / L$ in radians per second, thus if α and L are picked so that $\alpha / (2L) = 440$, the string will have a fundamental frequency of 440 Hz. If we set $L = 1$, which would be realistic for a piano, then $\alpha = 880$. We model the

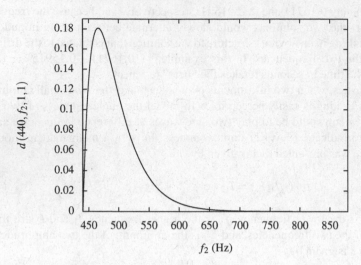

Fig. 11.42 Plot of the empirical dissonance function.

impact of the hammer as an initial condition where $1/100$ of the string is given an initial velocity of a magnitude of 100 over the segment of the string from $x = 0.09$ to $x = 0.1$.

Referring to the summary of the solution to the wave equation and using the parameter values from above, the solution is

$$u(x,t) = \sum_{n=1}^{\infty} \frac{200}{880n^2\pi^2} \left[\cos(0.1n\pi) - \cos(0.09n\pi)\right] \sin(n\pi x) \sin(880n\pi t).$$

Setting

$$l_{1,n} = \frac{200}{880n^2\pi^2} \left[\cos(0.1n\pi) - \cos(0.09n\pi)\right]$$

$$l_{2,m} = \frac{200}{2f_2n^2\pi^2} \left[\cos(0.1n\pi) - \cos(0.09n\pi)\right]$$

and $f_1 = 440$, then the sum of the dissonance for each pair of modes for each note is given by

$$d(f_2) = \sum_{n=1}^{\infty} \sum_{m=1}^{\infty} d\left((2n\pi)f_1, (2m\pi)f_2, l_{1,n}, l_{2,m}\right).$$

A plot of $d(f_2)$ versus f_2 is illustrated in Figure 11.43. Because the upper threshold of hearing for humans is near 20,000 Hz, $d(f_2)$ was only computed through the 46th mode, which has a frequency of approximately 20,240 Hz. Because of the very small segment of length over which the impact occurred, the magnitude of the higher modes decays very slowly, as illustrated in Figure 11.44, which makes the harmonics very important in computing the dissonance function. Note that many of the minima in the dissonance curve occur at ratios of the fundamental frequencies that are ratios of small integers.

- The minor third is at $(6/5)440 = 528$ Hz.
- The major third is at $(5/4)440 = 550$ Hz.
- The perfect fourth is at $(4/3)440 = 586.67$ Hz.
- The tritone is at $(7/5)440 = 616$ Hz.
- The perfect fifth is at $(3/2)440 = 660$ Hz.
- The major sixth is at $(5/3)440 = 733.33$ Hz.

This is because the harmonics are of similar ratios, and hence are unlikely to occur at intervals with high dissonance values.

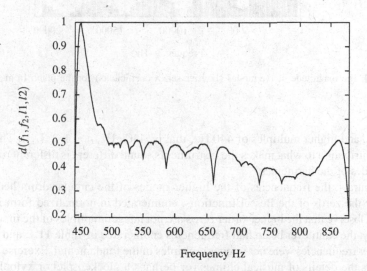

Fig. 11.43 The dissipation function for the piano mode in Example 11.18.

11.8.3 Percussion Instruments

The dynamics of a few percussion instruments have been studied, either in the main text or in the exercises at the end of the chapter. One feature to observe in many percussion instruments such as vibrating circular drum heads and oscillating beams is that the frequencies of the higher-order modes are not integer multiples of the fundamental frequency. In contrast, the harmonics of the one-dimensional wave equation, which models many wind instruments, pianos, guitars, and other stringed instruments, are integer multiples of the fundamental frequency.

For example, when a piano or guitar plays the note corresponding to Concert A, the fundamental frequency is at 440 Hz and all the harmonics are at 880 Hz,

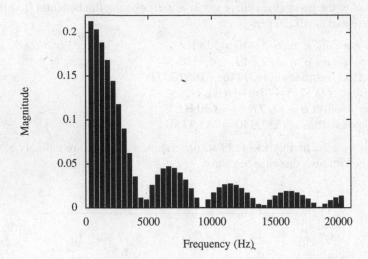

Fig. 11.44 The magnitude of the modes (Fourier series coefficients) for the piano from Example 11.18.

1320 Hz, and higher multiples of 440 Hz, that is, $440n$ Hz, $n = 1, 2, 3, \ldots$. The primary contribution to what makes the instruments sound different is different relative magnitudes of the harmonics.

In contrast, the frequencies of the higher modes of the circular drum head are given by the zeros of the Bessel functions, enumerated in normalized form in Table 5.2. Observe that the higher-order zeros are not integer multiples of the first zero. Similarly, the cantilever beam had frequencies enumerated in Table 11.1, and again the higher frequencies were not integer multiples of the fundamental. Exercise 11.26 considers the details of musical chimes (or perhaps a glockenspiel or xylophone), which have a similar characteristic. It is perhaps the nature of this relationship between the fundamental frequency and the harmonics that makes percussion instruments sound qualitatively different, and even less musical or more primitive, than other instruments.

11.9 Additional Topics

This chapter only considered solving the simplest types of partial differential equations and one approach to solving them, namely, separation of variables. The reader is cautioned not to conclude this is the only solution method for partial differential equations, and, in fact, entire courses exist that focus on subsets of the solution methods which exist for partial differential equations. A list of some other approaches, adapted from [14], is as follows.

1. Integral transforms may be used to reduce the number of independent variables by one. This would be useful in the case where there are two independent variables, because the transform would reduce the number of independent variables to one, which would then make the equation an ordinary differential equation.
2. A change of coordinates may be used to convert the partial differential equation to either a simpler partial differential equation or an ordinary differential equation.
3. A transformation of the dependent variable may also simplify the equation.
4. Numerical methods, the most basic approach of which is presented in Section 12.6.
5. Perturbation methods can transform a single nonlinear equation with a small nonlinearity into a sequence of linear equations.
6. Impulse-response methods may decompose initial and boundary conditions into a set of simple responses, which are then combined to form the solution.
7. A partial differential equation can sometimes be converted into an integral equation which may have known solution methods.
8. Calculus of variations may be used if the solution to the partial differential equation can be reformulated into an optimization problem.

11.10 Exercises

11.1. Compute the Fourier series for each of the following functions.

1. $f(x) = f(x+8\pi)$ and

$$f(x) = \begin{cases} 1, & 0 < x \leq \pi, \\ 0, & \pi < x \leq 7\pi, \\ -1, & 7\pi < x \leq 8\pi. \end{cases}$$

2. $f(x) = f(x+3)$ and

$$f(x) = \begin{cases} 0, & 0 < x \leq 1, \\ 1, & 1 < x \leq 2, \\ 2, & 2 < x \leq 3. \end{cases}$$

3. $f(x) = f(x+2)$ and

$$f(x) = \begin{cases} -2, & 0 < x \leq 1, \\ 8, & 1 < x \leq 2. \end{cases}$$

4. $f(x) = f(x+6)$ and

$$f(x) = \begin{cases} -1, & -3 < x \leq -1, \\ 1, & -1 < x \leq 1, \\ -1, & 1 < x \leq 3. \end{cases}$$

5. $f(x) = f(x+2)$ and

$$f(x) = \begin{cases} x, & -1 < x \leq 0, \\ x+1, & 0 < x \leq 1. \end{cases}$$

In each case, check your answer by plotting partial sums of the Fourier series to compare with the given function.

11.2. Consider the mass–spring–damper system illustrated in Figure 4.1. Let

$$f(t) = \begin{cases} 1, & 0 \leq t < 1 \\ 0, & 1 \leq t < 2, \end{cases}$$

and $f(t+2) = f(t)$. Also let $x(0) = \dot{x}(0) = 0$.

1. Determine the solution by considering each interval of time $t_n \leq t \leq t_n + 1$, where $t_n = 0, 1, 2, \ldots$ separately. Find the solution for $0 \leq t < 1$. Then using the value of that solution at $t = 1$ for the initial conditions for the next interval, determine the solution for $1 \leq t < 2$. Do this for the first several intervals.
2. Write a computer program to determine an approximate numerical solution for this system and compare the answer to the answer from part 1.
3. Expand $f(t)$ in a Fourier series and use the method of undetermined coefficients to find the solution. *Hint:* The solution will be a series. Plot the solution for including various numbers of terms in the series solution and compare it to the numerical solution in part 2.

11.3. Determine the solution to the one-dimensional wave equation with $u(0,t) = u(L,t) = 0$ and with the specified parameter values and initial conditions.

1. $\alpha = 2, L = 20, \partial u/\partial t(x,0) = 0$, and

$$u(x,0) = \begin{cases} \frac{x}{20}, 0 < x \leq 10, \\ \frac{20-x}{20}, 10 < x \leq 20. \end{cases}$$

2. $\alpha = 3, L = 10, \partial u/\partial t(x,0) = 0$, and

$$u(x,0) = \begin{cases} \frac{x}{3}, 0 < x \leq 3, \\ \frac{10-x}{7}, 3 < x \leq 10. \end{cases}$$

3. $\alpha = 1, L = 10, u(x,0) = 0$, and

$$\frac{\partial u}{\partial t}(x,0) = \begin{cases} 0, 0 < x \leq 3, \\ 1, 3 < x \leq 4, \\ 0, 4 < x \leq 10. \end{cases}$$

4. $\alpha = 2, L = 20, u(x,0) = 0$, and

$$\frac{\partial u}{\partial t}(x,0) = \begin{cases} 0, 0 < x \leq 3, \\ \sin 2\pi x, 3 < x \leq \frac{7}{2}, \\ 0, \frac{7}{2} < x \leq 20. \end{cases}$$

In each case, plot a partial sum of the solution including enough terms so that it is accurate for various times. Alternatively, make a movie.

11.4. Determine the solution to the one-dimensional heat condition equation with homogeneous boundary conditions and with the specified parameter values and initial condition.

1. $\alpha = 1, L = 4$, and

$$u(x,0) = \begin{cases} \frac{x}{2}, 0 < x \leq 2, \\ \frac{4-x}{2}, 2 < x \leq 4. \end{cases}$$

2. $\alpha = 3, L = 10$, and

$$u(x,0) = \begin{cases} \frac{x}{3}, 0 < x \leq 3, \\ \frac{10-x}{7}, 3 < x \leq 10. \end{cases}$$

3. $\alpha = 5, L = 3$, and

$$u(x,0) = \begin{cases} 0, 0 < x \leq 1, \\ 1, 1 < x \leq 2, \\ 0, 2 < x \leq 3. \end{cases}$$

4. $\alpha = 2, L = 20, u(x,0) = 0$, and

$$\frac{\partial u}{\partial t}(x,0) = \begin{cases} 0, 0 < x \leq 3, \\ \sin 2\pi x, 3 < x \leq \frac{7}{2}, \\ 0, \frac{7}{2} < x \leq 20. \end{cases}$$

In each case, plot a partial sum of the solution including enough terms to that it is accurate for various times. Alternatively, make a movie.

11.5. Repeat Exercise 11.4, but assume the following nonhomogeneous boundary conditions for the corresponding part from Exercise 11.4.

1. $u(0,t) = 100$ and $u(L,t) = 200$
2. $u(0,t) = 0$ and the end at $x = L$ insulated.
3. $u(0,t) = 10$ and $u(L,t) = 2$.
4. End at $x = 0$ insulated and $u(L,t) = 3$.

11.6. Show that the the eigenvalue for the one-dimensional heat conduction equation with homogeneous boundary conditions must be positive.

11.7. Referring to the one-dimensional heat equation given in Equation (11.30), explain what the effects of changing the thermal conductivity, the specific heat, and the density will have on the rate at which the temperature at a point will change and why those effects make sense.

11.8. Derive the solution given at the beginning of Section 11.3.3 for the one-dimensional heat equation with an insulated end.

11.9. Determine the solution to Laplace's equation

$$\frac{\partial^2 u}{\partial^2 x} + \frac{\partial^2 u}{\partial^2 y} = 0$$

where

1. $u(0,y) = f(y)$, $u(L_x,y) = 0$, $u(x,0) = 0$, and $u(x,L_y) = 0$
2. $u(0,y) = 0$, $u(L_x,y) = f(y)$, $u(x,0) = 0$, and $u(x,L_y) = 0$
3. $u(0,y) = 0$, $u(L_x,y) = 0$, $u(x,0) = f(x)$, and $u(x,L_y) = 0$

11.10. Derive the heat conduction equation in three dimensions in Cartesian coordinates.

11.11. Show that if the thermal conductivity is not uniform throughout the body, then the heat conduction equation in Cartesian coordinates in three dimensions is

$$\frac{\partial}{\partial x}\left(k\frac{\partial u}{\partial x}\right) + \frac{\partial}{\partial y}\left(k\frac{\partial u}{\partial y}\right) + \frac{\partial}{\partial z}\left(k\frac{\partial u}{\partial z}\right) = \rho c_p \frac{\partial u}{\partial t}.$$

11.12. Determine the solution to

$$\alpha^2 \frac{\partial^2 u}{\partial x^2} - \alpha^2 u = \frac{\partial^2 u}{\partial t^2},$$

where

$$u(0,t) = u(L,t) = 0,$$

and

$$u(x,0) = f(x), \quad \frac{\partial u}{\partial t}(x,0) = 0.$$

11.13. Show that Equation (11.52) is the solution to the wave equation in Cartesian coordinates in two dimensions.

11.14. By starting with the two-dimensional heat condition equation in Cartesian coordinates, show that the two-dimensional heat equation in polar coordinates is given by

$$\frac{\partial^2 u}{\partial r^2} + \frac{1}{r}\frac{\partial u}{\partial r} + \frac{1}{r^2}\frac{\partial^2 u}{\partial \theta^2} = \frac{\rho c_p}{k}\frac{\partial u}{\partial t}.$$

11.15. Determine the solution to the two-dimensional heat equation on a rectangular domain $x \in [0,3]$ and $y \in [0,4]$, with homogeneous boundary conditions where $\alpha = 4$ and the initial temperature everywhere is zero except for the rectangle $1 < x < 2$ and $1 < y < 2$ where the temperature is one. Plot the solution for various times.

11.16. Determine the solution to the two-dimensional wave equation on a rectangular domain $x \in [0,3]$ and $y \in [0,4]$, with homogeneous boundary conditions where $\alpha = 4$ and the initial displacement that is everywhere zero and an initial velocity that is everywhere zero except for the rectangle $1 < x < 2$ and $1 < y < 2$ where the velocity is one. Plot the solution for various times.

11.17. Model a drumstick impacting a drum by solving the two-dimensional wave equations in polar coordinates. Make up your own physical parameters for the system that seem realistic. For the initial conditions, let the initial displacement be zero, and make up an initial velocity profile that would be a reasonable model for the effect of a drumstick impacting the drum, such as having zero initial velocity everywhere except in a small region where the stick impacts the drum head. Plot the solution for various times, or alternatively, make a movie.

11.18. Drummers typically do not hit a drum head right in the center, but rather somewhat offset from the center. On a tympani, the mallet typically impacts the drum head near the edge. Repeat the previous problem, but solve it two times, once with an initial velocity profile that is zero everywhere except in a region in the center, and once with the velocity zero everywhere except a region offset from the center. Compare the solutions and spectrum of the response.

11.19. In this problem solve what is known as the *interior Dirichlet problem* for a circle. Use separation of variables to show that the solution to

$$\frac{\partial^2 u}{\partial r^2} + \frac{1}{r}\frac{\partial u}{\partial r} + \frac{1}{r^2}\frac{\partial^2 u}{\partial \theta^2} = 0$$
$$u(1,\theta) = g(\theta)$$

is

$$u(r,\theta) = \sum_{n=0}^{\infty} r^n \left(a_n \cos n\theta + b_n \sin n\theta\right).$$

Do not substitute this solution into the partial differential equation to show it is a solution (it is). Go through the separation of variables process to construct it. Also, provide formulae for the coefficients a_n and b_n.

11.20. Explain in words the physical interpretation of Equation (11.71).

11.21. Show that for the static deflection for the beam illustrated in Figure 11.45, the equivalent spring force is

$$k = \frac{12EI}{L^3},$$

or in the case of a rectangular cross-section,

$$k = \frac{Ewh^3}{L^3}.$$

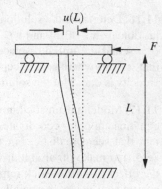

Fig. 11.45 A deflecting column.

11.22. When solving the static deflection problem for beams, note that because the equation is fourth-order, resulting in four coefficients in the solution, only four boundary conditions can be satisfied. Consider the simply-supported beam subjected to a point force at $x = x_F$ illustrated in Figure 11.46. The boundary conditions at $x = 0$ and $x = L$ require

$$u(0) = 0, \quad u''(0) = 0, \quad u(L) = 0, \quad u''(L) = 0.$$

Because a point force at location x is related to $u'''(x)$ this imposes another condition similar to a boundary condition. This gives a total of five conditions, which, in general, cannot be satisfied by a fourth-order equation.

The way to solve this problem is to determine two solutions, one, $u_1(x)$, from $x = 0$ to $x = x_F$, and the other, $u_2(x)$, from $x = x_F$ to $x = L$. In this case the boundary conditions are

$$u_1(0) = 0, \quad u_1''(0) = 0, \quad u_2(L) = 0, \quad u_2''(L) = 0.$$

The point force at $x = x_f$ requires

$$EIu_1(x_F) = F(L - x_F)/L, \quad EIu_2(x_F) = -Fx_F/L.$$

Finally, because the beam is continuous, there are two *matching conditions*, namely,

$$u_1(x_F) = u_2(x_F), \quad u_1'(x_F) = u_2'(x_F).$$

1. Explain why the matching conditions do not include equating the second derivatives of the two solutions.
2. Determine the two solutions.
3. Use some reasonable numerical values for the parameters of the system and plot the solution to check that your answer is reasonable. Plot it for $x_L = L/2$ and for an off-center force as well. *Hint:* for a point force of magnitude F at the midpoint of the beam, the relationship between the amount of static deflection of the midpoint and the applied force is given by

$$F = \frac{48EI}{L^3} u(L/2)$$

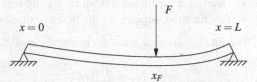

Fig. 11.46 Simply supported beam for
Exercise 11.22.

11.23. Show that for a beam cantilevered at both ends, the relationship between a force applied at the midpoint of the beam and the static displacement at the midpoint is given by $F = 192u(L/2)EI/L^3$.

11.24. Work through the whole separation of variables process to show that vibrations of a simply supported beam of length $L = 1$ which are solutions to

$$\frac{\partial^2 u}{\partial t^2} + \frac{\partial^4 u}{\partial x^4} = 0$$

($\alpha = 1$) with boundary conditions

$$u(0,t) = 0, \qquad \frac{\partial^2 u}{\partial x^2}(0,t) = 0, \qquad u(L,t) = 0, \qquad \frac{\partial^2 u}{\partial x^2}(L,t) = 0$$

and initial conditions given by

$$u(x,0) = f(x), \qquad \frac{\partial u}{\partial t}(x,0) = g(x)$$

are described by

$$u(x,t) = \sum_{n=1}^{\infty} \sin(n\pi x) \left[a_n \sin\left((n\pi)^2 t\right) + b_n \cos\left((n\pi)^2 t\right) \right],$$

where

$$a_n = \frac{2}{(n\pi)^2} \int_0^1 g(x)\sin(n\pi x)\,dx, \quad b_n = 2 \int_0^1 f(x)\sin(n\pi x)\,dx.$$

Hint: Use

$$X(x) = c_1 \cos\sqrt{\lambda}x + c_2 \sin\sqrt{\lambda}x + c_3 \cosh\sqrt{\lambda}x + c_4 \sinh\sqrt{\lambda}x.$$

Be sure to verify those four functions form a linearly independent set.

11.25. Verify the set made up of the four functions used in Equation (11.82), $\cos\sqrt{\lambda}x + \cosh\sqrt{\lambda}x, \cos\sqrt{\lambda}x - \cosh\sqrt{\lambda}x, \sin\sqrt{\lambda}x + \sinh\sqrt{\lambda}x, \sin\sqrt{\lambda}x - \sinh\sqrt{\lambda}x$, is linearly independent.

11.26. A *chime* is a musical instrument that may be modeled as beams with free ends. Consider the case where $\alpha = L = 1$.

- Determine an expression for the frequencies of the modes.
- Determine the numerical values for the first five modes.
- Plot the mode shapes for the first five modes.

If both ends are free, then the boundary conditions are

$$\frac{\partial^2 u}{\partial x^2}(0,t) = 0, \qquad \frac{\partial^3 u}{\partial x^3}(0,t) = 0, \qquad \frac{\partial^2 u}{\partial x^2}(L,t) = 0, \qquad \frac{\partial^3 u}{\partial x^3}(L,t) = 0.$$

11.27. In this exercise design your own six-string guitar. Assume that the length of all six strings is $L = 0.65$ m, and that the mass per unit length of each string is 0.1 kg/m (this is not realistic because most guitars have strings with different thickness and therefore mass per unit lengths). For each string on the guitar, determine the tension and location of the frets so that each string will play the notes indicated in Table 11.3 using equal temperament. Note that the frets go straight across the neck of the guitar, so the location of the frets that work for one string should work for them all.

String	Open Frequency (Hz)	Open Note	Fret 1 Note	Fret 2 Note	Fret 3 Note	Fret 4 Note	Fret 5 Note	Fret 6 Note	Fret 7 Note
1	329.60	E	F	$F^\sharp$	G	$G^\sharp$	A	$A^\sharp$	B
2	246.90	B	C	$C^\sharp$	D	$D^\sharp$	E	F	$F^\sharp$
3	196.00	G	$G^\sharp$	A	$A^\sharp$	B	C	$C^\sharp$	D
4	146.80	D	$D^\sharp$	E	F	$F^\sharp$	G	$G^\sharp$	A
5	110.00	A	$A^\sharp$	B	C	$C^\sharp$	D	$D^\sharp$	E
6	82.40	E	F	$F^\sharp$	G	$G^\sharp$	A	$A^\sharp$	B

Table 11.3 Table of frequencies and notes for Exercise 11.27

Chapter 12
Numerical Methods

This chapter deals with numerical methods for determining approximate solutions for differential equations. Unfortunately, most differential equations cannot be solved "by hand" to determine a solution that is expressed in terms of elementary functions. For example, most nonlinear differential equations are of this nature.

This chapter presents the derivation of the methods as well as analyses of the types of errors that are inherent in the methods. Section 12.1 presents Euler's method with more mathematical rigor than was considered in Section 1.6. Section 12.2 presents a method based upon Taylor series, which, actually, is the basis for all the methods we consider. The material in the section is not presented in many expositions on numerical methods, but is included for pedagogical purposes because it provides the theoretical foundation for the higher-order methods. Section 12.3 presents the ubiquitous Runge–Kutta method. Section 12.4 presents a derivation for an expression on the bound of the error for Euler's method with rather obvious extensions to higher-order methods. All the methods presented in Sections 12.1 through Section 12.3 work for a single, first-order ordinary differential equation. Section 12.5 extends these methods to systems of coupled first-order ordinary differential equations. Finally, Section 12.6 presents some basic techniques for determining approximate numerical solutions for partial differential equations.

12.1 Another Look at Euler's Method

In Section 1.6, Euler's method was derived as an approximation to the usual definition of the derivative. In particular, for a first-order, ordinary differential equation of the form

$$\dot{x} = f(x,t) \tag{12.1}$$

the derivative with respect to time is approximated by

$$\dot{x}(t) \approx \frac{x(t+\Delta t) - x(t)}{\Delta t}$$

B. Goodwine, *Engineering Differential Equations: Theory and Applications*, DOI 10.1007/978-1-4419-7919-3_12, © Springer Science+Business Media, LLC 2011

for $\Delta t \ll 1$. Consequently, solving for $x(t + \Delta t)$ gives

$$x(t + \Delta t) \approx x(t) + f(x(t), t)\Delta t \tag{12.2}$$

for $\Delta t \ll 1$.

In this section a slightly more sophisticated analysis is undertaken that allows for easy extensions to higher-order methods and error analyses. In particular, the analysis is based upon a Taylor series expansion of the form

$$x(t + \Delta t) = x(t) + \Delta t \frac{dx}{dt}(t) + \frac{(\Delta t)^2}{2!}\frac{d^2 x}{dt^2}(t) + \frac{(\Delta t)^3}{3!}\frac{d^3 x}{dt^3}(t) + \cdots . \tag{12.3}$$

Because the problem statement includes the fact that

$$\dot{x}(t) = f(x(t), t),$$

substituting this into Equation (12.3) gives

$$x(t + \Delta t) = x(t) + \Delta t f(x(t), t) + \frac{(\Delta t)^2}{2!}\frac{df}{dt}(x(t), t) + \frac{(\Delta t)^3}{3!}\frac{d^2 f}{dt^2}(x(t), t) + \cdots . \tag{12.4}$$

Clearly, Euler's method amounts to only using the first two terms in the series to approximate $x(t + \Delta t)$, and if $\Delta t \ll 1$, the *local truncation error* due to the fact that only a finite number of terms is used is proportional to $(\Delta t)^2$. In other words, if the time step is cut in half, the truncation error is reduced by $(1/2)^2 = 1/4$ and if Δt is reduced by an order of magnitude, the truncation error is reduced by $(1/10)^2 = 1/100$.

The following example illustrates the method as well as the effect of the time step on the error.

Example 12.1. Use Euler's method to determine an approximate solution to

$$\dot{x} = 5x,$$

where $x(0) = 1$ for $0 < t \le 2$.

Note that the exact solution is easy to compute and is $x(t) = e^{5t}$. At a given time, the equation to compute the value of the solution at the next time step is given by

$$x(t + \Delta t) = x(t) + f(x(t), t)\Delta t = x(t) + 5x(t)\Delta t.$$

A program listing in the C programming language for this problem appears in Appendix E.1.0.4. A program listing in FORTRAN for this problem appears in Appendix E.2.0.19. A plot of the solutions for two time steps as well as the exact solution is illustrated in Figure 12.1.

Remark 12.1. Until Section 12.4 which contains a detailed error analysis, when we refer to "overall error" it simply means an approximate measure of the error averaged over all the time considered in the problem.

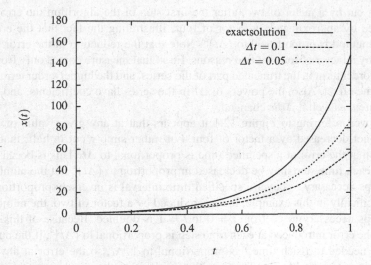

Fig. 12.1 Solution to Example 12.1 illustrating the fact that for Euler's method the overall error is proportional to Δt.

The first few steps of the results of the computations for the case where $\Delta t = 0.1$ are listed in Table 12.1. The first few steps of the results of the computations for the case where $\Delta t = 0.05$ are listed in Table 12.2.

t	$x(t)$	e^{5t}
0.000000	1.000000	1.000000
0.100000	1.500000	1.648721
0.200000	2.250000	2.718282

Table 12.1 First three steps in Euler's method for Example 12.1 with $\Delta t = 0.1$

t	$x(t)$	e^{5t}
0.000000	1.000000	1.000000
0.050000	1.250000	1.284025
0.100000	1.562500	1.648721

Table 12.2 First three steps in Euler's method for Example 12.1 with $\Delta t = 0.05$

After the first step when $\Delta t = 0.1$, the error in the approximate solution is $1.648721 - 1.5000000 = 0.14872$. When $\Delta t = 0.05$ the error after the first step is $1.284025 - 1.250000 = 0.034025$. The critical observation is that when the time step was cut by a factor of two, after the first step of the algorithm the error was decreased by approximately a factor of four, illustrating the fact that the error in Euler's method is proportional to $(\Delta t)^2$. Note that the reduction in the error is not exactly by a factor of four for two reasons. First, that measure comes only from the leading-order term in the truncated part of the series, and the higher-order terms will have some effect. Also, the powers of Δt in the series have coefficients, and those are different as well if Δt is changed.

However, referring to Figure 12.1 it appears that at any given value of t, the error is not decreased by a factor of four, but rather simply cut in half; that is, it appears that the error at a specified time is proportional to Δt. This is because the error at each time step may be decreased in proportion to $(\Delta t)^2$ but the number of time steps necessary to cover a specified time interval is inversely proportional to Δt. Specifically in this example, if Δt is reduced by a factor of two, the number of time steps necessary to go from $t = 0$ to $t = 1$ is doubled. Because of this, even though the error introduced at each time step is proportional to $(\Delta t)^2$, if the number of steps needed to reach time t is proportional to $1/\Delta t$, so the error at any time t will be proportional to Δt. This intuitive analysis of the error properties of the method is generally correct as we show in the more rigorous analysis developed in Section 12.4.

Another example helps flesh out the relationship between the changes in step size and the resultant error.

Example 12.2. Determine an approximate solution to

$$\dot{x} = -\sin t,$$

where $x(0) = 1$ using Euler's method. Not too much thought (or even less work) gives the exact solution as $x(t) = -\cos t$. A plot of the approximate solutions for $\Delta t = 1.0$ and $\Delta t = 0.5$ as well as the exact solution is illustrated in Figure 12.2.

Note that, as was the case in the previous example, decreasing the time step by a factor of two generally decreases the overall error by a factor of two as well. In other words, the overall error is proportional to the time step. A program listing in the C programming language for this problem appears in Appendix E.1.0.5. A program listing in FORTRAN for this problem appears in Appendix E.2.0.20.

12.2 Taylor Series Methods

If it is necessary to increase the accuracy of the approximate solution without the computational burden of an excessively small step size, the relatively obvious thing to do is to start including higher-order terms from the Taylor series expansion for $x(t + \Delta t)$ in Equation (12.4). Upon initially considering this notion, it may appear to be a rather trivial exercise. Although it is manageable to include the $(\Delta t)^2$ term, and even possibly the $(\Delta t)^3$ term, a quick review of multivariable calculus illustrates

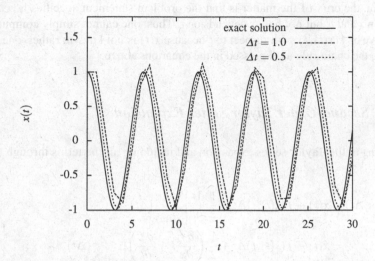

Fig. 12.2 Solution to Example 12.2 illustrating the fact that for Euler's method the overall error is proportional to Δt.

that the complexity of such an endeavor quickly becomes rather burdensome. This is because the function f depends upon both x and t, but x also depends upon t, and determining exactly that dependence of x on t is the whole point of the problem, that is, determining $x(t)$. From Equation (12.4), we need to compute the derivatives of f with respect to t.

From the chain rule, we have

$$\frac{df}{dt} = \frac{\partial f}{\partial x}\frac{dx}{dt} + \frac{\partial f}{\partial t} = \frac{\partial f}{\partial x}f + \frac{\partial f}{\partial t}.$$

or, including the dependence

$$\frac{df}{dt}(x(t),t) = \left(\frac{\partial f}{\partial x}\frac{dx}{dt} + \frac{\partial f}{\partial t}\right)\Bigg|_{(x(t),t)} = \left(\frac{\partial f}{\partial x}f + \frac{\partial f}{\partial t}\right)\Bigg|_{(x(t),t)}, \qquad (12.5)$$

where the notation

$$\frac{\partial f}{\partial x}\Bigg|_{(x(t),t)}$$

means, as usual, to compute the partial derivative of f with respect to x and then evaluate it at the values of $x(t)$ and t.[1]

---

[1] In this chapter, the manner in which the dependent variables are presented varies, depending on which way provides the most clarity. In fact, in Equation (12.5) two different ways are used; furthermore, the notation with the vertical bar used outside a group of terms with parentheses means that every term inside is evaluated at the indicated points.

Again, the crux of the matter is that the problem statement specifies how $\dot{x}$ depends on x and t, but not how x depends on t. Thus, one cannot simply compute the derivative of $f(x(t),t)$ with respect to t because $x(t)$ is not known; rather, one must resort to the chain rule as expressed in the equations above.

12.2.1 Second-Order Taylor Series Expansion

Returning to the Taylor series expansion and including all the terms through $(\Delta t)^2$ gives

$$
\begin{aligned}
x(t+\Delta t) &= x(t) + f(x(t),t)\Delta t + \frac{1}{2}\left.\frac{df}{dt}\right|_{(x(t),t)}(\Delta t)^2 + \cdots \\
&= x(t) + f(x(t),t)\Delta t + \frac{1}{2}\left(\frac{\partial f}{\partial x}f + \frac{\partial f}{\partial t}\right)\bigg|_{(x(t),t)}(\Delta t)^2 + \cdots .
\end{aligned}
\tag{12.6}
$$

Hence, keeping all terms through $(\Delta t)^2$, which should produce a step truncation error proportional to $(\Delta t)^3$ and an overall error proportional to $(\Delta t)^2$ is given by

$$
\boxed{x(t+\Delta t) = x(t) + f(x(t),t)\Delta t + \frac{(\Delta t)^2}{2}\left(\frac{\partial f}{\partial x}f + \frac{\partial f}{\partial t}\right)\bigg|_{(x(t),t)}.}
$$

Returning to Example 12.1 illustrates the fact that the error is indeed as would be expected.

Example 12.3. Use a second-order Taylor series expansion to determine an approximate solution to

$$\dot{x} = 5x,$$

where $x(0) = 1$ for $0 < t \le 2$.
   Because $f(x(t),t) = 5x(t)$,

$$\frac{\partial f}{\partial x} = 5, \quad \frac{\partial f}{\partial t} = 0.$$

Hence,

$$x(t+\Delta t) = x(t) + 5x(t)\Delta t + \frac{25}{2}x(t)(\Delta t)^2. \tag{12.7}$$

A program listing in the C programming language for this problem appears in Appendix E.1.0.6. A program listing in FORTRAN for this problem appears in Appendix E.2.0.21.
   The first few steps of the results of the computations for the case where $\Delta t = 0.1$ are given in Table 12.3 and and first few steps of the results of the computations

for the case where $\Delta t = 0.05$ are given in Table 12.4. Plots of the two approximate solutions and the exact solution are illustrated in Figure 12.3.

t	$x(t)$	e^{5t}
0.000000	1.000000	1.000000
0.100000	1.625000	1.648721
0.200000	2.640625	2.718282

Table 12.3 The first three time steps for the second-order Taylor series method for Example 12.3 with $\Delta t = 0.1$.

t	$x(t)$	e^{5t}
0.000000	1.000000	1.000000
0.050000	1.281250	1.284025
0.100000	1.641602	1.648721

Table 12.4 The first three time steps for the second-order Taylor series method for Example 12.3 with $\Delta t = 0.05$.

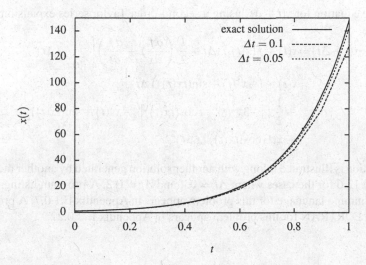

Fig. 12.3 Solution to Example 12.3. Note that for the second-order Taylor series, the error is generally proportional to $(\Delta t)^2$.

Observe that after the first time step, the error for $\Delta t = 0.1$ is $1.648721 - 1.625000 = 0.023721$, and the error for $\Delta t = 0.05$ is $1.284025 - 1.281250 = 0.0027750$. Because $0.023721/0.0027750 = 8.5 \approx 8$ it is clear that the error is proportional to $(\Delta t)^3$, which is the order of the first term truncated in the Taylor series, because the step size was reduced by a factor of two and the error was reduced by a factor of eight. With respect to the overall error, referring to Figure 12.3, it is clear that the overall error is proportional to $(\Delta t)^2$ inasmuch as the $\Delta t = 0.05$ curve has approximately $1/4$ the error of the $\Delta t = 0.1$ curve.

Also, when comparing the two methods, observe that for the case of $\Delta t = 0.1$ in Figures 12.1 and 12.3, the overall error decreases by an order of magnitude, which is consistent with the second-order Taylor series in the latter case including the $(\Delta t)^2$ in the expansion.

Because of the form of the partial derivatives in Equation (12.6), an example with the function $f(x,t)$ that includes both x and t may be helpful.

Example 12.4. Determine an approximate solution to

$$\dot{x} = -x^3 + \sin(tx),$$

where $x(0) = 1$ using a second-order Taylor series expansion.

In this problem

$$f(x,t) = -x^3 + \sin(tx)$$

Hence,

$$\frac{\partial f}{\partial x}(x(t),t) = -3x^2 + t\cos(tx), \quad \frac{\partial f}{\partial t}(x(t),t) = x\cos(tx).$$

Thus, the equation for $x(t + \Delta t)$ using a second-order Taylor series expansion is

$$\begin{aligned}
x(t+\Delta t) =& x(t) + f(x(t),t)\Delta t + \frac{1}{2}\left(\frac{\partial f}{\partial x}f + \frac{\partial f}{\partial t}\right)\Bigg|_{(x(t),t)}(\Delta t)^2 \\
=& x(t) + \left(-x^3(t) + \sin(tx(t))\right)\Delta t \\
&+ \frac{1}{2}\left\{\left[-3x^2(t) + t\cos(tx(t))\right]\left[-x^3(t) + \sin(tx(t))\right]\right. \\
&\left. + x(t)\cos(tx(t))\right\}(\Delta t)^2.
\end{aligned}$$

The solution is illustrated (along with another solution generated by another method) in Figure 12.6 for the cases where $\Delta t = 0.4$ and $\Delta t = 0.2$. A program listing in the C programming language for this problem appears in Appendix E.1.0.7. A program listing in FORTRAN for this problem appears in Appendix E.2.0.22.

12.2.2 Third-Order Taylor Series Expansion

The obvious thing to do at this point to improve the accuracy of the method is to try to include the third-order terms in the expansion, so, let us go for it. Starting with Equation (12.4),

$$x(t+\Delta t) = x(t) + f(x(t),t)\Delta t + \frac{1}{2!}\frac{df}{dt}(x(t),t)(\Delta t)^2 + \frac{1}{3!}\frac{d^2 f}{dt^2}(x(t),t)(\Delta t)^3 + \cdots.$$
(12.8)

As has already been stated, the dependence of f on x and t is specified, but not the dependence of x on t. Hence, as in the case of the second-order Taylor series, the chain rule must be used to expand the derivatives in terms of known quantities. In particular, as above

$$\frac{df}{dt} = \frac{\partial f}{\partial x}f + \frac{\partial f}{\partial t}.$$
(12.9)

So, to start to compute the next higher-order term,

$$\frac{d^2 f}{dt^2} = \frac{d}{dt}\left(\frac{df}{dt}\right) = \frac{d}{dt}\left(\frac{\partial f}{\partial x}f + \frac{\partial f}{\partial t}\right)$$

$$= \frac{d}{dt}\left(\frac{\partial f}{\partial x}\right)f + \frac{\partial f}{\partial x}\frac{d}{dt}(f) + \frac{d}{dt}\left(\frac{\partial f}{\partial t}\right),$$
(12.10)

where the second line is using the product rule for differentiating the $(\partial f/\partial x)f$ term. Recall that the need for the expansion in Equation (12.9) was the fact that f depended on both x and t, but x also depended on t, but only the derivative of x with respect to t is known. Similarly, $\partial f/\partial x$ and $\partial f/\partial t$ can depend on both x and t as well, so must be expanded similarly. Hence,

$$\frac{d}{dt}\left(\frac{\partial f}{\partial x}\right) = \frac{\partial^2 f}{\partial x^2}f + \frac{\partial^2 f}{\partial x \partial t}, \quad \frac{d}{dt}\left(\frac{\partial f}{\partial t}\right) = \frac{\partial^2 f}{\partial x \partial t}f + \frac{\partial^2 f}{\partial t^2}.$$

Using these two expressions as well as the one for df/dt in Equation (12.9) in Equation (12.10) gives

$$\frac{d^2 f}{dt^2} = \left(\frac{\partial^2 f}{\partial x^2}f + \frac{\partial^2 f}{\partial x \partial t}\right)f + \frac{\partial f}{\partial x}\left(\frac{\partial f}{\partial x}f + \frac{\partial f}{\partial t}\right) + \frac{\partial^2 f}{\partial x \partial t}f + \frac{\partial^2 f}{\partial t^2}.$$
(12.11)

Finally, substituting the terms from Equations (12.11) and (12.9) gives

$$\begin{aligned}
x(t+\Delta t) = {} & x(t) + f(x(t),t)\Delta t + \frac{(\Delta t)^2}{2}\left.\left(\frac{\partial f}{\partial x}f + \frac{\partial f}{\partial t}\right)\right|_{(x(t),t)} \\
& + \frac{(\Delta t)^3}{6}\left[\left(\frac{\partial^2 f}{\partial x^2}f + \frac{\partial^2 f}{\partial x \partial t}\right)f + \frac{\partial f}{\partial x}\left(\frac{\partial f}{\partial x}f + \frac{\partial f}{\partial t}\right)\right. \\
& + \left.\left.\frac{\partial^2 f}{\partial x \partial t}f + \frac{\partial^2 f}{\partial t^2}\right]\right|_{(x(t),t)}.
\end{aligned}$$
(12.12)

Remark 12.2.

1. It is theoretically possible to use Equation (12.12), however, as a practical matter it would be arduous to compute correctly all the partial derivatives, products, and the like.

2. If even greater accuracy is needed, including the fourth-order terms in Δt will result in an absolutely huge expansion because every term in Equation (12.12) depends on f, which will result in two partial derivative terms when expanded, as will all the terms that are products of two terms, which is every term except one. Clearly, an approach that gives higher-order accuracy without the hassle of such computations would be useful, which is exactly the point of the Runge–Kutta method in the next section.

Example 12.5. Use the first-, second-, and third-order Taylor series methods to determine an approximate numerical solution to

$$\dot{x} = 10x(1-x)$$

$$x(-1) = \frac{1}{1+e^{10}}$$

and compare it to the exact solution, which is

$$x(t) = \frac{1}{1+e^{-10t}}.$$

For this problem,

$$\dot{x} = f(x,t) = 10x(1-x)$$

so for the second- and third-order methods we need to compute

$$\frac{\partial f}{\partial x} = 10 - 20x, \quad \frac{\partial f}{\partial t} = 0, \quad \frac{\partial^2 f}{\partial x^2} = -20, \quad \frac{\partial^2 f}{\partial t^2} = 0, \quad \frac{\partial^2 f}{\partial t \partial x} = 0.$$

Thus, the equation for the first-order method (or Euler's method) is

$$x(t+\Delta t) = x(t) + f(x,t)\Delta t = x(t) + 10x(1-x)\Delta t.$$

The equation for the second-order method is

$$x(t+\Delta t) = x(t) + f(x,t)\Delta t + \frac{1}{2}\frac{df}{dt}(\Delta t)^2$$

$$= x(t) + f(x,t)\Delta t + \frac{1}{2}\left[\frac{\partial f}{\partial x}f + \frac{\partial f}{\partial t}\right](\Delta t)^2$$

$$= x(t) + (10x(1-x))\Delta t + \frac{1}{2}[(10-20x)(10x(1-x))](\Delta t)^2.$$

The equation for the third-order method is

$$x(t+\Delta t) = x(t) + f(x,t)\Delta t + \frac{1}{2}\frac{df}{dt}(\Delta t)^2 + \frac{1}{6}\frac{d^2f}{dt^2}(\Delta t)^3$$

$$= x(t) + f(x,t)\Delta t + \frac{1}{2}\left(\frac{\partial f}{\partial x}f + \frac{\partial f}{\partial t}\right)(\Delta t)^2$$

$$+ \frac{1}{6}\left[\left(\frac{\partial^2 f}{\partial x^2}f + \frac{\partial^2}{\partial x \partial t}\right) + \frac{\partial f}{\partial x}\left(\frac{\partial f}{\partial x}f + \frac{\partial f}{\partial t}\right) + \frac{\partial^2 f}{\partial x \partial t}f + \frac{\partial^2 f}{\partial t^2}\right](\Delta t)^3$$

$$= x(t) + (10x(1-x))\Delta t$$

$$+ \frac{1}{2}\left[(10-20x)(10x(1-x))\right](\Delta t)^2$$

$$+ \frac{1}{6}\left[-20(10x(1-x)) + (10-20x)^2(10x(1-x))\right](\Delta t)^3.$$

Clearly, as the order of the method increases, so does the complexity of the expression for $x(t+\Delta t)$.

12.3 The Runge–Kutta Method

The main idea behind the so-called Runge–Kutta methods is, instead of evaluating all the partial derivatives necessary in the Taylor series computations, to approximate the derivatives to the same order of accuracy using combinations of the function $f(x,t)$ evaluated not only at $x(t)$ and t, but other x and t values as well.

Consider the function $x(t)$ illustrated in Figure 12.4. The curve represents the unknown function $x(t)$. Assume that $x(t)$ is known exactly at two points, say at $t = 1.5$ and $t = 2.0$, so

$$\dot{x} \approx \frac{x(t+\Delta t) - x(t)}{\Delta t} = \frac{x(2.0) - x(1.5)}{0.5}.$$

In the figure, it is clear that the derivative of $x(t)$ at $t = 1.5$, which is the slope of the tangent line at that point, is approximately the same as the slope of the line connecting the values of $x(1.5)$ and $x(2.0)$ at times $t = 1.5$ and $t = 2.0$. Similarly, for that matter, the slope at $t = 2.0$ is approximately the same as well, as is the slope at any point between $t = 1.5$ and $t = 2.0$. Furthermore, the smaller the difference between the two points in time is, the better the approximation will be.

Now, consider the task of computing an approximation to the second derivative of x with respect to t. The second derivative is the derivative of the derivative, thus it is necessary to have an approximate computation for the derivative at two values for t. Hence, assume that the exact values for $x(t)$ are known for three points, say $t = 1.5$, $t = 2.0$, and $t = 2.5$, as is illustrated in Figure 12.5. The second derivative, then, is approximated by

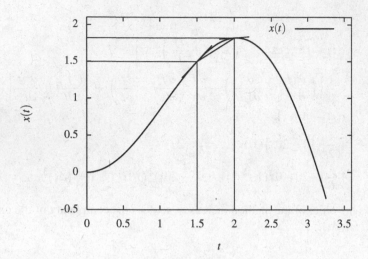

Fig. 12.4 Approximating derivatives of a function $x(t)$ by computing the slope of a line connecting two points.

$$\ddot{x} \approx \frac{\dot{x}(t+\Delta t)-\dot{x}(t)}{\Delta t} \approx \frac{\dot{x}(2.0)-\dot{x}(1.5)}{0.5} \approx \frac{\left(\frac{x(2.5)-x(2.0)}{0.5}\right)-\left(\frac{x(2.0)-x(1.5)}{0.5}\right)}{0.5}$$
$$= \frac{x(2.5)-2x(1.5)+x(1.5)}{(0.5)^2}$$

where the second to last term was obtained by substituting the equation for the approximate value of the derivative for each of $t=1.5$ and $t=2.0$. The main point is that the computation for an approximation for the second derivative of x with respect to t required that three points of $x(t)$ be known, which should make sense because, for a curve, the second derivative is a measure of the curvature, which cannot be captured by only two points.

So, to summarize, in order to approximate the derivative of x we needed to evaluate $x(t)$ at two points in time. In order to approximate the second derivative, we needed to compute $x(t)$ at three points in time. Clearly, to compute an approximate for the nth derivative, we need to evaluate $x(t)$ at $n+1$ points in time.

The main approach of the Runge–Kutta methods in this section is, in order to avoid all the complications associated with expanding the derivatives of $f(x(t),t)$ in a Taylor series, the higher-order derivatives are approximated by evaluating $f(x(t),t)$ at different x and t values to approximate the higher-order terms in the Taylor series.

So, in the case of attempting to compute approximate solutions to

$$\dot{x} = f(x(t),t)$$

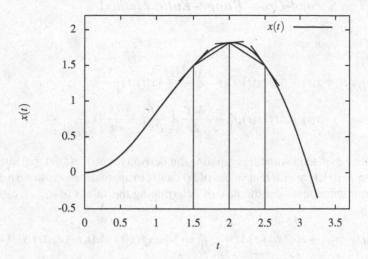

Fig. 12.5 Approximating derivatives of a function $x(t)$ by computing the slope of a line connecting two points.

there is a slight twist, which is that the first derivative of x is already given by the problem; namely, $f(x(t),t)$. So, the picture gets a little more abstract because the approximate derivatives that we are computing will not be for $x(t)$, but rather for $f(x(t),t)$; that is, we approximate the terms in Equation (12.4) instead of Equation (12.3). The one final conceptual complication is that the whole point of the problem is to determine $x(t)$; hence, these approximations for derivatives are not simple to compute because the $x(t)$ to plug into $f(x(t),t)$ is not known.

The approach is to approximate the $x(t + \Delta t)$ value that is used to evaluate the $f(x(t + \Delta t), t + \Delta t)$ values, that is used to determine approximations to the derivatives of $f(x(t),t)$ that appear in the Taylor series expansion of $x(t + \Delta t)$ in order to compute an approximation for $x(t + \Delta t)$.

12.3.1 The First-Order Runge–Kutta Method

Approximating

$$x(t+\Delta t) = x(t) + \Delta t f(x(t),t) + \frac{(\Delta t)^2}{2!}\frac{\mathrm{d}f}{\mathrm{d}t}(x(t),t) + \frac{(\Delta t)^3}{3!}\frac{\mathrm{d}^2 f}{\mathrm{d}t^2}(x(t),t) + \cdots$$

through the Δt term requires no derivative computations for $f(x(t),t)$. Hence, it is just Euler's method,

$$x(t+\Delta t) \approx x(t) + f(x(t),t)\Delta t.$$

12.3.2 The Second-Order Runge–Kutta Method

The goal is to compute

$$
\begin{aligned}
x(t+\Delta t) &= x(t)+\Delta t\, f(x(t),t)+\frac{(\Delta t)^2}{2!}\frac{\mathrm{d} f}{\mathrm{d} t}(x(t),t)+\cdots \\
&= x(t)+\Delta t\, f(x(t),t)+\frac{(\Delta t)^2}{2}\left(\frac{\partial f}{\partial x}f+\frac{\partial f}{\partial t}\right)\Bigg|_{(x(t),t)}+\cdots
\end{aligned}
\tag{12.13}
$$

through the $(\Delta t)^2$ term without computing the derivatives of $f(x(t),t)$, but rather by evaluating $f(x(t),t)$ at different values of $x(t)$ and t to approximate those derivatives. With that in mind, consider the task of determining the values of $c_1,\dots,c_4$ in the following

$$
x(t+\Delta t)=x(t)+c_1 f(x(t),t)\Delta t+c_2 f\big(x(t)+c_3 f(x(t),t)\Delta t,t+c_4\Delta t\big)\Delta t \tag{12.14}
$$

that makes it exactly equal to Equation (12.13) up to the $(\Delta t)^2$ term. Careful scrutiny of the second f term shows that this is the term where $f(x(t),t)$ is evaluated at different values for $x(t)$ and t; namely, $x(t)+c_3 f(x(t),t)\Delta t$ for the x-value and $t+c_4\Delta t$ for the t-value.

Although it is understandable that by this point the reader may be inclined to quit Taylor series for life, the way to determine the c_3 and c_4 constants is, obviously, to expand $f\big(x(t)+c_2 f(x(t),t)\Delta t,t+c_4\Delta t\big)$ in a Taylor series (be careful to match the parentheses correctly and identify what the two arguments to the function f are). In particular,

$$
\begin{aligned}
f\big(x(t)+c_3 f(x(t),t)\Delta t, t+c_4\Delta t\big) = \\
f(x(t),t)+\frac{\partial f}{\partial x}\bigg|_{(x(t),t)}[c_3 f(x(t),t)\Delta t]+\frac{\partial f}{\partial t}\bigg|_{(x(t),t)}c_4\Delta t+\cdots.
\end{aligned}
$$

Substituting this into Equation (12.14) gives

$$
\begin{aligned}
x(t+\Delta t) &= x(t)+c_1 f(x(t),t)\Delta t \\
&\quad +c_2\left(f(x(t),t)+\frac{\partial f}{\partial x}\bigg|_{(x(t),t)}(c_3 f(x(t),t)\Delta t)+\frac{\partial f}{\partial t}\bigg|_{(x(t),t)}c_4\Delta t+\cdots\right) \\
&= x(t)+(c_1+c_2)\,f(x(t),t)\Delta t \\
&\quad +\left(c_2 c_3\left(\frac{\partial f}{\partial x}f\right)\bigg|_{(x(t),t)}+c_2 c_4\frac{\partial f}{\partial t}\bigg|_{(x(t),t)}\right)(\Delta t)^2+\cdots.
\end{aligned}
\tag{12.15}
$$

Equating coefficients in Equations (12.13) and (12.15) gives

$$
c_1+c_2=1,\quad c_2 c_3=\frac{1}{2},\quad c_2 c_4=\frac{1}{2}.
$$

Clearly, there are multiple solutions, but

$$c_1 = \frac{1}{2}, \quad c_2 = \frac{1}{2}, \quad c_3 = 1, \quad c_4 = 1$$

is perhaps the most commonly used. Hence, substituting these values into Equation (12.14) gives

$$x(t + \Delta t) \approx x(t) + \frac{\Delta t}{2} \left[f(x(t), t) + f(x(t) + f(x(t), t)\Delta t, t + \Delta t) \right], \qquad (12.16)$$

which is known as either the *improved Euler* formula or *second-order Runge–Kutta* formula. It is worth emphasizing the following.

- The method summarized in Equation (12.16) holds for any $f(x, t)$ as long as the steps in the derivation are not violated, which basically require differentiability in both x and t.
- The main value is that the equation only involves f, and none of the derivatives of f, so very generic programs may be written where the only change needed for different problems is to change the line for f.

The following example illustrates a simple implementation of the method.

Example 12.6. Determine an approximate solution to

$$\dot{x} = 5x,$$

where $x(0) = 1$ for $0 < t \le 2$ using the second-order Runge–Kutta method.
  Because $f(x(t), t) = 5x$, then

$$f(x(t) + f(x(t), t)\Delta t, t + \Delta t) = 5(x(t) + f(x(t), t)\Delta t).$$

Hence, the second-order Runge–Kutta formula is

$$x(t + \Delta t) = x(t) + \frac{\Delta t}{2} \left[5(x(t)) + 5(x(t) + f(x(t), t)\Delta t) \right]$$

$$= x(t) + \frac{\Delta t}{2} \left[5(x(t)) + 5(x(t) + 5x(t)\Delta t) \right].$$

$$= x(t) + 5x(t)\Delta t + \frac{25}{2}x(t)(\Delta t)^2,$$

which happens to be identical to Equation (12.7). Hence, any computer program that computes an approximate solution will be the same as for Example 12.3. A program listing in the C programming language for this problem appears in Appendix E.1.0.8. A program listing in FORTRAN for this problem appears in Appendix E.2.0.23.

  Because of the rather complicated form of $f(x(t) + f(x(t), t)\Delta t, t + \Delta t)$, a slightly more complicated example is in order.

Example 12.7. Determine an approximate solution to

$$\dot{x} = -x^3 + \sin(tx),$$

where $x(0) = 1$ using the second-order Runge–Kutta formula (the improved Euler formula). This is the initial value problem from Example 12.4. Inasmuch as

$$f(x,t) = -x^3 + \sin(tx)$$

substituting $x(t) + f(x(t),t)\Delta t$ for x and $t + \Delta t$ for t gives

$$f\big(x(t) + f(x(t),t)\Delta t, t + \Delta t\big) = -\big(x(t) + f(x(t),t)\Delta t\big)^3$$
$$+ \sin\big((t + \Delta t)(x(t) + f(x(t),t)\Delta t)\big)$$

(verify this yourself; it is critical!). Hence, substituting for $f(x(t),t)$ and $f(x(t) + f(x(t),t)\Delta t, x + \Delta t)$ into Equation (12.16) gives

$$x(t + \Delta t) \approx x(t) + \frac{\Delta t}{2}\left[\big(x^3(t) + \sin(tx(t))\big)\right.$$
$$+ \left.\big((x(t) + f(x(t),t)\Delta t)^3 + \sin((t + \Delta t)(x(t) + f(x(t),t)\Delta t))\big)\right].$$

Figure 12.6 illustrates the approximate solution for the cases where $\Delta t = 0.2$ and $\Delta t = 0.4$. A program listing in the C programming language for this problem appears in Appendix E.1.0.8. A program listing in FORTRAN for this problem appears in Appendix E.2.0.23.

12.3.2.1 Comparison of Second-Order Runge–Kutta and Taylor Series Methods

As is clear by comparing the formulae in Examples 12.4 and 12.7, the second-order Taylor series method and the second-order Runge–Kutta method do not result in exactly the same approximate solution. However, both methods are accurate to the same order. Figure 12.6 illustrates an accurate solution (generated with a very small time step) and solutions from the two second-order approximate methods for two different time steps. Clearly, the two approximate solutions are not identical; however, they both demonstrated second-order accuracy.

12.3.2.2 Interpretation of the Second-Order Runge–Kutta Formula

Although the next two subsequent sections present the results of exactly this same approach carried out to third and fourth order, respectively, this approach yields a rather easy interpretation beyond the fact that it is the result of the above mathematical manipulations.

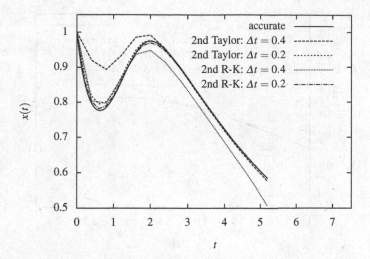

Fig. 12.6 A comparison of the solutions from different second-order methods from Examples 12.4 and 12.7. The second-order Taylor series and second-order Runge–Kutta do not give the same approximate solutions; however, both methods have the same order of accuracy.

One way to think of the second-order Runge–Kutta formula is that it is simply Euler's method using the average of the slopes of $x(t)$ at the two endpoints of the time interval, that is, the average of the slope at $x(t)$ and the slope at $x(t + \Delta t)$. Mathematically, this formula would be

$$x(t + \Delta t) = x(t) + \frac{1}{2} \left[f(x(t),t) + f(x(t + \Delta t), t + \Delta t) \right] \Delta t.$$

However, the term $x(t + \Delta t)$ appears on both sides of the equation and is exactly the term that is unknown. Also, unless the function $f(x(t),t)$ is of a very special form, it will generally be impossible to solve this equation for $x(t + \Delta t)$. The idea is to replace the $x(t + \Delta t)$ term that is on the right-hand side of the equation with an approximation for it, particularly, simply using Euler's formula for it on the right-hand side. Hence,

$$x(t + \Delta t) = x(t) + \frac{1}{2} \left[f(x(t),t) + f(x(t) + f(x(t),t)\Delta t), t + \Delta t) \right] \Delta t.$$

Initially, this approach may intuitively be no better than Euler's method because Euler's method was used on the right-hand side of the equation. However, inasmuch as it was used in a term already multiplied by Δt, the overall order of that term will be $(\Delta t)^2$ and hence an order better in accuracy.

This is conceptually illustrated in Figure 12.7 which illustrates the same function, $x(t)$ that was illustrated in Figures 12.4 and 12.5, but plotted over a much shorter time interval. In this figure, $t = 1.5$, $\Delta t = 0.5$, and $t + \Delta t = 2.0$. The slope of $x(t)$

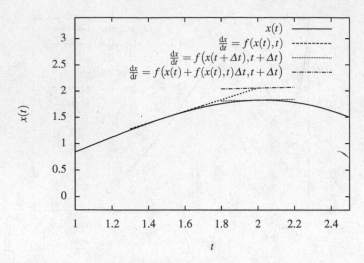

Fig. 12.7 Interpretation of the improved Euler method. It uses the average of the slopes at the values at the beginning of the time step and at the end of the time step, but with an $x(t + \Delta t)$ value computed using a first-order approximation.

is known and is $f(x(t),t)$. The value of $x(t + \Delta t)$ is not known, and hence $f(x(t + \Delta t), t + \Delta t)$ cannot be directly computed. However, if Δt is small, then $x(t + \Delta t) \approx x(t) + f(x(t),t)\Delta t$ and also then

$$f(x(t + \Delta t), t + \Delta t) \approx f(x(t) + f(x(t),t)\Delta t, t + \Delta t),$$

which is illustrated graphically in Figure 12.7.

12.3.3 The Third-Order Runge–Kutta Method

The third-order Runge–Kutta method (as well as the fourth-order method in the following subsection) is derived in exactly the same manner as the second-order Runge–Kutta method, except to third and fourth orders, respectively. Hence the goal is to compute Equation (12.12) through the $(\Delta t)^3$ term without explicitly computing the derivatives of $f(x(t),t)$ but rather approximating those derivatives to third order by evaluating $f(x(t),t)$ at different x and t values. In particular, equating

$$x(t+\Delta t) = x(t) + f(x(t),t)\Delta t + \frac{1}{2}\left(\frac{\partial f}{\partial x}f + \frac{\partial f}{\partial t}\right)\Bigg|_{(x(t),t)}(\Delta t)^2$$

$$+ \frac{1}{6}\left[\left(\frac{\partial^2 f}{\partial x^2}f + \frac{\partial^2 f}{\partial x \partial t}\right)f + \frac{\partial f}{\partial x}\left(\frac{\partial f}{\partial x}f + \frac{\partial f}{\partial t}\right)\right.$$

$$+ \left.\frac{\partial^2 f}{\partial x \partial t}f + \frac{\partial^2 f}{\partial t^2}\right]\Bigg|_{(x(t),t)}(\Delta t)^3 + \cdots$$

and

$$x(t+\Delta t) = x(t) + [c_1 f + c_2 f(x + c_3 f\Delta t, t + c_4\Delta t)$$
$$+ c_5 f(x + c_6 f\Delta t + c_7 f(x + c_8 f\Delta t, t + c_9\Delta t)\Delta t, t + c_{10}\Delta t)]\Delta t$$

(if no arguments to f are specified, it is evaluated at $(x(t),t)$) to third order gives

$$c_1 = \frac{1}{6}, \qquad c_2 = \frac{2}{3}, \qquad c_3 = \frac{1}{2}, \qquad c_4 = \frac{1}{2}, \qquad c_5 = \frac{1}{6},$$

$$c_6 = -1, \qquad c_7 = 2, \qquad c_8 = \frac{1}{2}, \qquad c_9 = \frac{1}{2}, \qquad c_{10} = 1.$$

See [26] for more details and a general derivation for the Runge–Kutta methods for various orders.

A more standard expression of this solution is

$$x(t+\Delta t) = x(t) + \frac{1}{6}(v_1 + 4v_2 + v_3), \tag{12.17}$$

where

$$v_1 = f(x(t),t)\Delta t$$
$$v_2 = f\left(x(t) + \frac{1}{2}v_1, t + \frac{1}{2}\Delta t\right)\Delta t \tag{12.18}$$
$$v_3 = f(x(t) + 2v_2 - v_1, t + \Delta t)\Delta t.$$

Example 12.8. Determine an approximate solution to

$$\dot{x} = -x^3 + \sin(tx)$$

where $x(0) = 1$ using the third-order Runge–Kutta method.

This is simply a matter of substituting into Equations (12.17) and (12.18) as follows,

$$x(t+\Delta t) = x(t) + \frac{1}{6}(v_1 + 4v_2 + v_3),$$

where

$$v_1 = \left(-(x(t))^3 + \sin(tx(t)) \right) \Delta t$$

$$v_2 = \left(-\left(x(t) + \frac{1}{2}v_1 \right)^3 + \sin\left(\left(t + \frac{1}{2}\Delta t \right) \left(x(t) + \frac{1}{2}v_1 \right) \right) \right) \Delta t$$

$$v_3 = \left(-(x(t) + 2v_2 - v_1)^3 + \sin\left((t + \Delta t)(x(t) + 2v_2 - v_1) \right) \right) \Delta t.$$

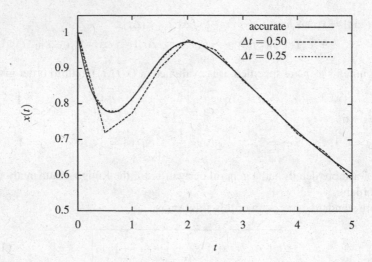

Fig. 12.8 Accurate and approximate solutions for Example 12.8. The third-order Runge–Kutta has a local truncation error proportional to $(\Delta t)^4$ and an overall error proportional to $(\Delta t)^3$.

An accurate solution determined with a very small time step as well as approximate solutions for $\Delta t = 0.5$ and $\Delta t = 0.25$ are illustrated in Figure 12.8. Note the substantial increase in accuracy when the time step is cut by a factor of two. A program listing in the C programming language for this problem appears in Appendix E.1.0.10. A program listing in FORTRAN for this problem appears in Appendix E.2.0.25.

12.3.4 The Fourth-Order Runge–Kutta Method

Again, the idea is exactly the same as the previous Runge–Kutta derivations. The famous fourth-order Runge–Kutta formula is

$$x(t + \Delta t) = x(t) + \frac{1}{6}(k_1 + 2k_2 + 2k_3 + k_4), \tag{12.19}$$

where

$$
\boxed{
\begin{aligned}
k_1 &= f\left(x(t),t\right)\Delta t \\
k_2 &= f\left(x(t)+\frac{1}{2}k_1,t+\frac{1}{2}\Delta t\right)\Delta t \\
k_3 &= f\left(x(t)+\frac{1}{2}k_2,t+\frac{1}{2}\Delta t\right)\Delta t \\
k_4 &= f\left(x(t)+k_3,t+\Delta t\right)\Delta t.
\end{aligned}
}
$$

(12.20)

Example 12.9. Determine an approximate solution to

$$
\dot{x} = -x^3 + \sin(tx),
$$

where $x(0) = 1$ using the fourth-order Runge–Kutta method.

This is simply a matter of substituting into Equations (12.19) and (12.20) as follows,

$$
x(t+\Delta t) = x(t) + \frac{1}{6}\left(k_1 + 2k_2 + 2k_3 + k_4\right),
$$

where

$$
\begin{aligned}
k_1 &= \left(-(x(t))^3 + \sin(tx(t))\right)\Delta t \\
k_2 &= \left(-\left(x(t)+\frac{1}{2}k_1\right)^3 + \sin\left(\left(t+\frac{1}{2}\Delta t\right)\left(x(t)+\frac{1}{2}k_1\right)\right)\right)\Delta t \\
k_3 &= \left(-\left(x(t)+\frac{1}{2}k_2\right)^3 + \sin\left(\left(t+\frac{1}{2}\Delta t\right)\left(x(t)+\frac{1}{2}k_2\right)\right)\right)\Delta t \\
k_4 &= \left(-(x(t)+k_3)^3 + \sin((t+\Delta t)(x(t)+k_3))\right)\Delta t.
\end{aligned}
$$

An accurate solution determined with a very small time step as well as approximate solutions for $\Delta t = 0.5$ and $\Delta t = 0.25$ is illustrated in Figure 12.9. Note the substantial increase in accuracy when the time step is cut by a factor of two and the generally better accuracy than the lower-order methods for the same time steps. A program listing in the C programming language for this problem appears in Appendix E.1.0.11. A program listing in FORTRAN for this problem appears in Appendix E.2.0.26.

12.4 Error Analysis

In each of the methods we have considered, we have explicitly accounted for a certain number of terms in the Taylor series expansion

$$
x(t+\Delta t) = x(t) + \Delta t f(x(t),t) + \frac{(\Delta t)^2}{2!}\frac{\mathrm{d}f}{\mathrm{d}t}(x(t),t) + \frac{(\Delta t)^3}{3!}\frac{\mathrm{d}^2 f}{\mathrm{d}t^2}(x(t),t) + \cdots.
$$

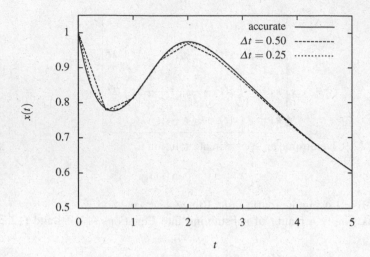

Fig. 12.9 Accurate and approximate solutions for Example 12.9. The fourth-order Runge–Kutta has a local truncation error proportional to $(\Delta t)^5$ and an overall error proportional to $(\Delta t)^4$.

The local truncation error is the error introduced at each time step that arises because only a finite number of terms in the Taylor series expansion are used, and we have dealt with that in a relatively intuitive manner so far. This section considers the issue of *convergence* of the numerical solution to the actual solution and considers two types of errors: the truncation error already considered as well as roundoff error. The latter is due to the fact that only a finite number of bits are used in a computer to represent a number, and whenever a computer makes a computation, error is introduced because of this fact. The analysis focuses on Euler's method, but the fundamental result, which is that there is a trade-off between truncation and roundoff error, holds generally.

Up to this point, $x(t)$ has been used to denote both the actual solution as well as for a representation of a numerical solution. In order to determine the error, we have to distinguish the two. In particular, for $\dot{x} = f(x, t)$, where an initial condition is specified at t_0, let $x(t)$ denote the actual solution and $x_n(t)$ denote the approximate solution[2] determined numerically.

At time $t + \Delta t$, the actual solution is given by

$$x(t + \Delta t) = x(t) + f(x(t), t)\,\Delta t + \frac{(\Delta t)^2}{2}\left.\frac{\mathrm{d}f}{\mathrm{d}t}\right|_{\hat{t}},$$

---

[2] In actuality, $x_n(t)$ is an abuse of notation because the numerical solution is only defined at a sequence of times; however, inasmuch as we only use it at the times at which it is defined, this is used to make it convenient to compare with the real solution.

where, by Taylor's theorem, $t \leq \hat{t} \leq t + \Delta t$. Note that this expression is exact, not truncated, but it can not be practically evaluated because the value for $\hat{t}$ is not known other than it belongs to the specified interval. For Euler's method, the numerical approximation is given by

$$x_n(t + \Delta t) = x_n(t) + \Delta t f(x_n(t), t) + \rho,$$

where ρ is the total roundoff error due to evaluating the expression.

Hence, an expression for the error at time $t + \Delta t$ is given by the difference between these two equations

$$err(t + \Delta t) = x(t + \Delta t) - x_n(t + \Delta t)$$

$$= \left[x(t) + \Delta t f(x(t), t) + \frac{(\Delta t)^2}{2} \left. \frac{df}{dt} \right|_{\hat{t}} \right] - [x_n(t) + \Delta t f(x_n(t), t) + \rho]$$

$$= [x(t) - x_n(t)] + \Delta t [f(x(t), t) - f(x_n(t), t)] + \frac{(\Delta t)^2}{2} \left. \frac{df}{dt} \right|_{\hat{t}} - \rho$$

$$= err(t) + \Delta t [f(x(t), t) - f(x_n(t), t)] + \frac{(\Delta t)^2}{2} \left. \frac{df}{dt} \right|_{\hat{t}} - \rho.$$

By the mean value theorem, the second term in the last line can be written as

$$f(x(t), t) - f(x_n(t), t) = \left. \frac{\partial f}{\partial x} \right|_{\hat{x}} (x(t) - x_n(t)),$$

where $\hat{x}$ is a value between $x(t)$ and $x_n(t)$. Hence,

$$x(t + \Delta t) = err(t) + err(t) \Delta t \left. \frac{\partial f}{\partial x} \right|_{\hat{x}} + \frac{(\Delta t)^2}{2} \left. \frac{df}{dt} \right|_{\hat{t}} - \rho$$

$$= \left(1 + \left. \frac{\partial f}{\partial x} \right|_{\hat{x}} \Delta t \right) err(t) + \frac{(\Delta t)^2}{2} \left. \frac{df}{dt} \right|_{\hat{t}} - \rho.$$

Where this is headed is next to replace $err(t)$ on the right-hand side with $err(t - \Delta t)$ and recursively work back to t_0. However, each time this is done the points $\hat{x}$ and $\hat{t}$ will belong to different ranges. Hence, define the region $S = (x, t)$ to be the domain of x and t in which we are interested in knowing whether the numerical solution converges to the actual solution. Then define M_1 and M_2 to be bounds on the terms in the expression for $err(t + \Delta t)$; that is,

$$M_1 \geq \left| \frac{\partial f}{\partial x}(x(t), t) \right|, \quad (x, t) \in S$$

$$M_2 \geq \left| \frac{\partial f}{\partial t}(x(t), t) \right|, \quad (x, t) \in S.$$

Because these are bounds on these terms, our error expression becomes an inequality of the form

$$|err(t+\Delta t)| \leq (1+M_1\Delta t)|err(t)| + \frac{1}{2}M_2(\Delta t)^2 + |\rho|. \tag{12.21}$$

Replacing t with $t - \Delta t$ in the above expression gives

$$|err(t)| \leq (1+M_1\Delta t)|err(t-\Delta t)| + \frac{1}{2}M_2(\Delta t)^2 + |\rho|, \tag{12.22}$$

and substituting this into the $err(t)$ term in Equation (12.21) and rearranging gives

$$|err(t+\Delta t)| \leq (1+M_1\Delta t)^2 |err(t-\Delta t)| + [1+(1+M_1\Delta t)]\left(\frac{1}{2}M_2(\Delta t)^2 + |\rho|\right).$$

Determining an expression for $err(t-2\Delta t)$ similar to the process for Equation (12.22) and substituting into this equation gives

$$|err(t+\Delta t)| \leq (1+M_1\Delta t)^3 |err(t-\Delta t)|$$
$$+ \left[1+(1+M_1\Delta t)+(t+M_1\Delta t)^2\right]\left(\frac{1}{2}M_2(\Delta t)^2 + |\rho|\right).$$

Let n be such that $t - t_0 = (n-1)\Delta t$. Then

$$|err(t+\Delta t)| \leq (1+M_1\Delta t)^n |err(t_0)|$$
$$+ \left[1+(1+M_1\Delta t)+\cdots+(t+M_1\Delta t)^{n-1}\right]\left(\frac{1}{2}M_2(\Delta t)^2 + |\rho|\right).$$

Inasmuch as

$$\left[1+(1+M_1\Delta t)+\cdots+(1+M_1\Delta t)^{n-1}\right]\frac{1-(1+M_1\Delta t)}{1-(1+M_1\Delta t)} = \frac{(1+M_1\Delta t)^n - 1}{M_1\Delta t}$$

then

$$|err(t+\Delta t)| \leq (1+M_1\Delta t)^n |err(t_0)|$$
$$+ \frac{(1+M_1\Delta t)^n - 1}{M_1\Delta t}\left(\frac{1}{2}M_2(\Delta t)^2 + |\rho|\right)$$
$$\leq (1+M_1\Delta t)^n |err(t_0)| + [(1+M_1\Delta t)^n - 1]\left(\frac{M_2}{2M_1}\Delta t + \frac{|\rho|}{M_1\Delta t}\right)$$
$$\leq (1+M_1\Delta t)^n \left(|err(t_0)| + \frac{M_2}{2M_1}\Delta t + \frac{|\rho|}{M_1\Delta t}\right).$$

Finally,

$$e^{M_1\Delta t} = 1 + (M_1\Delta t) + \sum_{i=2}^{\infty}\frac{(M_1\Delta t)^i}{i!}$$

and because $M_1 \Delta t$ is positive every term in the summation is positive, and hence

$$e^{M_1 \Delta t} \geq 1 + M_1 \Delta t \quad \Longrightarrow \quad e^{n M_1 \Delta t} \geq (1 + M_1 \Delta t)^n.$$

Using this, the definition of n, and assuming $|err(t_0)| = 0$, finally gives

$$\left| err(t + \Delta t) \right| \leq e^{M_1 (t + \Delta t - t_0)} \left(\frac{M_2}{2 M_1} \Delta t + \frac{|\rho|}{M_1 \Delta t} \right). \tag{12.23}$$

The error bound given by Equation (12.23) has three terms. The exponential term represents the fact that as time increases, the bound on the error grows exponentially. This exponential term also depends on $f(x,t)$ because the term M_1 appears in the exponent. The first term it multiplies depends on $f(x,t)$ as represented by M_1 and M_2 and is proportional to Δt. As the time step in increased, the contribution of this term to the error bound increases, which represents the fact that the truncation error at each step will be greater if the time step increases. The second term is due to the roundoff error and is inversely proportional to Δt. If the time step is decreased, then more steps are required to reach a given time, which accumulates more roundoff error. Figure 12.10 illustrates the qualitative relationship among the truncation error, the roundoff error, and the sum of them. An example which shows that the error does, in fact, increase if the step size is too small appears in the exercises.

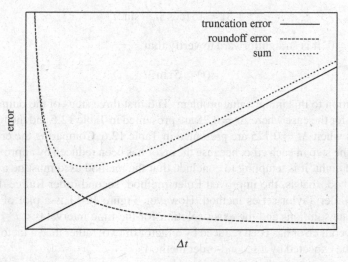

Fig. 12.10 Truncation error, roundoff error, and the sum of them as a function of the time step for Euler's method.

The analysis presented for Euler's method extends to higher-order methods as well. The results are listed in Table 12.5 for the different methods considered in this book.

Method	Local Truncation Error	Overall Error
Euler First-order Taylor Series First-order Runge–Kutta	$\mathcal{O}(\Delta t)^2$	$\mathcal{O}(\Delta t)$
Second-order Taylor Series	$\mathcal{O}(\Delta t)^3$	$\mathcal{O}(\Delta t)^2$
Second-order Runge–Kutta Improved Euler	$\mathcal{O}(\Delta t)^3$	$\mathcal{O}(\Delta t)^2$
Third-order Taylor Series	$\mathcal{O}(\Delta t)^4$	$\mathcal{O}(\Delta t)^3$
Third-order Runge–Kutta	$\mathcal{O}(\Delta t)^4$	$\mathcal{O}(\Delta t)^3$
Fourth-order Runge–Kutta	$\mathcal{O}(\Delta t)^5$	$\mathcal{O}(\Delta t)^4$

Table 12.5 Local truncation error and overall error for various numerical method schemes for $\Delta t \ll 1$. Equivalent methods are listed in the same row

It is worth emphasizing that the analysis presented is true in general, but that does not preclude the existence of somewhat pathological cases that seemingly defy the rules, such as the following example.

Example 12.10. Consider

$$\dot{x} + 3x = 15 \left(\cos 3t + \sin 3t \right)$$

with $x(0) = 0$. It is straightforward to verify that

$$x(t) = 5 \sin 3t$$

is the solution to this initial value problem. The first three steps of the output of the algorithm for the case where $\Delta t = 0.25$ are presented in Table 12.6 and the first three time steps when $\Delta t = 0.125$ are presented in Table 12.6. Comparing the error after the first time step in each case, because the error has been reduced by approximately a factor of eight, it is tempting to conclude that the method used must be a second-order method, that is, the improved Euler method, second-order Runge–Kutta or a second-order Taylor series method. However, Figure 12.11 is a plot of the two approximate solutions and the exact solution for the time interval $0 < t \le 1$. Note that the overall error has been reduced by a factor of two, rather than a factor of four as would be expected by a second-order method.

This apparent contradiction is resolved by studying the exact solution. Because

$$x(t) = 5 \sin 3t$$

the Taylor series for $x(t)$ is

t	$x(t)$	$5\sin 3t$	$5\sin 3t - x(t)$
0.000000	0.000000	0.000000	0.000000
0.250000	3.750000	3.408194	−0.341806
0.500000	6.237479	4.987475	−1.250004

Table 12.6 The first three time steps for Example 12.10 with $\Delta t = 0.25$

t	$x(t)$	$5\sin 3t$	$5\sin 3t - x(t)$
0.000000	0.000000	0.000000	0.000000
0.125000	1.875000	1.831363	−0.043637
0.250000	3.603338	3.408194	−0.195144

Table 12.7 The first three time steps for Example 12.10 with $\Delta t = 0.125$

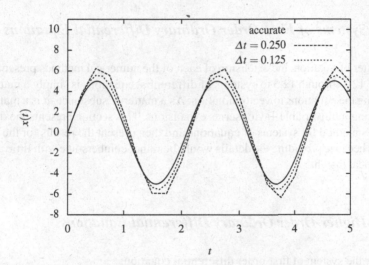

Fig. 12.11 Exact and approximate solutions for Example 12.10 exhibiting an overall error proportional to Δt.

$$x(t + \Delta t) = x(t) + \Delta t \frac{dx}{dt}\bigg|_t + \frac{(\Delta t)^2}{2} \frac{d^2 x}{dt^2}\bigg|_t + \cdots$$

$$= 5\sin 3t + 15\cos 3t\,(\Delta t) - \frac{45}{2}\sin 3t\,(\Delta t)^2 + \cdots.$$

Every other term contains $\sin 3t$, at $t = 0$, thus every other term is zero. Thus when comparing the local truncation error by examining the error after the first time step, a first-order method will look like a second-order method due to the fact that the coefficient of the $(\Delta t)^2$ term in the Taylor series is zero. Similarly, a third-order method will look like a fourth-order one, and so on. After the first time step, however, the relevant coefficients are nonzero, and hence the overall error behaves as expected. A program listing in the C programming language for this problem implementing Euler, second-order Runge–Kutta, a second-order Taylor series expansion, and fourth-order Runge–Kutta appears in Appendix E.1.0.12.

12.5 Numerical Methods for Higher-Order Systems

All the examples so far have been for first-order ordinary differential equations. This section presents the relatively easy extension to systems of first-order equations and higher-order ordinary differential equations and highlights the one subtlety with respect to computer implementation of the algorithms.

12.5.1 Systems of First-Order Ordinary Differential Equations

As a matter of notation, the extension of each of the numerical methods presented in Sections 12.1 through 12.3 to systems of differential equations is simply a matter of converting the equations to vector notation. As a matter of substance, it is a matter of considering multivariable Taylor series expansions. This section presents the details of Euler's method for systems of equations and then present the results for the other methods because providing the details would be rather cumbersome with little added pedagogical insight.

12.5.2 Higher-Order Ordinary Differential Equations

Consider the system of first-order differential equations

$$
\begin{aligned}
\dot{x}_1 &= f_1(x_1(t), x_2(t), \ldots, x_n(t), t) \\
\dot{x}_2 &= f_2(x_1(t), x_2(t), \ldots, x_n(t), t) \\
&\vdots = \vdots \\
\dot{x}_n &= f_1(x_1(t), x_2(t), \ldots, x_n(t), t).
\end{aligned}
\tag{12.24}
$$

Expanding each of the x_i in a Taylor series gives

$$x_1(t+\Delta t) = x_1(t) + \Delta t \frac{dx_1}{dt}(t) + \frac{(\Delta t)^2}{2}\frac{d^2x_1}{dt^2}(t) + \cdots$$

$$x_2(t+\Delta t) = x_2(t) + \Delta t \frac{dx_2}{dt}(t) + \frac{(\Delta t)^2}{2}\frac{d^2x_2}{dt^2}(t) + \cdots$$

$$\vdots = \vdots$$

$$x_n(t+\Delta t) = x_n(t) + \Delta t \frac{dx_n}{dt}(t) + \frac{(\Delta t)^2}{2}\frac{d^2x_n}{dt^2}(t) + \cdots,$$

or expressing the derivatives in terms of the functions f_i gives

$$x_1(t+\Delta t) = x_1(t) + \Delta t f_1\big(x_1(t),x_2(t),\ldots,x_n(t),t\big) + \frac{(\Delta t)^2}{2}\frac{df_1}{dt}(t) + \cdots$$

$$x_2(t+\Delta t) = x_2(t) + \Delta t f_2\big(x_1(t),x_2(t),\ldots,x_n(t),t\big) + \frac{(\Delta t)^2}{2}\frac{df_2}{dt}(t) + \cdots$$

$$\vdots = \vdots$$

$$x_n(t+\Delta t) = x_n(t) + \Delta t f_n\big(x_1(t),x_2(t),\ldots,x_n(t),t\big) + \frac{(\Delta t)^2}{2}\frac{df_n}{dt}(t) + \cdots.$$

12.5.3 Euler's Method

Euler's method for a single first-order equation was based upon keeping the terms in the Taylor series of $x(t)$ up through the Δt term, and the same is easily done in the case of a system of equations. In particular, Euler's method is

$$x_1(t+\Delta t) = x_1(t) + \Delta t f_1\big(x_1(t),x_2(t),\ldots,x_n(t),t\big)$$

$$x_2(t+\Delta t) = x_2(t) + \Delta t f_2\big(x_1(t),x_2(t),\ldots,x_n(t),t\big)$$

$$\vdots = \vdots$$

$$x_n(t+\Delta t) = x_n(t) + \Delta t f_n\big(x_1(t),x_2(t),\ldots,x_n(t),t\big).$$

$$(12.25)$$

Rewriting all of this in vector notation simplifies the expressions and furthermore makes the relationship between the methods for systems of equations and for a single first-order equation transparent. Let

$$\mathbf{x}(t) = \begin{bmatrix} x_1(t) \\ x_2(t) \\ \vdots \\ x_n(t) \end{bmatrix}$$

$$(12.26)$$

and

$$\mathbf{f}(x_1(t), x_2(t), \ldots, x_n(t), t) = \begin{bmatrix} f_1(x_1(t), x_2(t), \ldots, x_n(t), t) \\ f_2(x_1(t), x_2(t), \ldots, x_n(t), t) \\ \vdots \\ f_n(x_1(t), x_2(t), \ldots, x_n(t), t) \end{bmatrix}$$

or substituting the vector notation for $\mathbf{x}(t)$ from Equation (12.26)

$$\mathbf{f}(\mathbf{x}(t), t) = \begin{bmatrix} f_1(\mathbf{x}(t), t) \\ f_2(\mathbf{x}(t), t) \\ \vdots \\ f_n(\mathbf{x}(t), t) \end{bmatrix}.$$

Then the original system of equations expressed in Equation (12.24) becomes

$$\dot{\mathbf{x}} = \mathbf{f}(\mathbf{x}(t), t),$$

which looks remarkably like Equation (12.1) with a few of the terms in bold face font. Furthermore, expressing Equation (12.25) in this notation reduces the expression to

$$\boxed{\mathbf{x}(t + \Delta t) = \mathbf{x}(t) + \mathbf{f}(\mathbf{x}(t), t) \Delta t,} \qquad (12.27)$$

which, again, is exactly the same as the equation for Euler's method for a single first-order equation with the vector terms in boldface.

Example 12.11. Determine an approximate numerical solution to

$$\dot{x} = y$$
$$\dot{y} = (1 - x^2)y - x,$$

where $x(0) = 0.02$ and $y(0) = 0.0$ using Euler's method. Substituting into Equation (12.27) gives the system of equations

$$x(t + \Delta t) = x(t) + y(t)\Delta t$$
$$y(t + \Delta t) = y(t) + \left(\left(1 - (x(t))^2 \right) y(t) - x(t) \right) \Delta t.$$

Figure 12.12 illustrates both components of the solution for $0 < t < 20$. A program listing in the C programming language for this problem appears in Appendix E.1.0.13. A program listing in FORTRAN for this problem appears in Appendix E.2.0.27.

Observe that the right-hand side of Equation (12.27) is evaluated at time t. It is very easy to write a computer program that does not quite do that, as the following example illustrates.

Example 12.12. Consider the system from Example 12.11 and the following lines of code:

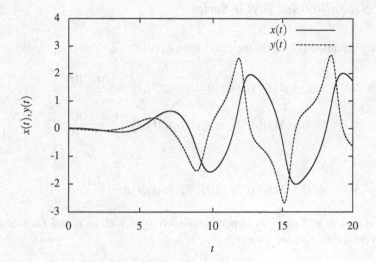

Fig. 12.12 Numerical solutions for the system of equations in Example 12.11 using Euler's method.

```
x = x + y*dt;
y = y + ((1.0 - x*x)*y - x)*dt;
```

This seemingly incorrectly implements Euler's method because the value for x, which appears on the right-hand side of the second y equation has already been changed from $x(t)$ to $x(t + \Delta t)$ by the first line.

Although this approach deviates from the exact expression for Euler's method (and does indeed result in a different approximate solution because the second equation uses the $x(t + \Delta t)$ values instead of $x(t)$), it is inconsequential with respect to the accuracy of the method. To see this consider the second equation using the "incorrect" method:

$$y(t) = y(t) + ((1 - x(t + \Delta t)x(t + \Delta t))y(t) - x(t + \Delta t))\Delta t$$
$$= y(t) + [(1 - (x(t) + f(x,y,t)\Delta t)(x(t) + f(x,y,t)\Delta t))y(t)$$
$$- (x(t) + f(x,y,t)\Delta t)]\Delta t$$
$$= y(t) + [(1 - x(t)x(t))y(t) - x(t)]\Delta t + \mathcal{O}\left((\Delta t)^2\right),$$

where the notation $\mathcal{O}\left((\Delta t)^2\right)$ means a collection of terms that multiply $(\Delta t)^2$.

The bottom line is that although this approach modifies the second equation and adds some extra terms to the expression for $y(t + \Delta t)$, these added terms are of a higher order than the accuracy of the method, and hence do not affect the order of accuracy of the approach. As shown subsequently, more care must be taken when using higher-order methods.

12.5.4 Second-Order Taylor Series

Extending Equation (12.25) to the $(\Delta t)^2$ term gives

$$x_1(t+\Delta t) = x_1(t) + \Delta t f_1\big(x_1(t),x_2(t),\ldots,x_n(t),t\big) + \frac{(\Delta t)^2}{2}\frac{\mathrm{d}f_1}{\mathrm{d}t}(t)$$

$$x_2(t+\Delta t) = x_2(t) + \Delta t f_2\big(x_1(t),x_2(t),\ldots,x_n(t),t\big) + \frac{(\Delta t)^2}{2}\frac{\mathrm{d}f_2}{\mathrm{d}t}(t) \tag{12.28}$$

$$\vdots = \vdots$$

$$x_n(t+\Delta t) = x_n(t) + \Delta t f_n\big(x_1(t),x_2(t),\ldots,x_n(t),t\big)\Delta t + \frac{(\Delta t)^2}{2}\frac{\mathrm{d}f_n}{\mathrm{d}t}(t).$$

Each component of $\mathbf{f}$ possibly depends on each x_i as well as t, and each of the x_i depends on t, therefore we have

$$\frac{\mathrm{d}f_i}{\mathrm{d}t} = \frac{\partial f_i}{\partial x_1}f_1 + \frac{\partial f_i}{\partial x_2}f_2 + \cdots + \frac{\partial f_i}{\partial x_n}f_n + \frac{\partial f_i}{\partial t}.$$

Expanding each derivative term in Equation (12.28) would be cumbersome, so to use a more compact notation, recall the definition of the Jacobian

$$\frac{\partial \mathbf{f}}{\partial \mathbf{x}} = \begin{bmatrix} \frac{\partial f_1}{\partial x_1} & \frac{\partial f_1}{\partial x_2} & \cdots & \frac{\partial f_1}{\partial x_n} \\ \frac{\partial f_2}{\partial x_1} & \frac{\partial f_2}{\partial x_2} & \cdots & \frac{\partial f_2}{\partial x_n} \\ \vdots & & \ddots & \vdots \\ \frac{\partial f_n}{\partial x_1} & \frac{\partial f_n}{\partial x_2} & \cdots & \frac{\partial f_n}{\partial x_n} \end{bmatrix}.$$

Using this, the Taylor series to second-order may be written in vector form as

$$\boxed{\mathbf{x}(t+\Delta t) = \mathbf{x}(t) + \mathbf{f}(\mathbf{x},t)\,\Delta t + \frac{(\Delta t)^2}{2}\left(\frac{\partial \mathbf{f}}{\partial \mathbf{x}}\mathbf{f} + \frac{\partial \mathbf{f}}{\partial t}\right).} \tag{12.29}$$

Obviously, computing all the n^2 partial derivatives would be a hassle. There is not much point to doing so because the same accuracy may be obtained by using the Runge–Kutta methods, as outlined subsequently.

12.5.5 Fourth-Order Runge–Kutta

Rather than provide the details for every method, this section skips right to the fourth-order Runge–Kutta method. From it, the generalizations necessary to implement the other methods for systems of equations should be obvious. Similar to the manner in which Euler's method generalized to the case of a system of equations, fourth-order Runge–Kutta may be expressed as the following. For

$$\dot{\mathbf{x}} = \mathbf{f}(\mathbf{x}(t), t)$$

let

$$\mathbf{x}(t + \Delta t) = \mathbf{x}(t) + \frac{1}{6}(\mathbf{k}_1 + 2\mathbf{k}_2 + 2\mathbf{k}_3 + \mathbf{k}_4), \tag{12.30}$$

where

$$
\begin{aligned}
\mathbf{k}_1 &= \mathbf{f}(\mathbf{x}(t), t)\,\Delta t \\
\mathbf{k}_2 &= \mathbf{f}\left(\mathbf{x}(t) + \frac{1}{2}\mathbf{k}_1, t + \frac{1}{2}\Delta t\right)\Delta t \\
\mathbf{k}_3 &= \mathbf{f}\left(\mathbf{x}(t) + \frac{1}{2}\mathbf{k}_2, t + \frac{1}{2}\Delta t\right)\Delta t \\
\mathbf{k}_4 &= \mathbf{f}(\mathbf{x}(t) + \mathbf{k}_3, t + \Delta t)\,\Delta t.
\end{aligned}
\tag{12.31}
$$

Note that $\mathbf{k}_1$ through $\mathbf{k}_4$ are vector quantities because $\mathbf{f}(\mathbf{x}(t), t)$ is a vector.

Example 12.13. Determine an approximate solution to

$$
\begin{aligned}
\dot{x} &= y \\
\dot{y} &= (1 - x^2)y - x\sin t,
\end{aligned}
$$

where $x(0) = 0.02$ and $y(0) = 0.0$ using the fourth-order Runge–Kutta method.

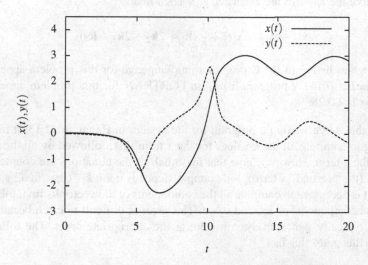

Fig. 12.13 Approximate numerical solutions for the system of equations from Example 12.13 using the fourth-order Runge–Kutta method.

Let

$$\mathbf{x}(t) = \begin{bmatrix} x(t) \\ y(t) \end{bmatrix} = \begin{bmatrix} x_1(t) \\ x_2(t) \end{bmatrix}$$

and

$$\mathbf{f}(\mathbf{x}(t),t) = \begin{bmatrix} x_2(t) \\ \left(1 - (x_1(t))^2\right) x_2(t) - x_1(t)\sin t \end{bmatrix}.$$

Then

$$\mathbf{k}_1 = \begin{bmatrix} k_{11} \\ k_{21} \end{bmatrix} = \begin{bmatrix} x_2 \\ \left(1 - (x_1)^2\right) x_2 - x_1\sin t \end{bmatrix} \Delta t$$

$$\mathbf{k}_2 = \begin{bmatrix} k_{12} \\ k_{22} \end{bmatrix}$$

$$= \begin{bmatrix} x_2 + \frac{1}{2}k_{21} \\ \left(1 - \left(x_1 + \frac{1}{2}k_{11}\right)^2\right)\left(x_2 + \frac{1}{2}k_{21}\right) - \left(x_1 + \frac{1}{2}k_{11}\right)\sin\left(t + \frac{1}{2}\Delta t\right) \end{bmatrix} \Delta t$$

$$\mathbf{k}_3 = \begin{bmatrix} k_{13} \\ k_{23} \end{bmatrix}$$

$$= \begin{bmatrix} x_2 + \frac{1}{2}k_{22} \\ \left(1 - \left(x_1 + \frac{1}{2}k_{12}\right)^2\right)\left(x_2 + \frac{1}{2}k_{22}\right) - \left(x_1 + \frac{1}{2}k_{12}\right)\sin\left(t + \frac{1}{2}\Delta t\right) \end{bmatrix} \Delta t$$

$$\mathbf{k}_4 = \begin{bmatrix} k_{14} \\ k_{24} \end{bmatrix} = \begin{bmatrix} x_2 + k_{23} \\ \left(1 - (x_1 + k_{13})^2\right)(x_2 + k_{23}) - (x_1 + k_{13})\sin(t + \Delta t) \end{bmatrix} \Delta t,$$

where all of the x_i terms are evaluated at t. Then finally

$$\mathbf{x}(t + \Delta t) = \mathbf{x}(t) + \frac{1}{6}\left(\mathbf{k}_1 + 2\mathbf{k}_2 + 2\mathbf{k}_3 + \mathbf{k}_4\right).$$

A program listing in the C programming language for this problem appears in Appendix E.1.0.14. A program listing in FORTRAN for this problem appears in Appendix E.2.0.28.

Note that when writing a program for the system in Example 12.13, it may be tempting to compute all the k values for the x term first, followed by all the k values for the y term. However, note that inasmuch as the equations are coupled, the k_{12} term (the "second" x term), for example, depends upon k_{21} (the "first" y term). Hence, it is necessary to compute all the components of the vector $\mathbf{k}_1$ first, followed by all the components of $\mathbf{k}_2$, and so on. This error is difficult to catch because the solution typically will converge, just not at the appropriate order. The following example illustrates this fact.

Example 12.14. Compute an approximate numerical solution for

$$\dot{x} = y$$
$$\dot{y} = -x,$$

where $x(0) = 0$ and $y(0) = 1$ using the fourth-order Runge–Kutta method. Compare the approximate solution when the terms in the algorithm are computed in the correct and incorrect order.

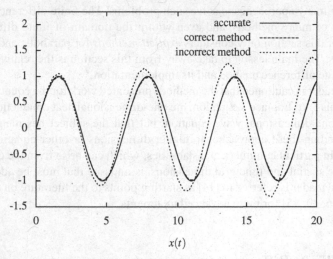

$x(t)$

Fig. 12.14 Comparison of approximate solutions from Example 12.14 when making a common error in implementing the fourth-order Runge–Kutta algorithm for systems of first-order differential equations.

Using the notation from Example 12.13, the correct order of computation for the k values is

$$k_{11}, k_{21}, k_{12}, k_{22}, k_{13}, k_{23}, k_{14}, k_{24}.$$

By comparison, the incorrect, but tempting, order is

$$k_{11}, k_{12}, k_{13}, k_{14}, k_{21}, k_{22}, k_{23}, k_{24}.$$

Figure 12.14 illustrates an accurate solution and compares the approximate solutions for $x(t)$ for both cases when $\Delta t = 0.4$. Clearly, the correct approach produces a much more accurate approximate solution.

12.6 Numerical Methods for Partial Differential Equations

This section considers the so-called *finite difference method* for partial differential equations. The main idea is (similar to the manner in which time was discretized for determining a numerical approximation to the time derivative for ordinary differential equations) that the spatial dimension(s) must be similarly discretized for partial differential equations with an independent variable corresponding to the spatial direction. Doing so results in a system of coupled *ordinary* differential equations. It would be tempting to think that it would then be a simple matter to use the methods covered previously in this chapter to solve the resulting system of ordinary differential equations, but unfortunately, doing so is not always so simple.

The subject of numerical methods for partial differential equations is an advanced one, with many approaches to solve such problems. The finite difference method is only one of many methods, and even within the domain of finite differences the subject of this section only considers *explicit methods* for parabolic and hyperbolic equations. The main lesson to take away from this section is the relative simplicity of the finite difference method and its implementation.

The reader is cautioned that the methods presented work for the equations considered in this text, Laplace's equation, the one-dimensional heat conduction equation, and the one-dimensional wave equation, but that the subject has many subtleties and extending these approaches to higher dimensions or other equations requires caution. In particular, numerical instabilities, which can arise from the coupling between the spatial and time grid dimensions, is an issue that must be addressed. An interested reader is referred to [14] as starting points to the literature on this subject, and to [26, 29, 45] for more advanced treatments.

12.6.1 Finite Difference Approximation

Consider, for example, the function $u(x,t)$ that describes the solution to the wave equation

$$\alpha^2 \frac{\partial^2 u}{\partial x^2} = \frac{\partial^2 u}{\partial t^2}. \tag{12.32}$$

Consider the fixed x values, $m\Delta x$ where $m = 0, 1, 2, \ldots, M$ (so $L = M\Delta x$) and $\Delta x \ll 1$. In a manner analogous to the situation for ordinary differential equations where an approximate solution is determined only at a finite number of discrete times, $x(n\Delta t)$, for a partial differential equation with one spatial dimension and one time dimension the approximate solution is determined at a set of discrete points in space and in time, that is, $u(m\Delta x, n\Delta t)$.[3] To clarify the presentation, $u(m,n)$ is written instead of $u(m\Delta x, n\Delta t)$. This is illustrated in Figure 12.15 for the string equation. The purpose of the finite difference method is to approximate the second-order spatial derivative on the left-hand side of Equation (12.32) with values of $u(x,t)$ at other fixed values of x.

---

[3] Writing $u(m\Delta x, n\Delta t)$ is actually an abuse of notation because it may imply that u is a function of x and t and the numerical approximation is only defined at discrete points in time. However this is adopted for purposes of clarity over rigor because using a more standard notation such as u_m^n, although technically more correct, may be confusing for those learning the subject for the first time.

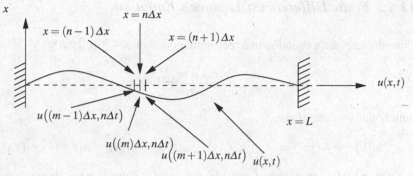

Fig. 12.15 Vibrating string.

In particular, consider approximating the first partial derivative with respect to x at $x = m\Delta x$. This is simple enough using the definition of the derivative

$$\frac{\partial u}{\partial x}(m\Delta x, t) \approx \frac{u(m\Delta x, t) - u\big((m-1)\Delta x, t\big)}{\Delta x}$$

if $\Delta x \ll 1$. Now, because the second derivative is the derivative of the derivative, then

$$\frac{\partial^2 u(m\Delta x, t)}{\partial x^2} \approx \frac{\frac{\partial u\big((m+1)\Delta x, t\big)}{\partial x} - \frac{\partial u(m\Delta x, t)}{\partial x}}{\Delta x}$$

$$\approx \frac{\left(\frac{u\big((m+1)\Delta x, t\big) - u(m\Delta x, t)}{\Delta x}\right) - \left(\frac{u(m\Delta x, t) - u\big((n-1)\Delta x, t\big)}{\Delta x}\right)}{\Delta x}$$

$$= \frac{u\big((m+1)\Delta x, t\big) - 2u(m\Delta x, t) + u\big((m-1)\Delta x, t\big)}{(\Delta x)^2}.$$

Finally, substituting this approximation back into Equation (12.32), using the same finite difference for the second-order time derivative, and dropping the explicit dependence on Δx and Δt terms u gives

$$\alpha^2 \frac{u(m+1, n) - 2u(m, n) + u(m-1, n)}{(\Delta x)^2} = \frac{u(m, n+1) - 2u(m, n) + u(m, n-1)}{(\Delta t)^2}.$$

(12.33)

The boundary conditions are expressed as fixed values at $u(0, t)$ and $u(M\Delta x, t)$. The initial conditions, which need a little more consideration are incorporated into the method as specified values for $u(m, 0)$ and $\dot{u}(m, 0)$ for all $m \in \{1, \ldots, M\}$.

The following sections implement the finite difference method on all three classes of linear partial differential equations: namely, Laplace's equation (elliptic), the one-dimensional heat conduction equation (parabolic), and the one-dimensional wave equation (hyperbolic).

12.6.2 Finite Differences: Laplace's Equation

Consider Laplace's equation in a rectangular domain, $x \in [0,a]$, $y \in [0,b]$

$$\frac{\partial^2 u}{\partial x^2} + \frac{\partial^2 u}{\partial y^2} = 0$$

with boundary conditions

$$u(0,y) = f_1(y), \quad u(a,y) = f_2(y), \quad u(x,0) = g_1(x), \quad u(x,b) = g_2(x).$$

In Section 11.4.1 separation of variables was used to solve this equation where three of the four boundary conditions were zero. First we implement the finite difference method to solve this equation and compare it to the solutions obtained previously in Section 11.4.1. Of course, the real value in a numerical method is the fact it can easily deal with complex geometry, so a second example implements the method on a problem that would be very difficult to solve by hand.

The finite difference approximation to Laplace's equation is

$$\frac{u\big((m-1)\Delta x, n\Delta y\big) - 2u\,(m\Delta x, n\Delta y) + u\big((m+1)\Delta x, n\Delta y\big)}{(\Delta x)^2}$$
$$+ \frac{u\,(m\Delta x, (n-1)\Delta y) - 2u\,(m\Delta x, n\Delta y) + u\,(m\Delta x, (n+1)\Delta y)}{(\Delta y)^2} = 0,$$

or dropping the cumbersome Δx and Δy terms

$$\frac{u\,(m-1,n) - 2u\,(m,n) + u\,(m+1,n)}{(\Delta x)^2}$$
$$+ \frac{u\,(m,n-1) - 2u\,(m,n) + u\,(m,n+1)}{(\Delta y)^2} = 0.$$

If $\Delta x = \Delta y = h$, then solving for $u(m,n)$ gives

$$\boxed{u(m,n) = \frac{1}{4}\left[u(m-1,n) + u(m+1,n) + u(m,n-1) + u(m,n+1)\right]} \qquad (12.34)$$

which has the very simple interpretation that the value for a grid point is the average of its four neighbors. This also makes complete intuitive physical sense when considering the fact that the solution to Laplace's equation can be interpreted as the steady-state temperature distribution in a two-dimensional solid.

A reasonable algorithm which would be the first way to attempt to find a solution would be to assign all the interior points a starting value, such as the average of all the boundary points, and then systematically working through all the interior points and setting the value of a grid point to the average of its neighbors. Such an approach does not have the fastest convergence, however, it is still workable, as the following two examples illustrate.

Example 12.15. This example uses the finite difference method to find an approximate solution to

$$\frac{\partial^2 u}{\partial x^2} + \frac{\partial^2 u}{\partial y^2} = 0$$

on the domain $x \in [0,4]$, $y \in [0,2]$ where

$$u(0,y) = 0, \qquad\qquad u(4,y) = 0,$$

$$u(x,0) = 0, \qquad\qquad u(x,2) = \begin{cases} x, & 0 < x \le 2, \\ 4 - x, & 2 < x \le 4. \end{cases}$$

A grid size of $\Delta x = \Delta y = 0.1$ is used. All the interior grid points are set to a value of $1/3$ initially. A FORTRAN program implementing the finite difference method for this problem appears in Appendix E.2.0.29. Plots of the numerical solution after 1, 10, 20 and 100 iterations appear in Figures 12.16, 12.17, and 12.18 respectively.

Note, these plots are not a depiction of the evolution of the system in time. Laplace's equation does not have time as an independent variable. The figures illustrate how the method *converges*. Hence, the first plots do not represent the solution, and the last one only does approximately.

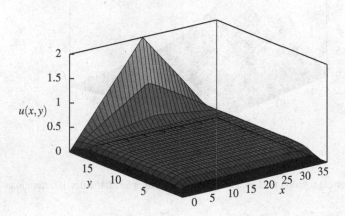

Fig. 12.16 Numerical solution for Laplace's equation in Example 12.15 after 1 iteration.

Of course, the real value in a numerical method is in determining a solution that would be difficult or impossible to find by hand. The following example has a

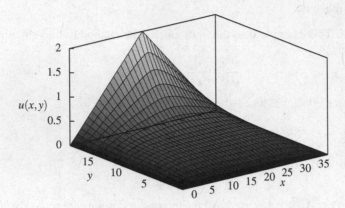

Fig. 12.17 Numerical solution for Laplace's equation in Example 12.15 after 20 iterations.

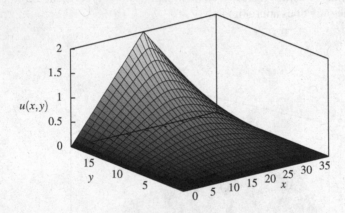

Fig. 12.18 Numerical solution for Laplace's equation in Example 12.15 after 100 iterations.

relatively complicated boundary, and the only realistic approach to finding a solution would be a numerical method.

Example 12.16. Determine an approximate numerical solution to Laplace's equation on the interior of the H-shaped domain illustrated in Figure 12.19 where all the "vertical" boundaries (constant x-value) have a value of zero and the "horizontal" boundaries (constant y-value) on the top have a value of one and on the bottom have a value of zero. The initial value for the interior points were set at $u = 0.5$. The solution after 100 iterations is illustrated in Figure 12.20. A FORTRAN program for this example is in Appendix E.2.0.30.

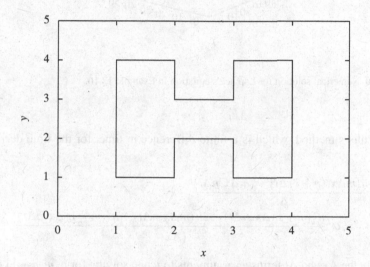

Fig. 12.19 Domain for Laplace's equation in Example 12.16.

12.6.3 Finite Differences: Heat Equation

Using finite differences for the spatial derivative in the heat equation

$$\frac{\partial u}{\partial t} = \alpha^2 \frac{\partial^2 u}{\partial x^2}$$

gives at the point $x = m\Delta x$,

$$\frac{\partial u}{\partial t}(m\Delta x, t) = \alpha^2 \frac{u((m+1)\Delta x, t) - 2u(m\Delta x, t) + u((m-1)\Delta x, t)}{(\Delta x)^2}.$$

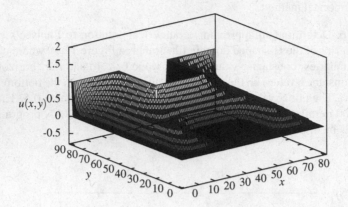

Fig. 12.20 Numerical solution for Laplace's equation in Example 12.16.

Using Euler's method, which is a finite difference in time, for the time derivative gives

$$\frac{u(m\Delta x,(n+1)\Delta t)-(m\Delta x,n\Delta t)}{\Delta t}$$
$$=\alpha^2\frac{u((m+1)\Delta x,n\Delta t)-2u(m\Delta x,n\Delta t)+u((m-1)\Delta x,n\Delta t)}{(\Delta x)^2}.$$

Dropping the Δx and Δt terms as arguments to u and solving for $u(m,n+1)$ gives

$$\boxed{u(m,n+1)=u(m,n)+\alpha^2\frac{u(m+1,n)-2u(m,n)+u(m-1,n)}{(\Delta x)^2}\Delta t.}\qquad(12.35)$$

Example 12.17. Use the finite-difference method to determine an approximate solution to

$$4\frac{\partial^2 u}{\partial x^2}=\frac{\partial u}{\partial t}$$

with

$$u(0,t)=u(10,t)=0$$

and

$$u(x,0)=\begin{cases}x, & 0<x\le 5,\\ 10-x, & 5<x\le 10.\end{cases}$$

This is the same problem that was considered in Example 11.8.

Using $\Delta x = 0.5$, the numerical solution with $\Delta t = 0.31$ is illustrated in Figure 12.21. A FORTRAN program that implements the method for this problem appears in Appendix E.2.0.31. Observe that the only significant difference between the numerical solution and the one obtained by separation of variables is due to the error in the series solution that was plotted because only the first 10 terms were included in the partial sum for the plot.

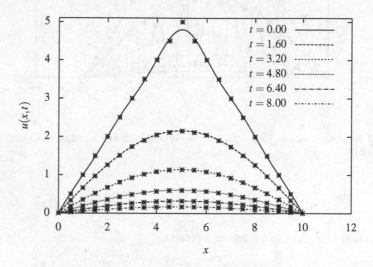

Fig. 12.21 Numerical solution (with $*$) compared with the partial sum solution using the first 10 terms obtained in Example 11.8, for the one-dimensional heat equation in Example 12.17 with $\Delta t = 0.031$ and $\Delta x = 0.5$.

At this point, everything seems rather straightforward. In the previous example, with a rather coarse grid and large time step, the numerical solution was very accurate. However, there is a subtlety lurking beneath the surface. With the finite difference method the spatial and temporal derivative approximations can interact in a manner that makes the numerical algorithm unstable. To illustrate this, the previous example is repeated with a slightly larger time step that will make the numerical solution unstable. The relationship between Δt and Δx for the algorithm to be stable is presented after the example.

Example 12.18. The approximate solution to the equation in Example 12.17 with $\Delta t = 0.32$ is illustrated in Figure 12.22, where obviously the solution no longer converges to the accurate solution. This instability is purely numerical, meaning that there is something in the relationship between the spatial and time derivative approximations that causes the approximate solution to diverge.

To develop some insight into the nature of this numerical instability, let us consider the heat equation with homogeneous boundary conditions

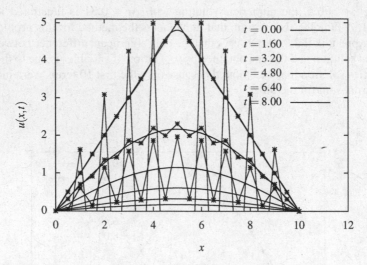

Fig. 12.22 Numerical solution (with ∗) for the one-dimensional heat equation in Example 12.17 with $\Delta t = 0.032$.

$$\frac{\partial u}{\partial t} = \alpha^2 \frac{\partial^2 u}{\partial x^2}, \quad u(0,t) = 0, \quad u(L,t) = 0, \quad u(x,0) = f(x).$$

From Section 11.3, the exact solution is given by

$$u(x,t) = \sum_{n=1}^{\infty} c_n e^{-(\alpha^2 n^2 \pi^2 t/L^2)} \sin \frac{n\pi x}{L}, \quad c_n = \frac{2}{L} \int_0^L f(x) \sin \frac{n\pi x}{L} dx.$$

The solution for $u(x,t)$ is the superposition of an infinite number of modes, therefore let us consider what happens to an individual mode when it is used in the finite-difference method. In particular, consider

$$u(x,t) = c_n e^{-(\alpha^2 n^2 \pi^2 t/L^2)} \sin \frac{n\pi x}{L}$$

for some integer n. Substituting this solution into Equation (12.35) gives

$$u(x,t+\Delta t) = u(x,t) + \left(\frac{\alpha}{\Delta x}\right)^2 (u(x+\Delta x,t) - 2u(x,t) + u(x-\Delta x,t))\Delta t$$

$$= c_n e^{-(\alpha^2 n^2 \pi^2 t/L^2)} \sin \frac{n\pi x}{L} + \left(\frac{\alpha}{\Delta x}\right)^2 c_n e^{-(\alpha^2 n^2 \pi^2 t/L^2)}.$$

$$\left(\sin \frac{n\pi(x+\Delta x)}{L} - 2\sin \frac{n\pi x}{L} + \sin \frac{n\pi(x-\Delta x)}{L}\right)\Delta t.$$

Expanding the $\sin(n\pi(x+\Delta x)/L)$ and $\sin(n\pi(x-\Delta x)/L)$ terms reduces this to

$$u(x,t+\Delta t) = c_n e^{-(\alpha^2 n^2 \pi^2 t/L^2)} \sin\frac{n\pi x}{L} + \left(\frac{\alpha}{\Delta x}\right)^2 c_n e^{-(\alpha^2 n^2 \pi^2 t/L^2)}.$$

$$\left(2\sin\frac{n\pi x}{L}\cos\frac{n\pi\Delta x}{L} - 2\sin\frac{n\pi x}{L}\right)\Delta t$$

$$= c_n e^{-(\alpha^2 n^2 \pi^2 t/L^2)} \sin\frac{n\pi x}{L}\left[1 + 2\Delta t\left(\frac{\alpha}{\Delta x}\right)^2\left(\cos\frac{n\pi\Delta x}{L} - 1\right)\right].$$

$$(12.36)$$

Keeping in mind that this analysis is looking at the effect of the algorithm on one of the modes out of many in the solution, it makes sense to consider whether the magnitude of the mode will increase or decrease in the time step. If $n = M$, the number of discretization points, then $M\Delta x = L$ and

$$\cos\frac{n\pi\Delta x}{L} - 1 = \cos\frac{M\pi\Delta x}{L} - 1 = \cos\frac{L\pi}{L} - 1 = -2,$$

which is the maximum magnitude that term can obtain. Then, if Δt is selected such that

$$2\Delta t\left(\frac{\alpha}{\Delta x}\right)^2 < 1$$

then the term in the square brackets in Equation (12.36) will have a magnitude less than one, which means that mode in the numerical solution will decrease in magnitude rather than increase that mode. In other words, the structure of the numerical solution method will excite the Mth mode if Δt is too large in relation to Δx, and it is purely an artifact of the numerical method, not the physics of the problem.

Referring to Figure 12.22, it is exactly the Mth mode that is going unstable. This may perhaps be seen the best on the second, $t = 1.60$, curve where the points are alternating above and below the actual solution. Recall that M is the number of discretization points, and hence the Mth mode in the series solution is the one that oscillates M times between $x = 0$ and $x = L$.

The above is intended to give some insight into the nature of the instability, but the reader is cautioned that it does not amount to a proof. The references at the beginning of this section provide a more rigorous treatment. Nonetheless, it is the case that the finite difference method for the one-dimensional heat equation requires that

$$\boxed{\Delta t < \frac{1}{2}\left(\frac{\Delta x}{\alpha}\right)^2}$$

$$(12.37)$$

for numerical stability. The difference between Examples 12.17 and 12.18 was that the former satisfied this stability bound and the latter did not.

12.6.4 Finite Differences: The Wave Equation

Normally, spending a lot of time discussing the wrong way to solve a problem is not of much value. However, in the case of the wave equation, the finite difference method, the way that is probably the most natural to use to solve it actually is problematic, so we will dispense with that first.

Because the wave equation is second order in time compared to the heat conduction equation which is first order, it is tempting to write

$$\dot{u}(1,t) = v(1,t)$$

$$\dot{v}(1,t) = (u(2,t) - 2u(1,t) + u(0,t))\left(\frac{\alpha^2}{\Delta x}\right)^2$$

$$\vdots = \vdots \tag{12.38}$$

$$\dot{u}(M-1,t) = v(M-1,t)$$

$$\dot{v}(M-1,t) = (u(M,t) - 2u(M-1,t) + u(M-2,t))\left(\frac{\alpha^2}{\Delta x}\right)^2$$

and use some standard method from earlier in the chapter to solve these. Motivated by the derivation of the stability condition for the heat equation given by Equation (12.37), let us consider what happens when the Mth mode of the solution to the wave equation is substituted into Equation (12.38). Recall M is selected so that $M\Delta x = L$.

The Mth mode of the solution to the wave equation is given by Equation (11.20),

$$u_M(x,t) = \sin\frac{M\pi x}{L}\left(a_M \sin\frac{\alpha M\pi t}{L} + b_M \cos\frac{\alpha M\pi t}{L}\right).$$

For simplicity we let $b_M = 0$. Computing the derivative with respect to time gives

$$v(x,t) = a_M\frac{M\pi t}{L}\sin\frac{M\pi x}{L}\cos\frac{\alpha M\pi t}{L}, \tag{12.39}$$

where $v(x,t) = \dot{u}(x,t)$. Using Euler's method and Equation (12.38) gives

$$v(x,t+\Delta t) = v(x,t) + \left(\frac{\alpha}{\Delta x}\right)^2 [u(x-\Delta x,t) - 2u(x,t) + u(x-\Delta x,t)]\Delta t.$$

Substituting for $v(x,t)$ from Equation (12.39) and using trigonometric identities to simplify the terms in the square brackets in a manner identical to what was done for the heat equation gives

$$v(x,t+\Delta t) = a_M \frac{\alpha M\pi}{L} \sin \frac{M\pi x}{L} \cos \frac{\alpha M\pi t}{L} \left[1 + 2\Delta t \left(\frac{\alpha}{\Delta x}\right)^2 \left(\cos \frac{M\pi \Delta x}{L} - 1\right)\right]$$

$$= a_M \frac{\alpha M\pi}{L} \sin \frac{M\pi x}{L} \cos \frac{\alpha M\pi t}{L} \left[1 - 4\Delta t \left(\frac{\alpha}{\Delta x}\right)^2\right].$$

Now, we use this to compute $u(x,t+2\Delta t)$. Euler's method is

$$u_M(x,t+2\Delta t) = u_M(x,t+\Delta t) + v(x,t+\Delta t)\Delta t,$$

and substituting for $u_M(x,t+\Delta t)$ and $v(x,t+\Delta t)$ gives

$$u_M(x,t+2\Delta t) = a_M \sin \frac{M\pi x}{L} \sin \frac{\alpha M\pi (t+\Delta t)}{L}$$
$$+ a_M \frac{\alpha M\pi}{L} \sin \frac{M\pi x}{L} \cos \frac{\alpha M\pi t}{L} \left[1 - 4\Delta t \left(\frac{\alpha}{\Delta x}\right)^2\right]\Delta t. \quad (12.40)$$

Using the usual trigonometric identity

$$\sin \frac{\alpha M\pi (t+\Delta t)}{L} = \sin \frac{\alpha M\pi t}{L} \cos \frac{\alpha M\pi \Delta t}{L} + \sin \frac{\alpha M\pi \Delta t}{L} \cos \frac{\alpha M\pi t)}{L}$$

and assuming a small Δt, which is reasonable inasmuch as we are considering convergence, gives

$$\sin \frac{\alpha M\pi (t+\Delta t)}{L} \approx \sin \frac{\alpha M\pi t}{L} + \frac{\alpha M\pi \Delta t}{L} \cos \frac{\alpha M\pi t}{L}.$$

Substituting this into Equation (12.40) and collecting terms results in

$$u_M(x,t+2\Delta t) = a_M \sin \frac{M\pi x}{L} \sin \frac{\alpha M\pi t}{L}$$
$$+ a_M \frac{\alpha M\pi}{L} \sin \frac{M\pi x}{L} \cos \frac{\alpha M\pi t}{L} \left[2 - 4\Delta t \left(\frac{\alpha}{\Delta x}\right)^2\right]\Delta t.$$

The magnitude of this solution is

$$|u_M(x,t+2\Delta t)| = a_M \sin \frac{M\pi x}{L} \sqrt{1 + \left(\frac{\alpha M\pi}{L} \left[2 - 4\Delta t \left(\frac{\alpha}{\Delta x}\right)^2\right]\Delta t\right)^2}$$

which is increasing, regardless of the relationship between Δx and Δt.

The proper way to use finite differences for the wave equation is to treat the time variable much like the spatial one and write

$$\alpha^2 \frac{u(m+1,n) - 2u(m,n) + u(m-1,n)}{(\Delta x)^2} = \frac{u(m,n+1) - 2u(m,n) + u(m,n-1)}{(\Delta t)^2}.$$

Thus, the value of u at time $t = (m+1)\Delta t$ depends not only on the value at the previous time step, but the time step before that as well. The only issue is at the initial time step, but because the initial velocity is also part of the problem specification, the first step can be approximated by the product of the velocity and the time step (basically ignoring the string physics for that one step). Then, starting with the second time step,

$$
u(m,n+1) = \left(\frac{\alpha\Delta t}{\Delta x}\right)^2 [u(m+1,n) - 2u(m,n) + u(m-1,n)]
$$
$$
+ 2u(m,n) - u(m,n-1). \tag{12.41}
$$

Example 12.19. Use the finite-difference method and Euler's method to determine an approximate solution to

$$
\alpha^2 \frac{\partial^2 u}{\partial x^2} = \frac{\partial^2 u}{\partial t^2},
$$

where $L = 3$ and $\alpha = 2$, subjected to the boundary conditions

$$
u(0,t) = 0, \quad u(L,t) = 0
$$

and initial conditions

$$
u(x,0) = \begin{cases} x, & x \le 1, \\ \frac{3-x}{2}, & 1 < x \le 3, \end{cases}
$$
$$
\left.\frac{\partial u}{\partial t}\right|_{t=0} = 0.
$$

This is the same system that was solved using separation of variables in Example 11.2. A plot of the approximate numerical solution is illustrated in Figure 12.23. This approximate solution is very close to the one obtained using separation of variables illustrated in Figure 11.6. A program listing in the C programming language for this problem which implements the finite difference method is in Appendix E.1.0.15.

12.7 Exercises

12.1. For each of the following initial value problems, write a computer program that uses Euler's method to determine an approximate numerical solution.

1. $\dot{x} + 5x = 0$, $x(0) = 1$.
2. $\dot{x} + 5x = 5$, $x(0) = 0$.
3. $\dot{x} + 5x = \sin t$, $x(0) = 0$.
4. $\dot{x} = t^2 + x$, $x(0) = 0$.

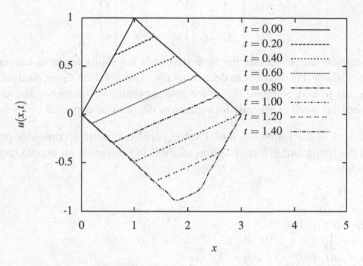

Fig. 12.23 Approximate numerical solution to the wave equation for Example 12.19. For comparison, the solution obtained using separation of variables is illustrated in Figure 11.6.

5. $\ddot{x} + x = 0$, $x(0) = 1$, $\dot{x}(0) = -1$.
6. $\ddot{x} + 5x = \sin t$, $x(0) = 1$, $\dot{x}(0) = 0$.
7. $\dot{x} + x + x^3 = 0$, $x(0) = 1$, $\dot{x}(0) = 1$.
8. $\dot{\xi} = A\xi$, where

$$A = \begin{bmatrix} -3 & 2 & 0 \\ -1 & -1 & 1 \\ 4 & -6 & 1 \end{bmatrix}, \quad \xi(0) = \begin{bmatrix} 1 \\ 3 \\ 5 \end{bmatrix}.$$

In every case be sure to reduce the time step until the solution seems to converge. In the cases where it is possible to determine the exact solution using methods from this book, do so and compare the exact and approximate solutions, and in these cases, if the time step is reduced, does the error decrease as expected?

12.2. For each of the following initial value problems, write a computer program that uses the second-order Taylor series method to determine an approximate numerical solution.

1. $\dot{x} + 5x = 0$, $x(0) = 1$.
2. $\dot{x} + 5x = 5$, $x(0) = 0$.
3. $\dot{x} + 5x = \sin t$, $x(0) = 0$.
4. $\ddot{x} + tx = 0$, $x(0) = 1$, $\dot{x}(0) = -1$.
5. $\ddot{x} + 5x = \sin t$, $x(0) = 1$, $\dot{x}(0) = 0$.
6. $\dot{x} + x + x^3 = 0$, $x(0) = 1$, $\dot{x}(0) = 1$.
7. $\dot{\xi} = A\xi$, where

$$A = \begin{bmatrix} -3 & 2 & 0 \\ -1 & -1 & 1 \\ 4 & -6 & 1 \end{bmatrix}, \quad \xi(0) = \begin{bmatrix} 1 \\ 3 \\ 5 \end{bmatrix}.$$

In every case be sure to reduce the time step until the solution seems to converge. In the cases where it is possible to determine the exact solution using methods from this book, do so and compare the exact and approximate solutions, and in these cases, if the time step is reduced, does the error decrease as expected?

12.3. For each of the following initial value problems, write a computer program that uses the fourth-order Runge–Kutta method to determine an approximate numerical solution.

1. $\dot{x} + 5x = 0$, $x(0) = 1$.
2. $\dot{x} + 5x = 5$, $x(0) = 0$.
3. $\dot{x} + 5x = \sin t$, $x(0) = 0$.
4. $\ddot{x} + tx = 0$, $x(0) = 1$, $\dot{x}(0) = -1$.
5. $\ddot{x} + 5x = \sin t$, $x(0) = 1$, $\dot{x}(0) = 0$.
6. $\dot{x} + x + x^3 = 0$, $x(0) = 1$, $\dot{x}(0) = 1$.
7. $\dot{\xi} = A\xi$, where

$$A = \begin{bmatrix} -3 & 2 & 0 \\ -1 & -1 & 1 \\ 4 & -6 & 1 \end{bmatrix}, \quad \xi(0) = \begin{bmatrix} 1 \\ 3 \\ 5 \end{bmatrix}.$$

In every case be sure to reduce the time step until the solution seems to converge. In the cases where it is possible to determine the exact solution using methods from this book, do so and compare the exact and approximate solutions, and in these cases, if the time step is reduced, does the error decrease as expected?

12.4. Consider

$$\ddot{x} + 5x = \sin t,$$

where $x(0) = 1$ and $\dot{x}(0) = 0$. If you completed Exercises 12.1–12.3 you have already written the programs for this problem. What is the largest time step that gives reasonable accuracy for $0 \le t \le 50$ for Euler's method, for the second-order Taylor series method and for the fourth-order Runge–Kutta method? If the answers are not the same, explain the reason why.

12.5. This problem considers some nonlinear differential equations that we encounter in the next chapter. Write a program that uses the fourth-order Runge–Kutta method to determine approximate numerical solutions for

1. $\ddot{x} + 0.2\dot{x} - x + x^2 = 0$, $x(0) = -1$, $\dot{x}(0) = 10$.
2. $\ddot{x} + 0.2\dot{x} - x + x^2 = 0.3\cos t$, $x(0) = 0$, $\dot{x}(0) = 0$.
3. $\ddot{x} + \dot{x} - 2x + x^2 = 0$, $x(0) = 1$, $\dot{x}(0) = 1$.

For each problem do the following.

• Continue to reduce the time step until the solution appears to converge.

- Plot the solutions versus t for a period of time that is sufficient to provide an accurate representation of the qualitative nature of the solution.
- Plot the solutions on a plot where the axes are $x(t)$ and $\dot{x}(t)$.

You may want to be sure to save your programs for the next chapter.

12.6. Consider

$$\ddot{x} + \dot{x} + x = \sin t$$

and

$$\ddot{x} + \dot{x} + x = \sin 30t,$$

where $x(0) = 0$ and $\dot{x}(0) = 0$. Write computer programs that use Euler's method to compute approximate numerical solutions for each one and run it using the following time steps, $\Delta t = 0.2$, $\Delta t = 0.175$, $\Delta t = 0.15$, $\Delta t = 0.125$, $\Delta t = 0.1$, $\Delta t = 0.05$, $\Delta t = 0.025$, $\Delta t = 0.0125$. Continue to reduce the time step until the numerical solution is a good approximation for the exact solution. Explain any unusual aspects of this problem. Would using a higher-order method eliminate these unusual aspects?

12.7. Consider $\dot{x} + x = \sin t$ where $x(0) = -1$.

1. Write a computer program to compute an approximate numerical solution for this differential equation using

   a. Euler's method
   b. The second-order Taylor series method
   c. The fourth-order Runge–Kutta method

   You may decide to write three separate programs or include all three methods in one program. Your program should also compute the exact solution for comparison purposes.
2. For each of the following time steps

   a. $\Delta t = 0.5$
   b. $\Delta t = 0.25$
   c. $\Delta t = 0.125$
   d. $\Delta t = 0.01$

   plot the exact solution and the approximate solution using each of the three methods for the time interval $t = 0$ to $t = 10$. Thus, there should be four plots and each plot should have four curves.
3. Plot the difference between the exact solution and the numerically computed solutions for the same time steps and time interval as in part 2 above. In each case indicate the factor by which the overall error changes as the time step changes and indicate whether such a factor would be expected for the global truncation error for the corresponding method.
4. What is the difference between the exact solution and the numerically computed solutions after the first time step for each method and time step size above? Determine the factor by which this error (the local truncation error) changes as the

time step changes and indicate whether such a factor corresponds to what is theoretically expected.

12.8. Write a computer program using a programming language that supports single precision to determine an approximate numerical solution for $\dot{x} = 2x$ where $x(0) = 1$ using Euler's method. For $\Delta t = 0.1$ compare the numerical solution to the exact solution at $t = 1$. Repeatedly run the program and reduce Δt and plot the error at $t = 1$ versus Δt. Continue to reduce the time step until the error at $t = 1$ begins to increase for several different values of Δt and explain the cause of this.

12.9. Write a computer program that determines an approximate numerical solution to

$$\ddot{x} + \dot{x} + 9x = \cos 3t$$

where $x(0) = 0$ and $\dot{x}(0) = 0$ using the fourth-order Runge–Kutta method. Plot the solution and exact solution for $0 \leq t \leq 15$. On a different plot, plot the error for the same range of time. Cut the time step in half and plot the error on the same plot as the error with the larger time step. By what factor was the error generally reduced? By what factor did you expect it to be reduced?

12.10. Consider $\dot{x} = 1$ where $x(0) = 0$. Of the three methods,

1. Euler's method
2. the third-order Taylor series method
3. the fourth-order Runge–Kutta method

for a given $\Delta t \ll 1$, which would determine an approximate numerical solution with the least error? Explain your answer.

12.11. Consider $\dot{x} = t$ where $x(0) = 1$. Write two computer programs to determine an approximate numerical solution for this equation. One program should use the second-order Runge–Kutta method and the other should use the fourth-order Runge–Kutta method. Compare each answer to the exact solution at $t = 3$. Then decrease the time step and compare to the exact answer again. Did the error in each case decrease by the factor that you expected? Explain any discrepancies you observe.

12.12. Write a computer program that uses the fourth-order Runge–Kutta method to solve the famous Lorenz equations:

$$\frac{dx}{dt} = \sigma(y - x), \quad \frac{dy}{dt} = x(\rho - z) - y, \quad \frac{dz}{dt} = xy - \beta z.$$

where $\sigma = 10$, $\beta = 8/3$, and $\rho = 28$ and the initial conditions are $x(0) = 10$, $y(0) = 10$, and $z(0) = 10$. Are these equations linear or nonlinear? When you do not have an exact solution to which to compare your numerical solution, you need to decrease the time step until the approximate numerical solutions converge to one solution. Create a three-dimensional plot of (x, y, z) parametrized by t where $0 \leq t \leq 50$. Also create a three-dimensional plot of (x, y, z) parametrized by t when the other values are the same except for $\rho = 99.96$.

12.13. Consider $\ddot{x} + \sin\left(xt^2\right)\dot{x} + x^3 = \cos(3\dot{x}t)$ where $x(0) = 2$ and $\dot{x}(0) = 3$. Write a computer program that uses the fourth-order Runge–Kutta method to determine an approximate numerical solution to this differential equation.

12.14. Reproduce the results from Example 12.14 for the initial value problem

$$\ddot{x} + 9x = 5,$$

where $x(0) = 0$ and $\dot{x}(0) = 0$. By reducing Δt, what order does the incorrect method seem to be?

12.15. Write a general-purpose computer program that computes an approximate numerical solution for

$$\dot{x} = f(x,t),$$

where $x(0) = x_0$. Write the program so that all you would have to change for a different problem would be the line that defines $f(x,t)$, the line that specifies the initial condition, and the line that specifies the time step.

12.16. Write a general-purpose computer program that computes an approximate numerical solution for

$$\dot{x}_1 = f_1(x_1, x_2, t)$$
$$\dot{x}_2 = f_2(x_1, x_2, t).$$

Write the program so that all you would have to change for a different problem would be the line that defines $f_1(x_1, x_2, t)$, the line that defines $f_2(x_1, x_2, t)$, the lines that specify the initial conditions, and the line that specifies the time step.

12.17. Write a computer program that uses the finite-difference method to determine an approximate solution to Laplace's equation given in part 1 of Exercise 11.9, where $L_x = 3$, $L_y = 4$, and $f(y) = \sin \pi y$.

12.18. Write a computer program that uses finite-differences to compute the solution to Laplace's equation on the L-shaped domain illustrated in Figure 12.24. Choose one of the boundary segments to have a value of one and the rest zero.

12.19. Write a computer program that uses the finite difference method to compute an approximate solution to the one-dimensional heat conduction equation given in part 3 of Exercise 11.5 (which incorporates part 3 from Exercise 11.4). Compare the approximate answer to that obtained using the separation of variables method. Be sure to reduce Δx and Δt until the solution appears to converge. Also, verify the stability condition given in Equation (12.37) by using a time step slightly too large.

12.20. Determine how to implement the finite difference for the one-dimensional heat conduction equation where the end at $x = L$ is insulated. Use it to determine an approximate numerical solution to the system from Example 11.10.

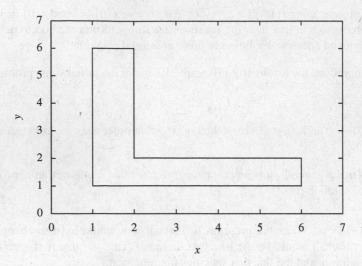

Fig. 12.24 Domain for Exercise 12.18.

12.21. Write computer programs to determine approximate numerical solutions to the one-dimensional wave given in parts 2 and 3 of Exercise 11.3. Compare the numerical answers to that obtained by using the separation of variables method.

12.22. Use the finite-difference method to determine an approximate numerical solution to

$$\alpha^2 \frac{\partial^2 u}{\partial x^2} - \alpha^2 u = \frac{\partial^2 u}{\partial t^2},$$

where $\alpha = 3$,

$$u(0,t) = u(10,t) = 0,$$

and

$$u(x,0) = f(x) = \sin \pi x, \quad \frac{\partial u}{\partial t}(x,0) = 0.$$

Chapter 13
Introduction to Nonlinear Systems

A quick review of the subject matter of this book up to this point confirms the fact that, with the exception of some specific first-order, ordinary differential equations which happen to be exact or separable, all the solution methods covered so far, other than numerical methods, have only been applicable to linear differential equations. Although the study of nonlinear differential equations is extremely interesting, it is also substantively difficult and rather advanced. The purpose of this chapter is to introduce some of the reasons why nonlinear systems are important, and why they are interesting, and to present a couple of the more basic analysis tools.

13.1 Motivation and Introduction

This section presents a couple of aspects of nonlinear systems which are features that are not present in linear systems. It also presents two of the most basic analysis tools; namely, the phase plane and Poincaré sections. The former are used extensively.

13.1.1 Multiple Equilibria and Chaos

By way of one example, this section illustrates the complexity of nonlinear systems and introduces the *phase plane*, which is one specific tool that is useful for two-dimensional systems. The example is the famous *Duffing's equation*, and both the forced and unforced cases are considered.

Example 13.1. Consider
$$\ddot{x} + b\dot{x} - x + x^3 = 0. \tag{13.1}$$

This equation is nonlinear in x, thus none of the methods considered previously in this text are applicable to determine a solution, so this example solves it numerically.

B. Goodwine, *Engineering Differential Equations: Theory and Applications*,
DOI 10.1007/978-1-4419-7919-3_13, © Springer Science+Business Media, LLC 2011

Using the fourth-order Runge–Kutta method with $\Delta t = 0.00001$ to determine an approximate solution to this equation with $b = 0.2$, $x(0) = -1$, and $\dot{x}(0)$ equal to 10.2 and 10.3, the solutions are illustrated in Figure 13.1.[1]

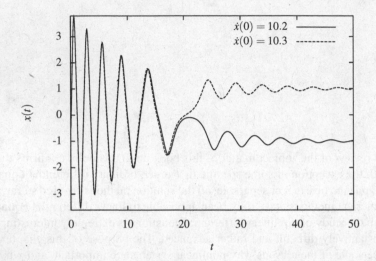

Fig. 13.1 Solutions for Equation (13.1) from Example 13.1 for two slightly different initial conditions.

The obvious feature of these two solutions is that although the initial condition was changed very slightly, and indeed the two solutions were nearly indistinguishable up until approximately $t = 20$, near that time the solutions rather radically diverged and ultimately appeared to approach different steady-state values. Although it is the case that such features are very sensitive to numerical errors, it is hoped to be clear subsequently that such a feature is inherent in this system and is actually fundamental feature of it.

This example illustrates the fact that solutions to nonlinear differential equations may be sensitive to initial conditions and furthermore may have multiple equilibria. In this case, the equilibria illustrated are the two steady-state solutions in Figure 13.1; namely, $(x(t), \dot{x}(t)) = (1,0)$ and $(x(t), \dot{x}(t)) = (-1,0)$.

The following example illustrates the fact that an additional complexity, namely, a time-varying inhomogeneous term, may result in a chaotic solution.

Example 13.2. Consider

$$\ddot{x} + b\dot{x} - x + x^3 = \gamma\cos\omega t, \tag{13.2}$$

---

[1] A reader is encouraged to reproduce these results. If the exact same method with the same time step is not used, however, the initial conditions which cause the solutions to diverge to different steady-state values may be slightly different and may require some trial and error to determine.

where

$$b = 0.2, \quad \gamma = 0.3, \quad \omega = 1.0.$$

A plot of a numerical solution to this equation with $x(0) = 0$ and $\dot{x}(0) = 0$ computed using the fourth-order Runge–Kutta method with a time step of $\Delta t = 0.0000001$ is illustrated in Figure 13.2.

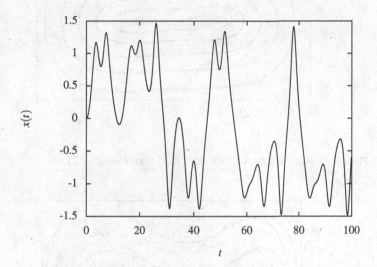

Fig. 13.2 Chaotic solution to Equation (13.2) from Example 13.2.

Although any precise definition of the term chaos is beyond the scope of this book, it is clear from the solution illustrated in Figure 13.2 that the numerical solution is "chaotic" at least in the sense of the common use of the term. At least for the time interval plotted, the solution does not appear to repeat; that is, it is nonperiodic, and seems to evolve in a rather unpredictable way.

13.1.2 The Phase Plane

For a second-order system, the *phase portrait* is a plot of the solution $x(t)$ versus $\dot{x}(t)$. The domain upon which the solution is plotted is often referred to as the *phase plane*.

Example 13.3. The phase portraits for the solutions from Examples 13.1 and 13.2 are illustrated in Figures 13.3 and 13.4, respectively.

Although arguably there is not much to be gained from the second figure, Figure 13.4, the first figure, Figure 13.3, is somewhat enlightening. Judging from the point where the two solutions diverge after closely tracking each other for quite a

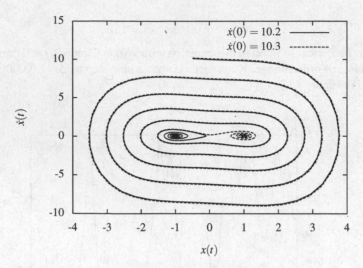

Fig. 13.3 Phase portraits for solutions to Equation (13.1) from Example 13.1.

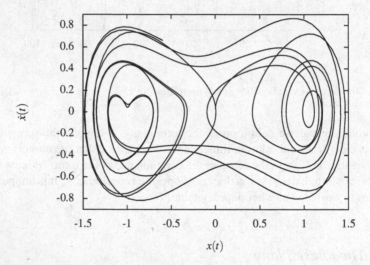

Fig. 13.4 Phase portrait for solutions to Equation (13.2).

long time, it is reasonable to infer that the geometric structure of the origin in the phase plane may be significant. In fact, as is developed subsequently, this is indeed the case.

13.1.3 Poincaré Sections

As indicated in Example 13.3, other than the chaotic nature of the solution, there is not much to observe from the solution to Equation (13.2) illustrated in Figure 13.4. However, one slight modification of the manner in which the data are illustrated reveals some very interesting structure to the solution. In particular, instead of plotting the complete solution curves $x(t)$ versus $\dot{x}(t)$, Figure 13.5 illustrates the discrete values of $x(t)$ versus $\dot{x}(t)$ for $t = 0, 2\pi, 4\pi \ldots$.

Example 13.4. Considering again the system in Equation (13.2) and computing an approximate numerical solution for the same parameter values and initial conditions, but for a much larger time range $0 \le t < 10000\pi$, a plot of the discrete values of $x(t)$ versus $\dot{x}(t)$ for the $t = 2m\pi$ is illustrated in Figure 13.5.

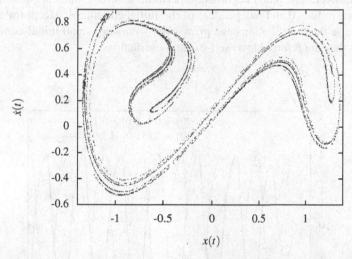

Fig. 13.5 Poincaré section for the forced Duffing equation in Example 13.2.

Note that a rather coherent structure becomes apparent when the data are presented in this manner. This topic is not pursued further in this text, but the reader should be at least made aware of its existence and its name: a *strange attractor*.

13.1.4 Limit Cycles

Another feature of nonlinear systems not present in solutions to linear differential equations is a *limit cycle*. Second-order linear systems can have periodic solutions either as homogeneous solutions when there is no damping, or as particular solutions when there is a periodic inhomogeneous term.

Similarly, for a nonlinear system, a limit cycle is a periodic solution to the nonlinear equation. In contrast to periodic solutions for linear systems, the limit cycle is independent of the initial conditions, which is not the case for the undamped homogeneous solutions. Furthermore, they are unforced. The following example illustrates a system with a limit cycle solution.

Example 13.5. Consider the *van der Pol equation* given by

$$\ddot{x} + (x^2 - 1)\dot{x} + x = 0. \tag{13.3}$$

One way to gain some insight into this equation is to observe that if $|x| < 1$, the system has negative damping and when $|x| > 1$, the system has positive damping. Hence, in the former case, solutions should increase in magnitude and in the latter case should decrease.

Figures 13.6 and 13.7 illustrate two solutions to Equation (13.3) with different initial conditions. Note that, in contrast to periodic solutions to linear equations, the solutions are unforced and independent of the initial conditions, except for a phase shift. Because these oscillations can grow even from very small initial conditions, sometimes they are referred to as self-excited oscillations.

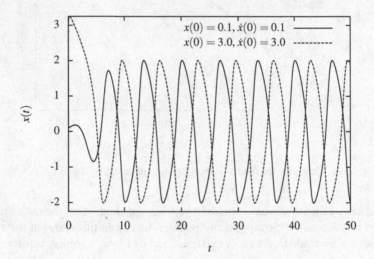

Fig. 13.6 Limit cycle solutions to the van der Pol equation from Example 13.5.

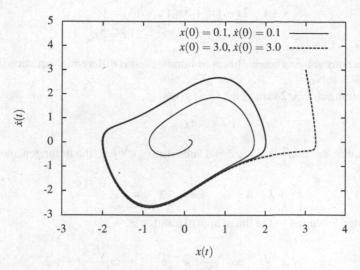

Fig. 13.7 Limit cycle solutions in the phase plane to the van der Pol equation from Example 13.5.

13.2 Linearization

One obvious approach to attempt to determine at least the basic features of a non-linear differential equation is to determine a differential equation that we can solve that is a good approximation to it. In the case where the nonlinear equation is homogeneous and has constant coefficients, if a good linear approximation can be determined, then, because it can be solved, at least some of the features of the solution of the nonlinear equation may be determined from the solution to the linear one.

The initial approach presented is simply to compute a Taylor series for all of the nonlinear terms about some point and keep only the first two terms from the Taylor series, which results in a linear differential equation. The following example illustrates this approach.

Example 13.6. Consider

$$\ddot{x} + \dot{x} - 2x + x^3 = 0. \qquad (13.4)$$

Determine a linear approximation for this equation by substituting a Taylor series for the x^3 term, where the Taylor series is computed about $x = \sqrt{2}$. The Taylor series for x^3 about $x = x_0$ is

$$x^3 = x_0^3 + 3x_0^2(x - x_0) + 6x_0(x - x_0)^2 + \cdots$$

thus keeping only the first two terms gives

$$\ddot{x} + \dot{x} - 2x + \left(x_0^3 + 3x_0^2(x - x_0)\right) = 0$$

$$\ddot{x} + \dot{x} + \left(3x_0^2 - 2\right)x = 2x_0^3 \tag{13.5}$$

which is a constant-coefficient, linear, inhomogeneous differential equation that we know how to solve.

Substituting $x_0 = \sqrt{2}$ Equation (13.5) gives

$$\ddot{x} + \dot{x} + 4x = 4\sqrt{2}. \tag{13.6}$$

The particular solution is $x_p = \sqrt{2}$ and substituting $e^{\lambda t}$ into the homogeneous equation gives

$$\lambda^2 + \lambda + 4 = 0 \quad \Longleftrightarrow \quad \lambda = \frac{-1}{2} \pm i\frac{\sqrt{15}}{2},$$

so the general solution to the linear approximation is

$$x(t) = c_1 e^{-(1/2)t} \cos\frac{\sqrt{15}}{2}t + c_2 e^{-(1/2)t} \sin\frac{\sqrt{15}}{2}t + \sqrt{2}. \tag{13.7}$$

Intuitively, in Example 13.6, the solutions to the linear approximation, Equation (13.6) are approximately the same as the solutions to Equation (13.4) as long as $x \approx \sqrt{2}$. Only the first two terms of the Taylor series were used, therefore the neglected terms, which were the higher powers of $(x - x_0)$ are only small if x stays near $\sqrt{2}$. The following example illustrates this fact.

Example 13.7. Figure 13.8 illustrates the solutions to Equation (13.4) and (13.6) for $x(0) = 1.4$ and $\dot{x}(0) = 0.2$. Note that the approximate solution closely tracks the solution to the nonlinear equation.

If the initial conditions are moved farther away from the point of linearization, say $x(0) = 1.0$ and $\dot{x}(0) = 0.2$, as is illustrated in Figure 13.9 the linear solution is not as good an approximation to the nonlinear solution as was the case illustrated in Figure 13.8.

If the initial conditions are even farther away from $x = \sqrt{2}$, say $x(0) = -1, \dot{x}(0) = 0.0$, then the two solutions are as illustrated in Figure 13.10. The solution is not near the point of linearization, thus the solution to the linearized differential equation is not even remotely a good approximation to the solution to the nonlinear equation.

Now, let us investigate what is happening near the origin.

Example 13.8. Determine the best linear approximation to

$$\ddot{x} + \dot{x} - 2x + x^3 = 0$$

for values of x near 0. Substituting $x_0 = 0$ into Equation (13.5) gives

$$\ddot{x} + \dot{x} - 2x = 0,$$

which is linear, constant-coefficient, homogeneous, and ordinary, so solutions are of the form $x = e^{\lambda t}$. Substituting gives the characteristic equation

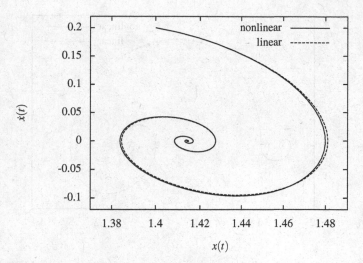

Fig. 13.8 Comparison of solutions of Equation (13.4) (nonlinear) with Equation (13.6) (linear approximation) with initial conditions near $x = \sqrt{2}$.

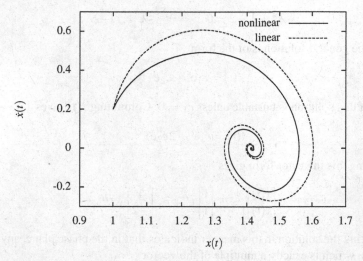

Fig. 13.9 Comparison of solutions of Equation (13.4) (nonlinear) with Equation (13.6) (linear approximation) with initial conditions slightly farther from $x = \sqrt{2}$ than in Figure 13.8.

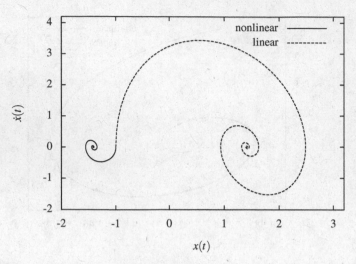

Fig. 13.10 Comparison of solutions of Equation (13.4) (nonlinear) with Equation (13.6) (linear approximation) with initial conditions far from $x = \sqrt{2}$.

$$\lambda^2 + \lambda - 2 = 0 \quad \Longleftrightarrow \quad \lambda = 1, -2.$$

Hence, the general solution is of the form

$$x(t) = c_1 e^t + c_2 e^{-2t}.$$

Note that this solution is unstable unless $c_1 = 0$. Computing $\dot{x}(t)$ gives

$$\dot{x}(t) = c_1 e^t - 2c_2 e^{-2t}.$$

Expressing this in vector form gives

$$\begin{bmatrix} x(t) \\ \dot{x}(t) \end{bmatrix} = c_1 \begin{bmatrix} 1 \\ 1 \end{bmatrix} e^t + c_2 \begin{bmatrix} 1 \\ -2 \end{bmatrix} e^{-2t}. \tag{13.8}$$

Considering the solution in this manner indicates that in the phase plane, any initial condition which is exactly a multiple of the vector

$$\begin{bmatrix} x(0) \\ \dot{x}(0) \end{bmatrix} = \alpha \begin{bmatrix} 1 \\ -2 \end{bmatrix}$$

results in $c_1 = 0$ and hence is stable; that is,

$$\lim_{t \to \infty} \begin{bmatrix} x(t) \\ \dot{x}(t) \end{bmatrix} = \begin{bmatrix} 0 \\ 0 \end{bmatrix}.$$

Any other initial condition will have a nonzero c_1 and hence will be unstable. If the initial condition is very close to the

$$\begin{bmatrix} x(0) \\ \dot{x}(0) \end{bmatrix} = \alpha \begin{bmatrix} 1 \\ -2 \end{bmatrix}$$

vector, then c_1 may be very small and the stable solution may initially dominate; however, due to the exponential term the solution will ultimately be unstable.

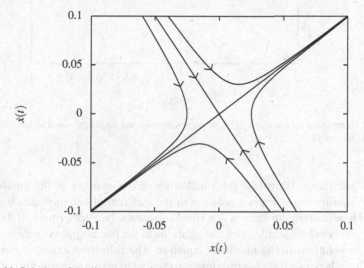

Fig. 13.11 Solution of the linear approximation to Equation (13.4) near the origin.

In Figure 13.11 of the six solutions plotted, three start in the upper left portion of the plot and initially move toward the origin. Similarly, the other three solutions start in the lower right portion of the graph and initially head to the origin as well. However, eventually the e^t term dominates and all six solutions ultimately move away from the origin and grow unbounded. Note that the solutions grow unbounded along the vector multiplying e^t in the vector form of the solution.

Figure 13.12 illustrates the solutions to the linear approximation near the origin (Equation (13.5) with $x_0 = 0$) and the solution to the nonlinear equation (Equation (13.4)) with the same initial conditions near the origin. Although initially the solutions are similar in nature, due to the instability of both solutions they both ultimately leave the domain in which the linear equation is a good approximation to the nonlinear equation.

In principle, it is appropriate to compute a Taylor series approximation for any nonlinear terms in a differential equation and keep only the linear terms to determine some of the features of the solution of the nonlinear equation near the point about which the linearization was computed. However, the main utility of linearization is to determine a linear approximation to a differential equation near an equilibrium

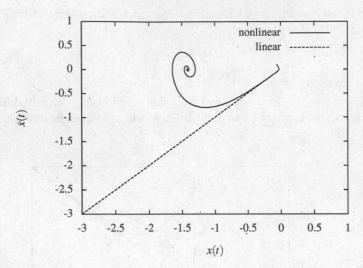

Fig. 13.12 Comparison of solutions of linear approximation and nonlinear solution for initial conditions near the origin.

point because that will provide information about the stability of the equilibrium point. An equilibrium point is a solution to the differential equation that is a constant, and how to compute them is described subsequently. Furthermore, if the equilibrium point is indeed stable, then the solutions to the linear approximation will be close to the solutions to the nonlinear equation. The following example illustrates the fact that whereas a linear approximation computed near a nonequilibrium point gives some information about the nonlinear solution, it is only transiently valid and furthermore, information regarding an equilibrium point of the linearized equation has nothing to do with the nonlinear system.

Example 13.9. Determine a linear approximation to

$$\ddot{x} + \dot{x} - 2x + x^3 = 0 \tag{13.9}$$

near $x_0 = 4$. Substituting $x_0 = 4$ into Equation (13.5) gives

$$\ddot{x} + \dot{x} + -2x\left(64 + 48\left(x - 4\right)\right) = 0$$
$$\ddot{x} + \dot{x} + 46x = 128 \tag{13.10}$$

which has the general solution

$$x(t) = c_1 e^{-(1/2)t} \cos \frac{\sqrt{183}}{2}t + c_2 e^{-(1/2)t} \sin \frac{\sqrt{183}}{2}t + \frac{64}{23}.$$

A plot of the the solution to the nonlinear equation and the linear approximation is illustrated in Figure 13.13.

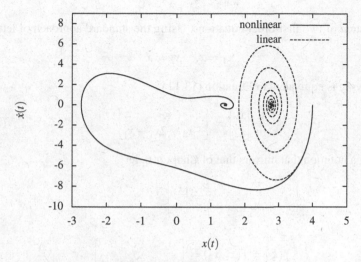

Fig. 13.13 Solutions to Equations (13.9) and (13.10).

Note that although the solutions stay near $x = 4$, they are nearly identical. However, as expected, as they diverge from $x = 4$ the linear approximation is increasingly less valid. Also note that the linearized equation has an equilibrium at $x = 64/23$ which is not an equilibrium for the nonlinear equation. The stability of the equilibrium point for the linearized equation at $x = 64/23$ has nothing to do with the stability or instability of the nonlinear equation near that point.

Typically, linear approximations to nonlinear differential equations are only computed about equilibrium points. This is due to the fact that the linear approximation about a nonequilibrium point in general will have an equilibrium that is not the same as the nonlinear equation. Furthermore, in applications such as feedback control, stabilizing a system to an equilibrium is typically the desired goal and hence it is desirable for the linearized approximation of the nonlinear equation to have an equilibrium in common.

The next section considers the more standard and systematic approach to doing this; namely, if necessary, converting a system of higher-order differential equations to a system of first-order equations and computing the Jacobian.

13.3 Jacobian Linearization

This is initially developed by mirroring the example from the previous section, Example 13.6.

Example 13.10. Convert

$$\ddot{x} + \dot{x} - 2x + x^3 = 0 \tag{13.11}$$

into a system of two first-order equations. Using the standard approach of letting

$$x_1 = x, \quad x_2 = \dot{x}$$

the following is equivalent to Equation (13.11)

$$\frac{d}{dt} \begin{bmatrix} x_1 \\ x_2 \end{bmatrix} = \begin{bmatrix} x_2 \\ -x_2 + 2x_1 - x_1^3 \end{bmatrix}. \tag{13.12}$$

Adopting a notation that mirrors that of Chapter 6, let

$$\xi = \begin{bmatrix} x_1 \\ x_2 \end{bmatrix}$$

and

$$f(\xi) = f(x_1, x_2) = \begin{bmatrix} x_2 \\ -x_2 + 2x_1 - x_1^3 \end{bmatrix}.$$

Observe carefully that in Example 13.10, both ξ and $f(\xi)$ are vectors and that the whole system may be represented by

$$\dot{\xi} = f(\xi).$$

Because it may be represented so compactly, even though it is, in general, a system of differential equations, it may simply be referred to as a or the differential equation.

Definition 13.1. A point ξ_0 is an *equilibrium point* of

$$\dot{\xi} = f(\xi)$$

if

$$f(\xi_0) = 0,$$

where the 0 on the right-hand side of the equation is a vector of zeros that is the same dimension as ξ and $f(\xi)$.

Note, that if ξ_0 is an equilibrium point, then

$$\dot{\xi} = f(\xi_0) = 0$$

so

$$\xi(t) = \xi_0$$

is a solution to

$$\dot{\xi} = f(\xi)$$

if

$$\xi(0) = \xi_0.$$

Example 13.11. Determine the equilibrium points for Equation (13.12). Clearly, $x_2 = 0$ is necessary to make the first component vanish. For the second component any of the three $x_1 = 0$ or $x_1 = \pm\sqrt{2}$. So, any of the three vectors

$$\xi_0 = \begin{bmatrix} 0 \\ 0 \end{bmatrix}, \quad \begin{bmatrix} \sqrt{2} \\ 0 \end{bmatrix}, \quad \begin{bmatrix} -\sqrt{2} \\ 0 \end{bmatrix}$$

when substituted into Equation (13.12) results in

$$f(\xi_0) = \begin{bmatrix} 0 \\ 0 \end{bmatrix}.$$

First we define the Jacobian, and then by referring to the previous examples show that it may be used to determine an equivalent linear approximation to a nonlinear equation near an equilibrium point.

Definition 13.2. For a vector-valued function, f of a vector

$$\xi = \begin{bmatrix} \xi_1 \\ \xi_2 \\ \vdots \\ \xi_n \end{bmatrix},$$

denoted by

$$f(\xi) = \begin{bmatrix} f_1(\xi_1, \xi_2, \ldots, \xi_n) \\ f_2(\xi_1, \xi_2, \ldots, \xi_n) \\ \vdots \\ f_m(\xi_1, \xi_2, \ldots, \xi_n) \end{bmatrix},$$

the *Jacobian matrix* for $f(\xi)$ is given by

$$\frac{\partial f}{\partial \xi} = \begin{bmatrix} \frac{\partial f_1}{\partial \xi_1} & \frac{\partial f_1}{\partial \xi_2} & \cdots & \frac{\partial f_1}{\partial \xi_n} \\ \frac{\partial f_2}{\partial \xi_1} & \frac{\partial f_2}{\partial \xi_2} & \cdots & \frac{\partial f_2}{\partial \xi_n} \\ \vdots & \vdots & \vdots & \vdots \\ \frac{\partial f_m}{\partial \xi_1} & \frac{\partial f_m}{\partial \xi_2} & \cdots & \frac{\partial f_m}{\partial \xi_n} \end{bmatrix},$$

where $f(\xi) \in \mathbb{R}^m$; that is, f is m elements "tall" and $\xi \in \mathbb{R}^n$; that is, ξ is n elements tall.

For a system of n first-order differential equations, the equations themselves typically only depend on the n state variables; hence, for systems of first-order equations that we consider in this book, the Jacobian matrix is always square. Now, we can define a linearization that is equivalent to the Taylor series method outlined previously.

Definition 13.3. For the system of n first-order, homogeneous differential equations

$$\dot{\xi} = f(\xi),\tag{13.13}$$

where $f(\xi_0) = 0$, the *linear approximation to Equation* (13.13) *about* ξ_0 is given by

$$\dot{\xi} = \left.\frac{\partial f}{\partial \xi}\right|_{\xi_0}(\xi - \xi_0).\tag{13.14}$$

Let us compare the results of using this linearization method with the linearization approximations determined using Taylor series in the previous examples.

Example 13.12. Consider

$$\ddot{x} + \dot{x} - 2x + x^3 = 0,$$

which, when converted to two first-order equations is given by

$$\frac{\mathrm{d}}{\mathrm{d}t}\begin{bmatrix} x_1 \\ x_2 \end{bmatrix} = \begin{bmatrix} x_2 \\ -x_2 + 2x_1 - x_1^3 \end{bmatrix},$$

where

$$\xi = \begin{bmatrix} x_1 \\ x_2 \end{bmatrix} = \begin{bmatrix} x \\ \dot{x} \end{bmatrix}.$$

The Jacobian for this system of equations is

$$\frac{\mathrm{d}f}{\mathrm{d}\xi} = \begin{bmatrix} \frac{\partial f_1}{\partial x_1} & \frac{\partial f_1}{\partial x_2} \\ \frac{\partial f_2}{\partial x_1} & \frac{\partial f_2}{\partial x_2} \end{bmatrix} = \begin{bmatrix} 0 & 1 \\ 2 - 3x_1^2 & -1 \end{bmatrix}.$$

Evaluated at the equilibrium point

$$\xi_0 = \begin{bmatrix} 0 \\ 0 \end{bmatrix}$$

and substituting into Equation (13.14) gives

$$\dot{\xi} = \frac{\mathrm{d}}{\mathrm{d}t}\begin{bmatrix} x_1 \\ x_2 \end{bmatrix} = \begin{bmatrix} 0 & 1 \\ 2 & -1 \end{bmatrix}\begin{bmatrix} x_1 \\ x_2 \end{bmatrix}.$$

From Example 13.8, the Taylor series linearization about $x_0 = 0$ was

$$\ddot{x} + \dot{x} - 2x = 0,$$

which, when converted to two first-order equations, gives the same result.
Similarly, at

$$\xi_0 = \begin{bmatrix} \sqrt{2} \\ 0 \end{bmatrix}$$

the linearization is

$$\frac{\mathrm{d}}{\mathrm{d}t}\begin{bmatrix} x_1 \\ x_2 \end{bmatrix} = \begin{bmatrix} 0 & 1 \\ -4 & -1 \end{bmatrix}\left(\begin{bmatrix} x_1 \\ x_2 \end{bmatrix} - \begin{bmatrix} \sqrt{2} \\ 0 \end{bmatrix}\right).$$

Referring back to Example 13.6, the Taylor series linearization resulted in

$$\ddot{x} + \dot{x} + 4x = 4\sqrt{2},$$

which is the same.

It is worth remarking that the general equation for the Taylor series of a vector valued function of several variables about the point ξ_0 is of the form

$$f(\xi) = f(\xi_0) + \left.\frac{\partial f}{\partial \xi}\right|_{\xi_0} (\xi - \xi_0) + \cdots.$$

Because ξ_0 was assumed to be an equilibrium point in the above development, what is obviously happening is that $f(\xi)$ is replaced by the first two terms in its Taylor series, but the first term $f(\xi_0)$ happens to be zero.

Finally, although there is nothing wrong with Equation (13.14), the inhomogeneous term resulting from the "$-\xi_0$" term in $(\xi - \xi_0)$ adds a bit of extra work that is easily avoided. By letting

$$\eta = \xi - \xi_0$$

then, because ξ_0 is a constant, $\dot{\eta} = \dot{\xi}$ and hence the linear approximation can be expressed simply as

$$\boxed{\dot{\eta} = \left.\frac{\partial f}{\partial \xi}\right|_{\xi_0} \eta.} \tag{13.15}$$

Clearly, the origin for η is the fixed point ξ_0, and the constant inhomogeneous term is eliminated by the coordinate transformation.

Example 13.13. Referring back to Example 13.12, determine the *homogeneous* linear approximation to

$$\ddot{x} + \dot{x} - 2x + x^3 = 0$$

about the fixed point

$$\xi_0 = \begin{bmatrix} \sqrt{2} \\ 0 \end{bmatrix}.$$

Letting

$$\eta = \begin{bmatrix} y_1 \\ y_2 \end{bmatrix} = \begin{bmatrix} x_1 \\ x_2 \end{bmatrix} - \begin{bmatrix} \sqrt{2} \\ 0 \end{bmatrix}$$

and using Equation (13.15), then

$$\frac{d}{dt} \begin{bmatrix} y_1 \\ y_2 \end{bmatrix} = \begin{bmatrix} 0 & 1 \\ -4 & -1 \end{bmatrix} \begin{bmatrix} y_1 \\ y_2 \end{bmatrix}.$$

13.4 Geometry and Stability of Equilibrium Points in the Phase Plane

This section first outlines the procedure to solve systems of two first-order, linear, homogeneous differential equations. This material is a bit of a review of the methods from Chapter 6, but is developed with the ultimate goal of gaining some insight to the relationship between the linear algebra (i.e., eigenvalue and eigenvectors), and the nature and geometry of solutions of systems near equilibrium points.

Examples 13.6 and 13.8 from Section 13.2 determined linear approximations to the nonlinear Duffing equation and solved them. In the case of Example 13.6, the solution to the linear approximation near the point $x_0 = \sqrt{2}$ is

$$x(t) = c_1 e^{-(1/2)t} \cos \frac{\sqrt{15}}{2} t + c_2 e^{-(1/2)t} \sin \frac{\sqrt{15}}{2} t + \sqrt{2},$$

and in the case of Example 13.8, the solution to the linear approximation near the point $x_0 = 0$ is

$$x(t) = c_1 e^t + c_2 e^{-2t}. \tag{13.16}$$

Note that both of these solutions are easily differentiated. In particular, for the solution of the linearization about the origin, we can write

$$\frac{d}{dt} \begin{bmatrix} x_1(t) \\ x_2(t) \end{bmatrix} = c_1 \begin{bmatrix} 1 \\ 1 \end{bmatrix} e^t + c_2 \begin{bmatrix} 1 \\ -2 \end{bmatrix} e^{-2t}, \tag{13.17}$$

where $x_1(t) = x(t)$ and $x_2(t) = \dot{x}(t) = \dot{x}_1(t)$ as usual. The second line is computed by simply differentiating the solution from Equation (13.16).

Observing the form of Equation (13.17), rather than computing it from the solution of the original scalar equation, it seems reasonable that an alternative solution method designed to determine the solution directly from the vector form of the equation would be reasonably useful.

In particular, note that Equation (13.15) is a system of first-order, homogeneous differential equations, which may be expressed in the form

$$\dot{\eta} = A\eta,$$

where $A \in \mathbb{R}^{2 \times 2}$ and

$$A = \left. \frac{\partial f}{\partial \xi} \right|_{\xi_0}.$$

Referring to Equation (13.17), it seems reasonable simply to assume a solution of the form

$$\eta(t) = \hat{\eta} e^{\lambda t},$$

where $\eta \in \mathbb{R}^2$; that is, it is a vector. Note that $\dot{\eta} = \lambda \eta e^{\lambda t}$. Substituting this into the differential equation gives

$$\lambda \hat{\eta} e^{\lambda t} = A\hat{\eta}.$$

Rearranging gives

$$A\hat{\eta}e^{\lambda t} - \lambda\hat{\eta}e^{\lambda t} = 0$$
$$A\hat{\eta}e^{\lambda t} - \lambda I\hat{\eta}e^{\lambda t} = 0$$
$$(A - \lambda I)\hat{\eta}e^{\lambda t} = 0$$
$$(A - \lambda I)\hat{\eta} = 0, \tag{13.18}$$

where I is the 2×2 identity matrix. Canceling the $e^{\lambda t}$ terms is justified because it may never be zero.

Hence, solutions of

$$\dot{\eta} = A\eta$$

are of the form

$$\eta(t) = \hat{\eta}e^{\lambda t}$$

where $\hat{\eta}$ and λ satisfy Equation (13.18). It is not a coincidence that Equation (13.18) is the equation for the eigenvalues and eigenvectors of the matrix A; in fact, one of the primary uses of eigenvalue and eigenvector computations is to solve systems of first-order, linear, constant-coefficient differential equations.

Recall that the procedure is to compute the λ values that satisfy Equation (13.18) by observing that the equation only has solutions for nonzero η if

$$\det(A - \lambda I) = 0.$$

Once the values for λ are determined, each value is substituted into Equation (13.18) and the corresponding eigenvector $\hat{\eta}$ is computed. Various procedures are necessary depending upon whether the eigenvalues are real or complex and whether they are repeated. A compete consideration of all these cases appears in Chapter 6, a summary of which is as follows.

Theorem 13.1. *For the linear, homogeneous, constant-coefficient system of n first-order ordinary differential equations*

$$\dot{\xi} = A\xi,$$

if λ_i are the eigenvalues of A and $\hat{\xi}^i$ are the corresponding eigenvectors, then the general solution $\xi(t)$ depends upon the nature of the eigenvalues as follows.

1. If the eigenvalues are distinct, then

$$\xi(t) = c_1\hat{\xi}^1 e^{\lambda_1 t} + c_1\hat{\xi}^2 e^{\lambda_2 t} + \cdots c_n\hat{\xi}^n e^{\lambda_n t}.$$

2. If there are any complex eigenvalues, say λ_i and $\lambda_{i+1} = \overline{\lambda}_i$ with complex conjugate eigenvectors $\hat{\xi}^i$ and $\hat{\xi}^{i+1} = \overline{\hat{\xi}^i}$, respectively, then the two terms in the general solution corresponding to λ_i and λ_{i+1} satisfy

$$c_i \hat{\xi}^i e^{\lambda_i t} + c_{i+1} \hat{\xi}^{i+1} e^{\lambda_{i+1} t}$$

$$= k_1 e^{\mu t} \left(\hat{\xi}^i_r \cos \omega t - \hat{\xi}^i_i \sin \omega t \right) + k_2 e^{\mu t} \left(\hat{\xi}^i_r \sin \omega t + \hat{\xi}^i_i \cos \omega t \right), \qquad (13.19)$$

where $\lambda_i = \mu + i\omega$ and $\hat{\xi}^i = \hat{\xi}^i_r + i\hat{\xi}^i_i$. *It is usually more convenient to have terms in the general solution to be in terms of the trigonometric functions instead of the complex exponentials; hence, it is preferable to replace the left-hand side of Equation (13.19) with the right-hand side for the corresponding terms in the general solution.*

3. *For any repeated eigenvalues, if λ_i is repeated m times, then there will be m solutions to*

$$(A - \lambda_i I)^m \hat{\xi} = 0. \qquad (13.20)$$

Then for each of the m solutions to Equation (13.20), a term of the form

$$\xi(t) = \left(\hat{\xi} + t (A - \lambda_i I) \hat{\xi} + \frac{t^2}{2!} (A - \lambda_i I)^2 \hat{\xi} + \cdots + \frac{t^{m-1}}{(m-1)!} (A - \lambda_i I)^{m-1} \hat{\xi} \right) e^{\lambda_i t}$$

will appear in the general solution and the general solution will be a linear combination of these.

Considering the special case where the system is two-dimensional, we have a limited number of possible combinations for the eigenvalues and can enumerate the possible forms of the solutions in the phase plane. For inferring the nature of solutions for nonlinear systems near an equilibrium point, the following are the cases that are useful.

1. If both eigenvalues are real, the equilibrium point is called a *node*.

   a. If both eigenvalues are real and negative, then all of the solutions are attracted to the equilibrium point, which is called a *stable node*.
   b. If both eigenvalues are real and positive, then all of the solutions are repelled by the equilibrium point, which is called a *unstable node*.
   c. If both eigenvalues are real and one is positive and one is negative, almost all solutions are repelled from the equilibrium point, which is called a *saddle node*.

2. If either of the eigenvalues is complex, then they both must be and they must be a complex-conjugate pair. Because the solutions contain sine and cosine terms, the solutions are characterized by spirals.

   a. If the real part of the eigenvalues is negative, then the solutions spiral into the equilibrium point, which is called a *stable spiral*.
   b. If the real part of the eigenvalues is positive, then the solutions spiral away from the equilibrium point, which is called an *unstable spiral*.

Remark 13.1. The only remaining possibilities for the two-dimensional case are where the real part of some or all of the eigenvalues is zero. These correspond to neutrally stable systems and in such a case it is not surprising that the linear approximation does not provide any useful information regarding whether the equilibrium is stable. This is because it is the higher-order nonlinear terms that dictate the stability, even near the equilibrium. These are explored further in the exercises. An interested reader is referred to *center manifold theory*, for example, in [54], which is a systematic way to convert a system with eigenvalues with zero real part into a form where an analysis of the nonlinear terms and their effect on stability is often possible. An equilibrium point that has a nonzero real part is called *hyperbolic*.

It is beyond the scope of this book to prove it, but what would be expected is true. Namely, sufficiently close to a hyperbolic fixed point, the solutions to the linear equation approximate the solutions to the nonlinear equation. If the fixed point is nonhyperbolic, then we cannot infer anything from the linearization.

Thus, in order to be able to qualitatively describe the nature of the nonlinear solutions near equilibria, we need to investigate the nature of the linear solutions for the possible combinations of eigenvalues for hyperbolic fixed points. Either in the following examples or the exercises, each of the possible cases is considered.

Example 13.14. This is an example of a stable node. Consider

$$\ddot{x} + 3\dot{x} + 2x = 0.$$

Converting to two first-order equations gives

$$\frac{d}{dt} \begin{bmatrix} x \\ \dot{x} \end{bmatrix} = \begin{bmatrix} 0 & 1 \\ -2 & -3 \end{bmatrix} \begin{bmatrix} x \\ \dot{x} \end{bmatrix}$$

which has eigenvalues $\lambda_1 = -1$ and $\lambda_2 = -2$ and associated eigenvectors

$$\hat{\xi}^1 = \begin{bmatrix} 1 \\ -1 \end{bmatrix}, \quad \hat{\xi}^2 = \begin{bmatrix} -1 \\ 2 \end{bmatrix},$$

so

$$\begin{bmatrix} x(t) \\ \dot{x}(t) \end{bmatrix} = c_1 \begin{bmatrix} 1 \\ -1 \end{bmatrix} e^{-t} + c_2 \begin{bmatrix} -1 \\ 2 \end{bmatrix} e^{-2t}.$$

Observe the following.

1. If the initial conditions are exactly along one of the eigenvectors, then the constant in the general solution corresponding to the other one will be zero, and hence the trajectory in the phase plane will stay on that eigenvector and go straight to the origin.
2. If the initial conditions are not exactly along one of the eigenvectors, then both c_1 and c_2 will be nonzero, and the solution will be a combination of the two terms. Considering the two eigenvectors are basis elements, then c_1 and c_2 can be thought of as how far off the other eigenvector the solution is. Because e^{-2t} decays faster than e^{-t}, the component of the solution corresponding to the eigenvector for $\lambda_2 = -2$ will shrink faster, and hence the solutions in the phase plane are curved toward the eigenvalue with the smaller magnitude.

These observations are both illustrated in the phase plane for the system, illustrated in Figure 13.14. An unstable node would have a similar geometry, but with the trajectories moving in the opposite direction, away from the origin.

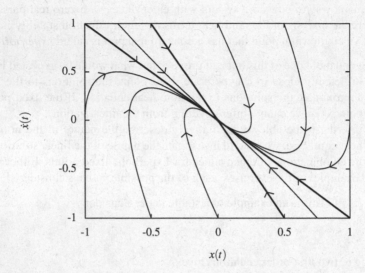

Fig. 13.14 Phase plane for the stable node in Example 13.14. The thicker lines are the eigenvectors.

Example 13.15. This is an example of a system with a saddle node at the origin. Consider

$$\ddot{x} + \dot{x} - 2x = 0,$$

which is equivalent to

$$\frac{d}{dt}\begin{bmatrix} x \\ \dot{x} \end{bmatrix} = \begin{bmatrix} 0 & 1 \\ 2 & -1 \end{bmatrix}\begin{bmatrix} x \\ \dot{x} \end{bmatrix}$$

which has eigenvalues of $\lambda_1 = 1$ and $\lambda_2 = -2$ and associated eigenvectors

$$\hat{\xi}^1 = \begin{bmatrix} 1 \\ 1 \end{bmatrix}, \quad \hat{\xi}^2 = \begin{bmatrix} -1 \\ 2 \end{bmatrix}.$$

Hence the solution is

$$\begin{bmatrix} x(t) \\ \dot{x}(t) \end{bmatrix} = c_1 \begin{bmatrix} 1 \\ 1 \end{bmatrix} e^t + c_2 \begin{bmatrix} -1 \\ 2 \end{bmatrix} e^{-2t}.$$

If the initial condition is exactly on the second eigenvector, then $c_1 = 0$ and the solution will go to the origin. For any other initial condition, $c_1 \neq 0$ and, due to the e^t term, that component will increase in magnitude. The phase plot is illustrated in Figure 13.15.

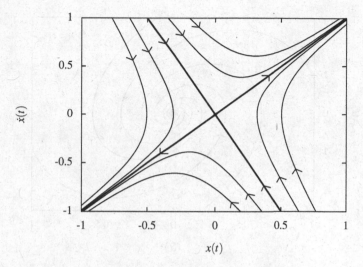

Fig. 13.15 Phase plane for the saddle node in Example 13.15. The thicker lines are the eigenvectors.

Example 13.16. This is an example of a stable spiral. Consider

$$\ddot{x} + 2\dot{x} + 17x = 0$$

which is equivalent to

$$\frac{d}{dt} \begin{bmatrix} x \\ \dot{x} \end{bmatrix} = \begin{bmatrix} 0 & 1 \\ -17 & -2 \end{bmatrix} \begin{bmatrix} x \\ \dot{x} \end{bmatrix},$$

which has eigenvalues $\lambda_1 = -1 + 4i$ and $\lambda_2 = \bar{\lambda}_1 = -1 - 4i$. Because the real part is negative, the solutions in the phase plane are stable spirals, as is illustrated in Figure 13.16. If the real part were positive, the spirals would still be oriented in the clockwise direction, but would spiral out from the origin instead of into it.

Now we can put all this to use to predict the behavior of a nonlinear system near its equilibrium points in the phase plane.

Example 13.17. Consider the second-order, nonlinear differential equation

$$\ddot{x} + \dot{x} - 2x + x^2 = 0, \tag{13.21}$$

which is equivalent to

$$\frac{d}{dt} \begin{bmatrix} x \\ \dot{x} \end{bmatrix} = \begin{bmatrix} \dot{x} \\ -\dot{x} + 2x - x^2 \end{bmatrix}. \tag{13.22}$$

The equilibrium points are where the right-hand side of Equation (13.22) is zero. In particular

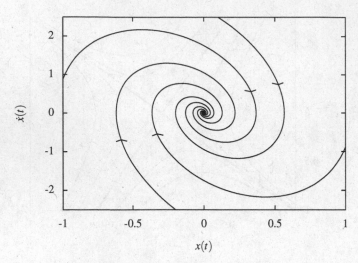

Fig. 13.16 Phase plane for a stable spiral in Example 13.16.

$$\begin{bmatrix} \dot{x} \\ -\dot{x} + 2x - x^2 \end{bmatrix} = \begin{bmatrix} 0 \\ 0 \end{bmatrix} \quad \Longleftrightarrow \quad \begin{bmatrix} x \\ \dot{x} \end{bmatrix} = \begin{bmatrix} 0 \\ 0 \end{bmatrix}, \begin{bmatrix} 0 \\ 2 \end{bmatrix}. \tag{13.23}$$

The Jacobian is

$$\frac{\partial f}{\partial x} = \begin{bmatrix} 0 & 1 \\ 2 - 2x & -1 \end{bmatrix}.$$

Near $(x, \dot{x}) = (0, 0)$,

$$\frac{\mathrm{d}}{\mathrm{d}t} \begin{bmatrix} x \\ \dot{x} \end{bmatrix} = \begin{bmatrix} 0 & 1 \\ 2 & -1 \end{bmatrix} \begin{bmatrix} x \\ \dot{x} \end{bmatrix}$$

which has eigenvalues $\lambda_1 = 1$ and $\lambda_2 = -2$ and associated eigenvectors

$$\hat{\xi}^1 = \begin{bmatrix} 1 \\ 1 \end{bmatrix}, \quad \hat{\xi}^2 = \begin{bmatrix} -1 \\ 2 \end{bmatrix}.$$

Thus, near the equilibrium point, near the vectors $\pm\hat{\xi}^1$ the trajectories will be moving away from the origin and near the vectors $\pm\hat{\xi}^2$ the trajectories will be moving toward the origin.

Near $(x, \dot{x}) = (2, 0)$

$$\frac{\mathrm{d}}{\mathrm{d}t} \begin{bmatrix} x \\ \dot{x} \end{bmatrix} = \begin{bmatrix} 0 & 1 \\ -2 & -1 \end{bmatrix} \begin{bmatrix} x \\ \dot{x} \end{bmatrix}$$

which has eigenvalues $\lambda = -1/2 \pm i\sqrt{7}/2$. Because the real part is negative, trajectories are clockwise spirals that spiral into the point $(x, \dot{x}) = (2, 0)$.

The phase portrait for the nonlinear system is illustrated in Figure 13.17, which is characterized by the features described above that are based on the linearization.

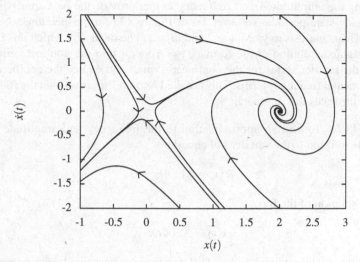

Fig. 13.17 Phase portrait for nonlinear system in Example 13.17.

13.5 Harmonic Balance and Describing Functions

The previous sections simplified nonlinear ordinary differential equations by determining a linear differential equation that approximated the nonlinear equation. The linear equation, of course, can be solved, which provided a means to determine certain aspects of the structure of the solution to the nonlinear equation, particularly near equilibrium points. The linearized equation was equivalent to considering a Taylor series for the nonlinear terms, and truncating the series after the linear terms.

In contrast, this section focuses on a similar idea, which is to consider a Fourier series for a potential periodic solution, but limiting the analysis to the first term, or fundamental frequency term, in the Fourier series. Because it is based on an analysis tool that is primarily applicable to periodic functions, this method is particularly useful for computing approximations to periodic solutions, which, in the case of nonlinear equations with no forcing, are limit cycles.

13.5.1 Harmonic Balance

The basic idea is to approximate a limit cycle solution with a harmonic solution. Referring back to Example 13.5 and the solutions illustrated in Figure 13.6, this method would be to use a function of the form

$$x(t) = A \cos \omega t$$

and finding the amplitude A and frequency ω that provide the best approximation to the steady-state periodic solution. From Figure 13.6, by inspection, we would expect such an analysis to yield $A \approx 2$ and $\omega \approx 1$. The means by which this is done is to substitute a solution of the form $x(t) = A \cos \omega t$ into the nonlinear equation, compute the Fourier series for the nonlinear terms,[2] and equate the coefficients of the fundamental frequency terms. This is called *harmonic balance* and the following example illustrates this approach.

Example 13.18. To find an approximation to the frequency and amplitude of the limit cycle solution to the van der Pol equation

$$\ddot{x} + \left(x^2 - 1\right)\dot{x} + x = 0 \tag{13.24}$$

assume a solution of the form

$$x(t) = A \cos \omega t$$

and substitute to determine A and ω. Differentiating the assumed solution gives

$$\dot{x} + -\omega A \sin \omega t, \quad \ddot{x} = -\omega^2 A \cos \omega t$$

and substituting into Equation (13.24) gives

$$-\omega^2 A \cos \omega t + \left(\left(A \cos \omega t\right)^2 - 1\right)\left(-\omega A \sin \omega t\right) + A \cos \omega t = 0,$$

and rearranging

$$A\left(1 - \omega^2\right)\cos \omega t + \omega A \sin \omega t = -\omega A^3 \left(\cos^2 \omega t\right) \sin \omega t. \tag{13.25}$$

If we are interested in harmonic solutions in terms of the fundamental frequency, the term on the right-hand side can be expanded in a Fourier series to express it as a linear combination of individual harmonic functions. In other words, if we express

$$-\omega A^3 \left(\cos^2 \omega t\right)\sin \omega t = a_1 \cos \omega t + b_1 \sin \omega t + a_2 \cos 2\omega t + b_2 \sin 2\omega t + \cdots \tag{13.26}$$

then, because $\cos \omega t$ and $\sin \omega t$ are linearly independent functions, we can equate the coefficients of the $\cos \omega t$ and $\sin \omega t$ terms to obtain expressions for A and ω.

To proceed, follow the usual approach for computing a Fourier series by multiplying both sides of Equation (13.26) by $\cos \omega t$ and integrating from 0 to $2\pi/\omega$ with respect to t to find the a_i coefficients and similarly by multiplying by $\sin \omega t$ and integrating to find the b_i. In detail, making use of the orthogonality property of sine and cosine functions, and hence dropping all terms in the Fourier series except

---

[2] Alternatively, and much more popular in nonlinear analysis texts, is to use trigonometric identities to manipulate the nonlinear terms to be of a form containing terms with the fundamental frequency and higher modes. This approach is not systematic and also does not apply to the types of nonlinearities we consider subsequently for typical nonlinear elements in control systems, therefore we use the Fourier series approach.

$\cos \omega t$, the left-hand side is

$$\int_0^{\frac{2\pi}{\omega}} \cos \omega t \left[-\omega A^3 \left(\cos^2 \omega t \right) \sin \omega t \right] dt = 0$$

and the right-hand side is

$$\int_0^{\frac{2\pi}{\omega}} a_1 \cos^2 \omega t \, dt = a_1 \frac{\pi}{\omega}.$$

Hence, $a_1 = 0$. Similarly, multiplying each side by $\sin \omega t$ and integrating gives, for the left-hand side

$$\int_0^{\frac{2\pi}{\omega}} \sin \omega t \left[-\omega A^3 \left(\cos^2 \omega t \right) \sin \omega t \right] dt = A^3 \frac{\pi}{4},$$

and for the left-hand side

$$\int_0^{\frac{2\pi}{\omega}} b_1 \sin^2 \omega t \, dt = b_1 \frac{\pi}{\omega}.$$

Hence, $b_1 = A^3 \omega / 4$.

Returning to Equation (13.25), and substituting the Fourier series for the right-hand side gives

$$A \left(1 - \omega^2 \right) \cos \omega t + \omega A \sin \omega t$$
$$= A^3 \frac{\omega}{4} \sin \omega t + a_2 \cos 2\omega t + b_2 \sin 2\omega t + a_3 \cos 3\omega t + b_3 \sin \omega t + \cdots$$

or dropping the higher harmonics

$$A \left(1 - \omega^2 \right) \cos \omega t + \omega A \sin \omega t \approx A^3 \frac{\omega}{4} \sin \omega t$$

which gives the two equations

$$A \left(1 - \omega^2 \right) = 0, \quad \omega A = A^3 \frac{\omega}{4},$$

which yield for ω and A,

$$\omega = 1, \quad A = 2,$$

which are exactly what we determined by inspection.

13.5.2 Describing Functions

In the chapters on controls, all the analysis was focused on systems composed entirely of linear components. In addition to plants that are nonlinear in nature, typical control components, especially actuators, may also be nonlinear. An example would be a control surface, such as an aileron, on an aircraft. Due to physical limitations, there is a limit on how far it may be deflected. Another example would be a motor that has a limit on the amount of torque it can produce, perhaps due to a limit on the current that may be provided by the circuit that drives it. Both of these are examples of *saturation*, which is a nonlinear phenomenon. Figure 13.18 illustrates a *saturation function* where the output is equal to the input for small values of the input, but is limited to some maximum value, which can be described by

$$sat(x,r) = \begin{cases} r, & x \geq r \\ x, & -r < x < r \\ -r, & x \leq -r \end{cases}$$

for a positive limit, $r > 0$.

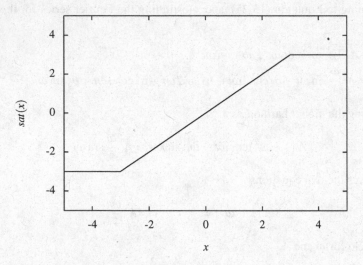

Fig. 13.18 Saturation function, $sat(x,r)$, with $r = 3$.

Another type of common nonlinearity would be a *dead zone* where there is a region around zero with no output. A simple example would be a force pushing on a mass that is resting on a surface, where forces with small magnitudes are less than the friction between the mass and the surface. A dead zone function would be of the form

$$dead(x,r) = \begin{cases} 0, & |x| < r, \\ x - r, & r \leq x, \\ x + r, & x \leq -r, \end{cases}$$

for a positive limit, $r > 0$, and is illustrated in Figure 13.19 for $r = .2$.

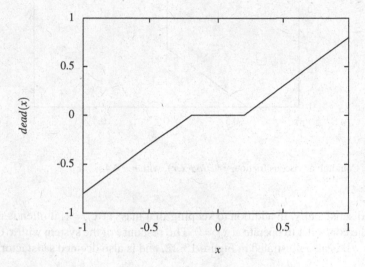

Fig. 13.19 Dead zone function, $dead(x,r)$, with $r = 0.2$.

A final type of nonlinearity we consider is illustrated in Figure 13.20. Considering the aileron example again, this could model the fact that when the desired position of the aileron is very close to zero, the input moving the aileron becomes very small, and is unable to overcome the internal friction in the mechanism. The function is given by

$$friction(x,r) = \begin{cases} x + r, & x \geq 0, \\ x - r, & x < 0, \end{cases}$$

for a positive limit, $r > 0$.

By way of example, let us now consider the effect of such a nonlinearity in a feedback system.

Example 13.19. Let us consider attempting to control the position of the mass illustrated in Figure 13.21 using PI control. Assume that the desired location x_d is a constant, and hence

$$f(t) = k_p(x_d - x(t)) + k_i \int_0^t (x_d - x(\tau)) \, d\tau.$$

Assume $m = 10$, $k = 4$, $b = 3$ and that a controller has already been designed with $k_p = 0.5$ and $k_i = 0.5$. Assume that the unit step response, illustrated in Figure 13.22

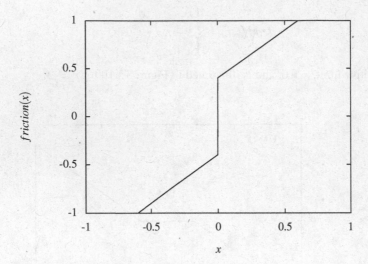

Fig. 13.20 Coulomb or viscous friction, $friction(x, r)$, with $r = 0.4$.

is deemed satisfactory. In addition to keeping that mass at $x_d = 1$, it often is necessary for the controller to operate at $x_d = 0$. The response of the system with $x(0) = 1$ and $\dot{x}(0) = 0$ is also illustrated in Figure 13.22, and is also deemed satisfactory.

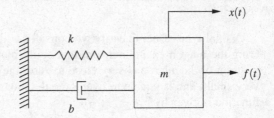

Fig. 13.21 System for Example 13.19.

However, now assume that the real system has an actuator with a response like that illustrated in Figure 13.20. If $r = 0.5$, then the unit step response and regulation to $x_d = 0$ are illustrated in Figure 13.23. The unit step response appears nearly identical; however, there are persistent steady-state oscillations about $x_d = 0$, which for nearly any application would be undesirable. This is a limit cycle, and we can use the method of harmonic balance to attempt to predict when such oscillations will exist.

Without the nonlinear friction element, the system in the previous example could be represented by the block diagram illustrated in Figure 13.24, where it is assumed that the actuator can provide whatever force is commanded by the controller. It is tempting to put a block between the controller $kG_c(s)$ and the plant $G_p(s)$ to represent the nonlinearity. However, if the definition and properties of the elements

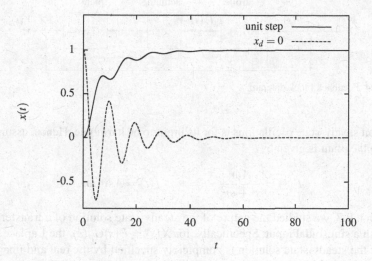

Fig. 13.22 Unit step response and regulation to zero for system in Example 13.19.

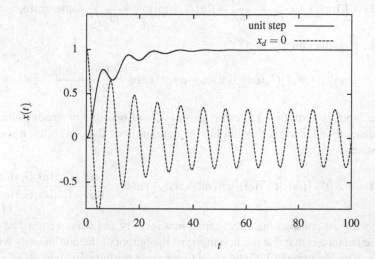

Fig. 13.23 Unit step response and regulation to zero for system in Example 13.19.

of block diagrams, which were developed in Section 8.6 are not to be violated, then we cannot do that, at least without a lot more work. At this point, consider the question of what is required for sustained oscillations, that is, a limit cycle solution, to exist when the input $R(s) = 0$ and there are some nonlinear actuator dynamics.

We have already considered this problem in the harmonic balance analysis above. Assume the plant is stable and the output to the plant has a limit cycle solution when the reference input $R(s) = 0$. The only way for the plant with all left half-plane poles

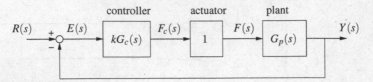

Fig. 13.24 Feedback block diagram.

to exhibit steady-state oscillations is for its input to be harmonic. Hence, assume the input to the plant is given by

$$F(s) = \frac{A\omega}{s^2 + \omega^2} \quad \Longleftrightarrow \quad f(t) = A\sin\omega t.$$

In Section 9.7, we studied the nature of the steady-state solution of a transfer function with a sinusoidal input. Specifically, for $X(s) = F(s)G_p(s)$, the Laplace transform of the steady-state solution is completely specified by the real and imaginary components of $G_p(i\omega)$. Specifically,

$$G_p(s) = A\left(\operatorname{Im}\left(G_p(i\omega)\right)\frac{\omega}{s^2 + \omega^2} + \operatorname{Re}\left(G_p(i\omega)\right)\frac{s}{s^2 + \omega^2}\right) + \text{other stable terms}.$$

and in the time domain

$$x_{ss}(t) = A\left|G_p(i\omega)\right|\sin(\omega t + \phi), \quad \tan\phi = \frac{\operatorname{Im}G_p(i\omega)}{\operatorname{Re}G_p(i\omega)}.$$

If the input is zero, then the error is $-X(s)$, and because in steady-state $X(s)$ is harmonic, then the same thing holds for the output of the controller, namely in steady-state

$$f_{c_{ss}}(t) = -A\left|G_p(i\omega)\right|\left|G_c(i\omega)\right|\sin(\omega t + \phi), \quad \tan\phi = \frac{\operatorname{Im}\left(-G_p(i\omega)G_c(i\omega)\right)}{\operatorname{Re}\left(-G_p(i\omega)G_c(i\omega)\right)}.$$
$$(13.27)$$

Now we must consider the effect of the nonlinearity and what we must have for harmonic balance is that that the first mode of the output of the nonlinearity with an input given by Equation (13.27) be equal to what we started with; namely,

$$f(t) = A\sin\omega t.$$

To proceed, we could compute the first mode with these specific inputs; alternatively and more generally, we compute the magnitude and phase of the fundamental mode of the output of the nonlinearity to an input of arbitrary magnitude and phase, and then match those to the properties of the system, which are manifested in the product $G_p(i\omega)G_c(i\omega)$ in the input.

Definition 13.4. For a nonlinear element, $f(x)$, if a_1 and b_1 are the coefficients of the fundamental cosine and sine terms, respectively, in the Fourier series of the output of $f(x)$ to a harmonic input with magnitude M and frequency, ω, that is,

$$f(M\sin\omega t) = a_1\cos\omega t + b_1\sin\omega t + a_2\cos 2\omega t + b_2\sin 2\omega t + \cdots$$

then the *describing function N* is given by

$$|N| = \frac{\sqrt{a_1^2 + b_1^2}}{M}, \quad \angle N = \tan^{-1}\frac{a_1}{b_1}.$$

Of course, the describing function, in general, depends on both M and ω. However, for all the nonlinearities considered in this text, the describing function only depends on M. The text, [42] contains an excellent exposition on describing functions and is recommended for further reading.[3] Thus, henceforth we write $N(M)$ to indicate the dependence of the describing function on the magnitude of the input.

Based on the discussion preceding the definition of a describing function, in order to have harmonic balance, we simply need

$$\left|G_p(i\omega)\,G_c(i\omega)\right| = \frac{1}{|N(M)|}$$

and

$$\angle\left(G_p(i\omega)\,G_c(i\omega)\right) = -\angle N(M),$$

which is identical to requiring that

$$\boxed{G_p(i\omega)\,G_c(i\omega) = -\frac{1}{N(M)}}$$

as complex numbers. Finally, observe that a Nyquist contour is a plot of the left-hand side of this equation. If we plot $-1/N$ on the same plot, then the terms are equal where the curves intersect. The magnitude of the limit cycle will be the value of M so that $N(M)$ is at the intersection point and the frequency is the value of ω so that the product $G_p(i\omega)\,G_c(i\omega)$ is at the intersection point.

Example 13.20. Consider the friction nonlinearity with $r = 0.5$, and $f_c(t) = M\cos\omega t$. Then

$$friction(f_c(t)) = friction(M\sin\omega t)$$

when $M = 0.25, 0.5$ and 1 with $\omega = 1$ is illustrated in Figure 13.25.

The input to the friction nonlinearity is

$$f_c(t) = M\sin\omega t$$

and we represent the output by the Fourier series

---

[3] Note that the citation is for the first edition of the text. Unfortunately, that material was apparently removed from subsequent editions.

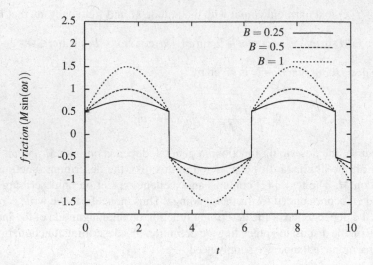

Fig. 13.25 Output of the friction function with $r = 0.5$ for various magnitude inputs and fixed frequency.

$$friction\,(\sin \omega t) = a_1 \cos \omega t + b_1 \sin \omega t + a_2 \cos 2\omega t + b_2 \sin 2\omega t + \cdots.$$

Multiplying both sides of the equation by $\cos \omega t$ and integrating with respect to t from 0 to $2\pi/\omega$ gives then

$$a_1 \frac{\omega}{\pi} = \int_0^{\frac{2\pi}{\omega}} (\cos \omega t)\,friction\,(M \sin \omega t)\,dt$$

$$= \int_0^{\frac{\pi}{\omega}} (\cos \omega t)\,(r + M \sin \omega t)\,dt + \int_{\frac{\pi}{\omega}}^{\frac{2\pi}{\omega}} (\cos \omega t)\,(-r + M \sin \omega t)\,dt$$

$$= 0.$$

Similarly

$$b_1 \frac{\omega}{\pi} = \int_0^{\frac{2\pi}{\omega}} (M \sin \omega t)\,friction\,(\sin \omega t)\,dt$$

$$= \int_0^{\frac{\pi}{\omega}} (\sin \omega t)\,(r + M \sin \omega t)\,dt + \int_{\frac{\pi}{\omega}}^{\frac{2\pi}{\omega}} (\sin \omega t)\,(-r + M \sin \omega t)\,dt$$

$$= \frac{4r + M\pi}{\omega}.$$

Hence,

$$a_1 = 0, \quad b_1 = M + \frac{4r}{\pi}.$$

Also,

$$\angle N(M) = \tan^{-1} \frac{0}{M + \frac{4r}{\pi}} = 0.$$

Hence, the describing function is

$$N(M) = 1 + \frac{4r}{M\pi},$$

which is a positive real number, so $\angle N(M) = 0$.

It is left as an exercise to show that the describing functions for the other nonlinearities are as follows.

- Saturation:

$$N = \frac{2}{\pi}\left(\sin^{-1}\frac{r}{M} + \frac{r}{M}\sqrt{1 - \left(\frac{r}{M}\right)^2}\right), \tag{13.28}$$

- Dead zone:

$$N = 1 - \frac{2}{\pi}\left(\sin^{-1}\frac{r}{M} + \frac{r}{M}\sqrt{1 - \left(\frac{r}{M}\right)^2}\right). \tag{13.29}$$

See [41] for more nonlinearities with slightly more general formulations.

Using the result from the previous example, we can plot the Nyquist contour for the plant and controller and find the intersection, if any, with $-1/N(M)$ to determine an approximate magnitude and frequency for the limit cycle.

Example 13.21. Returning to Example 13.19, the transfer function for the plant and controller are

$$G_p(s) = \frac{1}{10s^2 + 3s + 4}, \quad G_c(s) = 0.5\left(s + \frac{1}{s}\right).$$

The Nyquist contour for $G_p(s)G_c(s)$ is illustrated in Figure 13.26, along with the describing function, which is a straight line starting at the origin and ending at $s = -1$ as $M \to \infty$.

Because the curves intersect, there is a limit cycle. From the Nyquist contour, we need the frequency where $G_p(i\omega)G_c(i\omega)$ is negative and real, which can be determined by setting the imaginary component to zero

$$G_p(i\omega)G_c(i\omega) = 0.5\frac{i\omega + 1}{-10(i\omega)^3 + 3(i\omega)^2 + 4i\omega}$$

$$= 0.5\frac{i\omega + 1}{-3\omega^2 + i\omega(4 - 10\omega^2)}\left(\frac{-3\omega^2 - i\omega(4 - 10\omega^2)}{-3\omega^2 - i\omega(4 - 10\omega^2)}\right)$$

$$= \frac{\omega(\omega - 10\omega^3 + i(7\omega^2 - 4))}{2(-3\omega^2 - i\omega(4 - 10\omega^2))^2}.$$

Hence, for the imaginary part to be negative

$$7\omega^2 - 4 = 0 \quad \Longrightarrow \quad \omega = \sqrt{\frac{4}{7}} \approx 0.76,$$

which is the frequency of the limit cycle in Figure 13.23.

To determine the amplitude, note that the curves intersect at approximately $s = -0.3$. Hence, we need to compute the magnitude, M where $-1/N(M) = -0.3$. Thus

$$-\frac{1}{1 + \frac{4(0.5)}{\pi M}} = 0.3 \quad \Longrightarrow \quad M \approx 0.27,$$

which is the magnitude of the limit cycle in Figure 13.23.

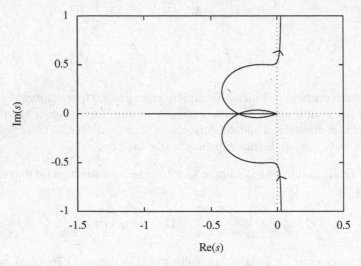

Fig. 13.26 Nyquist plot and describing function for Example 13.21.

13.6 Introduction to Local Bifurcation Analysis

Another phenomenon characteristic of nonlinear systems is that of a *bifurcation* which is a qualitative change in the nature of the solution to a differential equation as some parameter is varied. This section presents a catalog of typical bifurcations by way of examples. It is not intended to be a complete exposition on the subject; for that, interested readers are referred to [22, 54]. The examples are of the simplest type: namely, first-order and generally solvable.

13.6.1 Saddle-Node Bifurcations

Consider the first-order differential equation

$$\dot{x} = -x^2 + \mu. \tag{13.30}$$

The equilibria for this equation are $x_0 = \sqrt{\mu}$ if $\mu \geq 0$. If $\mu < 0$ there are no equilibria. These equilibrium values are plotted as a function of μ in Figure 13.27. The parameter μ is commonly called the *bifurcation parameter*.

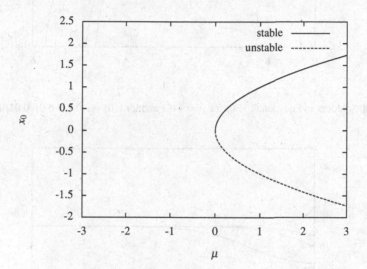

Fig. 13.27 Equilibrium values for Equation (13.30).

About x_0 the linear approximation is

$$\dot{x} + \left(x_0^2 + 2x_0 \left(x - x_0 \right) \right) - \mu = 0.$$

Substituting $x_0 = \pm\sqrt{\mu}$ gives

$$\dot{x} \pm 2\sqrt{\mu}x = 2\mu,$$

which has the general solution

$$x(t) = ce^{\mp 2\sqrt{\mu}t} + \sqrt{\mu},$$

which is stable for $+\sqrt{\mu}$ and unstable for $-\sqrt{\mu}$. These are indicated in Figure 13.27 with a solid line for the stable upper branch and a dashed line for the unstable lower branch.

Solutions for various initial conditions and $\mu = -0.1$ are illustrated in Figure 13.28. From Equation (13.30), if $\mu < 0$, $\dot{x}$ is always negative. Solutions for

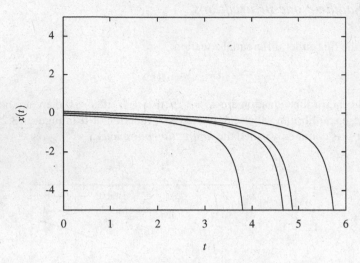

Fig. 13.28 Solutions to Equation (13.30) for $\mu = -0.1$ and for $x(0) = -0.1, -0.01, 0.01, 0.1$

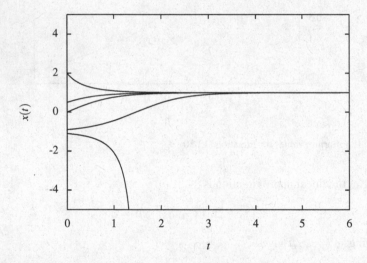

Fig. 13.29 Solutions to Equation (13.30) for $\mu = 0.5$ and for $x(0) = 2.0, 0.5, 0.0, -0.9, -1.1$.

various initial conditions and $\mu = 0.5$ are illustrated in Figure 13.29. Note that solutions that start with $x(0) > -\sqrt{\mu}$ are attracted to the upper stable equilibrium.
Solutions with $x(0) < -\sqrt{\mu}$ are unstable.

Of course the main point of much of mathematics is generalization, and for the
case at hand we would like to determine conditions on an equation of the form

$$\dot{x} = f(x, \mu) \tag{13.31}$$

that would guarantee the existence of a saddle-node bifurcation, which are as follows.

- The curve of equilibrium points in Figure 13.27 passed through the origin. This would require that

$$f(0,0) = 0. \tag{13.32}$$

- Referring to Figure 13.27, if we are able to solve for μ as a function of x, because the curve is tangent to the x-axis at the origin,

$$\frac{d\mu}{dx} = 0. \tag{13.33}$$

Also, because the curve is entirely to one side of the x-axis, it is not an inflection point, which requires that

$$\frac{d^2\mu}{dx^2} \neq 0. \tag{13.34}$$

In the simple example above, we were able to solve for μ as a function of x; however, if the equation is more complicated with many more higher-order terms, this may not be possible. Hence, we derive conditions on $f(x,\mu)$ that will ensure that the two conditions in Equations (13.33) and (13.34) are met. If we let $\mu = \mu(x)$ along the curve of equilibrium points, then

$$f(x, \mu(x)) = 0,$$

and

$$\frac{df}{dx} = \frac{\partial f}{\partial x} + \frac{\partial f}{\partial \mu} \frac{d\mu}{dx}. \tag{13.35}$$

This is zero inasmuch as $f(x, \mu(x)) = 0$, and solving for $d\mu/dx$ gives

$$\frac{d\mu}{dx} = -\frac{\frac{\partial f}{\partial x}}{\frac{\partial f}{\partial \mu}}.$$

Evaluating this at $(x,\mu) = (0,0)$ gives

$$\frac{d\mu}{dx}(0) = -\frac{\frac{\partial f}{\partial x}(0,0)}{\frac{\partial f}{\partial \mu}(0,0)}.$$

For this to be zero, we need

$$\frac{\partial f}{\partial x}(0,0) = 0, \tag{13.36}$$

and

$$\frac{\partial f}{\partial \mu}(0,0) \neq 0. \tag{13.37}$$

- Differentiating Equation (13.35) again lets us satisfy Equation (13.34). In particular,

$$\frac{d^2 f}{dx^2} = \frac{\partial^2 f}{\partial x^2} + 2\frac{\partial^2 f}{\partial x \partial \mu}\frac{d\mu}{dx} + \frac{\partial^2 f}{\partial \mu^2}\left(\frac{d\mu}{dx}\right)^2 + \frac{\partial f}{\partial \mu}\frac{d^2 \mu}{dx^2} = 0. \tag{13.38}$$

Because

$$\frac{d\mu}{dx}(0) = 0,$$

if we evaluate Equation (13.38) at $(x,\mu) = (0,0)$, we get

$$\frac{\partial^2 f}{\partial x^2}(0,0) + \frac{\partial f}{\partial \mu}(0,0)\frac{d^2 \mu}{dx^2}(0) = 0.$$

Hence,

$$\frac{d^2 \mu}{dx^2}(0) = -\frac{\frac{\partial^2 f}{\partial x^2}(0,0)}{\frac{\partial f}{\partial \mu}(0,0)}$$

which we need to be nonzero. Hence for a saddle-node bifurcation

$$\boxed{\frac{\partial^2 f}{\partial x^2}(0,0) \neq 0.} \tag{13.39}$$

The denominator was previously required to be nonzero.

Hence, any first order ordinary differential equation of the form of Equation (13.31) that satisfies Equations (13.32), (13.36), (13.37), and (13.39) will have the following features near the origin.

1. For one sign of μ all solutions will be unstable.
2. For the other sign of μ the parabola of equilibrium points will exist one of which will be stable and the other unstable. Between the branches the solutions will be attracted to the stable branch. Outside the branches, for values of x on the stable side, the solutions will be attracted to the stable branch; conversely, for values of x outside the unstable branch, solutions will be unstable.

The application of this result is shown in the following example.

Example 13.22. Consider the one-dimensional system

$$\dot{x} = 3\mu + 2x^2 + 6\mu x + x^4 + x^5. \tag{13.40}$$

We check each of the conditions for a saddle-node bifurcation.

- By direct substitution $f(0,0) = 0$, Equation (13.32) is satisfied.
- Checking Equation (13.36) gives

$$\frac{\partial f}{\partial x} = 4x + 6\mu + 4x^3 + 5x^4, \quad \implies \quad \frac{\partial f}{\partial x}(0,0) = 0.$$

- Checking Equation (13.37) gives

$$\frac{\partial f}{\partial \mu} = 3 + 6x, \quad \Longrightarrow \quad \frac{\partial f}{\partial \mu}(0,0) = 3 \neq 0.$$

- And finally, checking Equation (13.39) gives

$$\frac{\partial^2 f}{\partial x^2} = 4 + 12x^2 + 20x^3, \quad \Longrightarrow \quad \frac{\partial^2 f}{\partial x^2}(0,0) = 4 \neq 0.$$

Hence, we know that near the origin when μ switches sign on one side the system will have no fixed point and on the other it will have two fixed points, one stable and one unstable. Numerically solving Equation (13.40) for $\mu < 0$ for various initial conditions is illustrated in Figure 13.30. Note that the stable branch is negative and the unstable branch is positive. Figure 13.31 illustrates that all solutions for $\mu > 0$ are unstable. Note that both the sign for μ corresponding to the parabola of fixed points and the stable and unstable branches of fixed points are different from those in the example used to introduce the saddle-node bifurcation, which is permissible because nothing in the derivation of the conditions for the existence of this bifurcation required them to be one way or the other. Also, because of the higher-order terms, the structure is not exactly symmetric about $x = 0$.

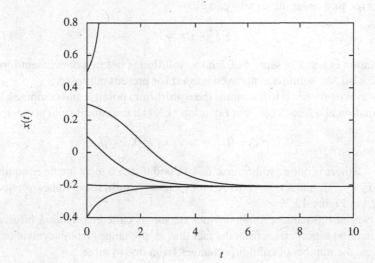

Fig. 13.30 Solutions for Example 13.22 for $\mu = -0.05$.

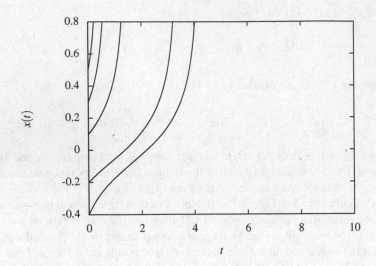

Fig. 13.31 Solutions for Example 13.22 for $\mu = 0.05$.

13.6.2 Pitchfork Bifurcations

Consider the first-order differential equation

$$\dot{x} + x^3 - \mu x = 0. \tag{13.41}$$

This equation is actually separable, but the solution is not in a convenient form for analysis. Also, the solution is not even needed for present purposes.

The way we proceed is to determine the equilibrium point(s) and compute a linear approximation about each one. For Equation (13.41), the equilibrium points satisfy

$$x_0^3 - \mu x_0 = 0 \quad \Longleftrightarrow \quad x_0 = 0, \pm\sqrt{\mu}.$$

So, if $\mu \leq 0$ there is one equilibrium at $x_0 = 0$, and if $\mu > 0$ there are three equilibria: $x_0 = 0$, $x_0 = \sqrt{\mu}$, and $x_0 = -\sqrt{\mu}$. A plot of these equilibrium values versus μ is illustrated in Figure 13.32.

This type of bifurcation, for an obvious reason, is called a *pitchfork* bifurcation. The bifurcation aspect arises from the fact that as μ changes from negative to positive values, the number of equilibria changes from one to three.

Now consider the stability of these equilibria. For any value of μ the linear approximation about the $x_0 = 0$ equilibrium is

$$\dot{x} - \mu x = 0,$$

which has solutions of the form

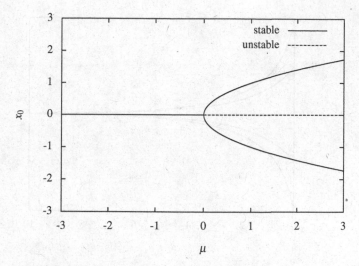

Fig. 13.32 Equilibrium values for Equation (13.41) versus μ.

$$x(t) = ce^{\mu t}.$$

Clearly, for $\mu < 0$ these solutions are stable and for $\mu > 0$ these solutions are unstable. About $x_0 = \pm\sqrt{\mu}$ for $\mu > 0$ the linearization is

$$\dot{x} + \left(\pm\mu^{\frac{3}{2}} + 3\mu\left(x \mp \sqrt{\mu}\right)\right) - \mu x = 0 \quad \Longrightarrow \quad \dot{x} + 2\mu x = \pm 2\mu^{3/2}. \tag{13.42}$$

Note the solution to Equation (13.42) is

$$x(t) = ce^{-2\mu} \pm \frac{1}{2}\sqrt{\mu}$$

which is stable regardless of the sign of $\pm\sqrt{\mu}$. Hence, referring back to Figure 13.32, the branches of the equilibrium solutions which are stable are indicated by solid lines and the unstable branch is indicated by dashed lines. Observe that the stability of the $x_0 = 0$ equilibrium switches from stable to unstable as μ switches from negative to positive. The two outer branches for positive μ are stable.

Plots of solutions of Equation (13.41) for $\mu = -0.5$ and for various initial conditions are illustrated in Figure 13.33. Note that the $x_0 = 0$ equilibrium point is stable. Solutions for $\mu = 0.5$ are illustrated in Figure 13.34. Note that the $x_0 = 0$ equilibrium point is unstable, whereas, the $x_0 = \pm 1/\sqrt{2}$ equilibria are stable.

A process that is similar to what we did for the saddle-node bifurcation gives the list of conditions for a general system of the form of Equation (13.31) to have a local pitchfork bifurcation.

1. $f(0,0) = 0$.
2. $\frac{\partial f}{\partial x}(0,0) = 0$.

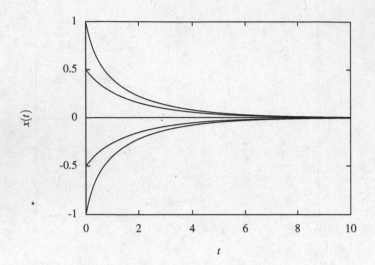

Fig. 13.33 Solutions to Equation (13.41) for $\mu = -0.5$ and for various initial conditions.

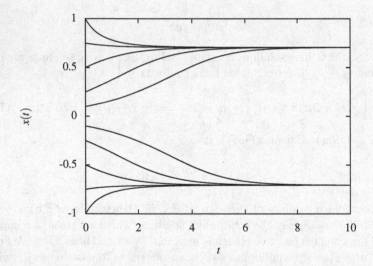

Fig. 13.34 Solutions to Equation (13.41) for $\mu = 0.5$ and for various initial conditions.

3. $\frac{\partial f}{\partial \mu}(0,0) = 0$.

4. $\frac{\partial^2 f}{\partial x^2}(0,0) = 0$.

5. $\frac{\partial^2 f}{\partial x \partial \mu}(0,0) \neq 0$.

6. $\frac{\partial^3 f}{\partial x^3}(0,0) \neq 0$.

13.6.3 Concluding Comments

Another type of one-dimensional bifurcation is called the *transcritical bifurcation*, which is characterized by two branches of equilibrium points, both of which exist for positive and negative values of the bifurcation parameter. One branch is stable and the other is unstable, and the stability properties switch as the bifurcation parameter changes sign. This type of bifurcation is explored in the exercises.

Observe that a condition for both the saddle-node bifurcation and the pitchfork bifurcation is that $\partial f / \partial x = 0$ when evaluated at $(x, \mu) = (0, 0)$. For a single, first order differential equation, this term is essentially the 1×1 Jacobian for the system, and its value is the eigenvalue.[4] It should make some sense that an eigenvalue must be approaching zero as the bifurcation parameter crosses zero, or else there would not be any qualitative change in the stability properties of the system. Considering this as a generic feature of bifurcations introduces the generic feature of Hopf bifurcations, which may occur when a complex conjugate pair of eigenvalues crosses the imaginary axis, which corresponds to having a zero real part at that point. Not surprisingly due to the nature of complex-conjugate eigenvalues, Hopf bifurcations are characterized by the existence of small-amplitude limit cycles.

13.7 Exercises

13.1. Consider $\ddot{x} + x - x^3 = 0$.

1. Write this as two first-order ordinary differential equations.
2. Determine all the equilibrium points.
3. Using the Jacobian, determine the differential equation that is the best linear approximation about the equilibrium that is farthest to the right, that is, about the equilibrium point that has the largest value.
4. Determine the general solution to the linear approximation.
5. Sketch the phase portrait near the equilibrium point. Include the eigenvectors of the Jacobian matrix evaluated at the equilibrium point in the sketch.

13.2. Consider

1. $\ddot{x} + \dot{x} - x + x^3 = 0$.
2. $\ddot{x} + \dot{x} + x - x^3 = 0$.
3. $\ddot{x} + \dot{x} - 2x - x^2 + x^3 = 0$.
4. $\ddot{x} + \dot{x} + 2x + x^2 - x^3 = 0$.

For each equation do the following.

- Find the equilibrium points.

---

[4] For a 1×1 matrix, the value in the matrix is also the eigenvalue because, when there is only one equation, $A - \lambda I$ must be zero for there to be a nonzero term such that $(A - \lambda I)\hat{\xi} = 0$.

- Determine the stability and nature of the solutions near each equilibrium point by computing the linear differential equations that approximate each equation near each equilibrium.
- Based on the analysis of the linear equations, sketch the phase plane for the system.
- Verify your sketch by using a computer package or program to compute approximate numerical solutions for various initial conditions for the nonlinear equation and plotting them on the phase plane. Note that some of the solutions may be unstable, and hence for those you may have to limit the time period that the solution method runs if the package does not automatically stop when a large number is reached.
- Plot $x(t)$ versus t for the nonlinear solution and the best linear approximation near $x_0 = 1$ for $x(0) = 0.75, x(0) = 0.5, x(0) = 0.25, x(0) = 0, x(0) = -0.25$, and $\dot{x}(0) = 1$ for all the cases. Make a different plot for each set of initial conditions. Each plot should contain two solutions, the nonlinear solution and the solution to the linear approximation, such as in Figure 13.9.

13.3. Repeat Exercise 13.2 for

$$\ddot{x} + \dot{x} + x(x-2)(x-1)(x+1)(x+2) = 0.$$

13.4. Consider

$$\ddot{x} + 0.2\dot{x} - x + x^2 = 0. \tag{13.43}$$

1. Determine a linear differential equation that is the best linear approximation to Equation (13.43) near an arbitrary point $x = x_0$.
2. For $x_0 = 1$ determine the general solution to the linear approximation. Make a plot in the phase plane for comparing the solution to the linear approximation and the nonlinear equation for various initial conditions. Include plots of solutions with initial conditions for which the solution to the linear equation is near the solution to the nonlinear equation as well as plots where the solution to the linear equation is not near the solution to the nonlinear equation.
3. Near $x_0 = 0$, determine the solution to the linear approximation. Write it in the form of

$$\frac{d}{dt}\begin{bmatrix} x_1(t) \\ x_2(t) \end{bmatrix} = c_1\hat{\xi}^1 e^{\lambda^1 t} + c_2\hat{\xi}^2 e^{\lambda^2 t},$$

where $\hat{\xi}^1$ and $\hat{\xi}^2$ are vectors. Plot the two vectors on the phase plane and also plot $-\hat{\xi}^1$ and $-\hat{\xi}^1$. Compare the solution to the linear approximation and the nonlinear approximation for initial conditions on either side of these vectors, that is, for initial conditions that are like the $\times$s illustrated in Figure 13.35.

  a. Plot all the solutions where the range for the axes is $x \in [-.5, .5]$ and $\dot{x} \in [-.5, .5]$.
  b. Plot all the solutions where the range for the axes is $x \in [-2, 2]$ and $\dot{x} \in [2, 2]$.

Note: The solutions in this problem are unstable. You may need to adjust the total time for the numerical solution for the nonlinear equation so that the solution does not exceed the maximum value allowable for the simulation and plotting program.

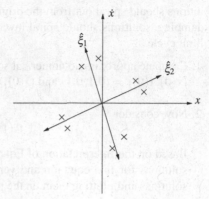

Fig. 13.35 Initial conditions for Exercise 13.4.

13.5. Consider $\ddot{x} - \sin(x) = 0$.

1. Determine the equilibrium points for this equation.
2. Determine a linear differential equation that is an approximation for this nonlinear equation near the equilibrium that is closest to the origin.
3. Determine the solution to the best linear approximation.
4. On the phase plane sketch the solutions near the equilibrium that is closest to the origin.

13.6. This exercise illustrates that when there is an eigenvalue of a system that is zero, the equilibrium may be stable, unstable, or neutrally stable.

1. Consider

$$\ddot{x} + \dot{x}^3 + x = 0.$$

By using either a Jacobian or a Taylor series for the $\dot{x}^3$ term, determine a linear approximation to this system near $\dot{x} = 0$. Does the linear approximation grow, decay, or is it neutrally stable? For $x(0) = 0.25$ and $\dot{x}(0) = 0$, plot the solution to the linear approximation and a numerical solution determined using a computer package. Does the equilibrium point at $(x, \dot{x}) = (0, 0)$ appear to be stable, unstable, or neutrally stable for the nonlinear system?

2. Repeat the previous step for

$$\ddot{x} + \dot{x}^2 + x = 0.$$

13.7. A feature of nonlinear equations that is not possible for a linear equation is a *limit cycle*. The *Van der Pol* equation,

$$\ddot{x} + \left(x^2 - 1\right)\dot{x} + x = 0 \tag{13.44}$$

exhibits a limit cycle. This exercise investigates that nature of limit cycles by numerically solving the nonlinear equation. Note that one way to interpret Equation (13.44) is if $|x| < 1$ thé system has negative damping; conversely, if $|x| > 1$ it has positive damping. Hence, in the phase plane for small initial conditions solutions should spiral out from the origin. For larger initial conditions, with positive damping, solutions should spiral inward. Where they meet is a solution called the limit cycle.

1. Compute approximate numerical solutions and plot them in the phase plane for $(x(0), \dot{x}(0)) = (0.1, 0.1)$ and $(x(0), \dot{x}(0)) = (3, 3)$ and plot these solutions versus t.
2. Now consider

$$\ddot{x} + \left(1 - x^2\right)\dot{x} + x = 0.$$

Based on the interpretation of Equation (13.44) above, predict the nature of the solutions for this equation and verify it by computing approximate numerical solutions and plotting them on the phase plane.

13.8. Another cool feature of nonlinear equations is the possibility of *subharmonic* and *superharmonic responses*. For a linear constant-coefficient ordinary differential equation, if the inhomogeneous term is harmonic, then the solution of the system has a component at the same frequency. Explain why this should be the case by considering the method of undetermined coefficients. A nonlinear system may have a harmonic solution that is at a different frequency than the inhomogeneous (forcing) term.

1. Determine an approximate numerical solution to

$$\ddot{x} + x + x^3 = 2\cos 6t$$

with $x(0) = 2$ and $\dot{x}(0) = 0$. Plot the solution versus t and on the same plot, plot the forcing term $2\cos 6t$. Explain why this is called the subharmonic response. Be sure to choose an appropriate interval of time over which to plot the response to clearly illustrate the nature of the phenomenon.
2. Determine an approximate numerical solution to

$$\ddot{x} + 0.005\dot{x} + x + 0.0025x^3 = 6\cos 0.38t$$

with $x(0) = 0$ and $\dot{x}(0) = 0$. Plot the solution versus t and on the same plot, plot the forcing term $2\cos 0.38t$. Explain why this is called the superharmonic response.

13.9. For some systems, it may be obvious that one or more of the variables are strongly stable, meaning that close to an equilibrium they decay to zero quickly. For example,

$$\dot{x} = x^2 y - x^5$$
$$\dot{y} = -y + x^2.$$

For these equations, as long as $|x| \ll 1$, then the linear approximation for the y equation is $\dot{y} = -y$, which has exponentially decaying solutions.

Because $y(t) \to 0$ quickly, it would be tempting to set $y = 0$ in the x equation and study $\dot{x} = -x^5$. This would lead us to conclude that $x(t) \to 0$ also because if $x > 0$, $\dot{x} < 0$ and vice versa. Such a simple approach is, unfortunately, incorrect. By using a computer package to compute an approximate numerical solution, verify that the origin is not a stable equilibrium point. Explain why the argument we made above that the origin is stable is incorrect.

13.10. Consider the system illustrated in Figure 13.21.

- Pick parameter values for m, b, and k. Design a controller that has a rise time less than one second, a 5% settling time less than one second, a percentage overshoot less than 25%, and steady-state error less than 0.05 for a unit step input.
- Assume that the spring is not exactly linear. In fact, assume that the spring is hardening in that the force required to compress the spring by an amount x is actually $f_s(x) = k\left(x + x^3/10\right)$. Use your controller on this nonlinear system and try step inputs with large and small magnitudes. Is the effect of the nonlinearity significant? Does it affect whether the specifications are still met?

13.11. Derive the describing functions for the dead zone and saturation nonlinearities provided by Equations (13.29) and (13.28), respectively.

13.12. For each of the following transfer functions

1. $kG_c(s)G_p(s) = \frac{50}{(s+1)(s+2)(s+3)}$,
2. $\frac{100}{s^2+2s+4}$,
3. $kG_c(s)G_p(s) = \frac{5}{s^3+6s^2+11s}$,

in each case for each of the saturation, dead zone, and friction nonlinearities, use a Nyquist plot and describing function to predict the existence of limit cycles. If there is a limit cycle, predict its magnitude and frequency. Verify your result numerically. For the saturation nonlinearity let $r = 0.5$, for the dead zone let $r = 0.25$, and for friction let $r = 0.33$.

13.13. The equation

$$\dot{x} = \mu x - x^2$$

exhibits a *transcritical bifurcation*. Construct the bifurcation diagram for this equation by plotting the fixed point(s) of the system versus μ. Also, analyze the stability of the fixed points and indicate the stable portions with solid lines and the unstable portions with dashed lines.

13.14. For each of the following first-order differential equations,

1. $\dot{x} = 3\mu + 2x^2 + 6\mu x + x^4 + x^5$;
2. $\dot{x} = 3\mu x + 2x^3 + x^3 \sin x + x^5$;
3. $\dot{x} = 3\mu x + 2x^3 + x^3 \sin x + x^5 + 1$;
4. $\dot{x} = 3\mu x^2 + 2x^3 + x^3 \sin x + x^5$;

5. $\dot{x} = -3\mu + 2x^2 + 6\mu x + x^4 + x^5$;
6. $\dot{x} = -3\mu - 2x^2 - 6\mu x + x^4 + x^5$;
7. $\dot{x} = 3\mu^2 + 2x + 6\mu x + x^4 + x^5$; and,
8. $\dot{x} = 3\mu - 3x^3 + 6x^5$;

determine whether the solutions exhibit the characteristics of a saddle-node bifurcation, pitchfork bifurcation, or neither as the parameter μ is varied. Verify your analysis by numerically solving the equations for different initial conditions and plotting the solutions. Remember that bifurcations are local so that you need to use small initial conditions and small values of μ for your numerical verifications.

13.15. This problem numerically investigates a Hopf bifurcation. Consider

$$\dot{x} = -x - y + \varepsilon x - z^3$$
$$\dot{y} = 2x + y + \varepsilon y - z^3$$
$$\dot{z} = x + 2y - z.$$

1. For $\varepsilon = 0.01$ and $\varepsilon = -0.01$, and small nonzero initial conditions, make a three-dimensional plot of an approximate numerical solution to show that the origin is an unstable equilibrium for one case and a stable equilibrium for the other.
2. For $\varepsilon = 0.01$, slowly increase the initial conditions until the solution no longer converges to the origin. Plot this solution on the same plot as for the second part of this problem. Print the plot and sketch by hand where you think the unstable limit cycle is. Keep the range for the plot to be between ± 0.4 for all three axes.

13.16. *Rayleigh's equation*

$$\ddot{x} + \left(\frac{1}{3}\dot{x}^2 - 1 \right)\dot{x} + x = 0$$

originated in the study of violin strings. Write a computer program to determine an approximate numerical solution for various initial conditions to show there is a limit cycle. Use the method of harmonic balance to determine the approximate magnitude and frequency of the limit cycle. Plot the limit cycle for the nonlinear equation on the phase plane as well as the approximation.

13.17. Consider

$$\ddot{x} + \left(x^2 + \dot{x}^2 - 1 \right)\dot{x} + x = 0.$$

Write a computer program to determine an approximate numerical solution for various initial conditions to show there is a limit cycle. Verify your answer using the method of harmonic balance.

13.18. If the van der Pol equation is changed to

$$\ddot{x} + \frac{1}{100}\left(1 - x^2 \right)\dot{x} + x = 0$$

it has negative damping for large $|x|$ and positive damping for small $|x|$, and so it should have an *unstable limit cycle*. Write a computer program to determine an approximate numerical solution for this system. Choose various initial conditions and try to pick some very close to the limit cycle. Note that because outside the limit cycle the system is unstable, it would be convenient to put a check in the program that makes it terminate once $x^2 + \dot{x}^2$ is large.

13.19. The equation of motion for a pendulum is

$$\ddot{\theta} + \frac{g}{l} \sin \theta = 0$$

if $\theta = 0$ when the pendulum is straight down.

1. Unlike a linear oscillator, the frequency of oscillation of the pendulum depends on its amplitude. Verify this by writing a computer program to determine an approximate numerical solution for this system and comparing the frequency of solutions for various initial conditions. Plot solutions to demonstrate that as the amplitude increases, the frequency decreases.
2. The method of harmonic balance can be used to find an approximate relationship between the magnitude of the oscillations and the frequency. If we assume $\theta(t) = \cos \omega t$, which, when substituted gives a term of the form $\sin(\cos \omega t)$, therefore it is easier to express

$$\sin \theta \approx \theta - \frac{1}{6} \theta^3.$$

Hence, use the method of harmonic balance on the equation

$$\ddot{\theta} + \frac{g}{l} \left(\theta - \frac{1}{6} \theta^3 \right) = 0$$

to show that an approximate relationship between the frequency ω and amplitude A is

$$\omega \approx \sqrt{\frac{g}{l} \left(1 - \frac{A^2}{8} \right)}.$$

3. Compare this relationship with the results from your numerical simulations. Is it better or worse than the frequency that would be obtained if the system were linearized about $\theta = 0$?

Appendix A
Some Complex Variable Theory

This appendix presents a very short overview of complex variable theory. An interested reader is referred to [10] for a complete exposition.

A.1 Complex Numbers

Historically, of course, imaginary numbers have a natural association with the square root of negative numbers. We develop the definitions of complex numbers in a more deductive manner and then show that the approach is consistent with the more historical view.

All readers should be familiar with the usual notion that a complex number has a real and imaginary component, where we may write

$$s = \sigma + i\omega,$$

where s is the complex number, σ is its real part and ω is the imaginary part. A complex number has two components, therefore it may naturally be considered an ordered pair of numbers. The only twist is to ensure that we define multiplication correctly.

Definition A.1. A complex number is an ordered pair of real numbers,

$$s = (a, b),$$

where for $s_1 = (a_1, b_1)$ and $s_2 = (a_2, b_2)$ addition is defined by

$$s_1 + s_2 = (a_1 + a_2, b_1 + b_2)$$

and multiplication is defined by

$$s_1 s_2 = (a_1 a_2 - b_1 b_2, a_1 b_2 + a_2 b_1).$$

B. Goodwine, *Engineering Differential Equations: Theory and Applications*,
DOI 10.1007/978-1-4419-7919-3, © Springer Science+Business Media, LLC 2011

This definition is consistent with the idea of using i because if we write $s_1 = a_1 + ib_1$ and $s_2 = a_2 + ib_2$ then

$$s_1 + s_2 = a_1 + ib_1 + a_2 + ib_2 = (a_1 + a_2) + i(b_1 + b_2)$$

and

$$\begin{aligned} s_1 s_2 &= (a_1 + ib_1)(a_2 + ib_2) \\ &= a_1 a_2 + ib_1 a_2 + a_1 ib_2 + ib_1 ib_2 \\ &= (a_1 a_2 - b_1 b_2) + i(a_1 b_2 + b_1 a_2). \end{aligned}$$

It follows from the definition of addition and multiplication that the additive inverse of $s = a + ib$ is $-s = -a - ib$ and the multiplicative inverse is

$$s^{-1} = \left(\frac{a}{a^2 + b^2}, -\frac{b}{a^2 + b^2} \right).$$

Using the multiplicative inverse, division may be defined as

$$\frac{s_1}{s_2} = s_1 s_2^{-1}.$$

An alternative representation is in polar coordinates where s is represented by a magnitude and phase which are the usual Euclidean norm and angle if the number is plotted in its Cartesian coordinates. Referring to Figure A.1, if $s = a + ib$, then

$$r = \sqrt{a^2 + b^2} = |s|$$

and

$$\theta = \tan^{-1}\left(\frac{b}{a} \right) = \angle s,$$

where the inverse tangent function is able to distinguish between quadrants. The number (angle) θ is called an *argument* of the complex number s. There are an infinite number of arguments that differ by a multiple of 2π. The *principal value* of the arguments is the unique value $\theta \in (-\pi, \pi]$.

The previous two equations relate the Cartesian to polar form. Going from polar to Cartesian form is simple geometry and is given by

$$s = r(\cos\theta + i\sin\theta).$$

The Cartesian form is easy to use for addition and subtraction because if $s_1 = a_1 + ib_1$ and $s_2 = a_2 + ib_2$, then

$$s_1 + s_2 = (a_1 + a_2) + i(b_1 + b_2).$$

However, multiplication is easier in polar form. In particular, if $s_1 = (r_1, \theta_1)$ and $s_2 = (r_2, \theta_2)$, then the product is

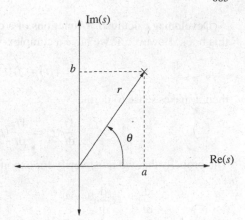

Fig. A.1 Cartesian, $s = a + ib$ and polar, $s = (r, \theta)$ forms of a complex number s.

$$s_1 s_2 = (r_1 r_2, \theta_1 + \theta_2)$$

and the quotient is

$$\frac{s_1}{s_2} = \left(\frac{r_1}{r_2}, \theta_1 - \theta_2 \right).$$

This multiplication rule is easily seen by writing $s_1 = r_1 (\cos \theta_1 + i \sin \theta_1)$ and $s_2 = r_2 (\cos \theta_2 + i \sin \theta_2)$. Taking the product

$$
\begin{aligned}
s_1 s_2 &= r_1 (\cos \theta_1 + i \sin \theta_1) \, r_2 (\cos \theta_2 + i \sin \theta_2) \\
&= r_1 r_2 [\cos \theta_1 \cos \theta_2 - \sin \theta_1 \sin \theta_2 + i (\sin \theta_1 \cos \theta_2 + \sin \theta_2 \cos \theta_1)] \\
&= r_1 r_2 [\cos (\theta_1 + \theta_2) + i \sin (\theta_1 + \theta_2)]
\end{aligned}
$$

so

$$s_1 s_2 = (r_1 r_2, \theta_1 + \theta_2).$$

A.2 Functions of a Complex Variable

The most important function of a complex variable is the exponential function due to the fact that exponentials are solutions to homogeneous, linear, constant-coefficient, ordinary differential equations.

Definition A.2. If $s = a + ib$, define

$$e^s = e^a (\cos b + i \sin b).$$

Remark A.1. Note that this is a *definition*, which we choose to adopt. It remains to determine whether it is a useful definition or whether it reduces to the usual form when the imaginary part of s is zero.

Developing calculus for functions of a complex variable is beyond the scope of this book. However, if we have a complex-valued function of a real variable,

$$f(t) = f_r(t) + i f_i(t)$$

then it makes sense to define

$$\frac{\mathrm{d}f}{\mathrm{d}t}(t) = \frac{\mathrm{d}f_r}{\mathrm{d}t}(t) + i\frac{\mathrm{d}f_i}{\mathrm{d}t}(t).$$

The property that we must verify for exponentials is

$$
\begin{aligned}
\frac{\mathrm{d}}{\mathrm{d}t}e^{st} &= \frac{\mathrm{d}}{\mathrm{d}t}e^{(a+ib)t} \\
&= \frac{\mathrm{d}}{\mathrm{d}t}\left(e^{at}\left(\cos bt + i\sin bt\right)\right) \\
&= ae^{at}\left(\cos bt + i\sin bt\right) + be^{at}\left(-\sin bt + i\cos bt\right) \\
&= ae^{at}\left(\cos bt + i\sin bt\right) + ibe^{at}\left(\cos bt + i\sin bt\right) \\
&= (a+ib)\left(e^{at}\left(\cos bt + i\sin bt\right)\right),
\end{aligned}
$$

so the usual rule for differentiating an exponential function holds.

A.3 Partial Fraction Decomposition

This subject is not limited to the field of complex variables, but because it appears in the process of solving for inverse Laplace transforms, it is included with the supplemental material, which is primarily complex-variable in nature.

The use of the partial fraction decomposition in this text is exclusively for the means of decomposing a rational function[1] into a linear combination of terms that appear in a Laplace transform table (Table 8.1). If it is possible to do this, then it completely avoids the rather arduous exercise of evaluating the inverse Laplace transform, which is given by Definition 8.3.

We use a partial fraction decomposition to reduce the degree of the polynomial appearing in the denominator of a rational function, which we always assume to be proper.[2] Reducing the degree of the denominator is useful because, referring to Table 8.1, the distinguishing features of different elements of the table are the denominators of the functions.

The approach is to express a rational function in the form

$$\frac{N(s)}{D(s)} = \frac{N_1(s)}{D_1(s)} + \frac{N_2(s)}{D_2(s)} + \cdots + \frac{N_n(s)}{D_n(s)},$$

---

[1] A *rational function* is a function that may be written as a ratio of polynomials.

[2] A rational function is proper if the degree of the numerator is less than the degree of the denominator.

where the denominators satisfy

$$D(s) = D_1(s)D_2(s)\cdots D_n(s)$$

and are of a desired form, that is, for our purposes of the form of the denominator of elements in Table 8.1. The $N_i(s)$, then are simply polynomials in s that make the equality hold. First we state a proposition that is helpful for computing the $N_i(s)$.

Proposition A.1. *If the function is proper, then the order of each $N_i(s)$ will be less than the order of the corresponding $D_i(s)$.*

Proof. If $P(s)$ is a polynomial in s, let $\mathcal{O}(P(s))$ denote the order of $P(s)$. Writing the decomposition and then putting it over a common denominator gives

$$
\begin{aligned}
\frac{N(s)}{D(s)} &= \frac{N_1(s)}{D_1(s)} + \frac{N_2(s)}{D_2(s)} + \cdots + \frac{N_n(s)}{D_n(s)} \\
&= \frac{(N_1(s)D_2(s)D_3(s)\cdots D_n(s)) + (N_2(s)D_1(s)D_3(s)\cdots D_n(s)) + \cdots}{D_1(s)D_2(s)\cdots D_n(s)} \\
&= \frac{\sum_{i=1}^{n}\prod_{j=1, j\neq i}^{n} N_i(s)D_j(s)}{\prod_{i=1}^{n} D_i(s)}.
\end{aligned}
\tag{A.1}
$$

At least one term in the sum in the numerator on the right in Equation (A.1) must have the same order as $N(s)$. Because $\mathcal{O}(N(s)) < \mathcal{O}(D(s))$, then each term in the sum in the numerator on the left hand side of Equation (A.1) has lower order than $D(s)$. Because

$$\mathcal{O}(P_1(s)P_2(s)) = \mathcal{O}(P_1(s)) + \mathcal{O}(P_2(s))$$

then

$$\mathcal{O}(D(s)) = \sum_{j=1}^{n} \mathcal{O}(D_j(s)).$$

Then, for any $i \in \{1,\ldots n\}$,

$$
\begin{aligned}
\mathcal{O}(D(s)) &> \mathcal{O}(N_i(s)D_1(s)D_2(s)\cdots D_{i-1}(s)D_{i+1}(s)\cdots D_n(s)) \\
&= \mathcal{O}(N_i(s)) + \mathcal{O}(D_1(s)) + \cdots \mathcal{O}(D_{i-1}(s)) \\
&\quad + \mathcal{O}(D_{i+1}(s)) + \cdots + \mathcal{O}(D_n(s)) \\
&= \mathcal{O}(N_i(s)) + \mathcal{O}(D(s)) - \mathcal{O}(D_i(s)).
\end{aligned}
$$

Hence

$$\mathcal{O}(D_i(s)) > \mathcal{O}(N_i(s)).$$

The proof was a bit detailed, but what it tells us is that if we need to assume a form for the numerator of one of the fractions $N_i(s)$, the largest its order can be is one less than the order of the denominator $D_i(s)$.

Example A.1. Compute the partial fraction decomposition for

$$G(s) = \frac{s+1}{(s+2)(s+3)}$$

so the result is a linear combination of terms appearing in Table 8.1.

Because they correspond to an entry in Table 8.1, we pick the two denominators to be $D_1(s) = s+2$ and $D_2(s) = s+3$. Both of these have order 1 in s, thus then each numerator must be of order 0, that is, a constant. Hence

$$\frac{s+1}{(s+2)(s+3)} = \frac{c_1}{s+2} + \frac{c_2}{s+3}.$$

The task now is to determine c_1 and c_2 so that the equality holds. We present two ways to do this.

1. One way would be to put the right hand side over the common denominator and equate the resulting numerators:

$$\frac{s+1}{(s+2)(s+3)} = \frac{c_1}{s+2} + \frac{c_2}{s+3} = \frac{c_1(s+3)+c_2(s+2)}{(s+2)(s+3)}.$$

Because the equality must hold, the numerators must be equal. So

$$s+1 = c_1(s+3)+c_2(s+2).$$

Also, for this to hold for any s, the coefficient of each power of s must be equal so

$$s+1 = (c_1+c_2)s+(3c_1+2c_2)$$

requires

$$c_1+c_2 = 1, \quad 3c_1+2c_2 = 1$$

which gives $c_1 = -1$ and $c_2 = 2$. Hence,

$$\frac{s+1}{(s+2)(s+3)} = \frac{-1}{s+2} + \frac{2}{s+3}.$$

2. Another way to determine an equation to compute the numerators is to multiply each side of the expression by the denominator corresponding to the numerator we want to compute and then take the limit as s approaches the value of the pole location for that term. So for

$$\frac{s+1}{(s+2)(s+3)} = \frac{c_1}{s+2} + \frac{c_2}{s+3}$$

to determine c_1, multiply both sides by $(s+2)$ and let $s \to -2$; that is,

$$\frac{s+1}{(s+2)(s+3)}(s+2) = \frac{c_1}{s+2}(s+2) + \frac{c_2}{s+3}(s+2)$$

or

$$\frac{s+1}{s+3} = c_1 + \frac{c_2}{s+3}(s+2).$$

Evaluating this as $s \to -2$, then

$$\frac{-2+1}{-2+3} = c_1 + \frac{c_2}{-2+3}(-2+2).$$

The last term is zero, therefore $c_1 = -1$. Similarly,

$$\frac{s+1}{(s+2)(s+3)}(s+3) = \frac{c_1}{s+2}(s+3) + \frac{c_2}{s+3}(s+3)$$

or

$$\frac{s+1}{(s+2)} = \frac{c_1}{s+2}(s+3) + c_2.$$

Evaluating this as $s \to -3$ gives $c_2 = 2$. Hence

$$\frac{s+1}{(s+2)(s+3)} = \frac{-1}{s+2} + \frac{2}{s+3}.$$

Either approach works for complex conjugate poles as well.

Example A.2. Compute the partial fraction decomposition for

$$G(s) = \frac{1}{(s+2)(s^2+2s+2)}$$

so the result is a linear combination of terms appearing in Table 8.1. The roots for the second term in the denominator are $s = -1 \pm i$. We could factor it, but the form $(s+1)^2 + 1$ is what appears in Table 8.1. Hence we wish to determine c_1, c_2, and c_3 such that

$$\frac{1}{(s+2)(s^2+2s+2)} = \frac{c_1}{s+2} + \frac{c_2 s + c_3}{(s+1)^2 + 1}. \tag{A.2}$$

1. Combining the terms on the right-hand side gives

$$\frac{1}{(s+2)(s^2+2s+2)} = \frac{c_1\left[(s+1)^2 + 1\right] + (c_2 s + c_3)(s+2)}{(s+2)(s^2+2s+2)},$$

and equating the numerators gives

$$1 = c_1\left(s^2 + 2s + 2\right) + (c_2 s + c_3)(s+2)$$
$$= (c_1 + c_2)s^2 + (2c_1 + 2c_2 + c_3)s + (2c_1 + 2c_3).$$

This equality must hold for all s, thus the coefficients of each power of s must be equal so

$$c_1 + c_2 = 0, \quad 2c_1 + 2c_2 + c_3 = 0, \quad 2c_1 + 2c_3 = 1,$$

which gives

$$c_1 = \frac{1}{2}, \quad c_2 = -\frac{1}{2}, \quad c_3 = 0.$$

Hence,

$$\frac{1}{(s+2)(s^2+2s+2)} = \frac{1}{2(s+2)} - \frac{s}{2\left[(s+1)^2+1\right]}.$$

2. Alternatively, multiplying both sides of Equation (A.2) by $s+2$ and computing the limit as $s \to -2$ gives

$$c_1 = \lim_{s \to -2} \frac{1}{s^2+2s+2} = \frac{1}{2}.$$

Similarly multiplying both sides of Equation (A.2) by s^2+2s+2 and computing the limit as $s \to -1+i$ gives

$$\lim_{s \to -1+i} (c_2 s + c_3) = \lim_{s \to -1+i} \frac{1}{s+2}$$

which gives

$$c_2(-1+i) + c_3 = \frac{1}{-1+i+2} = \frac{1}{1+i} \frac{1-i}{1-i} = \frac{1-i}{2}.$$

Equating the real and imaginary parts gives

$$-c_2 + c_3 = \frac{1}{2}, \quad c_2 = -\frac{1}{2},$$

and hence $c_3 = 0$, which is the same answer as before.

The only real complication is when there are repeated factors in the denominator. In that case, the assumed form for the decomposition must include separate terms for that factor of all orders up to its multiplicity, as is illustrated in the following example.

Example A.3. To compute a partial fraction decomposition for

$$G(s) = \frac{1}{(s+1)(s^2+2)^3},$$

we must assume

$$G(s) = \frac{c_1}{s+1} + \frac{c_2 s + c_3}{(s^2+2)} + \frac{c_4 s + c_5}{(s^2+2)^2} + \frac{c_5 s + c_6}{(s^2+2)^3}.$$

Computing all the coefficients gives

$$G(s) = \frac{1}{27} \frac{1}{s+1} - \frac{1}{27} \frac{s-1}{(s^2+2)} - \frac{1}{9} \frac{s-1}{(s^2+2)^2} - \frac{1}{3} \frac{s-1}{(s^2+2)^3}.$$

Appendix B
Linear Algebra Review

This appendix reviews some basic concepts from linear algebra. In particular, the definition of a linear vector space and transformations between them are considered.

B.1 Linear Vector Spaces

The most fundamental object in linear algebra is a *vector space*. A vector space is a generalization of the usual notion of a collection of vectors in Euclidean space and is useful too because such a generalized space will have all the properties that the set of vectors has. Instead of simply defining a vector space, let us present a list of those so-called useful properties and give examples of sets of objects other than vectors that also exhibit them or examples of objects that do not satisfy them.

B.1.1 Properties of Vector Operations in the Euclidean Plane

The notation is abandoned subsequently, however, for this introductory section the common practice of denoting vectors with bold letters is used. Although a vector space is fundamentally a set, the important properties that define it as a vector space are related to operations on these vectors, particularly, how they add and how they are scaled. Specifically, define vector addition in the usual "head to tail" manner (illustrated in Figure B.1) and define

Property B.1. Vector addition is *commutative*; that is, for vectors $\mathbf{x}_1$ and $\mathbf{x}_2$,

$$\mathbf{x}_1 + \mathbf{x}_2 = \mathbf{x}_2 + \mathbf{x}_1.$$

This property is illustrated in the usual way in Figure B.1.

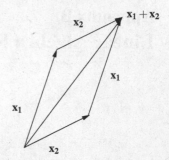

Fig. B.1 Vector addition is commutative.

Example B.1. An example of an operation that is not commutative is rigid body rotation. Consider the book illustrated on the left in Figure B.2 where the front cover is shaded and the top is indicated by arrows. If the book is first rotated about axis x_1 by an angle of $90°$ and then about axis x_2 by an amount $90°$, the final orientation is illustrated in the middle in Figure B.2. Positive directions of rotation are given by the right-hand rule. In contrast, if the body is rotated about axis x_2 by an amount $90°$ followed by a rotation about x_1 by an amount $90°$, the final orientation is illustrated on the right in Figure B.2.

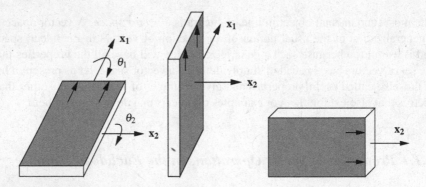

Fig. B.2 Two rotations of a book starting in the orientation in the left figure. The middle figure is after a rotation about x_1 and then x_2 by $90°$ each time. The right figure is a rotation about x_2 first and then x_1, which results in the book being in a different orientation.

If we use some sort of mathematical operation to represent these rotations, it may not commute, because one rotation followed by another rotation is, in general, not equal to the reverse order of rotations. It should not come as a surprise that rigid body rotations are often represented by matrices, and one rotation followed by another is represented by matrix multiplication, which does not commute.[1] An interested reader is referred to [12] for a complete exposition on rigid body kinematics and [35] for a more advanced treatment.

---

[1] Another popular representation for rotations is quaternions [53].

Property B.2. Vector addition is *associative*; that is, if $\mathbf{x}_1$, $\mathbf{x}_2$, and $\mathbf{x}_3$ are vectors, then

$$\mathbf{x}_1 + (\mathbf{x}_2 + \mathbf{x}_3) = (\mathbf{x}_1 + \mathbf{x}_2) + \mathbf{x}_3.$$

An example of a nonassociative operation is the cross-product in $\mathbb{R}^3$.

Example B.2. Let $\mathbf{i}$, $\mathbf{j}$, and $\mathbf{k}$ denote the usual coordinate axes in $\mathbb{R}^3$. Then observing that

$$\mathbf{i} \times \mathbf{j} = \mathbf{k}$$

and

$$\mathbf{i} \times \mathbf{i} = \mathbf{0},$$

then

$$\mathbf{i} \times (\mathbf{i} \times \mathbf{j}) = \mathbf{i} \times \mathbf{k} = -\mathbf{j}.$$

However,

$$(\mathbf{i} \times \mathbf{i}) \times \mathbf{j} = \mathbf{0} \times \mathbf{k} = \mathbf{0}.$$

Property B.3. Vector addition has an *identity element*; that is, there is a zero vector $\mathbf{0}$ such that for any vector $\mathbf{x}$,

$$\mathbf{x} + \mathbf{0} = \mathbf{x}.$$

In the case of vectors in Euclidean space, the zero vector has no length.

Property B.4. Vector addition has an *inverse*; that is, for each vector $\mathbf{x}$, there exists another vector denoted by $-\mathbf{x}$ such that

$$\mathbf{x} + (-\mathbf{x}) = \mathbf{0}.$$

In the case of vectors in Euclidean space, the additive inverse of a vector is a vector with the same length, but with the opposite orientation.

Property B.5. Scalar multiplication distributes over vector addition. So, for vectors $\mathbf{x}_1$ and $\mathbf{x}_2$, and a real number α,

$$\alpha (\mathbf{x}_1 + \mathbf{x}_2) = \alpha \mathbf{x}_1 + \alpha \mathbf{x}_2.$$

Example B.3. Considering the two vectors in Figure B.1 again, if we double the sum, it is equal to doubling the length of each vector first, and then adding them. This is illustrated on the left in Figure B.3.

Property B.6. Addition of two real numbers, a and b, distributes; that is,

$$(a + b)\mathbf{x} = a\mathbf{x} + b\mathbf{x}.$$

In other words, it does not matter if you add a and b first and then scale the vector, or if you multiply the vector individually by a and b and take the sum.

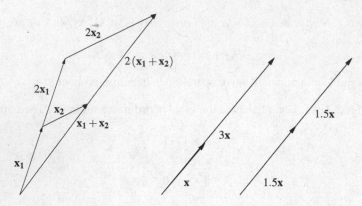

Fig. B.3 Distributive properties of vectors.

Example B.4. Considering the vector **x** illustrated on the right in Figure B.3, and the scalars $a = 1.5$ and $b = 1.5$, it is the case that

$$(1.5 + 1.5)\mathbf{x} = 3\mathbf{x},$$

as is illustrated on the right in Figure B.3.

Property B.7. Scalar multiplication of a vector is compatible with multiplication of real numbers; that is, for real numbers a and b

$$(ab)\mathbf{x} = a(b\mathbf{x}).$$

In other words, it does not matter if you multiply a and b together first and then scale the vector **x** or if you scale the vector by one of them followed by scaling by the other.

B.1.2 Definition and Examples of Vector Spaces

A vector space is simply any set where it is possible to add elements and scale them, as long as the manner in which they are added and scaled have properties similar to vector addition and scaling in Euclidean space. To add some degree of generality, we allow the vectors to be scaled by either real or complex numbers.

Definition B.1. Let the set[2] $\mathbb{F}$ be either $\mathbb{R}$ or $\mathbb{C}$ and let V be a set with

1. A mapping $V \times V \to V$ called *vector addition* and denoted by $\mathbf{x}_1 + \mathbf{x}_2$ for $\mathbf{x}_1$ and $\mathbf{x}_2 \in V$

---

[2] The set $\mathbb{F}$ is generally a *field*, but for the purposes of this book it is always the real or complex numbers.

2. A mapping $\mathbb{F} \times V \to V$ called *scalar multiplication* and denoted by $a\mathbf{x}$ for $a \in \mathbb{F}$ and $\mathbf{x} \in V$

where the mappings satisfy the following.

1. $\mathbf{x}_1 + \mathbf{x}_2 = \mathbf{x}_2 + \mathbf{x}_1$.
2. $(\mathbf{x}_1 + \mathbf{x}_2) + \mathbf{x}_3 = \mathbf{x}_1 + (\mathbf{x}_2 + \mathbf{x}_3)$.
3. there exists a $\mathbf{0} \in V$ such that $\mathbf{0} + \mathbf{x} = \mathbf{x}$ for all $\mathbf{x} \in V$.
4. for each $\mathbf{x} \in V$, there exists a $-\mathbf{x}$ such that $\mathbf{x} + (-\mathbf{x}) = \mathbf{0}$.
5. $(ab)\mathbf{x} = a(b\mathbf{x})$ for all $a, b \in \mathbb{F}$ and for all $\mathbf{x} \in V$.
6. $1\mathbf{x} = \mathbf{x}$ for all $\mathbf{x} \in V$.
7. $0\mathbf{x} = \mathbf{0}$ for all $\mathbf{x} \in V$.
8. $a(\mathbf{x}_1 + \mathbf{x}_2) = a\mathbf{x}_1 + b\mathbf{x}_2$ for all $a \in \mathbb{F}$ and for all $\mathbf{x}_1, \mathbf{x}_2 \in V$.
9. $(a + b)\mathbf{x} = a\mathbf{x} + b\mathbf{x}$ for all $a, b \in \mathbb{F}$ and for all $\mathbf{x} \in V$.

Example B.5. Consider the set of polynomials of the independent variable t with real coefficients and degree less than or equal to n. Denote this set by $P(t, n)$. Any element of $P(t, n)$ may be expressed as

$$\alpha_n t^n + \alpha_{n-1} t^{n-1} + \cdots + \alpha_1 t + \alpha_0 \in P(t, n).$$

If addition and scalar multiplication are defined in the usual manner, that is,

$$\begin{aligned}
&\left(\alpha_n t^n + \alpha_{n-1} t^{n-1} + \cdots + \alpha_1 t + \alpha_0\right) + \left(\beta_n t^n + \beta_{n-1} t^{n-1} + \cdots + \beta_1 t + \beta_0\right) \\
&= (\alpha_n + \beta_n) t^n + \cdots + (\alpha_1 + \beta_1) t + (\alpha_0 + \beta_0)
\end{aligned} \tag{B.1}$$

and

$$\beta\left(\alpha_n t^n + \alpha_{n-1} t^{n-1} + \cdots + \alpha_1 t + \alpha_0\right) = (\beta\alpha_n) t^n + \cdots + (\beta\alpha_1) t + (\beta\alpha_0), \tag{B.2}$$

then $P(t, n)$ is a vector space.

To actually prove this, we must verify each of the properties. This is generally a somewhat arduous exercise, but it is worth doing at least a few times.

1. For

$$p_1 = \alpha_n t^n + \alpha_{n-1} t^{n-1} + \cdots + \alpha_1 t + \alpha_0$$

and

$$p_2 = \beta_n t^n + \beta_{n-1} t^{n-1} + \cdots + \beta_1 t + \beta_0$$

we may write

$$p_1 + p_2 = (\alpha_n t^n + \cdots + \alpha_0) + (\beta_n t^n + \cdots + \beta_0) \tag{B.3}$$

$$= (\alpha_n + \beta_n) t^n + \cdots + (\alpha_1 + \beta_1) t + (\alpha_0 + \beta_0) \tag{B.4}$$

$$= (\beta_n + \alpha_n) t^n + \cdots + (\beta_1 + \alpha_1) t + (\beta_0 + \alpha_0) \tag{B.5}$$

$$= (\beta_n t^n + \cdots + \beta_0) + (\alpha_n t^n + \cdots + \alpha_0) \tag{B.6}$$

$$= p_2 + p_1. \tag{B.7}$$

These steps are justified as follows.

a. The step from Equation (B.3) to (B.4) is by the definition of vector addition in $P(t,n)$ given by Equation (B.1).

b. The step from Equation (B.4) to (B.5) is justified because the coefficients in the polynomial are real and real addition commutes.

c. The step from Equation (B.5) to (B.6) is by the definition of addition in $P(t,n)$.

Observe that, basically, addition of elements in $P(t,n)$ is defined in such a manner that the commutative property of addition of real numbers gives rise to the property that addition of two polynomials is also commutative.

2. For $p_1, p_2, p_3 \in P(t,n)$, where

$$p_1 = \alpha_n t^n + \cdots + \alpha_1 t + \alpha_0$$
$$p_2 = \beta_n t^n + \cdots + \beta_1 t + \beta_0$$
$$p_3 = \gamma_n t^n + \cdots + \gamma_1 t + \gamma_0$$

we have already that

$$(p_1 + p_2) = (\alpha_n + \beta_n) t^n + \cdots + (\alpha_1 + \beta_1) t + (\alpha_0 + \beta_0)$$

and

$$(p_2 + p_3) = (\beta_n + \gamma_n) t^n + \cdots + (\beta_1 + \gamma_1) t + (\beta_0 + \gamma_0).$$

Hence,

$$\begin{aligned}
(p_1 + p_2) + p_3 &= ((\alpha_n + \beta_n) + \gamma_n) t^n + \cdots + ((\alpha_0 + \beta_0) + \gamma_0) \\
&= (\alpha_n + (\beta_n + \gamma_n)) t^n + \cdots + (\alpha_0 + (\beta_0 + \gamma_0)) \\
&= p_1 + (p_2 + p_3).
\end{aligned}$$

Again, the associative property of vector addition basically follows from the definition of addition and the fact that real number addition is associative.

3. Define the zero polynomial as

$$p_0 = 0 t^n + \cdots + 0 t + 0.$$

Then for any other

$$p = \alpha_n t^n + \cdots + \alpha_1 t + \alpha_0$$

we have

$$\begin{aligned}
p_0 + p &= (0 + \alpha_n) t^n + \cdots + (0 + \alpha_1) t + (0 + \alpha_0) \\
&= \alpha_n t^n + \cdots + \alpha_1 t + \alpha_0 \\
&= p.
\end{aligned}$$

4. Because for any $\alpha \in \mathbb{R}$ there exists $-\alpha \in \mathbb{R}$, then for any

$$p = \alpha_n t^n + \cdots + \alpha_1 t + \alpha_0 \in P(t,n)$$

there exists a

$$(-p) = (-\alpha_n)t^n + \cdots + (-\alpha_1)t + (-\alpha_0) \in P(t,n)$$

such that

$$
\begin{aligned}
p + (-p) &= (\alpha_n t^n + \cdots + \alpha_1 t + \alpha_0) + ((-\alpha_n)t^n + \cdots + (-\alpha_1)t + (-\alpha_0)) \\
&= (\alpha_n - \alpha_n)t^n + \cdots + (\alpha_1 - \alpha_1)t + (\alpha_0 - \alpha_0) \\
&= 0t^n + \cdots + 0t + 0 \\
&= p_0.
\end{aligned}
$$

5. For

$$p = \alpha_n t^n + \cdots + \alpha_1 t + \alpha_0 \in P(t,n)$$

we have

$$
\begin{aligned}
(ab)p &= (ab)(\alpha_n t^n + \cdots + \alpha_1 t + \alpha_0) && \text{(B.8)} \\
&= ((ab)\alpha_n)t^n + \cdots + ((ab)\alpha_1)t + ((ab)\alpha_0) && \text{(B.9)} \\
&= (a(b\alpha_n))t^n + \cdots + (a(b\alpha_1))t + (a(b\alpha_0)) && \text{(B.10)} \\
&= a((b\alpha_n)t^n + \cdots + (b\alpha_1)t + (b\alpha_0)) && \text{(B.11)} \\
&= (a)(bp). && \text{(B.12)}
\end{aligned}
$$

The justification for each step is as follows.

a. The step from Equation (B.8) to (B.9) is the definition of scalar multiplication in $P(t,n)$ given by Equation (B.2).
b. The step from Equation (B.9) to (B.10) is justified because multiplication of real numbers is associative.
c. The step from Equation (B.10) to (B.11) is the definition of scalar multiplication in $P(t,n)$.

6. For

$$p = \alpha_n t^n + \cdots + \alpha_1 t + \alpha_0 \in P(t,n)$$

we have

$$
\begin{aligned}
1p &= 1(\alpha_n t^n + \cdots + \alpha_1 t + \alpha_0) \\
&= (1\alpha_n) + \cdots + (1\alpha_1)t + (1\alpha_0) \\
&= \alpha_n t^n + \cdots + \alpha_1 t + \alpha_0 \\
&= p.
\end{aligned}
$$

7. For

$$p = \alpha_n t^n + \cdots + \alpha_1 t + \alpha_0 \in P(t,n)$$

we have

$$0p = 0(\alpha_n t^n + \cdots + \alpha_1 t + \alpha_0)$$
$$= (0\alpha_n) + \cdots + (0\alpha_1)t + (0\alpha_0)$$
$$= 0t^n + \cdots + 0t + 0$$
$$= \mathbf{0}.$$

8. For

$$p_1 = \alpha_n t^n + \cdots + \alpha_1 t + \alpha_0, \quad p_2 = \beta_n t^n + \cdots + \beta_1 t + \beta_0$$

we have

$$a(p_1 + p_2) = a[(\alpha_n t^n + \cdots + \alpha_1 t + \alpha_0) + (\beta_n t^n + \cdots + \beta_1 t + \beta_0)] \tag{B.13}$$
$$= a[(\alpha_n + \beta_n)t^n + \cdots + (\alpha_0 + \beta_0)] \tag{B.14}$$
$$= [a(\alpha_n + \beta_n)t^n + \cdots + a(\alpha_0 + \beta_0)] \tag{B.15}$$
$$= [(a\alpha_n + a\beta_n)t^n + \cdots + (a\alpha_0 + a\beta_0)] \tag{B.16}$$
$$= ((a\alpha_n)t^n + \cdots + (a\alpha_1)t + (a\alpha_0) + (a\beta_n)t^n + \cdots$$
$$+ (a\beta_1)t + (a\beta_0)) \tag{B.17}$$
$$= ap_1 + ap_2. \tag{B.18}$$

Each step is justified as follows.

a. The step from Equation (B.13) to (B.14) is justified by the definition of addition in $P(t,n)$ given by Equation (B.1).
b. The step from Equation (B.14) to (B.15) is justified by the definition of scalar multiplication in $P(t,n)$ given by Equation (B.2).
c. The step from Equation (B.15) to (B.16) is justified by the fact that multiplication of real numbers distributes over addition of real numbers.
d. The step from Equation (B.16) to (B.17) is justified by the definition of addition in $P(t,n)$.

9. For

$$p = \alpha_n t^n + \cdots + \alpha_1 t + \alpha_0 \in P(t,n)$$

and $a, b \in \mathbb{R}$, we have

$$(a+b)p = (a+b)(\alpha_n t^n + \cdots + \alpha_1 t + \alpha_0) \tag{B.19}$$
$$= ((a+b)\alpha_n)t^n + \cdots + ((a+b)\alpha_0) \tag{B.20}$$
$$= (a\alpha_n + b\alpha_n)t^n + \cdots + (a\alpha_0 + b\alpha_0) \tag{B.21}$$
$$= (a\alpha_n)t^n + \cdots + (a\alpha_0) + (b\alpha_n)t^n + \cdots + (b\alpha_0) \tag{B.22}$$
$$= ap + bp. \tag{B.23}$$

Each step is justified as follows.

a. The step from Equation (B.19) to (B.20) is justified by the definition of scalar multiplication in $P(t,n)$ given by Equation (B.2).

b. The step from Equation (B.20) to (B.21) is justified by the fact that multiplication of real numbers distributes over addition of real numbers.

c. The step from Equation (B.21) to (B.22) is justified by the definition of vector addition in $P(t,n)$ given by Equation (B.1).

B.1.3 Linear Independence

Consider the set of vectors $\left\{\xi^1, \ldots, \xi^k\right\} \in \mathbb{R}^n$, that is, k vectors that are n elements "tall" such as

$$\xi^i = \begin{bmatrix} \xi_1^i \\ \xi_2^i \\ \vdots \\ \xi_n^i \end{bmatrix}.$$

Definition B.2 (Linear (in)dependence). The set $\left\{\xi^1, \ldots, \xi^k\right\}$ is *linearly dependent* if there exist scalars $\alpha_1, \ldots, \alpha_k$, where at least one $\alpha_i \neq 0$ such that

$$\alpha_1 \xi^1 + \alpha_2 \xi^2 + \cdots + \alpha_k \xi^k = \sum_{i=1}^{k} \alpha_i \xi^i = 0.$$

If the set is not linearly dependent, then it is *linearly independent*.

A simple example is in order.

Example B.6. Let $n = 3$ and

$$\xi^1 = \begin{bmatrix} 1 \\ 2 \\ 3 \end{bmatrix} \quad \xi^2 = \begin{bmatrix} 1 \\ 1 \\ 1 \end{bmatrix} \quad \xi^3 = \begin{bmatrix} 5 \\ 7 \\ 9 \end{bmatrix}.$$

Clearly, determining linear dependence or independence by inspection is not easy. So we try to solve

$$\alpha_1 \begin{bmatrix} 1 \\ 2 \\ 3 \end{bmatrix} + \alpha_2 \begin{bmatrix} 1 \\ 1 \\ 1 \end{bmatrix} + \alpha_3 \begin{bmatrix} 5 \\ 7 \\ 9 \end{bmatrix} = \begin{bmatrix} 0 \\ 0 \\ 0 \end{bmatrix}$$

or, as three scalar equations

$$\alpha_1 + \alpha_2 + 5\alpha_3 = 0$$
$$2\alpha_1 + \alpha_2 + 7\alpha_3 = 0$$
$$3\alpha_1 + \alpha_2 + 9\alpha_3 = 0.$$

A tedious calculation gives $\alpha_1 = 2$, $\alpha_2 = 3$, and $\alpha_3 = 1$, which determines that the set of vectors $\left\{\xi^1, \xi^2, \xi^3\right\}$ is linearly dependent.

An easier approach is to recall the following basic result from linear algebra [18].

Proposition B.1. *If $A \in \mathbb{R}^{n \times n}$ and if $\det(A) = 0$ then the set of vectors that are the columns of A are linearly dependent. Also, the set of vectors that are the rows of A are linearly dependent. If $\det(A) \neq 0$ then the columns and rows are linearly independent.*

Example B.7. Considering the system in Example B.6, an easy computation gives

$$\begin{vmatrix} 1 & 1 & 5 \\ 2 & 1 & 7 \\ 3 & 1 & 9 \end{vmatrix} = 0,$$

thus confirming the result from Example B.6 that the vectors are linearly dependent.

The primary utility of the notion of linear independence is that in an n-dimensional vector space, a set of n linearly independent vectors, $\{\xi^1, \ldots, \xi^n\}$, forms a *basis* for the vector space. Thus any vector in that space can be written as a linear combination: $\xi = \sum_{i=1}^{n} \alpha_i \xi^i$.

B.1.4 Eigenvalues and Eigenvectors

Given a matrix $A \in \mathbb{R}^{n \times n}$ and a vector $\xi \in \mathbb{R}^n$, the product $y = A\xi$ is simply another vector in $\mathbb{R}^n$. However, there are two classes of the vectors x that give a special result when multiplied into A. The first special case is then the resulting vector is all zeros and the second special case is when the resulting vector is just a scaled version of x. The following two definitions elaborate upon this.

Definition B.3. The *null space* of a matrix $A \in \mathbb{R}^{n \times n}$, denoted by $\mathcal{N}(A)$, is the set of all vectors $\xi \in \mathbb{R}^n$ such that

$$A\xi = 0.$$

In this case 0 is the vector in $\mathbb{R}^n$ full of n zeros.

Definition B.4. An *eigenvector* of a matrix $A \in \mathbb{R}^{n \times n}$ is a nonzero vector $\hat{\xi}$ such that

$$A\hat{\xi} = \lambda \hat{\xi}.$$

The number λ, which may be real or complex, is the associated *eigenvalue*.

To compute eigenvalues and eigenvectors, note that

$$A\hat{\xi} = \lambda \hat{\xi} \quad \Longrightarrow \quad A\hat{\xi} - \lambda \hat{\xi} = (A - \lambda I)\hat{\xi} = 0, \tag{B.24}$$

where I is the $n \times n$ identity matrix. By Cramer's rule, Equation (B.24) has a solution if and only if

$$\det(A - \lambda I) = 0. \tag{B.25}$$

Equation (B.25) is an nth-degree polynomial in λ with n solutions, and is called the *characteristic equation*. Thus, $A \in \mathbb{R}^{n \times n}$ has n eigenvalues. At this point, all we know is that there are n eigenvalues. Note that the eigenvalues may be all real and distinct, or some of them may be repeated or complex conjugate pairs.

To compute the eigenvalue associated with a particular eigenvalue λ, substitute the value for λ into Equation (B.24) and solve for each component of $\hat{\xi}$. As the following example illustrates, the eigenvector can only be determined up to a unique scaling factor.

Example B.8. Compute the eigenvalues and eigenvectors of

$$A = \begin{bmatrix} 1 & 2 \\ 1 & 3 \end{bmatrix}.$$

First, to compute the eigenvalues,

$$\det(A - \lambda I) = \det\left(\begin{bmatrix} 1 & 2 \\ 1 & 3 \end{bmatrix} - \lambda \begin{bmatrix} 1 & 0 \\ 0 & 1 \end{bmatrix}\right) = \det\left(\begin{bmatrix} 1-\lambda & 2 \\ 1 & 3-\lambda \end{bmatrix}\right)$$
$$= (1-\lambda)(3-\lambda) - 2 = \lambda^2 - 4\lambda + 1 = 0.$$

Thus,

$$\lambda = 2 \pm \sqrt{3}.$$

To compute the eigenvectors, substituting the two values for λ into Equation (B.25) gives

$$\left(A - \left(2+\sqrt{3}\right)I\right) = \begin{bmatrix} 1-2-\sqrt{3} & 2 \\ 1 & 3-2-\sqrt{3} \end{bmatrix} \begin{bmatrix} \xi_1 \\ \xi_2 \end{bmatrix}$$

which gives

$$\left(-1-\sqrt{3}\right)\xi_1 + 2\xi_2 = 0, \quad \xi_1 + \left(1-\sqrt{3}\right)\xi_2 = 0.$$

A quick computation shows that if we try to solve for one variable, say ξ_2, from one of the equations and substitute into the other equation, we will end up with the degenerate equation $0 = 0$. This is precisely due to the fact that we are trying to solve a system of linearly dependent equations. Thus there are an infinite number of solutions.

The most straightforward approach may be to set one of the variables equal to one and solve for the others. So, in this example, arbitrarily let $\xi_2 = 1$. Both equations then give $\xi_1 = \sqrt{3} - 1$, and hence the eigenvector corresponding to the eigenvalue $\lambda = 2 + \sqrt{3}$ is

$$\hat{\xi} = \begin{bmatrix} \sqrt{3} - 1 \\ 1 \end{bmatrix}.$$

Note that any vector of the form

$$\hat{\xi} = \alpha \begin{bmatrix} \sqrt{3} - 1 \\ 1 \end{bmatrix},$$

where α is a real or complex number is also an eigenvector corresponding to the eigenvalue $\lambda = 2 + \sqrt{3}$.

A similar computation (and again arbitrarily setting $\xi_2 = 1$) gives

$$\xi = \begin{bmatrix} -\sqrt{3} - 1 \\ 1 \end{bmatrix}$$

as an eigenvector corresponding to the eigenvalue $\lambda = 2 - \sqrt{3}$.

In order to be more systematic in the approach to computing eigenvectors recall that to solve a set of linear equations

$$Ax = b,$$

where $A \in \mathbb{R}^{n \times n}$, $b, x \in \mathbb{R}^n$ where A and b are given and x is to be determined, one approach is to construct the augmented matrix

$$[A|b]$$

and use row reduction operations to convert the left part of the augmented matrix to a convenient form (typically triangular form). In the case of determining eigenvectors, b will be a column of zeros, so the problem will be somewhat simpler. The details of the approach are illustrated by the following example.

Example B.9. Determine the eigenvalues and eigenvectors of

$$A = \begin{bmatrix} 1 & 0 & 0 & 0 \\ -1 & 2 & 0 & 0 \\ -1 & 0 & 1 & 1 \\ -1 & 0 & -1 & 3 \end{bmatrix}.$$

We have

$$\det(A - \lambda I) = (1 - \lambda) \begin{vmatrix} 2 - \lambda & 0 & 0 \\ 0 & 1 - \lambda & 1 \\ 0 & -1 & 3 - \lambda \end{vmatrix}$$

$$= (1 - \lambda)(2 - \lambda) \begin{vmatrix} 1 - \lambda & 1 \\ -1 & 3 - \lambda \end{vmatrix}$$

$$= (1 - \lambda)(2 - \lambda)((1 - \lambda)(3 - \lambda) + 1)$$

$$= (1 - \lambda)(2 - \lambda)(\lambda^2 - 4\lambda + 4)$$

$$= (1 - \lambda)(2 - \lambda)(\lambda^2 - 4\lambda + 4)$$

$$= (1 - \lambda)(2 - \lambda)^3.$$

So, $\lambda = 1$ is an eigenvalue and $\lambda = 2$ is an eigenvalue that is repeated three times.

Note that, in general, for matrices larger than two by two we are not able to do such computations by hand. It was only due to the particular structure of the way the zeros were arranged in A that allowed us to do it in this example. In general, for matrices larger than 2×2 using a computer program or calculator will be necessary to compute the eigenvalues, which are the roots of the characteristic equation.

Now, to compute the eigenvectors, substituting $\lambda = 1$ into $(A - \lambda I)\hat{\xi}^1 = 0$ gives

$$\begin{bmatrix} 0 & 0 & 0 & 0 \\ -1 & 1 & 0 & 0 \\ -1 & 0 & 0 & 1 \\ -1 & 0 & -1 & 2 \end{bmatrix} \begin{bmatrix} \hat{\xi}_1^1 \\ \hat{\xi}_2^1 \\ \hat{\xi}_3^1 \\ \hat{\xi}_4^1 \end{bmatrix} = \begin{bmatrix} 0 \\ 0 \\ 0 \\ 0 \end{bmatrix}$$

which, in augmented matrix form is

$$\begin{bmatrix} 0 & 0 & 0 & 0 & | & 0 \\ -1 & 1 & 0 & 0 & | & 0 \\ -1 & 0 & 0 & 1 & | & 0 \\ -1 & 0 & -1 & 2 & | & 0 \end{bmatrix} .$$

Interchanging the first and fourth rows gives

$$\begin{bmatrix} -1 & 0 & -1 & 2 & | & 0 \\ -1 & 1 & 0 & 0 & | & 0 \\ -1 & 0 & 0 & 1 & | & 0 \\ 0 & 0 & 0 & 0 & | & 0 \end{bmatrix}$$

and subtracting the first row from the second and third rows gives

$$\begin{bmatrix} -1 & 0 & -1 & 2 & | & 0 \\ 0 & 1 & 1 & -2 & | & 0 \\ 0 & 0 & 1 & -1 & | & 0 \\ 0 & 0 & 0 & 0 & | & 0 \end{bmatrix}$$

which is in upper triangular form. If we choose $\hat{\xi}_4^1 = 1$, then from the third row we have $\hat{\xi}_3^1 = 1$. Substituting both of these values into the second row gives $\hat{\xi}_2^1 = 1$ and finally the first row gives $\hat{\xi}_1^1 = 1$. Hence

$$\xi^1 = \begin{bmatrix} 1 \\ 1 \\ 1 \\ 1 \end{bmatrix} .$$

Now, for $\lambda = 2$ we have $\det(A - 2I)\hat{\xi} = 0$ as

$$\begin{bmatrix} -1 & 0 & 0 & 0 \\ -1 & 0 & 0 & 0 \\ -1 & 0 & -1 & 1 \\ -1 & 0 & -1 & 1 \end{bmatrix} \begin{bmatrix} \hat{\xi}_1^2 \\ \hat{\xi}_2^2 \\ \hat{\xi}_3^2 \\ \hat{\xi}_4^2 \end{bmatrix} = \begin{bmatrix} 0 \\ 0 \\ 0 \\ 0 \end{bmatrix} ,$$

or, without elaborating on all the details, the row reductions give

$$\begin{bmatrix} -1 & 0 & 0 & 0 & | & 0 \\ -1 & 0 & 0 & 0 & | & 0 \\ -1 & 0 & -1 & 1 & | & 0 \\ -1 & 0 & -1 & 1 & | & 0 \end{bmatrix} \iff \begin{bmatrix} -1 & 0 & 0 & 0 & | & 0 \\ 0 & 0 & 0 & 0 & | & 0 \\ -1 & 0 & -1 & 1 & | & 0 \\ 0 & 0 & 0 & 0 & | & 0 \end{bmatrix} \iff \begin{bmatrix} -1 & 0 & 0 & 0 & | & 0 \\ -1 & 0 & -1 & 1 & | & 0 \\ 0 & 0 & 0 & 0 & | & 0 \\ 0 & 0 & 0 & 0 & | & 0 \end{bmatrix}$$

$$\iff \begin{bmatrix} -1 & 0 & 0 & 0 & | & 0 \\ 0 & 0 & -1 & 1 & | & 0 \\ 0 & 0 & 0 & 0 & | & 0 \\ 0 & 0 & 0 & 0 & | & 0 \end{bmatrix}. \quad \text{(B.26)}$$

The procedure we adopt is the following.

1. Inspecting each row in the reduced matrix, we identify the variables as not free if they are the first nonzero term in any row. So, in the preceding matrix, the components $\hat{\xi}_1$ and $\hat{\xi}_3$ are not free.
2. The remaining variables are free. Choose one of the free variables to be equal to one and the rest of the free variables to be equal to zero and compute the remaining components. This gives one eigenvector.
3. To compute another linearly independent eigenvector, choose another of the free variables to be one and the rest to be zero, and compute the remaining components. Continue through the entire list of free variables. This results in a linearly independent set of eigenvectors.

Returning to the example, $\hat{\xi}_4^2$ and $\hat{\xi}_2^2$ are free. Choosing $\hat{\xi}_4^2 = 1$, $\hat{\xi}_2^2 = 0$, $\hat{\xi}_4^3 = 0$, and $\hat{\xi}_2^3 = 1$ gives

$$\hat{\xi}^2 = \begin{bmatrix} 0 \\ 0 \\ 1 \\ 1 \end{bmatrix}, \quad \hat{\xi}^3 = \begin{bmatrix} 0 \\ 1 \\ 0 \\ 0 \end{bmatrix}.$$

B.2 Matrix Computations

It is necessary to be able to compute matrix determinants and inverses. This section reviews how to do so. It is the way to do it by hand and is also the way that a computer program does it.

B.2.1 Computing Determinants

Define the determinant of a 2×2 matrix

$$A = \begin{bmatrix} a_{11} & a_{12} \\ a_{21} & a_{22} \end{bmatrix}$$

by

$$\det A = a_{11}a_{22} - a_{12}a_{21}.$$

For matrices larger than 2×2, we use the following theorem from [18].

Theorem B.1. *Let $A = (a_{ij})$ be an $n \times n$ matrix, where $n \geq 2$. Let A_{ij} be the $(n-1) \times (n-1)$ matrix formed by deleting row i and column j from A. Defining the cofactor*

$$c_{ij} = (-1)^{i+j} \det A_{ij}$$

we then have the expansion by row i:

$$\det A = a_{i1}c_{i1} + a_{i2}c_{i2} + \cdots + a_{in}c_{in}$$

and we have the expansion by column j:

$$\det A = a_{1j}c_{1j} + a_{2j}c_{2j} + \cdots + a_{nj}c_{nj}.$$

The 3×3 case should probably be memorized.

Corollary B.1. *For*

$$A = \begin{bmatrix} a_{11} & a_{12} & a_{13} \\ a_{21} & a_{22} & a_{23} \\ a_{31} & a_{32} & a_{33} \end{bmatrix}$$

applying Theorem B.1 gives

$$\det A = a_{11}(a_{22}a_{33} - a_{23}a_{32}) - a_{12}(a_{21}a_{33} - a_{23}a_{31}) + a_{13}(a_{21}a_{32} - a_{22}a_{31}).$$

Example B.10. Compute the determinant of

$$A = \begin{bmatrix} -2 & 0 & 1 & -3 \\ -1 & -1 & 1 & -3 \\ 2 & -2 & -3 & -1 \\ 0 & 0 & 0 & -4 \end{bmatrix}.$$

Clearly, it is best to expand on a row or column with the most zeros. Expanding across the fourth row gives

$$\det A = 0(-1)^{4+1}\det A_{41} + 0(-1)^{4+2}\det A_{42} + 0(-1)^{4+3}\det A_{43}$$
$$+ -4(-1)^{4+4}\det A_{44}$$

$$= -4 \begin{vmatrix} -2 & 0 & 1 \\ -1 & -1 & 1 \\ 2 & -2 & -3 \end{vmatrix} = -4[-2(3+2) - 0(4-2) + 1(2+2)] = 24.$$

B.2.2 Computing a Matrix Inverse

The following theorem is from [18] and is useful to compute matrix inverses by hand.

Theorem B.2. *Let A be an $n \times n$ matrix, with $n > 1$. Let A_{ij} be the $(n-1) \times (n-1)$ matrix formed by deleting row i and column j from A. Define the cofactor matrix*

$$\mathrm{cof} A = C = \left[(-1)^{i+j} \det A_{ij} \right] \qquad (i,j = 1,\ldots,n).$$

Let $\Delta = \det A$. Then if $\Delta \neq 0$

$$A^{-1} = \frac{1}{\Delta} C^T.$$

Example B.11. Compute the inverse of

$$A = \begin{bmatrix} -2 & 0 & 1 & -3 \\ -1 & -1 & 1 & -3 \\ 2 & -2 & -3 & -1 \\ 0 & 0 & 0 & -4 \end{bmatrix}.$$

From Example B.10 we know that $\Delta = 24$. The terms in the cofactor matrix are

$$C_{11} = (-1)^{1+1} \begin{vmatrix} -1 & 1 & -3 \\ -2 & -3 & -1 \\ 0 & 0 & -4 \end{vmatrix} = (-1)^{1+1} (-1)^{3+3} [-4(3+2)] = -20,$$

$$C_{12} = (-1)^{1+2} \begin{vmatrix} -1 & 1 & -3 \\ 2 & -3 & -1 \\ 0 & 0 & -4 \end{vmatrix} = -1[-1(12) - 1(-8) - 3(0)] = 4,$$

$$C_{13} = (-1)^{1+3} \begin{vmatrix} -1 & -1 & -3 \\ 2 & -2 & -1 \\ 0 & 0 & -4 \end{vmatrix} = (1)[-1(8) - (-1)(-8) + 0] = -16,$$

$$C_{14} = (-1)^{1+4} \begin{vmatrix} -1 & -1 & 1 \\ 2 & -2 & 3 \\ 0 & 0 & 0 \end{vmatrix} = 0,$$

and so on. Completing the tedious calculations gives

$$A^{-1} = \frac{1}{24} \begin{bmatrix} -20 & 4 & -16 & 0 \\ 8 & -16 & 16 & 0 \\ -4 & -4 & -8 & 0 \\ 22 & -14 & 26 & -6 \end{bmatrix}.$$

Appendix C
Detailed Computations

This appendix contains some of the important, but detailed or cumbersome computations, inclusion of which in the main text perhaps would be distracting.

C.1 Computations Related to Fourier Series

Proposition C.1. *The integral*

$$\int_0^L \sin\left(\frac{n\pi x}{L}\right) \sin\left(\frac{m\pi x}{L}\right) dx = \begin{cases} 0, & m \neq n, \\ \frac{L}{2}, & m = n, \end{cases}$$

$m, n \in \mathbb{N}$.

Proof. The case for $n \neq m$ is shown simply by integration by parts.

$$\int_0^L \sin\left(\frac{n\pi x}{L}\right) \sin\left(\frac{m\pi x}{L}\right) dx$$

$$= -\frac{L}{m\pi} \sin\left(\frac{n\pi x}{L}\right) \cos\left(\frac{m\pi x}{L}\right)\Big|_0^L + \frac{n}{m} \int_0^L \cos\left(\frac{n\pi x}{L}\right) \cos\left(\frac{m\pi x}{L}\right) dx$$

$$= \frac{n}{m} \int_0^L \cos\left(\frac{n\pi x}{L}\right) \cos\left(\frac{m\pi x}{L}\right) dx,$$

which is clearly asking us to integrate by parts again. So

$$\frac{n}{m} \int_0^L \cos\left(\frac{n\pi x}{L}\right) \cos\left(\frac{m\pi x}{L}\right) dx$$

$$= \frac{n}{m} \frac{L}{m\pi} \cos\left(\frac{n\pi x}{L}\right) \sin\left(\frac{m\pi x}{L}\right)\Big|_0^L + \left(\frac{n}{m}\right)^2 \int_0^L \sin\left(\frac{m\pi x}{L}\right) \sin\left(\frac{n\pi x}{L}\right) dx.$$

Thus

$$\int_0^L \sin\left(\frac{n\pi x}{L}\right) \sin\left(\frac{m\pi x}{L}\right) dx = \left(\frac{n}{m}\right)^2 \int_0^L \sin\left(\frac{n\pi x}{L}\right) \sin\left(\frac{m\pi x}{L}\right) dx,$$

and if $n \neq m$ the integral must be zero.

The case where $n = m$ is simply done by a trigonometric substitution. Recall that

$$\cos 2\theta = \cos\theta\cos\theta - \sin\theta\sin\theta = \cos^2\theta - \sin^2\theta = \left(1 - \sin^2\theta\right) - \sin^2\theta$$
$$= 1 - 2\sin^2\theta,$$

so $\sin^2\theta = (1 - \cos 2\theta)/2$. Hence

$$\int_0^L \sin^2\left(\frac{n\pi x}{L}\right) dx = \frac{1}{2} \int_0^L 1 - \cos\left(\frac{2n\pi x}{L}\right) dx = \frac{x}{2}\Big|_0^L - \frac{L}{4n\pi}\sin\left(\frac{2n\pi x}{L}\right)\Big|_0^L = \frac{L}{2}.$$

$\square$

Note that $\mathbb{N}$ is the set of *natural numbers*, $\mathbb{N} = \{1, 2, 3, \ldots\}$. The following two results are proved similarly.

Proposition C.2. *The integral*

$$\int_0^L \cos\left(\frac{n\pi x}{L}\right) \cos\left(\frac{m\pi x}{L}\right) dx = \begin{cases} 0, & m \neq n, \\ \frac{L}{2}, & m = n. \end{cases}$$

Proposition C.3. *The integral*

$$\int_0^L \sin\left(\frac{n\pi x}{L}\right) \cos\left(\frac{m\pi x}{L}\right) dx = 0$$

for all $m, n \in \mathbb{N}$.

Proof. The proofs to these two propositions obviously mirror the proof to Proposition C.1.

Appendix D
Some Theorems

D.1 Existence and Uniqueness Theorems

This section presents the basic theorem for first-order, ordinary differential equations which is adapted from [47]. As one would expect, for the first-order ordinary differential equation

$$\frac{\mathrm{d}x}{\mathrm{d}t}(t) = f(x(t), t)$$

whether a solution exists and is unique depends on the properties of the function f, and as long as f is not too crazy a unique solution probably exists. In fact, as the following theorem shows, this is the case.

Theorem D.1. *For the first-order ordinary differential equation with initial condition*

$$\dot{x} = f(x, t), \quad x(0) = x_0 \tag{D.1}$$

assume $f(x, t)$ is a piecewise continuous function of t and for each $T \in [0, \infty)$ there exist finite constants k_T and h_T such that for all $t \in [0, T]$,

$$\begin{aligned} |f(x, t) - f(\bar{x}, t)| &\leq k_T |x - \bar{x}| \\ |f(x_0, t)| &\leq h_T. \end{aligned} \tag{D.2}$$

Then Equation (D.1) has exactly one solution on the interval $[0, T]$ for all $T < \infty$.

Equation (D.2) is called the *Lipschitz condition*. The Lipschitz condition may not always be easy to check, so the following lemma is useful.

Lemma D.1. *A function that is continuously differentiable with a bounded first derivative satisfies the Lipschitz condition.*

Example D.1. Consider

$$\dot{x} = x^2 + 35t + \sin t$$

with $x(0) = 6$. For this equation

$$f(x,t) = x^2 + 35t + \sin t$$

and inasmuch as

$$\frac{\partial f}{\partial x} = 2x$$

is bounded, then this system has a unique solution.

Example D.2. Consider

$$\dot{x} = 3x^{\frac{2}{3}}, \quad x(0) = 0. \tag{D.3}$$

Because near the origin, the graph of $x^{2/3}$ is vertical it does not satisfy the Lipschitz condition. Hence we are not able to conclude that Equation (D.3) has a unique solution.

Be careful to properly interpret the theorem. It says that if certain properties on the right-hand side of the differential equation hold, then a unique solution exists. In other words it provides a *sufficient condition* for existence and uniqueness. However, the conditions are not *necessary*; in other words, even if the conditions are not satisfied it may be the case that a unique solution exists.

D.2 Series Solutions About a Singular Point

The following two theorems are from [8]. Theorem D.2 gives all the solutions to Euler's equation and Theorem D.3 gives solutions for all the cases for series solutions about a regular singular point.

Theorem D.2. *To determine the general solution to*

$$x^2 \frac{d^2 y}{dx^2}(x) + p_0 x \frac{dy}{dx}(x) + q_0 y(x) = 0 \tag{D.4}$$

for $x > 0$, assume a solution of the form

$$y(x) = x^m,$$

and substitute it into Equation (D.4). This results in a quadratic equation for m, with roots m_1 and m_2.

- *If the values determined for m are real and unequal, then the general solution is*

$$y(x) = c_1 x^{m_1} + c_2 x^{m_2}.$$

- *If the values determined for m are real and equal, $m_1 = m_2 = m$, then the general solution is*

$$y(x) = c_1 x^m + c_2 x^m \ln x.$$

- If the values determined for m are complex conjugates given by $m = \mu \pm i\omega$, then the general solution is given by

$$y(x) = c_1 x^\mu \cos(\omega \ln(x)) + c_2 x^\mu \sin(\omega \ln(x)).$$

Theorem D.3. *Let $x = 0$ be a regular singular point of*

$$x^2 \frac{d^2 y}{dx^2}(x) + x\hat{p}(x)\frac{dy}{dx}(x) + \hat{q}(x)y(x) = 0, \quad \text{(D.5)}$$

where $\hat{p}(x)$ and $\hat{q}(x)$ have Taylor series

$$\hat{p}(x) = p_0 + p_1 x + \sum_{n=2}^{\infty} \frac{p_n}{n!} x^n, \quad \hat{q}(x) = q_0 + q_1 x + \sum_{n=2}^{\infty} \frac{q_n}{n!} x^n,$$

that converge for $|x| < \rho$ and let m_1 and m_2 be the solutions to

$$m(m-1) + p_0 m + q_0 = 0.$$

Then, for either $-\rho < x < 0$ or $0 < x < \rho$, Equation (D.5) has two linearly independent solutions, $y_1(x)$ and $y_2(x)$ of the form:

1. *If $m_1 - m_2$ is neither an integer nor zero, then*

$$y_1(x) = |x|^{m_1}\left(1 + \sum_{n=1}^{\infty} a_n(m_1)x^n\right), \quad y_2(x) = |x|^{m_2}\left(1 + \sum_{n=1}^{\infty} a_n(m_2)x^n\right);$$

2. *If $m_1 = m_2$, then*

$$y_1(x) = |x|^{m_1}\left(1 + \sum_{n=1}^{\infty} a_n(m_1)x^n\right),$$

$$y_2(x) = y_1(x)\ln|x| + |x|^{m_1}\sum_{n=1}^{\infty} b_n(m_1)x^n;$$

3. *If $m_1 - m_2$ is a positive integer, then*

$$y_1(x) = |x|^{m_1}\left(1 + \sum_{n=1}^{\infty} a_n(m_1)x^n\right),$$

$$y_2(x) = ay_1(x)\ln|x| + |x|^{m_2}\left(1 + \sum_{n=1}^{\infty} c_n(m_2)x^n\right).$$

The coefficients $a_n(m_1)$, $a_n(m_2)$, $b_n(m_1)$, $c_n(m_2)$, and a can be determined by substituting the form of the series solutions for y into Equation (D.5) and equating the coefficients of the powers of x.

Appendix E
Example Programs

E.1 C Programs

Typically to compile a C program on a UNIX or LINUX platform, type gcc example.c -1m in a terminal and then type a.out to execute the program. In a terminal on a Windows or MacOS platform, the syntax is similar. In an integrated development environment, there are typically user-interface buttons to compile and execute the program.

E.1.0.1 Program for Example 1.28

```
/*  Example C program to determine an approximate solution to
 *
 *  x' = sin(2 t), x(0) = 3
 *
 *  using Euler's method.
 */

#include<stdio.h>
#include<math.h>

int main() {

  int n;
  float x,t,dt,f;
  FILE *fp;

  fp = fopen("eulerexample05.d","w");

  n = 0;
  dt = 0.5;
  x = 3.0;

  for(t=0;t<10;t+=dt) {
```

```
      f = sin(2.0*t);
      fprintf(fp,"%f \t %d \t %f \t %f \t ",t,n,x,f);
      x += f*dt;
      fprintf(fp,"%f \t %f\n",x,7.0/2.0 - cos(2.0*(t+dt))/2.0);
      n++;
   }
   fclose(fp);

   return 0;
}
```

E.1.0.2 Program for Example 1.29

```
/*  Example C program to determine an approximate solution to
 *
 *  x' = 75 x (1 - x)
 *
 *  using Euler's method.
 */

#include<stdio.h>
#include<math.h>

int main() {

   int n,points=1000,count=1;
   double x,t,dt,f,limit;
   FILE *fp;

   fp = fopen("output00001.d","w");

   n = 0;
   dt = 0.0001;
   limit = 1/dt/points;
   x = 1.0/(1.0+exp(75.0));

   for(t=-1;t<=1;t+=dt) {
      f = 75.0*x*(1-x);
      if(count > limit) {
         fprintf(fp,"%f\t%d\t%f\t%f\t%f\n ",t,n,x,f,x+f*dt);
         count = 1;
      } else {
         count++;
      }
      x += f*dt;
      n++;
   }
   fclose(fp);

   return 0;
}
```

E.1.0.3 Program for Example 1.30

```
/*  Example C program to determine an approximate solution to
 *
 *  x'' + sin(t) x' + cos(t) x = exp(-5*t)
 *
 *  using Euler's method.
 */

#include<stdio.h>
#include<math.h>

int main() {

  int n;
  float x[2],t,dt,f[2];
  FILE *fp;

  fp = fopen("output.d","w");

  n = 0;
  dt = 0.01;
  x[0] = 2;
  x[1] = 5;

  for(t=0;t<30;t+=dt) {
    f[0] = x[1];
    f[1] = exp(-5*t) - sin(t) * x[1] - cos(t)*x[0];
    fprintf(fp,"%f\t%d\t%f\t%f\t%f\t%f\t%f\t%f\n ",t,n,
    x[0],f[0],x[1],f[1],x[0]+f[0]*dt,x[1]+f[1]*dt);
    x[0]  += f[0]*dt;
    x[1]  += f[1]*dt;
    n++;
  }
  fclose(fp);

  return 0;
}
```

E.1.0.4 Program for Example 12.1

```
/*  Example C program to determine an approximate solution to
 *
 *  x' = 5 x, x(0) = 1
 *
 *  using Euler's method.  The exact solution is also printed to the
 *  data file for comparison purposes.
 */

#include<stdio.h>
#include<math.h>
```

```c
int main() {

  float x,t,dt;
  FILE *fp;

  fp = fopen("data.d","w");

  dt = 0.1;
  x = 1.0;

  for(t=0;t<=1;t+=dt) {
    fprintf(fp,"%f\t%f\t%f\n",t,x,exp(5.0*t));
    x += 5.0*x*dt;
  }
  fprintf(fp,"%f\t%f\t%f\n",t,x,exp(5.0*t));
  fclose(fp);

  return 0;
}
```

E.1.0.5 Program for Example 12.2

```c
/*  Example C program to determine an approximate solution to
 *
 *  x' = -sin t, x(0) = 1
 *
 *  using Euler's method.  The exact solution is also printed to the
 *  data file for comparison purposes.
 */

#include<stdio.h>
#include<math.h>

int main() {

  float x,t,dt;
  FILE *fp;

  fp = fopen("eulererroranalysis2.d","w");

  dt = 1.0;
  x = 1.0;

  for(t=0;t<30;t+=dt) {
    fprintf(fp,"%f\t%f\t%f\n",t,x,cos(t));
    x += -sin(t)*dt;
  }
  fprintf(fp,"%f\t%f\t%f\n",t,x,cos(t));
  fclose(fp);
```

```
   return 0;
}
```

E.1.0.6 Program for Example 12.3

```c
/*  Example C program to determine an approximate solution to
 *
 *  x' = 5 x, x(0) = 1
 *
 *  using a second order Taylor series expansion.  The exact solution
 *  is also printed to the data file for comparison purposes.
 */

#include<stdio.h>
#include<math.h>

int main() {

  float x,t,dt;
  FILE *fp;

  fp = fopen("secondordertaylor.d","w");

  dt = 0.1;
  x = 1.0;

  for(t=0;t<=1;t+=dt) {
    fprintf(fp,"%f\t%f\t%f\n",t,x,exp(5.0*t));
    x += 5.0*x*dt + 25.0/2.0*x*pow(dt,2);
  }
  fprintf(fp,"%f\t%f\t%f\n",t,x,exp(5.0*t));
  fclose(fp);

  return 0;
}
```

E.1.0.7 Program for Example 12.4

```c
/*  Example C program to determine an approximate solution to
 *
 *  x' = -x^3 + sin(t x), x(0) = 1
 *
 *  using a second order Taylor series expansion.
 */

#include<stdio.h>
#include<math.h>

int main() {
```

```
  float x,t,dt;
  FILE *fp;

  fp = fopen("secondordertaylor2.d","w");

  dt = 0.4;
  x = 1.0;

  for(t=0;t<5;t+=dt) {
    fprintf(fp,"%f\t%f\n",t,x);
    x += (-pow(x,3) + sin(t*x))*dt
      + 1.0/2.0*((-3*pow(x,2) + t*cos(t*x))*(-pow(x,3) + sin(t*x))
+ x*cos(t*x))*pow(dt,2);
  }
  fprintf(fp,"%f\t%f\n",t,x);
  fclose(fp);

  return 0;
}
```

E.1.0.8 Program for Example 12.6

```
/*  Example C program to determine an approximate solution to
 *
 *  x' = -x^3 + sin(t x), x(0) = 1
 *
 *  using the second order Runge-Kutta (or improved Euler) method. The
 *  exact solution is also printed to the data file for comparison
 *  purposes
 */

#include<stdio.h>
#include<math.h>

int main() {

  float x,t,dt;
  FILE *fp;

  fp = fopen("rk2.d","w");

  dt = 0.1;
  x = 1.0;

  for(t=0;t<=1;t+=dt) {
    fprintf(fp,"%f\t%f\t%f\n",t,x,exp(5.0*t));
    x += 5.0*x*dt;
  }
  fprintf(fp,"%f\t%f\t%f\n",t,x,exp(5.0*t));
  fclose(fp);
```

```
   return 0;
}
```

E.1.0.9 Program for Example 12.7

```
/*  Example C program to determine an approximate solution to
 *
 *  x' = -x^3 + sin(t x), x(0) = 1
 *
 *  using the second-order Runge-Kutta method.
 */


#include<stdio.h>
#include<math.h>

double f(double x, double t);

int main() {

  double x,t,dt;
  FILE *fp;

  fp = fopen("secondorderrk2.d","w");

  dt = 0.4;
  x = 1.0;

  for(t=0;t<5;t+=dt) {
    fprintf(fp,"%f\t%f\n",t,x);
    x += dt/2*(f(x,t) + f(x+f(x,t)*dt,t+dt));
  }
  fprintf(fp,"%f\t%f\n",t,x);
  fclose(fp);

  return 0;
}

double f(double x, double t) {
  return -pow(x,3) + sin(t*x);
}
```

E.1.0.10 Program for Example 12.8

```
/*  Example C program to determine an approximate solution to
 *
 *  x' = -x^3 + sin(t x), x(0) = 1
 *
```

```
*   using the third-order Runge-Kutta method.
*/


#include<stdio.h>
#include<math.h>

double f(double x, double t);

int main() {

  double x,t,dt;
  double v1,v2,v3;
  FILE *fp;

  fp = fopen("rk3.d","w");

  dt = 0.5;
  x = 1.0;

  for(t=0;t<5;t+=dt) {
    fprintf(fp,"%f\t%f\n",t,x);
    v1 = f(x,t)*dt;
    v2 = f(x+v1/2.0,t+dt/2.0)*dt;
    v3 = f(x+2.0*v2-v1,t+dt)*dt;
    x += 1.0/6.0*(v1+4.0*v2+v3);
  }
  fprintf(fp,"%f\t%f\n",t,x);
  fclose(fp);

  return 0;
}

double f(double x, double t) {
  return -pow(x,3) + sin(t*x);
}
```

E.1.0.11 Program for Example 12.9

```
/* Example C program to determine an approximate solution to
 *
 *   x' = -x^3 + sin(t x), x(0) = 1
 *
 *   using the fourth-order Runge-Kutta method.
 */


#include<stdio.h>
#include<math.h>

double f(double x, double t);
```

```
int main() {

  double x,t,dt;
  double k1,k2,k3,k4;
  FILE *fp;

  fp = fopen("rk4.d","w");

  dt = 0.5;
  x = 1.0;

  for(t=0;t<5;t+=dt) {
    fprintf(fp,"%f\t%f\n",t,x);
    k1 = f(x,t)*dt;
    k2 = f(x+k1/2.0,t+dt/2.0)*dt;
    k3 = f(x+k2/2.0,t+dt/2.0)*dt;
    k4 = f(x+k3,t+dt)*dt;

    x += 1.0/6.0*(k1 + 2.0*k2 + 2.0*k3 + k4);
  }
  fprintf(fp,"%f\t%f\n",t,x);
  fclose(fp);

  return 0;
}

double f(double x, double t) {
  return -pow(x,3) + sin(t*x);
}
```

E.1.0.12 Program for Example 12.10

```
/*  This is from the file C/subtleerror.c
 *
 *  Example C program to determine an approximate solution to
 *
 *  x' + 3 x = 15(cos(3 t) + sin(3 t)), x(0) = 1
 *
 *  using Euler's method, 2nd order RK, a 2nd order Taylor series
 *  expansion, and 4th order RK.  The exact solution is also printed to
 *  the data file for comparison purposes.
 */

#include<stdio.h>
#include<math.h>

double f(double x, double t);

main() {
```

```c
   double xe,xie,t,dt=0.25;
   double xts,x4rk;
   double w1,w2,w3,w4;
   double t_final=5;
   double exact;
   FILE *fp;

   fp = fopen("subtledata.d","w");
   xe = 0.0;                           /* euler's method */
   xie = 0.0;                          /* 2nd order RK */
   xts = 0.0;                          /* 2nd order TS */
   x4rk = 0.0;                         /* 4th order RK */
   exact = 0.0;

   for(t=0;t<=t_final;t+=dt) {
     fprintf(fp,"%f\t%f\t%f\t%f\t%f\t%f\t%f\t%f\t%f\t%f\n",
     t,xe,xie,xts,x4rk,exact,exact-xe,exact-xie,exact-xts,exact-x4rk);

     xe += f(xe,t)*dt;
     xie += (f(xie,t)+f(xie+f(xie,t)*dt,t+dt))*dt/2.0;
     xts += f(xts,t)*dt + pow(dt,2)/2.0*
        (-3.0*f(xts,t)+45.0*(-sin(3.0*t) + cos(3.0*t)));
     w1 = f(x4rk,t)*dt;
     w2 = f(x4rk+w1/2.0,t+dt/2.0)*dt;
     w3 = f(x4rk+w2/2.0,t+dt/2.0)*dt;
     w4 = f(x4rk+w3,t+dt)*dt;
     x4rk += 1.0/6.0*(w1 + 2.0*w2 + 2.0*w3 + w4);
     exact = 5.0*sin(3.0*(t+dt));
   }
     fclose(fp);
}


double f(double x, double t) {
   return 15.0*(cos(3.0*t)+sin(3.0*t)) - 3.0*x;
}
```

E.1.0.13 Program for Example 12.11

```c
/*  Example C program to determine an approximate solution to
 *
 *  x' = y, y' = (1 - x^2)y - x
 *  x(0) = 0.0 2, y(0) = 0.0
 *
 *  using Euler's method.
 */

#include<stdio.h>
#include<math.h>

int main() {
```

```
  double x[2],t,dt;
  int i;
  FILE *fp;

  fp = fopen("systemeuler.d","w");

  dt = 0.001;
  x[0] = 0.02;
  x[1] = 0.0;

  for(t=0;t<=20;t+=dt) {
    fprintf(fp,"%f\t%f\t%f\n",t,x[0],x[1]);
    x[0] += x[1]*dt;
    x[1] += ((1.0-pow(x[0],2))*x[1]-x[0])*dt;
  }
  fclose(fp);

  return 0;
}
```

E.1.0.14 Program for Example 12.13

```
/* Example C program to determine an approximate solution to
 *
 * x' = y, y' = (1 - x^2)y - x sin(t)
 * x(0) = 0.02, y(0) = 0.0
 *
 * using the fourth order Runge-Kutta method.
 */

#include<stdio.h>
#include<math.h>

double f(double x, double y, double t);
double g(double x, double y, double t);

int main() {

  double x,y,t,dt;
  double v1,v2,v3,v4,w1,w2,w3,w4;
  FILE *fp;

  fp = fopen("systemrk4.d","w");

  dt = 0.01;
  x = 0.02;
  y = 0.0;

  for(t=0;t<=20;t+=dt) {
    fprintf(fp,"%f\t%f\t%f\n",t,x,y);
```

```
    v1 = f(x, y, t)*dt;
    w1 = g(x, y, t)*dt;
    v2 = f(x+v1/2.0, y+w1/2.0, t+dt/2.0)*dt;
    w2 = g(x+v1/2.0, y+w1/2.0, t+dt/2.0)*dt;
    v3 = f(x+v2/2.0, y+w2/2.0, t+dt/2.0)*dt;
    w3 = g(x+v2/2.0, y+w2/2.0, t+dt/2.0)*dt;
    v4 = f(x+v3, y+w3, t+dt)*dt;
    w4 = g(x+v3, y+w3, t+dt)*dt;

    x += (v1 + 2.0*v2 + 2.0*v3 + v4)/6.0;
    y += (w1 + 2.0*w2 + 2.0*w3 + w4)/6.0;
  }
  fclose(fp);

  return 0;
}

double f(double x, double y, double t) {
  return y;
}

double g(double x, double y, double t) {
  return (1.0 - pow(x,2))*y - x*sin(t);
}
```

E.1.0.15   Program for Example 12.19

```
#include<stdio.h>
#include<math.h>

#define N 100

main() {

  double u[N+1][3],udic[N+1];
  double t,dt,dx,L=3.0,alpha=2.0;
  double tfinal=2.0*2.0*L/alpha;       /* gives two cycles for first mode
  int i,j,n,m;
  FILE *fp;

  fp = fopen("wavedata.d","w");

  dx = L/N;
  dt = 0.01;
  printf("%f\t%f\n",dx,dt);


  /* initial conditions */
  for(n=0;n<=N+1;n++) {
    if(n < N/3) {
```

```
     u[n][0] = dx*n;
   } else {
     u[n][0] = (3.0-dx*n)/2.0;
   }
     udic[n] = 0.0;
}

for(n=0;n<=N;n++) {
   fprintf(fp,"%f\t",u[n][0]);
}
fprintf(fp,"\n");

/* use the initial condition for velocity to compute u[1] */

for(n=0;n<=N;n++) {
   u[n][1] = u[n][0] + udic[n]*dt;
}
for(n=0;n<=N;n++) {
   fprintf(fp,"%f\t",u[n][1]);
}
fprintf(fp,"\n");

u[0][2] = 0.0;
u[N][2] = 0.0;
for(t=0;t<tfinal;t+=dt) {
   printf("t = %f\n",t);
   fprintf(fp,"%f\t",u[0][2]);
   for(n=1;n<N;n++) {
     u[n][2] = pow(alpha*dt/dx,2)*(u[n+1][1] - 2.0*u[n][1] + u[n-1][1])
+ 2.0*u[n][1] - u[n][0];
     fprintf(fp,"%f\t",u[n][2]);
   }
   fprintf(fp,"%f\n",u[N][2]);

   for(n=0;n<=N;n++) {
     u[n][0] = u[n][1];
     u[n][1] = u[n][2];
   }
}

   fclose(fp);
}
```

E.2 FORTRAN Programs

To compile FORTRAN programs on a UNIX or LINUX machine, type f77 eulerexample.f in a terminal and then type a.out to execute it. The syntax is similar on Windows and MacOS terminals. If an integrated development environment is used, a user must consult the documentation.

E.2.0.16 Program for Example 1.28

```
      program eulerexample

c     This is a sample FORTRAN program that solves the differential
c     equation
c
c     x' = sin(2*t)
c
c     using Euler's method.

      real x,t,dt,f
      integer n


      open(unit=13,file="output.d")

      n = 0
      dt = 0.01
      x = 3.0

c 100  format(f3.5,i4,f3.5,f3.5,f3.5,f3.5)
      do 10 t = 0, 10, dt
         f = sin(2.0*t)
         write(13,*) t,n,x,f,x+f*dt, 7.0/2.0 - cos(2.0*(t+dt))/2.0
         x = x + f*dt
         n = n + 1
 10     continue
      stop
      end
```

E.2.0.17 Program for Example 1.29

```
      program eulerexample

c     This is a sample FORTRAN program that solves the differential
c     equation
c
c     x' = 1/(1 + exp(-10*(t-5)))
c
c     using Euler's method.

      double precision x,t,dt,f
      integer n

      open(unit=13,file="output.d")

      n = 0
      dt = 0.00001
      x = 1/(1+exp(75.0))
```

```
        do 10 t = -1, 1, dt
          f = 75*x*(1-x)
          write(13,*) t,n,x,f,x+f*dt
          x = x + f*dt
          n = n + 1
 10     continue
        stop
        end
```

E.2.0.18 Program for Example 1.30

```
        program eulerexample

c       This is a sample FORTRAN program that solves the differential
c       equation
c
c       x'' + sin(t) x' + cos(t) x = exp(-5*t)
c       x(0) = 2, x'(0) = 5
c
c       using Euler's method.

        double precision x(2),t,dt,f(2)
        integer n

        open(unit=13,file="output.d")

        n = 0
        dt = 0.02
        x(1) = 2.0
        x(2) = 5.0

        do 10 t = 0, 30, dt
          f(1) = x(2)
          f(2) = exp(-5.0*t) - sin(t)*x(2) - cos(t)*x(1)
          write(13,*) t,x(1),x(2)
          x(1) = x(1) + f(1)*dt
          x(2) = x(2) + f(2)*dt
          n = n + 1
 10     continue
        stop
        end
```

E.2.0.19 Program for Example 12.1

```
        program eulererroranalysis

c       This is a sample FORTRAN program that solves the differential
```

```
c       equation
c
c       x' = 5 x, x(0) = 1
c
c       using Euler's method.

        double precision x,t,dt

        open(unit=13,file="fortrandata.d")

        dt = 0.1
        x = 1.0

        do 10 t = 0, 1, dt
           write(13,*) t,x,exp(5*t)
           x = x + 5*x*dt
 10     continue
        write(13,*) t,x,exp(5*t)
        stop
        end
```

E.2.0.20 Program for Example 12.2

```
        program eulererroranalysis2

c       This is a sample FORTRAN program that solves the differential
c       equation
c
c       x' = -sin(t),   x(0) = 1
c
c       using Euler's method.

        double precision x,t,dt;

        open(unit=13,file="fortrandata.d")

        dt = 1
        x = 1.0

        do 10 t = 0, 30, dt
           write(13,*) t,x,cos(t)
           x = x - sin(t)*dt
 10     continue
        write(13,*) t,x,cos(t)
        stop
        end
```

E.2.0.21 Program for Example 12.3

```
      program secondordertaylor

c     This is a sample FORTRAN program that solves the differential
c     equation
c
c     x' = 5 x, x(0) = 1
c
c     using a second order Taylor series expansion

      double precision x,t,dt;

      open(unit=13,file="secondordertaylor.d")

      dt = 0.1
      x = 1.0

      do 10 t = 0, 1, dt
         write(13,*) t,x,exp(5*t)
         x = x + 5*x*dt + 25/2*x*dt**2
 10   continue
      write(13,*) t,x,exp(5*t)
      stop
      end
```

E.2.0.22 Program for Example 12.4

```
      program secondordertaylor2

c     This is a sample FORTRAN program that solves the differential
c     equation
c
c     x' = -x^3 + sin(t x), x(0) = 1
c
c     using a second order Taylor series expansion

      double precision x,t,dt;

      open(unit=13,file="secondordertaylor2a.d")

      dt = 0.2
      x = 1.0

      do 10 t = 0, 5, dt
         write(13,*) t,x
         x = x + (-x**3 + sin(t*x))*dt
c     + 1.0/2.0*((-3*x**2 + t*cos(t*x))*(-x**3 + sin(t*x)) +
c     x*cos(t*x))*dt**2
 10   continue
      write(13,*) t,x
```

```
      stop
      end
```

E.2.0.23 Program for Example 12.6

```
      program rk2

c     This is a sample FORTRAN program that solves the differential
c     equation
c
c     x' = 5*x, x(0) = 1
c
c     using the second order Runge-Kutta method.

      double precision x,t,dt;

      open(unit=13,file="secondorderrk.d")

      dt = 0.1
      x = 1.0

      do 10 t = 0, 1, dt
         write(13,*) t,x,exp(5.0*t)
         x = x + dt/2.0*(f(x,t) + f(x+f(x,t)*dt,t+dt))
 10   continue
      write(13,*) t,x,exp(5.0*t)
      stop
      end

      double precision function f(x,t)
      double precision x,t

      f = 5*x

      return
      end
```

E.2.0.24 Program for Example 12.7

```
      program secondorderrk

c     This is a sample FORTRAN program that solves the differential
c     equation
c
c     x' = -x^3 + sin(t x), x(0) = 1
c
c     using the second order Runge-Kutta method.

      double precision x,t,dt;
```

```
         open(unit=13,file="secondorderrk2a.d")

         dt = 0.2
         x = 1.0

         do 10 t = 0, 5, dt
            write(13,*) t,x
            x = x + dt/2.0*(f(x,t) + f(x+f(x,t)*dt,t+dt))
 10      continue
         write(13,*) t,x
         stop
         end

         double precision function f(x,t)
         double precision x,t

         f = -x**3 + sin(t*x)

         return
         end
```

E.2.0.25 Program for Example 12.8

```
         program rk3

 c       This is a sample FORTRAN program that solves the differential
 c       equation
 c
 c       x' = -x^3 + sin(t x), x(0) = 1
 c
 c       using the third order Runge-Kutta method.

         double precision x,t,dt;
         double precision v1,v2,v3;

         open(unit=13,file="rk3.d")

         dt = 0.25
         x = 1.0

         do 10 t = 0, 5, dt
            write(13,*) t,x
            v1 = f(x,t)*dt
            v2 = f(x+0.5*v1,t+0.5*dt)*dt
            v3 = f(x+2.0*v2-v1,t+dt)*dt
            x = x + (v1 + 4*v2 + v3)/6.0
 10      continue
         write(13,*) t,x
         stop
         end
```

```
        double precision function f(x,t)
        double precision x,t

        f = -x**3 + sin(t*x)

        return
        end
```

E.2.0.26 Program for Example 12.9

```
        program rk4

c       This is from the file FORTRAN/rk4.f
c
c       This is a sample FORTRAN program that solves the differential
c       equation
c
c       x' = -x^3 + sin(t x), x(0) = 1
c
c       using the fourth order Runge-Kutta method.

        double precision x,t,dt;
        double precision k1,k2,k3,k4

        open(unit=13,file="rk4.d")

        dt = 0.25
        x = 1.0

        do 10 t = 0, 5, dt
           write(13,*) t,x
           k1 = f(x,t)*dt
           k2 = f(x+k1/2.0,t+dt/2.0)*dt
           k3 = f(x+k2/2.0,t+dt/2.0)*dt
           k4 = f(x+k3,t+dt)*dt
           x = x + (k1 + 2.0*k2 + 2.0*k3 + k4)/6.0
 10     continue
        write(13,*) t,x
        stop
        end

        double precision function f(x,t)
        double precision x,t

        f = -x**3 + sin(t*x)

        return
        end
```

E.2.0.27 Program for Example 12.11

```
      program systemeuler

c     This is a sample FORTRAN program that solves the differential
c     equation
c
c     x' = y, y' = (1 - x^2)y - x
c     x(0) = 0.02, y(0) = 0.0
c
c     using Euler's method.

      double precision x(2),t,dt
      double precision copy(2)

      open(unit=13,file="systemfortran.d")

      dt = 0.001
      x(1) = 0.02
      x(2) = 0.0

      do 10 t = 0, 20, dt
         write(13,*) t,x(1),x(2)
         copy(1) = x(1)
         copy(2) = x(2)
         x(1) = x(1) + (copy(2))*dt
         x(2) = x(2) + ((1.0 - copy(1)**2)*x(2) - x(1))*dt
 10   continue
      stop
      end
```

E.2.0.28 Program for Example 12.13

```
      program systemrk4

c     This is a sample FORTRAN program that solves the differential
c     equation
c
c     x' = y, y' = (1 - x^2)y - x sin(t)
c     x(0) = 0.02, y(0) = 0.0
c
c     using the fourth order Runge-Kutta method.

      double precision x,y,t,dt
      double precision v1,v2,v3,v4,w1,w2,w3,w4

      open(unit=13,file="systemrk4f.d")

      dt = 0.001
      x = 0.02
      y = 0.0
```

```
      do 10 t = 0, 20, dt
         write(13,*) t,x,y

         v1 = f(x, y, t)*dt
         w1 = g(x, y, t)*dt

         v2 = f(x+v1/2.0, y+w1/2.0, t+dt/2.0)*dt
         w2 = g(x+v1/2.0, y+w1/2.0, t+dt/2.0)*dt

         v3 = f(x+v2/2.0, y+w2/2.0, t+dt/2.0)*dt
         w3 = g(x+v2/2.0, y+w2/2.0, t+dt/2.0)*dt

         v4 = f(x+v3, y+w3, t+dt)*dt
         w4 = g(x+v3, y+w3, t+dt)*dt

         x = x + (v1 + 2.0*v2 + 2.0*v3 + v4)/6.0
         y = y + (w1 + 2.0*w2 + 2.0*w3 + w4)/6.0
10    continue
      stop
      end

      double precision function f(x,y,t)
      double precision x,y,t

      f = y

      return
      end

      double precision function g(x,y,t)
      double precision x,y,t

      g = (1.0-x**2)*y - x*sin(t)

      return
      end
```

E.2.0.29 Program for Example 12.15

```
      program laplace
      implicit none

c     This is an example FORTRAN program that solves Laplace's equation
c     on a rectangular domain using the finite difference method.

      double precision u(0:40,0:20)
      integer i,j,n,m

c     set the initial values
      data u /861*0.3/
```

```
      open(unit=13,file="output.d")

c     set the boundary conditions
      do 10 m=0,20
         u(0,m)=0
         u(40,m)=0
 10   continue

      do 20 n=0,40
         u(n,0)=0
         if(n .LE. 20) then
            u(n,20) = n/10.0
         else
            u(n,20) = (40-n)/10.0
         endif
 20   continue

c     iterate
      do 30 i=1,100
         do 40 n=1,39
            do 50 m=1,19
               u(n,m) = (u(n-1,m)+u(n+1,m)+u(n,m-1)+u(n,m+1))/4.0
 50         continue
 40      continue
 30   continue
      write(13,100) ((u(i,j),j=0,20),i=0,40)
 100  format (41(21(F6.3, 1X),/))

      stop
      end
```

E.2.0.30 Program for Example 12.16

```
      program laplace2
      implicit none

c     This is an example FORTRAN program that solves Laplace's equation
c     on an H-shaped domain using the finite difference method.

      double precision u(0:90,0:90)
      integer i,j,n,m,whereami,iteration

c     set the initial values
      data u /8281*0.5/

      open(unit=13,file="output.d")
```

```fortran
c       set the boundary conditions

        do 10 n=0,90
           do 20 m=0,90
              if(whereami(n,m).EQ.0) THEN
                 u(n,m) = 0
              elseif(whereami(n,m).EQ.1) THEN
                 u(n,m) = 0.5
              elseif(whereami(n,m).EQ.3) THEN
                 u(n,m) = 0
              else
                 u(n,m) = 1
              endif
20         continue
10      continue
        write(*,100) ((u(i,j),j=0,90),i=0,90)
        do 30 iteration=1,100
           do 40 n=1,89
              do 50 m=1,89
                 if(whereami(n,m).EQ.1) THEN
                    u(n,m) = (u(n-1,m)+u(n+1,m)+u(n,m-1)+u(n,m+1))/4.0
                 endif
50            continue
40         continue
30      continue

        write(13,100) ((u(i,j),j=0,90),i=0,90)
        write(*,100) ((u(i,j),j=0,90),i=0,90)
100     format (91(91(F5.1, 1X),/))

        stop
        end

c       0 = exterior
c       1 = interior
c       2 = constant boundary points with u=1
c       3 = constant boundary points with u=0
        integer function whereami(n,m)
        integer n,m

        if(((n.GT.30).AND.(n.LT.60)).AND.((m.GT.60).OR.(m.LT.30)))THEN
           whereami = 0
        elseif((n.GT.0).AND.(n.LT.30).AND.((m.gt.0).AND.(m.lt.90))) THEN
           whereami = 1
        elseif((n.GT.60.AND.n.LT.90).AND.((m.gt.0).AND.(m.lt.90))) THEN
           whereami = 1
        elseif((n.GE.30.AND.n.LE.60).AND.(m.gt.30.AND.m.lt.60))THEN
           whereami = 1
        elseif(((n.LE.30).OR.(n.GE.60)).AND.(m.EQ.90)) THEN
           whereami = 2
        elseif(((n.GT.30).AND.(n.LT.60)).AND.(m.EQ.60)) THEN
           whereami = 2
        elseif(((n.EQ.30).OR.(n.EQ.60)).AND.(m.GE.60)) THEN
           whereami = 2
```

```
         else
            whereami = 3
         endif
         end
```

E.2.0.31 Program for Example 12.16

```
         program heat
         implicit none

c        This program solves the one-dimensional heat conduction equation
c        with alpha=2, L=10.  The initial heat distribution is triangular
c        with a maximum at x=5 and a maximum value of 5.

         integer i,j,k,steps,dim
         parameter(dim=20)
         double precision u(0:dim),un(0:dim),t,dt,dx,alpha

         open(unit=13,file="heathomogeneous.d")


         dx = 10.0/dim
         dt = 0.031
         alpha=2
         steps=8/dt+1
         write(*,*) 'dt should be less than ',0.5*(dx/alpha)**2
         write(*,*) 'dt is ',dt

c        set the initial conditions
         t=0
         do 10 i=0,dim
            if(i .LT. dim/2) then
               u(i) = 2.0*5.0*i/dim
            else
               u(i) = 5.0-2.0*5.0*(i-dim/2.0)/dim
            endif
 10      continue

         write(13,100) (u(i),i=0,dim)

         do 20 k=1,steps
            do 30 j=1,dim-1
               un(j) = u(j)+alpha**2*((u(j-1)-2*u(j)+u(j+1))/dx**2)*dt
 30         continue
            do 40 j=1,dim-1
               u(j)=un(j)
 40         continue
            t=t+dt

            if(abs(t-1.6).LT.dt/2) then
               write(*,*) t
```

```
            write(13,100) (u(i),i=0,dim)
        elseif(abs(t-3.2).LT.dt/2) then
            write(*,*) t
            write(13,100) (u(i),i=0,dim)
        elseif(abs(t-4.8).LT.dt/2) then
            write(*,*) t
            write(13,100) (u(i),i=0,dim)
        elseif(abs(t-6.4).LT.dt/2) then
            write(*,*) t
            write(13,100) (u(i),i=0,dim)
        elseif(abs(t-8).LT.dt/2) then
            write(*,*) t
            write(13,100) (u(i),i=0,dim)
        endif
20      continue

        if(abs(t-8).LT.dt/2) then
            write(*,*) t
            write(13,100) (u(i),i=0,dim)
        endif
        write(*,*) t

100     format(201(F10.3, 1X))

        stop
        end
```

References

1. Abraham, R., Marsden, J.E.: Foundations of Mechanics, 2nd edn. Addison Wesley, Reading, MA: (1978)
2. Anderson, B.D.O., Moore, J.B.: Optimal Control: Linear Quadratic Methods. Prentice Hall, Upper Saddle River, NJ: (1990)
3. Antsaklis, P.J., Michel, A.M.: Linear Systems. McGraw Hill, New York (1997)
4. Arnold, V.: Mathematical Methods of Classical Mechanics, 2nd edn. Springer, New York (1997)
5. Åström, K.J., Murray, R.M.: Feedback Systems: An Introduction for Scientists and Engineers. Princeton University Press, Princeton, NJ (2008)
6. Barbour, J.M.: Tuning and Temperament: A Historical Survey. Dover, New York (2005)
7. Bloom, B.: Taxonomy of Educational Objectives. Longmans, New York (1956)
8. Boyce, W.E., DiPrima, R.C.: Elementary Differential Equations and Boundary Value Problems, 7th edn. John Wiley & Sons, Hoboken, NJ (2001)
9. Braun, M.: Differential Equations and Their Applications, 4th edn. Springer-Verlag, New York (1993)
10. Churchill, R.V., Brown, J.W., Verhey, R.F.: Complex Variables and Applications, 3rd edn. McGraw-Hill, New York (1974)
11. Corben, H., Stehle, P.: Classical Mechanics. Wiley, New York (1950)
12. Craig, J.J.: Introduction to Robotics, 2nd edn. Addison-Wesley, Reading, MA (1989)
13. Dorf, R.C., Bishop, R.H.: Modern Control Systems, 10th edn. Prentice Hall, New York (2005)
14. Farlow, S.J.: Partial Differential Equations for Scientists and Engineers. Dover, New York (1993)
15. Felder, R.: Reaching the second tier: Learning and teaching styles in college science education. Journal of College Science Teaching 23(5), 286–290 (1993)
16. Felder, R., Silverman, L.: Learning and teaching styles in engineering eduction. Engineering Education 78(7), 674–681 (1988)
17. Franklin, G., Powell, J., Emami-Naeini, A.: Feedback Control of Dynamic Systems, 5th edn. Prentice Hall, Upper Saddle River, NJ (005)
18. Franklin, J.N.: Matrix Theory. Prentice-Hall, Inc., Upper Saddle River, NJ (1968)
19. Goldstein, H.: Classical Mechanics, 2nd edn. Addison-Wesley, Reading, MA (1980)
20. Hartog, J.P.D.: Mechanics. McGraw-Hill, New York (1948)
21. Hartog, J.P.D.: Mechanical Vibrations, 4th edn. McGraw-Hill, New York (1956)
22. Hirsch, M.W., Smale, S.: Differential Equations, Dynamical Systems, and Linear Algebra. Academic Press, New York (1974)
23. Howland, R.A.: Intermediate Dynamics: A Linear Algebraic Approach. Springer, New York (2005)
24. Hungerford, T.W.: Algebra. Springer-Verlag, New York (1974)
25. Incropera, F.P., DeWitt, D.P.: Fundamentals of Heat and Mass Transfer, 2nd edn. Wiley, New York (1895)
26. Isaacson, E., Keller, H.B.: Analysis of Numerical Methods. John Wiley & Sons, New York (1966)
27. Kofler, M.: Maple: An Introduction and Reference. Addison-Wesley Longman, Boston, MA (1997)
28. Kuo, B.C.: Automatic Control Systems, 7th edn. Prentice Hall, Upper Saddle River, NJ (1995)
29. LeVeque, R.J.: Numerical Methods for Conservation Laws. Birkhäuser, Boston (1992)
30. Levine, W.S. (ed.): The Control Handbook. CRC Press, Boca Raton, FL (1996)
31. Levitin, D.J.: This is Your Brain on Music: The Science of a Human Obsession. Penguin Group, London (2006)
32. Mancini, R.: Feedback amplifier analysis tools. Tech. rep., Texas Instruments (2001)
33. McKeachie, W.: Improving lectures by understanding students' information processing. In: W. McKeachie (ed.) New Directions for Teaching and Learning, 2, chap. Learning, Cognition and College Teaching, p. 32. Jossey-Bass, San Francisco, CA (1980)

34. Meriam, J.L., Kraige, L.G.: Engineering Mechanics , Dynamics, 5th edn. Wiley, Hoboken, NJ (2001)
35. Murray, R.M., Li, Z.X., Sastry, S.S.: A Mathematical Introduction to Robotic Manipulation. CRC Press, Boca Raton, FL (1994)
36. Myint-U, T.: Ordinary Differential Equations. North-Holland, Amsterdam (1978)
37. National Institute of Standards and Technology: The International System of Units (SI). NIST Special Publication 330. United States Department of Commerce (2001)
38. Naylor, A.W., Sell, G.R.: Linear Operator Theory in Engineering and Science. Springer-Verlag, New York (1982)
39. Netwon, I.: Philosophiae naturalis principia mathematica (1687)
40. Newton, I.: The Principia. Prometheus, Amherst, NY (1995)
41. Ogata, K.: Modern Control Engineering. Prentice Hall, Englewood Cliffs, NJ (1970)
42. Ogata, K.: Modern Control Engineering, 4th edn. Prentice Hall, Upper Saddle River, NJ (2002)
43. Pars, L.A.: A Treatise on Analytic Dynamics. Ox Bow Press, Woodbridge, CT (1979)
44. Polking, J., Boggess, A., Arnold, D.: Differential Equations with Boundary Value Problems, 2nd edn. Prentice Hall, Upper Saddle River, NJ (2005)
45. Richtmyer, R.D., Morton, K.W.: Difference Methods for Initial-Value Problems, 2nd edn. John Wiley & Sons, New York (1967)
46. Routh, E.: A Treatise on the Stability of a Given State of Motion. Macmillan, London (1877)
47. Sastry, S.: Nonlinear Systems: Analyis, Stability and Control. Springer, New York (1999)
48. Schwartz, L.: Theorie des Distributions. Hermann, Paris (1966)
49. Sethares, W.A.: Tuning, Timbre, Spectrum, Scale, 2nd edn. Springer, New York (2005)
50. Tenenbaum, M., Pollard, H.: Ordinary Differential Equations. Dover Publications, New York (1963)
51. Timoshenko, S., Young, D.H.: Advanced Dynamics. McGraw-Hill, New York (1940)
52. White, F.M.: Fluid Mechanics. McGraw-Hill, New York (1979)
53. Whittaker, E.: A Treatise on the Analytical Dynamics of Particles and Rigid Bodies. Cambridge at the University Press (1927)
54. Wiggins, S.: Introduction to Applied Nonlinear Dynamical Systems and Chaos. Springer, New York (1990)
55. Wolfram, S.: The Mathematica Book, 4th edn. Cambridge University Press, New York (1999)
56. Zadeh, L.: Fuzzy sets. Information and Control **8**, 338–353 (1965)

Index

Printed in the United States
By Bookmasters